ANATOMICAL CHART COMPANY

# ATLAS of HUMAN ANATOMY

ANATOMICAL CHART COMPANY

# ATLAS of HUMAN ANATOMY

**SPRINGHOUSE**
Springhouse, Pennsylvania

## Staff

**Senior Publisher**
Donna O. Carpenter

**Clinical Director**
Marguerite Ambrose, RN, MSN, CCRN, CS

**Creative Director**
Jake Smith

**Executive Editor**
H. Nancy Holmes

**Clinical Project Manager**
Patricia Kardish Fischer, RN, BSN

**Art Director**
Elaine Kasmer Ezrow (design project manager)

**Editors**
Audrey Selena Hughes (editorial project manager), Jennifer P. Kowalak

**Copy Editors**
Catherine B. Cramer, Dolores Connors Matthews

**Clinical Editors**
Jill M. Curry, RN, CCRN; Lori Musolf Neri, RN, MSN, CRNP

**Designers**
Arlene Putterman (associate design director), Joseph John Clark, Jacalyn B. Facciolo, Linda J. Franklin, Jan Greenberg, Donald G. Knauss, Donna S. Morris, Susan Hopkins Rodzewich

**Illustrators**
Anatomical Chart Company: Dawn Gorski, Marguerite Aitken, Peter Bachin, Liana Bauman, Ernie Beck, Carl Clingman, Birck Cox, Ruth Daly, Leonard Dank, Brian Evans, Robert Fletcher, Claudia Grosz, Fred Harwin, William Jacobson, Keith Kasnot, Jeanne Koelling, Lik Kwong, Lena Lyons, Kimberly Martins, Marcelo Oliver, Linda Warren, William Westwood, Christine Young

**Electronic Production Services**
Diane Paluba (manager), Joyce Rossi Biletz

**Manufacturing**
Patricia K. Dorshaw (manager), Otto Mezei (book production manager)

**Editorial Assistants**
Carol A. Caputo, Arlene Claffee, Beth Janae Orr

**Indexer**
Barbara E. Hodgson



The clinical procedures described and recommended in this publication are based on research and consultation with nursing, medical, and legal authorities. To the best of our knowledge, these procedures reflect currently accepted practice; nevertheless, they can't be considered absolute and universal recommendations. For individual application, all recommendations must be considered in light of the patient's clinical condition and, before administration of new or infrequently used drugs, in light of the latest package-insert information. The authors and the publisher disclaim responsibility for any adverse effects resulting directly or indirectly from the suggested procedures, from any undetected errors, or from the reader's misunderstanding of the text.



Printed in the United States of America.
AHA-D N O S A J J M A
03 02 01 10 9 8 7 6 5 4 3 2 1

**Library of Congress Cataloging-in-Publication Data**

Atlas of human anatomy.
p. ; cm.
Includes index.
ISBN 1-58255-108-1 (alk. paper)
1. Human anatomy — Atlases. 1. Springhouse Corporation.
[DNLM: 1. Anatomy — Atlases. QS 17 A8813 2001]
QM25 .A798 2001
611'.0022'2--dc21

2001020027

# Contents

# Consultants

**James Agostinucci,** ScD, OTR
Associate Professor
Physical Therapy Program
University of Rhode Island
Kingston

**Darlene Nebel Cantu,** RNC, MSN
Director
Baptist Health System
San Antonio, Texas

**Barbara J. Davis,** PhD
Assistant Professor in Neurobiology and Anatomy
University of Rochester (N.Y.)
School of Medicine and Dentistry

**Martin Dym,** PhD
Chairman
Department of Anatomy and Cell Biology
Georgetown University Medical Center
Washington, D.C.

**Ellie Z. Franges,** RN, MSN, CNRN
Director of Neuroscience Services
Sacred Heart Hospital
Allentown, Pennsylvania

**David J. Lash,** PA-C, MPAS
Head, Branch Medical Annex
Camp H.M. Smith
United States Navy
Hawaii

**Henry R. Lemke,** MMS, PA-C
Director
Physician Assistant Studies
University of North Texas Health Science Center
Fort Worth

**Carol A. Livingston,** RN, MS, NP
Adjunct Faculty
Regis University
Denver
Red Rocks Community College
Lakewood, Colorado

**Roger M. Morrell,** MD, PhD, FACP (P.C.), ABPN
Clinical Neurologist in private practice
Southfield, Michigan

**Joanne M. Orth,** PhD
Professor of Anatomy and Cell Biology
Temple University School of Medicine
Philadelphia

**Mary Clare A. Schafer,** MS, RN, ONC
Orthopedic Clinical Specialist
Infection Control Nurse
Graduate Hospital
Philadelphia

**Mary A. Stahl,** RN, CS, MSN, CCRN
Clinical Nurse Specialist
Saint Luke's Hospital
Kansas City, Missouri

**Lawrence C. Zoller,** PhD
Associate Professor of Anatomy
Department of Anatomy and Neurobiology
Boston University School of Medicine

# Foreword

With the introduction of the World Wide Web and the Internet over a decade ago and the amazing proliferation of computers in homes and practices, the promise of information at our fingertips seemed close to realization. Just by entering a key word, we expected we could depend on sophisticated search engines to deliver multiple Websites filled with relevant, accurate, easy-to-understand, and up-to-date information.

That was the dream. The *reality* is familiar to most health care providers – indeed to anyone attempting research on the Web. Finding a relevant Website that presents visually appealing material at an appropriate level of detail is so rare as to be a surprise. The more common result is a list of hundreds of Websites that are not remotely useful – and a search for anatomical information turns up unsavory sites generated by the "entertainment" industry. The expenditure of time and effort produces only enormous frustration.

No matter how many electronic gadgets we use, nothing currently can replace the solid presence of reputable reference books on our shelves. A good book is a known and trustworthy entity. For health care providers, a good book on anatomy is priceless, and *Atlas of Human Anatomy* is just such a book.

## A reliable tool for you

*Atlas of Human Anatomy* is at once a trusted reference, a handy review, and a valuable patient-teaching aid. For the most part, the body of anatomical knowledge is stable. Once you have a good grasp of the topic, details fall easily into place and a word or well-designed graphic can prompt adequate recall.

*Atlas of Human Anatomy* is the perfect tool for stimulating recall. This book has successfully applied a formula that realizes the promise of relevant, accurate, and understandable information at your fingertips. The full-color graphics are clear and detailed and the accompanying text concise without being telegraphic. The bulk of the atlas supplies anatomical drawings and text organized by body *region*, while the last part summarizes anatomical and physiological information for each body *system*.

The book's layout is perfectly suited to the busy professional who needs to quickly check anatomical details or review a major concept. New topics always appear on a left-hand page, and graphics on the right.

Throughout the book, color logos call your attention to key physiology concepts, age-related changes to normal anatomy, and clinical tips to aid in physical examination.

Helpful appendices include a chart of skeletal muscles, with their areas of origin and insertion, as well as charts on cranial and autonomic nerves. You'll also find memory-refresher lists of suffixes, prefixes, abbreviations, and symbols and a glossary of common anatomical terms.

## ... and for your patients

In addition to serving as a personal work reference, *Atlas of Human Anatomy* is well suited to a role in patient education. As more and more patients arrive armed with questions and misconceptions generated from their Web searches, or with information of questionable quality or utility, health care providers have an increasing need for solid, reliable resources to assist in patient education.

Explanations of illness are often best understood when based in the concrete reality of anatomy. With *Atlas of Human Anatomy*, you have a marvelous tool to promote patient understanding. Its myriad illustrations in full color and its overall visual attractiveness ensure its use as the perfect prop to assist you in answering questions, interpreting level of understanding, and correcting misconceptions.

I see this as a reference book that you will want to keep within arm's reach, one that has the elegance usually associated with more detailed texts, yet a delivery style perfectly suited to the time-challenged practitioner. *Atlas of Human Anatomy*'s concise, clear illustrations and functional layout make it a workhorse of a reference book and a priceless addition to every health care provider's library.

**Kathleen A. Mulligan,** PhD
Senior Lecturer
Department of Biological Structure
University of Washington
Seattle

# PART I

# General overview

The practice of any health care profession requires a basic understanding of anatomy and physiology. *Anatomy* is the study of the body structure and the relationships of body parts to one another. (See *Branches of anatomy*.) *Physiology* is the study of how body parts function in the organism; it includes both chemical and physical processes.

## Branches of anatomy

Branches of anatomy include gross anatomy, histology, developmental anatomy, applied anatomy, and pathologic anatomy.
- *Gross anatomy* (also called macroscopic anatomy): the study of anatomic structures visible to the unaided eye.
- *Histology*: the microscopic study of the structure of cells, tissues, and organs in relation to their function. *Cytology* is the study of the origin, structure, function, and pathology of isolated cells.
- *Developmental anatomy*: the study of structural changes from conception through old age. *Embryology* is the study of the origin, growth, development, and function of an organism from fertilization to birth. *Gerontology* is the study of the aging process and all aspects of the problems of aging.
- *Applied anatomy*: the use of anatomic findings to diagnose and treat medical disorders.
- *Pathologic anatomy*: the study of diseased, abnormal, or injured tissue.

## Anatomic terms

Anatomic terms describe the locations of structures (directional terms) in the body's planes, cavities, and regions.

### Directional terms

Directional terms help to describe the exact location of a structure in relation to the body, a body part, or another structure:
- *superior* — toward the head
- *inferior* — toward the feet
- *anterior* or *ventral* — toward the front
- *posterior* or *dorsal* — toward the back
- *medial* — toward the midline
- *lateral* — away from the midline
- *proximal* — toward to the trunk
- *distal* — away from the trunk
- *superficial* — at or toward the body surface
- *deep* — beneath or farthest from the body surface.

### Reference planes

Imaginary planes divide the body and its organs into sections, making it possible to locate body parts and structures in three dimensions. The four major reference planes are the sagittal, frontal, transverse, and oblique planes. (See *Body reference planes*.) The first three lie at right angles to one another:
- *sagittal* — along the body's longitudinal axis (vertically); divides the body into right and left regions
  - *median sagittal* or *midsagittal* — exactly on the midline
  - *parasagittal* — to either side of the midline
- *frontal* (coronal) — longitudinal but at a right angle to a sagittal plane; divides the body into anterior and posterior regions
- *transverse* — horizontal at a right angle to the vertical axis; divides the body into superior and inferior regions
- *oblique* — at an angle that lies between a horizontal plane and a vertical plane.

## Body reference planes

Body reference planes are used to indicate the locations of body structures. Shown here are the median sagittal, frontal, and transverse planes, which lie at right angles to one another. An oblique plane (not shown) is a slanted plane that lies between a horizontal plane and a vertical plane.

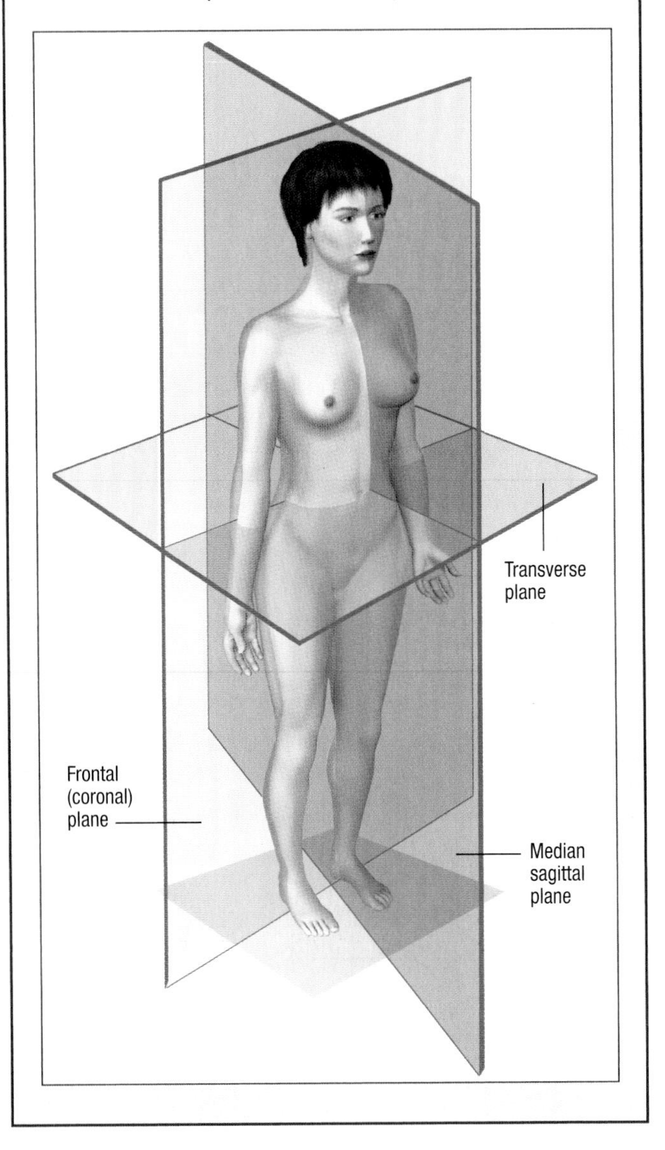
Transverse plane
Frontal (coronal) plane
Median sagittal plane

### Body cavities

Body cavities are spaces in the body that contain the internal organs. The dorsal and ventral cavities are the two major closed cavities. (See *Body cavities*.)

### Dorsal cavity

The dorsal cavity consists of the cranial cavity and the vertebral canal. The *cranial cavity* (skull) encases the brain. The *vertebral canal* (also called the *spinal* or *vertebral cavity*) is formed by portions of the bones (vertebrae) that form the spine. It encloses the spinal cord.

## Body cavities

The dorsal cavity, in the posterior region of the body, is divided into the cranial cavity and the vertebral canal (vertebral cavity). The ventral cavity, in the anterior region, is divided into the thoracic and abdominopelvic cavities.

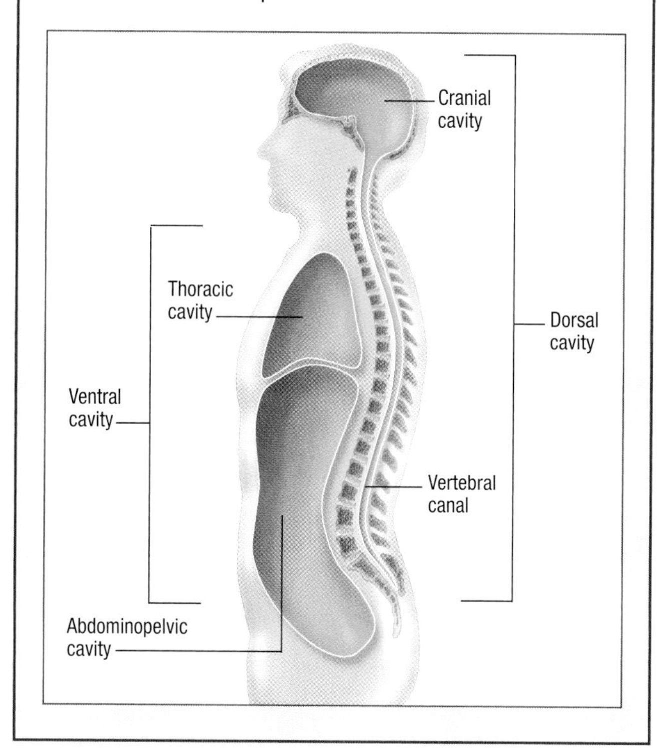
Cranial cavity
Thoracic cavity
Dorsal cavity
Ventral cavity
Vertebral canal
Abdominopelvic cavity

### Ventral cavity

The ventral cavity consists of the *thoracic cavity* and the *abdominopelvic cavity.* The two communicate through an opening in the diaphragm called the *hiatus.* A thin membrane, the *serosa,* lines the ventral cavity and the outer surfaces of its organs. The *parietal serosa* lines the walls; the *visceral serosa* covers the organs.

#### *Thoracic cavity*

The thoracic cavity is surrounded by the ribs and chest muscles. It's subdivided into the *pleural cavities*, each of which contains a lung, and the *mediastinum*, which contains the heart, large vessels of the heart, trachea, esophagus, thymus, lymph nodes, and other blood vessels and nerves. The mediastinum is further divided into four areas. The *middle mediastinum* contains the heart and pericardial sac; the *anterior, posterior,* and *superior* areas are named according to their positions relative to the middle mediastinum.

The serosa in the pleural cavities is called the *pleura.* In the middle mediastinum, the serosa surrounds the heart and is called the *pericardium.* The outer layer of the pericardium is made of fibrous tissue and is lined by parietal serosa. The visceral serosa covers the surface of the heart. The *pericardial cavity*, between the visceral and parietal layers of the serous pericardium, contains a thin film of fluid.

#### *Abdominopelvic cavity*

The abdominopelvic cavity has two regions: the *abdominal cavity* and the *pelvic cavity*. The abdominal cavity contains the stomach, intestines, spleen, liver, and other internal organs. The pelvic cavity, inferior to the abdominal cavity, contains the bladder, some reproductive structures, and the rectum. The serosa in these cavities is called the *peritoneum*.

*Note*: A plane defined by the rim of the pelvis divides the abdominal and pelvic cavities. No muscles or membranes physically separate them.

### Other cavities

The body also contains an oral cavity (the mouth), a nasal cavity (in the nose), orbital cavities (which contain the eyes), middle ear cavities (which contain the small bones of the middle ear), and synovial cavities (enclosed in the capsules surrounding certain joints).

## Body regions

Body regions are universally accepted designations for specific areas that have a special nerve or vascular supply or that perform a special function. Like other anatomical terms, the names of body regions describe the locations of various structures. (See *Regions of the human body*, page 4, and *Abdominal regions, anterior view,* below.)

## Abdominal regions, anterior view

This illustration shows the abdominal regions from the front in greater detail.

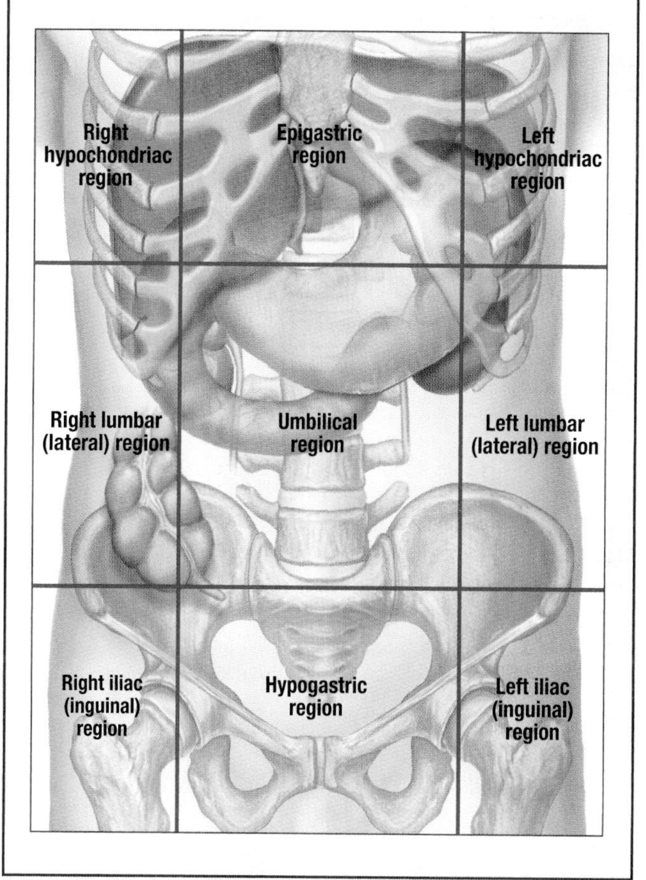
Right hypochondriac region
Epigastric region
Left hypochondriac region
Right lumbar (lateral) region
Umbilical region
Left lumbar (lateral) region
Right iliac (inguinal) region
Hypogastric region
Left iliac (inguinal) region

## Regions of the human body

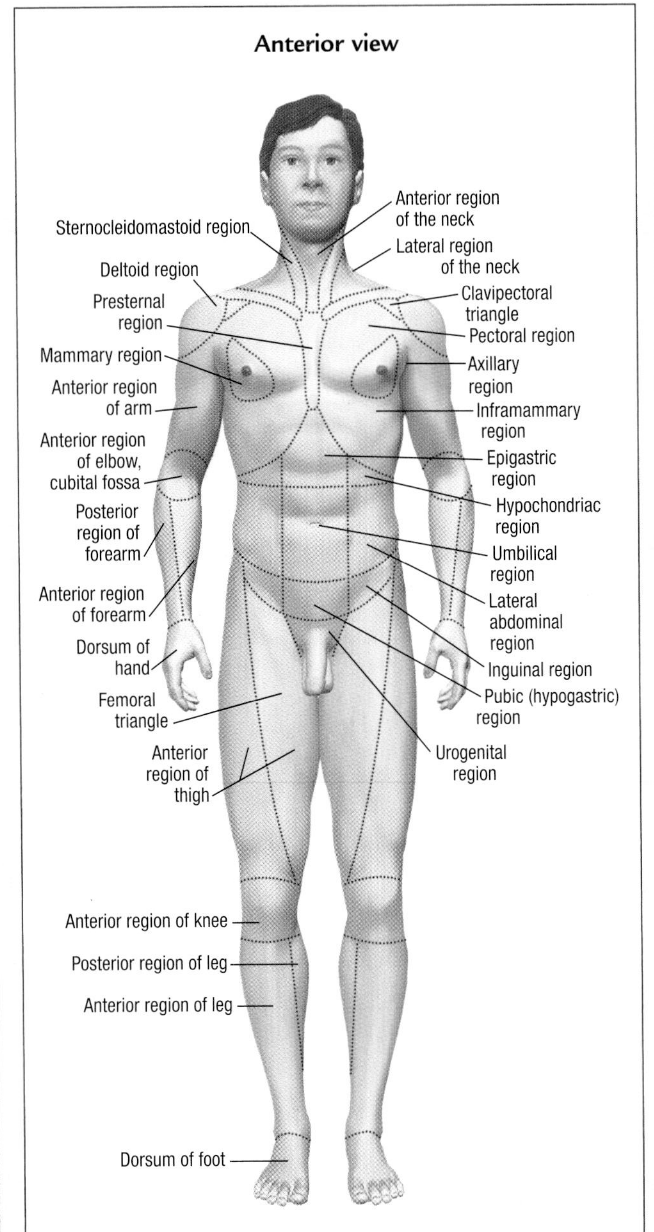

Anterior view
Sternocleidomastoid region
Anterior region of the neck
Lateral region of the neck
Deltoid region
Presternal region
Clavipectoral triangle
Pectoral region
Mammary region
Axillary region
Anterior region of arm
Inframammary region
Anterior region of elbow, cubital fossa
Epigastric region
Hypochondriac region
Posterior region of forearm
Umbilical region
Anterior region of forearm
Lateral abdominal region
Dorsum of hand
Inguinal region
Femoral triangle
Pubic (hypogastric) region
Anterior region of thigh
Urogenital region
Anterior region of knee
Posterior region of leg
Anterior region of leg
Dorsum of foot


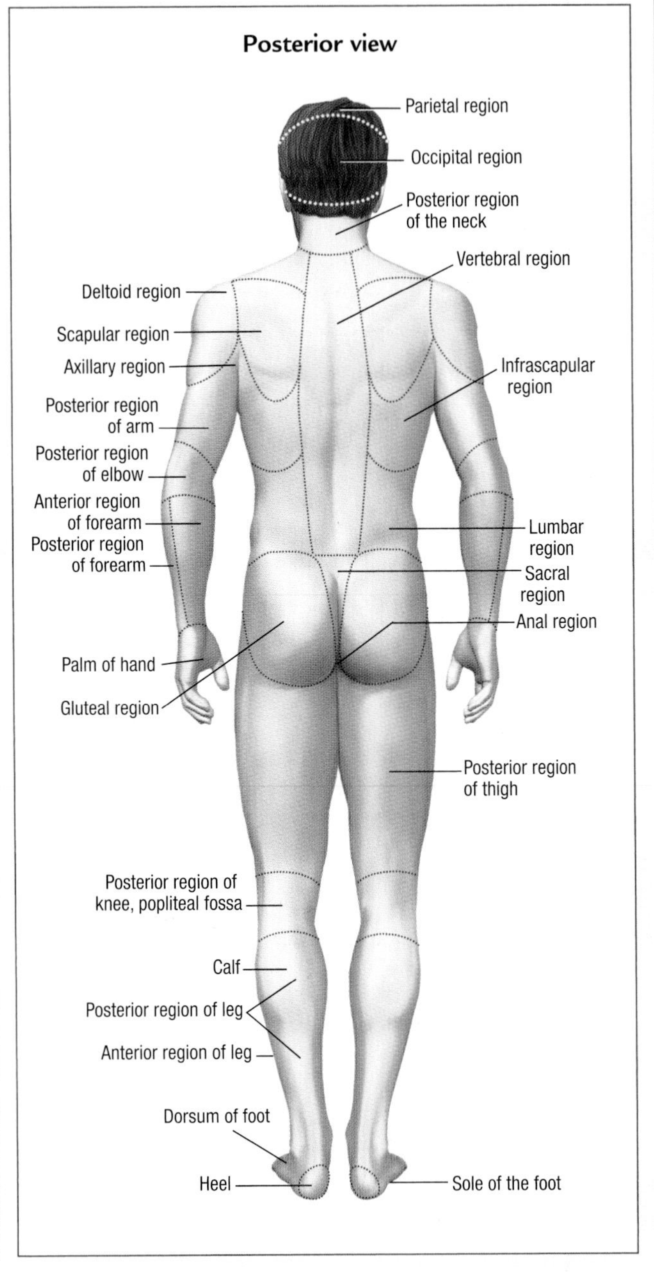

Posterior view
Parietal region
Occipital region
Posterior region of the neck
Vertebral region
Deltoid region
Scapular region
Axillary region
Infrascapular region
Posterior region of arm
Posterior region of elbow
Anterior region of forearm
Lumbar region
Posterior region of forearm
Sacral region
Anal region
Palm of hand
Gluteal region
Posterior region of thigh
Posterior region of knee, popliteal fossa
Calf
Posterior region of leg
Anterior region of leg
Dorsum of foot
Heel
Sole of the foot


### Abdominal regions

The following list of terms widely used to describe abdominal regions identifies their locations and contents:

- *epigastric* – immediately below the diaphragm and superior to the umbilical region; contains portions of the pancreas and stomach and the liver, inferior vena cava, abdominal aorta, and duodenum
- *right* and *left hypochondriac* – lateral to the epigastric region; contain portions of the diaphragm, kidneys, and stomach, the spleen, and part of the pancreas
- *umbilical* – around the umbilicus; includes portions of the small and large intestines, inferior vena cava, and abdominal aorta
- *right* and *left lumbar* (lateral) – lateral to umbilical region; contain portions of the small and large intestines and portions of the right and left kidneys
- *hypogastric* (pubic) – inferior to the umbilical region; contains the bladder, ureters, and portions of the sigmoid colon and small intestine
- *right* and *left iliac* (inguinal) – lateral to the hypogastric region; contain portions of the small and large intestines.

## Inside the cell

The fundamental unit of life, the cell consists of a nucleus, cytoplasm, and organelles. Within the nucleus are the blueprints of heredity – ribonucleic acid and deoxyribonucleic acid.

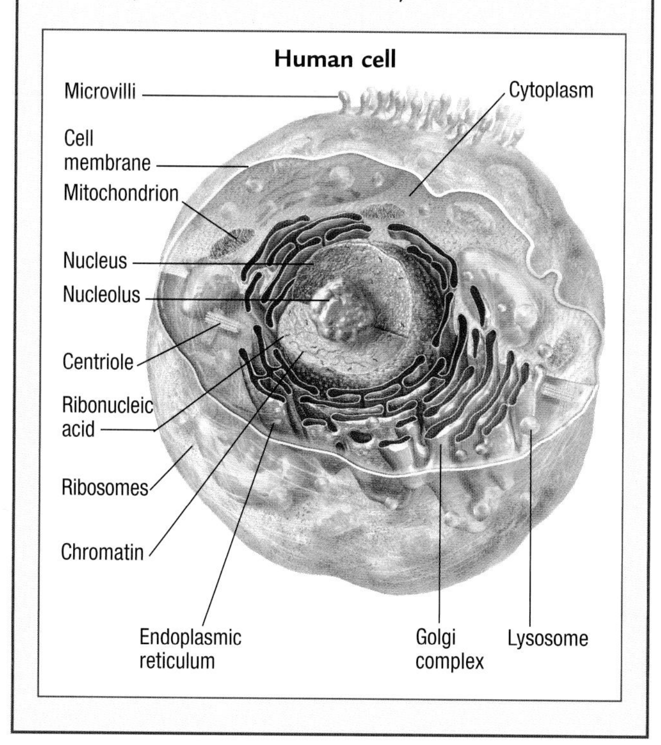
Human cell
Microvilli
Cytoplasm
Cell membrane
Mitochondrion
Nucleus
Nucleolus
Centriole
Ribonucleic acid
Ribosomes
Chromatin
Endoplasmic reticulum
Golgi complex
Lysosome

## Learning about organelle functions

- *Mitochondria* are threadlike structures that provide most of the body's adenosine triphosphate (ATP). ATP is an energy-storage molecule that fuels many cellular activities.
- *Ribosomes* are the sites of protein synthesis.
- The *endoplasmic reticulum* is an extensive network of membrane-enclosed tubules. Rough endoplasmic reticulum, covered with ribosomes, produces certain proteins; smooth endoplasmic reticulum contains enzymes that synthesize lipids.
- Each *Golgi apparatus* synthesizes carbohydrate molecules, which combine with protein produced by rough endoplasmic reticulum to form secretory products such as lipoproteins.
- *Lysosomes* are digestive bodies that break down nutrient material as well as foreign or damaged material in cells. A membrane surrounding each lysosome separates its digestive enzymes from the rest of the cytoplasm. The enzymes digest nutrient matter brought into the cell by means of endocytosis, in which a portion of the cell membrane surrounds and engulfs matter to form a membrane-bound intracellular vesicle. The membrane of the lysosome fuses with the membrane of the vesicle surrounding the endocytosed material. The lysosomal enzymes then digest the engulfed material. Lysosomes digest the foreign matter ingested by white blood cells by a similar process called *phagocytosis*.
- *Peroxisomes* contain *oxidases,* enzymes capable of reducing oxygen to hydrogen peroxide and hydrogen peroxide to water.
- *Cytoskeletal elements* form a network of protein structures.
- *Centrosomes* contain *centrioles,* short cylinders adjacent to the nucleus that take part in cell division.

# Cells

The *cell* is the basic unit of living matter. Human cells vary widely, ranging from the simple squamous epithelial cell to the highly specialized neuron. Generally, the simpler the cell, the shorter its life span and the greater its power to regenerate; the more specialized, the longer its life span and the weaker its regenerative power.

## Cell structure

The three basic components of a typical cell are the plasma membrane, cytoplasm, and nucleus.

### Plasma membrane

The semipermeable *plasma membrane* (cell membrane) serves as the cell's external boundary, separating it from other cells and from the external environment. Roughly 75Å (3/10 millionths of an inch) thick, the cell membrane is a double layer of phospholipids and associated protein molecules.

### Cytoplasm

A viscous, translucent material, *cytoplasm* is the primary component of plant and animal cells. It contains water, inorganic ions (such as potassium, calcium, magnesium, and sodium), and naturally occurring organic molecules (such as proteins, lipids, and carbohydrates). (See *Inside the cell.*)

Inorganic ions perform specific functions, such as regulating acid-base balance and the amount of intracellular water. When atoms – sodium, chlorine, potassium, or hydrogen, for example – gain or lose electrons, they acquire a positive or negative electrical charge, respectively, and are called *ions* or *electrolytes*.

*Nucleoplasm*, the cytoplasm in the cell nucleus, plays a part in reproduction. The cytoplasm surrounding the nucleus is the site of most synthesizing activity and conversion of raw materials to energy. The cytoplasm consists of cytosol, organelles, and inclusions.

#### *Cytosol*

Cytosol is a viscous, semitransparent fluid that's 70% to 90% water. It contains proteins, salts, and sugars.

#### *Organelles*

Organelles are the cell's metabolic units, and each performs a specific function to maintain the life of the cell. Organelles include mitochondria, ribosomes, the endoplasmic reticulum, Golgi apparatus, lysosomes, peroxisomes, cytoskeletal elements, and centrosomes. (See *Learning about organelle functions*.)

#### *Inclusions*

These nonmoving units are usually the products of cellular metabolic activity. Examples include the pigment *melanin* in epithelial cells, the stored nutrient *glycogen* in liver cells, and lipid droplets that contain stored cholesterol.

### Nucleus

The cell's control center, the *nucleus,* regulates cell growth, metabolism, and reproduction. The nucleus of a human somatic cell contains 46 *chromosomes* and one or more *nucleoli.* Chromosomes contain genes that, when activated, control cellular ac-

## More about DNA

DNA contains four different nucleotide bases:
- Adenine and guanine are double-ring molecules classified as purines
- Thymine and cytosine are single-ring molecules classified as pyrimidines.

Nucleotides are joined into long chains by chemical bonds between the phosphate group of the nucleotide and a carbon atom in the deoxyribose molecule of the adjacent nucleotide. Existing in pairs, DNA chains are held together by weak chemical attractions between the nitrogen atoms on adjacent chains. Because of the chemical shape of the bases, adenine bonds only with thymine, and guanine bonds only with cytosine. Bases that can link with each other are called *complementary.*

Linked DNA chains form a spiral structure, or *double helix,* that resembles a spiral staircase; the deoxyribose and phosphate groups form the railings, and the nitrogen-containing base pairs form the steps. The sequence of nucleotide bases in DNA chains forms a series of coded messages called the *genetic code.*

tivities and direct protein synthesis through ribosomes in the cytoplasm. *Nucleoli* synthesize *ribonucleic acids (RNA),* complex polynucleotides that control protein synthesis.

#### *Chromosomes*

Chromosomes appear as a network of chromatin granules in the nondividing cell; they assume recognizable form only when the cell begins to divide. Except in the *gametes* (germ cells), chromosomes exist in pairs. One chromosome from each pair comes from the male germ cell (*spermatozoon*); the other, from the female germ cell (*ovum*).

Normal human cells contain 23 pairs of chromosomes. In these cells, 22 pairs are called *homologous chromosomes*; both members of each pair contain genetic information that controls the same characteristics or functions. The 23rd pair (the X and Y chromosomes) determines gender: XX produces a genetic female; XY, a genetic male.

In the female, the genetic activity of both X chromosomes is essential only during the first few weeks after conception. Later development requires just one functional X chromosome. The other X chromosome is inactivated and appears as a dense chromatin mass – called a *Barr body,* or sex chromatin – attached to the nuclear membrane. In the cells of a normal male, who has only one functional X chromosome, the Barr body is absent.

#### *Deoxyribonucleic acid*

*Deoxyribonucleic acid (DNA)* is a large molecule that carries genetic information and provides the blueprint for protein synthesis. Its basic structural unit, the *nucleotide*, is a phosphate group linked to a five-carbon sugar, deoxyribose, joined to a nitrogen-containing compound called a nucleotide base. (See *More about DNA.*)

#### *Ribonucleic acid*

Like DNA, RNA consists of nucleotide chains, but some of its components differ from those of DNA. RNA transfers genetic information from nuclear DNA to ribosomes in the cytoplasm, where protein synthesis occurs. This process involves three types of RNA – transfer, messenger, and ribosomal – with specific functions.
- *Transfer RNA* consists of short nucleotide chains, each of which is specific for an individual amino acid. Transfer RNA carries the genetic code to messenger RNA.
- *Messenger RNA* directs the arrangement of amino acids to make proteins at the ribosomes. Its single strand of nucleotides is complementary to a segment of the DNA chain that contains instructions for protein synthesis. It moves from the nucleus into the cytoplasm, where it attaches to ribosomes.
- *Ribosomal RNA* makes ribosomes in the endoplasmic reticulum of the cytoplasm, where the cell produces proteins in accordance with information carried in the genetic code.

## Tissues

Tissues are groups of cells with the same general function. The human body contains four basic types of tissue – epithelial, connective, muscle, and nerve. (See *Types of body tissue.*) Blood is also considered a tissue, although anatomists debate whether it's a distinct tissue in itself or a type of connective tissue.

### Epithelial tissue

Epithelial tissue is a sheet of tightly opposed cells that covers the body's surface (*epithelium*), lines body cavities (*endothelium*), and forms certain glands. For example, *endothelium* is a single layer of squamous epithelium that lines the heart and blood vessels.

#### Types of epithelial tissue

Epithelial tissue is classified according to number of cell layers, shape of the cells on its surface, and other characteristics.
- Number of cell layers:
  - *simple* – one layer
  - *stratified* – two or more
  - *pseudostratified* – simple but appears stratified because all cells touch the basement membrane; however, some don't extend to the surface
- Shape of surface cells:
  - *squamous* – flat
  - *columnar* – tall, cylindrical, prism-shaped
  - *cuboidal* – cube-shaped
- Tissue-specific characteristics that contribute to organ function:
  - *intestinal lumen* – columnar epithelial cells with many small projections, called *villi,* that increase surface area and enhance absorption of nutrients and other materials
  - *renal tubules – brush border:* tiny, brushlike structures (*microvilli*) on the apical surface of columnar cells; amplify surface area to enhance absorption
  - *lining of epididymis – stereociliated epithelial cells* have tall microvilli, originally and erroneously called cilia; propel sperm along ducts
  - *luminal surfaces of many organs, such as the upper respiratory tract – ciliated epithelium* have surface *cilia,* fine hairlike protuberances, larger than microvilli, that move fluid and particles along the surface.

#### *Glandular epithelium*

Organs that produce secretions (glands) consist of a special type of epithelium called *glandular epithelium.* Many glands are enclosed in a dense capsule of connective tissue; these capsules

## Types of body tissue

The human body contains four basic types of tissue — epithelial, connective, muscle, and nerve. Epithelial tissue may be simple, stratified, or pseudostratified; and squamous, columnar, or cuboidal. Types of connective tissue include bone, cartilage, and adipose (fatty) tissue. Muscle tissue may be striated (skeletal), cardiac, or smooth. Nervous tissue consists of neurons and neuroglia. Two types of tissue are illustrated below.

**Striated muscle**

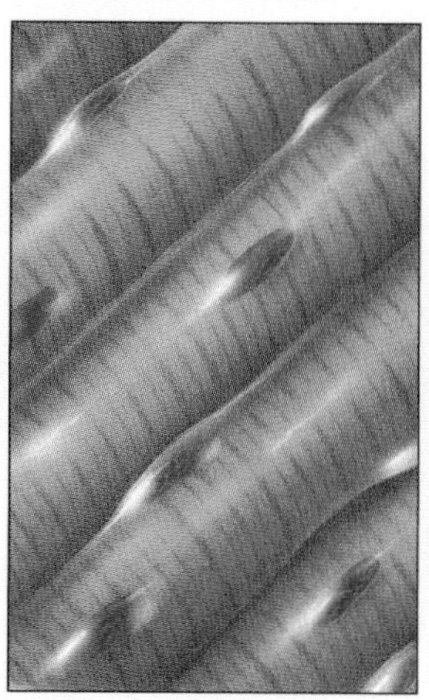

**Stratified squamous**

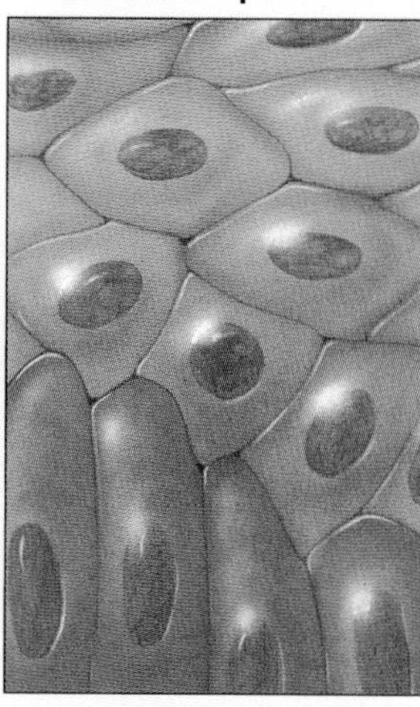

## Endocrine and Exocrine glands

Depending on how it secretes its products, a gland is classified as either *endocrine* or *exocrine*. *Secretion* — the process of producing a specific substance — may involve separation of an element of the blood, or production of a totally new chemical substance (such as the urine secreted by the kidneys). *Excretion,* in contrast, is the elimination of a product from the body. The urinary bladder, for example, excretes urine.

- *Endocrine glands* lack ducts and release their secretions directly into the blood or lymph. For instance, the medulla of the adrenal gland secretes epinephrine and norepinephrine into the bloodstream.
- *Exocrine glands* secrete into ducts. Simple exocrine glands have only one duct; compound glands, more than one. Examples are sweat and sebaceous glands of the skin.
- *Mixed glands* contain both endocrine and exocrine cells. The pancreas, a mixed gland, contains alpha and beta cells (in the islets of Langerhans); these endocrine cells produce glucagon and insulin, respectively. The pancreas also contains acinar cells, exocrine cells that secrete digestive juices.

are divided into *lobes*, then into smaller units called *lobules*. (See *Endocrine and exocrine glands*.)

## Connective tissue

Connective tissue — a category that includes bone, cartilage, and adipose (fatty) tissue — binds and supports body structures. Connective tissue is made up of collagenous, reticular, or elastin fibers, cells, and matrix.

*Collagenous* fibers are soft, flexible, white fibers made of the protein *collagen*; they are present in all types of connective tissue and are highly resistant to pulling forces. *Reticular* fibers are very thin fibers that occur in organs that change shape and volume such as the uterus. *Elastin* fibers are yellowish fibers with an elastic quality that are found in organs that are stretched, such as the aorta.

Connective tissue cells may be fixed or wandering. *Fixed cells* are typical cells that remain in place, whereas *wandering cells* may move from one site to another.

### Types of connective tissue

Connective tissue is classified as loose or dense.

- Loose (areolar) connective tissue consists predominantly of cells or matrix.
- Dense connective tissue, which provides structural support, has greater fiber concentration. Dense tissue is further subdivided into dense regular and dense irregular connective tissue.
  - Dense regular connective tissue consists of tightly packed fibers arranged in a consistent pattern. It comprises tendons, ligaments, and aponeuroses (flat fibrous sheets that surround and attach muscles to bones or other tissues).
  - Dense irregular connective tissue has tightly packed fibers arranged in an inconsistent pattern. It's found in the dermis, submucosa of the GI tract, fibrous capsules, and fascia.

#### *Adipose tissue*

Commonly called fat, *adipose tissue* is a specialized type of loose connective tissue in which a single lipid (fat) droplet occupies most of each cell. It cushions internal organs and acts as a reserve supply of energy. Adipose tissue is widely distributed subcutaneously.

**Clinical tip**

The distribution of adipose tissue varies with sex and age. When assessing for the presence of adipose tissue, keep the following in mind:

- In men, subcutaneous fat appears mainly in the nape of the neck, the regions overlying the seventh cervical vertebra, the deltoid and triceps muscles, the lumbosacral region, and the buttocks.
- In women, subcutaneous fat occurs chiefly in the breasts, buttocks, and thighs. In both sexes, fat accumulates extensively in the abdominal region.

### Special properties

Connective tissue in some parts of the body has special properties.

- Mucous connective tissue of the umbilical cord (Wharton's jelly) is a temporary tissue that supports the umbilical cord until after birth.
- Elastic connective tissue of the vocal cords makes speech possible.
- Reticular connective tissue of the spleen forms a soft skeleton for support of other cells. Some cells of this tissue are phagocytic, protecting the body against foreign cells and substances.
- An abundance of collagen (white, fibrous connective tissue) gives the sclera of the eyeball its white color.

## Muscle tissue

Muscle tissue consists of muscle cells and has a generous blood supply. Measuring up to several centimeters long, muscle cells have an elongated shape that enhances their *contractility* (ability to contract).

### Types of muscle tissue

The three basic types of muscle tissue are striated skeletal, striated cardiac, and smooth.

#### *Striated skeletal*

Striated muscle tissue gets its name from its striped, or *striated,* appearance. Its cells, called *fibers,* which contain masses of protoplasm and many nuclei, receive stimulation from cerebrospinal nerves. Striated muscle tissue that is capable of voluntary contraction is called *skeletal* muscle tissue. Striated muscles contain specialized *myofibrils,* bundles of fine fibers made up of thin and thick filaments. Thin filaments contain the contractile protein *actin*; thick filaments contain the contractile protein *myosin*.

#### *Striated cardiac*

The striated muscle tissue of the heart, cardiac muscle tissue is sometimes classified as striated, rather than placed in a separate category. However, it differs from skeletal muscle in two ways:

- Its fibers are separate cellular units, which don't contain many nuclei.
- It contracts involuntarily.

#### *Smooth*

Lacking the striped pattern of striated muscle, smooth muscle consists of long, spindle-shaped cells. Smooth muscle contraction is stimulated by the autonomic nervous system; it is not under voluntary control.

Smooth muscle lines the walls of many internal organs and other structures. (See *Locations of smooth muscle.*)

### Locations of smooth muscle

Smooth muscle occurs throughout the body, including the sites below:

- walls of internal organs such as those of the GI tract from the middle of the esophagus to the internal anal sphincter
- walls of the repiratory passages from the trachea to the alveolar ducts
- urinary and genital ducts
- walls of arteries and veins
- walls of larger lymphatic trunks
- hair follicles
- iris and ciliary body of the eye.

In the skin, smooth-muscle fibers form the *arrectores pilorum,* tiny muscles whose contraction causes the hair to stand erect. In the mammary glands, smooth muscle causes the nipples to become erect; in the scrotum, it wrinkles the skin to help raise the testes.

Smooth muscle in the ciliary body of the eye plays a part in accommodation (focusing the eye for clear vision at various distances). In the iris, smooth-muscle contraction constricts the pupil.

## Nerve tissue

The main function of nerve tissue is communication. Its primary properties are *irritability* (reaction to various physical and chemical agents) and *conductivity* (transmittal of the resulting reaction from one point to another). Nerve tissue cells may be neurons or neuroglia. Highly specialized cells, *neurons* generate and conduct nerve impulses. A typical neuron consists of a cell body with cytoplasmic extensions – numerous dendrites on one pole and a single axon on the other. These extensions allow the neuron to conduct impulses over long distances. *Neuroglia* form the support structure of nervous tissue, insulating and protecting neurons.

# PART II

# HEAD AND NECK

# Head and neck bones

The bones of the head and neck are part of the axial skeleton – bones that form the longitudinal axis of the body. The head and neck bones include the 22 bones of the skull, the seven cervical vertebrae, and three small bones in each ear, the auditory ossicles. Finally, the hyoid bone is a U-shaped bone suspended by the stylohyoid ligament from the styloid processes of the temporal bones, between the mandible and the larynx.

## The skull

The skull, also called the cranium, consists of 8 cranial bones and 14 facial bones. The cranial bones cover the cranial cavity, which contains the brain.

### Functions of the skull

- Protects the brain
- Protects the sensory organs, such as eyes, ears, and nose
- Provides for passage of air and food through its openings
- Contains teeth and jaws for mastication
- Contains openings for passage of nerves, arteries, and veins.

### Cranial bones

The eight cranial bones consist of the following:
- parietal bones (2)
- temporal bones (2)
- frontal bone
- occipital bone
- sphenoid bone
- ethmoid bone.

The calvaria, or skull cap, is the superior portion of the cranium, where the frontal, parietal, and occipital bones meet. The cranial base is formed by the occipital, ethmoid, and sphenoid bones.

### Facial bones

The 14 facial bones consist of the following:
- nasal bones (2)
- maxillae (2)
- lachrymal bones (2)
- zygomatic bones (2)
- palatine bones (2)
- nasal conchae (2)
- vomer
- mandible.

### Skull bone characteristics

Most skull bones are *flat* bones, which consist of two layers of compact bone separated by spongy bone and marrow.

The cranial bones are *irregular* bones, consisting of a spongy layer between internal and external tables (layers) of compact bone. Internally, bony ridges divide the cranial bones at the base of the skull into three fossae, or depressed regions, called the anterior, middle, and posterior cranial fossae.

The bones of the skull are joined together except for the mandible, which is movable. Small holes in the skull, called foramina, allow room for blood vessels and nerves.

## Cervical vertebrae

The cervical vertebrae, designated C1 through C7, constitute the skeleton of the neck. They are smaller than the other vertebrae of the spine. They articulate with the skull at the base of the cranium, where the spinal cord enters through the largest hole in the skull, the foramen magnum.
- C1, called the *atlas*, is highly specialized. Unlike the other cervical vertebrae, it is simply a ring of bone; it lacks a body or spinous process.
- C2, called the *axis*, has a toothlike process called the *dens* that projects superiorly from the vertebral body. The dens forms a specialized articulation with the atlas, the atlantoaxial joint.
- The remaining cervical vertebrae, C3 through C7, consist of a body (the discoid, weight-bearing portion) anteriorly and a vertebral arch posteriorly. The body and arch form the *vertebral foramen*; collectively, these openings form the vertebral canal, which houses the spinal cord. Each of these vertebrae has two *transverse processes*, lateral projections from the vertebral arch; and one *spinous process*, a posterior projection in the midline. The bony connections between the body and the transverse processes are called *pedicles*; the bony connections between the transverse and spinous processes are called *laminae*. (See also Part III, Back and spine, for additional illustrations of specific cervical vertebrae and for more information on the spine.)

Clinical tip

### Finding cervical spine reference points

To get a frame of reference when you are assessing the cervical spine area, instruct your patient to flex his neck. A bony prominence will bulge at the nape of the neck. That is the last cervical vertebra, C7, sometimes called the *vertebra prominens*.

## HEAD AND NECK BONES

### Anterior view

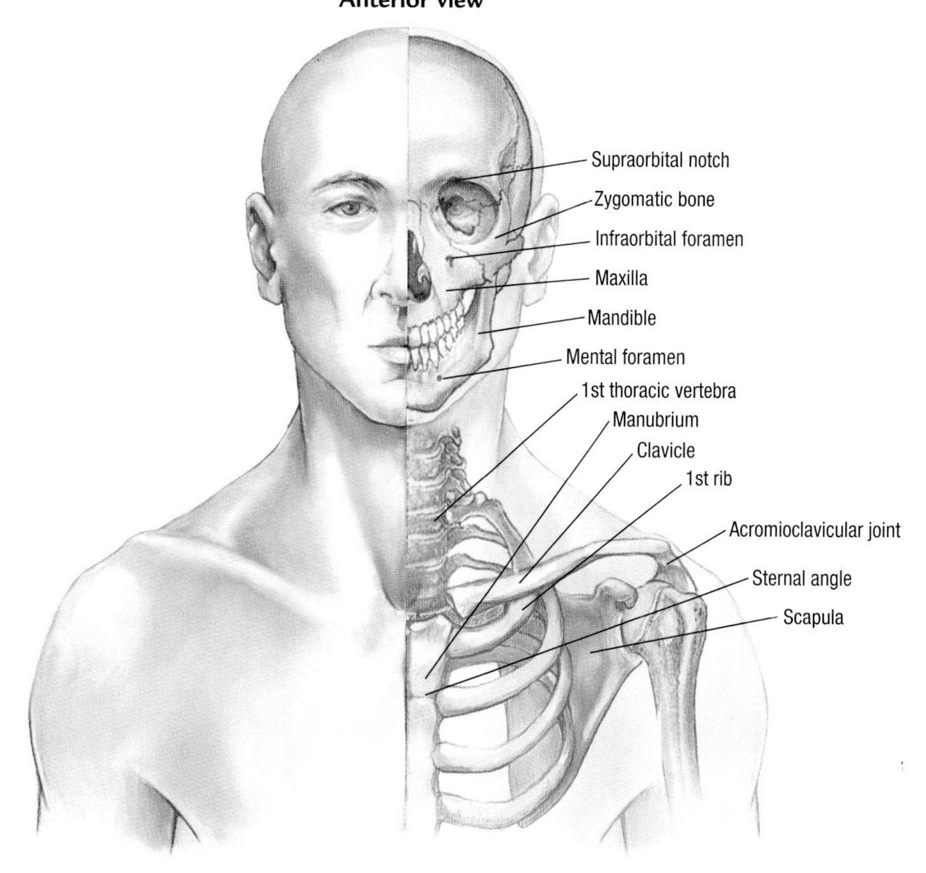

Supraorbital notch
Zygomatic bone
Infraorbital foramen
Maxilla
Mandible
Mental foramen
1st thoracic vertebra
Manubrium
Clavicle
1st rib
Acromioclavicular joint
Sternal angle
Scapula


### Posterior view

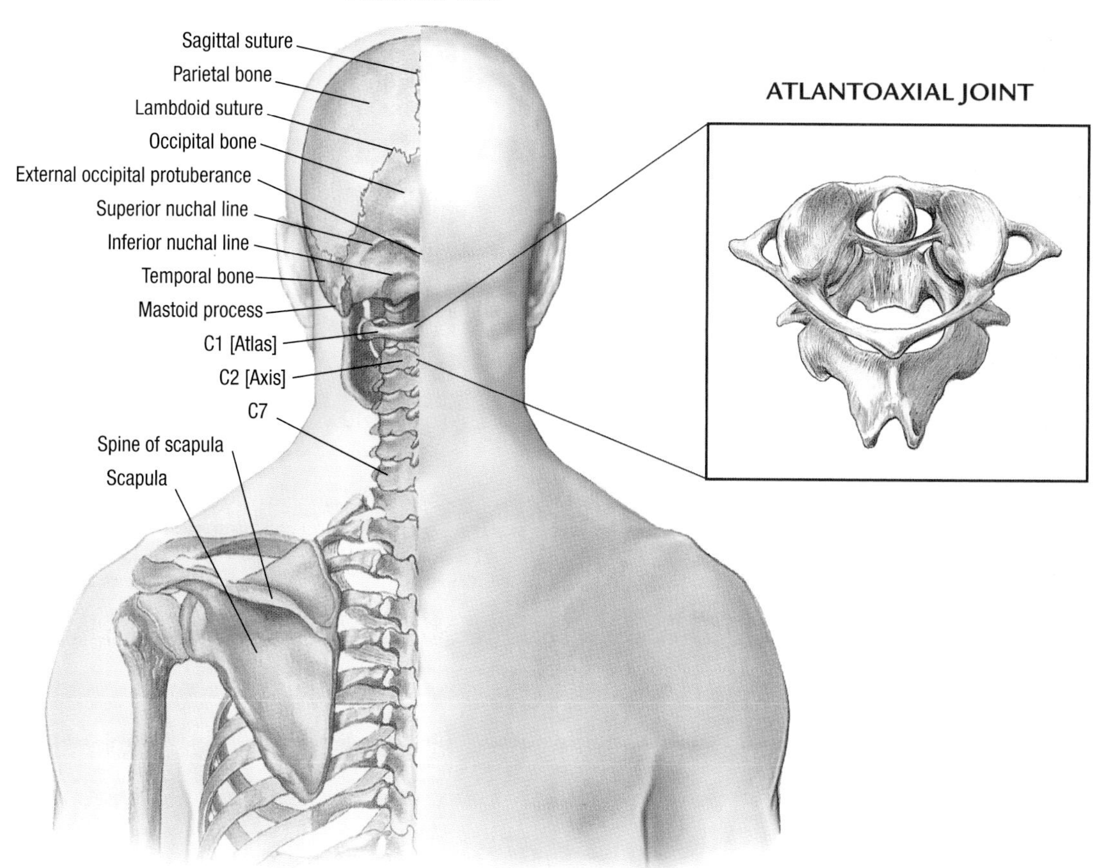

Sagittal suture
Parietal bone
Lambdoid suture
Occipital bone
External occipital protuberance
Superior nuchal line
Inferior nuchal line
Temporal bone
Mastoid process
C1 [Atlas]
C2 [Axis]
C7
Spine of scapula
Scapula
ATLANTOAXIAL JOINT

# Skull features

Numerous features of the skull can be identified when viewed from different aspects.

## Anterior aspect

In an anterior view, prominent skull features include the following:

- frontal bone – forms the forehead and articulates with the parietal bones
- nasion – depression at the root of the nose, formed by the intersection of the frontal bone and the two nasal bones; between the nasal bones lies the internasal suture, the line where the bones join
- superciliary arch – elevation that extends laterally from above the nasion and between the eyebrows
- orbits – bony cavities that house the eyes, including the inferior and superior orbital fissures
- zygomatic or malar bones – prominences of the cheeks
- piriform aperture – opening formed by the nasal bones and maxillae
- maxillae and mandible – upper and lower jaws, respectively
- paranasal sinuses – four pairs of sinuses named for the bones in which they lie: frontal, ethmoid, sphenoid, and maxillary.

## Superior aspect

The most significant structures in the superior view are the suture lines, the points where two bones are joined. Sutures have closely united opposing surfaces. (See *Adult skull* and *Neonate skull,* page 14.) In very young infants, the sutures are open. In adults, sutures are immovable. The four sutures are:

- coronal suture
- sagittal suture
- lambdoid suture
- squamous suture.

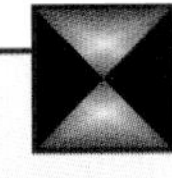

**Age-related changes**

### Fontanels

The cranial bones are not completely ossified at birth. Thus the head can change shape as it passes through the birth canal. The membranous, unossified areas, called *fontanels,* are at junctures of future suture lines. The fontanels close gradually, and closure is complete by about age 18 months. The neonate has six fontanels:

- Anterior. At the juncture of the future coronal, frontal, and sagittal sutures, this diamond-shaped fontanel is the largest. It remains open until age 12 to 18 months.
- Posterior. This triangular fontanel is at the juncture of the sagittal and lambdoidal sutures.
- Anterolateral (2). Located at each side of the juncture of the frontal, parietal, sphenoid, and temporal bones.
- Posterolateral (2). Located at each side of the juncture of the temporal, parietal, and occipital bones.

## Posterior aspect

In the posterior view, the paired parietal bones join the occipital and squamous parts of the temporal bones. The mastoid processes are prominent bilaterally. Each mastoid process is a conical projection of the caudal, posterior portion of the temporal bone to which several muscles are attached, including the sternocleidomastoid. A hollow section contains air cells that communicate with the tympanic cavity of the middle ear.

## Lateral aspect

When the skull is viewed from a lateral aspect, one of the most prominent features is the zygomatic arch, where the zygomatic process of the temporal bone articulates with the temporal process of the zygomatic bone. The sphenoid bone forms the vault of the cranium. It joins with the frontal and occipital bones. (See *Skull, lateral view,* page 15.)

The mandible is connected to the rest of the skull by paired synovial joints that articulate with the temporal bone. (See Temporomandibular joint, page 66.)

Other notable features of the lateral aspect include:

- coronoid process of the mandible
- mastoid and styloid processes of the temporal bone
- mental foramen – the opening for passage of the mental nerve and vessels located in the mandible
- pterion – the junction of the sphenoid, frontal, parietal, and temporal bones
- frontal, parietal, and occipital bones, superiorly
- ethmoid, sphenoid, and temporal bones, inferiorly.

## Sectional views

Some features of the skull are visible only in sections.

### Sagittal section

A sagittal section of the skull reveals the following: (See *Skull, sagittal section through skull,* page 15.)

- Frontal sinus, one of a pair of small paranasal cavities in the frontal bone that communicate with the nasal cavity, measuring approximately 3 cm (1¼″) in height, 2.5 cm (1″) in width, and 2.5 cm in depth
- The paranasal sinuses, which are lined with mucous membrane and open into the nasal cavity
- Perpendicular plate of the ethmoid bone
- Occipital condyles, the rounded projection at the ends of the occipital bones, which articulate with the atlas, allowing the motion of nodding
- Internal acoustic meatus, the internal auditory canal
- Vomer, the flat bone forming the posterior and inferior parts of the nasal septum
- Sella turcica, a saddle-shaped depression that crosses the midline of the superior surface of the body of the sphenoid bone, and contains the pituitary gland.

### Coronal section

Prominent structures visible in the coronal section include the ethmoid and sphenoid bones. (See *Skull, coronal section through anterior skull,* page 16.)

# SKULL

## Anterior view

Supraorbital notch
Supraorbital margin
Parietal bone
Temporal bone
**Greater wing of sphenoid bone**
Temporal surface
Orbital surface
Zygomatic arch
Zygomatic bone
Infraorbital foramen
Nasal bone
Inferior nasal concha
Nasal septum
Maxilla
Intermaxillary suture
Frontal bone
Coronal suture
Nasion
Lesser wing of sphenoid bone
Superior orbital fissure
Optic canal
Nasolacrimal canal
Inferior orbital fissure
Zygomatico-maxillary suture
Vomer
Anterior nasal spine
Mandible
Mental foramen

## Posterior view

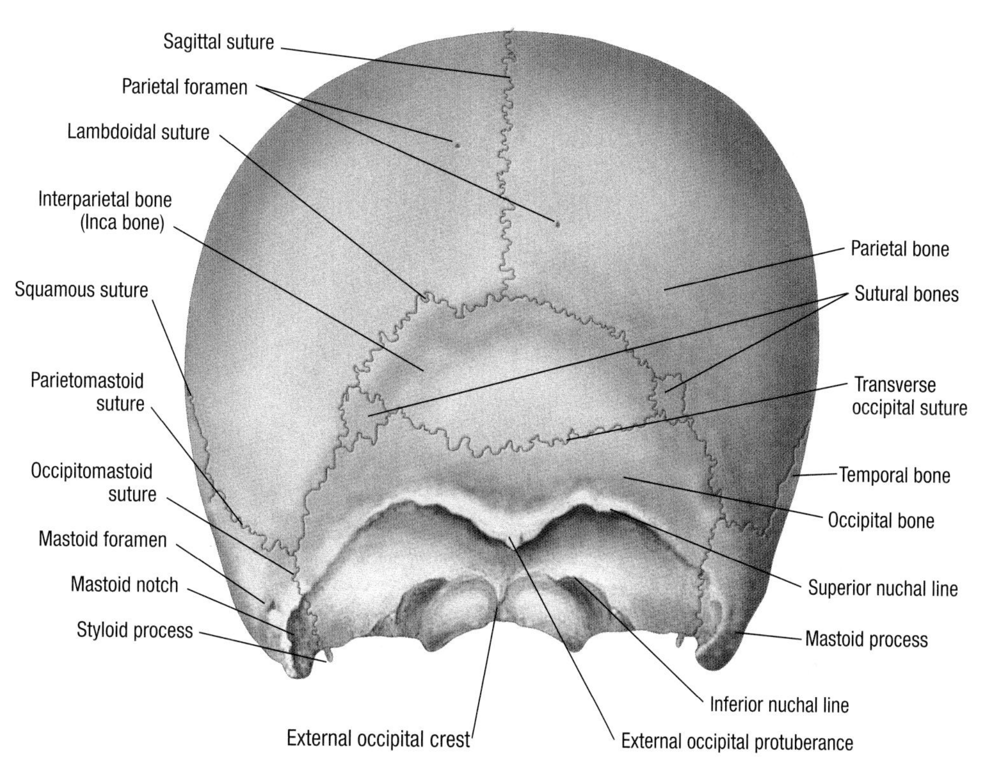
Sagittal suture
Parietal foramen
Lambdoidal suture
Interparietal bone (Inca bone)
Squamous suture
Parietomastoid suture
Occipitomastoid suture
Mastoid foramen
Mastoid notch
Styloid process
External occipital crest
External occipital protuberance
Inferior nuchal line
Mastoid process
Superior nuchal line
Occipital bone
Temporal bone
Transverse occipital suture
Sutural bones
Parietal bone

## ADULT SKULL
**Superior view**

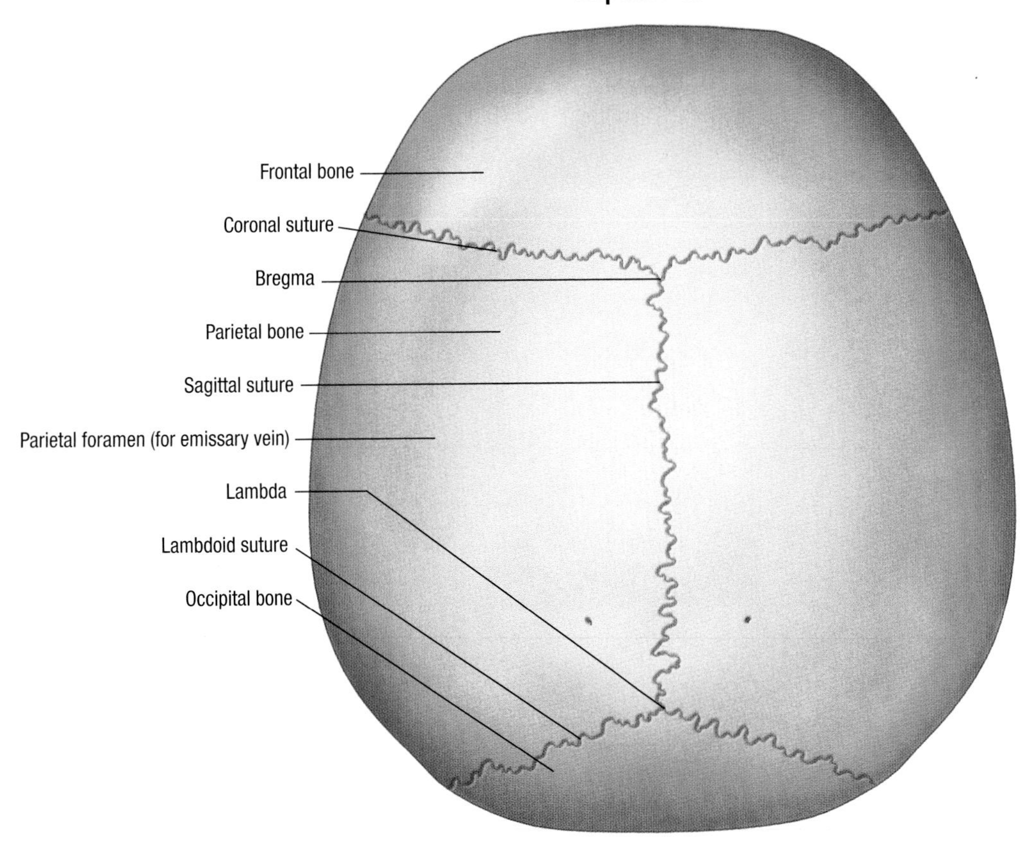

Frontal bone
Coronal suture
Bregma
Parietal bone
Sagittal suture
Parietal foramen (for emissary vein)
Lambda
Lambdoid suture
Occipital bone


## NEONATE SKULL
**Superior view**

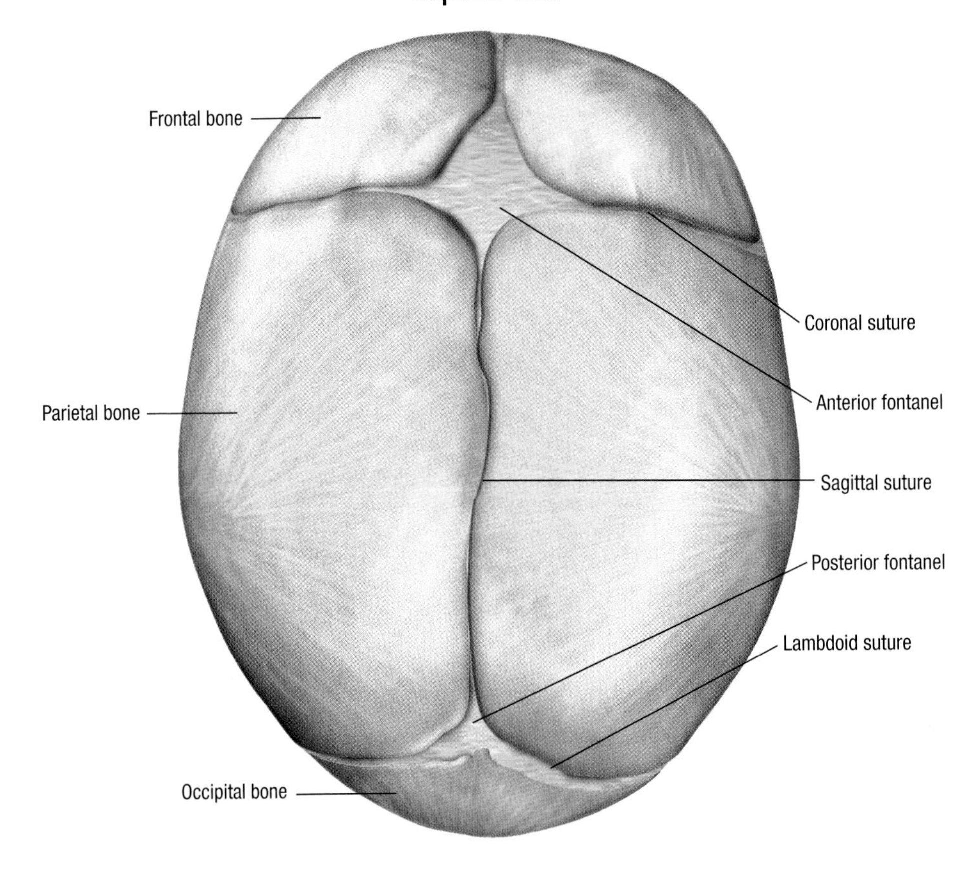

Frontal bone
Parietal bone
Occipital bone
Coronal suture
Anterior fontanel
Sagittal suture
Posterior fontanel
Lambdoid suture

# SKULL

## Lateral view

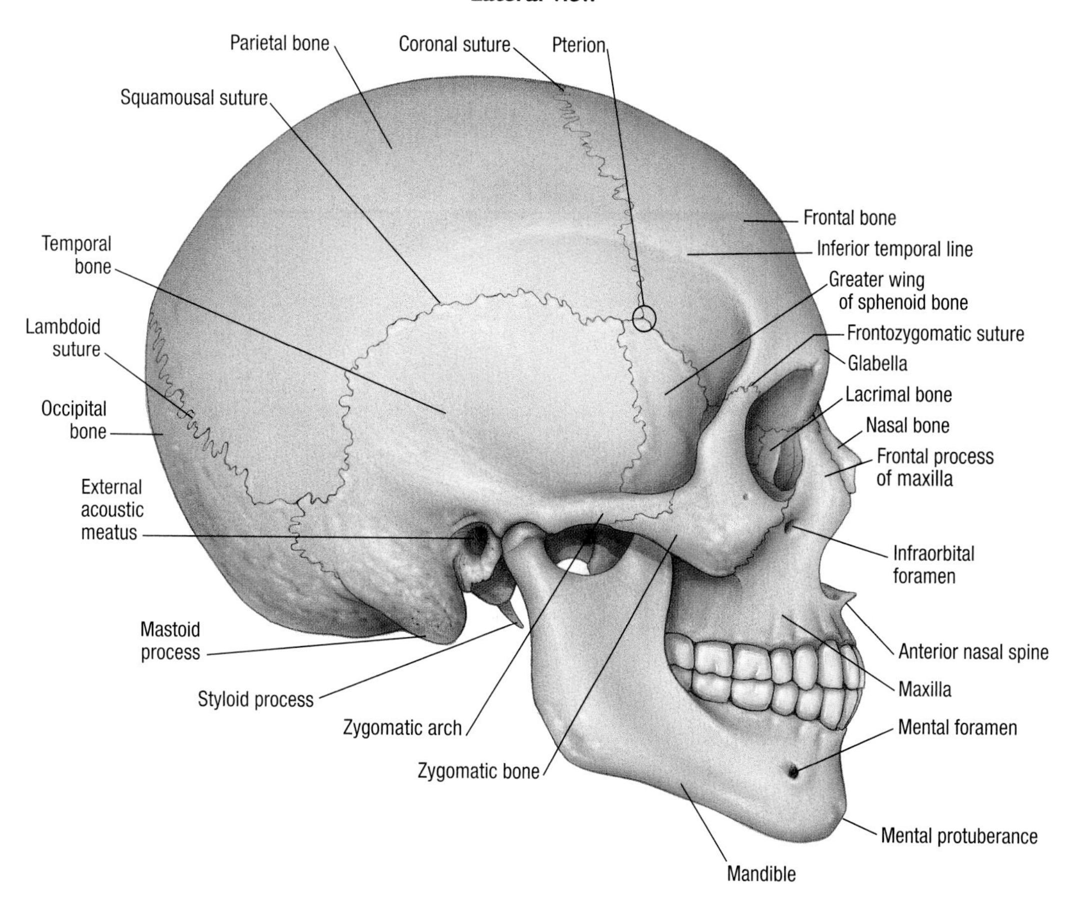

Parietal bone
Coronal suture
Pterion
Squamousal suture
Frontal bone
Inferior temporal line
Temporal bone
Greater wing of sphenoid bone
Lambdoid suture
Frontozygomatic suture
Glabella
Occipital bone
Lacrimal bone
Nasal bone
Frontal process of maxilla
External acoustic meatus
Infraorbital foramen
Mastoid process
Anterior nasal spine
Styloid process
Maxilla
Zygomatic arch
Mental foramen
Zygomatic bone
Mental protuberance
Mandible


## Sagittal section through skull

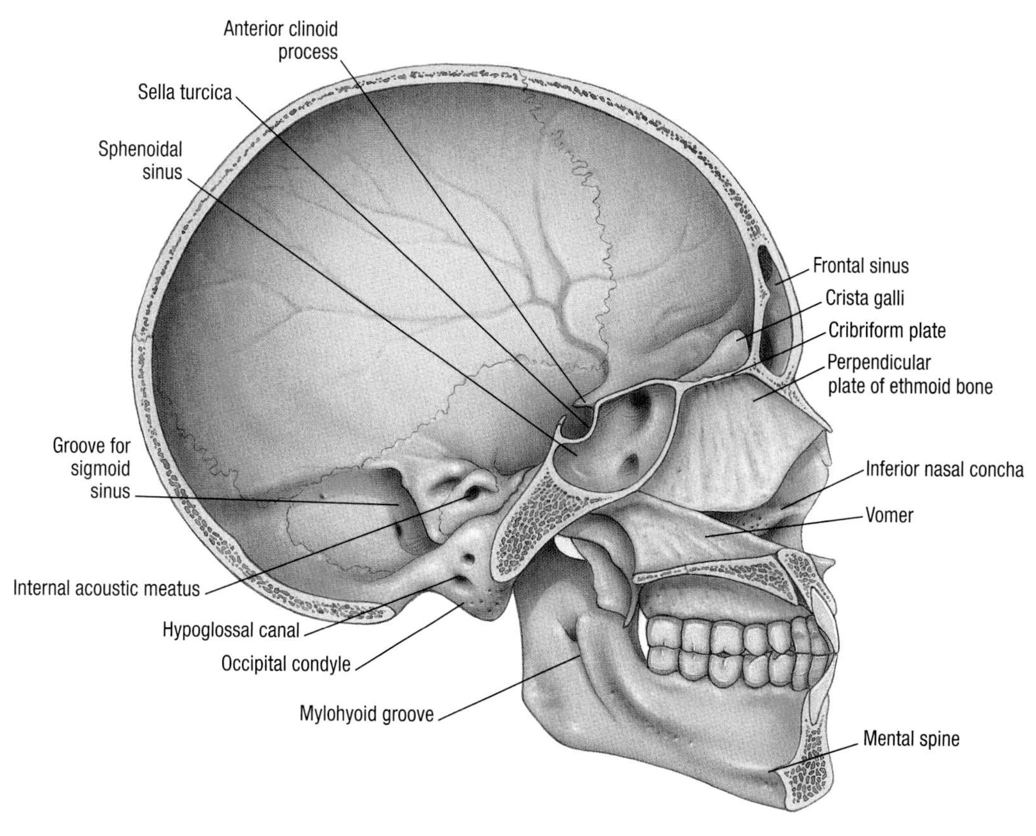

Anterior clinoid process
Sella turcica
Sphenoidal sinus
Frontal sinus
Crista galli
Cribriform plate
Perpendicular plate of ethmoid bone
Groove for sigmoid sinus
Inferior nasal concha
Vomer
Internal acoustic meatus
Hypoglossal canal
Occipital condyle
Mylohyoid groove
Mental spine

## SKULL
### Coronal section through anterior skull

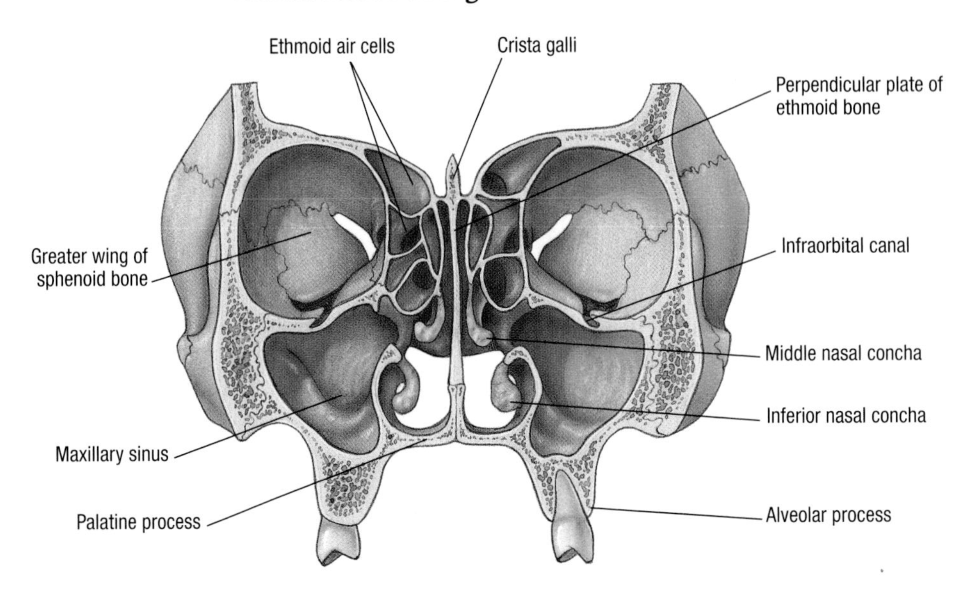
Ethmoid air cells
Crista galli
Perpendicular plate of ethmoid bone
Greater wing of sphenoid bone
Infraorbital canal
Middle nasal concha
Inferior nasal concha
Maxillary sinus
Palatine process
Alveolar process

## ETHMOID BONE
### Anterior and superior views

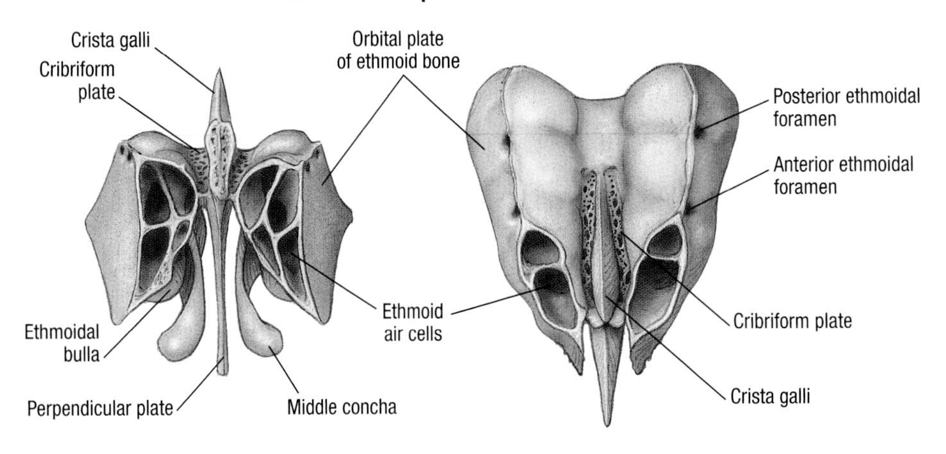
Crista galli
Cribriform plate
Orbital plate of ethmoid bone
Posterior ethmoidal foramen
Anterior ethmoidal foramen
Ethmoid air cells
Cribriform plate
Ethmoidal bulla
Perpendicular plate
Middle concha
Crista galli

## SPHENOID BONE
### Anterior view

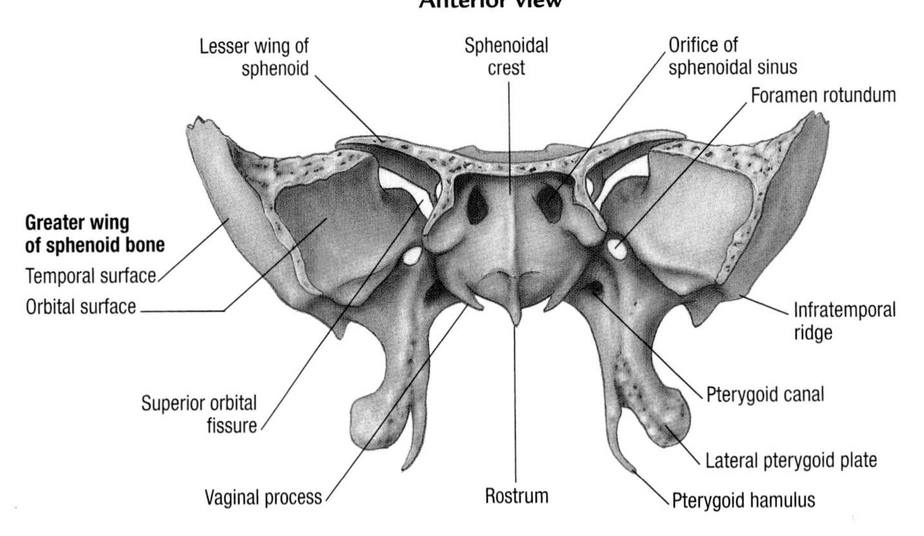
Lesser wing of sphenoid
Sphenoidal crest
Orifice of sphenoidal sinus
Foramen rotundum
Greater wing of sphenoid bone
Temporal surface
Orbital surface
Infratemporal ridge
Pterygoid canal
Superior orbital fissure
Lateral pterygoid plate
Vaginal process
Rostrum
Pterygoid hamulus

## LEFT NASAL CAVITY
### Lateral wall

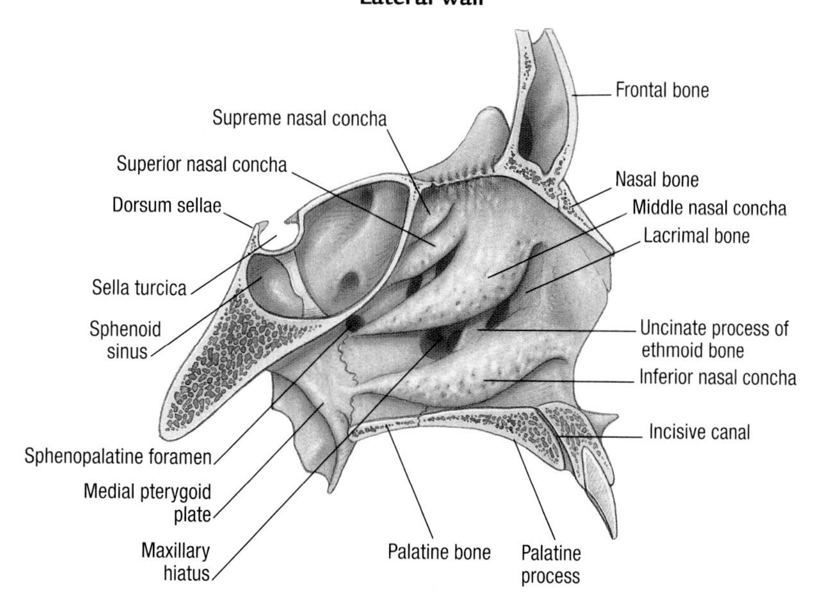
Frontal bone
Supreme nasal concha
Superior nasal concha
Nasal bone
Middle nasal concha
Dorsum sellae
Lacrimal bone
Sella turcica
Sphenoid sinus
Uncinate process of ethmoid bone
Inferior nasal concha
Incisive canal
Sphenopalatine foramen
Medial pterygoid plate
Maxillary hiatus
Palatine bone
Palatine process

## HORIZONTAL SECTION THROUGH MAXILLA
### Superior view

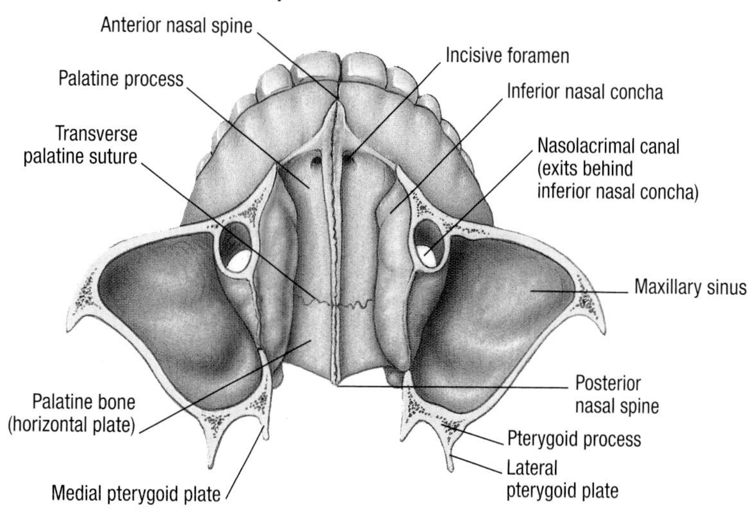
Anterior nasal spine
Incisive foramen
Palatine process
Inferior nasal concha
Transverse palatine suture
Nasolacrimal canal (exits behind inferior nasal concha)
Maxillary sinus
Posterior nasal spine
Palatine bone (horizontal plate)
Pterygoid process
Lateral pterygoid plate
Medial pterygoid plate

## MANDIBLE

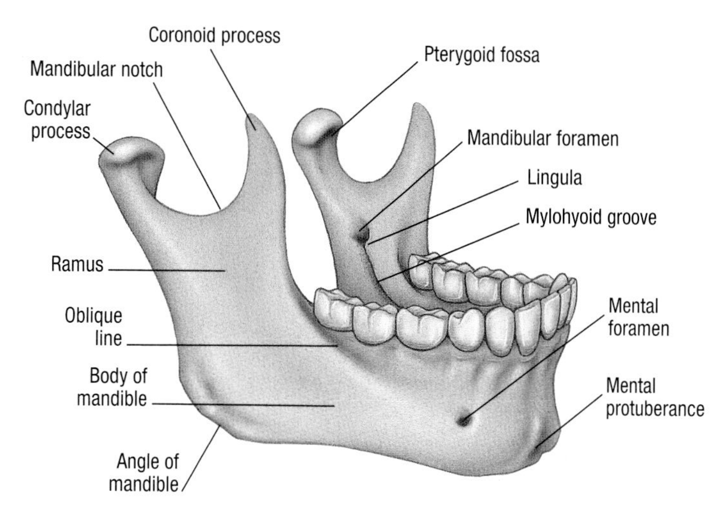
Coronoid process
Pterygoid fossa
Mandibular notch
Condylar process
Mandibular foramen
Lingula
Mylohyoid groove
Ramus
Oblique line
Mental foramen
Body of mandible
Mental protuberance
Angle of mandible

# Base of skull

Several structures and openings that can't be fully viewed from other aspects can be seen from the base of the skull.

## Bones

Bones that are only partly visible in views of other aspects of the human skull can be more clearly seen when the base of the skull from its inner surface are viewed. These include:
- lesser and greater wings of the sphenoid bone in the context of the whole skull
- palatine process of the maxilla and the palatine bone, which form the roof of the mouth and the floor of the nose (hard palate); the palatine bone also is a small part of the orbit
- vomer, which is the bony part of the septum.

## Openings

Openings in the skull permit the passage of blood vessels and nerves. The largest one provides passage for the spinal cord.

### Foramen magnum

Also called the great foramen, the foramen magnum is the passage in the occipital bone that connects the vertebral column and the cranial cavity. The spinal cord passes through the foramen magnum to connect with the medulla oblongata of the brain. On each side are occipital condyles, which articulate with the atlas of the cervical vertebral column.

**Clinical tip**

**Assessing for basilar skull injuries**

A basilar fracture may be undetected on skull X-rays. This can be a fatal injury if untreated, possibly causing injury to nearby cranial nerves and the brain stem as well as to blood vessels and the meninges.

When evaluating your patient for basilar skull injuries or fractures, look for ecchymosis over the mastoid process of the temporal bone — a sign of a basilar skull fracture. This is called Battle's sign. Force great enough to fracture the base of the skull damages supporting tissues of the mastoid area, causing ecchymosis. The sign may also result from seepage of the blood from the fracture site to the mastoid process.

Other signs of a basilar skull fracture include:
- periorbital ecchymosis
- conjunctival hemorrhage
- nystagmus
- ocular deviation
- epistaxis
- ansomia
- bulging tympanic membrane (from cerebrospinal fluid or blood accumulation)
- visible fracture line on the external auditory canal
- facial paralysis
- complaints about tinnitus, hearing difficulty, or vertigo.

### Other openings

Other openings and the structures that pass through them include the following:
- foramen ovale — mandibular nerve
- internal acoustic meatus — facial and vestibulocochlear nerves en route to the inner and middle ears
- hypoglossal canal — hypoglossal nerve
- foramen lacerum — greater petrosal nerve
- optic canals — optic nerves and ophthalmic arteries
- jugular foramen — internal jugular nerve and cranial nerves IX, X, XI
- superior orbital fissure — cranial nerves III, IV, V, VI.

## Cranial fossae

The inner surface of the base of the skull has a series of depressions, called cranial fossae. These depressions, classified by position, are the anterior, middle, and posterior cranial fossae.

### Anterior

The anterior cranial fossa supports the anterior and inferior parts of the frontal lobes. In this fossa, the *crista galli* projects from the *cribriform plate* and is attached to the *falx cerebri*, a sickle-shaped fold of dura mater that extends into and follows along the longitudinal fissure separating the cerebral hemispheres. Olfactory nerve filaments pass through small openings, or perforations, in the cribriform plate. The anterior cranial fossa is composed of three bones:
- frontal bone — orbital section
- ethmoid bone — cribriform plate
- sphenoid bone — lesser wing.

### Middle

The middle cranial fossa supports the temporal lobes of the brain. The hypophyseal fossa and the clinoid process form the sella turcica. The bones that make up the middle cranial fossa are:
- sphenoid bone — greater wings
- temporal bone — lateral (squamous) and posterior (petrous) portions.

### Posterior

The posterior cranial fossa, the deepest of the three, supports the cerebellum, pons, and medulla oblongata. It's composed of:
- occipital bone
- sphenoid bone — greater wing
- temporal bone — petrous and mastoid portions.

## BASE OF SKULL
### Outer surface

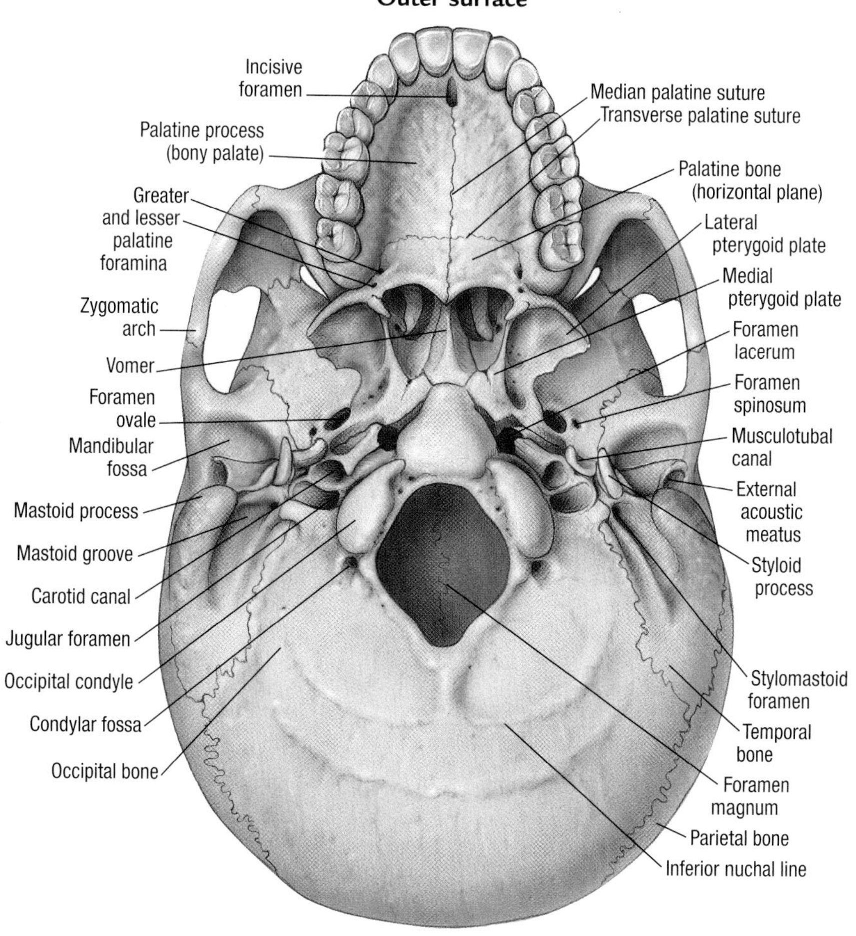

Incisive foramen
Palatine process (bony palate)
Greater and lesser palatine foramina
Zygomatic arch
Vomer
Foramen ovale
Mandibular fossa
Mastoid process
Mastoid groove
Carotid canal
Jugular foramen
Occipital condyle
Condylar fossa
Occipital bone
Median palatine suture
Transverse palatine suture
Palatine bone (horizontal plane)
Lateral pterygoid plate
Medial pterygoid plate
Foramen lacerum
Foramen spinosum
Musculotubal canal
External acoustic meatus
Styloid process
Stylomastoid foramen
Temporal bone
Foramen magnum
Parietal bone
Inferior nuchal line


## BASE OF SKULL
### Inner surface

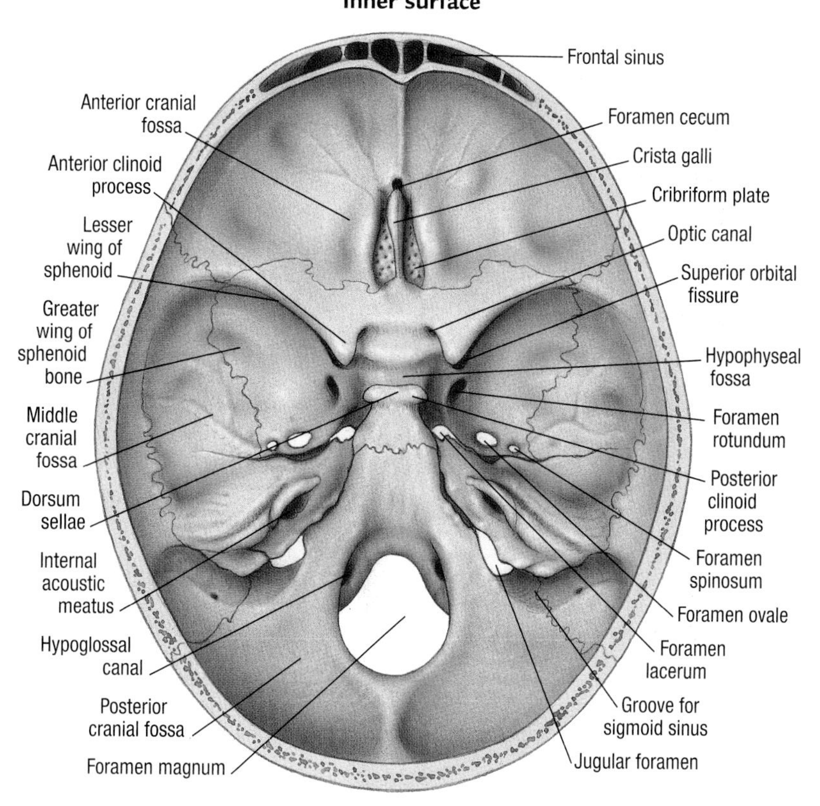

Anterior cranial fossa
Anterior clinoid process
Lesser wing of sphenoid
Greater wing of sphenoid bone
Middle cranial fossa
Dorsum sellae
Internal acoustic meatus
Hypoglossal canal
Posterior cranial fossa
Foramen magnum
Frontal sinus
Foramen cecum
Crista galli
Cribriform plate
Optic canal
Superior orbital fissure
Hypophyseal fossa
Foramen rotundum
Posterior clinoid process
Foramen spinosum
Foramen ovale
Foramen lacerum
Groove for sigmoid sinus
Jugular foramen

# Overview of head and neck structures

The head and neck contain many vital, life-sustaining structures. Many muscles, nerves, and blood vessels connect structures in the head and neck with the thorax and limbs. Each plays an important role in the function of the systems throughout the body.

## Nervous system

The central nervous system — the brain and part of the spinal cord — is in the head and neck. In addition, parts of the peripheral nervous system, such as the cranial nerves and the cervical spinal nerves, originate here.

## Respiratory system

The upper respiratory tract consists primarily of the following structures:

- nose
- mouth
- nasopharynx
- oropharynx
- laryngopharynx
- larynx.

## Sensory organs

Sensory stimulation allows the body to interact with the environment. The brain receives stimulation from the sensory organs as follows:

- eyes — vision
- ears — hearing and balance
- nose — smell
- tongue — taste
- skin — touch, pain.

## Gastrointestinal system

Nutrients enter the body through the mouth, pharynx, and esophagus, the superior structures of the alimentary canal. Muscles assist with mastication of food, and saliva moistens and softens it, preparing the food for the digestive process.

## Endocrine glands

Four endocrine glands are located in the head and neck:

- pituitary gland, at the base of the brain
- pineal gland, at the back of the third ventricle of the brain
- thyroid and parathyroid glands, in the neck.

## Musculoskeletal system

The vertebral column supports the weight of the head and protects the spinal cord from the brain to the sacrum. Functions of muscles include support and movement of the head, changes of facial expression, and mastication. The skull protects and surrounds the brain.

## Cardiovascular system

A network of blood vessels crosses the surfaces and runs through the deep aspects of the head and neck, providing the needed blood supply for daily function. Major vessels include:

- circle of Willis, formed by branches of the anterior, middle, and posterior cerebral arteries
- external and internal carotid arteries
- external and internal jugular veins
- facial artery and vein
- vertebral artery and associated branches
- superior sagittal sinus
- sigmoid sinus
- ophthalmic artery.

## Lymphatic system

In general, lymphatic vessels parallel blood vessels in the head and neck. (See *Head and neck lymphatics*, page 23.) The many lymph nodes of the head and neck include the following:

- preauricular nodes
- occipital nodes
- postauricular nodes
- tonsillar nodes
- superior deep cervical nodes
- submandibular nodes
- submental nodes
- anterior cervical chain
- deep cervical chain
- supraclavicular nodes.

**Clinical tip**

### Palpating lymph nodes

When palpating lymph nodes, expect the following characteristics:

- mobile
- soft
- less than 1 cm (½") in size
- nontender.

## HEAD AND NECK
### Median section

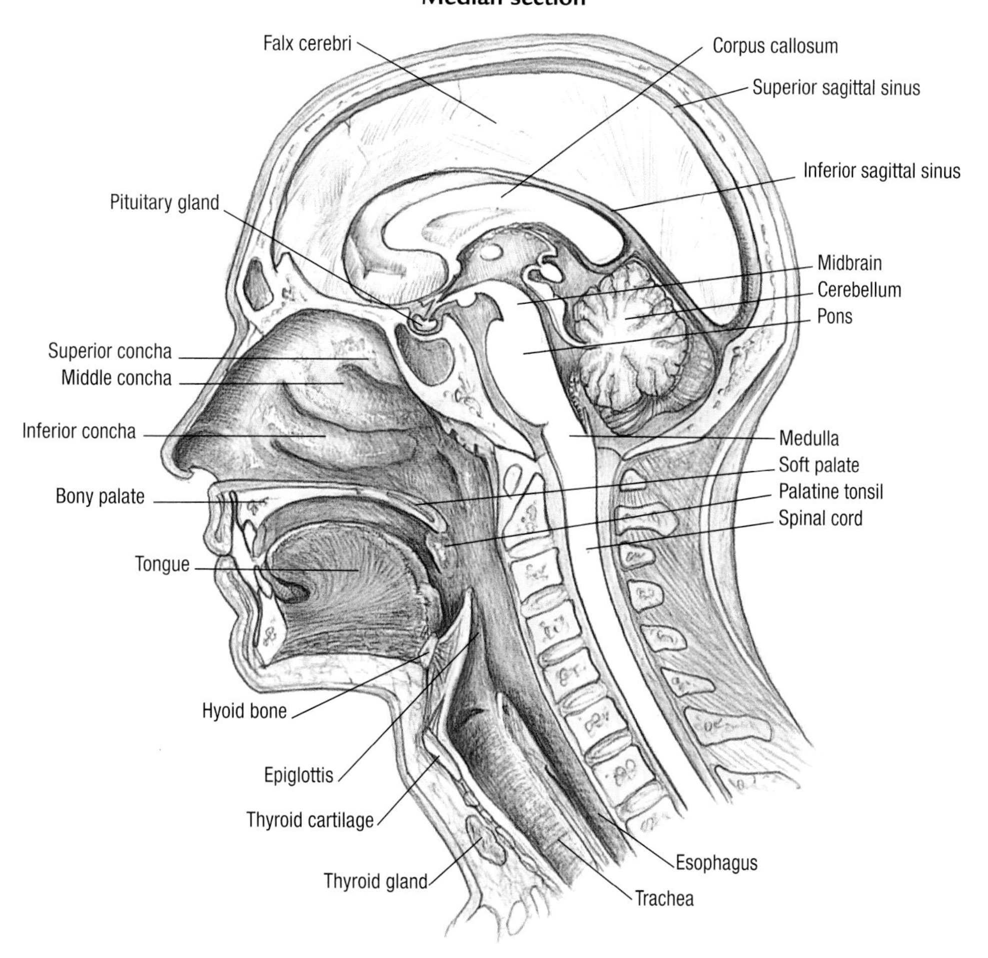
Falx cerebri
Corpus callosum
Superior sagittal sinus
Inferior sagittal sinus
Pituitary gland
Midbrain
Cerebellum
Pons
Superior concha
Middle concha
Inferior concha
Medulla
Soft palate
Palatine tonsil
Spinal cord
Bony palate
Tongue
Hyoid bone
Epiglottis
Thyroid cartilage
Thyroid gland
Esophagus
Trachea

## HEAD
### Horizontal section

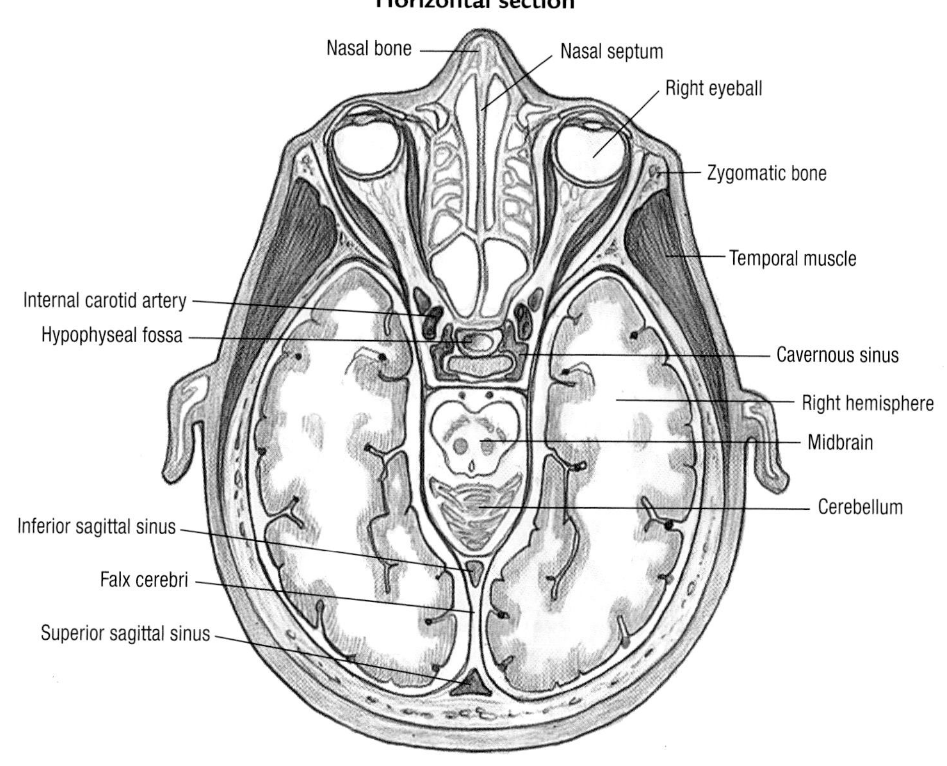
Nasal bone
Nasal septum
Right eyeball
Zygomatic bone
Temporal muscle
Internal carotid artery
Hypophyseal fossa
Cavernous sinus
Right hemisphere
Midbrain
Cerebellum
Inferior sagittal sinus
Falx cerebri
Superior sagittal sinus

## DEEP STRUCTURES OF THE NECK

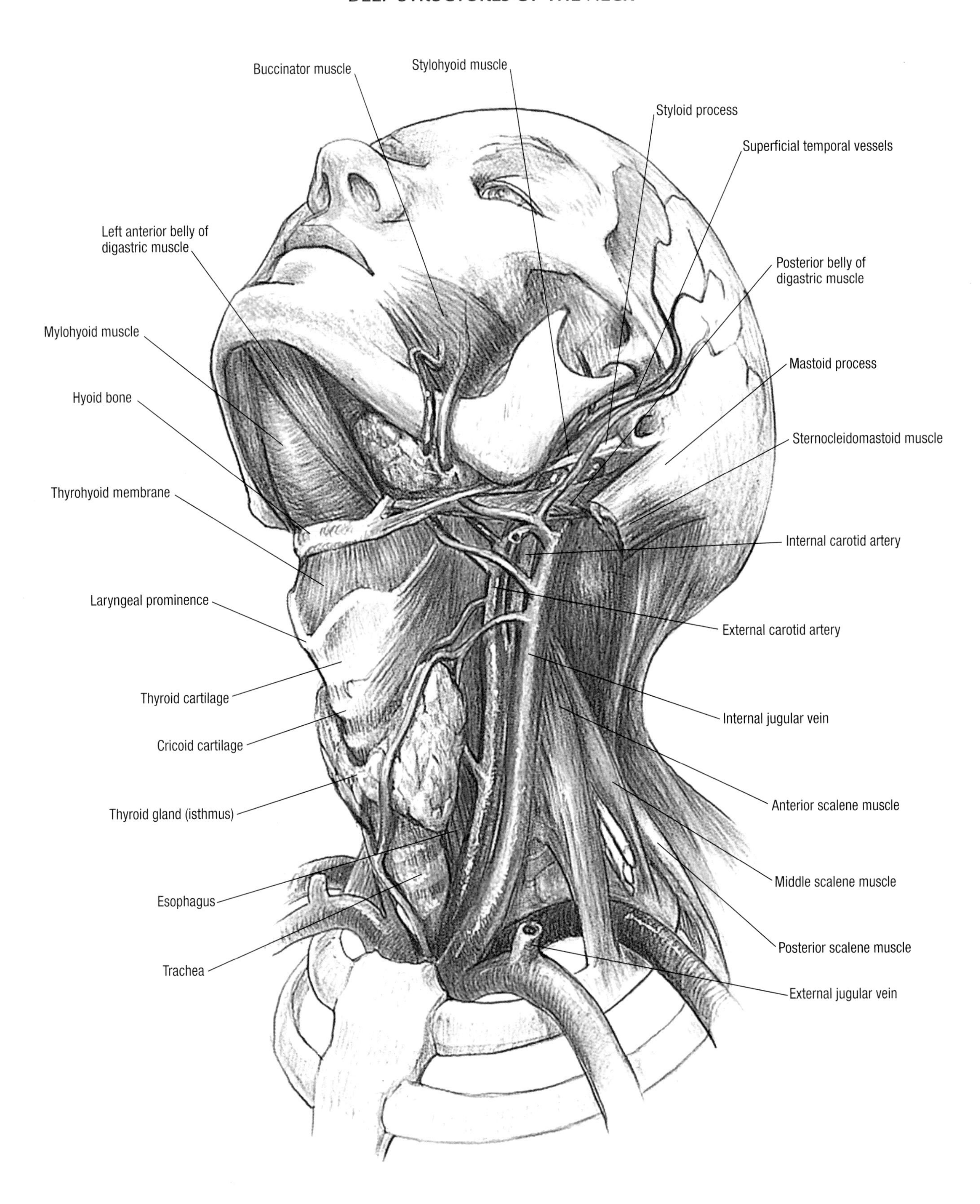

Buccinator muscle
Stylohyoid muscle
Styloid process
Superficial temporal vessels
Left anterior belly of digastric muscle
Posterior belly of digastric muscle
Mylohyoid muscle
Mastoid process
Hyoid bone
Sternocleidomastoid muscle
Thyrohyoid membrane
Internal carotid artery
Laryngeal prominence
External carotid artery
Thyroid cartilage
Internal jugular vein
Cricoid cartilage
Anterior scalene muscle
Thyroid gland (isthmus)
Middle scalene muscle
Esophagus
Posterior scalene muscle
Trachea
External jugular vein

## HEAD AND NECK LYMPHATICS

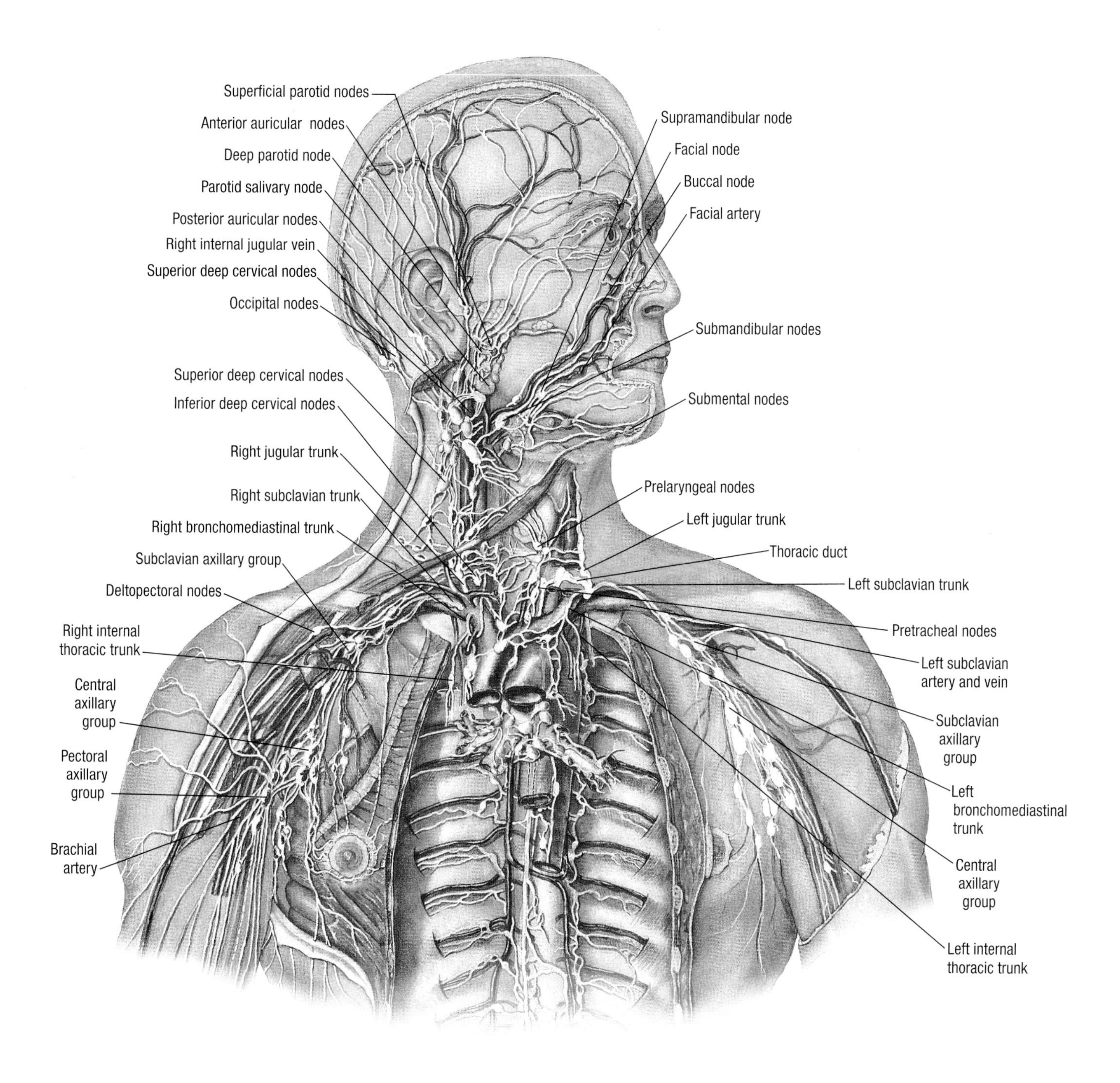

Superficial parotid nodes
Anterior auricular nodes
Deep parotid node
Parotid salivary node
Posterior auricular nodes
Right internal jugular vein
Superior deep cervical nodes
Occipital nodes
Superior deep cervical nodes
Inferior deep cervical nodes
Right jugular trunk
Right subclavian trunk
Right bronchomediastinal trunk
Subclavian axillary group
Deltopectoral nodes
Right internal thoracic trunk
Central axillary group
Pectoral axillary group
Brachial artery
Supramandibular node
Facial node
Buccal node
Facial artery
Submandibular nodes
Submental nodes
Prelaryngeal nodes
Left jugular trunk
Thoracic duct
Left subclavian trunk
Pretracheal nodes
Left subclavian artery and vein
Subclavian axillary group
Left bronchomediastinal trunk
Central axillary group
Left internal thoracic trunk

# Head and neck muscles

Muscles of the head and neck include those of the face, eyeballs, tongue, and neck. They provide facial expression, support mastication, and move the head.

## Facial expression

Many muscles attached to the facial skin work together to provide facial expression. (See *Muscles of facial expression, lateral view*, page 26.) The most significant muscles and their functions are the following:

- buccinator – compresses the cheek and pulls the corner of the mouth
- corrugator supercilii – draws the eyebrows together
- depressor anguli oris – pulls the mouth downward
- depressor labii inferioris – pulls the lower lip downward
- epicranius frontalis – raises the eyebrows and wrinkles the forehead skin
- epicranius occipitalis – draws the scalp backward
- levator labii superioris – raises the upper lip and opens the nostrils
- mentalis – raises the lower lip
- orbicularis oculi – closes the eyelids and tightens the forehead skin
- orbicularis oris – closes the lips
- platysma – depresses the jaw and tightens the skin of the neck
- procerus – wrinkles the skin between the eyebrows
- risorius – pulls the mouth backward
- zygomaticus major – raises the angle of the mouth.

## Mastication

Four muscles take part in mastication. The temporalis and the masseter close the jaws. The medial pterygoids close the jaws and help move them sideways. The lateral pterygoids open the jaws and help move them sideways. (See *Muscles of mastication*, page 27.)

## Neck muscles

Muscles of the neck include the sternocleidomastoid, suprahyoid, and infrahyoid.

- The sternocleidomastoid flexes the vertebral column and rotates the head to the opposite side.

**Clinical tip**

### Checking neck range of motion

Instruct the patient to do the following:

- Attempt to touch the right ear to the right shoulder and the left ear to the left shoulder. Expect a range of motion of 40 degrees on each side.
- Touch the chin to the chest and then point the chin toward the ceiling. The neck should flex 45 degrees and extend backward 55 degrees.
- Turn the head to each side without moving the trunk of the body. The chin should be parallel to the shoulders.
- Move the head in a circle. Expect a normal rotation of 70 degrees.

- Four suprahyoid muscles move the hyoid bone, a single bone in the neck:
  - digastric – raises the hyoid and helps open the jaws
  - stylohyoid – raises the hyoid and pulls it backward
  - mylohyoid – raises the hyoid and floor of the mouth
  - geniohyoid – pulls the hyoid forward.
- Four infrahyoid muscles, sometimes referred to as the strap muscles, support swallowing, speech, and chewing:
  - sternohyoid – pulls the hyoid downward
  - sternothyroid – pulls the larynx downward
  - thyrohyoid – pulls the hyoid downward and raises the larynx
  - omohyoid – pulls the hyoid downward.

Other key muscles of the head and neck include:

- splenius capitis and splenius cervicis – act together to extend the neck and act singly to laterally flex the neck and rotate the head toward the same side
- scalenus muscles – flex and rotate the neck. They also support inhalation because the are attached to the ribs.

## DEEP MUSCLES AND SENSORY NERVES OF THE HEAD

## HEAD AND NECK MUSCLES
**Posterior view**

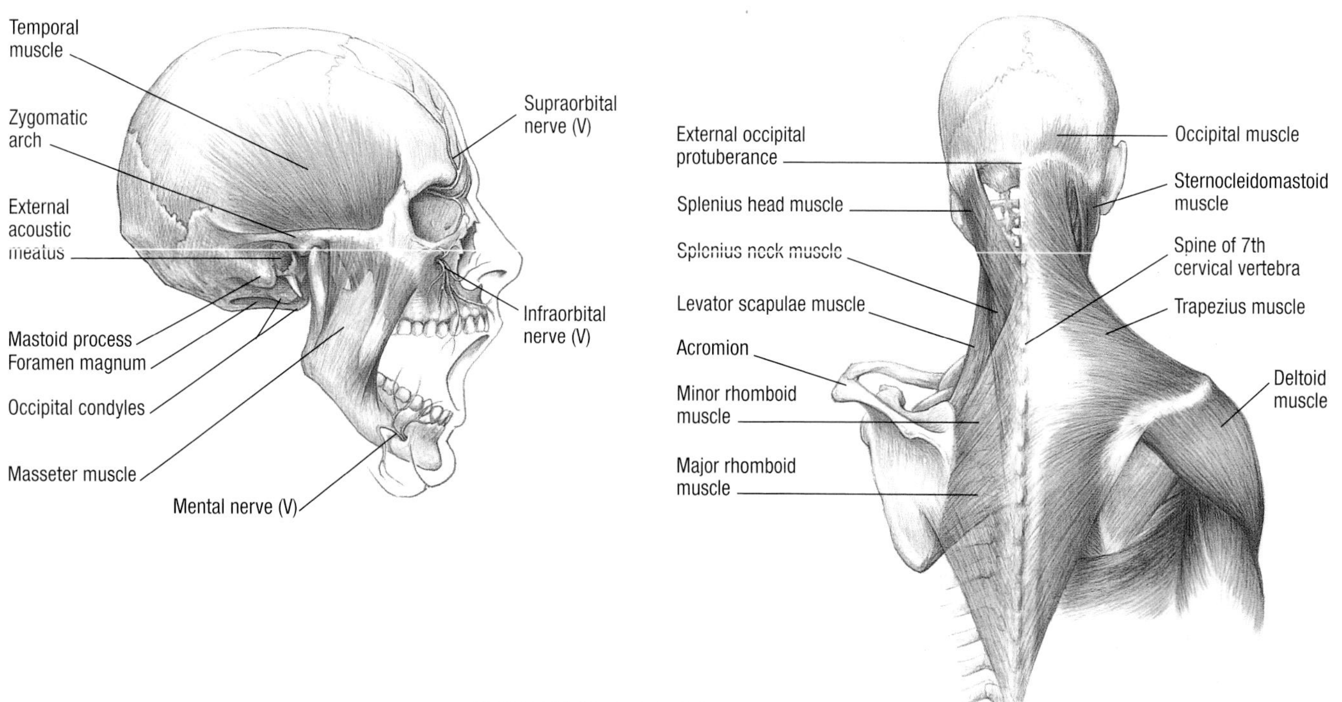
Temporal muscle
Zygomatic arch
External acoustic meatus
Mastoid process
Foramen magnum
Occipital condyles
Masseter muscle
Mental nerve (V)
Supraorbital nerve (V)
Infraorbital nerve (V)
External occipital protuberance
Splenius head muscle
Splenius neck muscle
Levator scapulae muscle
Acromion
Minor rhomboid muscle
Major rhomboid muscle
Occipital muscle
Sternocleidomastoid muscle
Spine of 7th cervical vertebra
Trapezius muscle
Deltoid muscle

## SUPERFICIAL MUSCLES AND NERVES OF THE HEAD AND NECK

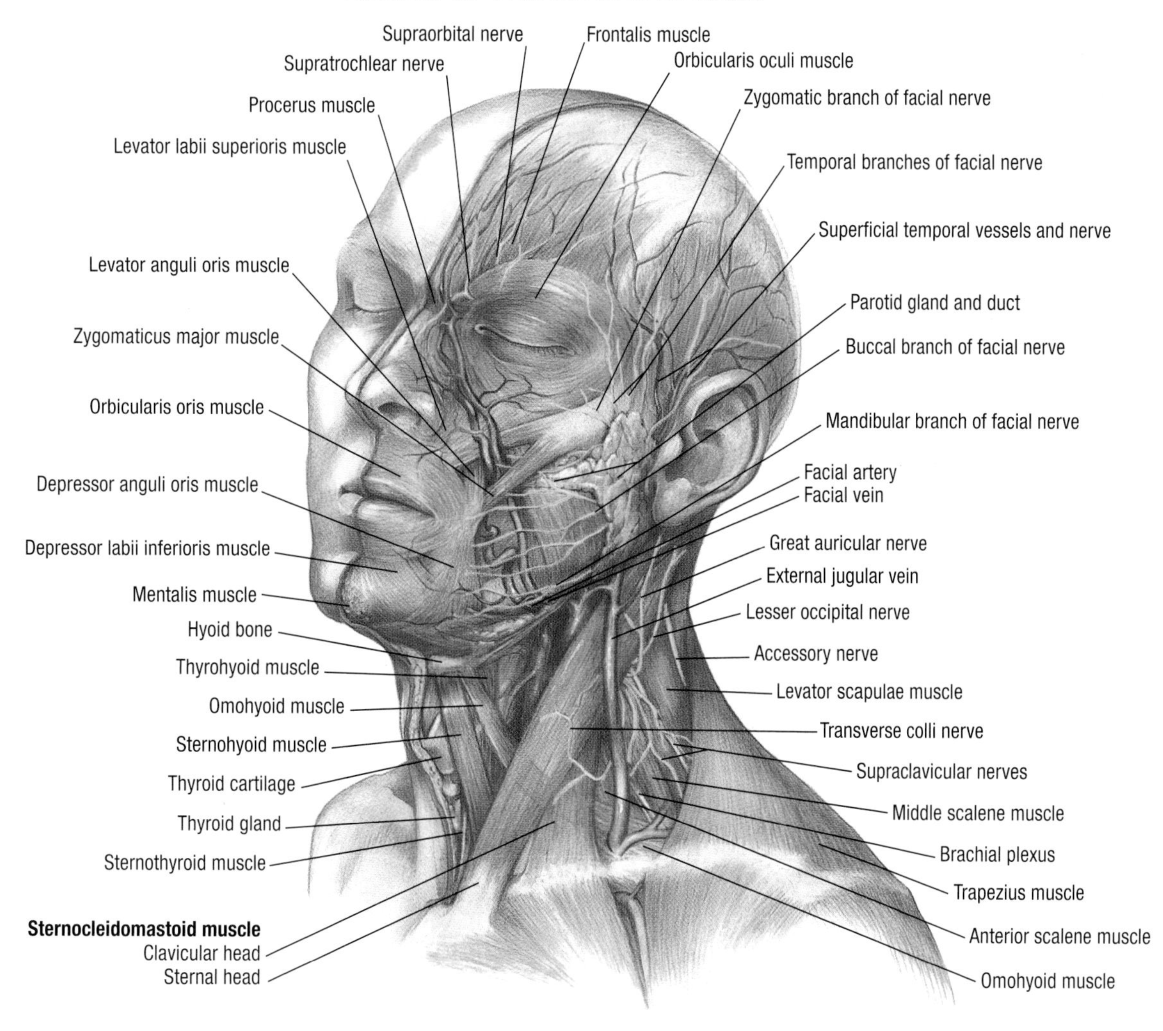
Supraorbital nerve
Supratrochlear nerve
Procerus muscle
Levator labii superioris muscle
Levator anguli oris muscle
Zygomaticus major muscle
Orbicularis oris muscle
Depressor anguli oris muscle
Depressor labii inferioris muscle
Mentalis muscle
Hyoid bone
Thyrohyoid muscle
Omohyoid muscle
Sternohyoid muscle
Thyroid cartilage
Thyroid gland
Sternothyroid muscle
**Sternocleidomastoid muscle**
Clavicular head
Sternal head
Frontalis muscle
Orbicularis oculi muscle
Zygomatic branch of facial nerve
Temporal branches of facial nerve
Superficial temporal vessels and nerve
Parotid gland and duct
Buccal branch of facial nerve
Mandibular branch of facial nerve
Facial artery
Facial vein
Great auricular nerve
External jugular vein
Lesser occipital nerve
Accessory nerve
Levator scapulae muscle
Transverse colli nerve
Supraclavicular nerves
Middle scalene muscle
Brachial plexus
Trapezius muscle
Anterior scalene muscle
Omohyoid muscle

# MUSCLES OF FACIAL EXPRESSION
**Lateral view**

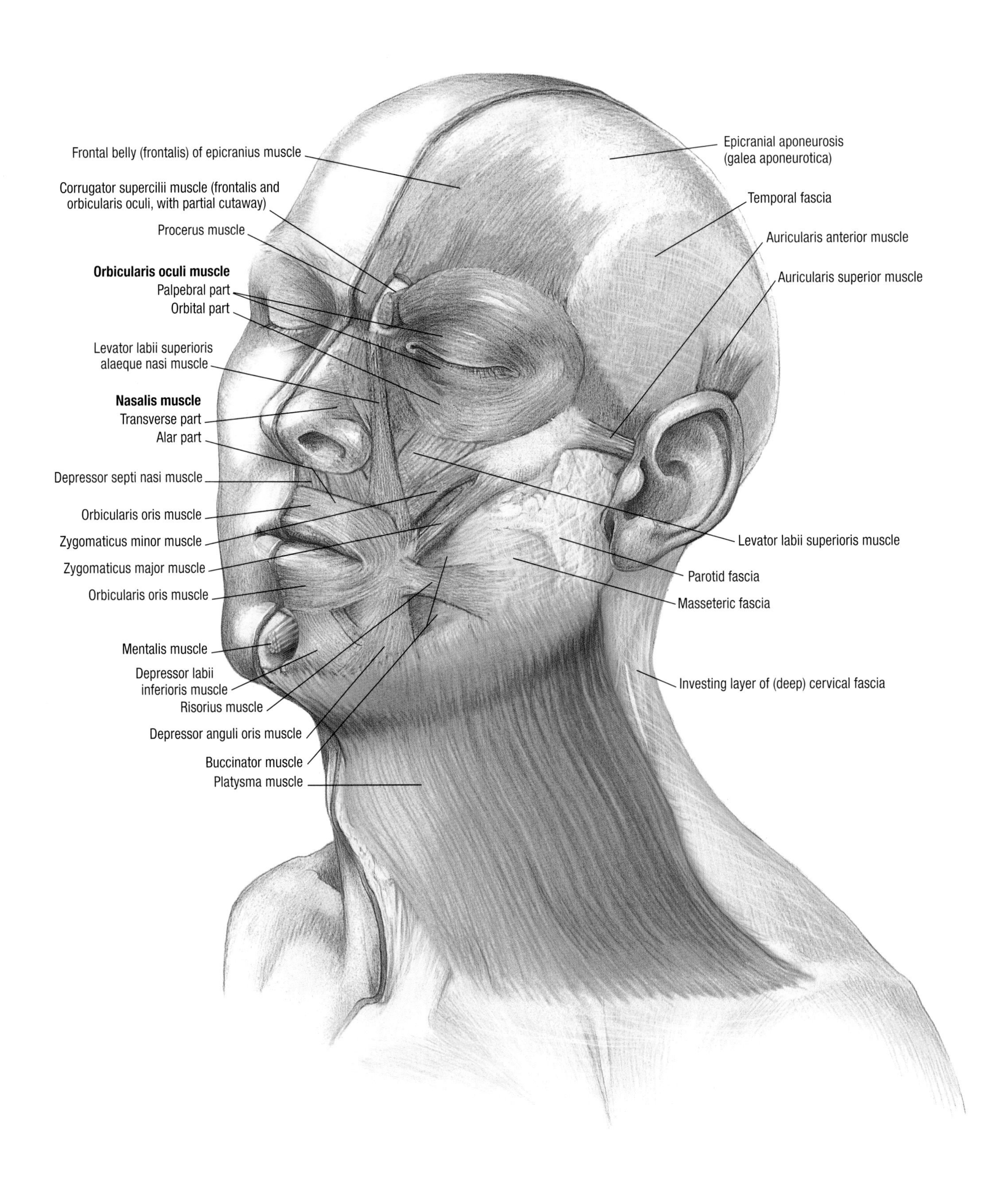
Frontal belly (frontalis) of epicranius muscle
Corrugator supercilii muscle (frontalis and orbicularis oculi, with partial cutaway)
Procerus muscle
Orbicularis oculi muscle
Palpebral part
Orbital part
Levator labii superioris alaeque nasi muscle
Nasalis muscle
Transverse part
Alar part
Depressor septi nasi muscle
Orbicularis oris muscle
Zygomaticus minor muscle
Zygomaticus major muscle
Orbicularis oris muscle
Mentalis muscle
Depressor labii inferioris muscle
Risorius muscle
Depressor anguli oris muscle
Buccinator muscle
Platysma muscle
Epicranial aponeurosis (galea aponeurotica)
Temporal fascia
Auricularis anterior muscle
Auricularis superior muscle
Levator labii superioris muscle
Parotid fascia
Masseteric fascia
Investing layer of (deep) cervical fascia

## Oblique view

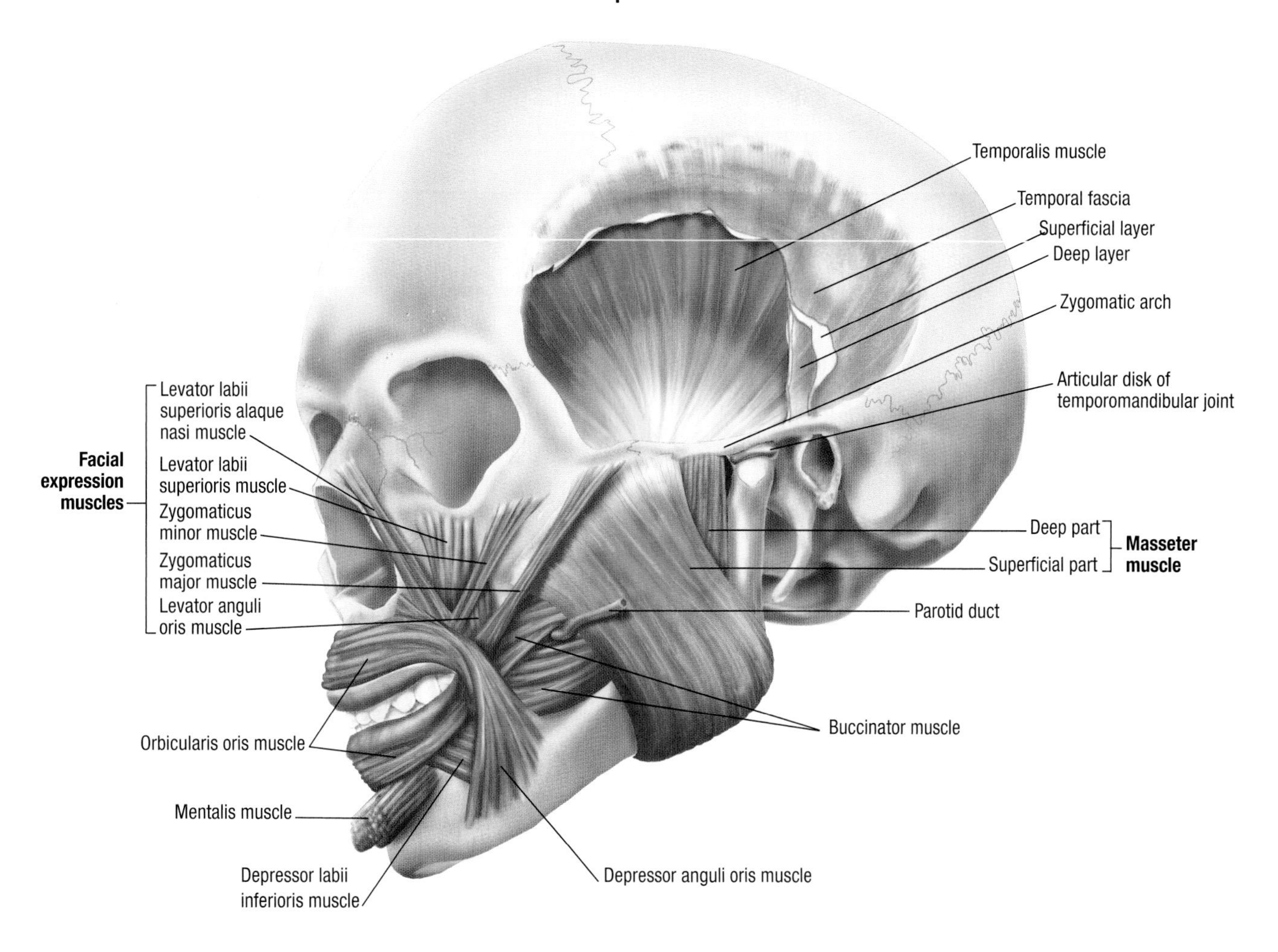
Temporalis muscle
Temporal fascia
Superficial layer
Deep layer
Zygomatic arch
Articular disk of temporomandibular joint
Deep part
Superficial part
Masseter muscle
Parotid duct
Buccinator muscle
Depressor anguli oris muscle
Facial expression muscles
Levator labii superioris alaque nasi muscle
Levator labii superioris muscle
Zygomaticus minor muscle
Zygomaticus major muscle
Levator anguli oris muscle
Orbicularis oris muscle
Mentalis muscle
Depressor labii inferioris muscle

## Posterior view

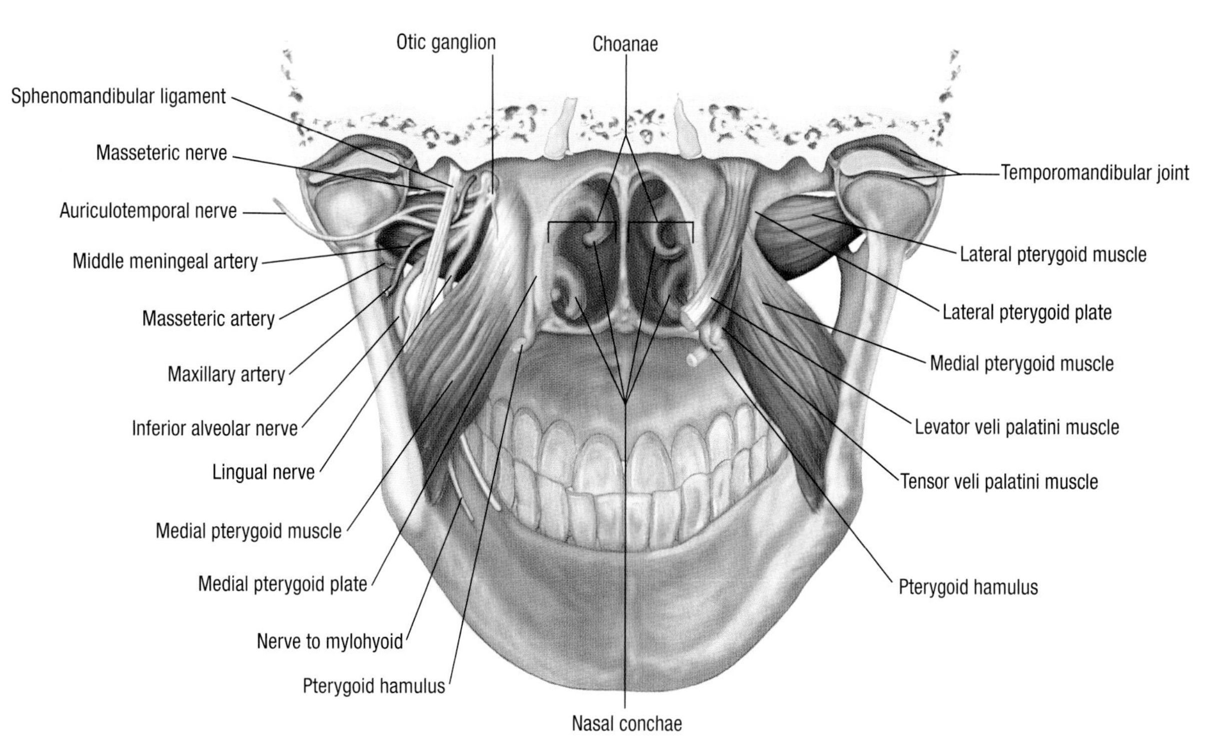
Otic ganglion
Choanae
Sphenomandibular ligament
Masseteric nerve
Auriculotemporal nerve
Middle meningeal artery
Masseteric artery
Maxillary artery
Inferior alveolar nerve
Lingual nerve
Medial pterygoid muscle
Medial pterygoid plate
Nerve to mylohyoid
Pterygoid hamulus
Nasal conchae
Temporomandibular joint
Lateral pterygoid muscle
Lateral pterygoid plate
Medial pterygoid muscle
Levator veli palatini muscle
Tensor veli palatini muscle
Pterygoid hamulus

# Brain

The brain consists of the cerebrum, cerebellum, brain stem, and primitive structures that lie deep in the cerebrum – the diencephalon, basal ganglia, limbic system, and reticular activating system. (See *Brain, coronal section*, page 30.) Brain and spinal cord make up the central nervous system (CNS). The blood-brain barrier separates CNS tissue from the bloodstream; it guards against invasion by disease-causing organisms and excludes most blood-borne chemicals from the brain.

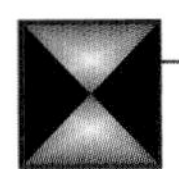

**AGE-RELATED CHANGES**

- After about age 50, the number of neurons decreases at a rate of about 1% a year. Clinical effects usually aren't noticeable until aging is more advanced.
- Nerve transmission typically slows down. An older person may react sluggishly to stimuli.
- Diminished cerebral blood flow increases the risk of stroke.

## CEREBRUM

The cerebrum, the largest region of the brain, occupies the superior portion of the cranial cavity. The cerebrum consists of right and left hemispheres. The right hemisphere controls the left side of the body; the left hemisphere, the right. The corpus callosum is a mass of nerve fibers connecting the hemispheres. The surface of the cerebrum is made up of convolutions (gyri) and creases or fissures (sulci). The median longitudinal fissure separates the hemispheres. The lateral fissure separates the temporal lobe from the rest of the cerebrum. The thin surface layer, the cerebral cortex, consists of gray matter (neuronal cell bodies and dendrites). Within the cerebrum lie white matter (myelinated axons) and islands of internal gray matter.

### Lobes of the brain

Each cerebral hemisphere is divided into four lobes, based on anatomical landmarks and functional differences. The lobes are named for the cranial bones that overlie them.

- frontal – contains the primary motor (movement) area and influences personality, judgment, abstract reasoning, social behavior, and language expression
- temporal – controls hearing, language comprehension, and storage and recall of memories
- parietal – interprets and integrates sensations, including pain, temperature, and touch; interprets size, shape, distance, and texture; important for awareness of body shape
- occipital – functions mainly to interpret visual stimuli.

## CEREBELLUM

The second largest brain region, the cerebellum lies posterior and inferior to the cerebrum. It has two hemispheres, each with an outer cortex of gray matter and an inner core of white matter that contains islands of gray matter (deep cerebellar nuclei). The cerebellum functions to maintain muscle tone, coordinate muscle movement, and control balance.

## BRAIN STEM

The brain stem lies immediately inferior to the cerebrum, just anterior to the cerebellum. It's continuous with the cerebrum superiorly and with the spinal cord inferiorly. (See *Midbrain, medulla oblongata, and cervical spinal cord, posterior view*, page 31.)

Composed of the midbrain, pons, and medulla oblongata, the brain stem relays messages between the parts of the nervous system. It has three main functions:

- It produces the reflex autonomic responses necessary for survival, such as increasing heart rate and stimulating the adrenal medulla to produce epinephrine.
- It provides pathways for nerve fibers between higher and lower neural centers.
- It serves as the origin for 10 of the 12 pairs of cranial nerves.

**PHYSIOLOGY**

A diffuse network of hyperexcitable neurons, the reticular activating system (RAS) fans out from the brain stem through the cerebral cortex. After screening all incoming sensory information, the RAS directs it primarily to the thalamus. RAS activity maintains consciousness.

### Midbrain

The midbrain connects dorsally with the cerebellum. It contains large voluntary motor nerve tracts running between the brain and the spinal cord. The substantia nigra in the midbrain helps to control movement. Lesions of substantia nigra cause Parkinson's disease.

### Pons

The pons connects the cerebellum with the cerebrum and links the midbrain to the medulla oblongata. Besides housing one of the brain's respiratory centers, the pons acts as a pathway for conduction tracts between brain centers and the spinal cord, and serves as the exit point for cranial nerves V, VI, VII, and VIII.

### Medulla oblongata

The most inferior portion of the brain stem, the medulla oblongata is a small, cone-shaped structure that joins the spinal cord at the level of the foramen magnum. The medulla oblongata serves as an autonomic reflex center to maintain homeostasis, regulating respiratory, vasomotor, and cardiac functions. It's the site of origin for cranial nerves IX, X, XI, and XII.

## CEREBRAL HEMISPHERES

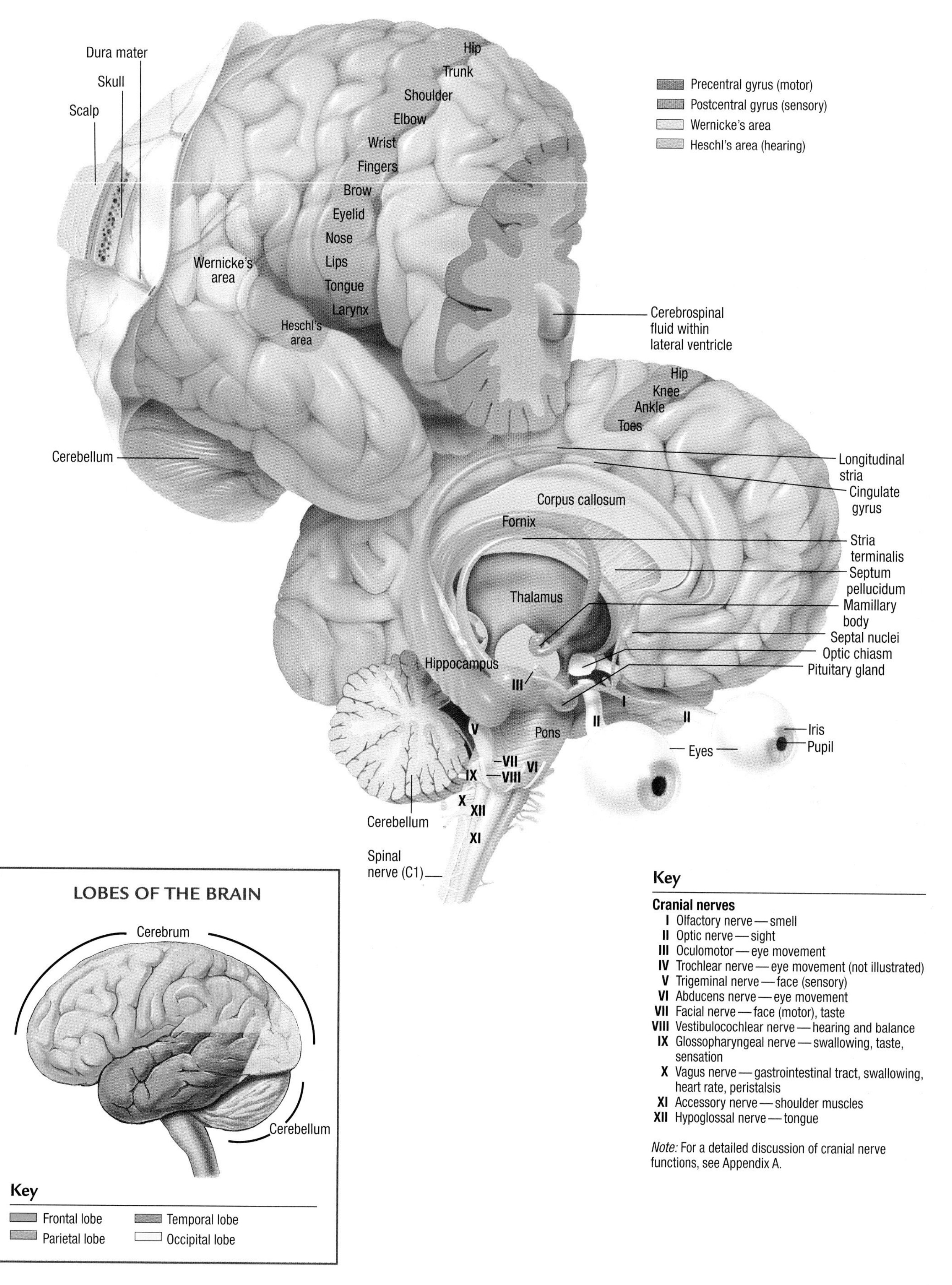
Dura mater
Skull
Scalp
Hip
Trunk
Shoulder
Elbow
Wrist
Fingers
Brow
Eyelid
Nose
Lips
Tongue
Larynx
Wernicke's area
Heschl's area
Precentral gyrus (motor)
Postcentral gyrus (sensory)
Wernicke's area
Heschl's area (hearing)
Cerebrospinal fluid within lateral ventricle
Hip
Knee
Ankle
Toes
Cerebellum
Longitudinal stria
Cingulate gyrus
Corpus callosum
Fornix
Stria terminalis
Septum pellucidum
Thalamus
Mamillary body
Septal nuclei
Optic chiasm
Pituitary gland
Hippocampus
III
I
II
II
V
Pons
Iris
Pupil
Eyes
VII
VI
IX
VIII
X
XII
XI
Cerebellum
Spinal nerve (C1)
LOBES OF THE BRAIN
Cerebrum
Cerebellum
Key
Frontal lobe
Parietal lobe
Temporal lobe
Occipital lobe
Key
Cranial nerves
I Olfactory nerve — smell
II Optic nerve — sight
III Oculomotor — eye movement
IV Trochlear nerve — eye movement (not illustrated)
V Trigeminal nerve — face (sensory)
VI Abducens nerve — eye movement
VII Facial nerve — face (motor), taste
VIII Vestibulocochlear nerve — hearing and balance
IX Glossopharyngeal nerve — swallowing, taste, sensation
X Vagus nerve — gastrointestinal tract, swallowing, heart rate, peristalsis
XI Accessory nerve — shoulder muscles
XII Hypoglossal nerve — tongue
Note: For a detailed discussion of cranial nerve functions, see Appendix A.

## BRAIN
### Coronal section

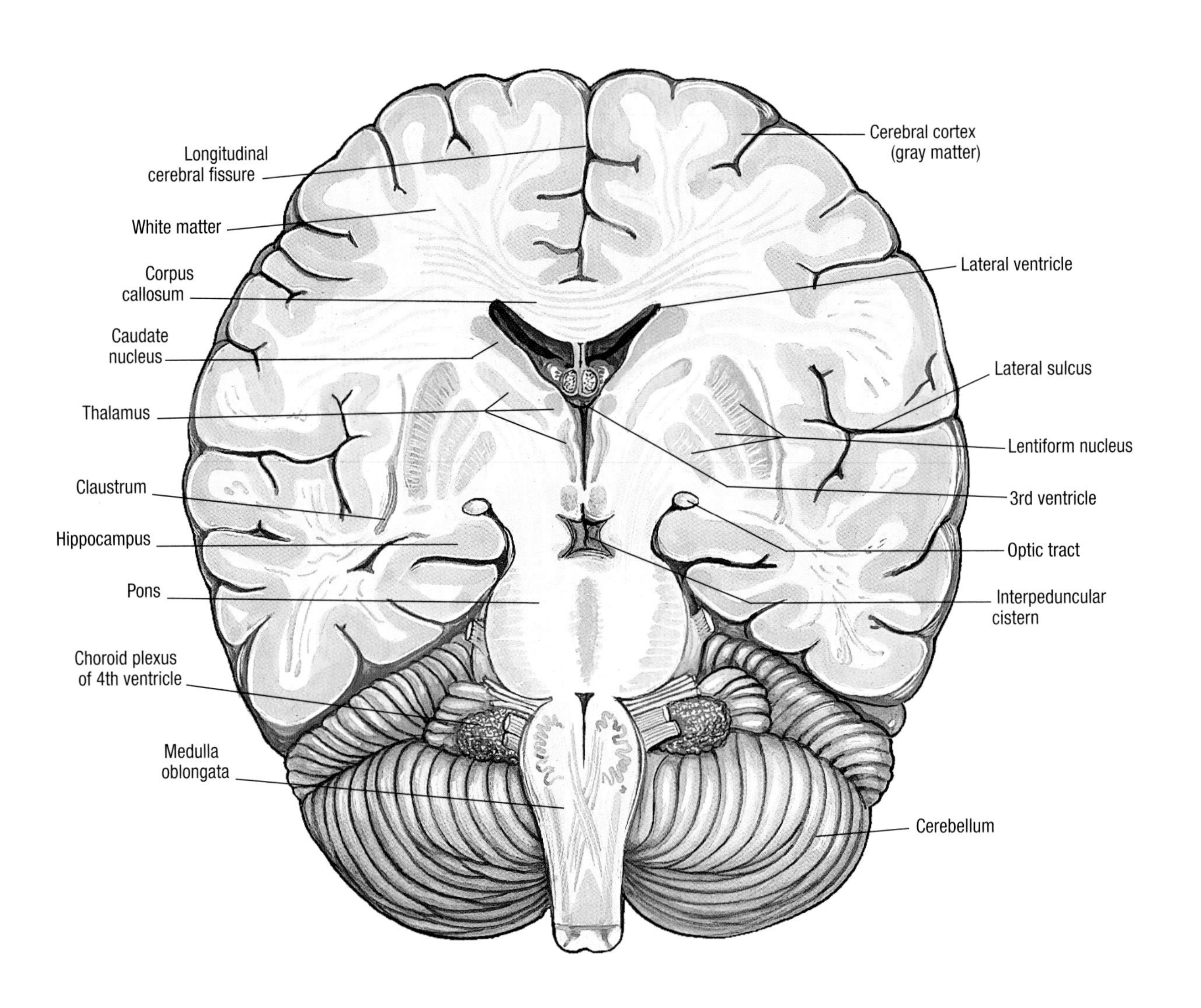

Longitudinal cerebral fissure
White matter
Corpus callosum
Caudate nucleus
Thalamus
Claustrum
Hippocampus
Pons
Choroid plexus of 4th ventricle
Medulla oblongata
Cerebral cortex (gray matter)
Lateral ventricle
Lateral sulcus
Lentiform nucleus
3rd ventricle
Optic tract
Interpeduncular cistern
Cerebellum

## MIDBRAIN, MEDULLA OBLONGATA AND CERVICAL SPINAL CORD
### Posterior view

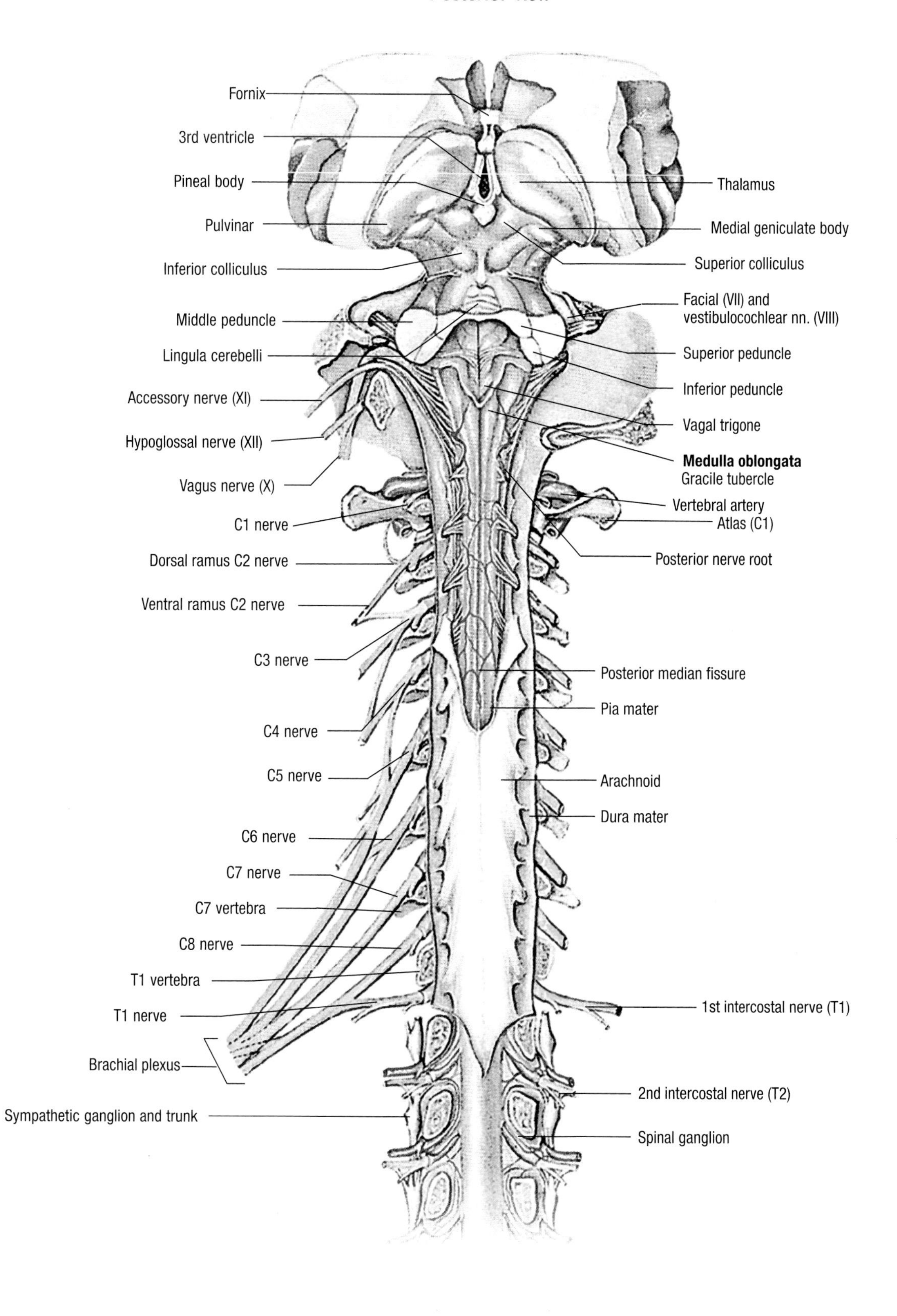
Fornix
3rd ventricle
Pineal body
Pulvinar
Inferior colliculus
Middle peduncle
Lingula cerebelli
Accessory nerve (XI)
Hypoglossal nerve (XII)
Vagus nerve (X)
C1 nerve
Dorsal ramus C2 nerve
Ventral ramus C2 nerve
C3 nerve
C4 nerve
C5 nerve
C6 nerve
C7 nerve
C7 vertebra
C8 nerve
T1 vertebra
T1 nerve
Brachial plexus
Sympathetic ganglion and trunk
Thalamus
Medial geniculate body
Superior colliculus
Facial (VII) and vestibulocochlear nn. (VIII)
Superior peduncle
Inferior peduncle
Vagal trigone
Medulla oblongata
Gracile tubercle
Vertebral artery
Atlas (C1)
Posterior nerve root
Posterior median fissure
Pia mater
Arachnoid
Dura mater
1st intercostal nerve (T1)
2nd intercostal nerve (T2)
Spinal ganglion

# Primitive structures and ventricles

Primitive structures of the brain – the diencephalon and limbic system – occupy its innermost area. The third ventricle (brain cavity) is located within the primitive structures. Three other ventricles are located in other areas of the brain.

## Diencephalon

The diencephalon forms the central core of the cerebral hemispheres and connects the cerebrum with the brain stem. It surrounds the third ventricle. The diencephalon consists of the thalamus, hypothalamus, and epithalamus.

### Thalamus

The thalamus forms the superolateral walls of the third ventricle; it serves as a relay station for sensory information. The thalamus relays all sensory stimuli (except olfactory). Its functions include primitive awareness of pain, screening of incoming stimuli, and focusing of attention.

### Hypothalamus

The hypothalamus extends from the optic chiasm to the posterior margin of the mamillary bodies. The hypothalamus controls body temperature, appetite, water balance, pituitary secretions, reproductive activity, emotions, and autonomic functions, including sleep-wake cycles.

**Age-related changes**

As a person ages, the hypothalamus becomes less effective at regulating body temperature.

### Epithalamus

The epithalamus is located dorsally, forming the floor of the third ventricle. It contains the pineal gland.

## Limbic system

The limbic system includes deep structures in the temporal, frontal, and parietal lobes, and parts of the thalamus and hypothalamus. Besides initiating basic drives – hunger, aggression, and emotional and sexual arousal – the limbic system screens all sensory messages traveling to the cerebral cortex.

## Ventricles

The brain contains four ventricles or small cavities, in which the cerebrospinal fluid is formed.

The right and left lateral ventricles are in the right and left cerebral hemispheres, respectively. They communicate with the narrow third ventricle in the diencephalon through a small opening, the interventricular foramen (foramen of Monro). The third ventricle is continuous with the fourth ventricle via the cerebral aqueduct that traverses the midbrain. The fourth ventricle is located dorsal to the pons and medulla, and ventral to the cerebellum. A single median aperture and a pair of lateral apertures provide continuity between the fourth ventricle and the subarachnoid space.

## Pineal and pituitary glands

### Pineal gland

The tiny pineal gland – only about ¼" (8 mm) in diameter – lies at the back of the third ventricle of the brain. It produces the hormone melatonin, which has widespread effects and may play a role in the neuroendocrine reproductive axis.

### Pituitary gland

Also called the hypophysis, or master gland, the pituitary rests in the sella turcica, a depression in the sphenoidal bone at the base of the brain. This pea-sized gland, which weighs less than 1¼ oz (less than 0.75 g), connects with the hypothalamus via the infundibulum, through which it receives chemical and neural stimuli.

The pituitary has two main regions. The larger region, the anterior pituitary (adenohypophysis), produces at least six hormones:

- growth hormone (GH), or somatotropin
- thyrotropin, or thyroid-stimulating hormone (TSH)
- corticotropin, or adrenocorticotropic hormone (ACTH)
- follicle-stimulating hormone (FSH)
- luteinizing hormone (LH)
- prolactin.

The posterior pituitary, which makes up about 25% of the gland, serves as a storage area for antidiuretic hormone (ADH, vasopressin) and oxytocin, which are produced by the hypothalamus.

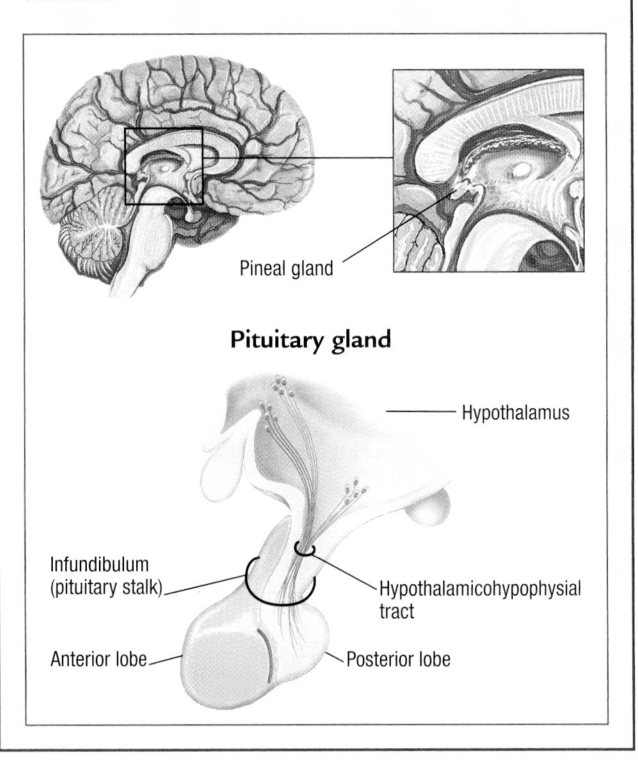
Pineal gland
Pituitary gland
Hypothalamus
Infundibulum (pituitary stalk)
Hypothalamicohypophysial tract
Anterior lobe
Posterior lobe

## LIMBIC SYSTEM

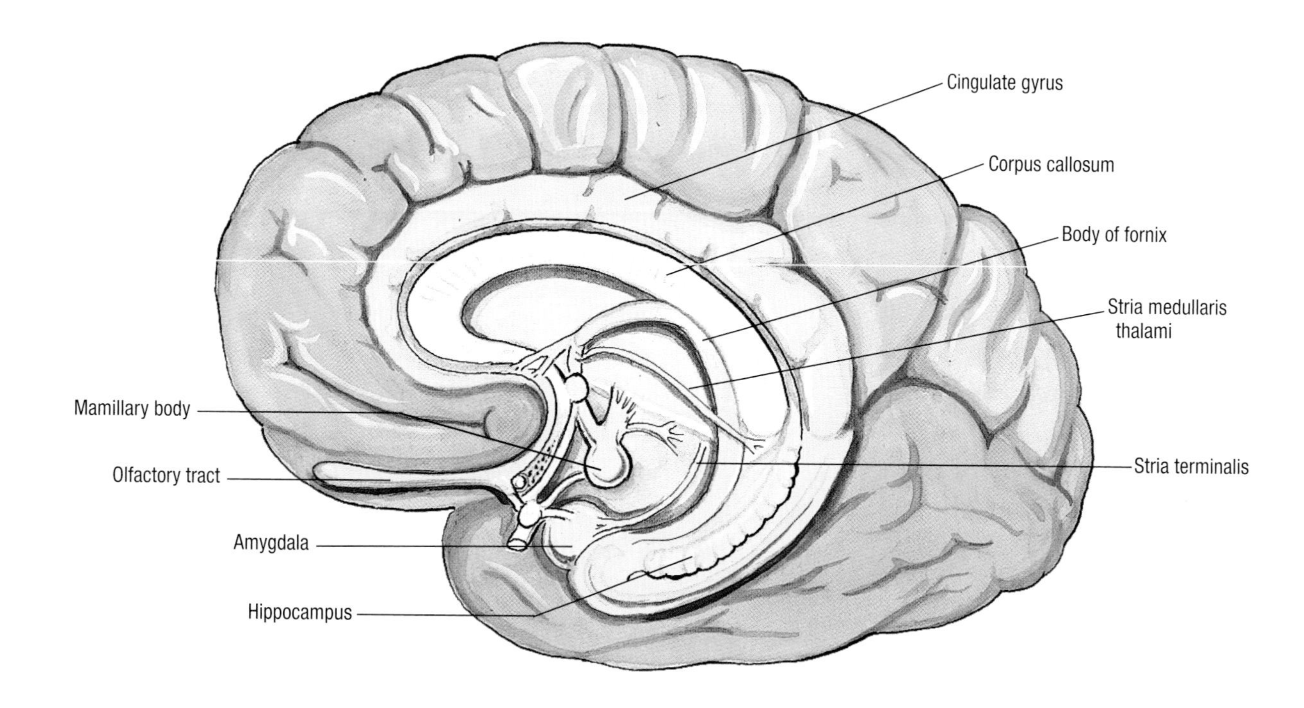
Cingulate gyrus
Corpus callosum
Body of fornix
Stria medullaris thalami
Mamillary body
Olfactory tract
Stria terminalis
Amygdala
Hippocampus

## VENTRICLES

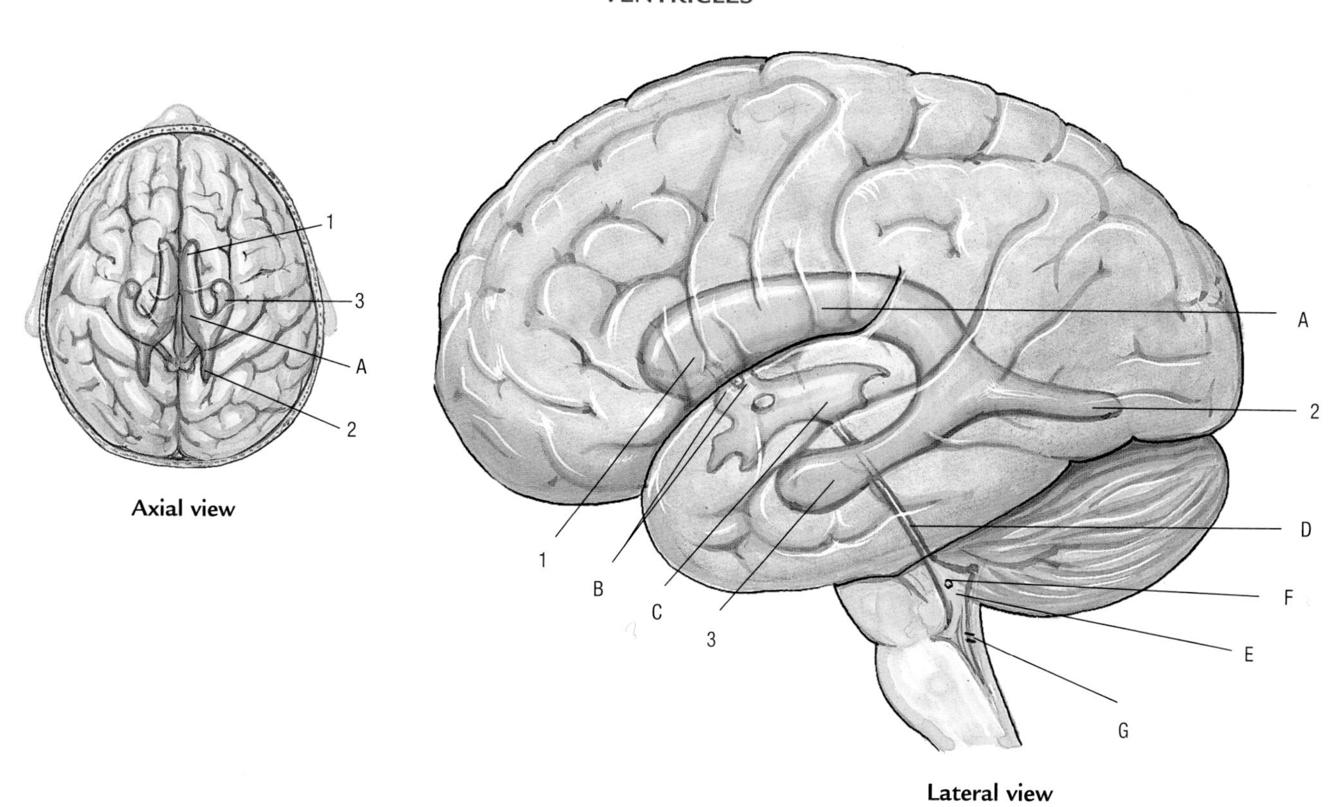
1
3
A
2
Axial view
A
2
D
F
E
G
1
B
C
3
Lateral view

**Key**

| | | |
|---|---|---|
| **A** Lateral ventricle | **B** Interventricular foramen (Monro) | **E** Lateral aperture (Luschka) |
| **1** Anterior horn | **C** 3rd ventricle | **F** 4th ventricle |
| **2** Posterior horn | **D** Cerebral aqueduct | **G** Median aperture (Magendie) |
| **3** Inferior horn | | |

# Protective structures

The brain and spinal cord are protected from trauma and infection by bones, meninges, and cerebrospinal fluid (CSF).

## Bones

The skull, formed of cranial bones, completely surrounds the brain. It opens at the foramen magnum, from which the spinal cord exits.

The vertebral column protects the spinal cord. Each vertebra is separated from its neighbors by intervertebral disks. Articulation of the vertebrae provides flexibility.

## Meninges

The meninges cover and protect the cerebral cortex and spinal column. They consist of three layers of connective tissue: dura mater, arachnoid membrane, and pia mater.

### Dura mater

The dura mater is a fibrous membrane that lines the skull and forms folds, or reflections, that descend into the brain's fissures and provide stability. The dural folds are the following:

- *falx cerebri* – lies in the longitudinal fissure and separates the cerebral hemispheres
- *tentorium cerebelli* – separates the cerebrum and the cerebellum
- *falx cerebelli* – separates the two lobes of the cerebellum.

Cerebrospinal fluid drains into the venous circulation through the arachnoid villi – projections of the arachnoid into the dural venous sinuses.

### Arachnoid membrane

A fragile, fibrous layer of moderate vascularity, the arachnoid membrane lies between the dura mater and the pia mater. Injury to its blood vessels during lumbar or cisternal puncture may cause hemorrhage.

### Pia mater

Extremely thin, the pia mater has a rich blood supply. It closely covers the brain's surface and extends into its fissures.

### Spaces

The subarachnoid space, filled with CSF, separates the arachnoid membrane and the pia mater.

In addition, the meningeal area has two potential spaces:

- epidural space – over the dura mater; becomes a real space in the presence of pathology, such as accumulation of blood from a torn meningeal artery (an epidural hematoma)
- subdural space – a closed space with no egress between the dura mater and the arachnoid membrane: often the site of hemorrhage after head trauma.

**Clinical tip**

#### Two key signs of meningitis

When evaluating your patient for signs of inflammation of the meninges, be sure to include these two diagnostic tests.

- Brudzinki's sign — While the patient is in a recumbent position, put your hands behind the neck. Attempt to bring the head forward to the chest. If meningitis is present, there will be pain and resistance to the movement, and the hips and knees will flex.
- Kernig's sign — With the patient in a recumbent position, flex the leg at the hip to a 90-degree angle, and then straighten the knee. If meningitis is present, the patient will resist and experience pain caused by inflammation of the meninges and spinal roots.

Because early recognition and treatment are so important, be sure you are familiar with the early warning signs of meningitis:

- sudden restlessness
- subtle changes in behavior, speech, or orientation
- deterioration in Glasgow Coma Scale score
- fever above 101.5°F (38.6°C)
- sluggish pupil reaction
- unilateral hippus (abnormally exaggerated rhythmic contraction and dilation of pupil)
- seizures
- increased resistance to passive range of movement
- inability to stretch out arms, or arms that tremble when outstretched
- elevated blood pressure
- bradycardia
- widened pulse pressure
- decreased sensory function
- vomiting
- loss of mental function
- severe headache
- stiff neck.

## MENINGES OF THE BRAIN

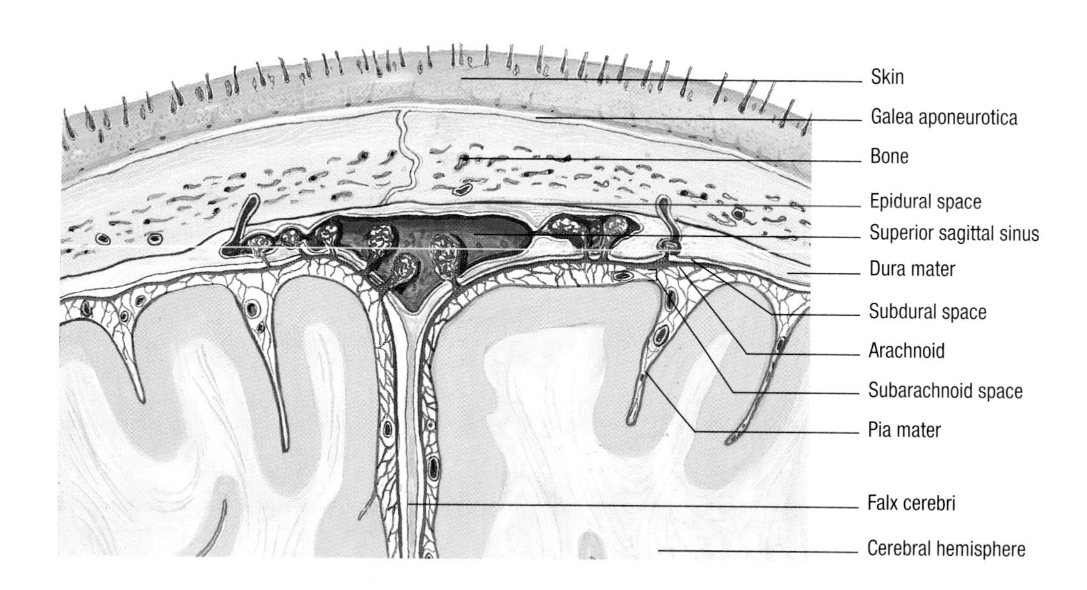

Skin
Galea aponeurotica
Bone
Epidural space
Superior sagittal sinus
Dura mater
Subdural space
Arachnoid
Subarachnoid space
Pia mater
Falx cerebri
Cerebral hemisphere


## MENINGES AND VENOUS SINUSES

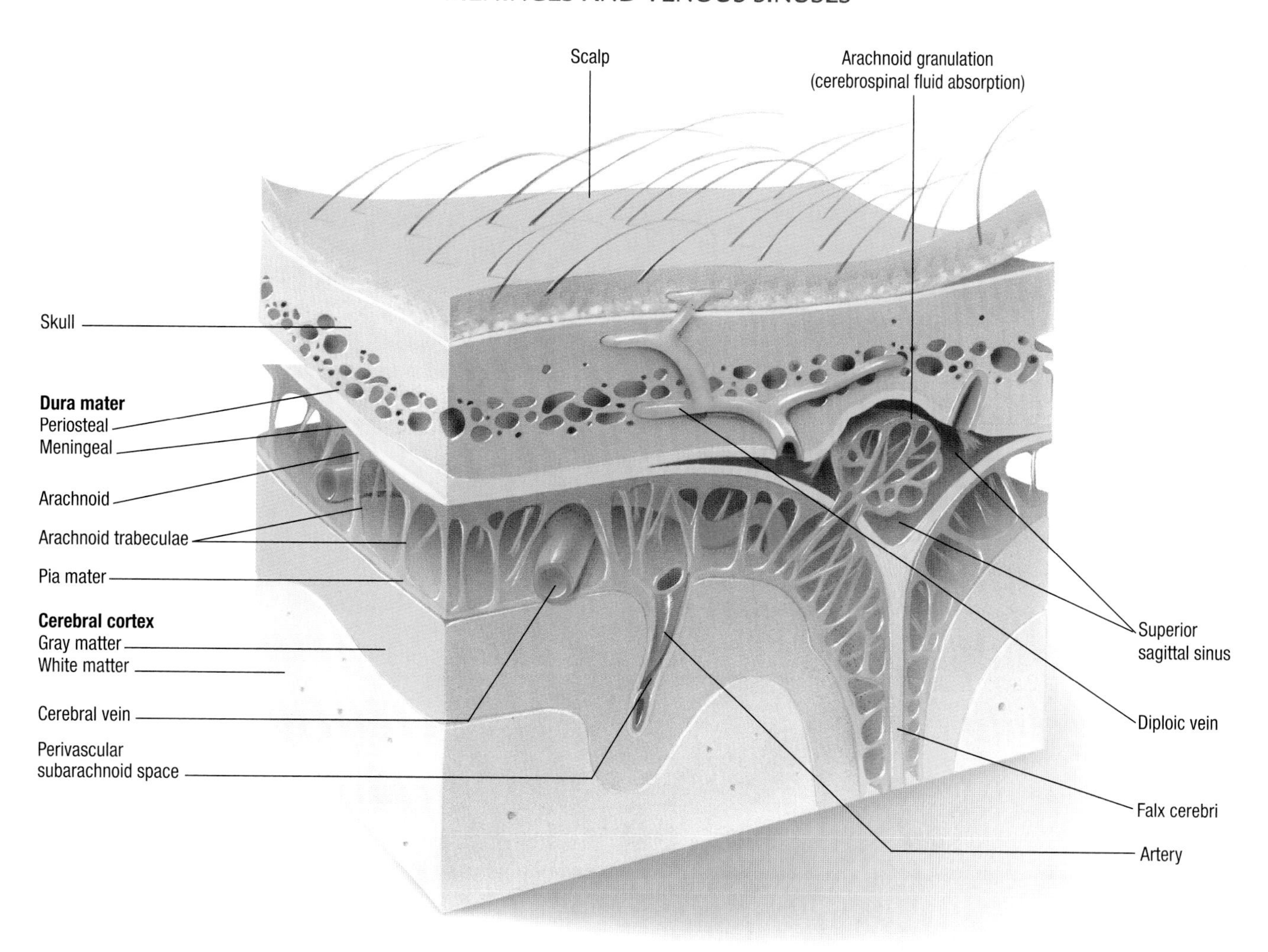

Scalp
Arachnoid granulation (cerebrospinal fluid absorption)
Skull
**Dura mater**
Periosteal
Meningeal
Arachnoid
Arachnoid trabeculae
Pia mater
**Cerebral cortex**
Gray matter
White matter
Cerebral vein
Perivascular subarachnoid space
Superior sagittal sinus
Diploic vein
Falx cerebri
Artery

# Cerebrospinal fluid

Cerebrospinal fluid (CSF) is a colorless, thin fluid found in the ventricles of the brain, the subarachnoid space, and the central canal of the spinal cord. CSF is produced from blood plasma. CSF cushions the brain and spinal cord, nourishes cells, and transports metabolic waste.

## Formation

CSF forms continuously in clusters of capillaries, called choroid plexuses, in the roof of each ventricle. The choroid plexuses produce approximately 500 ml (17 oz)/day of CSF. CSF flows from the choroid plexuses into the ventricles of the brain, each of which adds small quantities of CSF. (For additional information on the ventricles of the brain, see page 32.)

## Pathway

From the lateral ventricles, CSF flows through the interventricular foramen (foramen of Monro) to the third ventricle of the brain. It passes from the third to the fourth ventricle through the cerebral aqueduct, which traverses the midbrain. From the fourth ventricle, it reaches the subarachnoid space by passing through the medial and lateral apertures. CSF then passes under the base of the brain, upward over the brain's upper surfaces, and down around the spinal cord. Eventually, it reaches the arachnoid villi, where it's reabsorbed into venous blood at the venous sinuses. Normally, the rates of CSF production and absorption are balanced so the CSF volume remains constant and the CSF pressure remains normal. Blockage of any part of the pathway – by congenital defects, brain tumors, or head injury, for example – can disrupt CSF flow and raise intracranial pressure.

### Drainage

External drainage is done to reduce CSF pressure to a desired level, and then maintain it at that level. Fluid can be withdrawn from the lateral ventricle (ventriculostomy) or lumbar subarachnoid space, depending on the indication and the desired outcome. Ventricular drainage is used to reduce intracranial pressure (ICP), whereas lumbar drainage is used to aid healing of the dura mater. External CSF drainage is used most commonly to reduce ICP and to facilitate spinal or cerebral dural healing after traumatic injury or surgery. In either case, CSF is drained by a catheter or ventriculostomy tube into a sterile, closed collection system.

**Clinical tip**

### Spotting a CSF leak

Some skull fractures can cause a CSF leak. Often such leakage is manifested by a clear fluid draining from the nares or from the ear. A glucose test differentiates CSF from other bodily fluids in nasal drainage. If glucose isn't present, the fluid isn't CSF; if the fluid tests positive for glucose, be highly suspicious that it is CSF. Beware! A positive result will also occur if blood is mixed with another body fluid; the test may not be able to differentiate between CSF and the other fluid.

Another CSF indicator is the halo or double-ring sign. When a patient with CSF leakage rests his head on a paper- or cloth-covered pillow, the blood accumulates at the center and a halo of clear CSF surrounds it.

## Characteristics

CSF has a similar composition to blood plasma. It contains white blood cells, proteins, glucose, urea, and salts. However, CSF contains less protein than plasma, fewer calcium and potassium ions, and more sodium, magnesium, chloride, and hydrogen ions. About 125 ml or 4 oz of CSF circulates through the entire central nervous system.

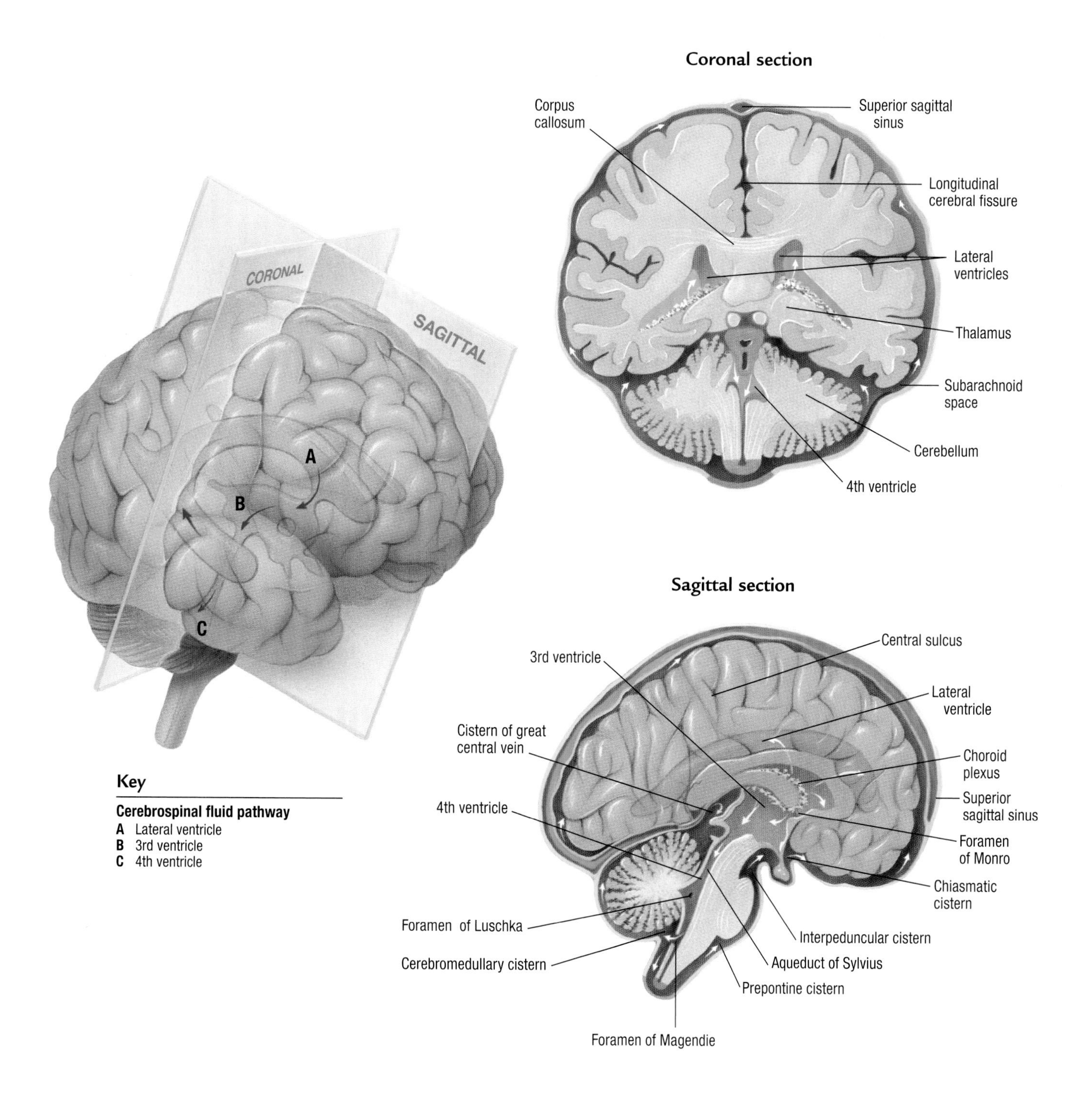
Coronal section
Corpus callosum
Superior sagittal sinus
Longitudinal cerebral fissure
Lateral ventricles
Thalamus
Subarachnoid space
Cerebellum
4th ventricle
CORONAL
SAGITTAL
A
B
C
Sagittal section
Central sulcus
3rd ventricle
Lateral ventricle
Cistern of great central vein
Choroid plexus
4th ventricle
Superior sagittal sinus
Foramen of Monro
Chiasmatic cistern
Foramen of Luschka
Interpeduncular cistern
Cerebromedullary cistern
Aqueduct of Sylvius
Prepontine cistern
Foramen of Magendie

## Key

**Cerebrospinal fluid pathway**

**A** Lateral ventricle
**B** 3rd ventricle
**C** 4th ventricle

# Head and neck blood supply

The vertebral and carotid arteries supply oxygenated blood to the neck, head, and brain. The external and internal jugular veins and the venous sinuses provide venous drainage.

## Arteries

Two vertebral and two carotid arteries and their branches supply the head, neck, and brain.

### Vertebral arteries

The two vertebral arteries branch off of the subclavian arteries. They ascend posteriorly through the transverse foramina of the cervical vertebrae. Branches of the vertebral arteries provide blood to deep neck muscles, parts of the spinal cord, and the medulla oblongata.

After passing through the foramen magnum, the vertebral arteries merge centrally to form the basilar artery, which supplies blood to the posterior brain – the cerebellum and the pons. The basilar artery separates into two posterior cerebral arteries, which supply blood to the occipital lobes, the thalamus, and the hypothalamus.

### Carotid arteries

The common carotids give rise to two pairs of blood vessels that provide an arterial blood supply to the head and neck – the internal and external carotid arteries.

#### Internal carotid arteries

The two internal carotids pass through the carotid foramina at the base of the brain to supply blood to the anterior and middle parts of the brain. They divide further into the anterior and middle cerebral arteries. These arteries interconnect through the circle of Willis, an anastomosis at the base of the brain that ensures continual circulation to the brain despite interruption of any of the brain's major vessels. The internal carotids and their branches also supply blood to the eyeball and orbital cavity, the nose, and the forehead.

#### External carotid arteries

Another major branch of each common carotid is the external carotid artery. The external carotid arteries have many branches providing blood to many areas in the head and neck.

At its terminal end, each external carotid artery branches to form the superficial temporal and maxillary arteries. The superficial temporal artery can be observed pulsating in front of the ear and is a convenient point for feeling the pulse. The maxillary artery provides the blood supply to deep structures of the face and jaw; it branches to form the middle meningeal artery, which supplies the dura mater.

### Carotid sinus

The carotid sinus is a dilated portion of the proximal part of the internal carotid artery, near the bifurcation of the common carotid artery. Changes in blood pressure stimulate vagal nerve endings in the wall of the carotid sinus to send signals along the vagus nerve to slow the heart rate; this response is referred to as the carotid sinus reflex.

**Clinical tip**

**When assessing the carotid arteries for bruits, use the bell of the stethoscope to elicit high-pitched sounds. Also, to avoid mistaking air movement sounds for a bruit, ask your patient to hold his breath while you are auscultating.**

## Circle of Willis

The arterial network, called the circle of Willis, is at the base of the brain. (See *Circle of Willis*, page 42.) It is formed by the following:

- internal carotid arteries, branches of the common carotids
- anterior cerebral arteries, branches of the internal carotids
- anterior communicating arteries, branches of the anterior cerebral arteries
- posterior cerebral arteries, branches of the basilar arteries
- posterior communicating arteries, branches of the posterior cerebral arteries.

## Venous drainage

The venous sinuses provide drainage from the cranial cavities and orbital structures and, eventually, into the internal jugular veins. (See *Venous sinuses*, page 43.) Most sinuses are single structures, but some occur in pairs.

### Jugular veins

From the subclavian, two pairs of veins arise that assist in the drainage of blood from the head and neck – the internal and external jugular veins. The internal jugular, typically the largest vein in the neck, collects blood from the following areas:

- brain
- eyes
- neck
- deep and superficial areas of the face.

At the end of the inferior jugular vein, the vein dilates to form the inferior bulb of the internal jugular vein. This area is unique because it contains a bicuspid valve, which prevents backflow while still allowing blood to flow to the heart should the area become inverted, such as when a person stands on his head.

The external jugular vein collects blood from the superficial areas of the face and neck and the exterior cranium.

## VESSELS OF HEAD AND NECK

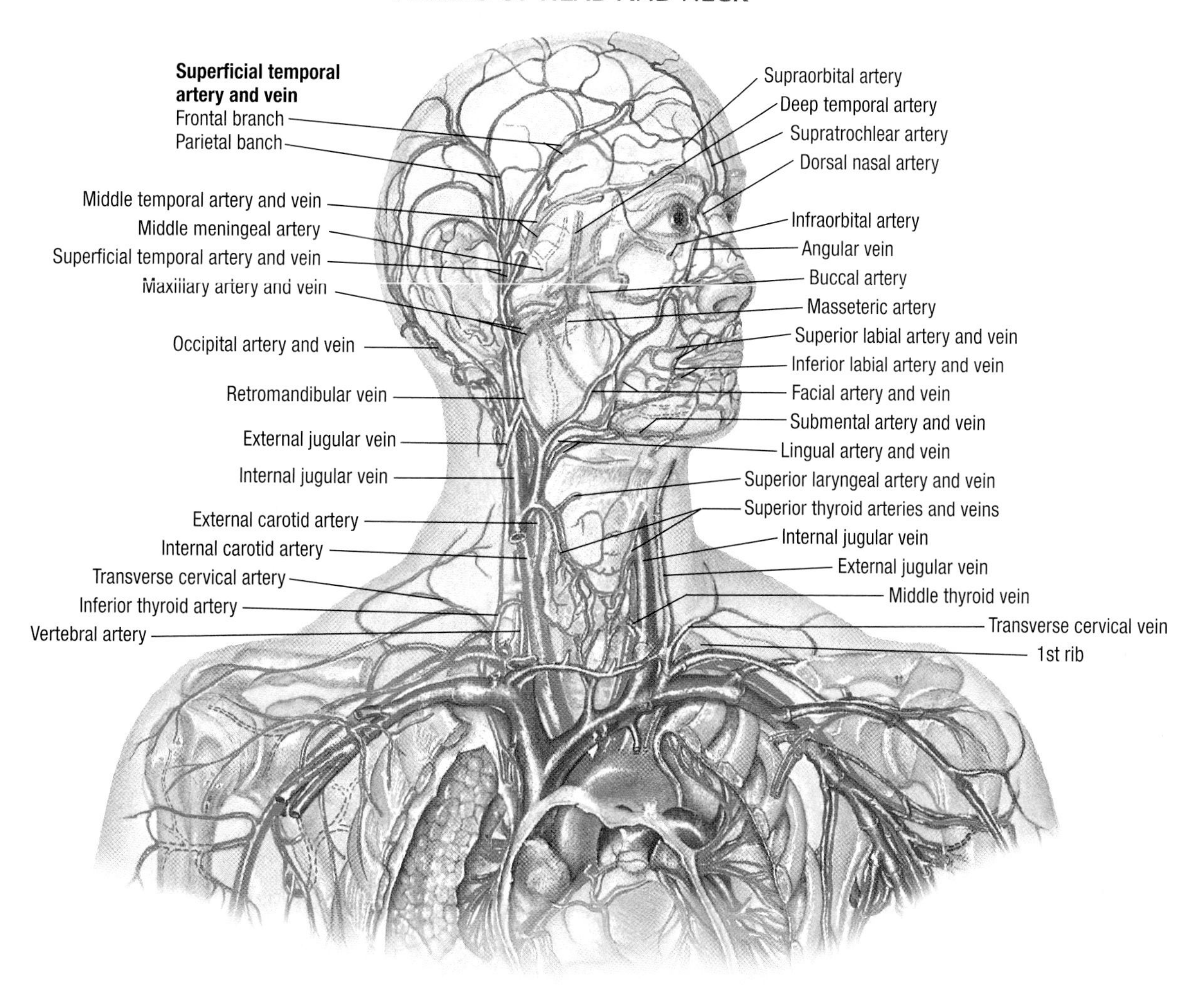

**Superficial temporal artery and vein**
Frontal branch
Parietal banch
Middle temporal artery and vein
Middle meningeal artery
Superficial temporal artery and vein
Maxillary artery and vein
Occipital artery and vein
Retromandibular vein
External jugular vein
Internal jugular vein
External carotid artery
Internal carotid artery
Transverse cervical artery
Inferior thyroid artery
Vertebral artery
Supraorbital artery
Deep temporal artery
Supratrochlear artery
Dorsal nasal artery
Infraorbital artery
Angular vein
Buccal artery
Masseteric artery
Superior labial artery and vein
Inferior labial artery and vein
Facial artery and vein
Submental artery and vein
Lingual artery and vein
Superior laryngeal artery and vein
Superior thyroid arteries and veins
Internal jugular vein
External jugular vein
Middle thyroid vein
Transverse cervical vein
1st rib


## INTERNAL CAROTID AND VERTEBRAL ARTERIES

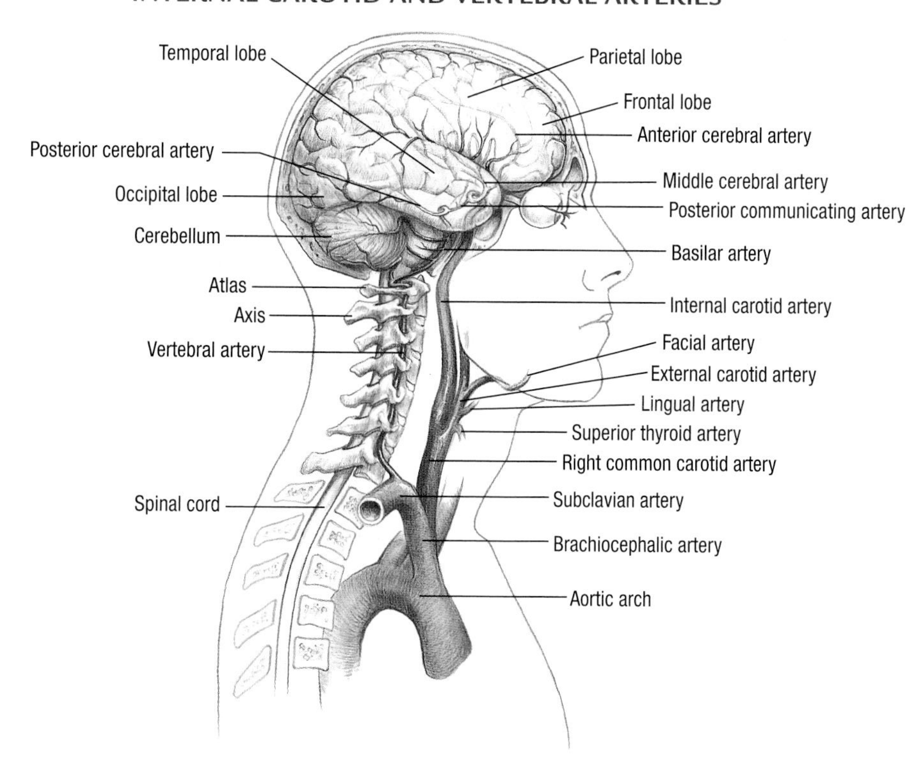

Temporal lobe
Posterior cerebral artery
Occipital lobe
Cerebellum
Atlas
Axis
Vertebral artery
Spinal cord
Parietal lobe
Frontal lobe
Anterior cerebral artery
Middle cerebral artery
Posterior communicating artery
Basilar artery
Internal carotid artery
Facial artery
External carotid artery
Lingual artery
Superior thyroid artery
Right common carotid artery
Subclavian artery
Brachiocephalic artery
Aortic arch

## BRAIN ARTERIES
### Sagittal section

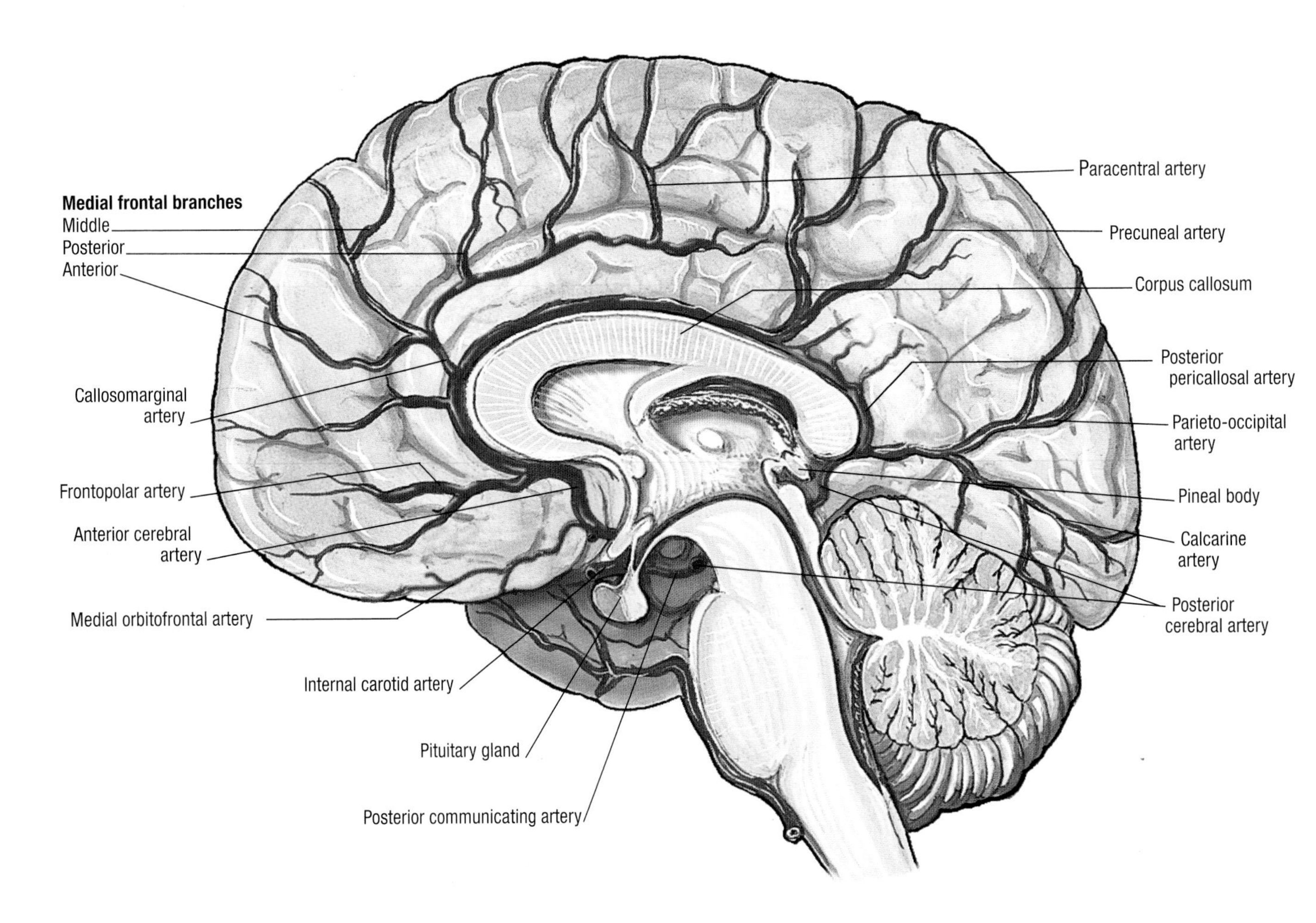

Medial frontal branches
Middle
Posterior
Anterior
Callosomarginal artery
Frontopolar artery
Anterior cerebral artery
Medial orbitofrontal artery
Internal carotid artery
Pituitary gland
Posterior communicating artery
Paracentral artery
Precuneal artery
Corpus callosum
Posterior pericallosal artery
Parieto-occipital artery
Pineal body
Calcarine artery
Posterior cerebral artery

## BRAIN ARTERIES
### Lateral view

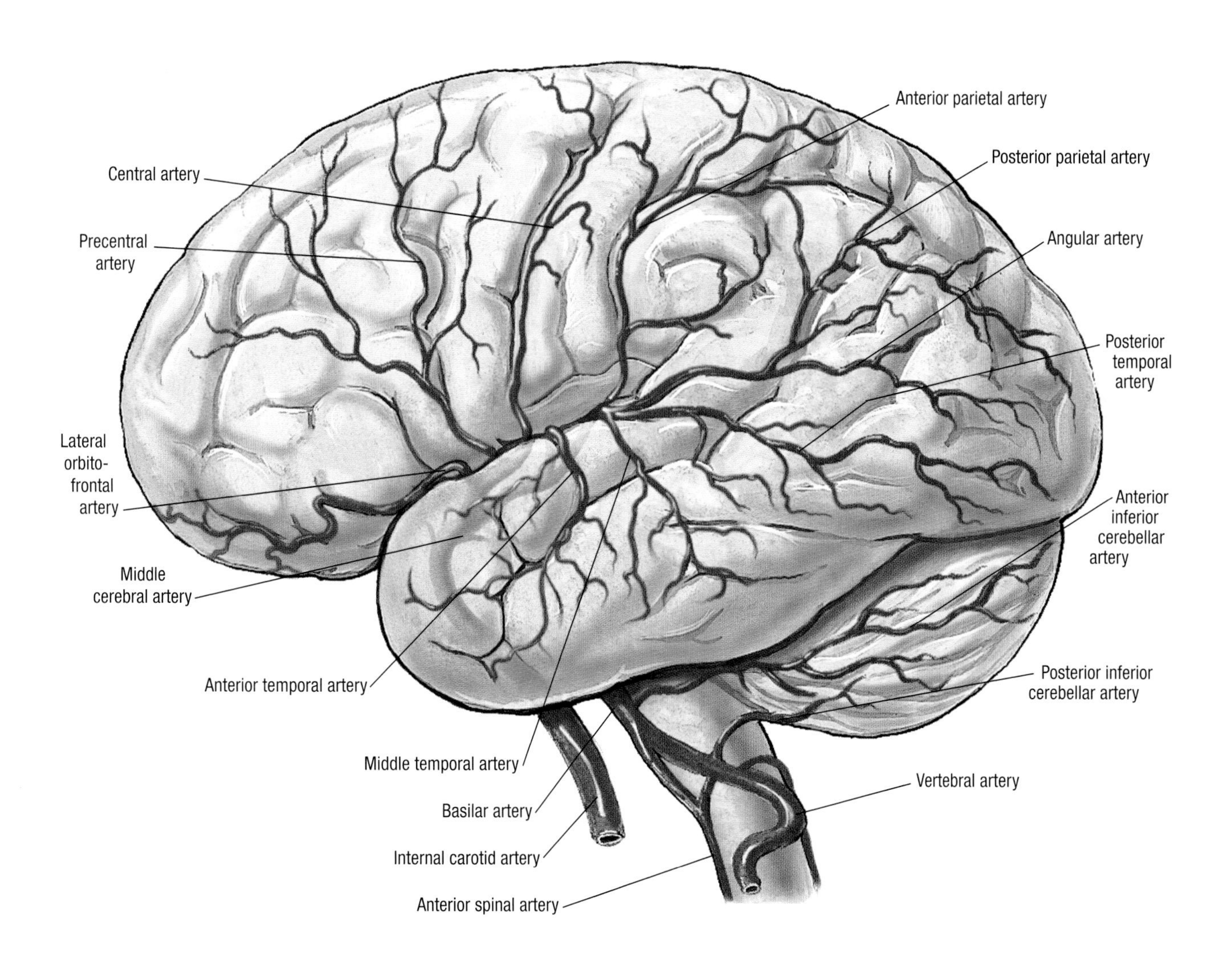
Anterior parietal artery
Posterior parietal artery
Central artery
Precentral artery
Angular artery
Posterior temporal artery
Lateral orbito-frontal artery
Anterior inferior cerebellar artery
Middle cerebral artery
Anterior temporal artery
Posterior inferior cerebellar artery
Middle temporal artery
Vertebral artery
Basilar artery
Internal carotid artery
Anterior spinal artery

## VESSELS OF THE BRAIN
### Inferior view

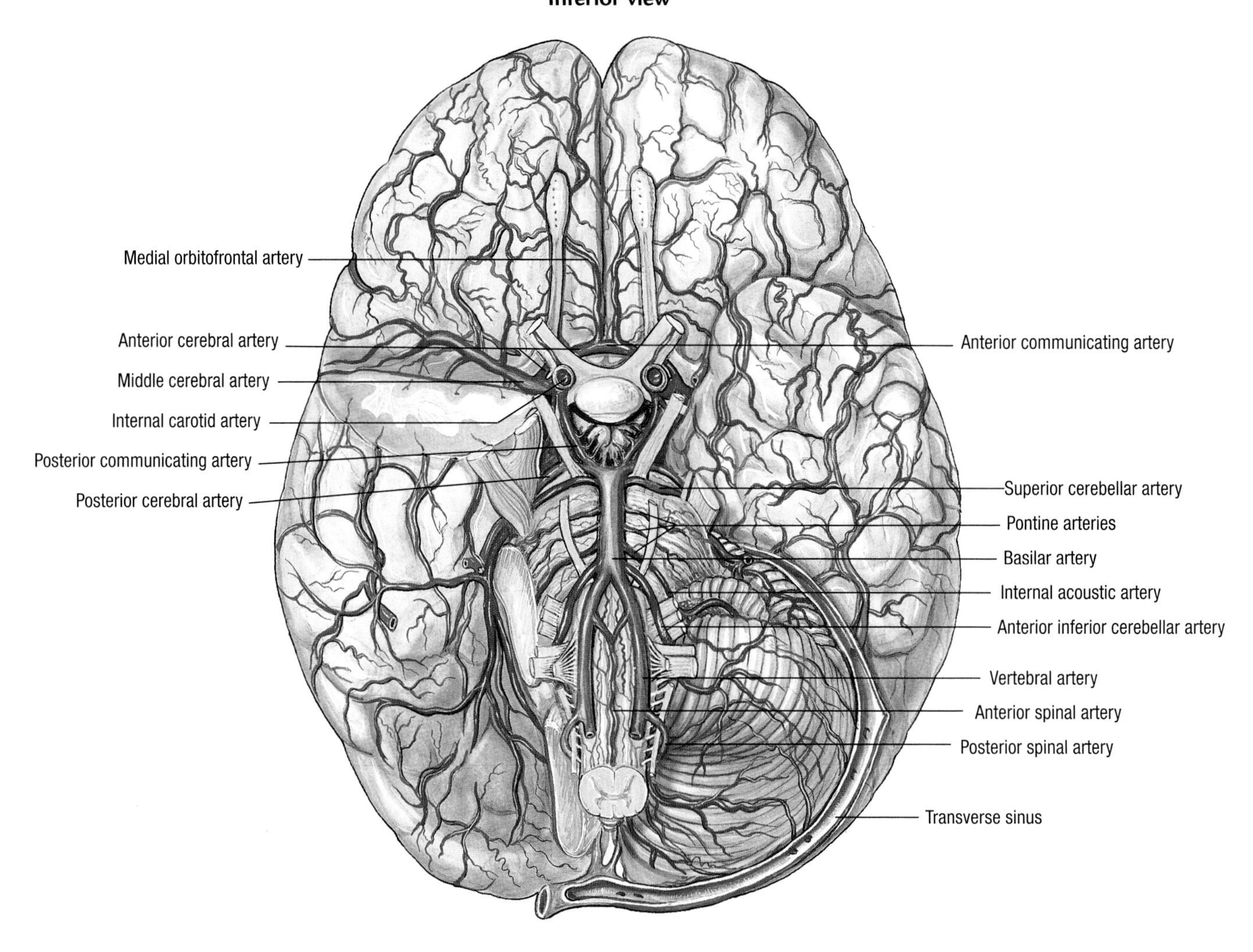
Medial orbitofrontal artery
Anterior cerebral artery
Middle cerebral artery
Internal carotid artery
Posterior communicating artery
Posterior cerebral artery
Anterior communicating artery
Superior cerebellar artery
Pontine arteries
Basilar artery
Internal acoustic artery
Anterior inferior cerebellar artery
Vertebral artery
Anterior spinal artery
Posterior spinal artery
Transverse sinus

## CIRCLE OF WILLIS

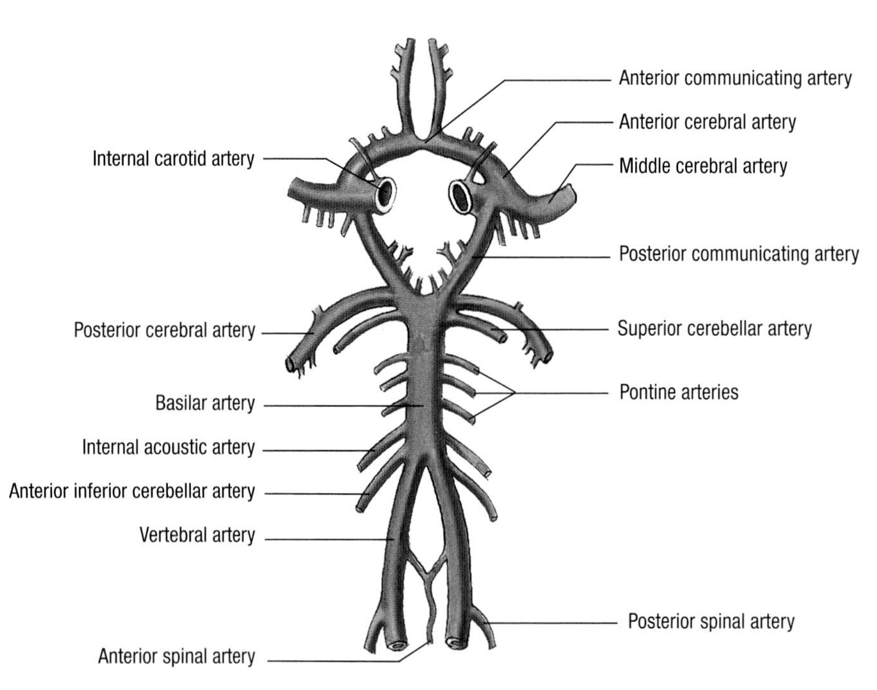
Anterior communicating artery
Anterior cerebral artery
Internal carotid artery
Middle cerebral artery
Posterior communicating artery
Posterior cerebral artery
Superior cerebellar artery
Pontine arteries
Basilar artery
Internal acoustic artery
Anterior inferior cerebellar artery
Vertebral artery
Posterior spinal artery
Anterior spinal artery

## VENOUS SINUSES

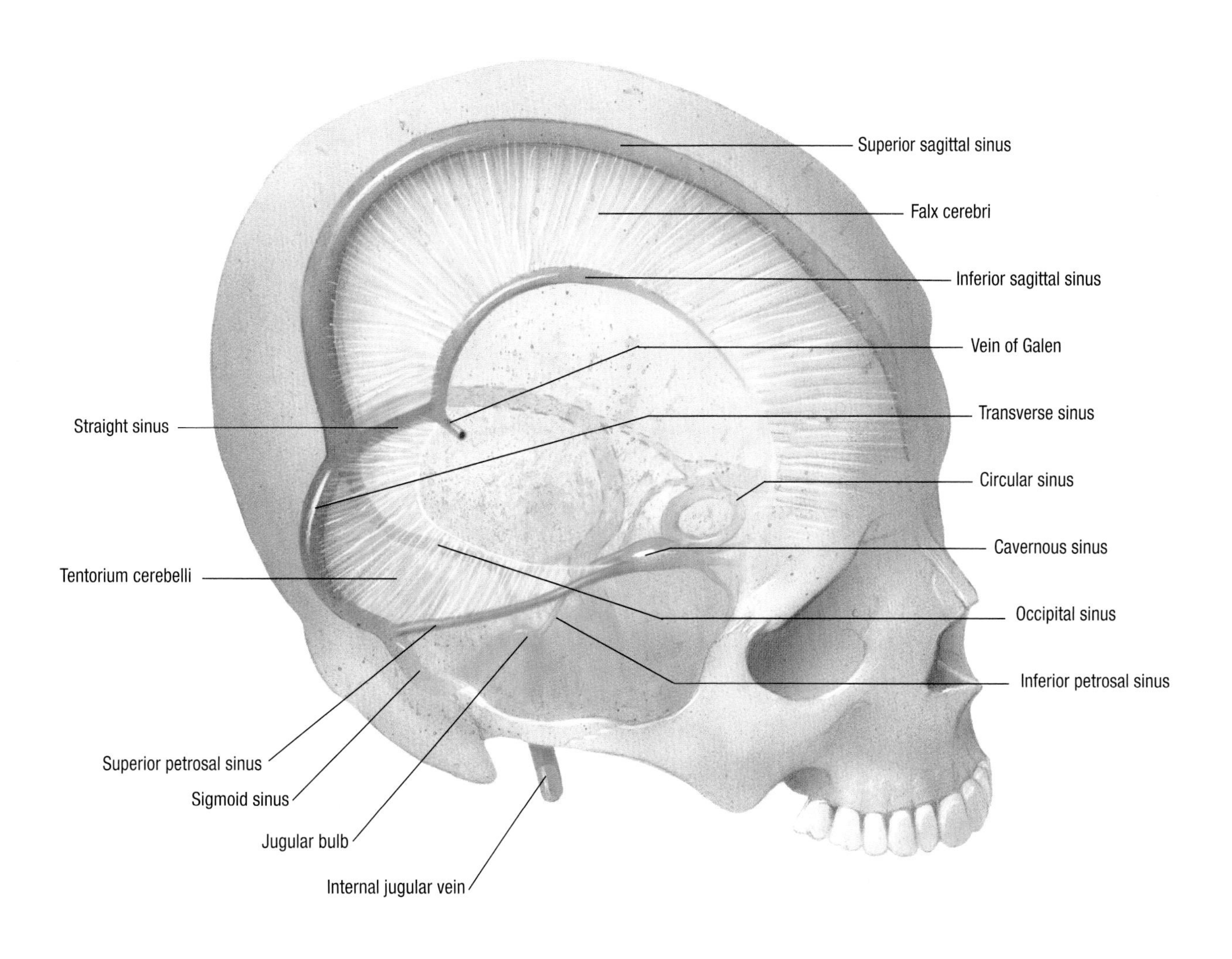
Superior sagittal sinus
Falx cerebri
Inferior sagittal sinus
Vein of Galen
Transverse sinus
Circular sinus
Cavernous sinus
Occipital sinus
Inferior petrosal sinus
Straight sinus
Tentorium cerebelli
Superior petrosal sinus
Sigmoid sinus
Jugular bulb
Internal jugular vein

# Head and neck nerves

The cranial nerves and cervical spinal nerves are part of the peripheral nervous system. These nerves serve as a link between the central nervous system (brain and spinal cord) and the rest of the body. (For more information on the nervous system, see Part IX, Body systems.)

## Cranial nerves

The 12 pairs of cranial nerves transmit motor or sensory messages or both, primarily between the brain or brain stem and the head and neck. All cranial nerves except the olfactory and optic nerves originate in the midbrain, pons, or medulla oblongata of the brain stem. The olfactory and optic nerves originate in the forebrain. (See *Cranial nerves and their functions*, Appendix A, and *Head and neck nerves*, page 46.)

### Branches of the cranial nerves

Most of the cranial nerves have many branches providing innervation to key areas. Branches of cranial nerves (CN) VII and X (facial and vagus, respectively) are particularly important.

#### Facial nerve (CN VII)

This is the only nerve supplying the muscles of facial expression. Its posterior auricular branch – a sensory nerve – forms just before the facial nerve enters the parotid salivary gland. (See *Facial nerve and parotid gland*, page 47.) Within the gland, the facial nerve divides to form the following 5 branches, which supply the muscles of facial expression:

- temporal
- zygomatic
- buccal
- mandibular
- cervical.

The parotid gland is encapsulated. When it becomes inflamed, there is little room for swelling. Compression of the facial nerve causes the painful syndrome called Bell's palsy. Care must be taken during surgery on the parotid gland to avoid injuring the facial nerve.

**Clinical tip**

When evaluating the motor component of CN VII, observe the patient's face for symmetry at rest and while smiling, frowning, and raising eyebrows. If a weakness is caused by a stroke or other condition that damages the cerebral cortex, the patient will be able to raise his eyebrows and wrinkle his forehead, but his lower face will be weak. If the weakness is due to an interruption of the facial nerve, the entire side of his face will be immobile.

#### Vagus nerve (CN X)

The vagus nerve has motor, sensory, and parasympathetic functions. It's notable for innervating more than the head and neck region. It originates in the medulla; passes through the skull, neck, thorax, and diaphragm; and extends into the abdomen. (See *Vagus nerve distribution*, page 47.) Key branches of the vagus nerve include:

- pharyngeal
- laryngeal
- esophageal
- cardiac
- pulmonary
- gastric
- gallbladder
- pancreatic
- intestinal (to the left colic flexure).

## Cervical spinal nerves

Eight pairs of cervical nerves originate in the spinal cord. The ventral rami (anterior branches) of the first four cervical nerves form the cervical plexus; this plexus connects with:

- skin and muscles of the head
- neck
- upper part of the shoulders
- cranial nerves XI and XII
- phrenic nerves.

The ventral rami of spinal nerves C5 to C8 and T1 (with contributions from C4 and T2) form the brachial plexus. This plexus provides the entire nerve supply to the upper extremities and to several neck and shoulder muscles.

## CRANIAL NERVES
### Inferior view

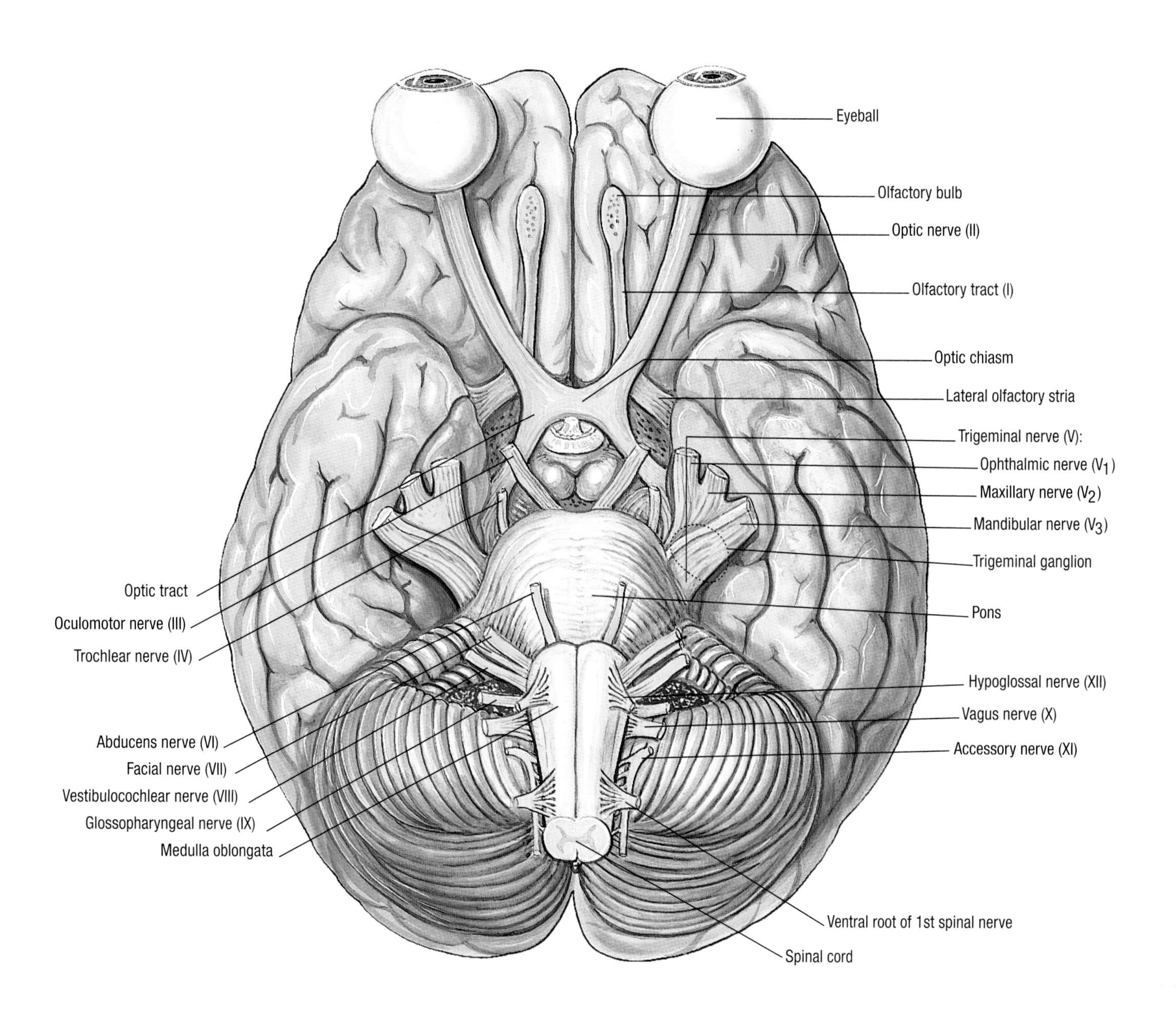

Eyeball
Olfactory bulb
Optic nerve (II)
Olfactory tract (I)
Optic chiasm
Lateral olfactory stria
Trigeminal nerve (V):
Ophthalmic nerve ($V_1$)
Maxillary nerve ($V_2$)
Mandibular nerve ($V_3$)
Trigeminal ganglion
Pons
Hypoglossal nerve (XII)
Vagus nerve (X)
Accessory nerve (XI)
Ventral root of 1st spinal nerve
Spinal cord
Optic tract
Oculomotor nerve (III)
Trochlear nerve (IV)
Abducens nerve (VI)
Facial nerve (VII)
Vestibulocochlear nerve (VIII)
Glossopharyngeal nerve (IX)
Medulla oblongata

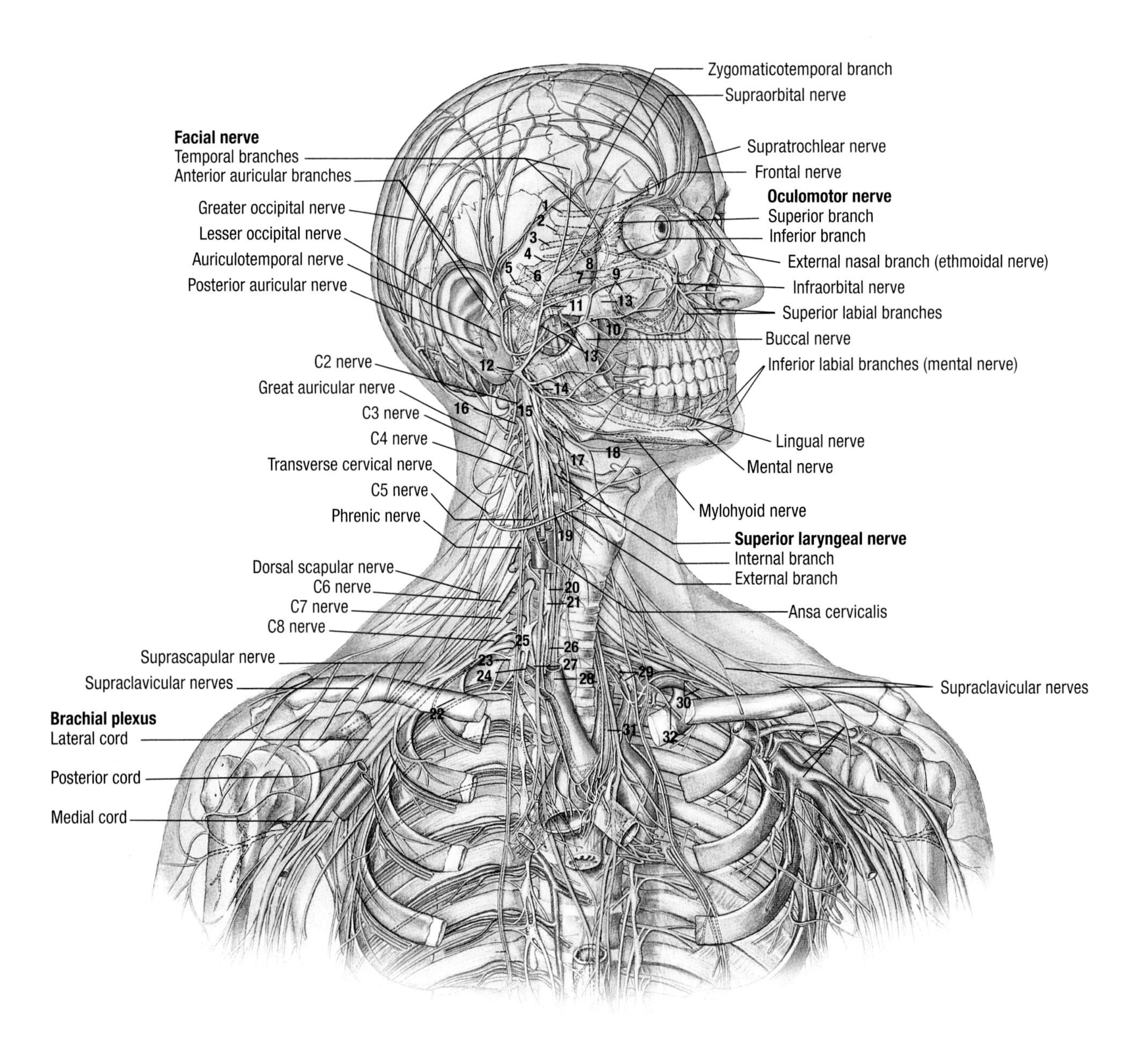
Zygomaticotemporal branch
Supraorbital nerve
Facial nerve
Temporal branches
Anterior auricular branches
Greater occipital nerve
Lesser occipital nerve
Auriculotemporal nerve
Posterior auricular nerve
Supratrochlear nerve
Frontal nerve
Oculomotor nerve
Superior branch
Inferior branch
External nasal branch (ethmoidal nerve)
Infraorbital nerve
Superior labial branches
Buccal nerve
Inferior labial branches (mental nerve)
C2 nerve
Great auricular nerve
C3 nerve
C4 nerve
Transverse cervical nerve
C5 nerve
Phrenic nerve
Lingual nerve
Mental nerve
Mylohyoid nerve
Superior laryngeal nerve
Internal branch
External branch
Dorsal scapular nerve
C6 nerve
C7 nerve
C8 nerve
Ansa cervicalis
Suprascapular nerve
Supraclavicular nerves
Supraclavicular nerves
Brachial plexus
Lateral cord
Posterior cord
Medial cord

## Key

**Head**

- **1** Olfactory tract (I)
- **2** Optic nerve (II)
- **3** Oculomotor nerve (III)
- **4** Trochlear nerve (IV)
- **5** Abducens nerve (VI)
- **6** Trigeminal nerve (V)
- **7** Maxillary nerve (V2)
- **8** Ophthalmic nerve (V1)
- **9** Superior alveolar branches
- **10** Palatine nerves
- **11** Mandibular nerve (V3)
- **12** Facial nerve (VII)
- **13** Zygomatic branches (Facial nerve)
- **14** Glossopharyngeal nerve (IX)
- **15** Vagus nerve (X)
- **16** Accessory nerve (XI)
- **17** Pharyngeal branch (Vagus nerve)
- **18** Hypoglossal nerve (XII)

**Neck**

- **19** Superior cardiac nerve
- **20** Sympathetic trunk
- **21** Middle cervical sympathetic ganglion
- **22** 1st intercostal nerve
- **23** 1st thoracic ganglion
- **24** Ansa subclavia
- **25** Inferior cervical sympathetic ganglion
- **26** Middle cardiac nerve
- **27** Inferior cardiac nerve
- **28** Right recurrent nerve
- **29** Vertebral plexus
- **30** Subclavian plexus
- **31** Superior cardiac branch (Vagus nerve)
- **32** Subclavian nerve

## FACIAL NERVE AND PAROTID GLAND

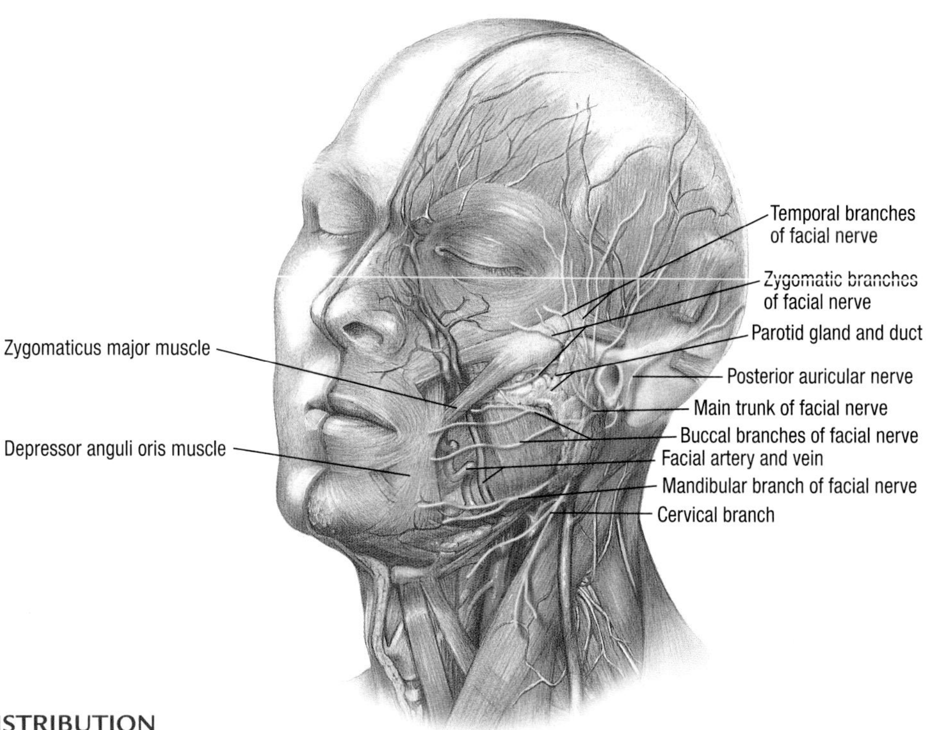
Temporal branches of facial nerve
Zygomatic branches of facial nerve
Parotid gland and duct
Posterior auricular nerve
Main trunk of facial nerve
Buccal branches of facial nerve
Facial artery and vein
Mandibular branch of facial nerve
Cervical branch
Zygomaticus major muscle
Depressor anguli oris muscle

## VAGUS NERVE DISTRIBUTION

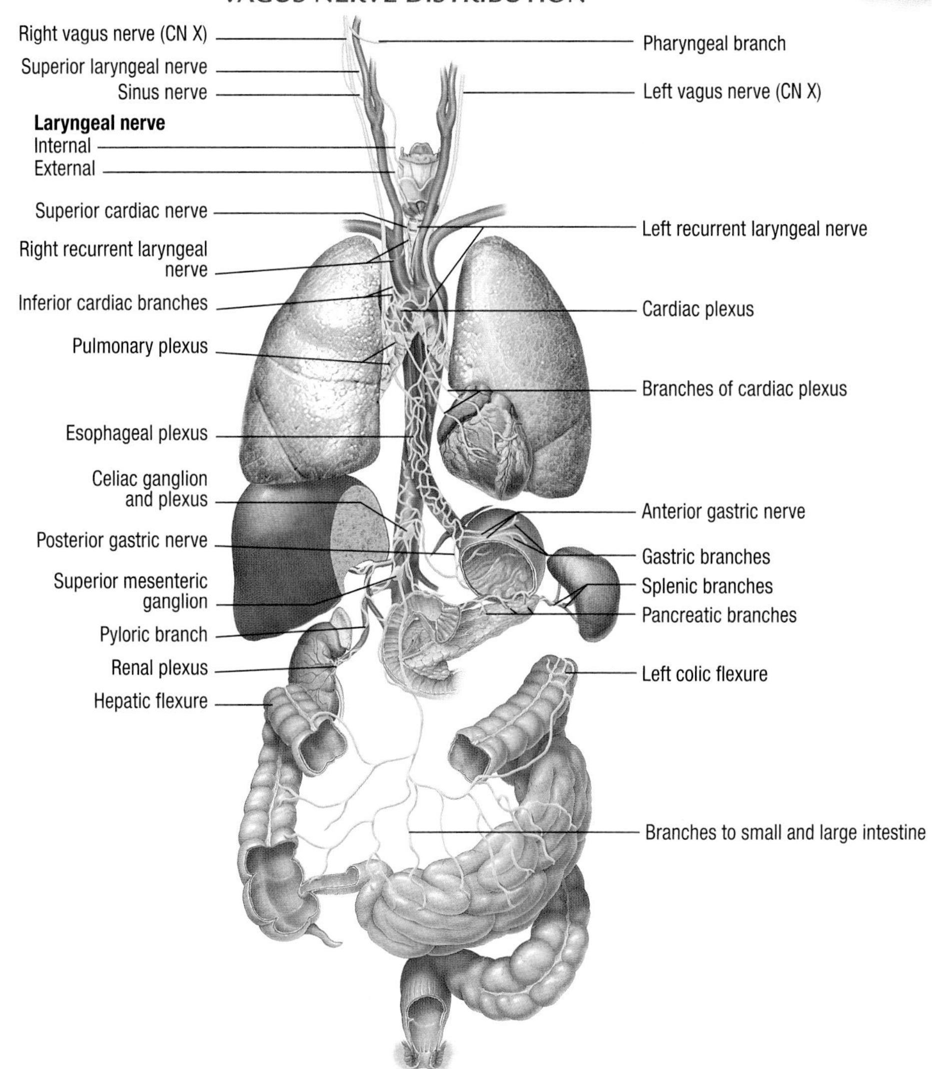
Right vagus nerve (CN X)
Superior laryngeal nerve
Sinus nerve
Laryngeal nerve
Internal
External
Superior cardiac nerve
Right recurrent laryngeal nerve
Inferior cardiac branches
Pulmonary plexus
Esophageal plexus
Celiac ganglion and plexus
Posterior gastric nerve
Superior mesenteric ganglion
Pyloric branch
Renal plexus
Hepatic flexure
Pharyngeal branch
Left vagus nerve (CN X)
Left recurrent laryngeal nerve
Cardiac plexus
Branches of cardiac plexus
Anterior gastric nerve
Gastric branches
Splenic branches
Pancreatic branches
Left colic flexure
Branches to small and large intestine

# Extraocular structures

The organ of vision, the eye contains about 70% of the body's sensory receptors. Although the eye measures about 1" (2.5 cm) in diameter, only its anterior surface is visible.

## Muscles

Extraocular muscles hold the eyes in place and control their movement. Their coordinated action keeps both eyes parallel and creates binocular vision. These muscles have mutually antagonistic actions: As one muscle contracts, its opposing muscle relaxes. Cranial nerves III (oculomotor), IV (trochlear), and VI (abducens) control eye movement.

### Muscle action

Each extraocular muscle has the following specific actions:
- *superior rectus* — rotates the eye upward and adducts and rotates the eye medially
- *inferior rectus* — rotates the eye downward and adducts and rotates the eye medially
- *lateral rectus* — turns the eye laterally
- *medial rectus* — turns the eye medially
- *superior oblique* — turns the eye downward and abducts and rotates the eye laterally
- *inferior oblique* — turns the eye upward and adducts and turns the eye laterally.

**Clinical tip**

When assessing eye movement, have the patient follow your finger through the six cardinal positions of gaze:
- left superior
- left lateral
- left inferior
- right superior
- right lateral
- right inferior.

Pause slightly before moving from one position to the next. This helps you to evaluate for nystagmus, or involuntary eye movement, and the ability to hold the gaze in that particular position.

## Structures

Extraocular structures include the eyelids, conjunctivae, and lacrimal apparatus. With the extraocular muscles, these structures support and protect the eyes.

### Eyelids

Also called the palpebrae, the eyelids are loose folds of skin covering the anterior portion of the eye. The eyelids protect the eye from foreign bodies, regulate the entrance of light, and distribute tears over the eye by blinking. The lid margins contain hair follicles, which in turn contain eyelashes and sebaceous glands. The levator palpebrae superioris muscle is connected to the upper lid by a thin tendon. It helps keep the upper eyelid from closing from its own weight, and it elevates the lid during upward looking or the last part of a blink.

When closed, the upper and lower eyelids cover the eye completely. When open, the upper eyelid extends beyond the limbus (the junction of the cornea and the sclera) and covers a small portion of the iris.

The palpebral fissure is the opening between the margins of the upper and lower lids; it should be equal in both eyes. The juncture of the upper and lower lids is called the canthus.

The eyelids contain three types of glands:
- meibomian glands — sebaceous glands that secrete sebum, an oily substance that prevents evaporation of tears and keeps the eye lubricated
- glands of Zeis — modified sebaceous glands connected to the follicles of the eyelashes
- Moll's glands — ordinary sweat glands.

### Conjunctivae

Conjunctivae are thin mucous membranes that line the inner surface of each eyelid and the anterior portion of the sclera, guarding the eye from invasion by foreign matter. The palpebral conjunctiva — the portion that lines the inner surface of the eyelids — appears shiny pink or red. The bulbar conjunctiva, which joins the palpebral portion, contains many small, normally visible blood vessels. (For an illustration of the conjunctivae, see *Eye,* page 51.)

### Lacrimal apparatus

The structures of the lacrimal apparatus — lacrimal gland, punctum, lacrimal sac, and nasolacrimal duct — lubricate and protect the cornea and conjunctivae by producing and absorbing tears.

The lacrimal gland occupies a shallow fossa beneath the superior temporal orbital rim on the lateral side of the orbit; it secretes tears, which flow through excretory ducts. Besides keeping the cornea and conjunctiva moist, tears contain lysozyme, an enzyme that protects against bacterial invasion.

As the eyelids blink, they direct the flow of tears to the inner canthus, where the tears pool and then drain through the punctum, a tiny opening at the medial junction of the upper and lower eyelids. From there, tears flow through the lacrimal canals, into the lacrimal sac, then through the nasolacrimal duct into the nose.

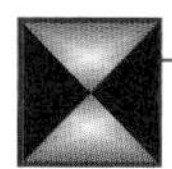

**Age-related changes**

With age, the eyes sit deeper in their sockets and the eyelids lose their elasticity, becoming baggy and wrinkled. The conjunctivae become thinner and yellow, and pingueculae — fat pads under the conjunctiva — may form. As the lacrimal apparatus gradually loses fatty tissue, the quantity of tears decreases and evaporation occurs more quickly.

## LACRIMAL APPARATUS

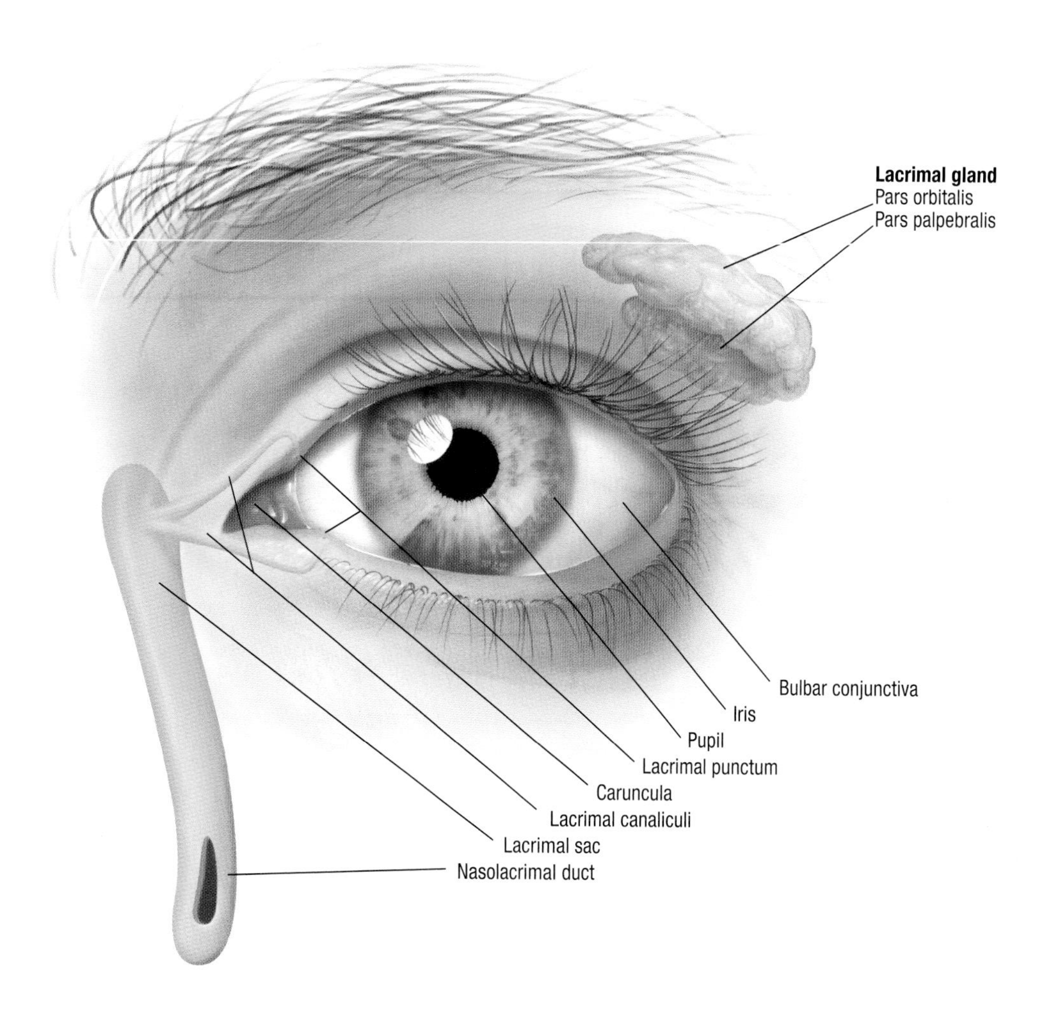

**Lacrimal gland**
Pars orbitalis
Pars palpebralis
Bulbar conjunctiva
Iris
Pupil
Lacrimal punctum
Caruncula
Lacrimal canaliculi
Lacrimal sac
Nasolacrimal duct


## EYE MUSCLES

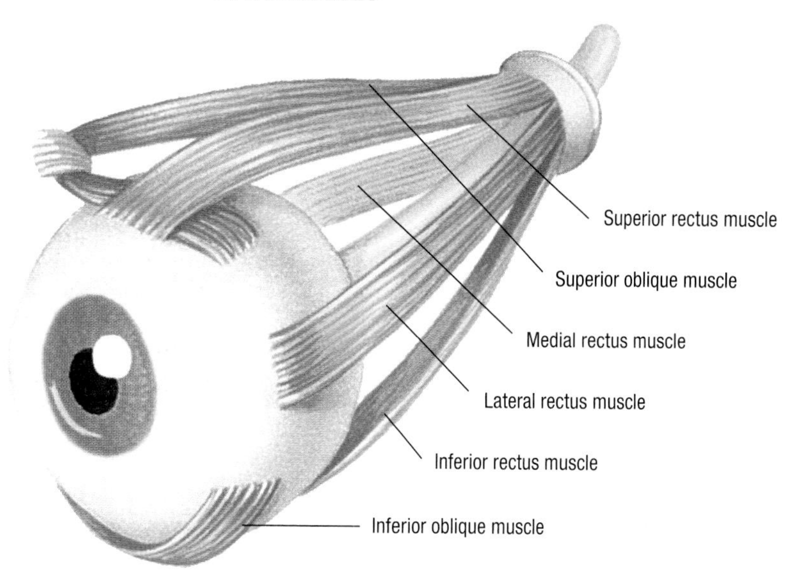

Superior rectus muscle
Superior oblique muscle
Medial rectus muscle
Lateral rectus muscle
Inferior rectus muscle
Inferior oblique muscle

# Intraocular structures

Intraocular structures are directly involved with vision. The eye has three layers of tissue. The outermost layer (fibrous tunic) contains the sclera and cornea. The middle layer (uvea or vascular tunic) has three regions — choroid, ciliary body, and iris. The innermost layer (sensory tunic) of the eye is the retina.

## Sclera and cornea

The white sclera coats four-fifths of the outside of the eyeball, maintaining its size and form. The transparent cornea, which is continuous with the sclera at the limbus, reveals the pupil and iris. A smooth, transparent tissue, the cornea has no blood supply. The corneal epithelium merges with the bulbar conjunctiva at the limbus. Kept moist by tears, the cornea is highly sensitive to touch.

## Iris

The iris is a circular contractile disk containing circular and radial smooth muscles. It has an opening in the center, the pupil. Eye color depends on the amount of pigment in the endothelial cells of the iris.

## Pupil

Pupils should be equal in size and round; depending on the patient's age, they should measure ⅛" to ¼" (3 to 7 mm) in diameter. Pupil size is controlled by involuntary dilatory and sphincter muscles in the posterior portion of the iris that regulate light entry.

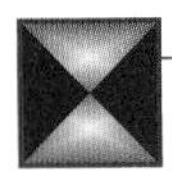

**Age-related changes**

Aging causes a number of changes in intraocular structures.

- The sclera becomes thick and rigid, and fat deposits cause yellowing.
- The cornea loses its luster and flattens.
- The iris fades or develops irregular pigmentation.
- Increased formation of connective tissue may cause sclerosis of the sphincter muscles.
- The pupil becomes smaller, reducing the amount of light that reaches the retina.
- The vitreous can degenerate, revealing opacities and floating debris, and can also detach from the retina.
- The lens enlarges and loses transparency. Impaired lens elasticity (presbyopia) can decrease accommodation.
- Reabsorption of intraocular fluid can diminish, predisposing a patient to glaucoma.

## Anterior chamber

The anterior chamber, a cavity bounded anteriorly by the cornea and posteriorly by the lens and iris, is filled with a clear, watery fluid called aqueous humor. (See *Anterior chamber*, page 52.) This fluid flows from the posterior chamber to the anterior chamber through the pupil. From there, it flows peripherally and filters through a network of connective tissue called the trabecular meshwork to Schlemm's canal, a venous sinus deep within the sclera. From this canal, the fluid ultimately enters the venous circulation.

## Lens

The lens, situated directly behind the iris at the pupillary opening, acts like a camera lens, refracting and focusing light onto the retina. It's composed of transparent fibers in an elastic membrane called the lens capsule. (See *Lens*, page 53.)

## Ciliary body

The ciliary body (three muscles and the iris, which make up the anterior part of the vascular uveal tract) controls lens thickness. Its action, coordinated with muscles in the iris, regulates the amount of light focused through the lens onto the retina.

## Posterior chamber

The posterior chamber, a small space directly posterior to the iris but anterior to the lens, is filled with aqueous humor. (See *Posterior chamber*, page 53.)

The vitreous humor — a thick, gelatinous material — fills the area behind the lens. The vitreous holds the retina in place and maintains the spherical shape of the eyeball.

## Posterior sclera

The posterior sclera, a white, opaque, fibrous layer, covers the posterior portion of the eyeball, continuing back to the dural sheath, which covers the optic nerve.

## Choroid

The choroid lines the inner aspect of the eyeball beneath the retina (adjacent to the sclera); it contains many small arteries and veins. Posteriorly, it ends at the point where the optic nerve enters the eye. Anteriorly, the choroid is continuous with the ciliary body. (For a detailed discussion of the retina, see page 54.)

## EYE

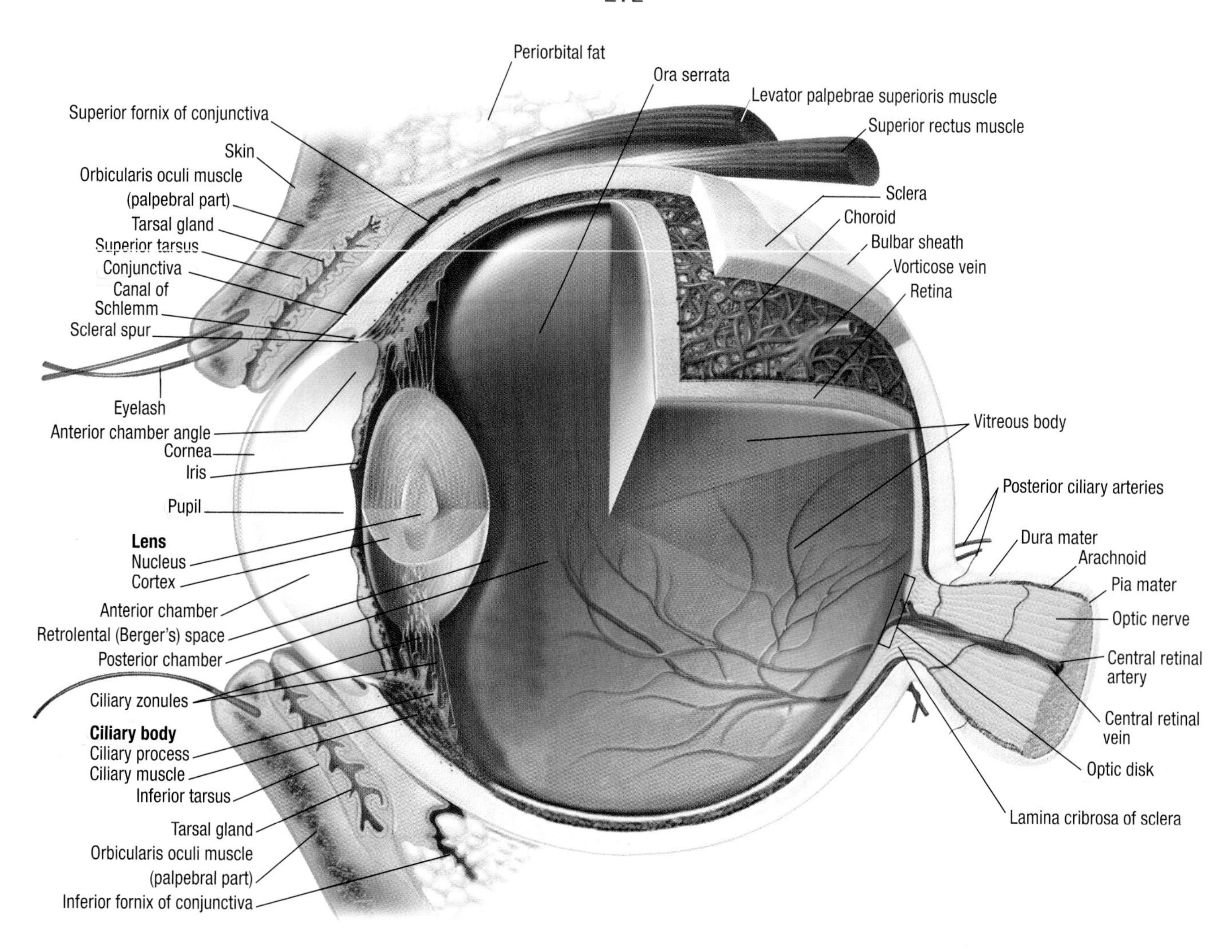

Periorbital fat
Ora serrata
Levator palpebrae superioris muscle
Superior rectus muscle
Superior fornix of conjunctiva
Skin
Orbicularis oculi muscle
(palpebral part)
Tarsal gland
Superior tarsus
Conjunctiva
Canal of
Schlemm
Scleral spur
Sclera
Choroid
Bulbar sheath
Vorticose vein
Retina
Eyelash
Anterior chamber angle
Cornea
Iris
Pupil
Vitreous body
Posterior ciliary arteries
Dura mater
Arachnoid
Pia mater
Optic nerve
Lens
Nucleus
Cortex
Anterior chamber
Retrolental (Berger's) space
Posterior chamber
Central retinal
artery
Ciliary zonules
Ciliary body
Ciliary process
Ciliary muscle
Inferior tarsus
Central retinal
vein
Optic disk
Tarsal gland
Orbicularis oculi muscle
(palpebral part)
Inferior fornix of conjunctiva
Lamina cribrosa of sclera


## CILIARY PROCESS

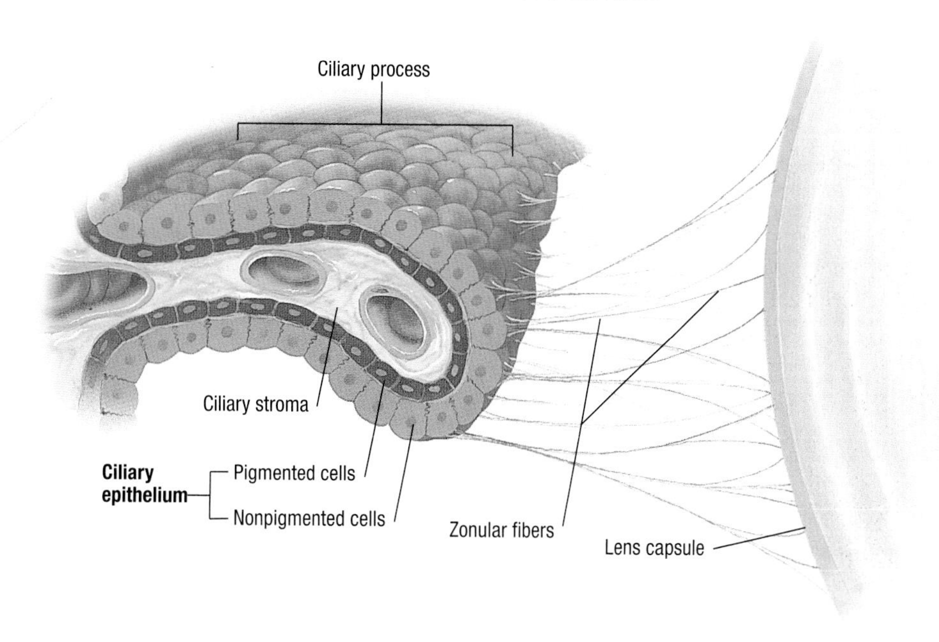

Ciliary process
Ciliary stroma
Ciliary
epithelium
Pigmented cells
Nonpigmented cells
Zonular fibers
Lens capsule

## EYE
### Sagittal section

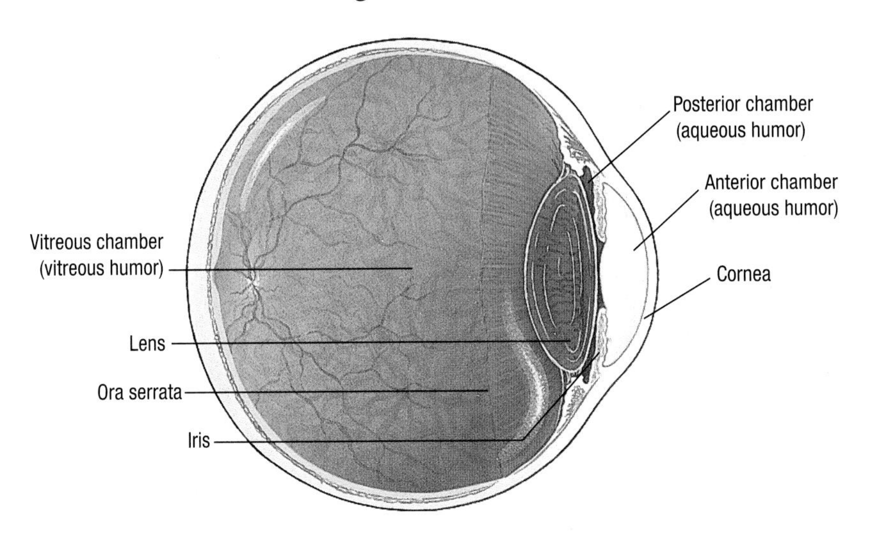
Posterior chamber (aqueous humor)
Anterior chamber (aqueous humor)
Cornea
Vitreous chamber (vitreous humor)
Lens
Ora serrata
Iris

## ANTERIOR CHAMBER

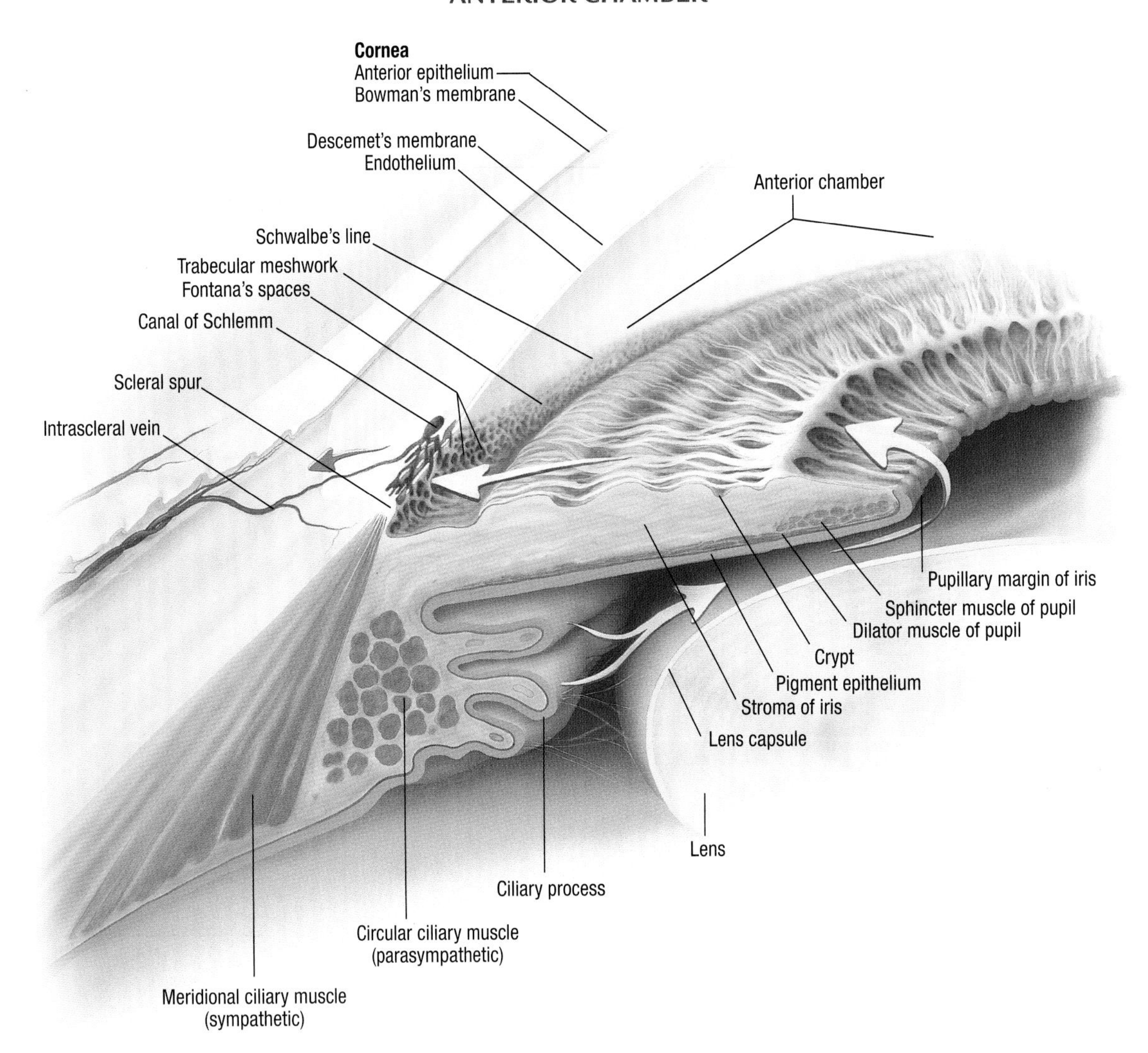
Cornea
Anterior epithelium
Bowman's membrane
Descemet's membrane
Endothelium
Anterior chamber
Schwalbe's line
Trabecular meshwork
Fontana's spaces
Canal of Schlemm
Scleral spur
Intrascleral vein
Pupillary margin of iris
Sphincter muscle of pupil
Dilator muscle of pupil
Crypt
Pigment epithelium
Stroma of iris
Lens capsule
Lens
Ciliary process
Circular ciliary muscle (parasympathetic)
Meridional ciliary muscle (sympathetic)

## LENS

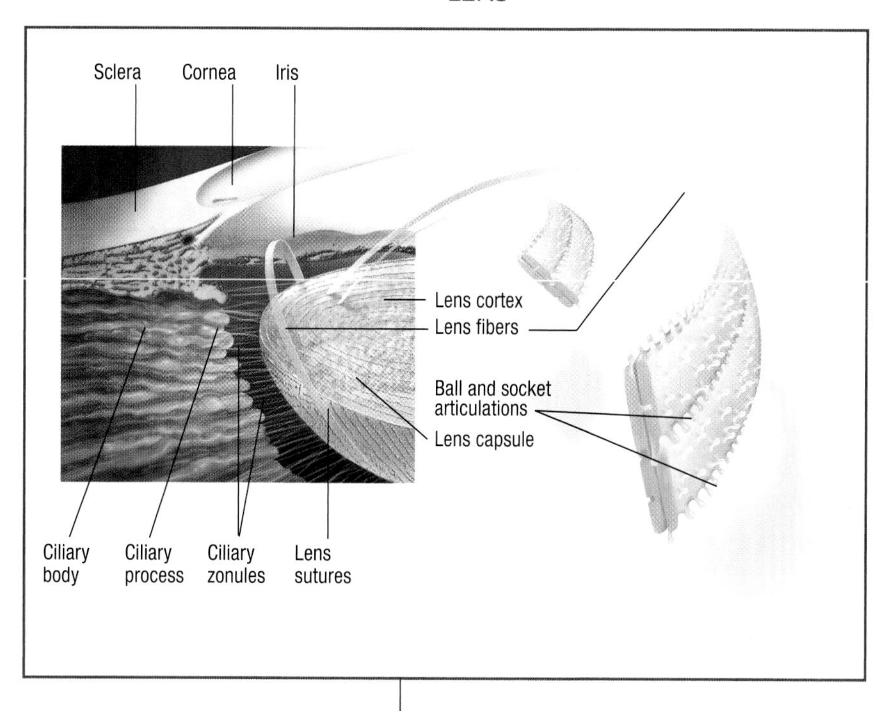
Sclera
Cornea
Iris
Lens cortex
Lens fibers
Ball and socket articulations
Lens capsule
Ciliary body
Ciliary process
Ciliary zonules
Lens sutures

## POSTERIOR CHAMBER

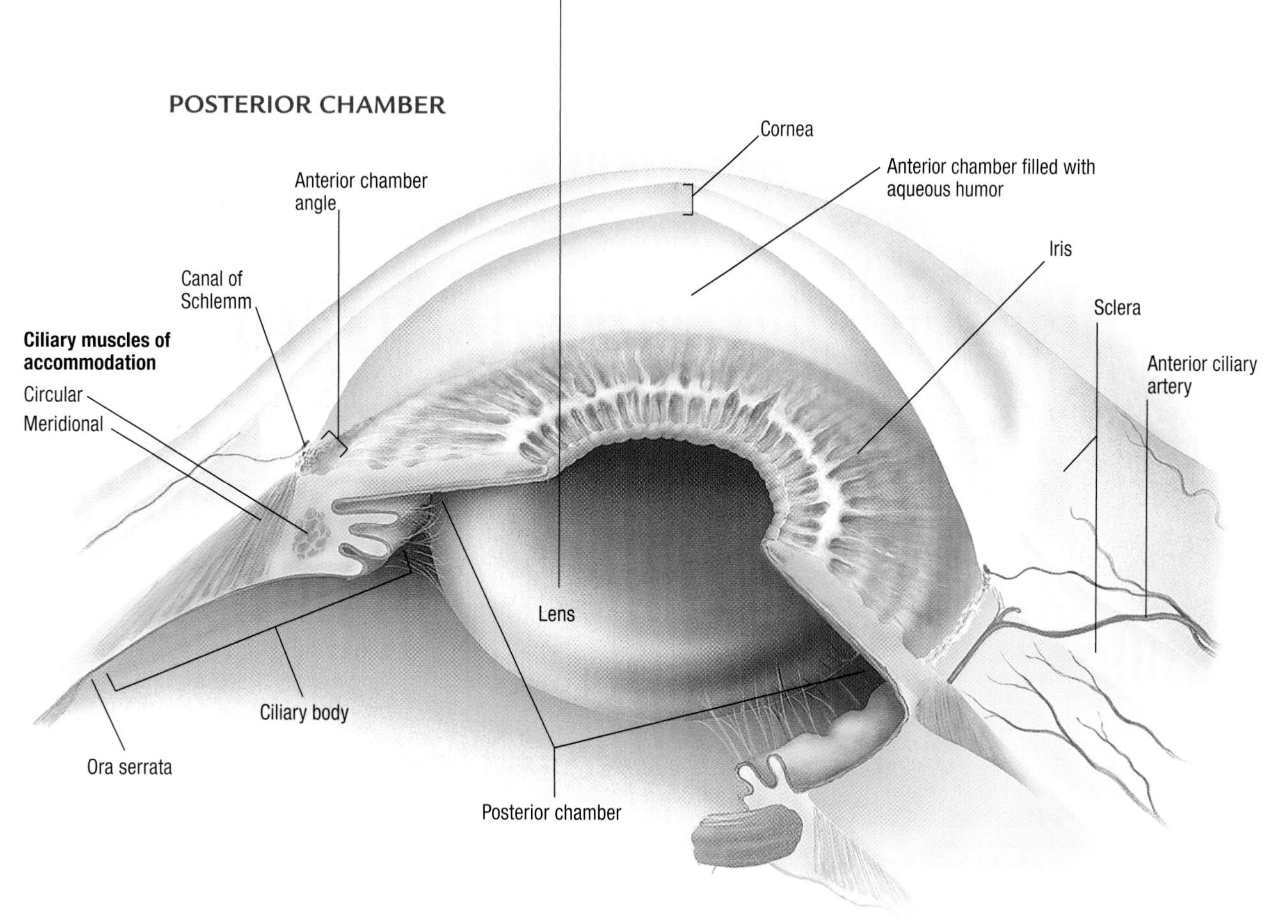
Cornea
Anterior chamber filled with aqueous humor
Anterior chamber angle
Canal of Schlemm
Ciliary muscles of accommodation
Circular
Meridional
Iris
Sclera
Anterior ciliary artery
Lens
Ciliary body
Ora serrata
Posterior chamber

# Retina

The innermost coat of the eyeball, the retina receives visual stimuli and sends them to the brain.

## Retinal structures

Each of the four sets of retinal vessels contains a transparent arteriole and a venule. Arterioles and veins become progressively thinner as they leave the optic disk.

### Optic disk

The optic disk is a well-defined, 1.5-mm (less than ⅛"), round or oval area in the nasal portion of the retina. The optic nerve head enters the retina at the creamy yellow to pink optic disk as the nerve head. A whitish to grayish crescent of scleral tissue may be present on the lateral side of the disk. The physiologic cup is a light-colored depression in the optic disk on the temporal side; it covers one-third of the center of the disk.

**Clinical tip**

When you use an ophthalmoscope, a normal vitreous body appears transparent. The appearance of a dark ring in front of the optic disk indicates that the vitreous has detached from the retina.

### Rods and cones

Photoreceptor cells called rods and cones make vision possible. Rods, concentrated toward the periphery of the retina, respond to low-intensity light and shades of gray. Cones, concentrated in the central fovea of the retina, respond to bright light and color.

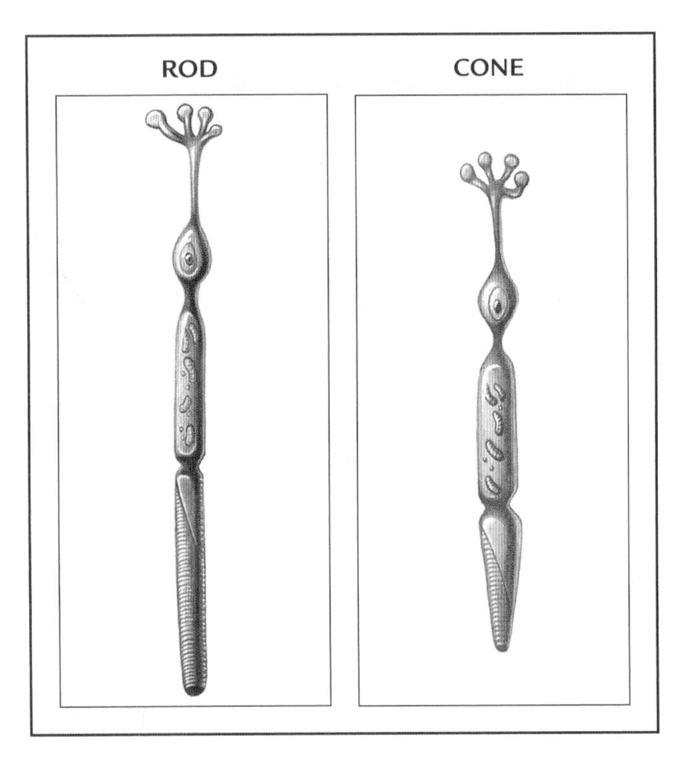
ROD
CONE

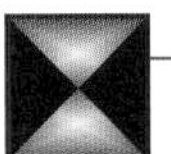

**Age-related changes**

Older adults need about three times as much light as younger people need to see objects clearly. Many older adults experience impaired color vision, especially in the blue and green ranges, because cones in the retina deteriorate.

### Macula lutea

The macula, lateral to the optic disk, is slightly darker than the rest of the retina and has no visible blood vessels. The central fovea, a slight depression in the center of the macula, contains the heaviest concentration of cones and is the area that provides sharpest vision and color perception.

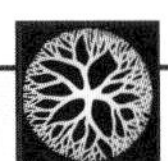

**Physiology**

#### Vision and image perception

**Vision pathway**

Intraocular structures perceive and form images, then send them to the brain for interpretation. To interpret these images, the brain relies on structures along the vision pathway, including the retina, optic nerve, optic chiasm, and optic tract, to create the proper visual fields. In the optic chaism, fibers from the nasal aspects of both retinas cross to opposite sides, and fibers from the temporal portions remain uncrossed. Both crossed and uncrossed fibers form the optic tracts. Injury to one of the optic nerves can cause blindness in the corresponding eye; injury or lesion in the optic chiasm can cause partial vision loss (for example, loss of the two temporal visual fields).

**Image perception and formation**

Image formation begins when the cornea, aqueous humor, lens, and vitreous humor refract light rays from an object, focusing them on the central fovea as an inverted and reversed image. In the retina, rods and cones transduce this projected image into nerve impulses and transmit them to the two optic nerves.

The impulses travel to the optic chiasm where the optic nerves unite, split again into two optic tracts, and then continue into the optic region of the cerebral cortex. There, the inverted and reversed images from the retina change back to their original form and upright orientation.

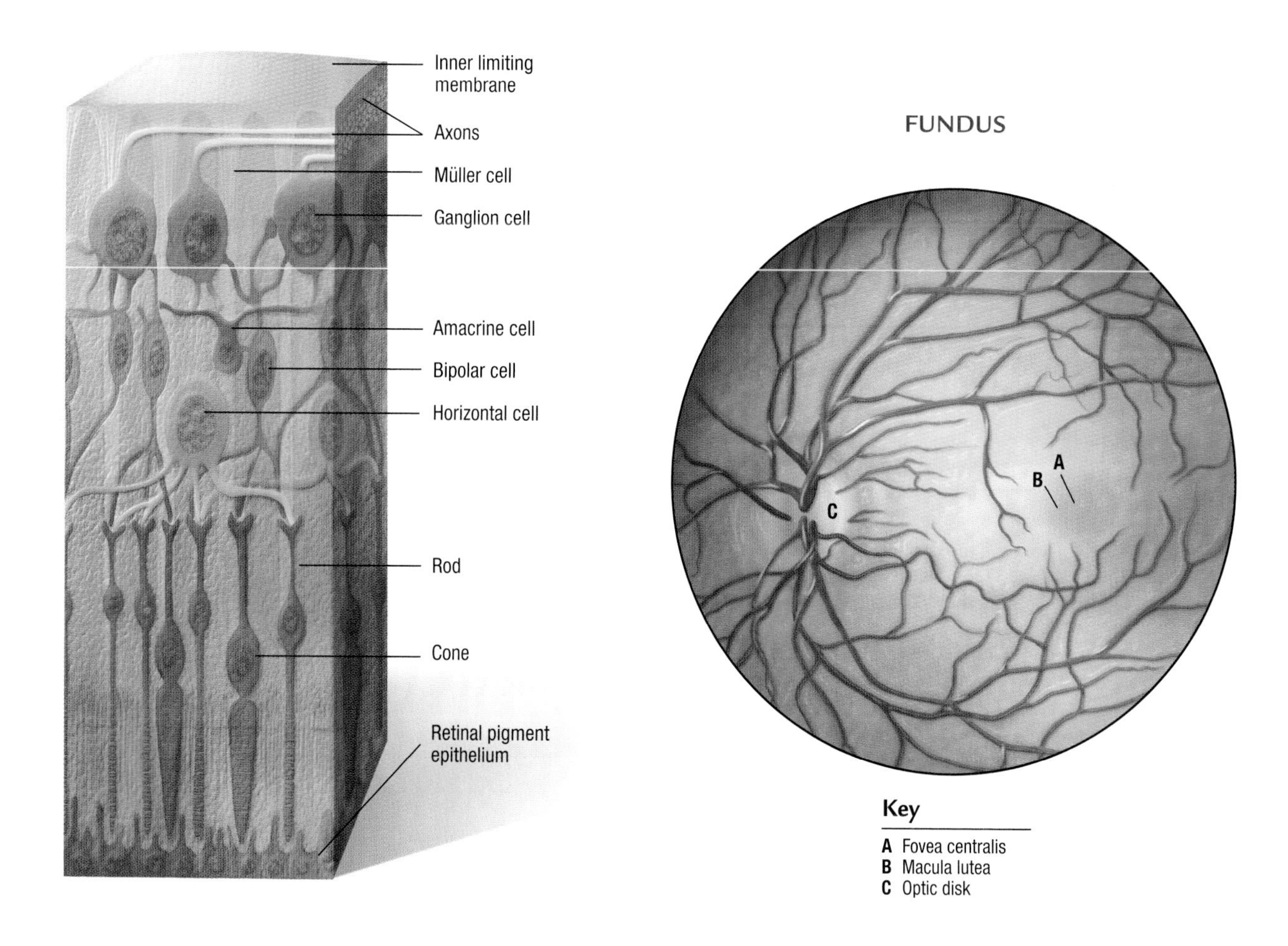
RETINA
Inner limiting membrane
Axons
Müller cell
Ganglion cell
Amacrine cell
Bipolar cell
Horizontal cell
Rod
Cone
Retinal pigment epithelium
FUNDUS
A
B
C
Key
A Fovea centralis
B Macula lutea
C Optic disk

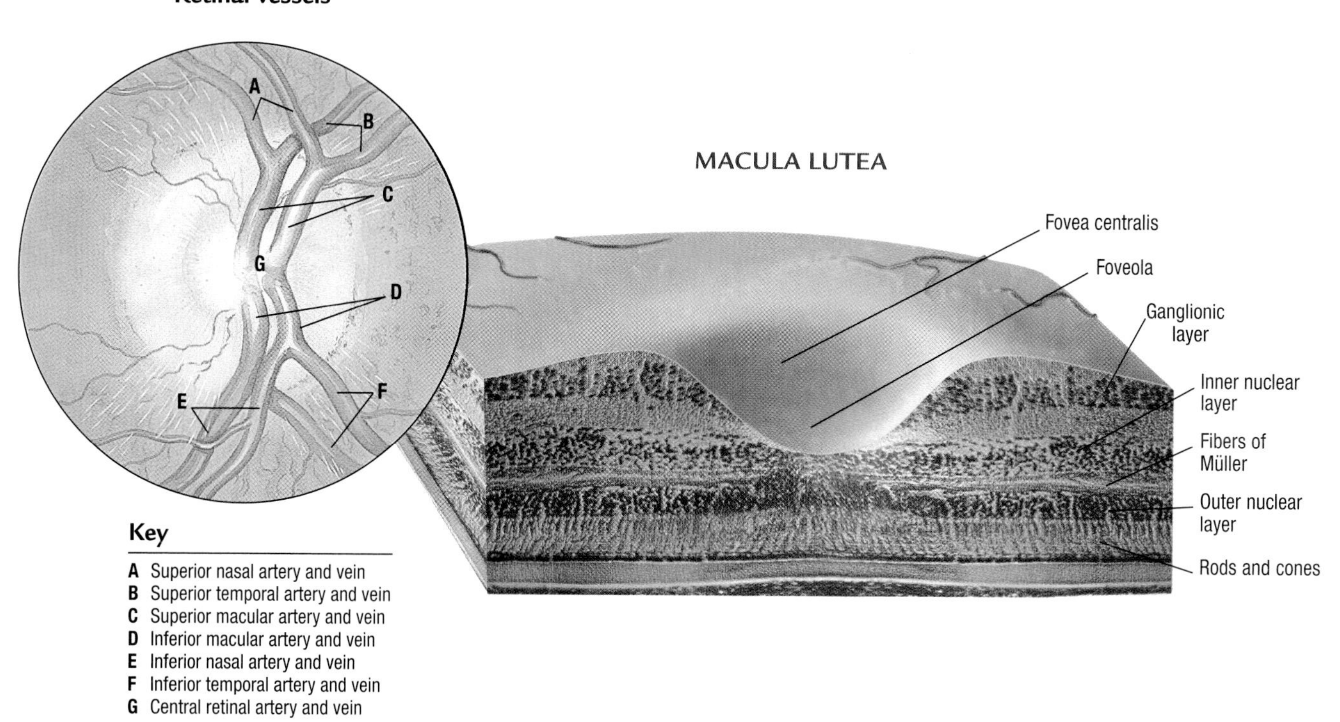
OPTIC DISK
Retinal vessels
A
B
C
G
D
E
F
Key
A Superior nasal artery and vein
B Superior temporal artery and vein
C Superior macular artery and vein
D Inferior macular artery and vein
E Inferior nasal artery and vein
F Inferior temporal artery and vein
G Central retinal artery and vein
MACULA LUTEA
Fovea centralis
Foveola
Ganglionic layer
Inner nuclear layer
Fibers of Müller
Outer nuclear layer
Rods and cones

# Ear

The ears gather sound waves and transmit them as nerve impulses to the brain; the brain interprets these impulses as hearing. The ear is divided into three main parts:
- external ear
- middle ear
- inner ear.

## External ear

The external or outer ear consists of the auricle (pinna) and the external auditory canal. (See *Ear structures*, page 58.)

### Auricle

The auricle consists of cartilaginous anthelix, crux of the helix, lobule, tragus, and concha. Although not part of the external ear, the mastoid process of the temporal bone is an important bony landmark behind the lower part of the auricle.

### External auditory canal

The external auditory canal is a narrow chamber measuring about 1″ (2.5 cm) long. This canal connects the auricle with the tympanic membrane in the middle ear. Thin, sensitive skin covers the cartilage that forms the outer third of the external auditory canal; bone covered by thin skin forms the inner two-thirds. Ceruminous glands, modified apocrine sweat glands, in this thin skin secrete cerumen, a brownish-yellow earwax.

**Clinical tip**

Two external signs of otitis externa are tenderness of the helix when the ear is pulled back and tenderness of the tragus, the cartilaginous projection anterior to the external opening of the ear, when it is palpated.

## Middle ear

Also called the tympanic cavity, the middle ear is an air-filled cavity within the petrous (hard) portion of the temporal bone. It contains three small bones that transmit sound. Lined with mucosa, it's bounded laterally by the tympanic membrane and medially by the oval and round windows. (See *Middle ear*, page 59, and *Middle ear development*, page 57.)

### Tympanic membrane

The tympanic membrane, consisting of layers of skin, fibrous tissue, and mucous membrane, transmits sound vibrations to the internal ear. (See *Right tympanic membrane*, page 59.) Through an otoscope it appears pearly gray, shiny, and translucent. The center of the tympanic membrane, called the umbo, covers the long process of the malleus. Around the outer border of the membrane is a pale-white, fibrous ring called the annulus.

### Oval and round windows

The oval window (fenestra ovalis) is an opening in the wall between the middle and inner ear into which the footpiece of the stapes (a tiny bone of the middle ear) fits; it transmits vibrations to the inner ear. The round window (fenestra cochleae), another opening in the same wall, is enclosed by the secondary tympanic membrane. Like the oval window, the round window transmits vibrations to the inner ear.

**Clinical tip**

**Otoscopic examination**

When examining an adult, before inserting the speculum of the otoscope into the external canal, tilt the patient's head away from you. Then grasp the superior posterior auricle with your thumb and index finger and gently pull it up and back to straighten the canal. If your patient is a child under age 3, pull the auricle down to get a good view of the tympanic membrane. Remember that everyone's ear is different and it may be necessary to vary the angle of the speculum until you can see the tympanic membrane.

Expect the following findings:
- tympanic membrane pearl gray and glistening
- annulus white and denser than the rest of the membrane
- no evidence of bleeding, bulging, retraction, lesions, or perforations, especially on the periphery
- light reflex between 4 and 6 o'clock in the right ear and between 6 and 8 o'clock in the left. If displaced or absent, the tympanic membrane may be bulging, inflamed, or retracted.
- bony landmarks: The *malleus* appears as a dense, white streak at the 12 o'clock position, and the *umbo*, the inferior point of the malleus, can be seen at the top of the light reflex.

### Middle ear bones

The middle ear contains three small bones called *ossicles*:
- malleus (hammer)
- incus
- stapes.

The malleus attaches to the tympanic membrane. The stapes attaches to the oval window. The incus lies between them. The three bones transmit vibratory motion — sound waves — from the eardrum to the inner ear, where the vibration excites receptor nerve endings. (See *Auditory ossicles*, page 59.)

### Eustachian tube

The eustachian tube connects the middle ear to the nasopharynx. This tube equalizes air pressure on either side of the tympanic membrane. A normally functioning eustachian tube keeps the middle ear free from contaminants from the nasopharynx. Upper respiratory tract infections can affect the tube by obstructing middle ear drainage and causing otitis media or effusion.

## EXTERNAL EAR

### Auricle

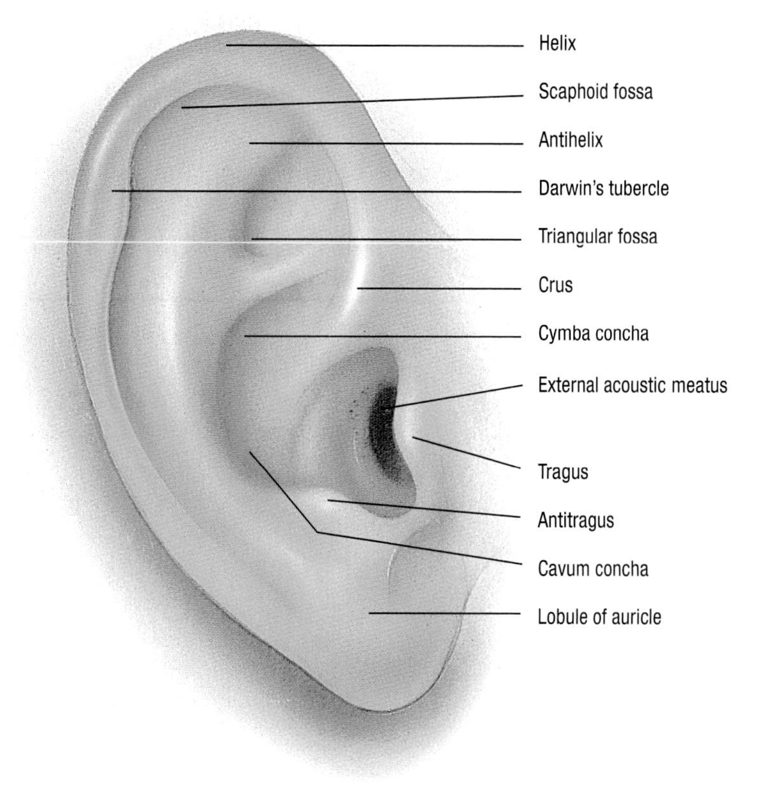
Helix
Scaphoid fossa
Antihelix
Darwin's tubercle
Triangular fossa
Crus
Cymba concha
External acoustic meatus
Tragus
Antitragus
Cavum concha
Lobule of auricle

**AGE-RELATED CHANGES**

The following illustration depicts the changes to the middle ear from infancy to adulthood.

### MIDDLE EAR DEVELOPMENT

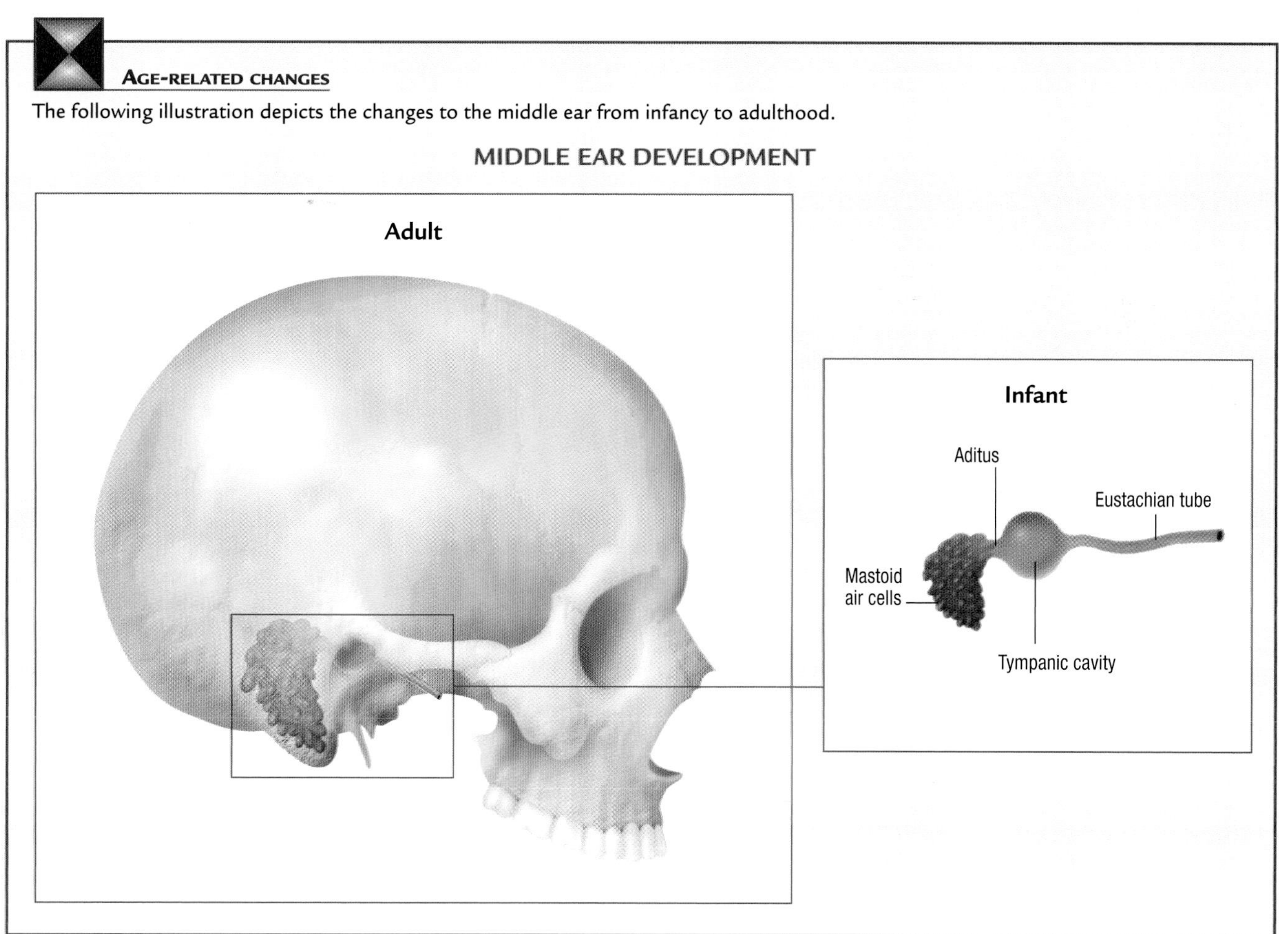
Adult
Infant
Aditus
Eustachian tube
Mastoid air cells
Tympanic cavity

## EAR STRUCTURES

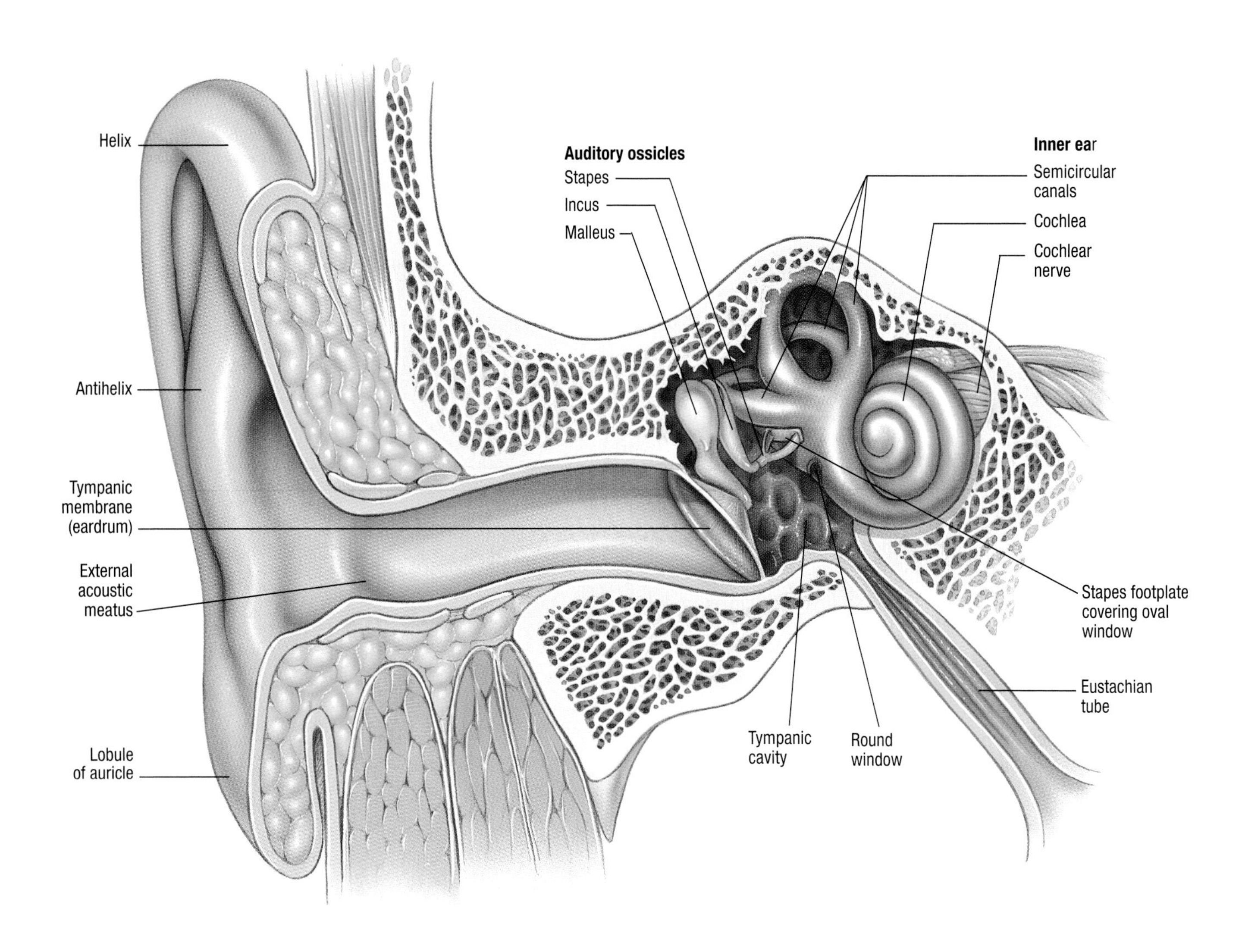

Helix
Antihelix
Tympanic membrane (eardrum)
External acoustic meatus
Lobule of auricle
Auditory ossicles
Stapes
Incus
Malleus
Inner ear
Semicircular canals
Cochlea
Cochlear nerve
Stapes footplate covering oval window
Eustachian tube
Tympanic cavity
Round window

## RIGHT TYMPANIC MEMBRANE

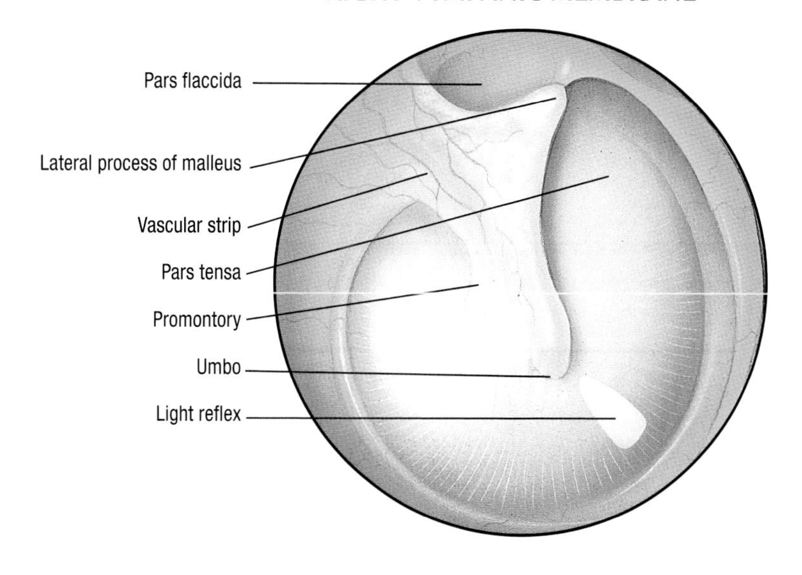
Pars flaccida
Lateral process of malleus
Vascular strip
Pars tensa
Promontory
Umbo
Light reflex

## MIDDLE EAR

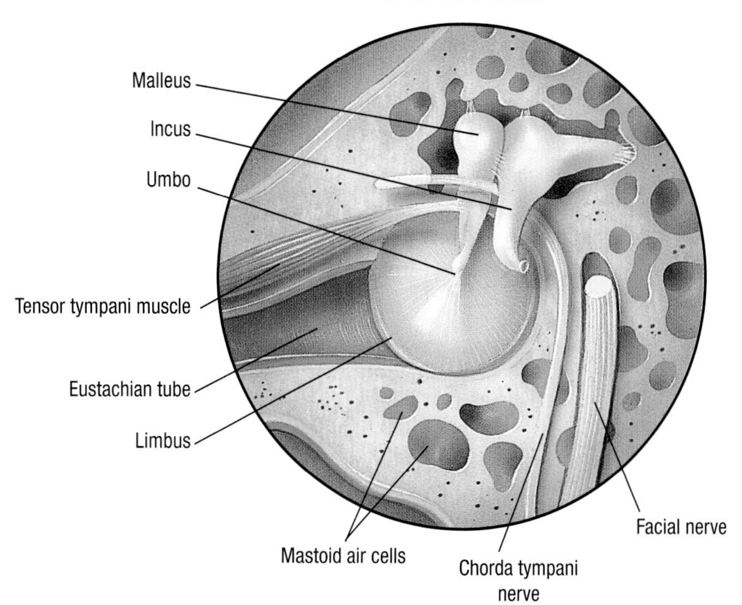
Malleus
Incus
Umbo
Tensor tympani muscle
Eustachian tube
Limbus
Mastoid air cells
Chorda tympani nerve
Facial nerve

## AUDITORY OSSICLES

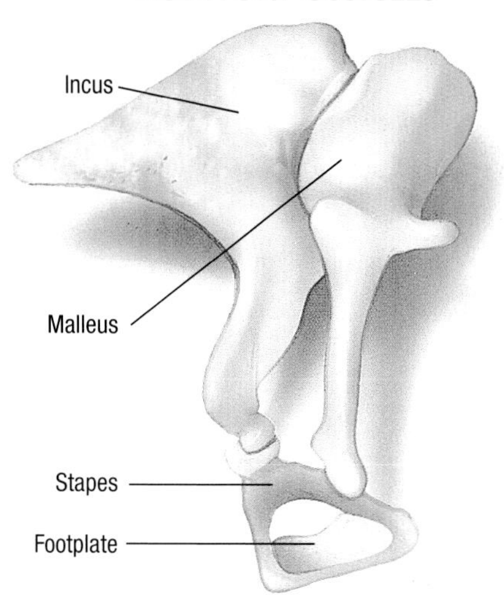
Incus
Malleus
Stapes
Footplate

# Inner ear and ear function

The inner ear is the third part of the auditory apparatus. It contains key structures important for hearing and equilibrium.

## Inner ear

The inner ear consists of closed, fluid-filled spaces within the temporal bone. It is a bony labyrinth, which includes three connected structures – the vestibule, the semicircular canals, and the cochlea. These structures are lined with a serous membrane that forms the membranous labyrinth. A fluid called perilymph fills the space between the bony labyrinth and the membranous labyrinth.

### Vestibule

The vestibule, the entrance to the inner ear, lies posterior to the cochlea and anterior to the semicircular canals. It houses two membranous sacs, the saccule and utricle, which sense gravity changes and linear and angular acceleration.

### Semicircular canals

The three semicircular canals project from the posterior aspect of the vestibule. Each canal is oriented in one of three planes – superior, posterior, or lateral. The semicircular duct, which traverses the canals, is continuous with the utricle anteriorly. At the juncture of each canal and the utricle lies a membranous ampulla. Within its center, the crista ampullaris, are tiny hair cells that respond to movement of endolymph in the canals, resulting in sensations of turning or spinning.

MEMBRANOUS AMPULLA
**Balance**

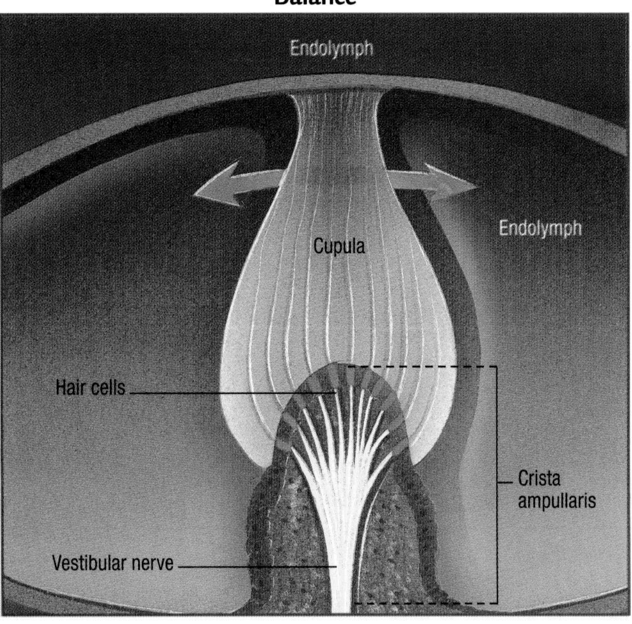
Endolymph
Endolymph
Cupula
Hair cells
Crista ampullaris
Vestibular nerve

### Cochlea

The cochlea, a bony, spiraling cone, extends from the anterior part of the vestibule. Within it lies the cochlear duct, a triangular, membranous structure housing the organ of Corti. The receptor organ for hearing, the organ of Corti transmits sound to the cochlear branch of the acoustic (CN VIII) nerve.

## Hearing

All three parts of the auditory apparatus play a role in hearing:
- external ear – *collects* sound
- middle ear – *conducts* sound
- inner ear – contains structures that *transmit* sound waves.

For hearing to occur, sound waves travel through the ear by two pathways – air conduction and bone conduction. Air conduction occurs when sound waves travel in the air through the external and middle ear to the inner ear. Bone conduction occurs when sound waves travel through bone to the inner ear.

Vibrations transmitted through air and bone stimulate nerve impulses in the inner ear. The cochlear branch of the acoustic nerve transmits these vibrations to the auditory area of the cerebral cortex, which then interprets the sound.

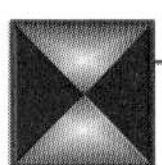

**Age-related changes**

Many older people lose some degree of hearing. Possible causes include:
- gradual buildup of cerumen
- slow, progressing deafness of aging, called presbycusis or senile deafness. This irreversible, bilateral sensorineural hearing loss usually starts at middle age, slowly worsens, and affects more men than women.

Presbycusis appears in four forms. The most common form is sensory presbycusis, caused by atrophy of the organ of Corti and the auditory nerve. The accompanying hearing loss occurs mostly in the high pitch ranges. By age 60, most adults have difficulty hearing above 4,000 Hz (normal speech is 500 – 2,000 Hz). Older adults can't easily distinguish the high-pitched consonants: s, z, t, f, and g.

## Equilibrium

The sense of equilibrium, or balance, maintains stability during movement or at rest. The vestibular apparatus of the inner ear controls the sense of equilibrium and position. It consists of:
- membranous labyrinth, or semicircular canals, which respond to motion
- utricle and the saccule, which respond to position changes.

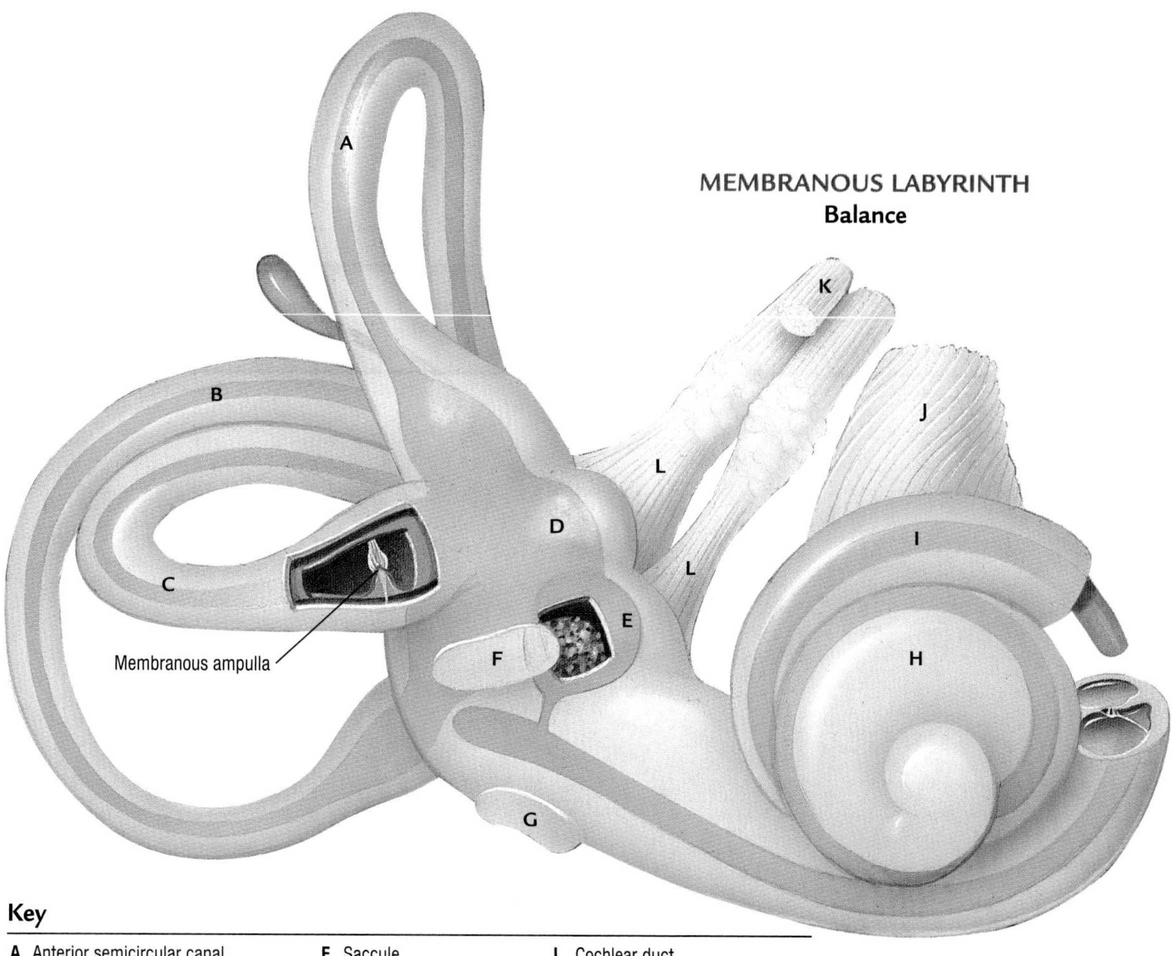
MEMBRANOUS LABYRINTH
Balance
A
B
C
D
E
F
G
H
I
J
K
L
L
Membranous ampulla

### Key

**A** Anterior semicircular canal
**B** Posterior semicircular canal
**C** Lateral semicircular canal
**D** Utricle
**E** Saccule
**F** Oval window
**G** Round window
**H** Cochlea
**I** Cochlear duct
**J** Cochlear nerve
**K** Facial nerve
**L** Vestibular nerve

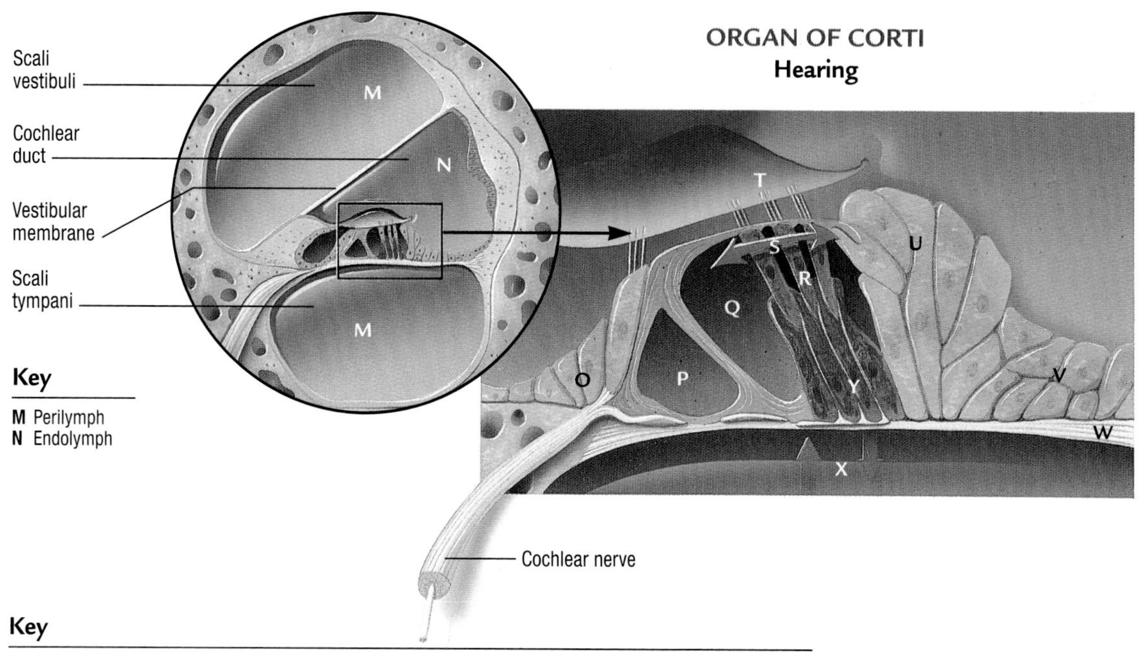
ORGAN OF CORTI
Hearing
Scali vestibuli
Cochlear duct
Vestibular membrane
Scali tympani
M
N
M
O
P
Q
R
S
T
U
V
W
X
Y
Cochlear nerve

### Key

**M** Perilymph
**N** Endolymph

### Key

**O** Inner hair cells
**P** Tunnel of Corti
**Q** Nuel's space
**R** Outer hair cells
**S** Movement of hair cells by sound waves
**T** Tectorial membrane
**U** Cells of Hensen
**V** Cells of Claudius
**W** Basilar membrane
**X** Vibration in basilar membrane
**Y** Cells of Deiters

# Nose

The nose is formed superiorly by the nasal and frontal bones, laterally and internally by the ethmoid bones, and inferiorly by flexible plates of hyaline cartilage (lateral and alar plates).

## Functions

The nose has four main functions, as follows:
- Acts as the usual site of inspiration (inhalation) and expiration (exhalation)
- Filters, warms, and moistens air
- Functions in the sensation of smell by containing olfactory (smell) receptors
- Acts as a resonating chamber for speech sounds.

## Nasal structures

The lower two-thirds of the external nose consists of flexible cartilage, and the upper one-third is rigid bone. The anterior portion of the nose, called the vestibule, is lined with coarse hair that helps filter out dust. The nose has two nasal cavities, which open on the face through the anterior nasal apertures (nares); the nasal septum separates the cavities. The anterior portion of the nasal septum is composed of hyaline cartilage. (See *Nasal septum* and *Lateral wall of nose,* page 64.)

**Age-related changes**

**Nasal cartilage continues to grow as people age, resulting in nose enlargement. At birth and in early life, the nasal septum is straight. The septum becomes slightly deviated or deformed in almost every adult.**

The posterior portion of the nose is composed of a flat bone called the vomer and the perpendicular plate of the ethmoid bone. (For an illustration of these bones, see pages 15 and 16.) Posteriorly, the nasal cavities join the pharynx through the internal nares (choanae).

### Conchae

Each nasal cavity has three mucosa-covered passageways, called the inferior, middle, and superior nasal conchae, also referred to as the nasal turbinates. The curved bony conchae and their mucosal covering ease breathing by warming, filtering, and humidifying inhaled air.

The conchae form four passages in the nose, as follows:
- sphenoethmoid recess – above the superior concha in front of the sphenoid body
- superior nasal meatus – below the superior concha
- middle nasal meatus – below the middle concha
- inferior nasal meatus – below the inferior concha

The conchae are lined with cilia (microscopic hairlike structures) that trap dust and foreign particles before they reach the lungs. Epithelial cells lining the nasal cavities secrete mucus, which collects foreign particles. Ciliated cells of the nasal mucosa move mucus toward the pharynx, where it is removed by swallowing, sneezing, or spitting.

### Paranasal sinuses

Four pairs of paranasal sinuses open into the internal nose. (See *Paranasal sinuses*, page 65). The sinuses are named according to their associated bones, as follows:
- maxillary sinuses – in the upper jaw below the eyes
- frontal sinuses – above the eyebrows
- ethmoidal and sphenoidal sinuses – behind the orbits and nasal cavities.

The small openings between the sinuses and the nasal cavity can become obstructed easily because they are lined with mucous membranes that can become inflamed and swollen. The maxillary and frontal sinuses are easily assessed. However, the ethmoid and sphenoid sinuses are not readily accessible.

**Clinical tip**

**Transillumination of the sinuses helps detect obstructions and tumors. All you need is a penlight and a darkened room.**
- **Have the patient close his eyes.**
- **Place the penlight on the supraorbital ring, and direct the light upward to illuminate the frontal sinuses just above the eyebrow.**
- **Place the penlight on the patient's cheekbone just below the eye and ask him to open his mouth. The light should transilluminate the maxillary sinuses easily and equally.**

## Sense of smell

The mucosal epithelium that lines the uppermost portion of the nasal cavity contains receptors for fibers of the olfactory (CN I) nerves, called olfactory (smell) receptors. Air must pass through the nose to stimulate the olfactory receptors; therefore, a person can't smell if the nostrils are blocked by mucus, polyps, or other substances.

### Olfactory receptors

Each olfactory receptor has a dendrite that terminates in several long cilia called olfactory hairs. These hairs contain the receptor proteins that transduce odorants into nerve impulses.

**Physiology**

**After smelling an odor for a short time, a person is no longer able to perceive it as intensely. This loss of odor sensitivity, called adaptation, results from reduced responsiveness of the olfactory receptors and diminished perception of odor in the olfactory portion of the cerebral cortex.**

**Sagittal section**

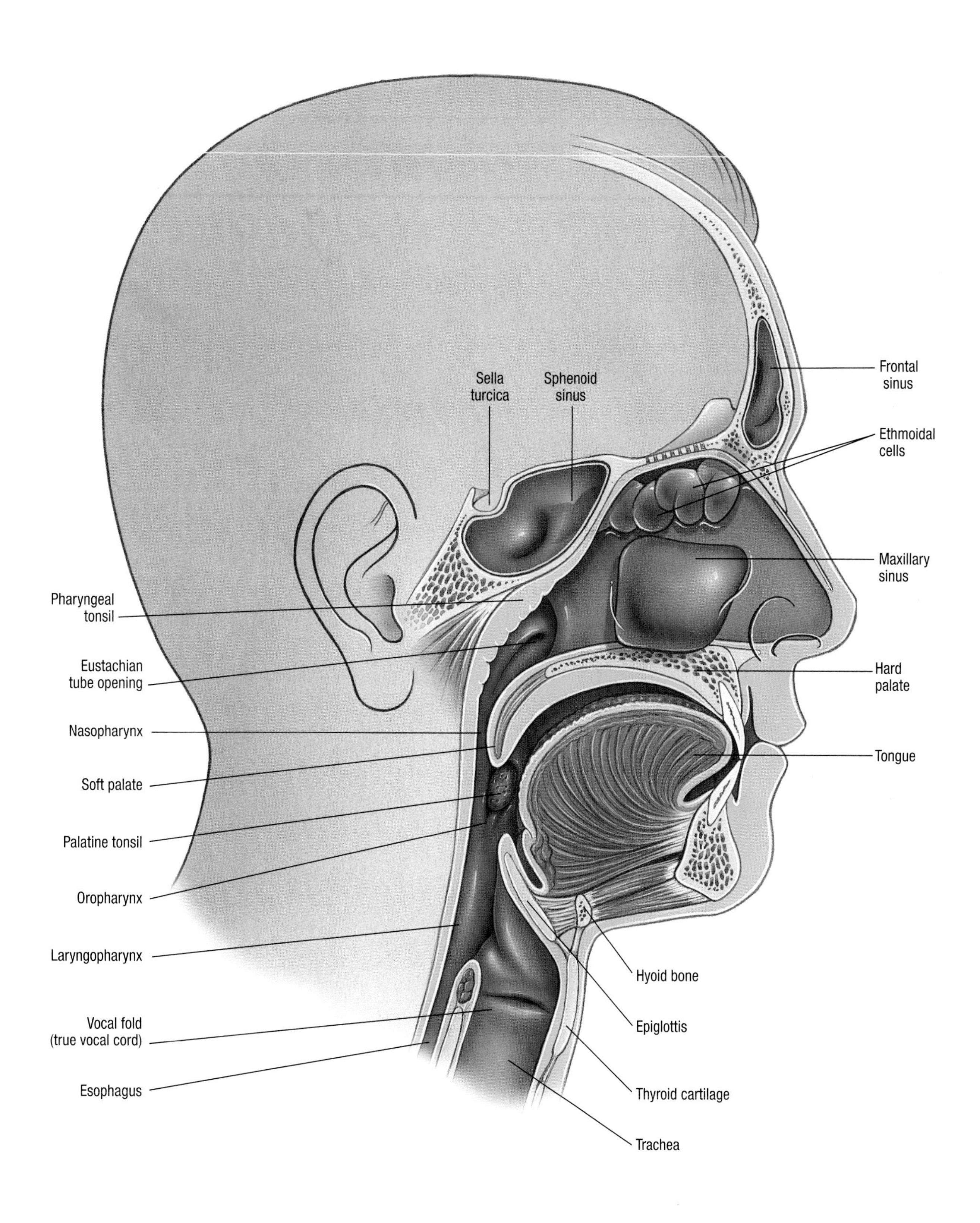
Sella turcica
Sphenoid sinus
Frontal sinus
Ethmoidal cells
Maxillary sinus
Pharyngeal tonsil
Eustachian tube opening
Hard palate
Nasopharynx
Tongue
Soft palate
Palatine tonsil
Oropharynx
Laryngopharynx
Hyoid bone
Epiglottis
Vocal fold (true vocal cord)
Esophagus
Thyroid cartilage
Trachea

## NASAL SEPTUM

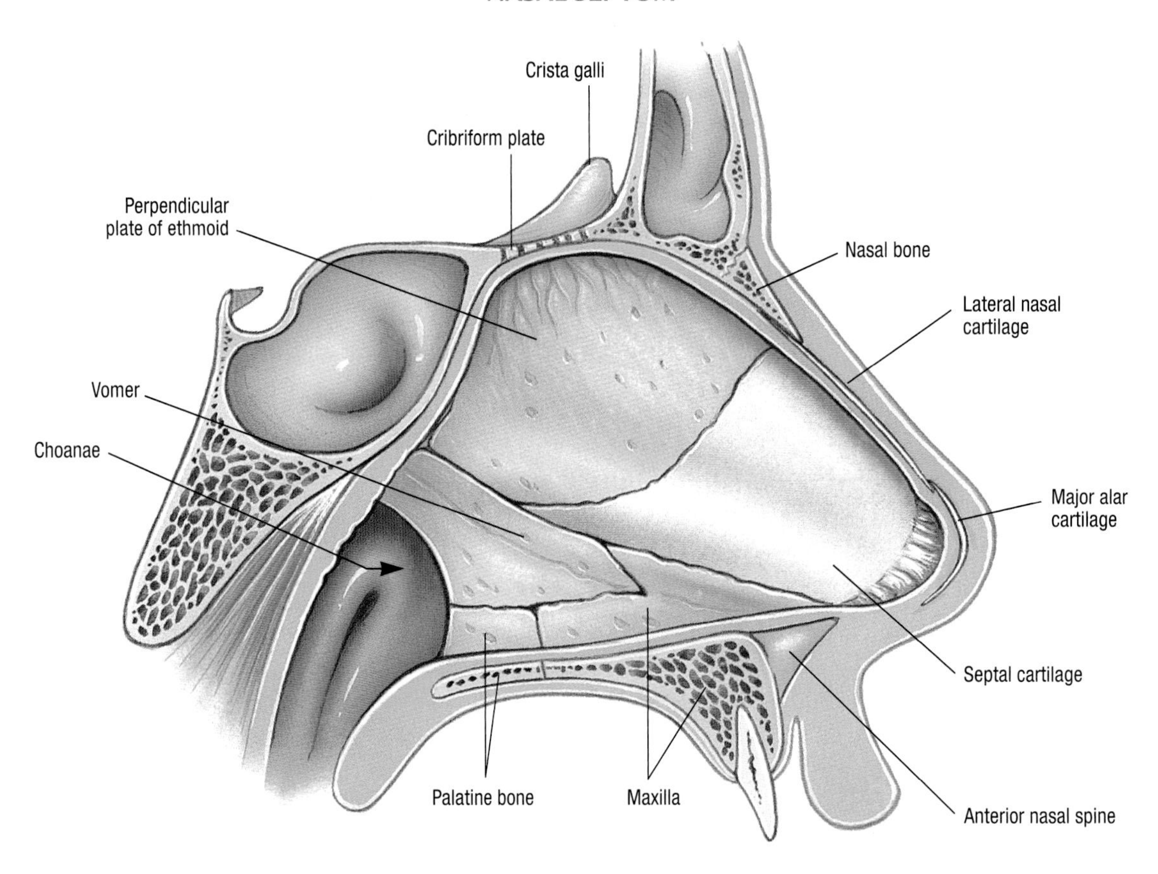

Crista galli
Cribriform plate
Perpendicular plate of ethmoid
Nasal bone
Lateral nasal cartilage
Vomer
Choanae
Major alar cartilage
Septal cartilage
Palatine bone
Maxilla
Anterior nasal spine


## LATERAL WALL OF NOSE

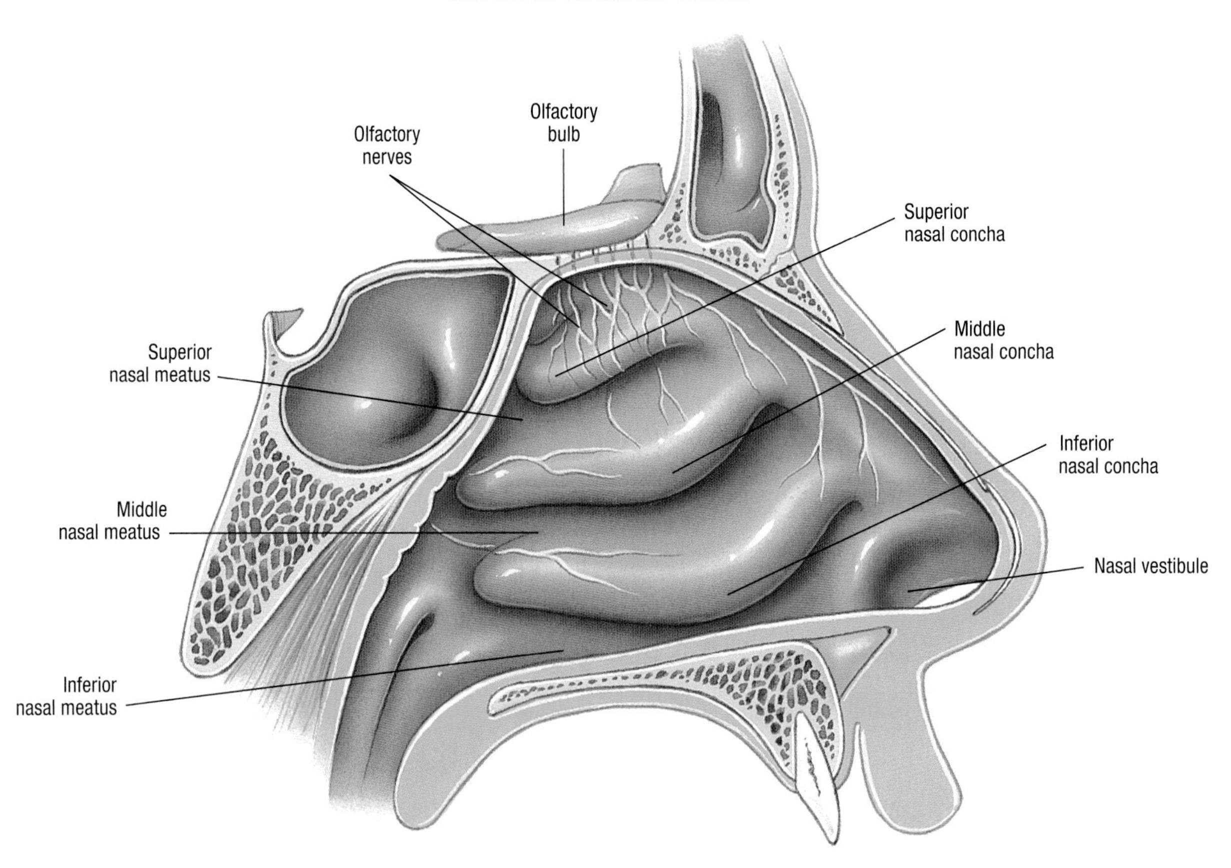

Olfactory bulb
Olfactory nerves
Superior nasal concha
Superior nasal meatus
Middle nasal concha
Inferior nasal concha
Middle nasal meatus
Nasal vestibule
Inferior nasal meatus

## PARANASAL SINUSES

### Anterior view

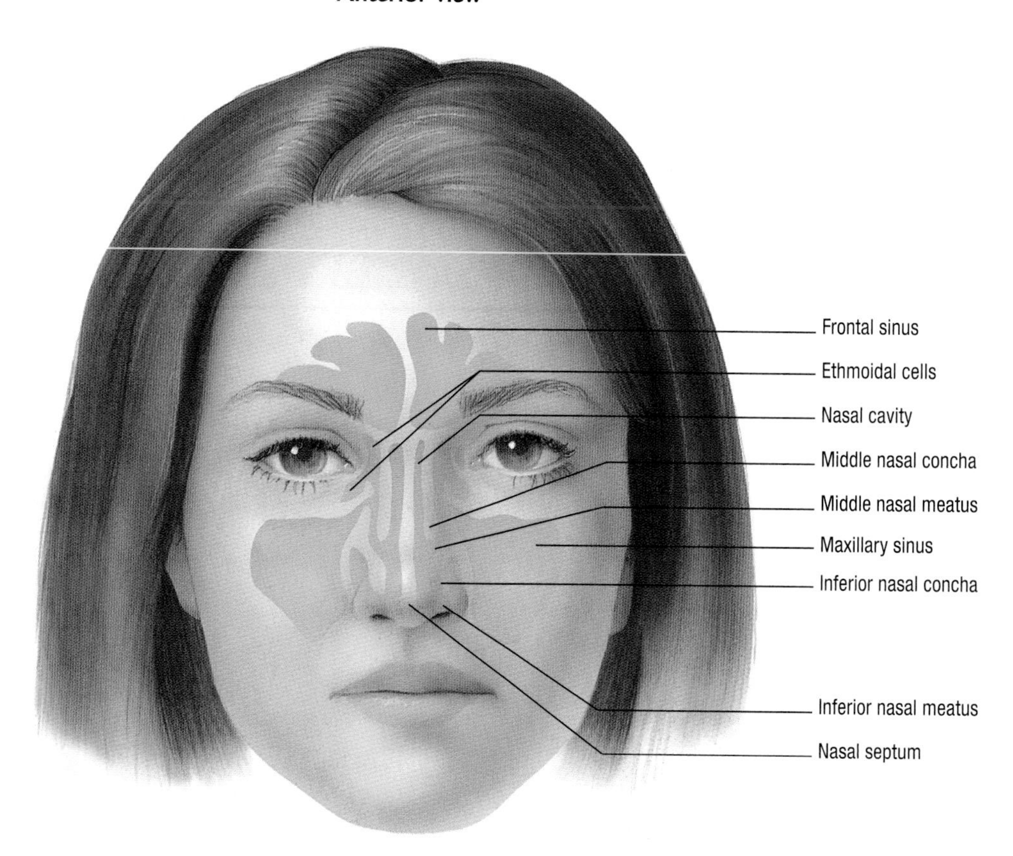
Frontal sinus
Ethmoidal cells
Nasal cavity
Middle nasal concha
Middle nasal meatus
Maxillary sinus
Inferior nasal concha
Inferior nasal meatus
Nasal septum

### Lateral view (Conchae removed)

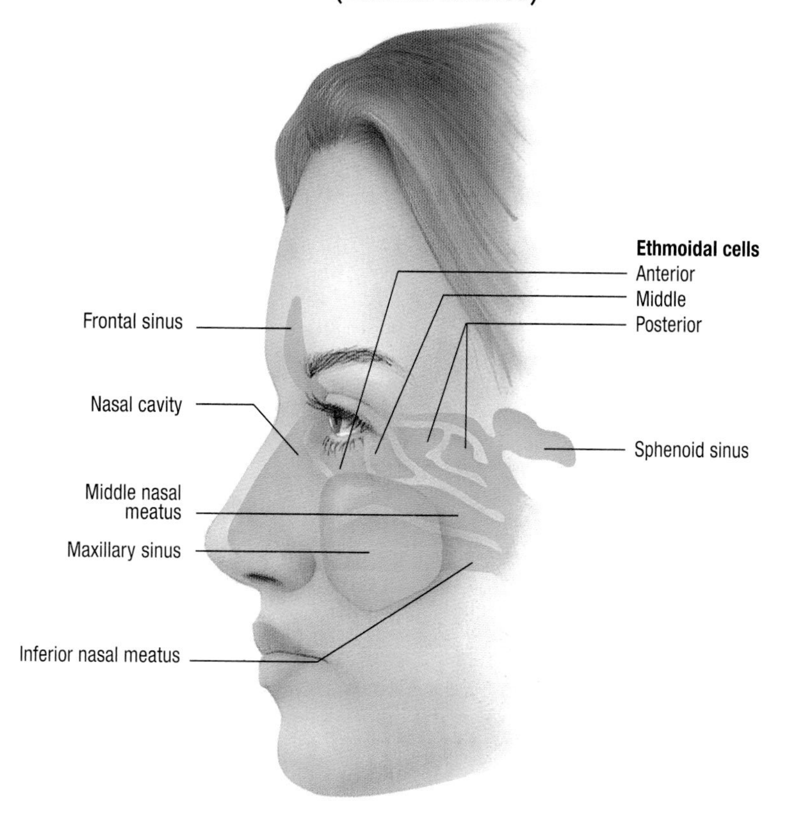
Ethmoidal cells
Anterior
Middle
Posterior
Frontal sinus
Nasal cavity
Sphenoid sinus
Middle nasal meatus
Maxillary sinus
Inferior nasal meatus

# Temporomandibular joint

The temporomandibular joint (TMJ) is a combined hinge and gliding joint that connects the mandible to the temporal bone. The only movable joint in the skull, it allows the mandible to move up and down, and from side to side.

## Components

The TMJ is formed by the following:
- parts of the mandibular fossa of the temporal bone
- articular tubercles
- mandibular condyles
- right and left bicondylar articulation.

The fibrous capsule of the joint attaches to the temporal bone and the mandible.

Clinical tip

### Checking range of motion

An easy way to evaluate the range of motion in the temporomandibular joint is to place the tips of your first two or three fingers in front of the middle of the ear. Ask the patient to open and close his mouth. Then place your fingers into the depressed area over the joint. Note the movement of the mandible. The patient should be able to open and close his jaw and protract and retract his mandible easily without pain or tenderness. An action such as a wide yawn or taking a large bite can actually dislocate the mandibular condyle.

Signs of possible TMJ disorder include:
- limited movement or locking of the jaw
- radiating pain in the face, neck, or shoulders
- painful clicking, popping, or grating sounds during opening or closure of the mouth
- sudden, major change in the way the upper and lower teeth fit together
- swelling
- pain.

### Condyles

The rounded ends of the mandible are called condyles. When the mouth is opened, the condyles glide along the joint socket of the temporal bone. The condyles slide back to their original position when the mouth is closed.

### Articular disk

The cartilagenous articular disk lies between the condyle and the temporal bone, dividing the area into two compartments. Synovial membrane lines the capsules superior and inferior to the disk. The disk cushions the bones of the TMJ when the mouth is opening and closing. When the mouth opens, the mandibular condyle rotates on a horizontal axis. At the same time, the condyle and disk glide forward and downward on the articular eminence. The articular disk functions to absorb the forces produced by chewing and other movements.

## Innervation and blood supply

Numerous branches of the facial (CN VII) and trigeminal (CN V) nerves are in the TMJ area. Offshoots of the mandibular branch of the trigeminal nerve, the auricotemporal and masseteric nerves, innervate the TMJ region. The superficial temporal and maxillary branches of the external carotid artery provide the blood supply.

## Muscles and ligaments

Muscles attached to and surrounding the jaw joint control position and movement.
- Opening – lateral pterygoid and suprahyoid (digastric) muscles
- Closing – temporalis, masseter, and medial pterygoid muscles
- Protrusion – lateral and medial pterygoid, and masseter muscles
- Retrusion – temporalis and masseter muscles.

The superior head of the lateral pterygoid muscle has its insertion at the intra-articular disk of the TMJ and pulls it laterally. Ligaments in the TMJ region include the sphenomandibular, stylomandibular, and lateral ligaments.

## OPEN JAW

## CLOSED JAW

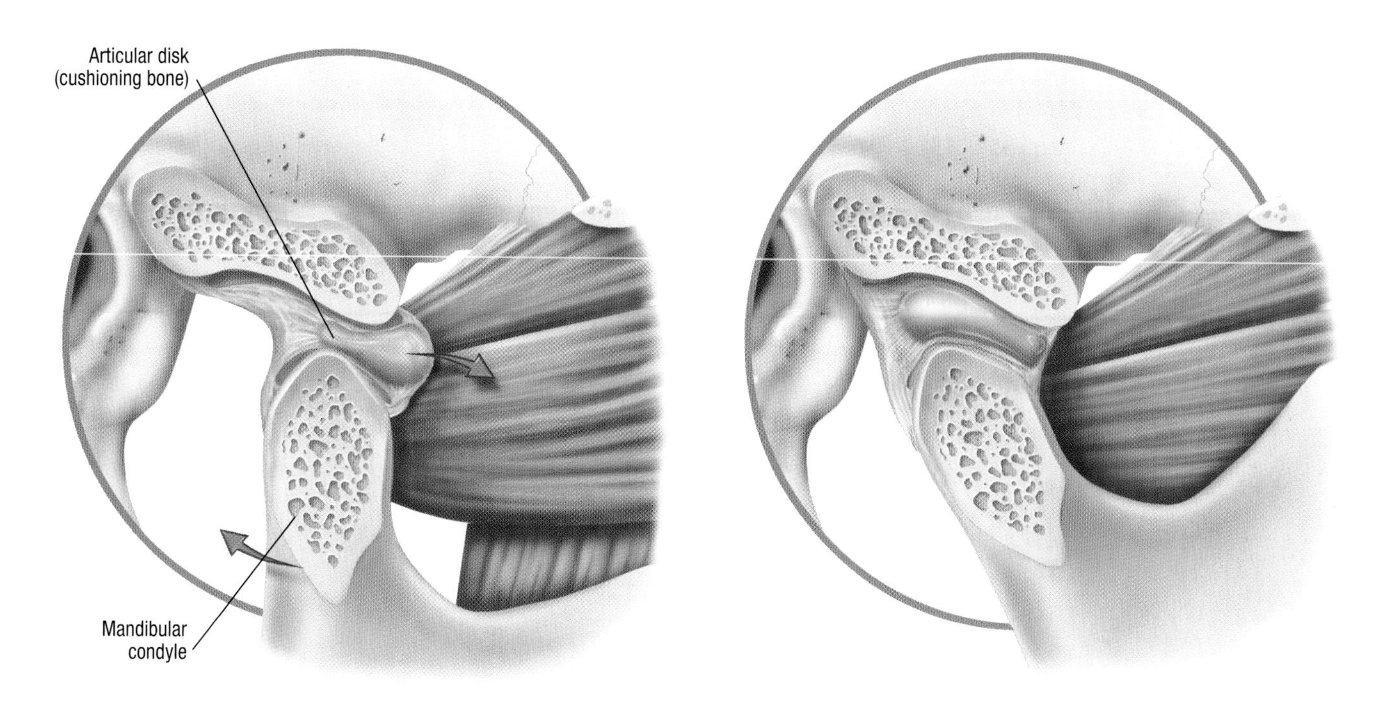

Articular disk
(cushioning bone)
Mandibular
condyle


## NERVES OF THE TEMPOROMANDIBULAR REGION

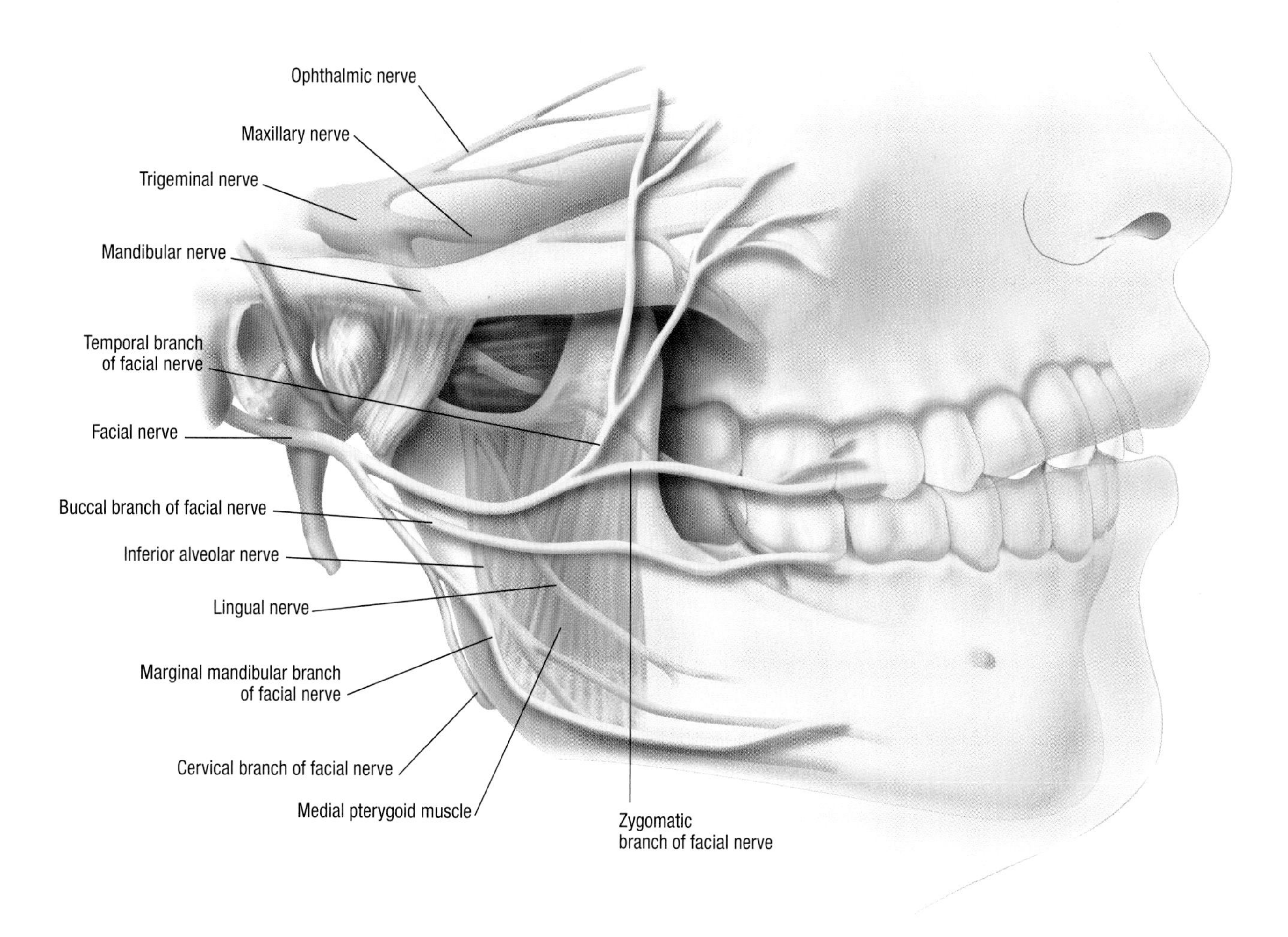

Ophthalmic nerve
Maxillary nerve
Trigeminal nerve
Mandibular nerve
Temporal branch
of facial nerve
Facial nerve
Buccal branch of facial nerve
Inferior alveolar nerve
Lingual nerve
Marginal mandibular branch
of facial nerve
Cervical branch of facial nerve
Medial pterygoid muscle
Zygomatic
branch of facial nerve

# Oral cavity

Also called the buccal cavity or mouth, the oral cavity receives ingested food and fluids and prepares them for swallowing.

## Role in digestion

The oral cavity is continuous with other gastrointestinal structures. Teeth cut and grind food into small particles. The salivary glands secrete saliva into the mouth. The food mixes with saliva to form a pliable mass (bolus) that can be easily swallowed. Saliva contains the enzymes amylase, which begins starch digestion, and lysozyme, which kills bacteria.

## Structures

The oral cavity is bounded by the lips (labia), cheeks, palate, and tongue and in adults contains 32 teeth. Gingiva, supportive tissue for the teeth, cover the crowns of unerupted teeth and encircle the necks of erupted teeth. For more information on the teeth, see page 70.

The oral cavity is lined with mucous membranes and is divided into two parts.

- The oral vestibule is bounded by the teeth, buccal gingiva, lips, and cheeks.
- The oral cavity proper is bounded laterally and anteriorly by the teeth, posteriorly by the pharynx, and inferiorly by the tongue. The oral cavity joins the pharynx at a junction called the fauces.

The nerve supply for the oral cavity is primarily from the trigeminal (CN V) nerve; some areas are innervated by the facial (CN VII, for taste sensation) and glossopharyngeal (CN IX) nerves. The blood supply is provided by branches of the external carotid artery.

### Lips and cheeks

The lips are covered by skin on the outside and are lined by mucous membranes. Muscles of the lips are the orbicularis oris and the superior and inferior labial muscles. The cheeks are composed of striated muscle tissue, primarily the buccinator muscle.

### Salivary glands

Ducts connect the mouth with three major pairs of salivary glands:

- parotid
- submandibular
- sublingual.

The parotid gland is the largest of the three main pairs of salivary glands. It's situated at the side of the face in front of and below the external ear. The capsule that surrounds it provides little room for swelling if it becomes inflamed. Five terminal branches of the facial (CN VII) nerve and blood vessels pass through the parotid gland. Swelling of one or both of the parotid glands is apparent when a person has mumps, an acute infectious disease caused by a paramyxovirus.

The submandibular glands lie in the submandibular triangle, which extends from the digastric muscles to the stylomandibular ligament. The sublingual glands, the smallest of the salivary glands, are under the tongue.

### Tongue

The tongue has three sets of skeletal muscles. The genioglossus protracts, retracts, and depresses the tongue. The hyoglossus depresses the tongue and draws its sides downward. The styloglossus retracts and elevates the tongue. These muscles are innervated by the hypoglossal (CN XII) nerve. The trigeminal (CN V) nerve provides the sensory input to the anterior two-thirds of the tongue; the glossopharyngeal (CN IX), the posterior portion.

**Physiology**

The tongue and the roof of the mouth contain most of the receptors for the taste nerve fibers in branches of the facial (CN VII), glossopharyngeal (CN IX), and vagus nerves (CN X). Located on taste cells in the taste buds, these receptors are stimulated by chemicals. They respond to four taste sensations perceived by specific areas on the tongue:

- sweet — on the tip
- sour — along the sides
- bitter — on the back
- salty — on the tip and sides.

All the other flavors a person senses result from a combination of taste-bud stimulation and olfactory-receptor stimulation by molecules in air passing through the nose. If the sense of smell is not functioning properly (for example, if the nose is congested from a cold), the taste perception may be diminished or unusual.

### Palate

The palate separates the oral and nasal cavities and forms the roof of the mouth. It is divided into the soft (or fleshy) palate at the back of the mouth and the hard (or rigid) palate at the front of the mouth. The hard palate is formed by the palatine process of the maxillae and the horizontal plate of the palatine bone.

### Tonsils

Lingual tonsils, lymphatic tissues embedded into the mucous membrane at the base of the tongue, protrude into the oropharynx. For more information on tonsils and an illustration, see pages 72 and 75.

## ORAL CAVITY

### Lateral view

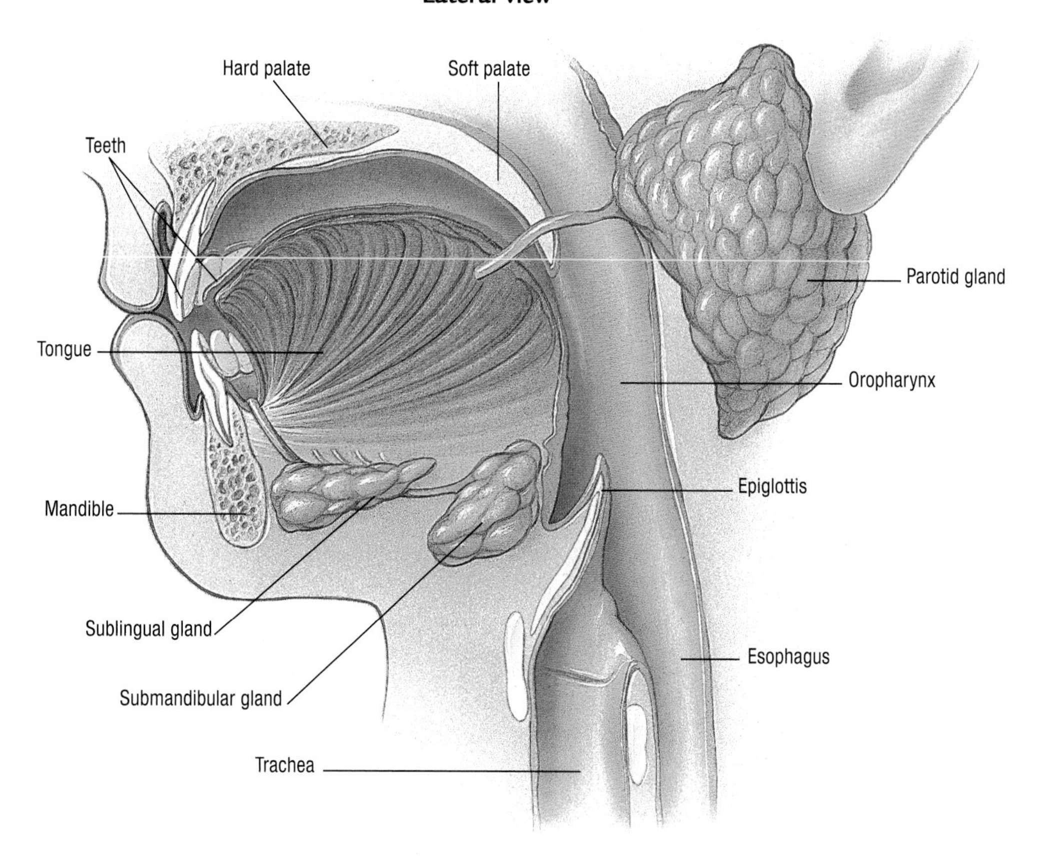

Hard palate
Soft palate
Teeth
Parotid gland
Tongue
Oropharynx
Epiglottis
Mandible
Sublingual gland
Esophagus
Submandibular gland
Trachea


### Anterior view

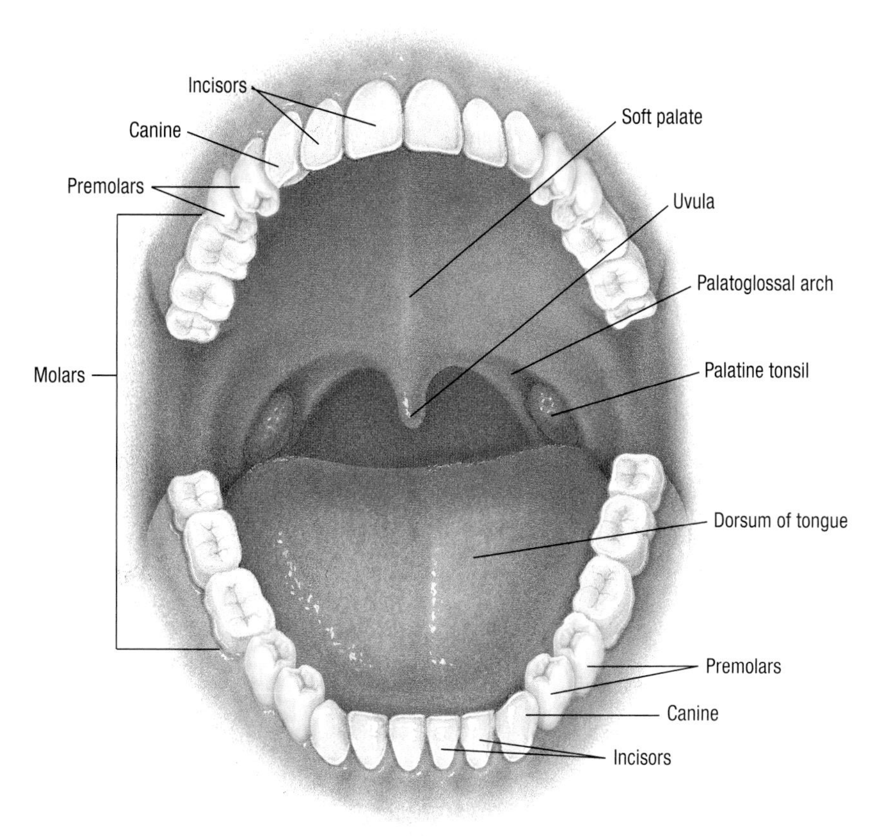

Incisors
Canine
Soft palate
Premolars
Uvula
Palatoglossal arch
Molars
Palatine tonsil
Dorsum of tongue
Premolars
Canine
Incisors

# Teeth

The most important function of the teeth is mastication. To prepare food for digestion, the teeth break it up into pieces small enough to swallow. Teeth are considered accessory digestive organs. Upper teeth are anchored in the alveoli (sockets) of the left and right maxillae; lower teeth, in the alveoli of the mandible.

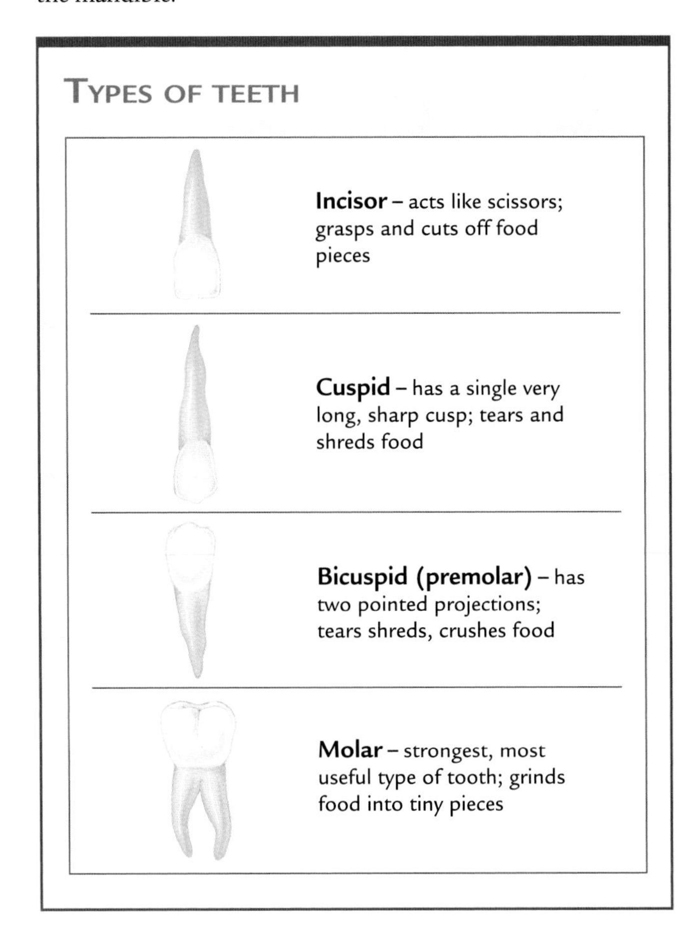

TYPES OF TEETH
Incisor – acts like scissors; grasps and cuts off food pieces
Cuspid – has a single very long, sharp cusp; tears and shreds food
Bicuspid (premolar) – has two pointed projections; tears shreds, crushes food
Molar – strongest, most useful type of tooth; grinds food into tiny pieces


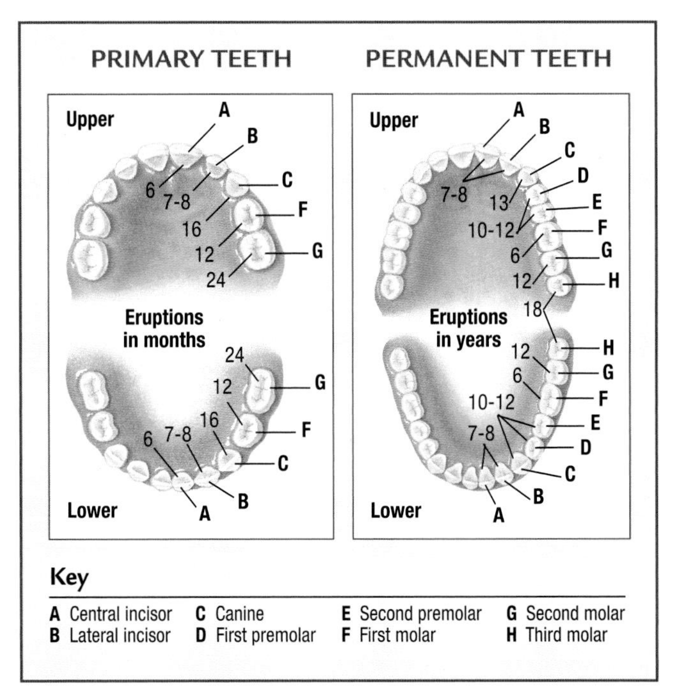

PRIMARY TEETH
PERMANENT TEETH
Upper
Lower
Eruptions in months
Eruptions in years
Key
A Central incisor
B Lateral incisor
C Canine
D First premolar
E Second premolar
F First molar
G Second molar
H Third molar


## Composition

The tooth consists of enamel, dentin, and pulp.

### Enamel

All exposed surfaces of the teeth are covered with enamel, the hardest tissue of the body. Enamel protects the underlying layers from food acids, heat, and cold. It's a shiny, hard, nonliving tissue that cannot repair itself once damaged.

### Dentin

Dentin is the second-hardest tissue in the body. It's the yellow substance under tooth enamel. Millions of tiny canals contain nerve fibers and odontoblast processes (the cells that form dentin). Dentin has a slight flexibility that protects teeth from breaking during chewing.

### Pulp

Pulp is the innermost part of the tooth. It holds tiny nerves and blood vessels. The root canal is a conduit for nerves and vessels between the tooth socket and the pulp area. A thin protective layer of cementum covers each tooth root. Cementum is similar to bone; it's alive and can repair itself.

## Ligaments

Periodontal ligaments anchor the teeth to the sockets and provide sensory information about the movements of the teeth. The ligaments provide a strong attachment for each tooth, while allowing movement during chewing and normal wear.

## Innervation and blood supply

The superior alveolar nerves are branches of the maxillary nerve. The inferior alveolar nerve is a branch of the mandibular nerve. The maxillary and mandibular nerves are branches of the trigeminal nerve (CN V). Each of the alveolar nerves forms a dental plexus from which branches leave to go to the roots of each tooth. The superior and inferior alveolar arteries and nerves innervate and provide the blood supply to the teeth. The superior and inferior alveolar veins provide venous drainage. The superior and inferior arteries are branches of the maxillary artery.

**Age-related changes**

Humans develop two sets of teeth over a lifetime. The 20 primary, or deciduous, teeth are replaced by 32 permanent teeth. A developing permanent tooth forms directly under the primary tooth it will replace. Eruption of the new tooth causes resorption of the primary tooth's root. Eventually, the primary tooth falls out.

## TOOTH ANATOMY

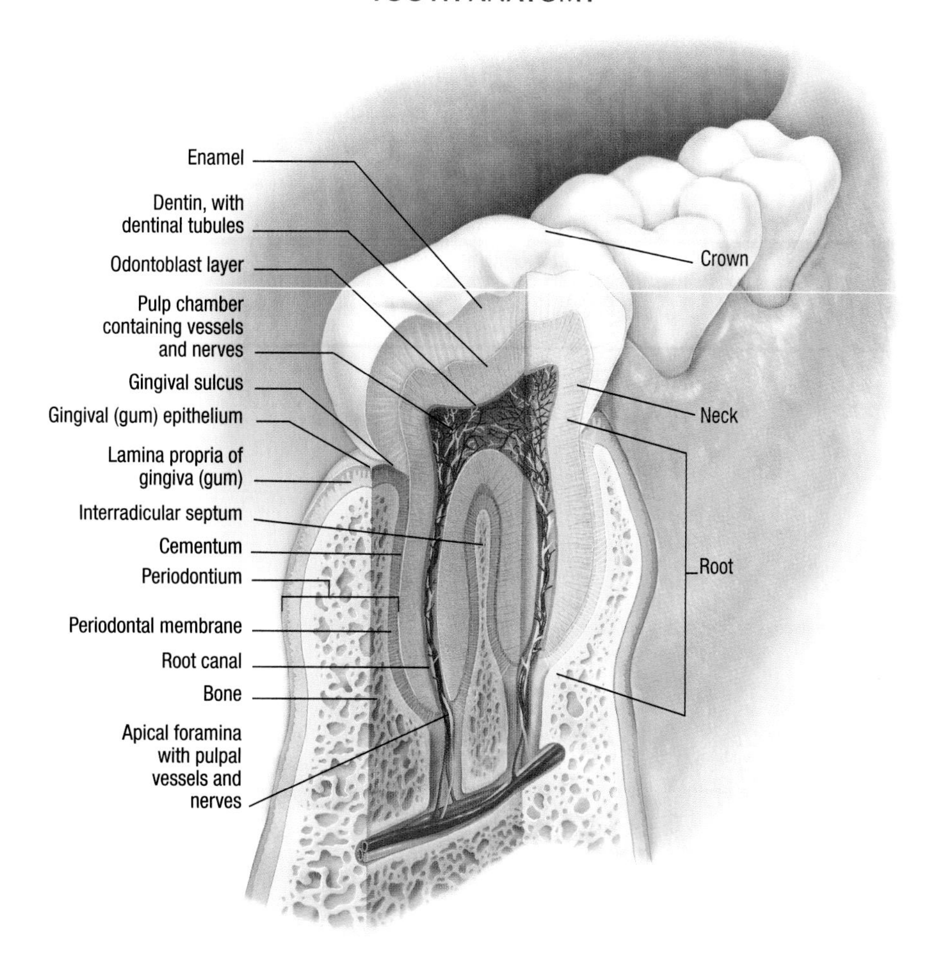

Enamel
Dentin, with dentinal tubules
Odontoblast layer
Pulp chamber containing vessels and nerves
Gingival sulcus
Gingival (gum) epithelium
Lamina propria of gingiva (gum)
Interradicular septum
Cementum
Periodontium
Periodontal membrane
Root canal
Bone
Apical foramina with pulpal vessels and nerves
Crown
Neck
Root


## TOOTH INNERVATION AND BLOOD SUPPLY

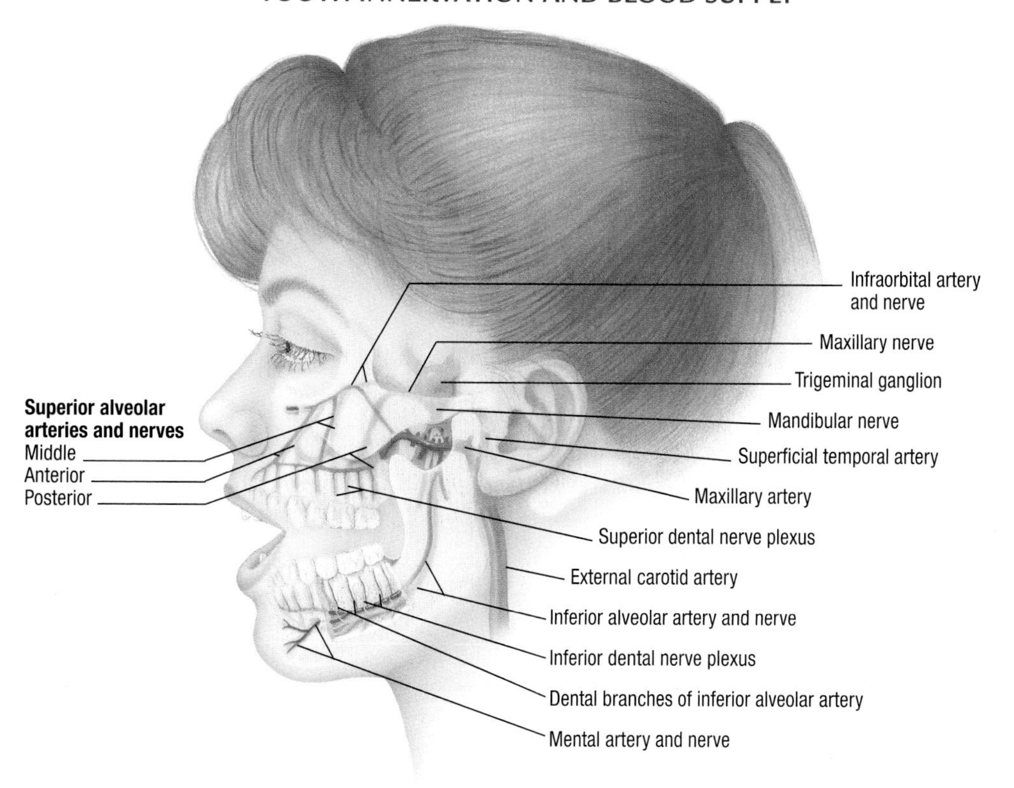

Superior alveolar arteries and nerves
Middle
Anterior
Posterior
Infraorbital artery and nerve
Maxillary nerve
Trigeminal ganglion
Mandibular nerve
Superficial temporal artery
Maxillary artery
Superior dental nerve plexus
External carotid artery
Inferior alveolar artery and nerve
Inferior dental nerve plexus
Dental branches of inferior alveolar artery
Mental artery and nerve

# Pharynx

The pharynx is a cavity that extends from the base of the skull to the esophagus (at the level of the sixth cervical vertebra). It extends from the junction of the nasal and oral cavities and splits into the larynx and the esophagus. The entire pharynx is about 13 cm (5 inches) long and is composed of striated muscle and lined with mucous membrane.

## Major functions

- Passageway for air entering from the nose
- Passageway for food entering the gastrointestinal tract
- Key role in speech, changing shape to allow phonation of vowel sounds.

## Divisions

The pharynx has three divisions: nasopharynx, oropharynx, and laryngopharynx.

### Nasopharynx

The nasopharynx is the most superior division of the pharynx. It is inferior to the sphenoid bone and lies at the level of the soft palate. The nasopharynx is lined with ciliated pseudostratified epithelium (respiratory epithelium). The pharyngeal tonsils (adenoids) are located on the posterior wall of the nasopharynx. The nasopharynx has four openings:

- two eustachian tubes – each opening out of a lateral wall and connecting with the middle ear (tympanic cavity)
- two openings of posterior nares (choanae).

### Oropharynx

The oropharynx, the middle division of the pharynx, is continuous with the posterior oral cavity and is lined with stratified squamous epithelium. It extends inferiorly from the soft palate to the hyoid bone. The opening into the oropharynx from the mouth is called the fauces. The lingual tonsils protrude into the oropharynx from the oral cavity at the base of the tongue. The anterolateral walls of the oropharynx support the palatine tonsils. (See also *Oral cavity*, page 68.)

### Laryngopharynx

The most inferior division of the pharynx, the laryngopharynx extends from the hyoid bone to the opening of the esophagus. The laryngopharynx is lined with stratified squamous epithelium.

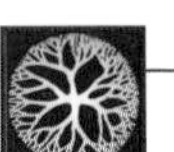

**Physiology**

Food mixes with saliva to form a pliable mass (bolus) for swallowing. A voluntary action propels food from the mouth into the oropharynx. The pharynx aids swallowing by grasping food and moving it toward the esophagus. The remaining actions of swallowing are involuntary. Normally, food does not pass through the nasopharynx because elevation of the soft palate closes it during swallowing. Coordinated contractions of the pharyngeal sphincters (superior, middle, and inferior) propel the food into the esophagus.

### Tonsils

The tonsils are lymphatic tissues embedded in the mucous membrane. They are located in the oral cavity and pharynx. (See *Tonsils*, page 75.)

- Lingual tonsils are at the base of the tongue and protrude into the oropharynx
- Palatine tonsils are supported by the anterolateral walls of the oropharynx
- Pharyngeal tonsils (adenoids) are on the posterior wall of the nasopharynx.

**Clinical tip**

Discomfort in part of the pharynx can range from feelings of scratchiness to severe pain. It can result from infection, trauma, allergy, neoplasms, and certain systemic disorders.
Signs of viral pharyngitis are a slight swelling and inflammation of the pharynx. Bacterial (streptococcal) pharyngitis causes significant swelling and inflammation of the pharynx, exudate from the tonsils, and lymph node enlargement and tenderness. Patients with infectious mononucleosis have enlarged inguinal and axillary lymph nodes in addition to exudative pharyngitis.

## PHARYNX
### Sagittal section

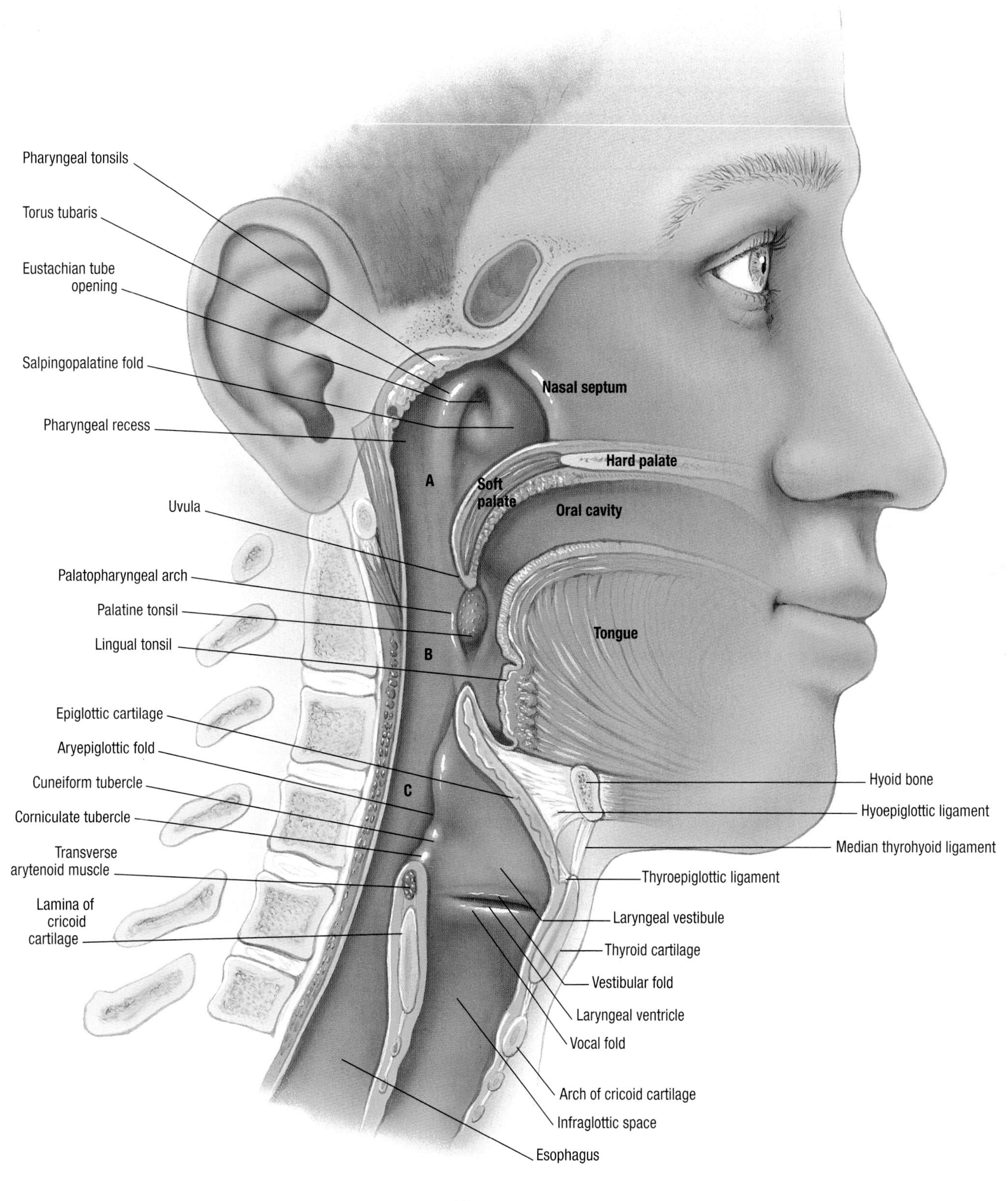
Pharyngeal tonsils
Torus tubaris
Eustachian tube opening
Salpingopalatine fold
Pharyngeal recess
Uvula
Palatopharyngeal arch
Palatine tonsil
Lingual tonsil
Epiglottic cartilage
Aryepiglottic fold
Cuneiform tubercle
Corniculate tubercle
Transverse arytenoid muscle
Lamina of cricoid cartilage
Nasal septum
Hard palate
Soft palate
Oral cavity
Tongue
A
B
C
Hyoid bone
Hyoepiglottic ligament
Median thyrohyoid ligament
Thyroepiglottic ligament
Laryngeal vestibule
Thyroid cartilage
Vestibular fold
Laryngeal ventricle
Vocal fold
Arch of cricoid cartilage
Infraglottic space
Esophagus

**Key**

**A** Nasopharynx
**B** Oropharynx
**C** Laryngopharynx

## Posterior view

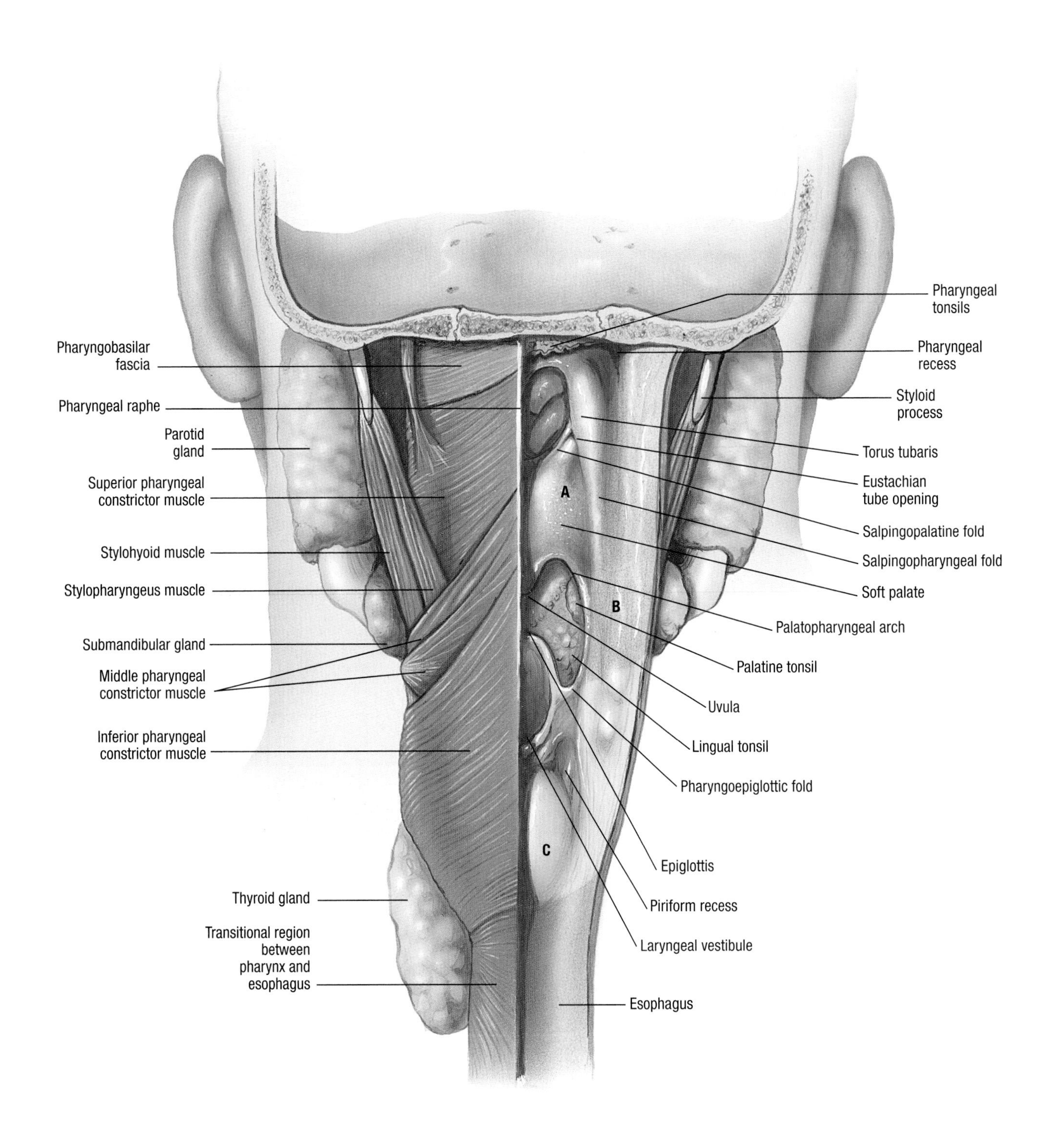
Pharyngeal tonsils
Pharyngobasilar fascia
Pharyngeal recess
Styloid process
Pharyngeal raphe
Parotid gland
Torus tubaris
Eustachian tube opening
Superior pharyngeal constrictor muscle
A
Salpingopalatine fold
Stylohyoid muscle
Salpingopharyngeal fold
Soft palate
Stylopharyngeus muscle
B
Palatopharyngeal arch
Submandibular gland
Palatine tonsil
Middle pharyngeal constrictor muscle
Uvula
Inferior pharyngeal constrictor muscle
Lingual tonsil
Pharyngoepiglottic fold
C
Epiglottis
Thyroid gland
Piriform recess
Transitional region between pharynx and esophagus
Laryngeal vestibule
Esophagus

**Key**

**A** Nasopharynx
**B** Oropharynx
**C** Laryngopharynx

## PHARYNX
### Deep lateral view

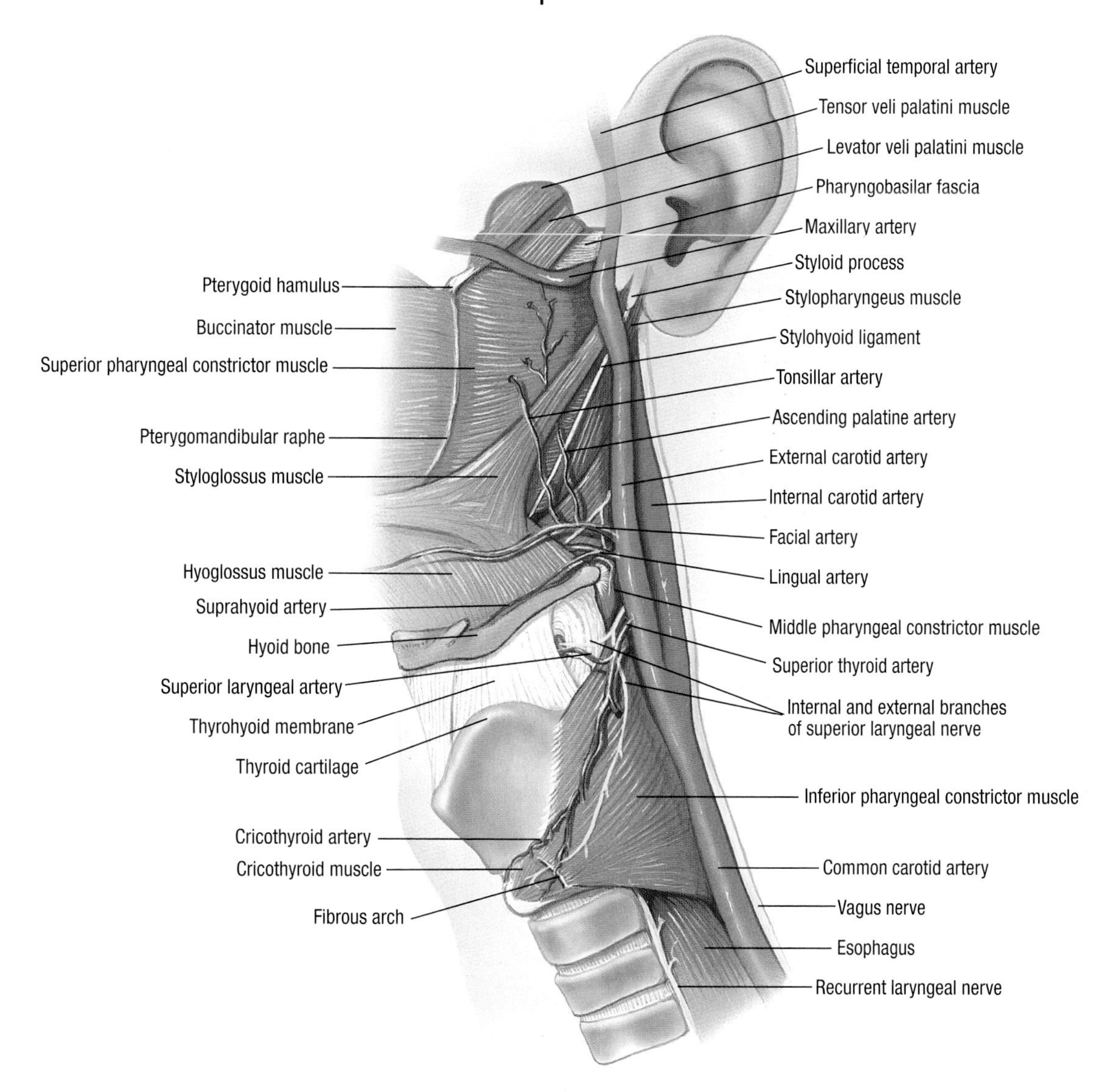

Superficial temporal artery
Tensor veli palatini muscle
Levator veli palatini muscle
Pharyngobasilar fascia
Maxillary artery
Styloid process
Stylopharyngeus muscle
Stylohyoid ligament
Tonsillar artery
Ascending palatine artery
External carotid artery
Internal carotid artery
Facial artery
Lingual artery
Middle pharyngeal constrictor muscle
Superior thyroid artery
Internal and external branches of superior laryngeal nerve
Inferior pharyngeal constrictor muscle
Common carotid artery
Vagus nerve
Esophagus
Recurrent laryngeal nerve
Pterygoid hamulus
Buccinator muscle
Superior pharyngeal constrictor muscle
Pterygomandibular raphe
Styloglossus muscle
Hyoglossus muscle
Suprahyoid artery
Hyoid bone
Superior laryngeal artery
Thyrohyoid membrane
Thyroid cartilage
Cricothyroid artery
Cricothyroid muscle
Fibrous arch


## TONSILS

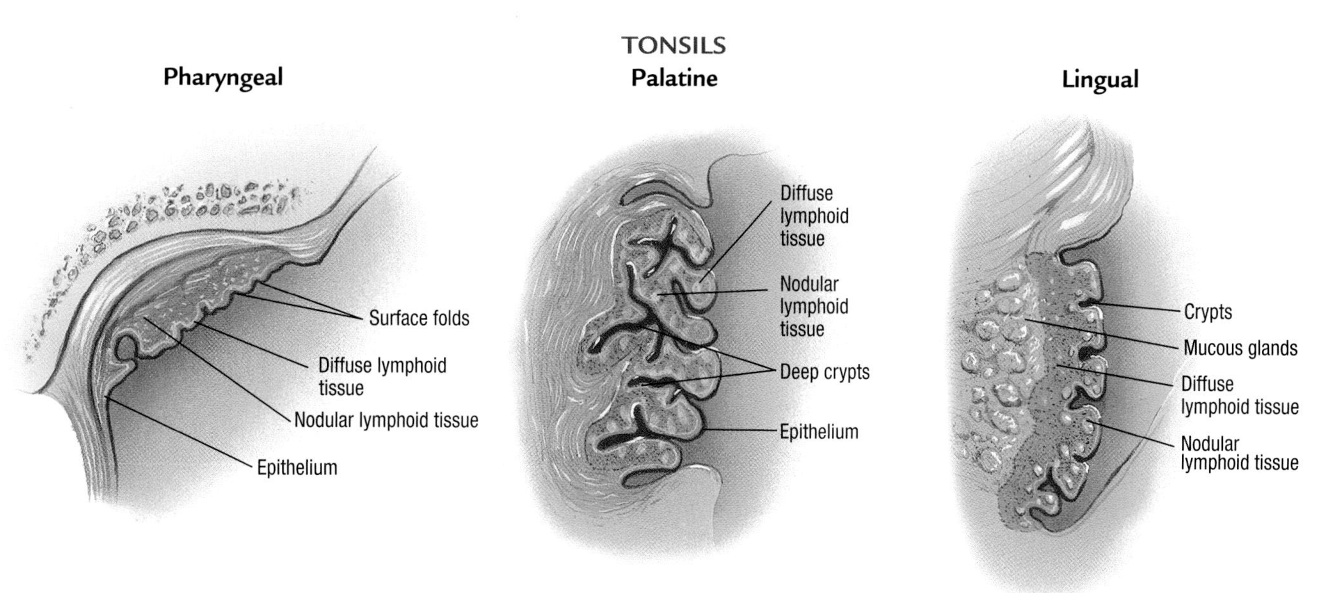

Pharyngeal
Surface folds
Diffuse lymphoid tissue
Nodular lymphoid tissue
Epithelium
Palatine
Diffuse lymphoid tissue
Nodular lymphoid tissue
Deep crypts
Epithelium
Lingual
Crypts
Mucous glands
Diffuse lymphoid tissue
Nodular lymphoid tissue

# Larynx

The larynx, which contains the vocal cords, connects the inferior part of the pharynx with the trachea. Triangular in shape, the larynx is located just below and in front of the most inferior part of the pharynx. It extends from the level of the fourth to the sixth cervical vertebra, attaching to the hyoid bone.

The main functions of the larynx are to provide a permanent airway to the lungs and to aid in vocalization. (See *Laryngeal function*, page 79.)

## Structures

The larynx is composed of nine cartilages, which are connected by ligaments and controlled by muscles.

### Cartilages

Three of the cartilages are single, and six form three pairs. The single cartilages are:

- cricoid – at the lower end of the larynx; often described as a signet ring with the signet portion facing posteriorly; attaches inferiorly to the first cartilaginous ring of the trachea
- epiglottic – a flexible cartilage above the thyroid cartilage
- thyroid – the largest laryngeal cartilage; shaped like a shield. (In males, the ventral edges of the cartilage are at a more acute angle than in females, and it forms a laryngeal prominence sometimes referred to as the Adam's apple.)

The paired cartilages are:

- arytenoids – above the cricoid cartilage attached to vocal cord ligaments
- corniculates – cover the apex of each arytenoid
- cuneiforms – anterior to the corniculates in the aryepiglottic folds.

### Muscles and ligaments

The laryngeal muscles are extrinsic or intrinsic. The extrinsic muscles move the entire larynx; the intrinsic muscles move specific laryngeal parts. Extrinsic muscles include the suprahyoid, infrahyoid, and stylopharyngeus muscles. Intrinsic muscles include the cricothyroid and cricoarytenoids.

The larynx has numerous ligaments, including the median thyrohyoid and cricothyroid ligaments. An emergency tracheostomy is performed through the cricothyroid ligament, which connects the thyroid and cricoid cartilages.

## Innervation and blood supply

Branches of the superior and inferior laryngeal nerves innervate the larynx. The laryngeal nerves are branches of the vagus nerve (CN X). The superior nerve divides into internal and external branches, a communicating branch between the superior and inferior branches, and a muscular branch of the inferior laryngeal nerve. The inferior laryngeal nerve divides into anterior and posterior branches, which supply most of the intrinsic laryngeal muscles. Most of the superior laryngeal nerve carries sensory impulses from the laryngeal mucosa (important in the cough reflex).

The superior laryngeal artery supplies the internal surface of the superior part of the larynx. The inferior laryngeal artery supplies mucous membranes and muscles of the inferior part of the larynx. The superior laryngeal vein empties into the internal jugular vein. The inferior laryngeal vein merges with the inferior thyroid vein of the anterior venous plexus of the thyroid gland.

## True vocal cords

The true vocal cords are a pair of horizontal folds that project into the laryngeal cavity. They are attached to the angle of the thyroid cartilage anteriorly and to the vocal process of the arytenoid cartilage posteriorly. Each vocal cord has a vocal ligament and a vocalis muscle. Vestibular folds between the thyroid and arytenoid cartilages are referred to as the false vocal cords.

The glottis includes the vocal folds and processes. The space between the vocal cords, which varies with vocal cord position, is called the rima glottidis. The glottis and rima glottidis serve as the phonation apparatus.

**Physiology**

Vibration of the vocal cords by expired air causes the sounds made in phonation. Movement of the arytenoid cartilage on the cricoid cartilages changes the length and depth of the vocal cord for sound production.

## LARYNX

### Anterior view

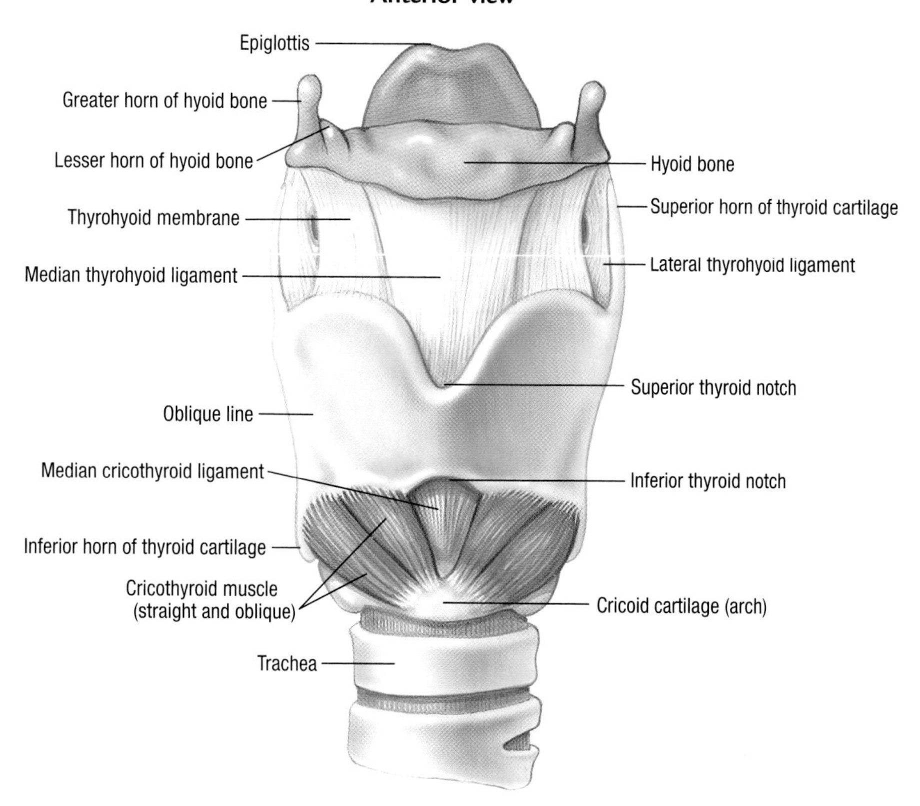

Epiglottis
Greater horn of hyoid bone
Lesser horn of hyoid bone
Thyrohyoid membrane
Median thyrohyoid ligament
Oblique line
Median cricothyroid ligament
Inferior horn of thyroid cartilage
Cricothyroid muscle
(straight and oblique)
Trachea
Hyoid bone
Superior horn of thyroid cartilage
Lateral thyrohyoid ligament
Superior thyroid notch
Inferior thyroid notch
Cricoid cartilage (arch)


### Posterior view

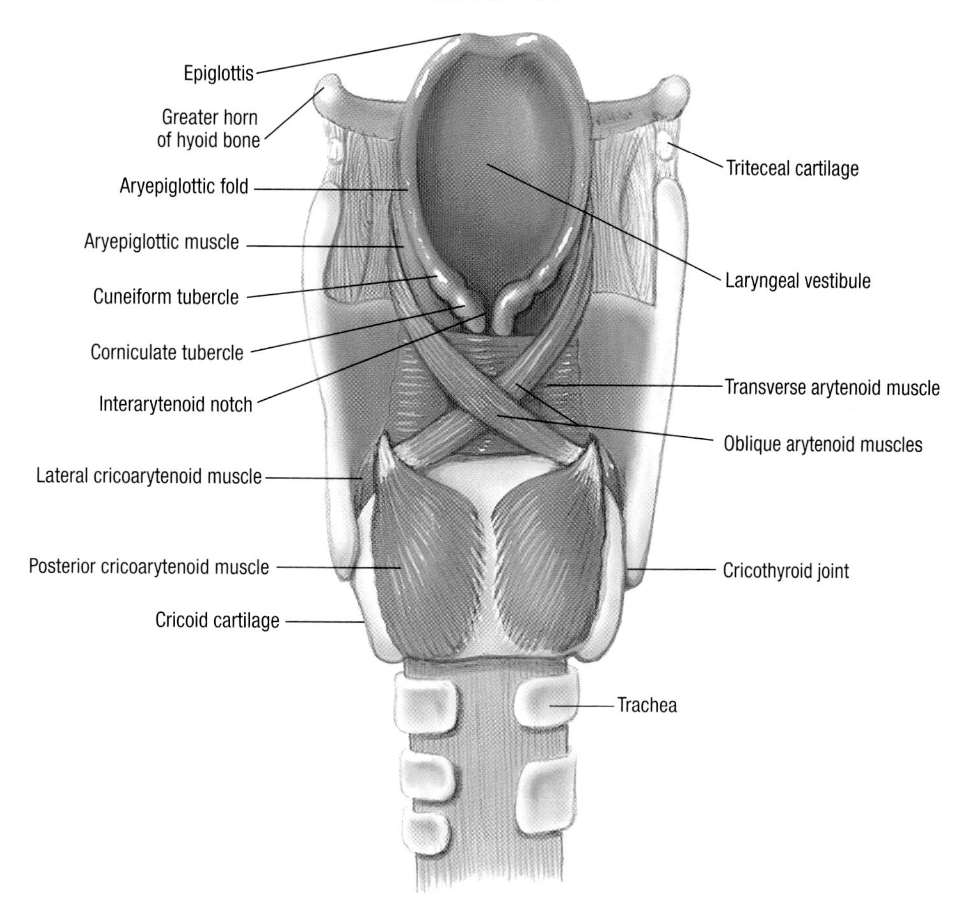

Epiglottis
Greater horn
of hyoid bone
Aryepiglottic fold
Aryepiglottic muscle
Cuneiform tubercle
Corniculate tubercle
Interarytenoid notch
Lateral cricoarytenoid muscle
Posterior cricoarytenoid muscle
Cricoid cartilage
Triteceal cartilage
Laryngeal vestibule
Transverse arytenoid muscle
Oblique arytenoid muscles
Cricothyroid joint
Trachea

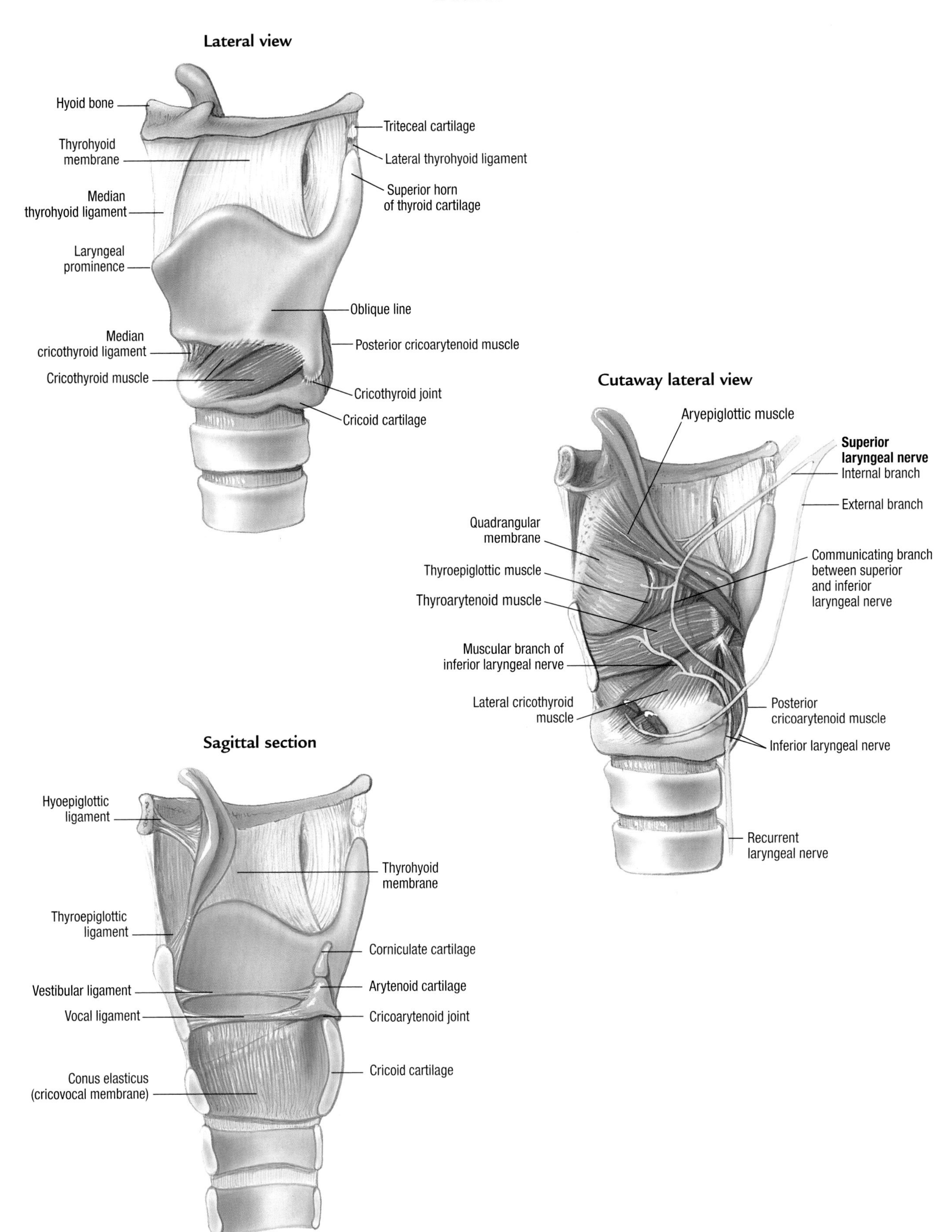
Lateral view
Hyoid bone
Thyrohyoid membrane
Median thyrohyoid ligament
Laryngeal prominence
Median cricothyroid ligament
Cricothyroid muscle
Triteceal cartilage
Lateral thyrohyoid ligament
Superior horn of thyroid cartilage
Oblique line
Posterior cricoarytenoid muscle
Cricothyroid joint
Cricoid cartilage
Cutaway lateral view
Aryepiglottic muscle
Superior laryngeal nerve
Internal branch
External branch
Quadrangular membrane
Thyroepiglottic muscle
Thyroarytenoid muscle
Muscular branch of inferior laryngeal nerve
Lateral cricothyroid muscle
Communicating branch between superior and inferior laryngeal nerve
Posterior cricoarytenoid muscle
Inferior laryngeal nerve
Recurrent laryngeal nerve
Sagittal section
Hyoepiglottic ligament
Thyroepiglottic ligament
Vestibular ligament
Vocal ligament
Conus elasticus (cricovocal membrane)
Thyrohyoid membrane
Corniculate cartilage
Arytenoid cartilage
Cricoarytenoid joint
Cricoid cartilage

## LARYNX
### Superior view
### (Epiglottis removed)

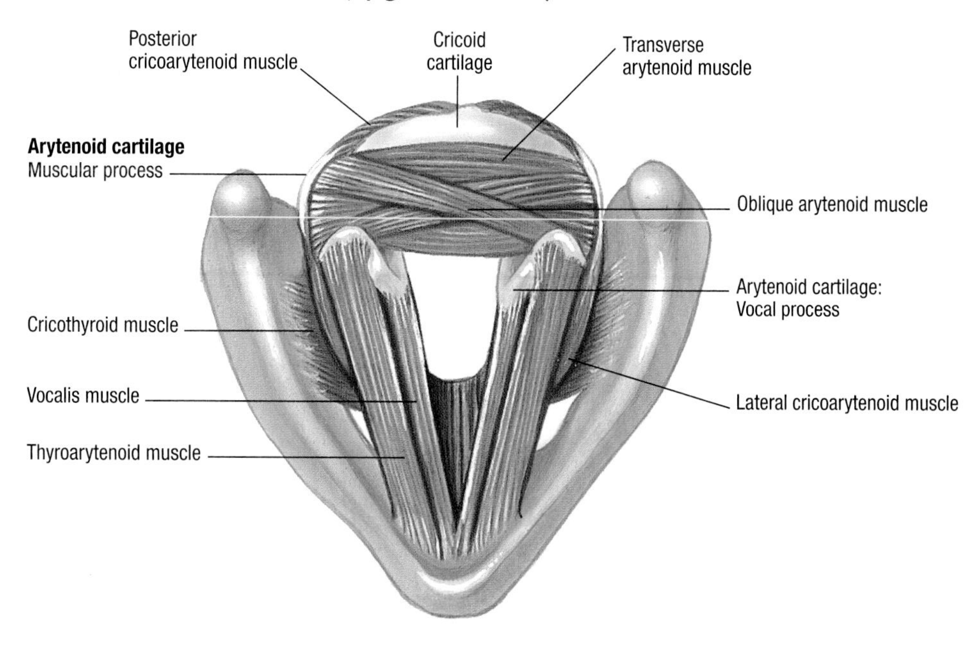
Posterior cricoarytenoid muscle
Cricoid cartilage
Transverse arytenoid muscle
Arytenoid cartilage
Muscular process
Oblique arytenoid muscle
Arytenoid cartilage: Vocal process
Cricothyroid muscle
Vocalis muscle
Lateral cricoarytenoid muscle
Thyroarytenoid muscle

## LARYNGEAL FUNCTION

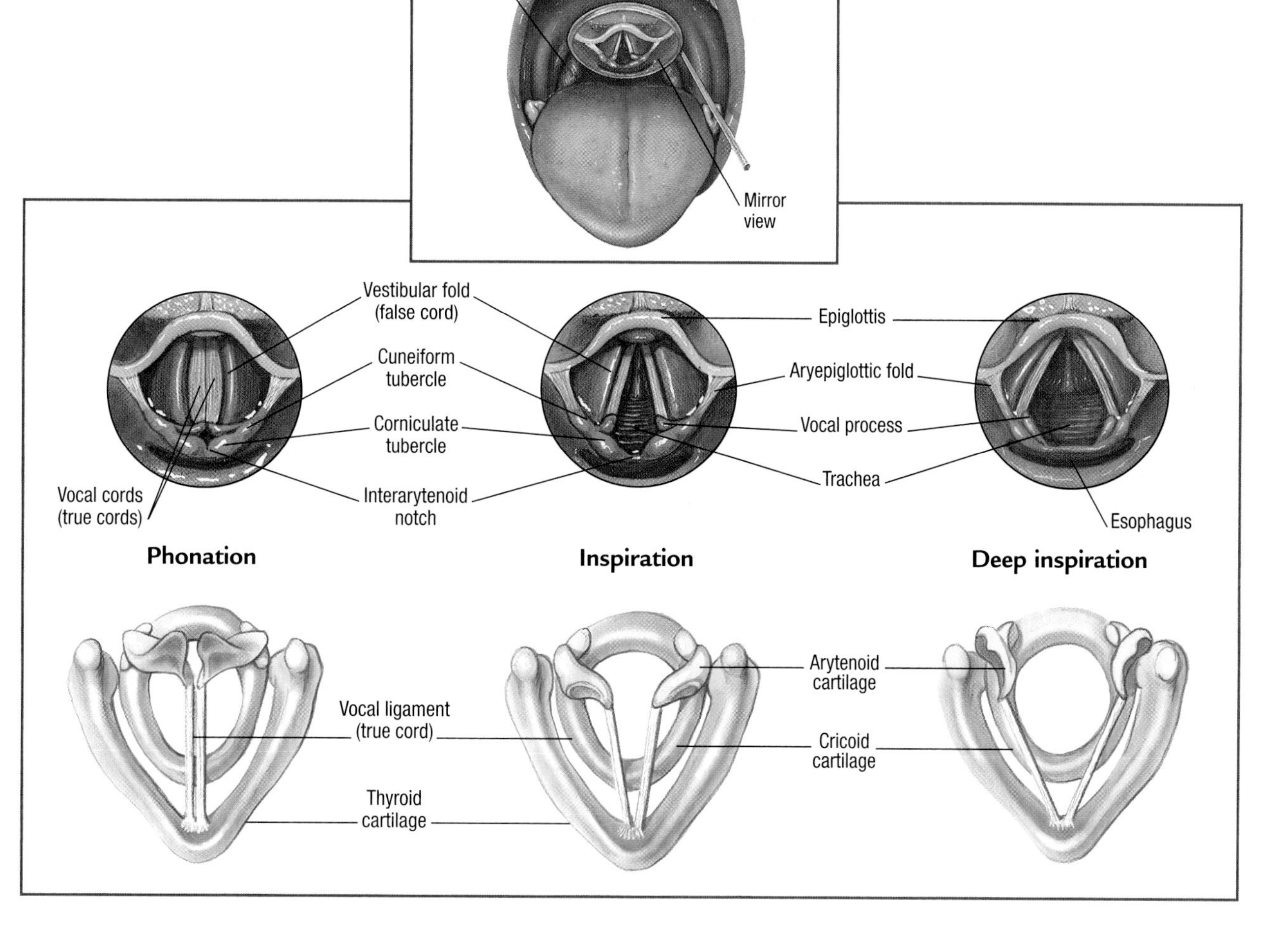
Palatine tonsil
Mirror view
Vestibular fold (false cord)
Cuneiform tubercle
Corniculate tubercle
Interarytenoid notch
Vocal cords (true cords)
Epiglottis
Aryepiglottic fold
Vocal process
Trachea
Esophagus
Phonation
Inspiration
Deep inspiration
Vocal ligament (true cord)
Thyroid cartilage
Arytenoid cartilage
Cricoid cartilage

# Thyroid and parathyroid glands

The thyroid and parathyroid glands have a close physical relationship. The parathyroid glands are embedded beneath the posterior surface of the thyroid gland.

## Thyroid gland

The thyroid gland lies in the lower anterior neck, overlying the inferior border of the larynx. The gland is fixed to the anterior surface of the upper trachea by loose connective tissue.

The thyroid consists of two lateral lobes – one on either side of the trachea – joined by a narrow tissue bridge, the isthmus. The corners of the lobes are irregular, extending farther than the main body of the lobe. This shape, along with the isthmus, gives the gland its butterfly shape. The thyroid gland is controlled by hormones from the pituitary gland.

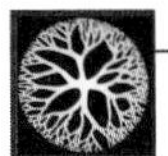

**Physiology**

The two lobes of the thyroid gland function as a unit to produce the hormones thyroxine ($T_4$), triiodothyronine ($T_3$), and calcitonin. Collectively referred to as thyroid hormone, $T_4$ and $T_3$ are the body's major metabolic hormones, regulating metabolism by stimulating cellular respiration. $T_3$ has several times the biologic activity of $T_4$. Calcitonin maintains the blood calcium level by inhibiting release of calcium from bone. Its secretion is controlled by the calcium concentration in the blood.

### Blood vessels

The arterial blood supply is provided by the superior and inferior thyroid arteries. The superior artery is a branch of the external carotid artery, and the inferior artery is a branch of the thyrocervical trunk, which originates from the subclavian artery. Three pairs of veins drain the thyroid gland:

- superior thyroid vein – drains the superior poles
- middle thyroid vein – drains the middle of the lobes
- inferior thyroid vein – drains the inferior poles.

## Parathyroid glands

The body's smallest known endocrine glands, the parathyroid glands are embedded beneath the posterior surface of the thyroid, one in each corner. Most people have four parathyroid glands, although the occurrence of more than four parathyroid glands has been noted. Branches of the inferior thyroid arteries provide the predominant blood supply to the parathyroid gland; however, other arteries in the neck area, such as the superior thyroid or tracheal arteries, may make a contribution.

**Physiology**

Working together as a single gland, the parathyroid glands produce parathyroid hormone (PTH). The main function of PTH is to help regulate the calcium balance by elevating low blood calcium levels. This hormone adjusts the rate at which calcium and magnesium ions are removed from bone and reabsorbed by the kidneys. It also increases the movement of phosphate ions from the blood to the urine for excretion.

**Clinical tip**

### Thyroid assessment

Keep these symptoms in mind when assessing for thyroid dysfunction.

Hypothyroidism:

- dry, cold skin
- coarse hair
- tired, cold, weak feeling
- swelling around the eyes (myxedema)
- neck swelling
- weight gain
- constipation
- difficulty with memory and concentration
- heavy menstrual blood flow.

Hyperthyroidism:

- nervous, irritable feelings
- warm or sweaty feeling
- breathing difficulty
- tired, weak feeling
- shakiness
- weight loss
- hunger
- frequent bowel movements
- decreased menstrual flow
- bulging, staring eyes (exophthalmos).

## THYROID AND PARATHYROID GLANDS

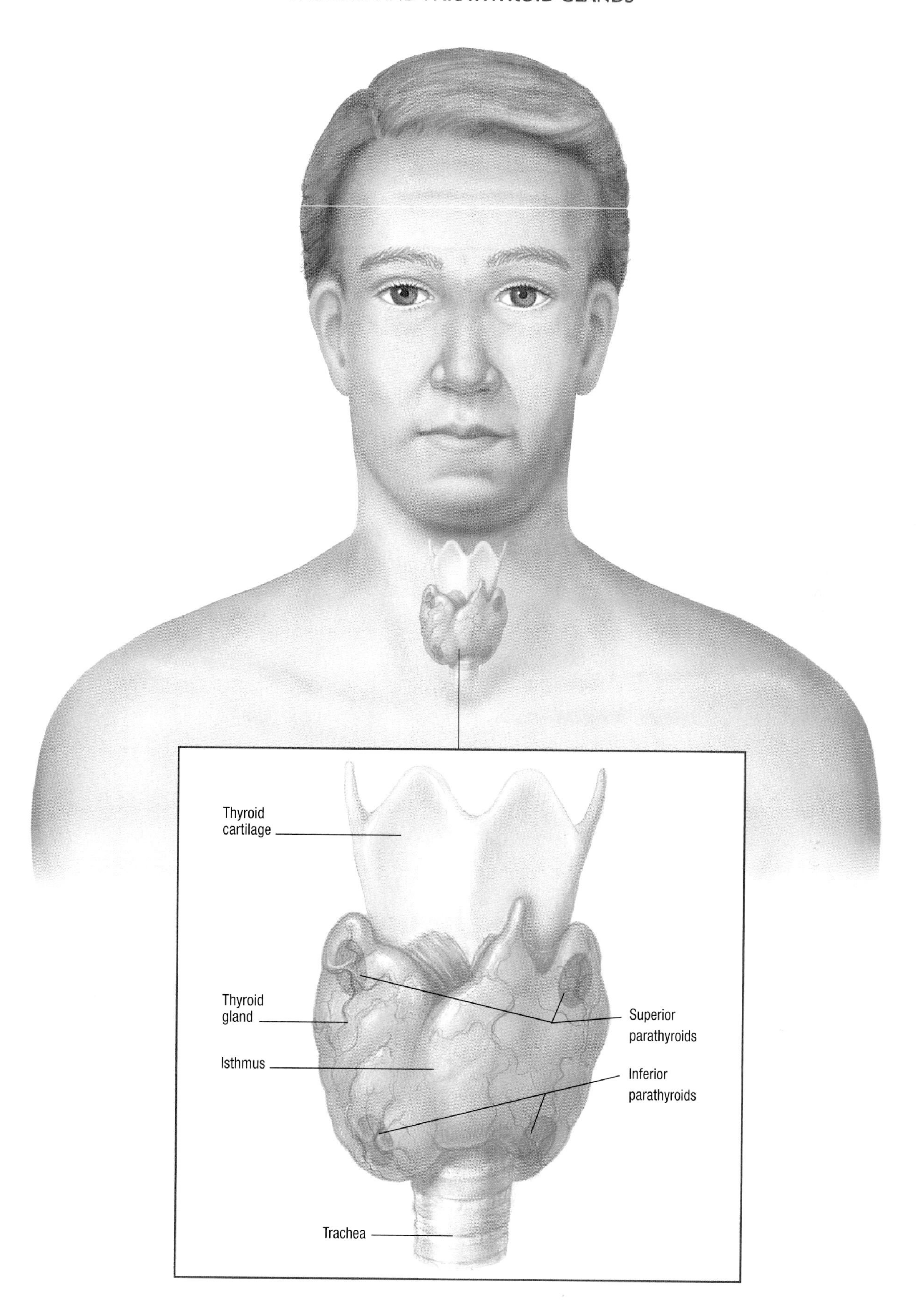

Thyroid cartilage
Thyroid gland
Isthmus
Superior parathyroids
Inferior parathyroids
Trachea

# PART III

# BACK AND SPINE

# Back structures

The back is the posterior area of the body between the neck and the buttocks. (See *Exterior landmarks of the back,* page 87.) In close proximity to the upper limbs and the neck, it contributes to the function and actions of each. The vertebral column (spine), the scapulae, the posterior portion of the bony thorax, and numerous muscles constitute the musculoskeletal system of the back.

## Bone function

The bones of the back and spine perform the following functions:
- enclose and protect the spinal cord
- support the head
- anchor the ribs
- articulate with the upper limbs, ribs, and pelvis (sacroiliac joint)
- provide the mechanical basis for movement.

For detailed information on the vertebral column, see page 88.

## The bony thorax

The bony thorax, or rib cage, forms a protective enclosure for the heart, lungs, and great vessels of the thoracic cavity. It is part of the axial skeleton. The sternum, part of the anterior bony thorax, supports the shoulder girdle by articulating with the clavicle at the sternoclavicular joint. Each of the first 10 ribs has two facets on each head to articulate with its associated vertebra and the vertebra superior to it. The eleventh and twelfth ribs, also called floating ribs, end at the posterior abdominal musculature without joining the sternum. More detail about ribs and the thorax can be found in Part V, Thorax.

## Scapula

The scapula (shoulder blade) is a triangular flat bone that articulates with the humerus, making arm movement possible. It is part of the appendicular skeleton. The dorsal section lies between the second and seventh ribs. The medial borders of the scapulae are approximately 5 cm from the vertebral column. Superiorly, the scapula has two prominent bony projections – the coracoid process in front and the acromion process in back. The spine of the scapula runs diagonally across the bone's superior portion.

## Muscles of the back

The major muscles of the back move the pectoral (shoulder) girdle, head, and upper limbs. (See *Back muscles,* page 86.) The back muscles that support and move the spine are described on page 88. The muscles of the back include the following:
- trapezius
- deltoid
- latissimus dorsi
- levator scapulae
- rhomboideus minor
- supraspinatus
- rhomboideus major
- infraspinatus
- serratus anterior
- serratus posterior
- teres minor
- teres major.

In addition to the actions discussed previously, back muscles support and move the vertebral column. For more information on vertebral column muscles, see page 88. For more information on neck muscles, see pages 24 and 25.

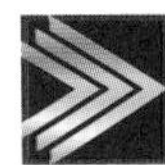

**Clinical tip**

These tips may help to distinguish pain secondary to back conditions from that caused by other medical conditions.
- Pain referred from the viscera is typically unaffected by activity and rest.
- Deep lumbar pain unaffected by activity suggests a dissecting abdominal aortic aneurysm; evaluate for a pulsing epigastric mass.
- Pain originating in the vertebral joints may worsen with activity and improve with rest.
- Back pain of neoplastic origin frequently is relieved by walking and worsens at night.

# BACK BONES

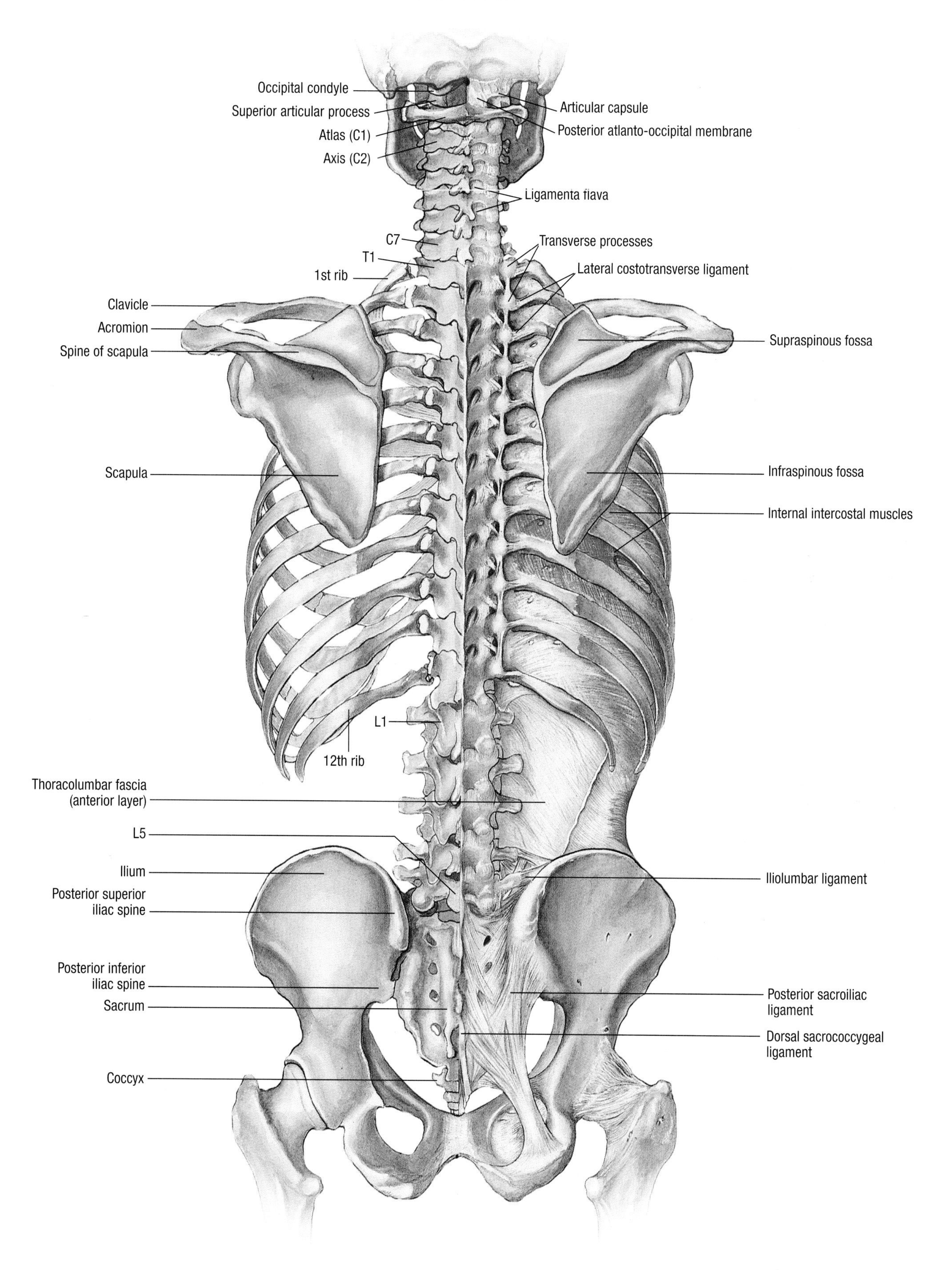

Occipital condyle
Superior articular process
Atlas (C1)
Axis (C2)
Articular capsule
Posterior atlanto-occipital membrane
Ligamenta flava
C7
T1
1st rib
Transverse processes
Lateral costotransverse ligament
Clavicle
Acromion
Spine of scapula
Supraspinous fossa
Scapula
Infraspinous fossa
Internal intercostal muscles
L1
12th rib
Thoracolumbar fascia
(anterior layer)
L5
Ilium
Iliolumbar ligament
Posterior superior
iliac spine
Posterior inferior
iliac spine
Sacrum
Posterior sacroiliac
ligament
Dorsal sacrococcygeal
ligament
Coccyx

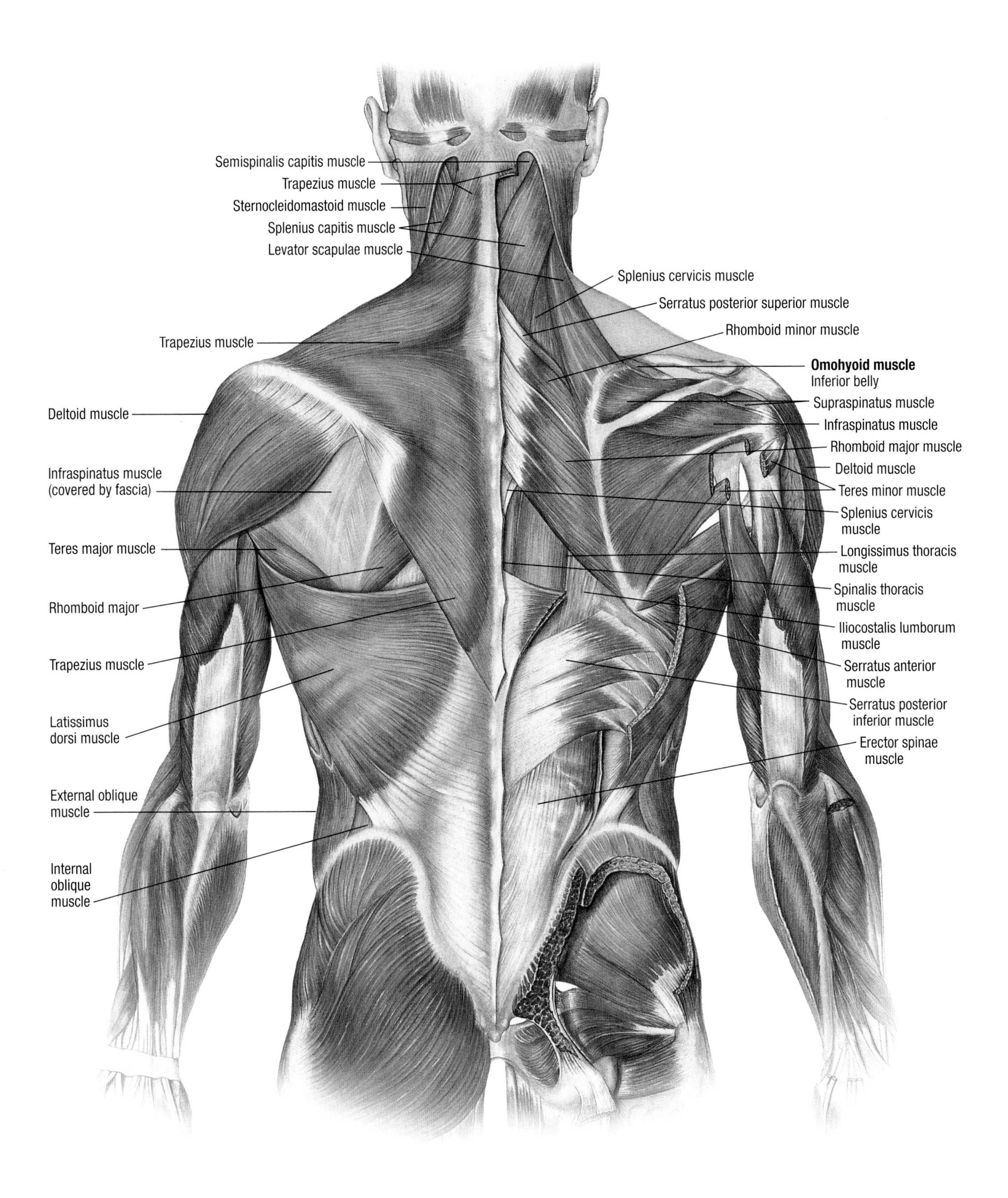
Semispinalis capitis muscle
Trapezius muscle
Sternocleidomastoid muscle
Splenius capitis muscle
Levator scapulae muscle
Splenius cervicis muscle
Serratus posterior superior muscle
Rhomboid minor muscle
Trapezius muscle
Omohyoid muscle
Inferior belly
Deltoid muscle
Supraspinatus muscle
Infraspinatus muscle
Rhomboid major muscle
Deltoid muscle
Infraspinatus muscle
(covered by fascia)
Teres minor muscle
Splenius cervicis
muscle
Teres major muscle
Longissimus thoracis
muscle
Spinalis thoracis
muscle
Rhomboid major
Iliocostalis lumborum
muscle
Trapezius muscle
Serratus anterior
muscle
Serratus posterior
inferior muscle
Latissimus
dorsi muscle
Erector spinae
muscle
External oblique
muscle
Internal
oblique
muscle

## EXTERIOR LANDMARKS OF THE BACK

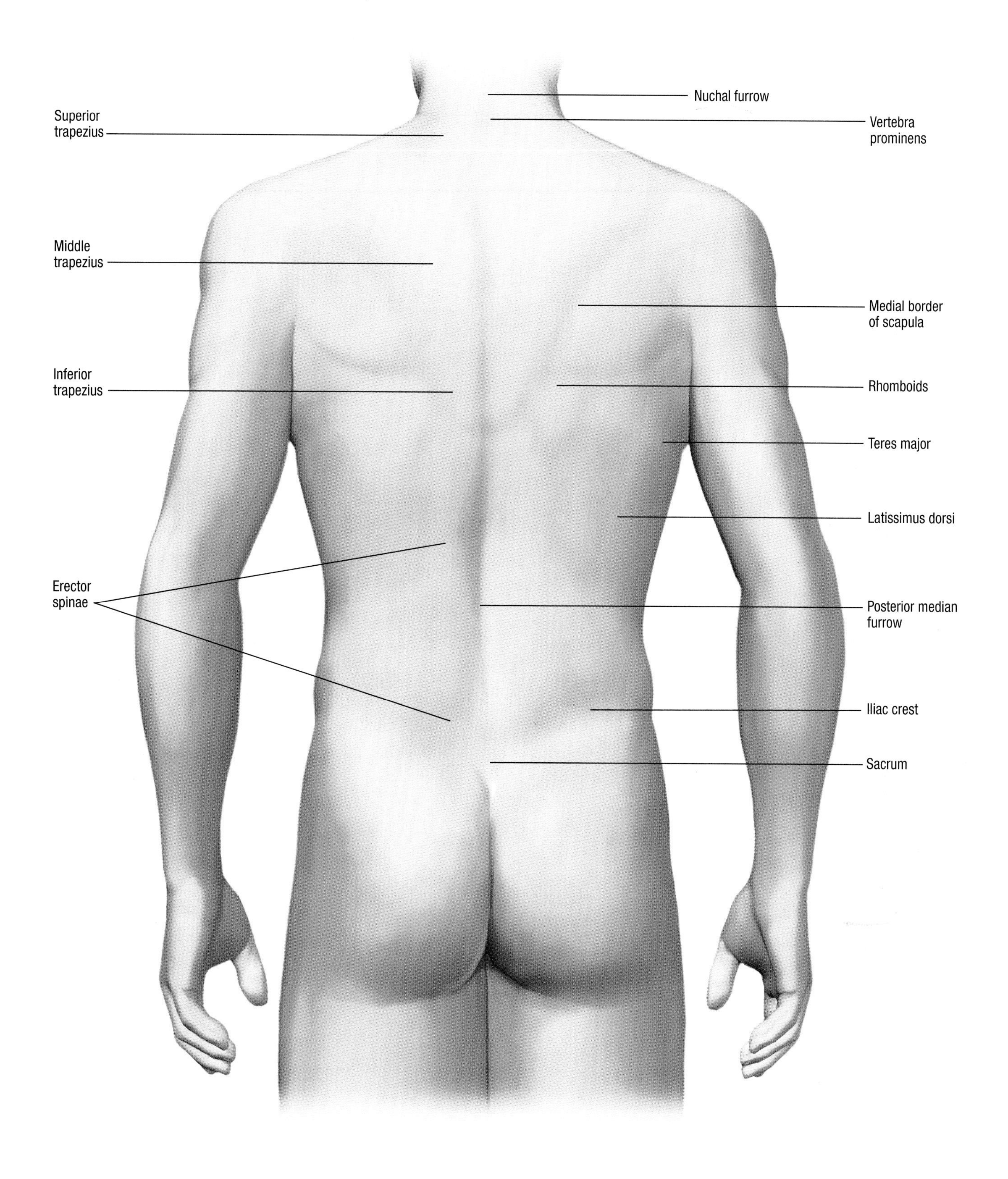

Nuchal furrow
Superior trapezius
Vertebra prominens
Middle trapezius
Medial border of scapula
Inferior trapezius
Rhomboids
Teres major
Latissimus dorsi
Erector spinae
Posterior median furrow
Iliac crest
Sacrum

# Vertebral column

The vertebral column (spine) extends from the skull to the pelvis. It provides the primary axial support for the body and protects the spinal cord. It consists of 24 movable, irregular bones (vertebrae) and two that represent fusions of several bones. An intervertebral disk between each two movable vertebrae provides flexibility and shock absorption.

## Vertebrae

Of the 33 vertebrae, 24 are movable and the rest are fused, as follows:

- cervical (7)
- thoracic (12)
- lumbar (5)
- sacrum (formed by the fusion of 5 vertebrae)
- coccyx (formed by the fusion of 4 vertebrae).

The vertebrae are discussed in detail on page 94.

## Curves

The vertebral column has four sagittal curves that provide strength, resilience, balance, and flexibility. From superior to inferior, they are the following:

- cervical
- thoracic
- lumbar
- sacral.

The posteriorly convex thoracic and sacral curves are called the *primary* curves because they form during fetal development. The anteriorly convex cervical and lumbar curves are called *secondary* curves because they develop after birth. (See *Abnormal spinal curves,* page 90.)

**Age-related changes**

**During gestation, the spine of the fetus has only a single C-shaped anterior curve. The cervical and lumbar curves develop during childhood, and the original C curve persists as the thoracic and sacral curves. The result is the S curve of the adult spine.**

## Ligaments

Several ligaments connect vertebrae. (See *Lumbar vertebral ligaments,* page 91.)

- anterior longitudinal – a broad sheath of connective tissue along the anterior surface of the vertebral bodies
- posterior longitudinal – narrower than the anterior; lies along the posterior surface of the vertebral bodies inside the vertebral canal
- flaval – lies on the anterior aspect of the posterior arch (lamina) of the vertebral column
- interspinous and supraspinous – connect spinous processes.

## Muscles

Muscles along the spine move the vertebral column. They are innervated by the dorsal rami of the spinal nerves and are enclosed by fascia. (See *Vertebral column muscles,* pages 92 and 93.) The deep back muscles are commonly divided into four groups, as follows:

- superficial – together extend the head and neck; singly abduct and rotate the head ipsilaterally (toward the same side)
  - splenius capitis
  - splenius cervicis
- intermediate erector spinae – the major extensors of the vertebral column; a series of overlapping muscles
  - iliocostalis lumborum, iliocostalis thoracis, iliocostalis cervicis – extend the vertebral column and bend it to one side
  - longissimus thoracis, longissimus cervicis, longissimus capitis – extend vertebral column and head; rotate head ipsilaterally
  - spinalis thoracis, spinalis cervicis – extend vertebral column.
- deep transversospinalis
  - semispinalis thoracis, semispinalis cervicis, semispinalis capitis – extend and rotate vertebral column and head
  - multifidus – extend and rotate the vertebral column
  - rotators – extend and rotate the vertebral column
- minor deep group
  - interspinalis – extend the vertebral column
  - intertransversarii – abduct the vertebral column; participate in lateral bending
  - levator costorum – participate in lateral bending; raise ribs during inspiration
  - semispinalis thoracis, semispinalis cervicis, and semispinalis capitis – extend and rotate vertebral column and head.

**Clinical tip**

**To assess spinal movement, measure the degrees of flexion, extension, and lateral bending. Then measure patient between the neck and the waist while he is standing straight up. Then have the patient bend forward at the waist, while you hold the tape. The length of the spine from the neck to the waist usually increases by at least 2″ (5 cm) when the patient bends forward. If it doesn't, the patient's mobility may be impaired.**

## VERTEBRAL COLUMN LOCATION

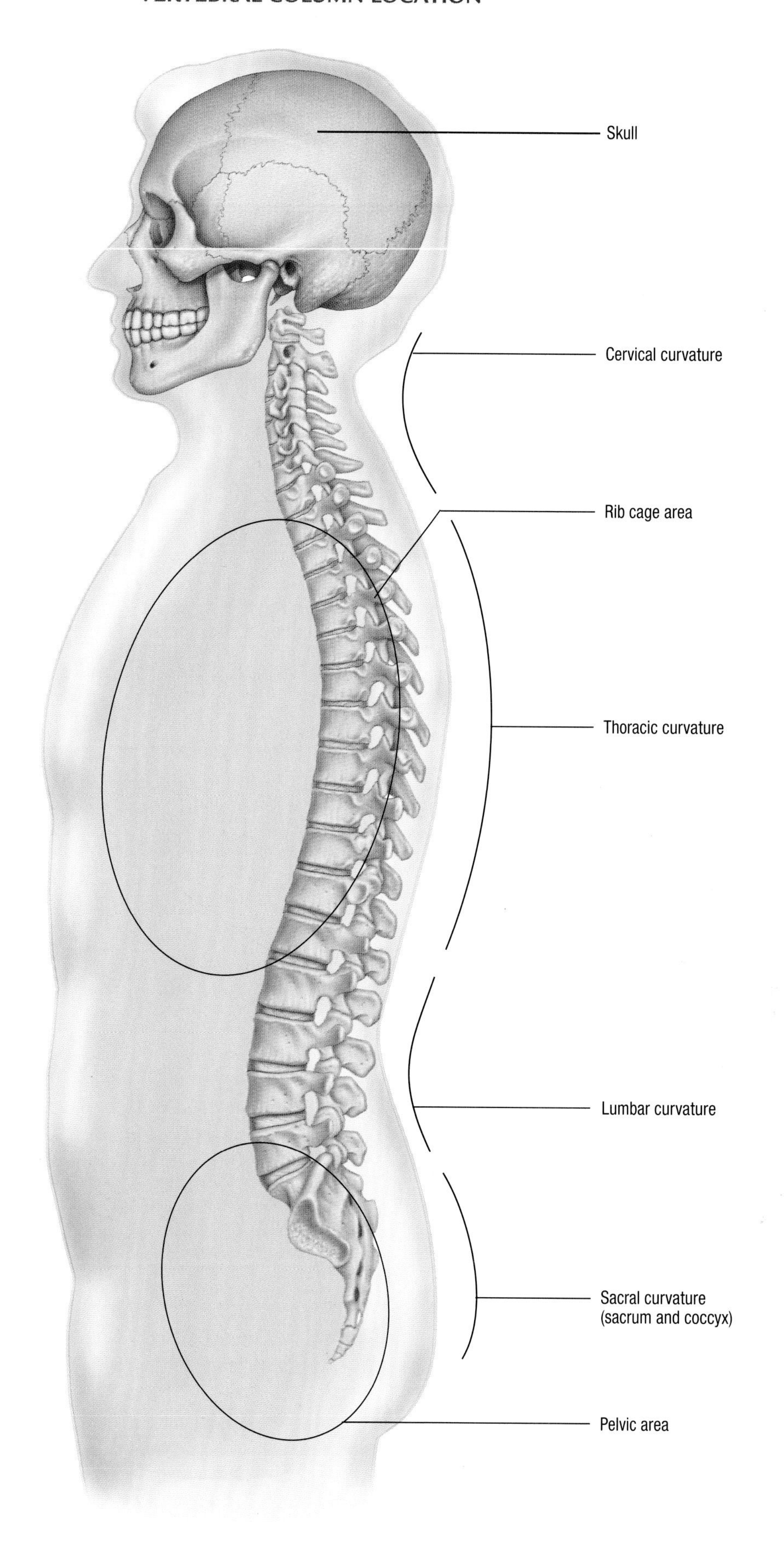

Skull
Cervical curvature
Rib cage area
Thoracic curvature
Lumbar curvature
Sacral curvature
(sacrum and coccyx)
Pelvic area

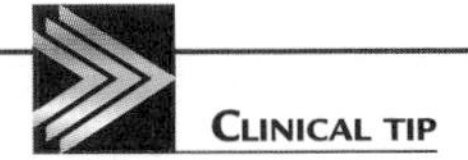

Clinical tip

## Abnormal spinal curves

**Lordosis** is an increased anterior concavity in the curvature of the lumbar spine, observed when the patient is viewed from the side. It is normal in young children and pregnant women.

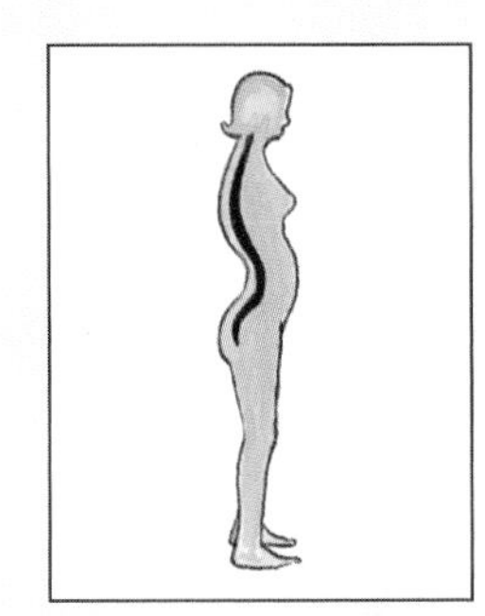

**Kyphosis** is an exaggeration of the posterior convexity of the thoracic vertebral column. It may be caused by absence of a vertebral body, malformation by incomplete segmentation of vertebral bodies, absence of a corner, or flattening by compression. This is common in elderly women who have osteoarthritis or osteoporosis.

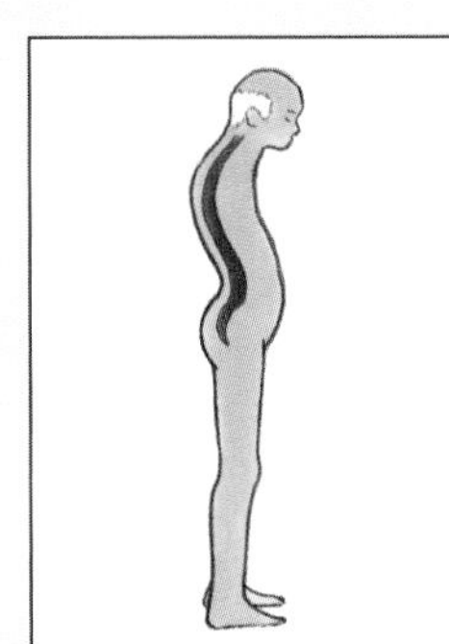

**Scoliosis** is an appreciable lateral curve of the spine, which can result from numerous causes, including congenital malformations of the spine, muscle paralysis, poliomyelitis, sciatica, and unequal leg length.

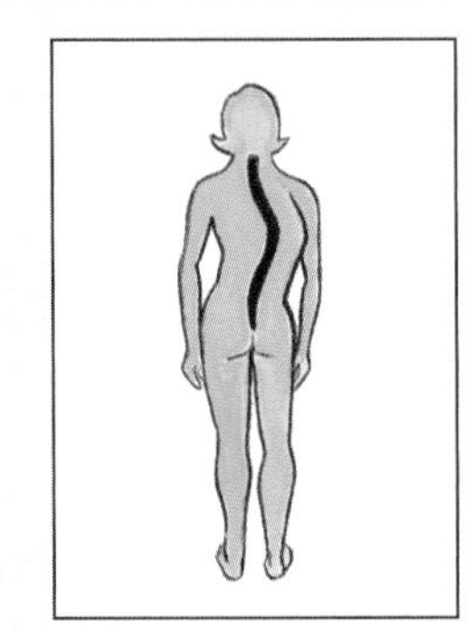

### Signs of scoliosis

When evaluating for scoliosis, have the patient remove his shirt and stand as straight as possible with his back toward you.

Look for:

- Uneven shoulder height and shoulder blade prominence
- Unequal distance between the arms and the body
- Asymmetrical waistline
- Uneven hip height
- Sideways lean

Also have the patient bend forward, keeping the head down and palms together.

Look for:

- Asymmetrical thoracic spine or prominent rib cage on either side
- Asymmetrical waistline

**VERTEBRAL COLUMN**
**Sagittal view**

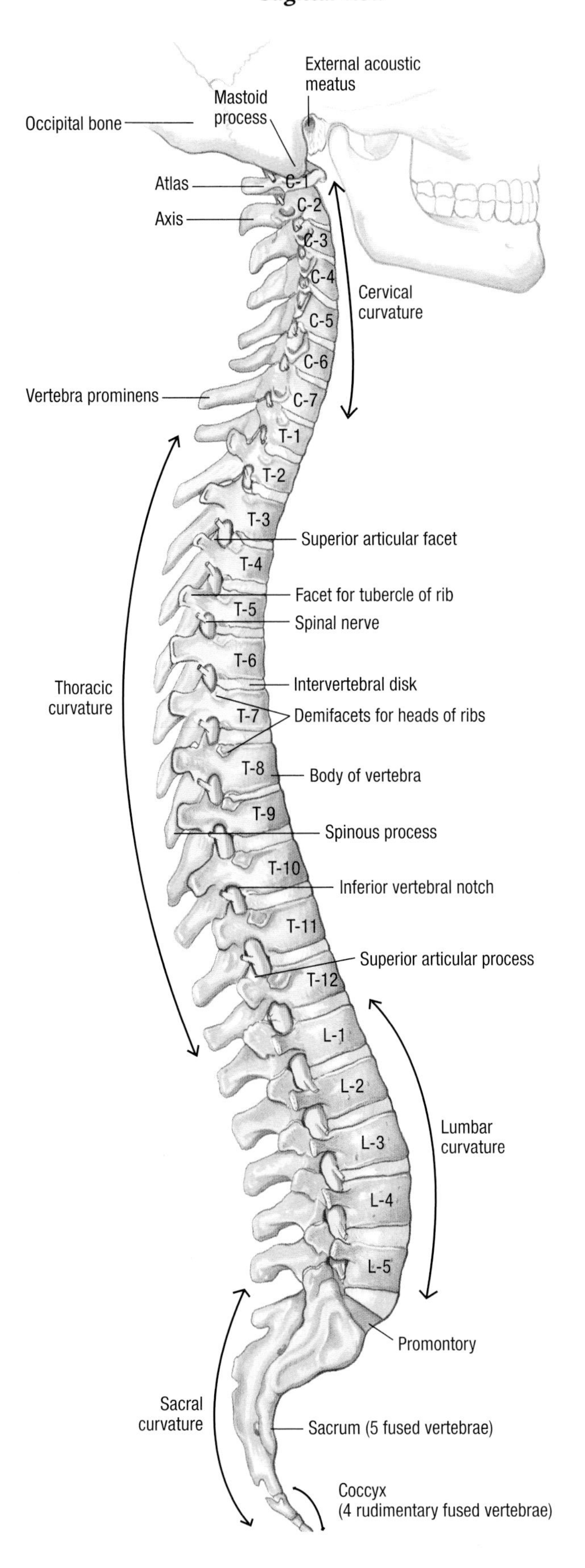
External acoustic meatus
Mastoid process
Occipital bone
Atlas
Axis
C-1
C-2
C-3
C-4
C-5
C-6
C-7
Cervical curvature
Vertebra prominens
T-1
T-2
T-3
T-4
T-5
T-6
T-7
T-8
T-9
T-10
T-11
T-12
Superior articular facet
Facet for tubercle of rib
Spinal nerve
Intervertebral disk
Demifacets for heads of ribs
Body of vertebra
Spinous process
Inferior vertebral notch
Superior articular process
Thoracic curvature
L-1
L-2
L-3
L-4
L-5
Lumbar curvature
Promontory
Sacral curvature
Sacrum (5 fused vertebrae)
Coccyx
(4 rudimentary fused vertebrae)

## LUMBAR VERTEBRAL LIGAMENTS
**Left lateral view**

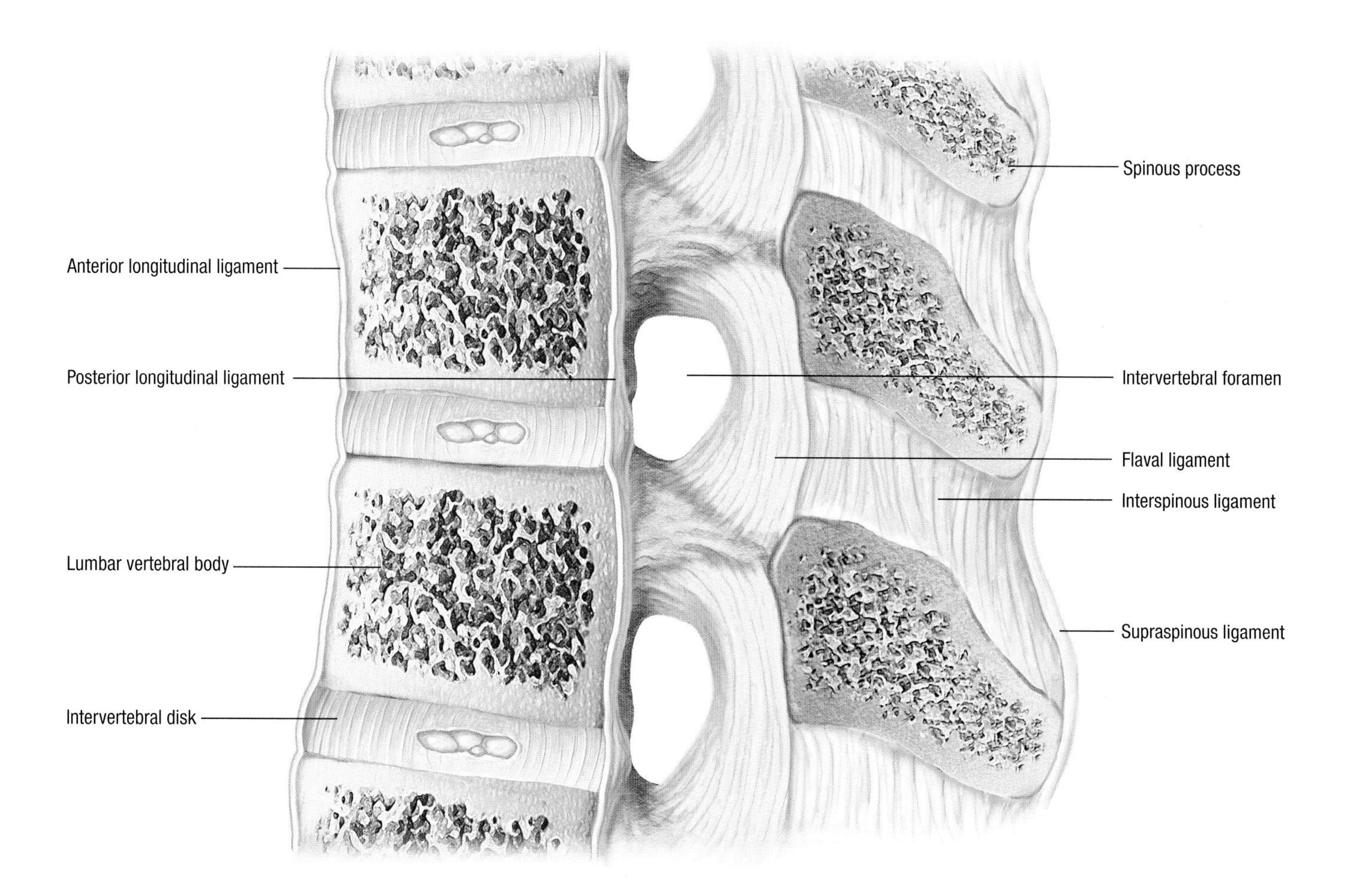
Spinous process
Anterior longitudinal ligament
Posterior longitudinal ligament
Intervertebral foramen
Flaval ligament
Interspinous ligament
Lumbar vertebral body
Supraspinous ligament
Intervertebral disk

## VERTEBRAL COLUMN MUSCLES
### Posterior view

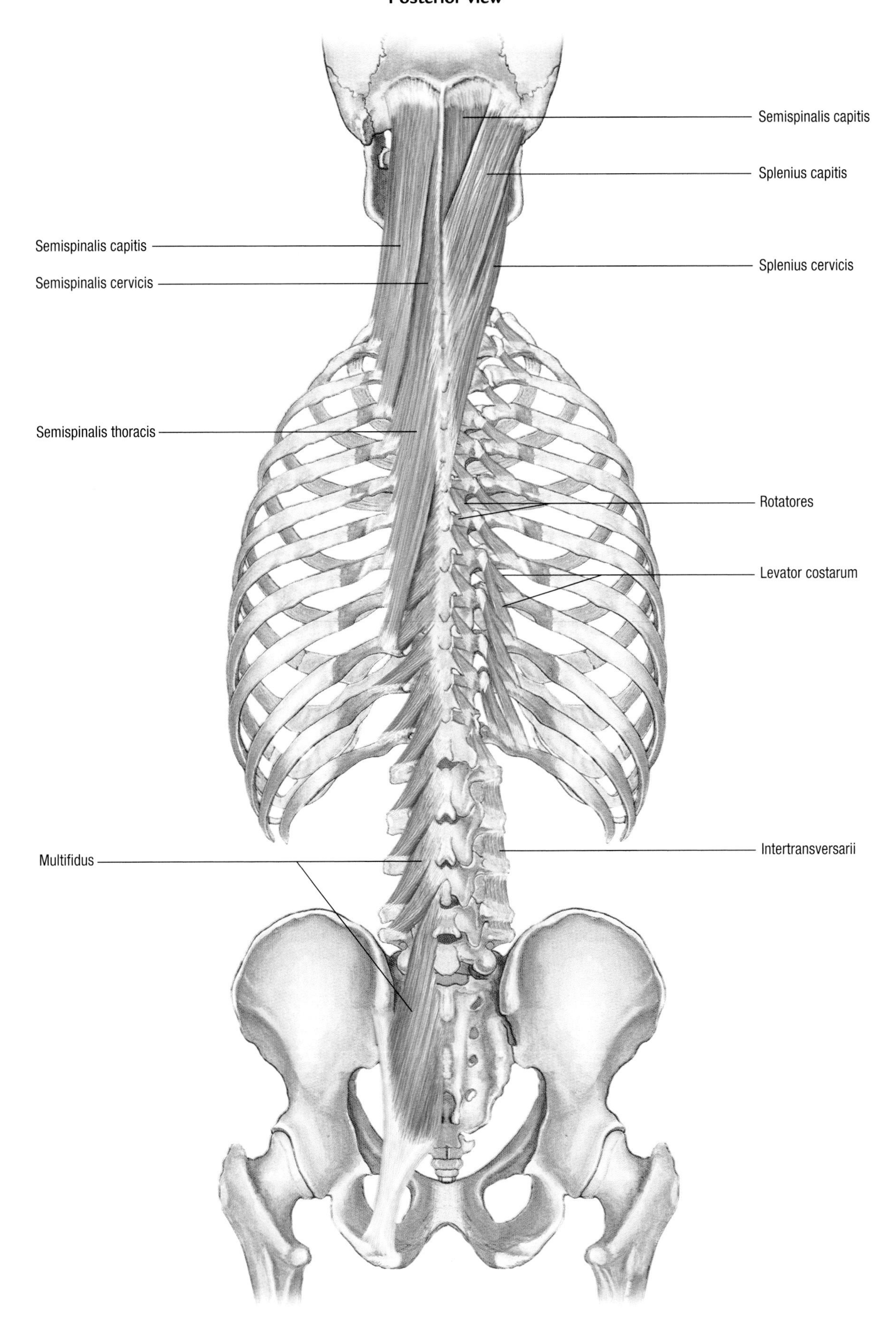

Semispinalis capitis
Splenius capitis
Semispinalis capitis
Semispinalis cervicis
Splenius cervicis
Semispinalis thoracis
Rotatores
Levator costarum
Multifidus
Intertransversarii

## VERTEBRAL COLUMN MUSCLES
### Lateral view

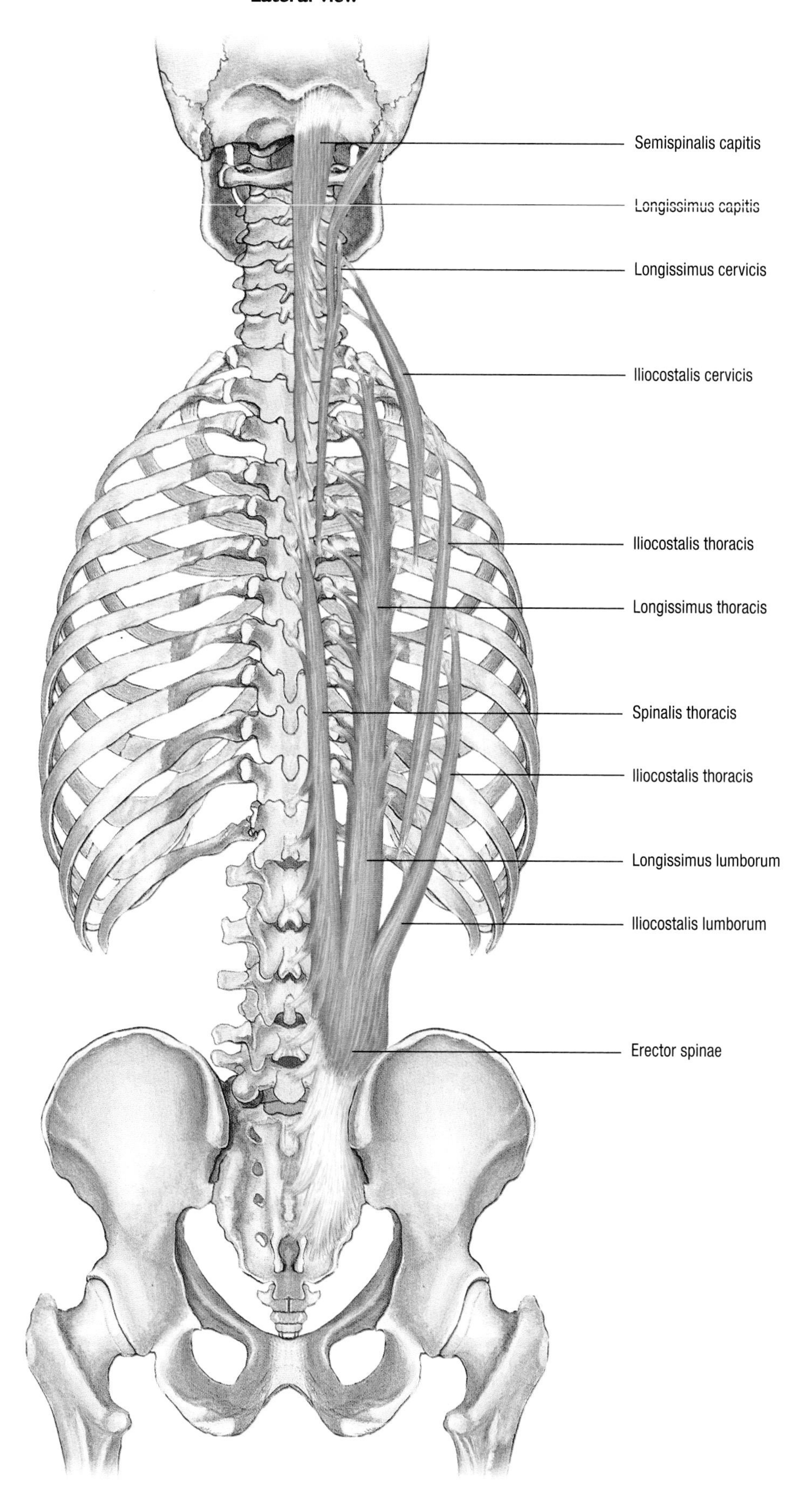
Semispinalis capitis
Longissimus capitis
Longissimus cervicis
Iliocostalis cervicis
Iliocostalis thoracis
Longissimus thoracis
Spinalis thoracis
Iliocostalis thoracis
Longissimus lumborum
Iliocostalis lumborum
Erector spinae

# Vertebrae

Vertebrae are the bones (both movable and fused) of the spinal column. A typical vertebra consists of:
- an ovoid body
- a vertebral arch (a portion of which is called the lamina)
- two transverse processes
- two superior articular processes
- two inferior articular processes
- one spinous process.

Muscles attach at the spinous process and the two transverse processes. The superior and inferior articular processes articulate with the vertebrae immediately above and below them. The openings between two adjacent vertebrae – called the *intervertebral* foramina – allow passage of spinal nerves. As vertebrae articulate, alignment of the *vertebral* foramina forms the vertebral canal (spinal canal).

## Cervical vertebrae

The cervical vertebrae (designated C1 to C7) constitute the skeleton of the neck. (See *Atlas and axis,* page 96, and *5th cervical vertebra,* page 97.) They are smaller than the other vertebrae. See Head and neck bones, pages 10 and 11, for a detailed discussion of the cervical vertebrae.

## Thoracic vertebrae

The thoracic vertebrae (T1 to T12) are located between the cervical and lumbar vertebrae. (See *7th thoracic vertebra,* page 98.)
- Each thoracic vertebra articulates bilaterally with ribs.
- The first 10 have facets on the transverse processes that articulate with the tubercles of the ribs.

## Lumbar vertebrae

The lumbar vertebrae (L1 to L5) are located between the thoracic vertebrae and the sacrum. They are the largest and strongest vertebrae. (See *2nd lumbar vertebra,* page 99.)

## Sacrum and coccyx

The sacrum, a triangular bone formed by the fusion of five sacral vertebrae, lies in the posterior portion of the pelvic girdle between the hip bones. The coccyx, a triangular bone formed by the fusion of the four coccygeal vertebrae, articulates superiorly with the sacrum. (See *Sacrum and coccyx,* pages 100 and 101.)

**FUNCTION OF INTERVERTEBRAL DISKS**

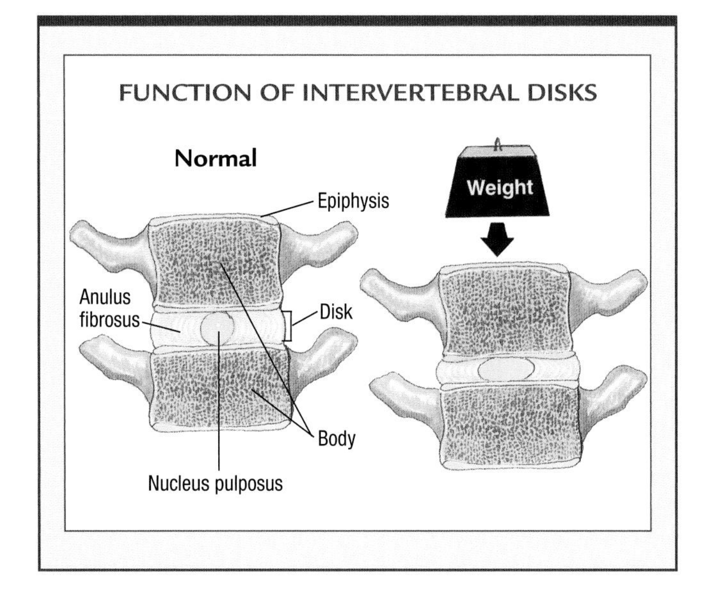
Normal
Epiphysis
Weight
Anulus fibrosus
Disk
Body
Nucleus pulposus

**Clinical tip**

Some spinal vertebrae are notable on visual inspection of the back. The most observable are those at the base of the neck, C7, T1, and T2. C7 is called the vertebra prominens because it's the most prominent vertebra when the neck is bent forward.

## Intervertebral disks

Intervertebral disks lie between two adjacent vertebrae. The nucleus pulposus, the central gelatinous part of the disk, is enclosed in several layers of fibrocartilaginous laminae. The disk functions as a cushion between the vertebrae and facilitates movement.

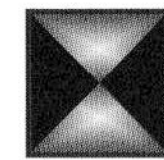

**Age-related changes**

With age, the nucleus pulposus hardens and becomes less resilient. Narrowing of the disks and compression of individual vertebrae result in loss of height with age. Further degeneration of disk tissue may cause bulging without herniation.

## TYPICAL VERTEBRA
### Superior aspect

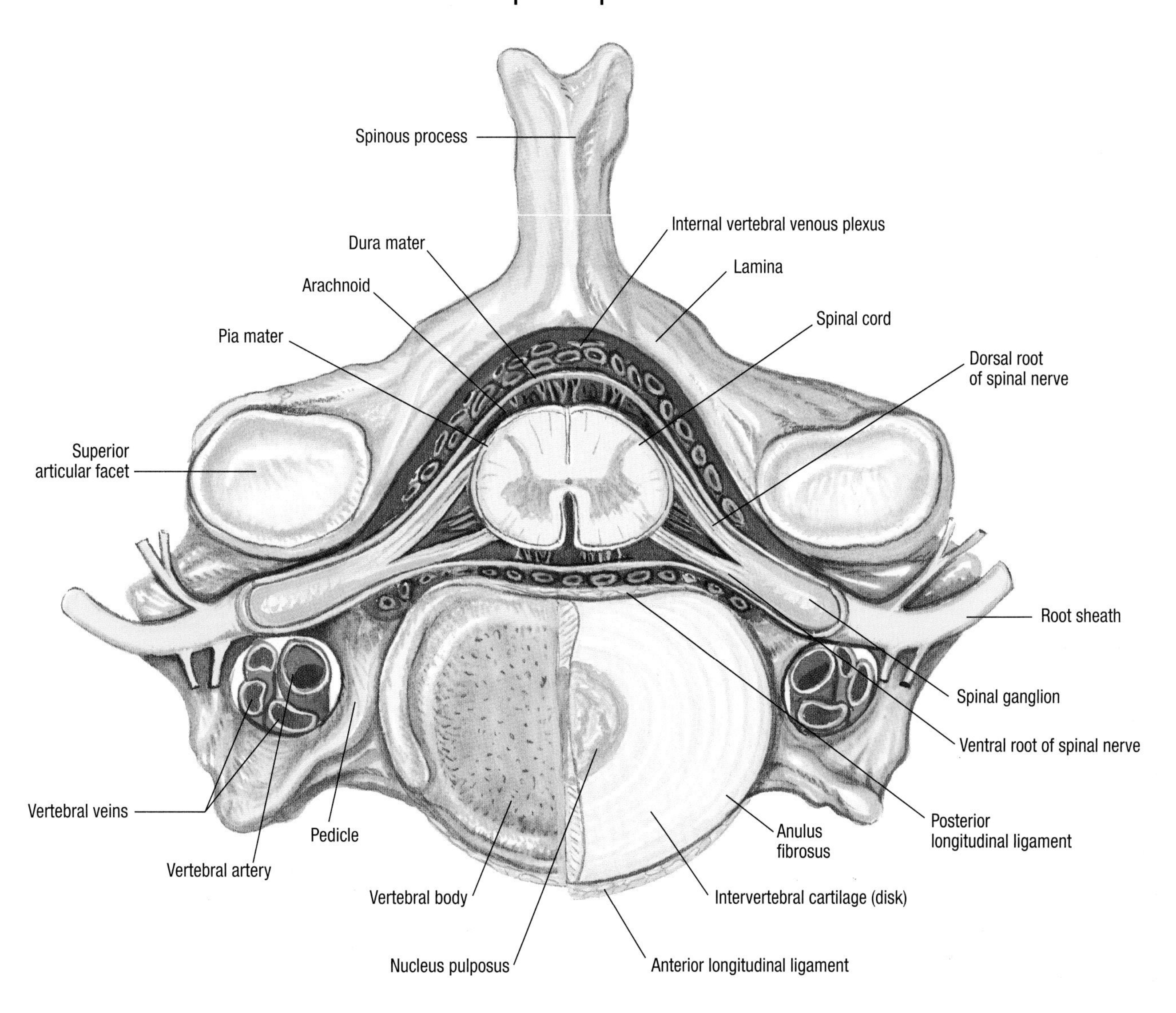

Spinous process
Internal vertebral venous plexus
Dura mater
Lamina
Arachnoid
Spinal cord
Pia mater
Dorsal root
of spinal nerve
Superior
articular facet
Root sheath
Spinal ganglion
Ventral root of spinal nerve
Vertebral veins
Pedicle
Posterior
longitudinal ligament
Anulus
fibrosus
Vertebral artery
Vertebral body
Intervertebral cartilage (disk)
Nucleus pulposus
Anterior longitudinal ligament


## STRUCTURAL FEATURES OF AN INTERVERTEBRAL DISK

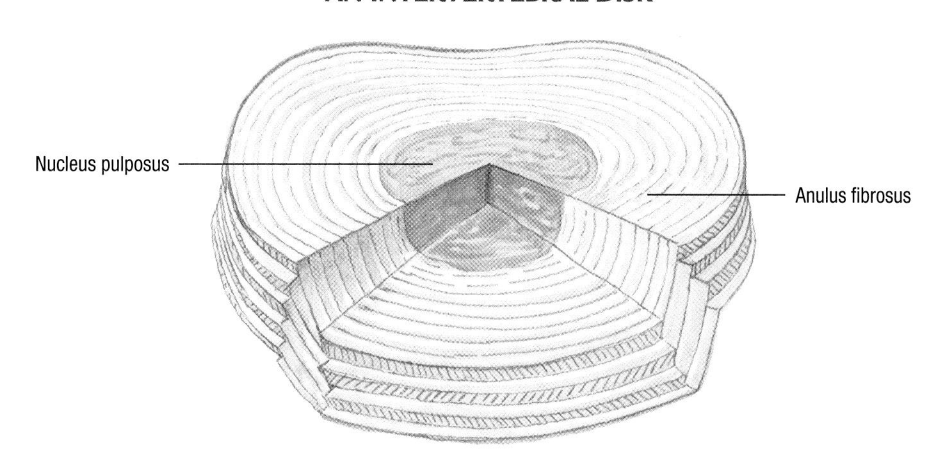

Nucleus pulposus
Anulus fibrosus

## ATLAS AND AXIS, C1 AND C2

### Articulated right lateral view

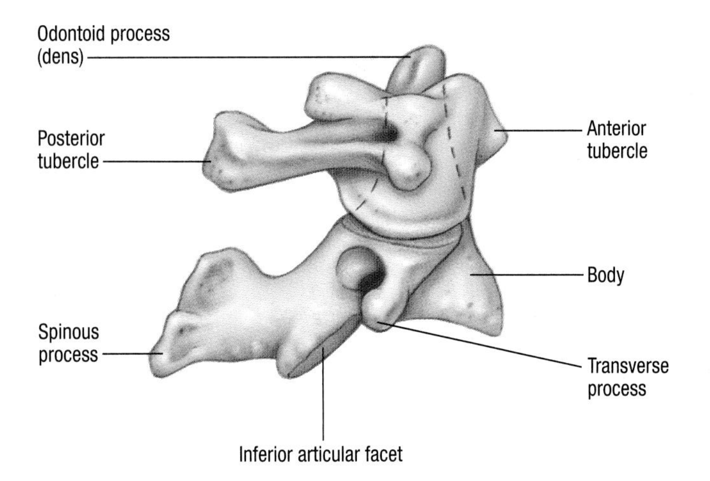
Odontoid process (dens)
Posterior tubercle
Anterior tubercle
Body
Spinous process
Transverse process
Inferior articular facet

## ATLAS, C1

### Disarticulated anterior view

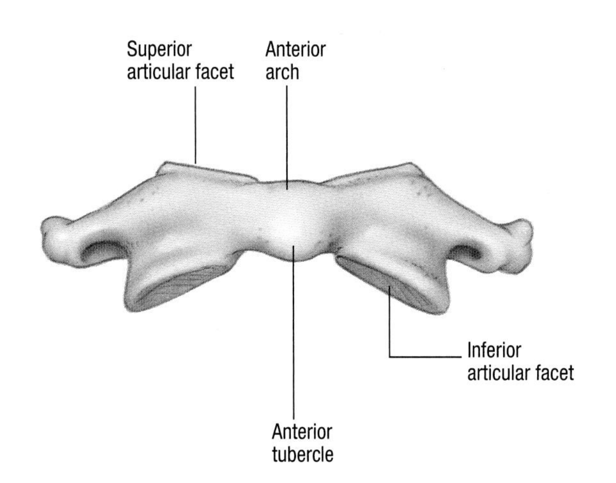
Superior articular facet
Anterior arch
Inferior articular facet
Anterior tubercle

## ATLAS AND AXIS, C1 AND C2

### Articulated posterior view

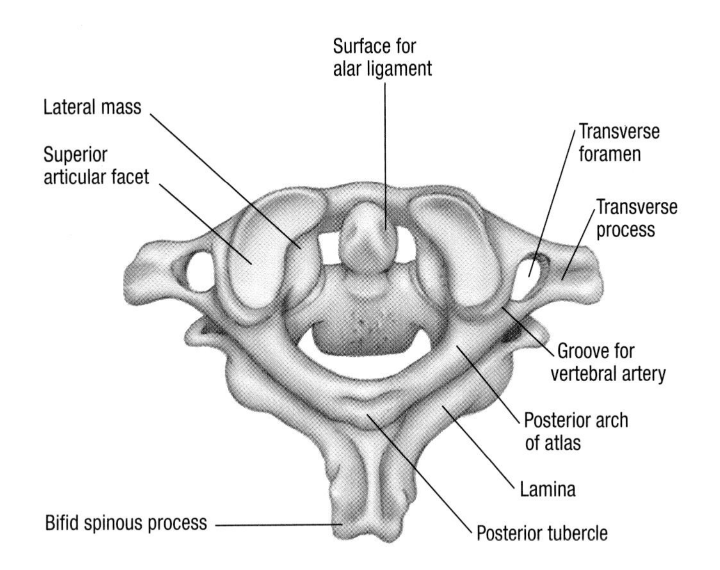
Surface for alar ligament
Lateral mass
Superior articular facet
Transverse foramen
Transverse process
Groove for vertebral artery
Posterior arch of atlas
Lamina
Bifid spinous process
Posterior tubercle

## AXIS, C2

### Disarticulated anterior view

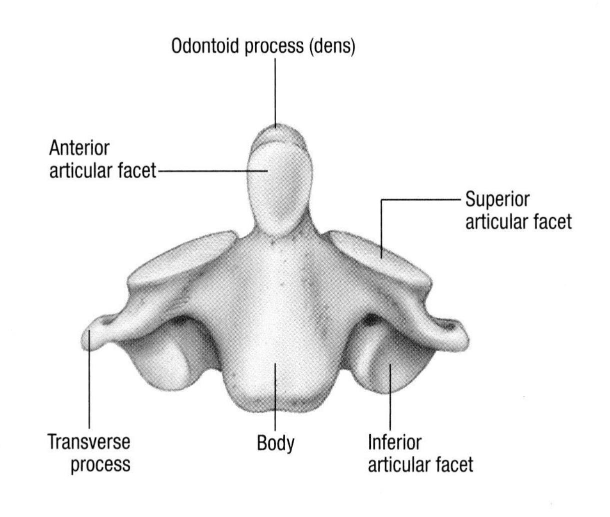
Odontoid process (dens)
Anterior articular facet
Superior articular facet
Transverse process
Body
Inferior articular facet

# 5TH CERVICAL VERTEBRA

## Lateral view

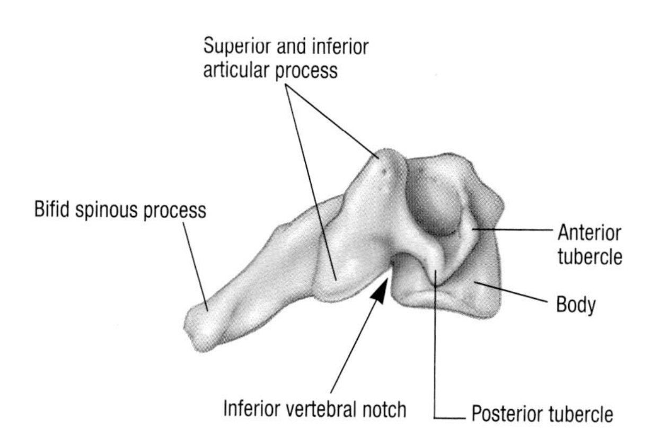
Superior and inferior articular process
Bifid spinous process
Anterior tubercle
Body
Inferior vertebral notch
Posterior tubercle

## Anterior view

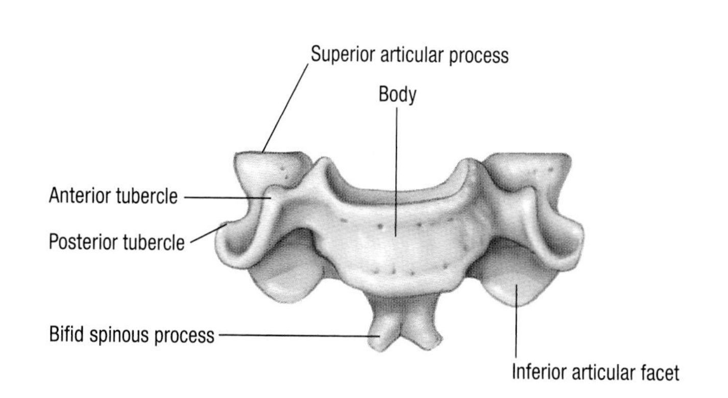
Superior articular process
Body
Anterior tubercle
Posterior tubercle
Bifid spinous process
Inferior articular facet

## Superior view

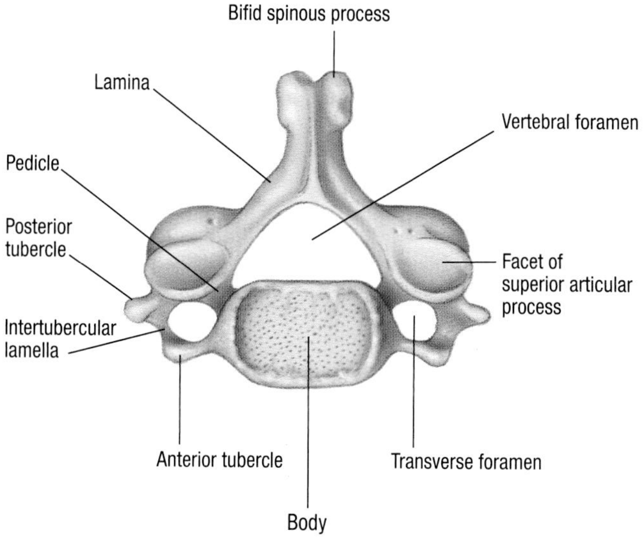

Bifid spinous process
Lamina
Vertebral foramen
Pedicle
Posterior tubercle
Facet of superior articular process
Intertubercular lamella
Anterior tubercle
Transverse foramen
Body

## Posterolateral oblique view

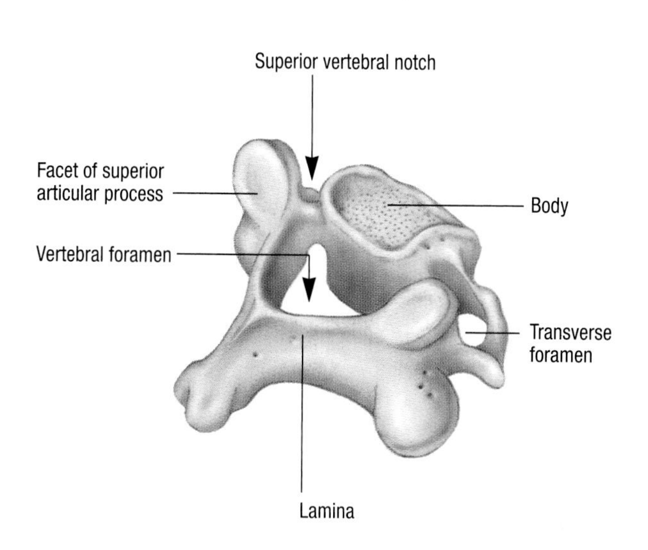
Superior vertebral notch
Facet of superior articular process
Body
Vertebral foramen
Transverse foramen
Lamina

Superior view

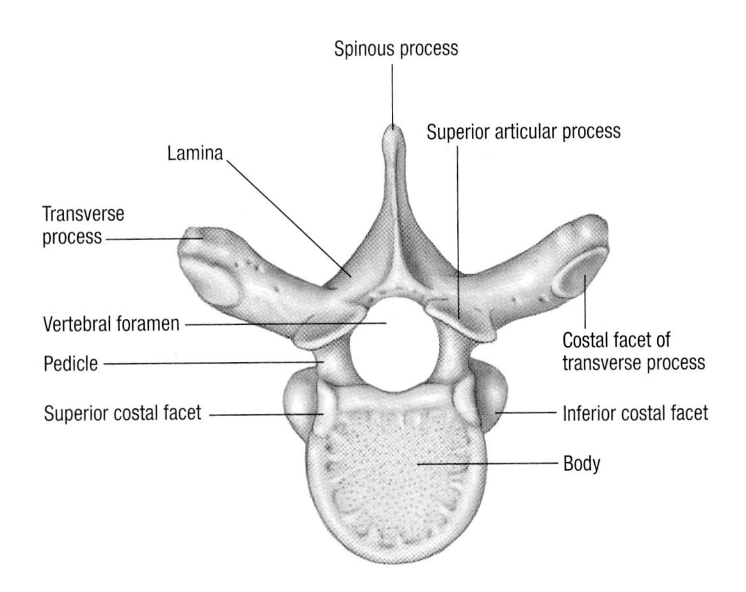
Spinous process
Superior articular process
Lamina
Transverse process
Vertebral foramen
Costal facet of transverse process
Pedicle
Superior costal facet
Inferior costal facet
Body

Lateral view

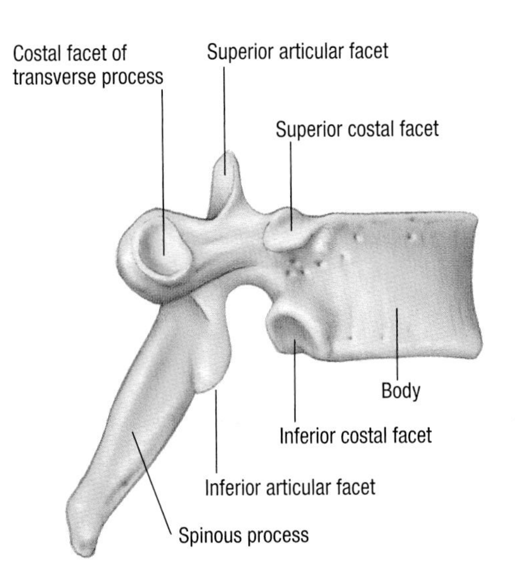
Costal facet of transverse process
Superior articular facet
Superior costal facet
Body
Inferior costal facet
Inferior articular facet
Spinous process

Posterolateral oblique view

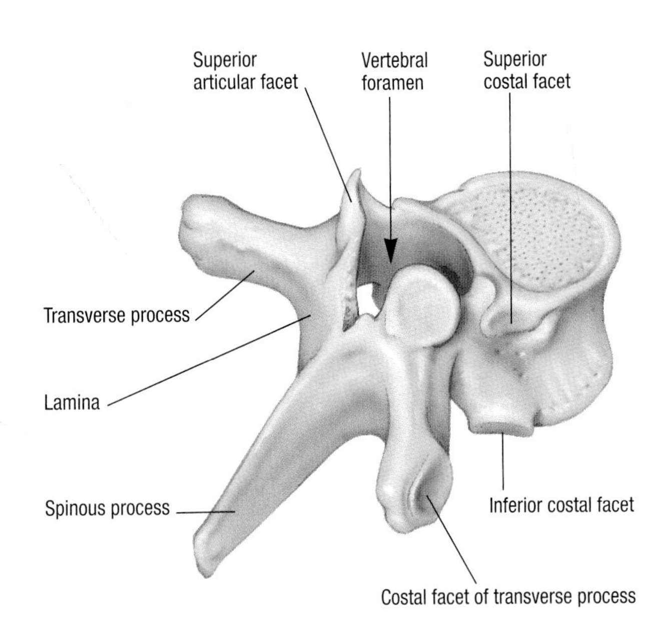
Superior articular facet
Vertebral foramen
Superior costal facet
Transverse process
Lamina
Spinous process
Inferior costal facet
Costal facet of transverse process

# 2ND LUMBAR VERTEBRA

## Superior view

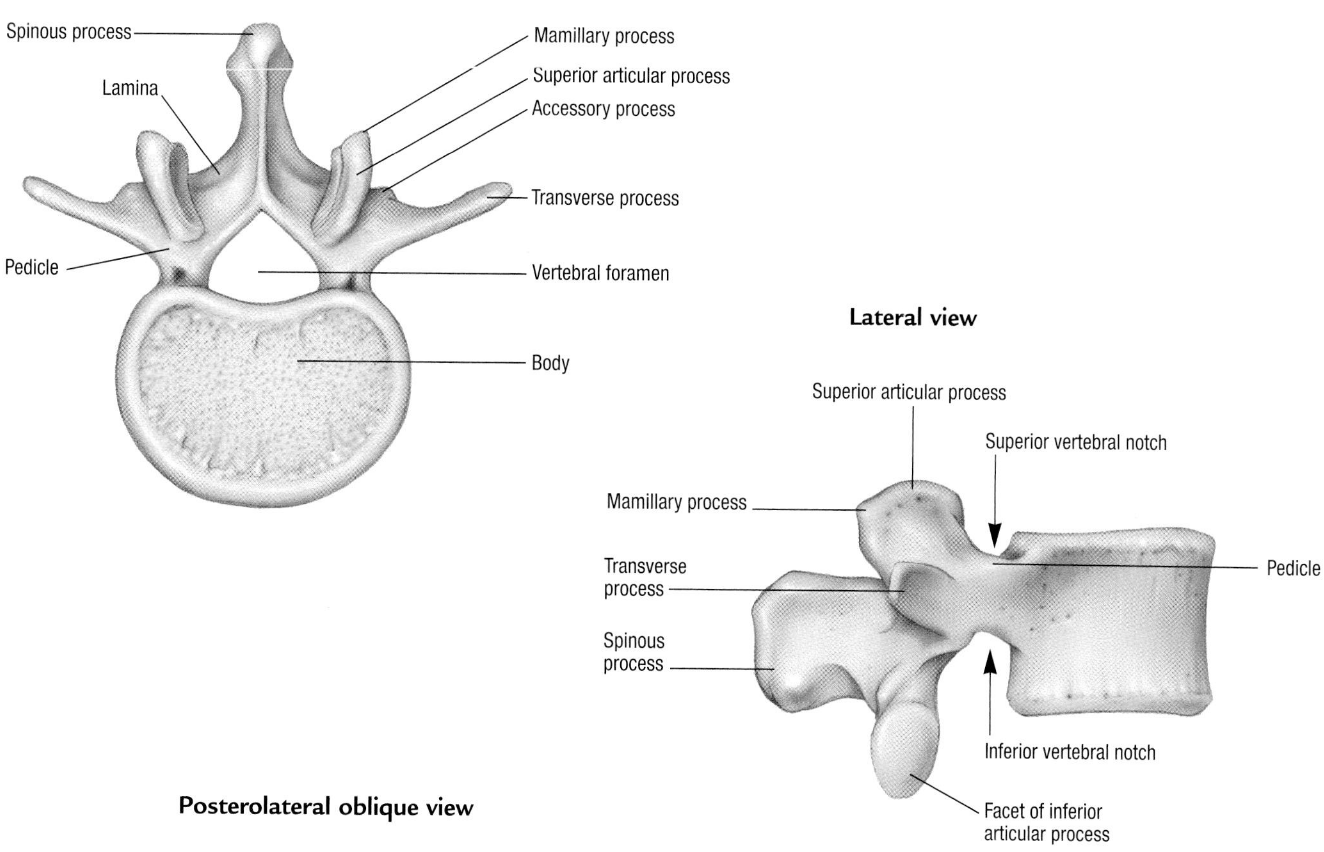

Spinous process
Lamina
Pedicle
Mamillary process
Superior articular process
Accessory process
Transverse process
Vertebral foramen
Body
Lateral view
Superior articular process
Superior vertebral notch
Mamillary process
Transverse process
Spinous process
Pedicle
Inferior vertebral notch
Facet of inferior articular process


## Posterolateral oblique view

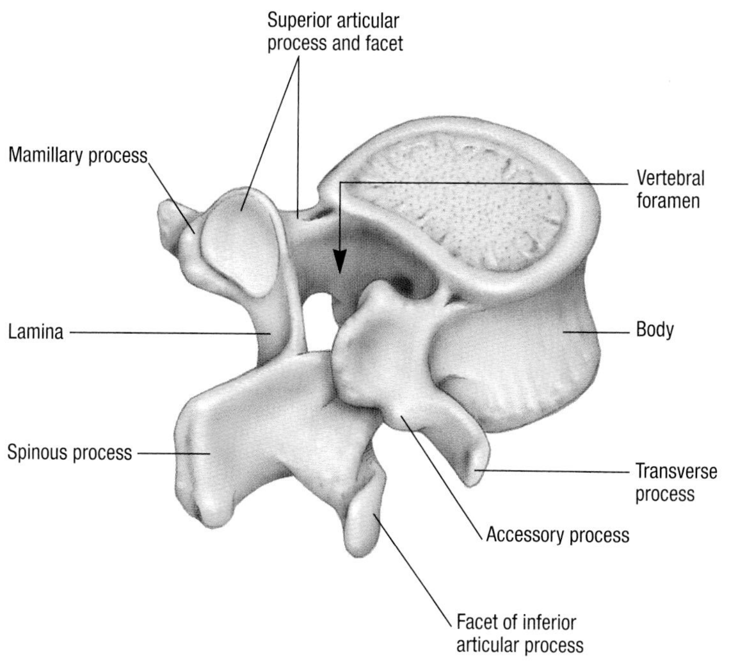

Superior articular process and facet
Mamillary process
Vertebral foramen
Lamina
Body
Spinous process
Transverse process
Accessory process
Facet of inferior articular process

## Dorsal surface

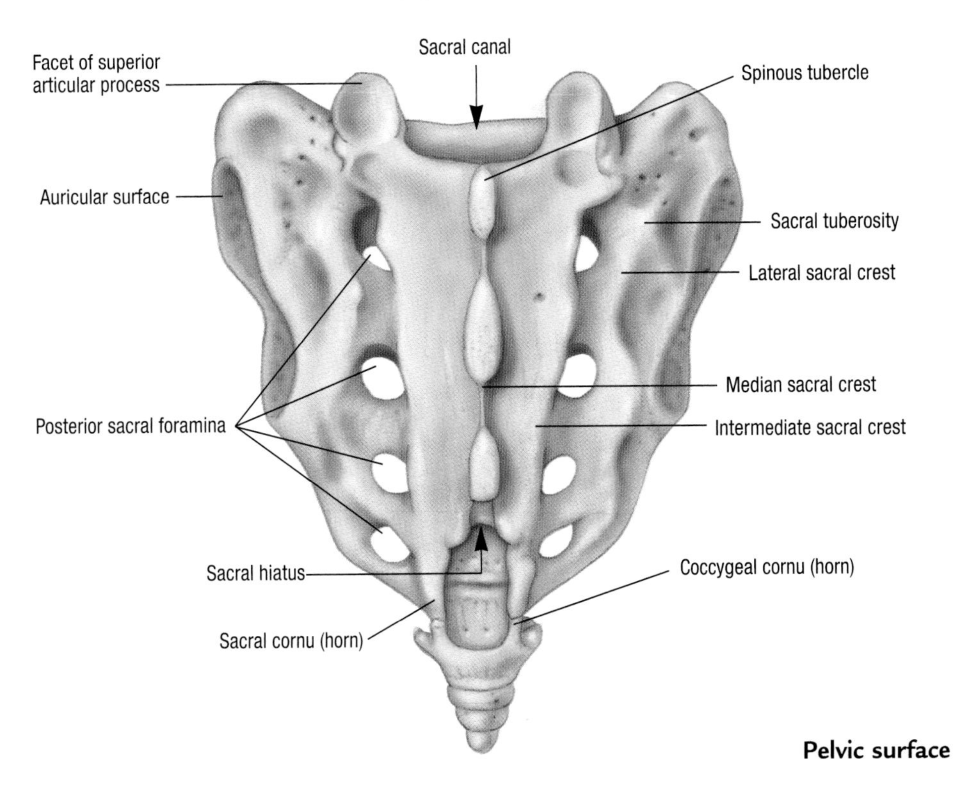
Sacral canal
Facet of superior articular process
Spinous tubercle
Auricular surface
Sacral tuberosity
Lateral sacral crest
Median sacral crest
Posterior sacral foramina
Intermediate sacral crest
Sacral hiatus
Coccygeal cornu (horn)
Sacral cornu (horn)

## Pelvic surface

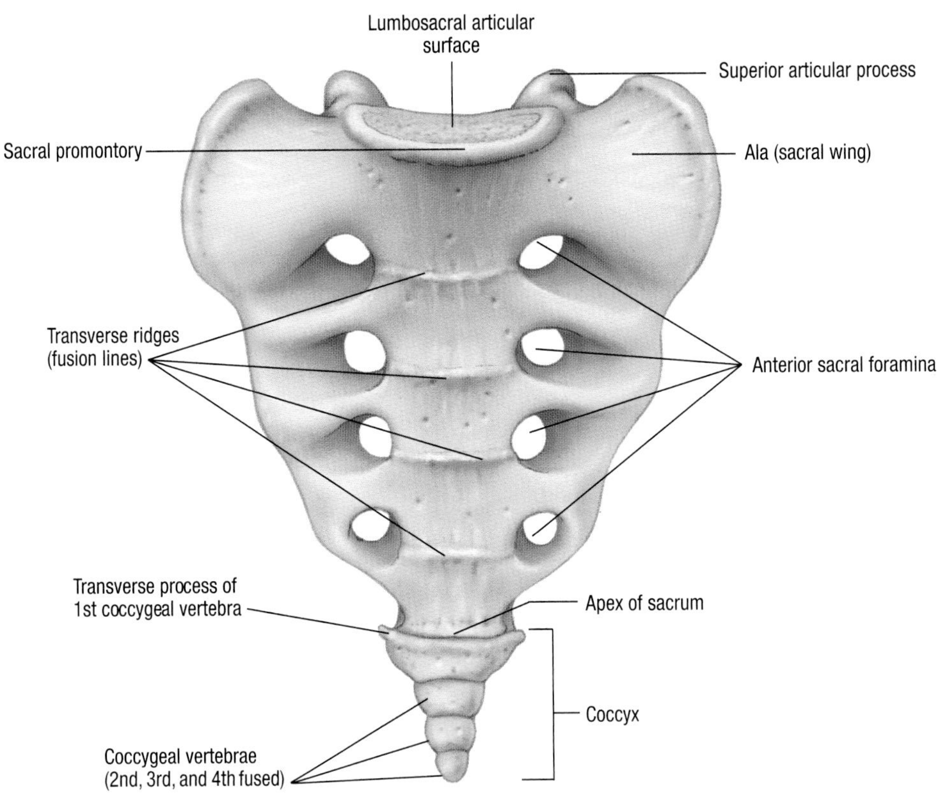
Lumbosacral articular surface
Superior articular process
Sacral promontory
Ala (sacral wing)
Transverse ridges (fusion lines)
Anterior sacral foramina
Transverse process of 1st coccygeal vertebra
Apex of sacrum
Coccyx
Coccygeal vertebrae (2nd, 3rd, and 4th fused)

## SACRUM AND COCCYX
**Lateral view**

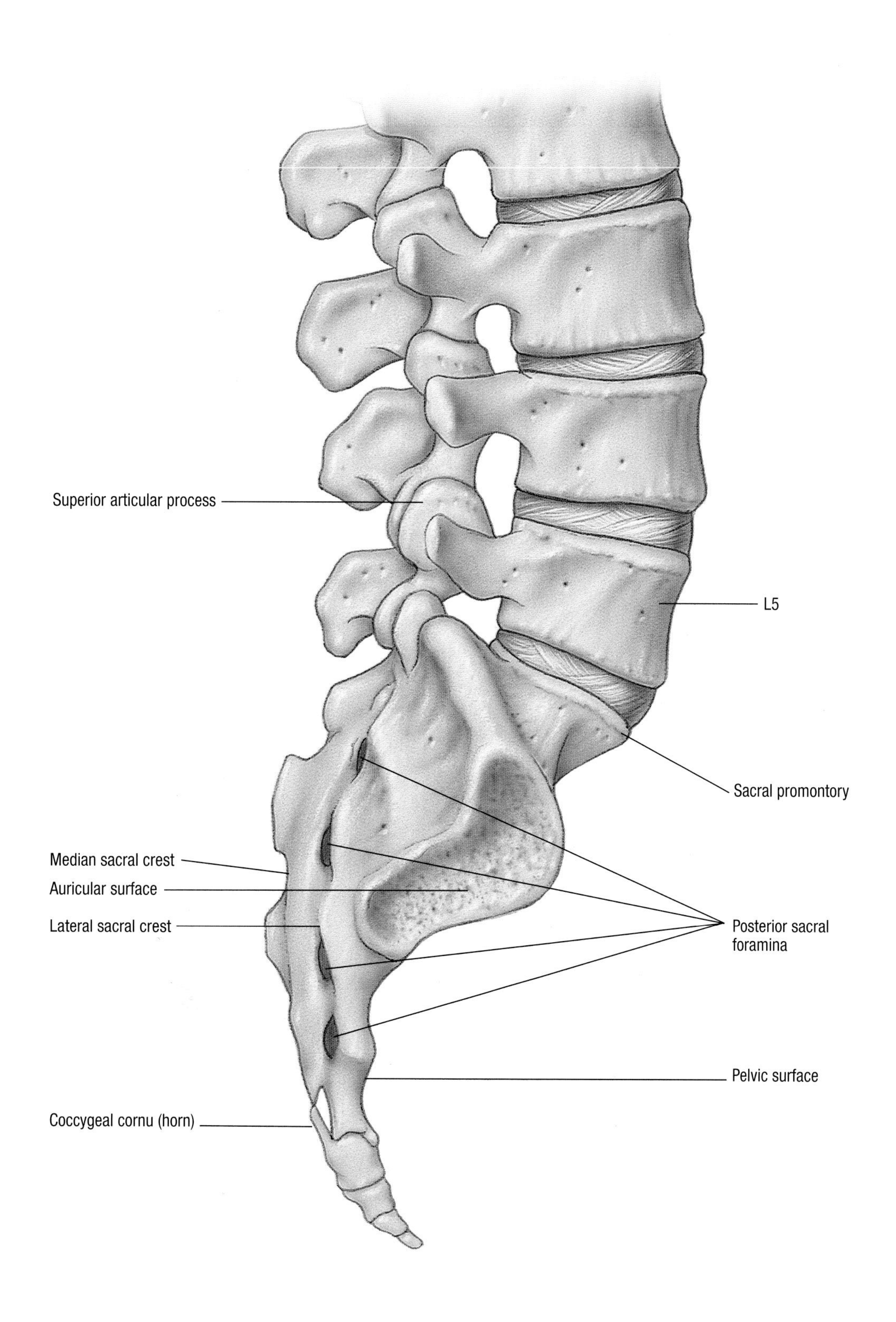
Superior articular process
L5
Sacral promontory
Median sacral crest
Auricular surface
Lateral sacral crest
Posterior sacral foramina
Pelvic surface
Coccygeal cornu (horn)

# Spinal cord and spinal nerves

A cylindrical structure in the vertebral canal, the spinal cord extends from the foramen magnum at the base of the skull to the upper lumbar region of the vertebral column. The brain and spinal cord are protected from shock and infection by bones, several cushioning layers called the meninges, and cerebrospinal fluid (CSF). See also Part II, Head and neck.

## Gray and white matter

Within the spinal cord, the H-shaped mass of gray matter is divided into horns, which consist mainly of neuronal cell bodies.
- Cell bodies in the posterior (dorsal) horn primarily relay sensations.
- Cell bodies in the anterior (ventral) horn play a part in voluntary and reflex motor activity.

White matter surrounding the outer part of these horns consists of myelinated nerve fibers grouped functionally in vertical columns, or tracts.

## Spinal nerves

Thirty-one pairs of spinal nerves arise from the cord and leave the vertebral column through openings called the intervertebral foramina. (See *Spinal nerves,* page 104, and *Thoracic spinal cord section with spinal nerves,* page 105.) The *cauda equina,* a collection of lumbar and sacral nerve roots that fills the caudal end of the spinal cord, resembles the tail of a horse. The nerves of the cauda equina typically exit the cord between the twelfth thoracic and third lumbar vertebrae.

The pairs of spinal nerves are designated from top to bottom as C1 through S5 as follows: 8 cervical, 12 thoracic, 5 lumbar, 5 sacral, and 1 coccygeal.

Each spinal nerve contains a dorsal (sensory or afferent) root and a ventral (motor or efferent) root.
- Cell bodies of the afferent (sensory) fibers lie outside the spinal cord in the dorsal root ganglion
- Cell bodies of efferent fibers lie in the ventral gray column of the cord.

The spinal cord provides pathways for nerve impulses to and from the brain; it performs sensory, motor, and reflex functions.

### Spinal nerve branches

Spinal nerves divide into branches called rami. Ventral primary rami exit anteriorly, and dorsal primary rami, posteriorly.

A nerve plexus is a network of adjacent nerves that join together. The name of each plexus describes the area its nerves supply. The major nerve plexuses and the areas they supply are:
- cervical – head, neck, shoulders, diaphragm
- brachial – upper limbs and some neck and shoulder muscles
- lumbar – part of the abdominal wall, lower limbs, and external male genitalia
- sacral – perineum, buttocks, and most of the lower limbs
- pudendal – external female genitalia.

Most spinal nerves innervate specific areas of the skin called *dermatomes.* These areas are variable and tend to overlap. (See *Dermatomes,* page 311.)

### Spinal pathways

Sensory impulses travel along the sensory (afferent, or ascending) neural pathways to the sensory cortex in the parietal lobe of the brain where they're interpreted.

Motor impulses travel from the brain to the muscles along the motor (efferent, or descending) pathways. These impulses originate in the motor cortex of the frontal lobe and travel along upper motor neurons to the peripheral nervous system.

Upper motor neurons originate in the brain and form two major systems, the pyramidal system and the extrapyramidal system.

### Reflex responses

Reflex responses occur automatically, without any required brain involvement, to protect the body. The brain receives information that a spinal reflex has been activated. Spinal nerves, which have both sensory and motor portions, mediate:
- Deep tendon reflexes (involuntary contractions of a muscle after brief stretching caused by tendon percussion)
- Superficial reflexes (for example, withdrawal reflexes elicited by noxious or tactile stimulation of the skin, cornea, or mucous membranes).

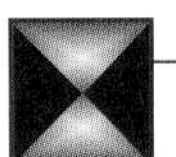

**Age-related changes**

The following reflexes tend to diminish with age: ankle, knee-jerk, and abdominal.

#### Knee-jerk reflex arc

A simple spinal reflex like the knee-jerk (patellar) reflex arc requires a sensory (afferent) neuron and a motor (efferent) neuron. In this reflex, the following events occur:
- A sensory receptor detects the mechanical stimulus produced by the reflex hammer striking the patellar tendon.
- The sensory neuron carries the impulse along its axon via a spinal nerve to the dorsal root, where it enters the spinal cord.
- In the anterior horn of the cord, the sensory neuron synapses with a motor neuron, which carries the impulse along its axon via a spinal nerve to the muscle.
- The motor neuron transmits the impulse to muscle fibers via stimulation of the motor end plate, a band of terminal fibers of the motor nerves of skeletal muscles. In response, the muscle contracts and the leg extends.

Other neurons and neural pathways (gamma afferent and efferent systems) also participate in spinal reflexes.

**Clinical tip**

If the brain can't send a message to the leg after a severe spinal cord injury, a stimulus can still cause the knee-jerk reflex as long as the spinal cord remains intact at the level of this reflex.

## MIDBRAIN, MEDULLA OBLONGATA, AND SPINAL CORD

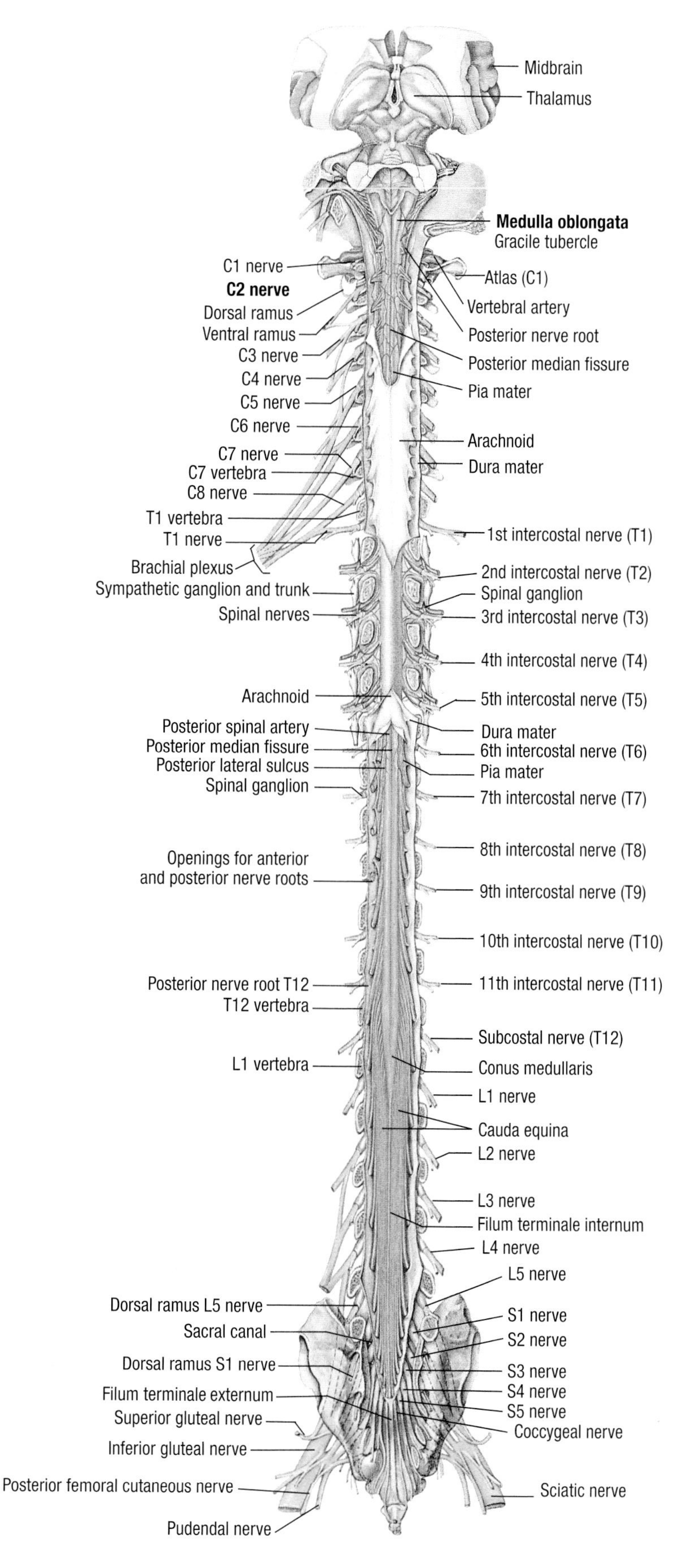

Midbrain
Thalamus
Medulla oblongata
Gracile tubercle
C1 nerve
C2 nerve
Dorsal ramus
Ventral ramus
C3 nerve
C4 nerve
C5 nerve
C6 nerve
C7 nerve
C7 vertebra
C8 nerve
T1 vertebra
T1 nerve
Brachial plexus
Sympathetic ganglion and trunk
Spinal nerves
Atlas (C1)
Vertebral artery
Posterior nerve root
Posterior median fissure
Pia mater
Arachnoid
Dura mater
1st intercostal nerve (T1)
2nd intercostal nerve (T2)
Spinal ganglion
3rd intercostal nerve (T3)
4th intercostal nerve (T4)
Arachnoid
5th intercostal nerve (T5)
Posterior spinal artery
Posterior median fissure
Posterior lateral sulcus
Spinal ganglion
Dura mater
6th intercostal nerve (T6)
Pia mater
7th intercostal nerve (T7)
8th intercostal nerve (T8)
Openings for anterior
and posterior nerve roots
9th intercostal nerve (T9)
10th intercostal nerve (T10)
Posterior nerve root T12
T12 vertebra
11th intercostal nerve (T11)
Subcostal nerve (T12)
L1 vertebra
Conus medullaris
L1 nerve
Cauda equina
L2 nerve
L3 nerve
Filum terminale internum
L4 nerve
L5 nerve
Dorsal ramus L5 nerve
Sacral canal
Dorsal ramus S1 nerve
Filum terminale externum
Superior gluteal nerve
Inferior gluteal nerve
Posterior femoral cutaneous nerve
Pudendal nerve
S1 nerve
S2 nerve
S3 nerve
S4 nerve
S5 nerve
Coccygeal nerve
Sciatic nerve

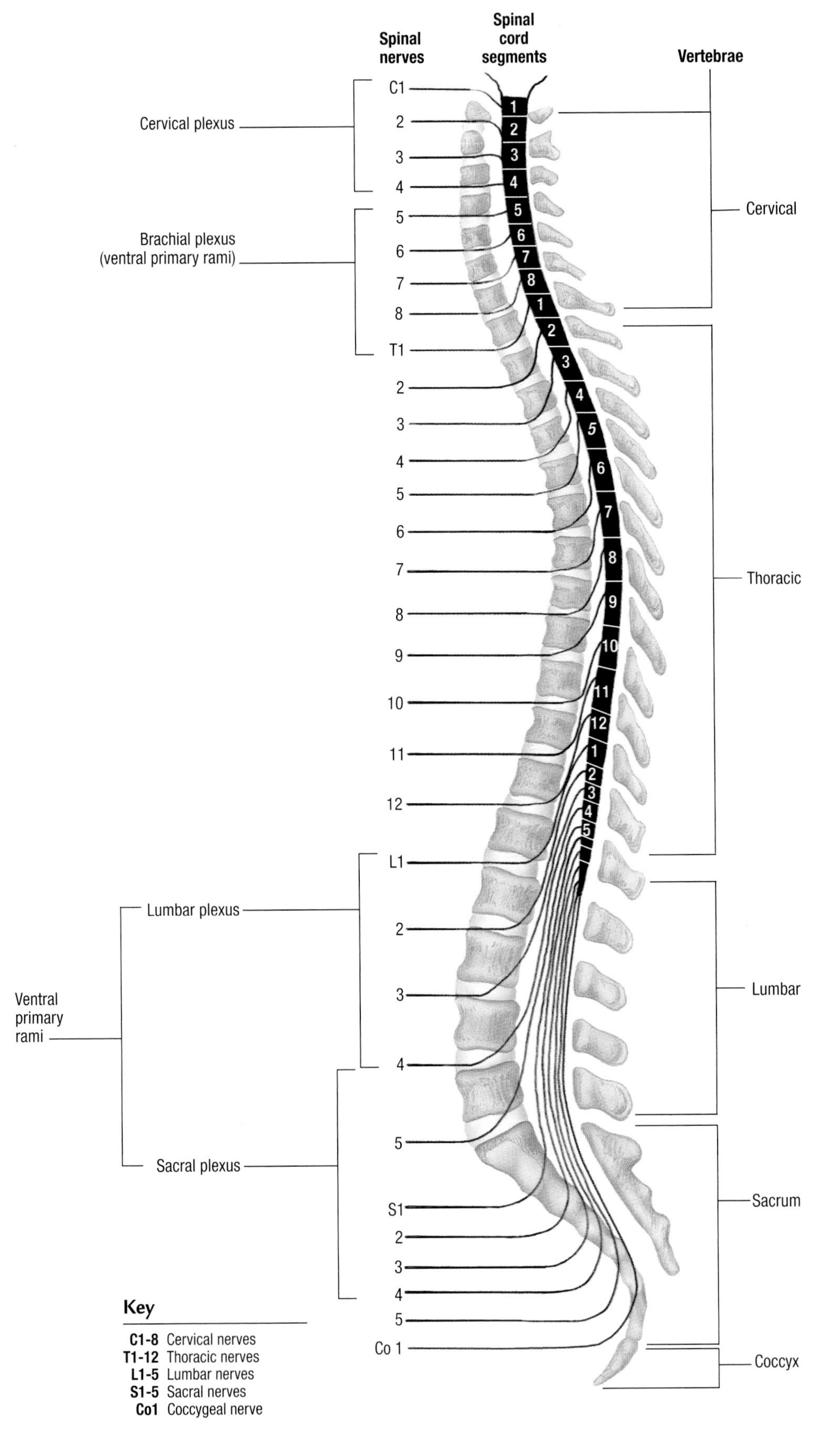
Spinal nerves
Spinal cord segments
Vertebrae
Cervical plexus
Brachial plexus (ventral primary rami)
Cervical
Thoracic
Lumbar plexus
Ventral primary rami
Lumbar
Sacral plexus
Sacrum
Coccyx
C1
T1
L1
S1
Co 1

**Key**

**C1-8** Cervical nerves
**T1-12** Thoracic nerves
**L1-5** Lumbar nerves
**S1-5** Sacral nerves
**Co1** Coccygeal nerve

## THORACIC SPINAL CORD SECTION WITH SPINAL NERVES

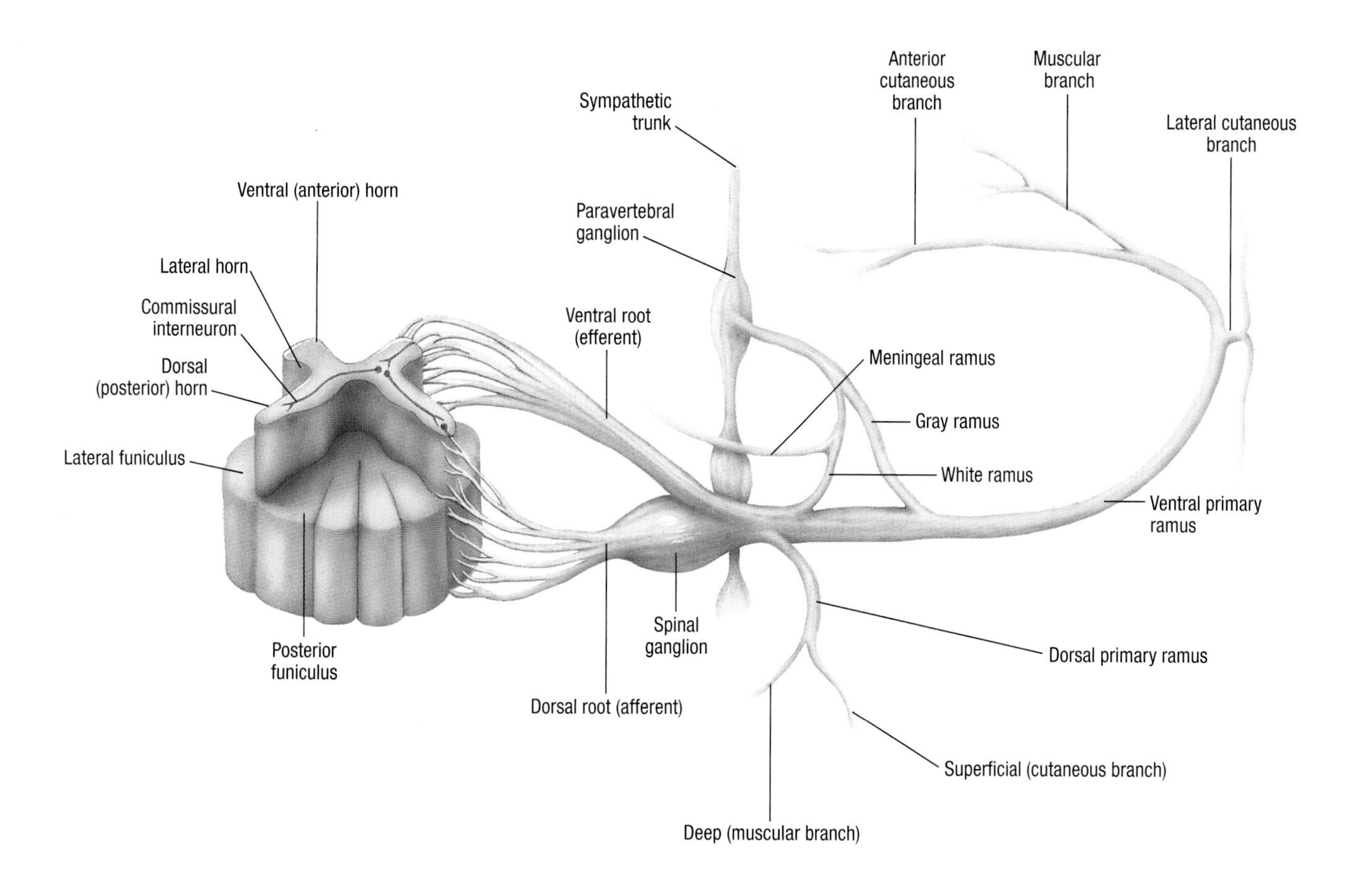

Anterior cutaneous branch
Muscular branch
Sympathetic trunk
Lateral cutaneous branch
Ventral (anterior) horn
Paravertebral ganglion
Lateral horn
Commissural interneuron
Ventral root (efferent)
Dorsal (posterior) horn
Meningeal ramus
Gray ramus
Lateral funiculus
White ramus
Ventral primary ramus
Spinal ganglion
Posterior funiculus
Dorsal primary ramus
Dorsal root (afferent)
Superficial (cutaneous branch)
Deep (muscular branch)

# PART IV

# Upper limb

# Upper limb structures

The upper limb, including the pectoral girdle, is part of the appendicular skeleton. Unique features of the upper limb are the shoulder joint, which can move in almost any plane, and the opposable thumb, which can move against other digits of the hand. This section provides a brief summary of upper limb structures; it is followed by detailed descriptions of structures in each region of the upper limb: shoulder and upper arm, elbow, forearm, and wrist and hand.

## Bones

Each upper limb consists of 30 bones: humerus (1), ulna (1), radius (1), and hand bones (27).

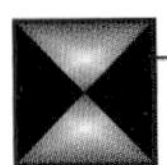

**Age-related changes**

Many of the upper limb bones ossify between birth and young adulthood – namely the clavicle, humerus, radius, ulna, metacarpals, and phalanges. X-rays are frequently used to assess skeletal age because they can distinguish between cartilage and bone. Wrist and hand X-rays can be used to evaluate the degree of ossification of the carpal bones. By preadolescence (about 11 years of age), the carpal bones should be fully ossified. Childhood injuries, such as fractures to the epiphyseal area (growth plate), are closely monitored through skeletal maturity. Growth plate injuries in children can cause limb-length discrepancy, joint incongruity, or angular deformity. The clavicular epiphysis is the last to ossify; cartilage may persist for longer than 30 years.

## Joints and ligaments

Joints, ligaments, muscles, and tendons make both gross and fine motor movements possible. Joints of the upper limb and their anatomical types include the following:

- shoulder (2) – ball-and-socket
- elbow (2) – hinge
- interphalangeals (18) – hinge
- metacarpophalangeals (10) – hinge
- first metacarpals (2) – saddle.

Ligaments in the upper limb and their attachments are the following:

- Coracoid and trapezoid – form the coracoclavicular ligament, which connects the coracoid surface (the curved projection on the superior border) of the scapula to the lateral end of the clavicle
- Coracohumeral – supports the superior part of the shoulder joint capsule
- Coracoacromial – forms an arch above the shoulder joint in conjunction with the coracoid process and the acromion; prevents upward displacement of the humeral head
- Ulnar collateral – attaches to the medial epicondyle
- Radial collateral – attaches to the lateral epicondyle.

## Bursae

Bursal sacs are present in some joints of the upper limb. Bursitis, inflammation of the bursa, is a common cause of shoulder pain. It can result from external injury or from recurrent trauma caused by increased friction between tissue layers, as may occur after inflammation. The elbow joint has a bursa between the olecranon process and the skin. The shoulder area contains the following three bursae, named according to location:

- subacromial
- subdeltoid
- subscapular.

## Muscles

Details about specific muscles of the upper limb appear on page 124.

## Nerves

The brachial plexus provides most of the nerve supply to the arms, and to several neck and shoulder muscles. It is formed by the ventral rami of spinal nerves C5 to C8 and T1 (with contributions from C4 and T2). (See *Upper limb nerves*, page 112.)

### Brachial plexus branches

The network of nerves branching off the brachial plexus innervate the muscles and skin of the chest, shoulder, and arms. (See *Brachial plexus and branches*, page 113.) The roots of the brachial plexus (ventral rami) and contributions from spinal nerves C4 and T2 join to form three trunks – superior, middle, and inferior. Each trunk separates into an anterior and a posterior division. The divisions then join to form three cords – lateral, medial, and posterior – from which the following nerves branch:

- lateral cord – lateral pectoral nerve
- medial cord – medial pectoral nerve, medial cutaneous nerve of arm, and medial cutaneous nerve of forearm
- posterior cord – upper subscapular and thoracodorsal (middle subscapular) nerves.

At the distal ends of the cords are the terminal branches of the brachial plexus, the major nerves of the upper limb. (See *Brachial plexus terminal branches*, page 113.)

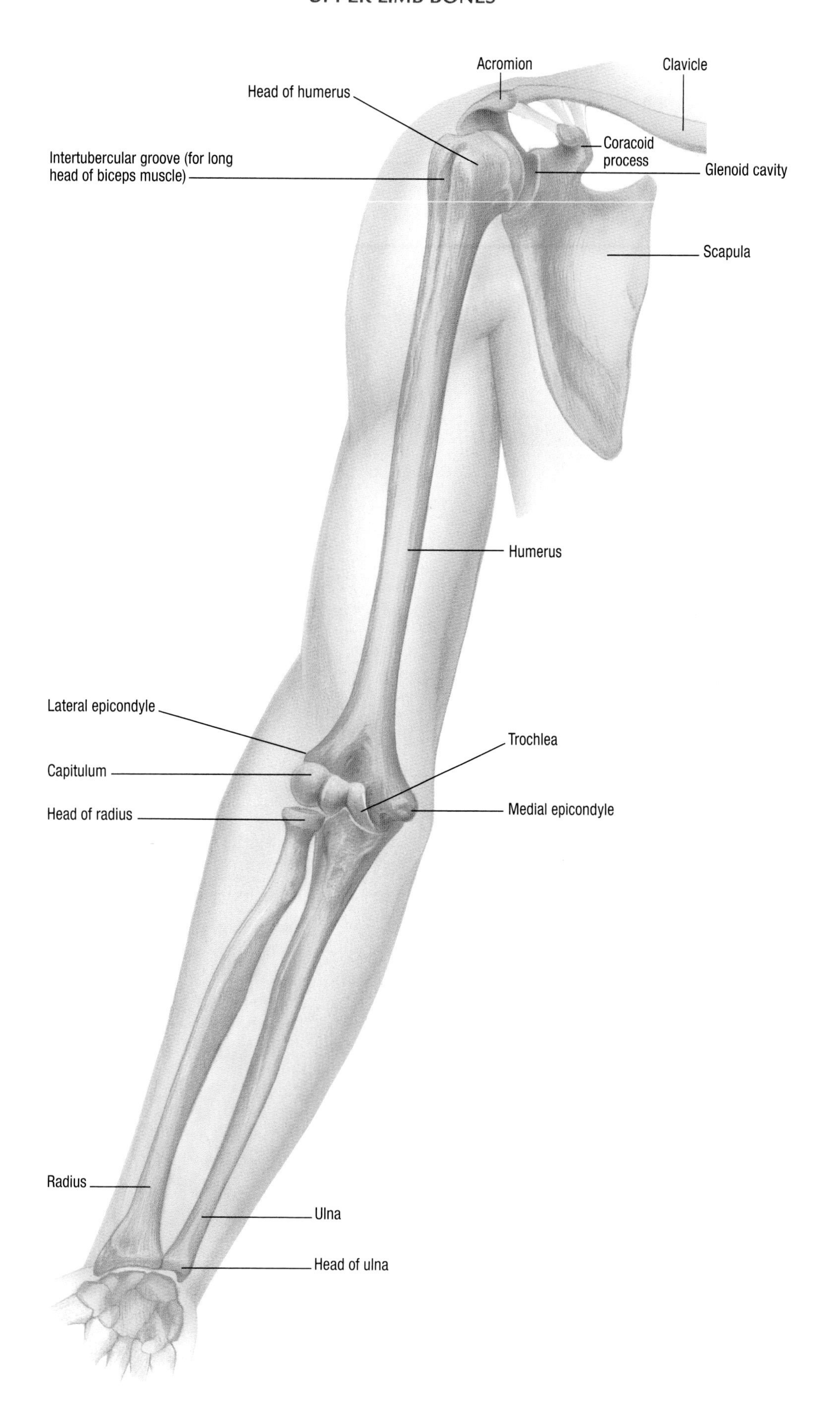
Acromion
Clavicle
Head of humerus
Coracoid process
Intertubercular groove (for long head of biceps muscle)
Glenoid cavity
Scapula
Humerus
Lateral epicondyle
Trochlea
Capitulum
Medial epicondyle
Head of radius
Radius
Ulna
Head of ulna

## UPPER LIMB BONES AND LIGAMENTS
### Anterior view

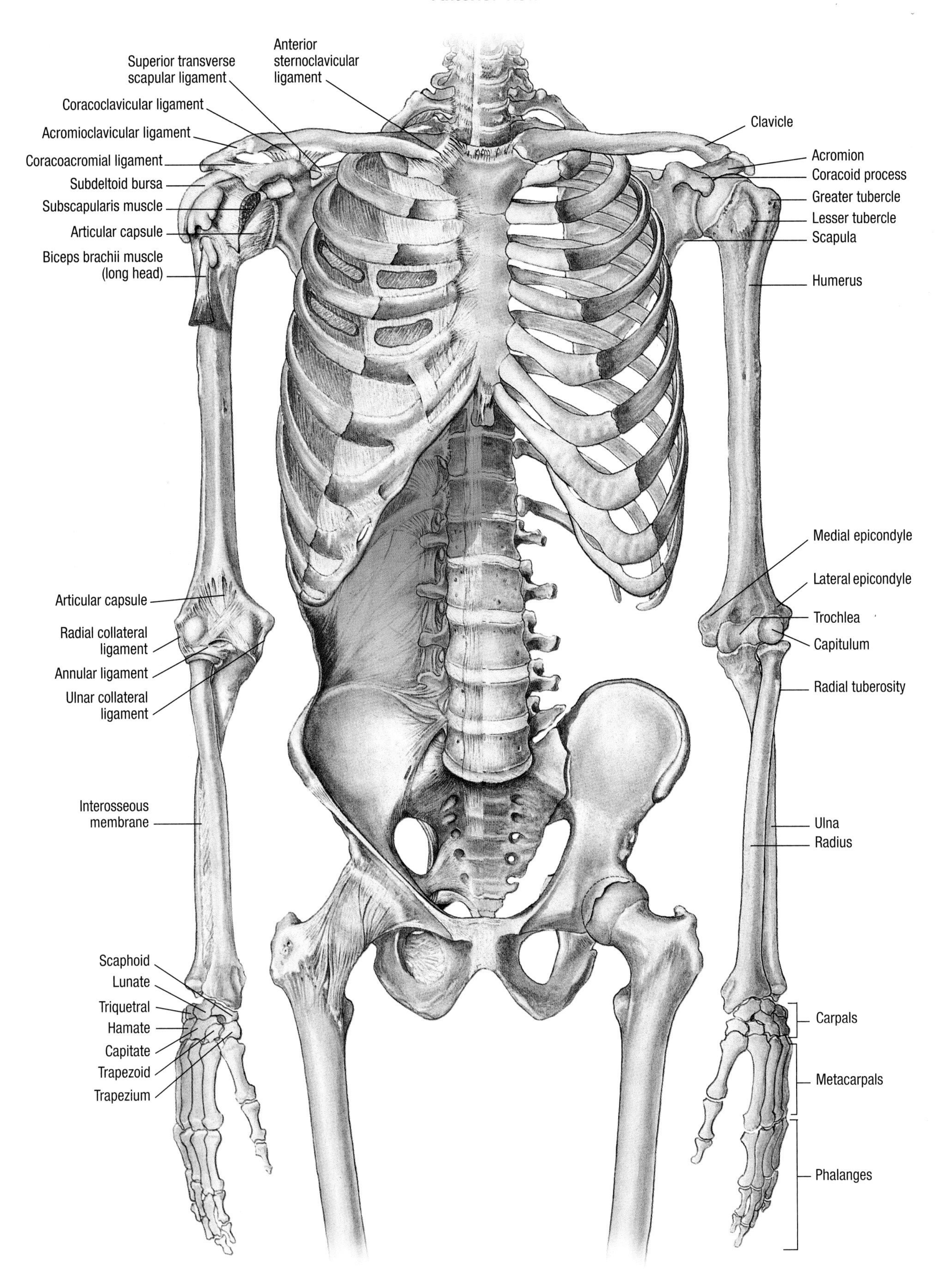

Superior transverse scapular ligament
Anterior sternoclavicular ligament
Coracoclavicular ligament
Acromioclavicular ligament
Coracoacromial ligament
Subdeltoid bursa
Subscapularis muscle
Articular capsule
Biceps brachii muscle (long head)
Clavicle
Acromion
Coracoid process
Greater tubercle
Lesser tubercle
Scapula
Humerus
Medial epicondyle
Lateral epicondyle
Trochlea
Capitulum
Radial tuberosity
Articular capsule
Radial collateral ligament
Annular ligament
Ulnar collateral ligament
Interosseous membrane
Ulna
Radius
Scaphoid
Lunate
Triquetral
Hamate
Capitate
Trapezoid
Trapezium
Carpals
Metacarpals
Phalanges

# UPPER LIMB BONES AND LIGAMENTS
## Posterior view

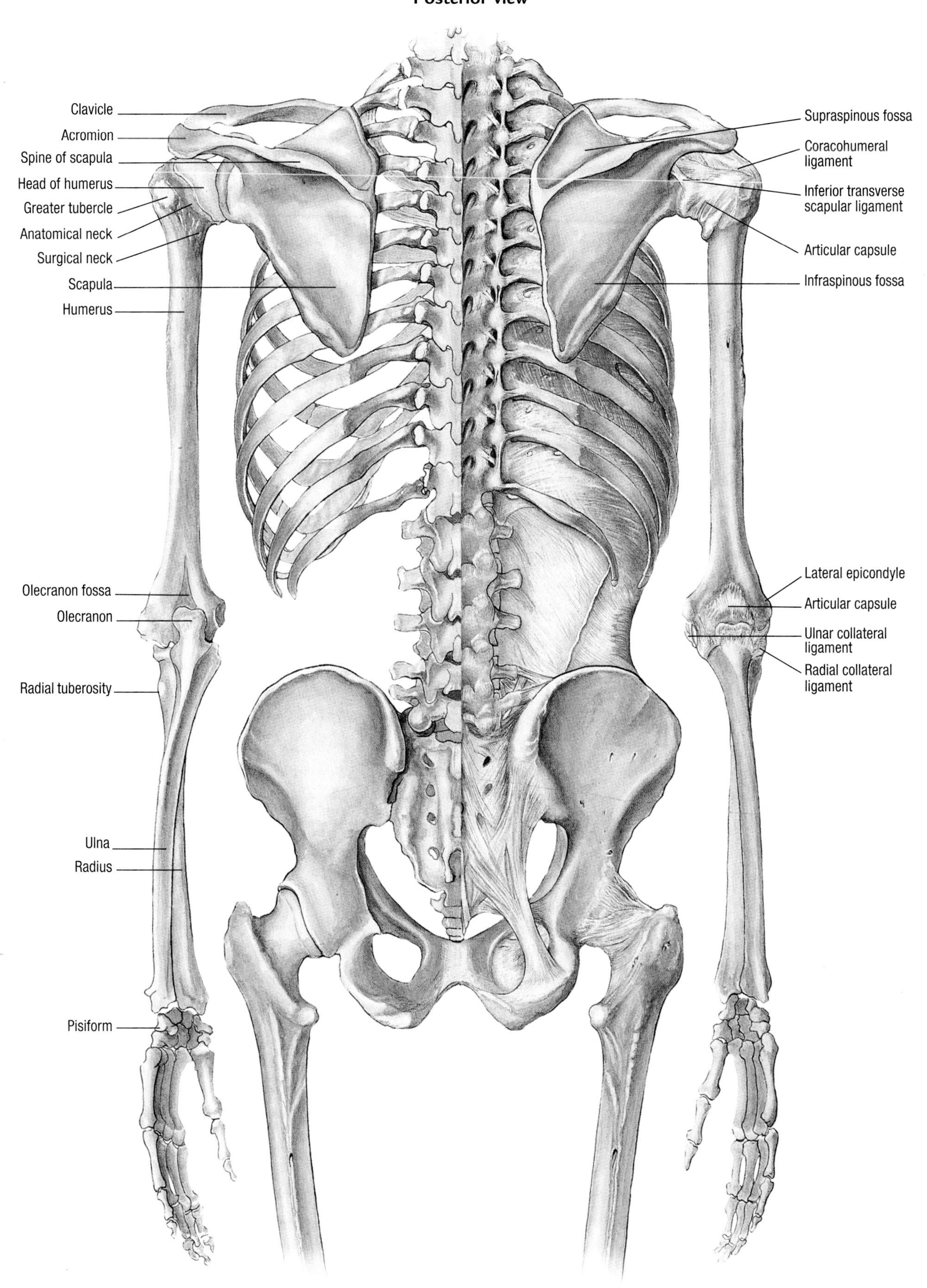
Clavicle
Acromion
Spine of scapula
Head of humerus
Greater tubercle
Anatomical neck
Surgical neck
Scapula
Humerus
Olecranon fossa
Olecranon
Radial tuberosity
Ulna
Radius
Pisiform
Supraspinous fossa
Coracohumeral ligament
Inferior transverse scapular ligament
Articular capsule
Infraspinous fossa
Lateral epicondyle
Articular capsule
Ulnar collateral ligament
Radial collateral ligament

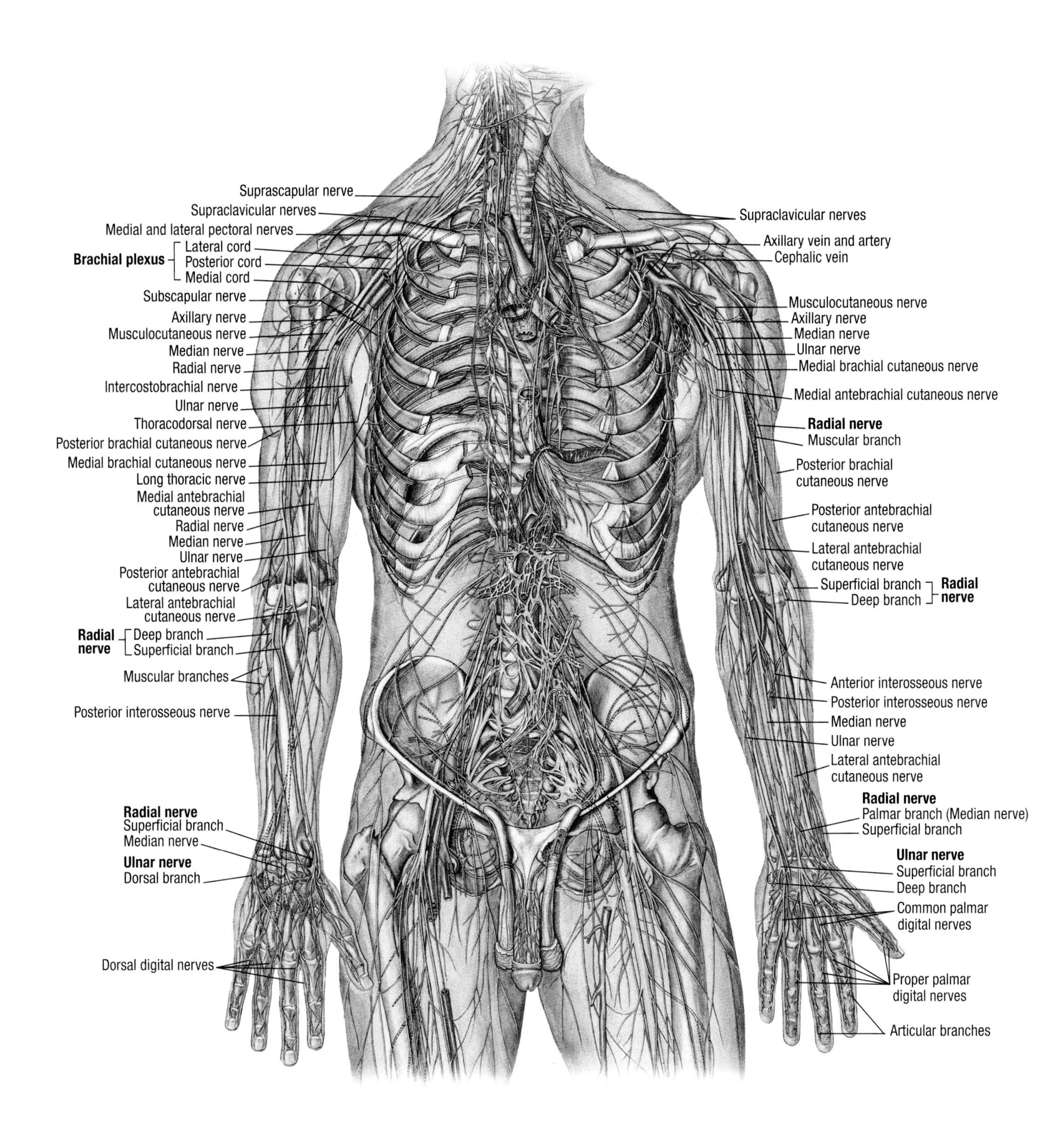

Suprascapular nerve
Supraclavicular nerves
Medial and lateral pectoral nerves
Brachial plexus
Lateral cord
Posterior cord
Medial cord
Subscapular nerve
Axillary nerve
Musculocutaneous nerve
Median nerve
Radial nerve
Intercostobrachial nerve
Ulnar nerve
Thoracodorsal nerve
Posterior brachial cutaneous nerve
Medial brachial cutaneous nerve
Long thoracic nerve
Medial antebrachial cutaneous nerve
Radial nerve
Median nerve
Ulnar nerve
Posterior antebrachial cutaneous nerve
Lateral antebrachial cutaneous nerve
Radial nerve
Deep branch
Superficial branch
Muscular branches
Posterior interosseous nerve
Radial nerve
Superficial branch
Median nerve
Ulnar nerve
Dorsal branch
Dorsal digital nerves
Supraclavicular nerves
Axillary vein and artery
Cephalic vein
Musculocutaneous nerve
Axillary nerve
Median nerve
Ulnar nerve
Medial brachial cutaneous nerve
Medial antebrachial cutaneous nerve
Radial nerve
Muscular branch
Posterior brachial cutaneous nerve
Posterior antebrachial cutaneous nerve
Lateral antebrachial cutaneous nerve
Superficial branch
Deep branch
Radial nerve
Anterior interosseous nerve
Posterior interosseous nerve
Median nerve
Ulnar nerve
Lateral antebrachial cutaneous nerve
Radial nerve
Palmar branch (Median nerve)
Superficial branch
Ulnar nerve
Superficial branch
Deep branch
Common palmar digital nerves
Proper palmar digital nerves
Articular branches

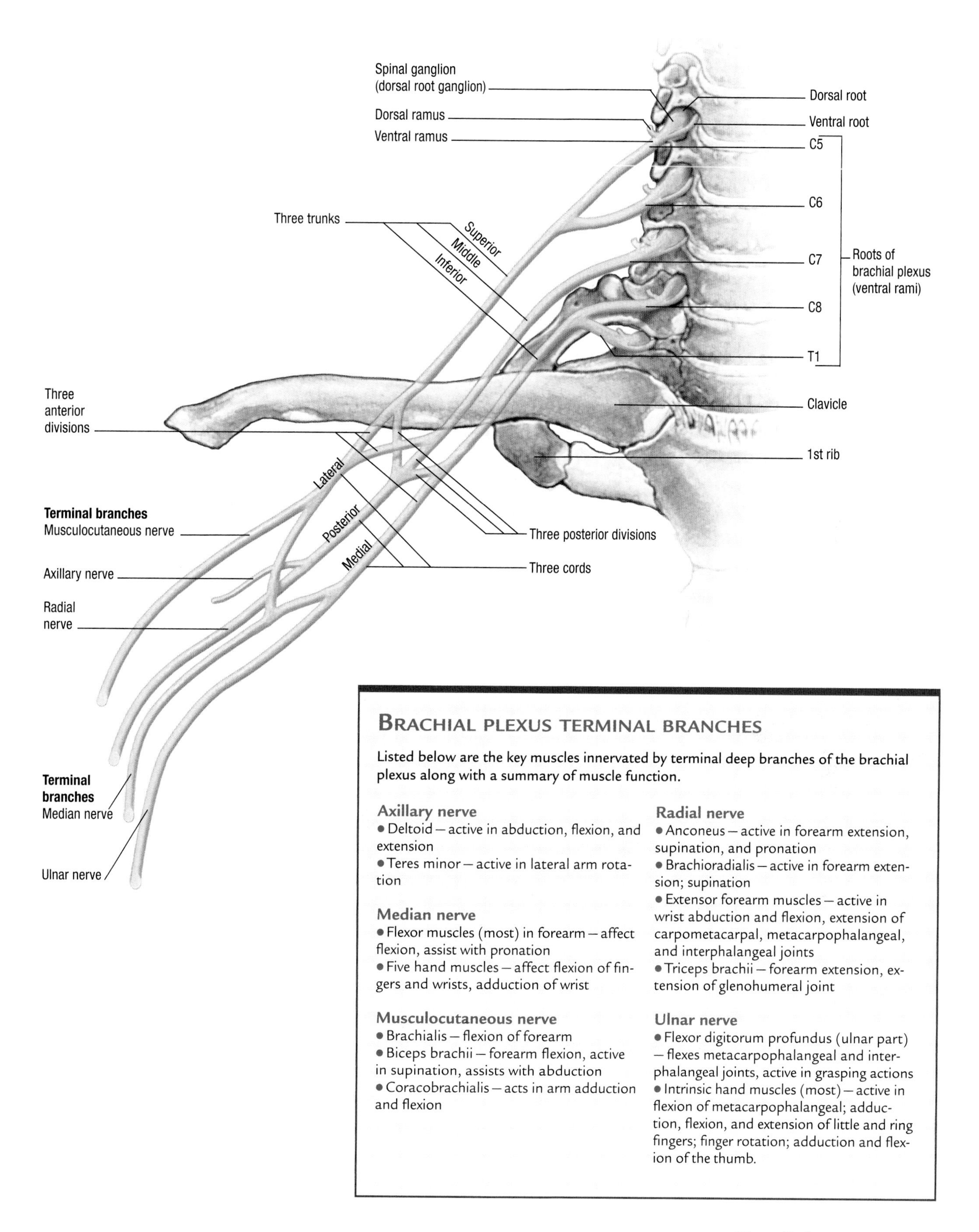
Spinal ganglion
(dorsal root ganglion)
Dorsal ramus
Ventral ramus
Dorsal root
Ventral root
C5
C6
C7
C8
T1
Roots of
brachial plexus
(ventral rami)
Three trunks
Superior
Middle
Inferior
Clavicle
1st rib
Three
anterior
divisions
Lateral
Posterior
Medial
Three posterior divisions
Three cords
Terminal branches
Musculocutaneous nerve
Axillary nerve
Radial
nerve
Terminal
branches
Median nerve
Ulnar nerve

## Brachial plexus terminal branches

Listed below are the key muscles innervated by terminal deep branches of the brachial plexus along with a summary of muscle function.

### Axillary nerve

- Deltoid – active in abduction, flexion, and extension
- Teres minor – active in lateral arm rotation

### Median nerve

- Flexor muscles (most) in forearm – affect flexion, assist with pronation
- Five hand muscles – affect flexion of fingers and wrists, adduction of wrist

### Musculocutaneous nerve

- Brachialis – flexion of forearm
- Biceps brachii – forearm flexion, active in supination, assists with abduction
- Coracobrachialis – acts in arm adduction and flexion

### Radial nerve

- Anconeus – active in forearm extension, supination, and pronation
- Brachioradialis – active in forearm extension; supination
- Extensor forearm muscles – active in wrist abduction and flexion, extension of carpometacarpal, metacarpophalangeal, and interphalangeal joints
- Triceps brachii – forearm extension, extension of glenohumeral joint

### Ulnar nerve

- Flexor digitorum profundus (ulnar part) – flexes metacarpophalangeal and interphalangeal joints, active in grasping actions
- Intrinsic hand muscles (most) – active in flexion of metacarpophalangeal; adduction, flexion, and extension of little and ring fingers; finger rotation; adduction and flexion of the thumb.

# Blood supply and lymphatic drainage

Blood and lymphatic vessels form networks throughout the upper limb.

## Blood supply

The main blood supply for the upper limb originates from the subclavian artery. After the subclavian crosses the lateral border of the first rib, it is called the axillary artery, and after it crosses the lower border of the teres major tendon – passing through the axilla – it becomes the brachial artery. The brachial artery branches at the cubital fossa into the radial and ulnar arteries, which then branch further to supply the lower portion of the upper limb, as follows:

- Major branches of the radial artery:
  - radial recurrent
  - superficial radial
  - first dorsal metacarpal
  - dorsal carpal (branches to dorsal metacarpal and dorsal digital)
  - princeps pollicis
  - radialis indicis.
- Major branches of the ulnar artery:
  - ulnar recurrent
  - common interosseous (with anterior and posterior branches)
  - dorsal carpal
  - superficial ulnar (and superficial palmar arch and palmar digital branch)
  - deep ulnar
  - palmar metacarpal.

## Venous drainage

Deep veins accompany the major arteries of the upper limb. In addition, two superficial veins – the cephalic and basilic – provide venous drainage. (See *Upper limb veins*, page 116.) Perforating veins allow the superficial veins to communicate through the deep fascia to the deep veins.

The cephalic vein ascends the upper limb from the radial side of the wrist and communicates with the median cubital vein, which forms a channel between the cephalic and brachial veins across the anterior aspect of the elbow. The cephalic vein continues upward along the deltopectoral groove, enters the deltopectoral triangle through the deep fascia, and empties into the axillary vein. The basilic vein ascends parallel to the brachial artery and merges with one of the parallel veins to form the axillary vein.

## Lymphatic vessels

The upper limb contains two types of lymphatic vessels :

- superficial lymphatics – originate in the lymphatic plexus of the hand, accompany superficial veins
- deep lymphatics – accompany major veins.

Lymph nodes in the upper limb, named according to location, include cubital, deltopectoral, and apical, central, and humeral axillary nodes. (See *Lymphatic drainage of the upper limb*, page 117.)

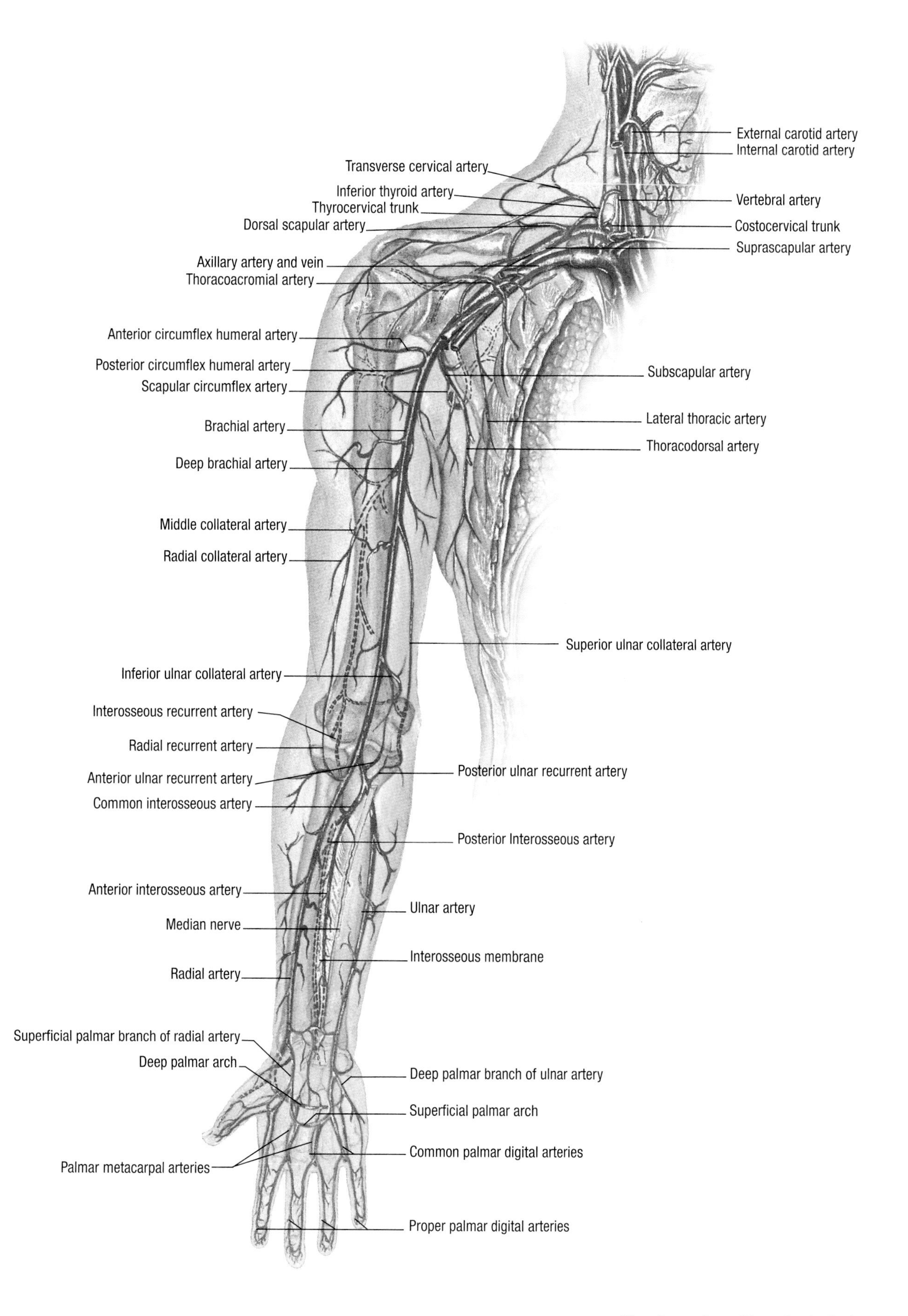
External carotid artery
Internal carotid artery
Transverse cervical artery
Inferior thyroid artery
Thyrocervical trunk
Dorsal scapular artery
Vertebral artery
Costocervical trunk
Suprascapular artery
Axillary artery and vein
Thoracoacromial artery
Anterior circumflex humeral artery
Posterior circumflex humeral artery
Scapular circumflex artery
Subscapular artery
Brachial artery
Lateral thoracic artery
Thoracodorsal artery
Deep brachial artery
Middle collateral artery
Radial collateral artery
Superior ulnar collateral artery
Inferior ulnar collateral artery
Interosseous recurrent artery
Radial recurrent artery
Anterior ulnar recurrent artery
Posterior ulnar recurrent artery
Common interosseous artery
Posterior Interosseous artery
Anterior interosseous artery
Ulnar artery
Median nerve
Interosseous membrane
Radial artery
Superficial palmar branch of radial artery
Deep palmar arch
Deep palmar branch of ulnar artery
Superficial palmar arch
Common palmar digital arteries
Palmar metacarpal arteries
Proper palmar digital arteries

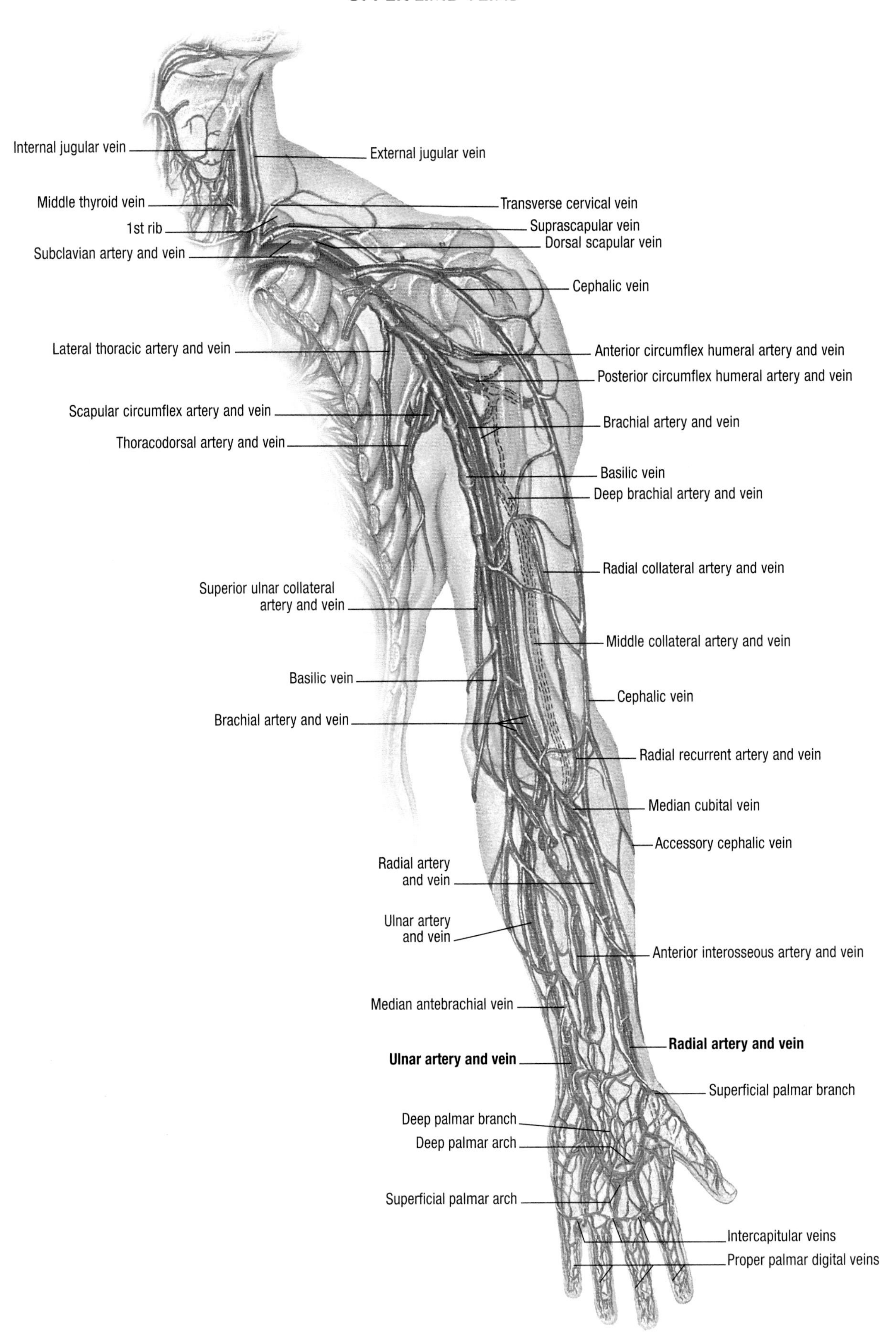
Internal jugular vein
External jugular vein
Middle thyroid vein
Transverse cervical vein
1st rib
Suprascapular vein
Subclavian artery and vein
Dorsal scapular vein
Cephalic vein
Lateral thoracic artery and vein
Anterior circumflex humeral artery and vein
Posterior circumflex humeral artery and vein
Scapular circumflex artery and vein
Brachial artery and vein
Thoracodorsal artery and vein
Basilic vein
Deep brachial artery and vein
Radial collateral artery and vein
Superior ulnar collateral artery and vein
Middle collateral artery and vein
Basilic vein
Cephalic vein
Brachial artery and vein
Radial recurrent artery and vein
Median cubital vein
Accessory cephalic vein
Radial artery and vein
Ulnar artery and vein
Anterior interosseous artery and vein
Median antebrachial vein
Radial artery and vein
Ulnar artery and vein
Superficial palmar branch
Deep palmar branch
Deep palmar arch
Superficial palmar arch
Intercapitular veins
Proper palmar digital veins

## LYMPHATIC DRAINAGE OF THE UPPER LIMB

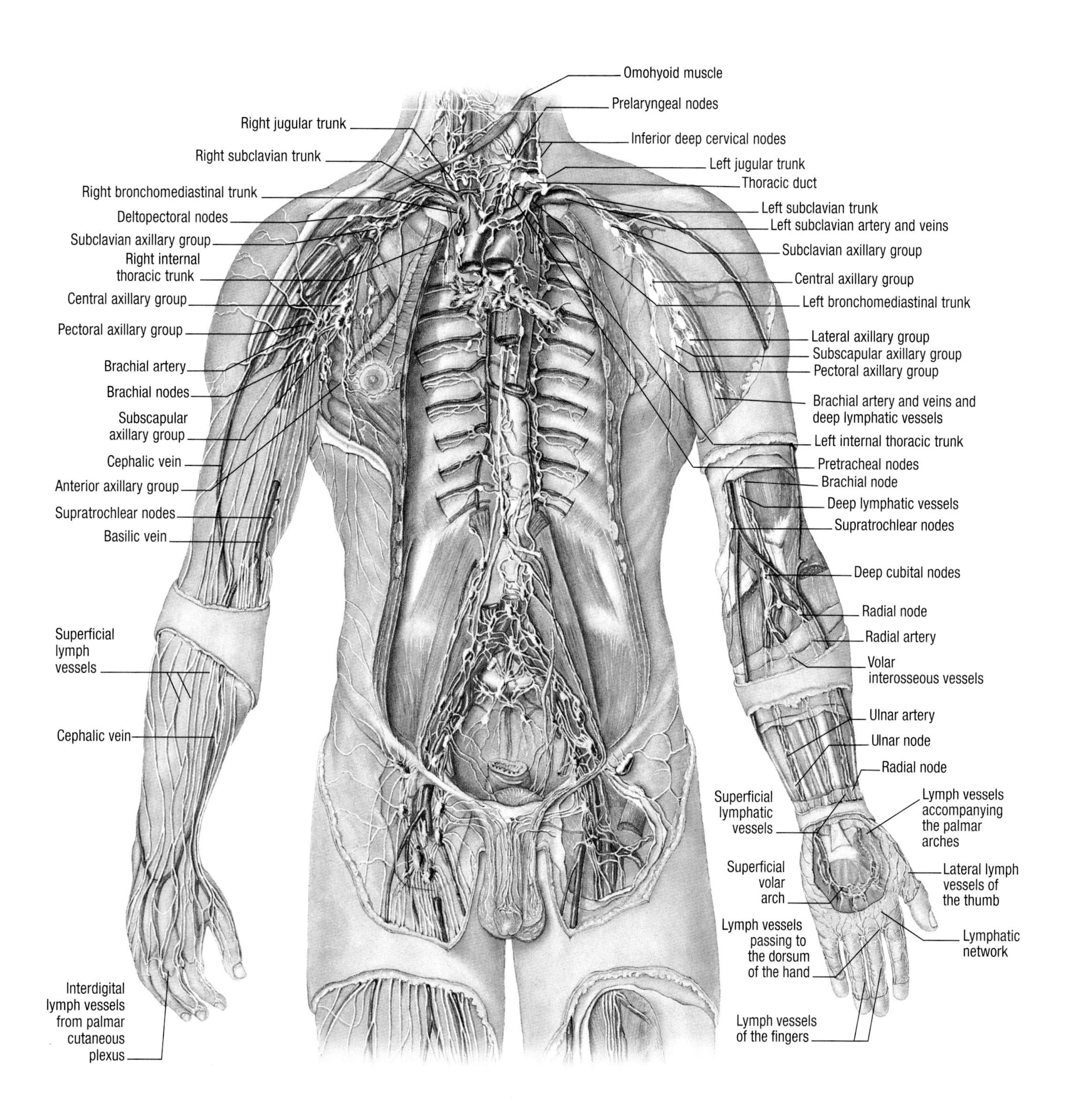
Omohyoid muscle
Prelaryngeal nodes
Right jugular trunk
Inferior deep cervical nodes
Right subclavian trunk
Left jugular trunk
Thoracic duct
Right bronchomediastinal trunk
Left subclavian trunk
Deltopectoral nodes
Left subclavian artery and veins
Subclavian axillary group
Subclavian axillary group
Right internal thoracic trunk
Central axillary group
Central axillary group
Left bronchomediastinal trunk
Pectoral axillary group
Lateral axillary group
Subscapular axillary group
Brachial artery
Pectoral axillary group
Brachial nodes
Brachial artery and veins and deep lymphatic vessels
Subscapular axillary group
Left internal thoracic trunk
Cephalic vein
Pretracheal nodes
Anterior axillary group
Brachial node
Deep lymphatic vessels
Supratrochlear nodes
Supratrochlear nodes
Basilic vein
Deep cubital nodes
Radial node
Radial artery
Superficial lymph vessels
Volar interosseous vessels
Ulnar artery
Cephalic vein
Ulnar node
Radial node
Superficial lymphatic vessels
Lymph vessels accompanying the palmar arches
Superficial volar arch
Lateral lymph vessels of the thumb
Lymph vessels passing to the dorsum of the hand
Lymphatic network
Interdigital lymph vessels from palmar cutaneous plexus
Lymph vessels of the fingers

# Shoulder and upper limb bones

The ball-and-socket joint of the shoulder has the capability of great flexibility. However, the anatomy that provides flexibility also increases the risk of instability; thus, the shoulder is one of the most common joints to be dislocated.

## Pectoral girdle

The pectoral (shoulder) girdle is the area formed by the clavicle and the scapula. Together they form an incomplete circle – open posteriorly – between the scapula and the vertebral column. (See *Shoulder bones*, pages 119 and 120.)

### Clavicle

The clavicle (collar bone) is a slender, curved, flat bone that forms the anterior portion of the shoulder girdle. Although classified as a long bone, it lacks the typical marrow cavity of a long bone. Rather, hard bone surrounds a cancellous (spongy) center. The clavicle is just above the first rib anteriorly; posteriorly, it overlaps portions of the second through seventh ribs.

The clavicles connect the trunk of the body to the upper limbs. At the medial end, the clavicle articulates with the manubrium of the sternum. At the distal end, it articulates with the scapula, forming the sternoclavicular joint. Two major functions of the clavicle are:
- bracing the shoulder
- protecting the nerves as they enter the upper arm.

**Clinical tip**

Suspect a possible clavicular fracture when a patient has a history of a recent fall, even if there is no evidence of a direct injury to the area. When a person tries to avoid injury by using the hand or upper limb as a brace, the force of the fall may fracture the clavicle. Signs of a clavicular fracture include:
- drooping shoulder on the affected side on visual exam
- deformity if the bones are not aligned properly
- swelling and tenderness at the fracture site
- pain radiating down the arm.

### Scapula

The scapula (shoulder blade) is a triangular flat bone. Laterally, it contains a rounded, shallow cavity called the glenoid fossa, which articulates with the head of the humerus to form the shoulder joint. (See *Shoulder joint socket*, page 122.) Superiorly, the scapula has two prominent bony projections – the coracoid process in front and the acromion process in back.

The coracoid process is a strong, curved projection of the superior border of the scapula, to which the pectoralis minor muscle is attached. It overhangs the shoulder joint.

The acromion process is the large triangular projection of the spine of the scapula that articulates with the clavicle and forms the point of the shoulder. The deltoid and trapezius muscles attach to it. Posteriorly, the scapula attaches only to muscle, not to the bony thorax.

## Humerus

The humerus is the largest and longest bone of the upper limb. (See *Left humerus,* page 121.) The rounded head of the humerus articulates medially with the glenoid cavity of the scapula.

Directly below the head of the humerus is the anatomical neck of the humerus. Near the area where the head and the neck of the humerus join the main shaft of the bone are two eminences, the greater and lesser tubercles, which are points of muscle attachment. The intertubercular groove is where the tendon of the long head of the biceps attaches.

Just distal to the tubercles is the surgical neck of the humerus, which has a smaller diameter than that of the anatomic neck and is a common fracture site. Distally, the humerus articulates with the radius at the capitulum, and with the ulna at the trochlea.

The lower end of the humerus has two hollows:
- Coronoid fossa, located anteriorly, where the coronoid process of the ulna fits when the forearm is flexed
- Olecranon fossa, located posteriorly, where the olecranon of the ulna fits when the forearm is extended.

On either side of the lower end of the humerus are two projections, called the medial and lateral epicondyles. The ulnar nerve passes into the forearm through a groove on the posterior aspect of the medial epicondyle. The radial nerve crosses the anterior aspect of the lateral epicondyle.

## SHOULDER BONES
**Posterior view**

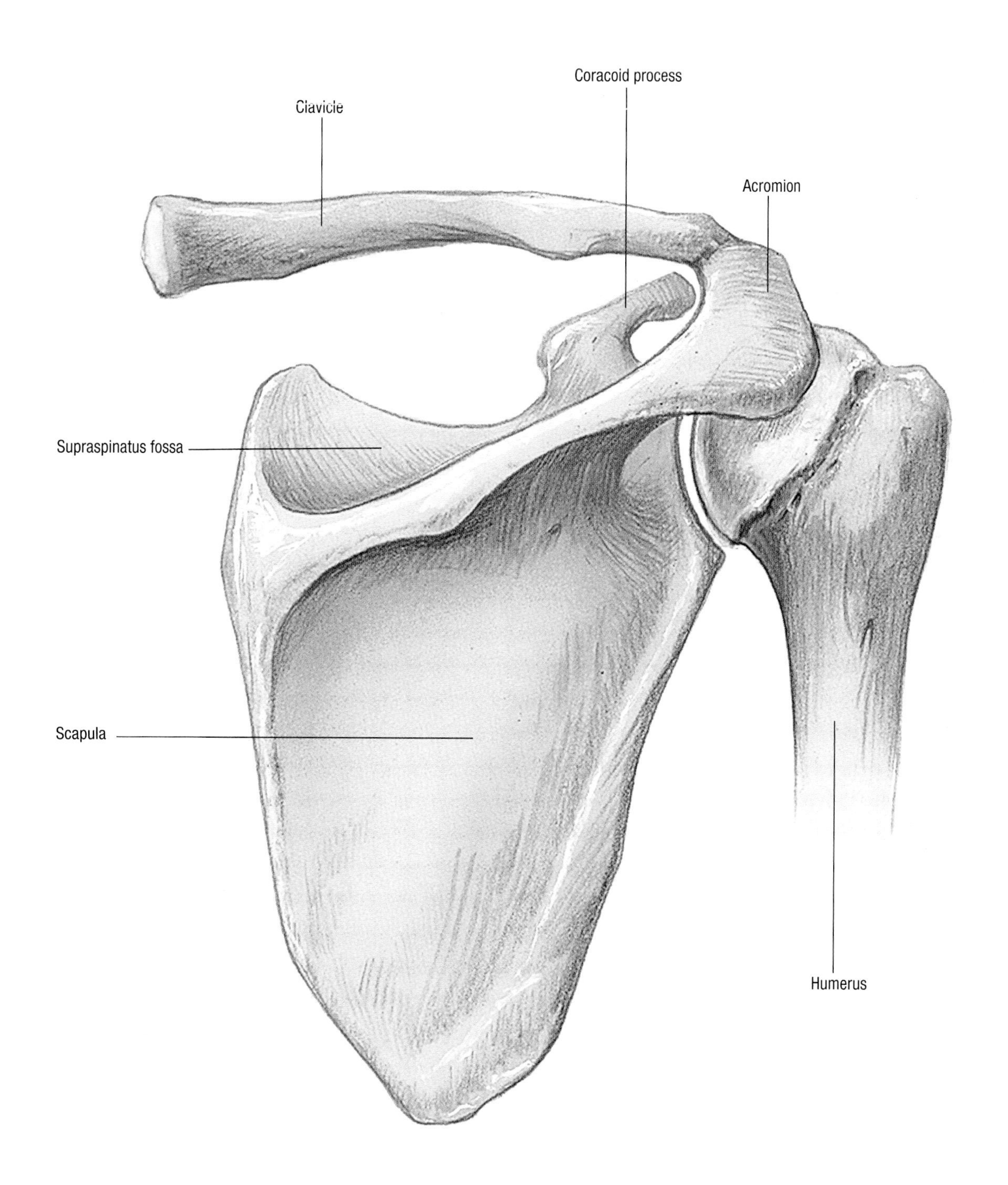

Coracoid process
Clavicle
Acromion
Supraspinatus fossa
Scapula
Humerus

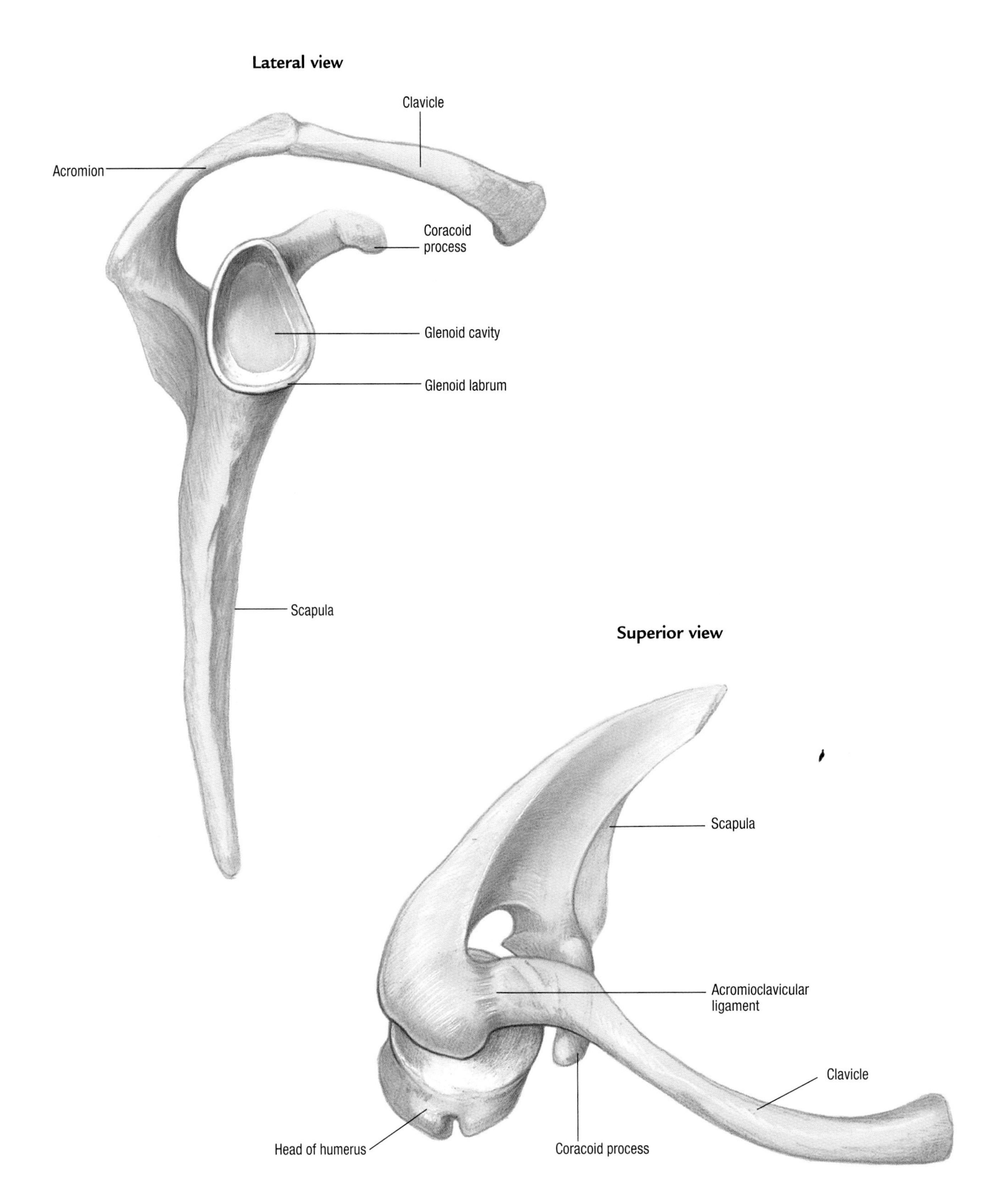
Lateral view
Clavicle
Acromion
Coracoid process
Glenoid cavity
Glenoid labrum
Scapula
Superior view
Scapula
Acromioclavicular ligament
Clavicle
Head of humerus
Coracoid process

## LEFT HUMERUS

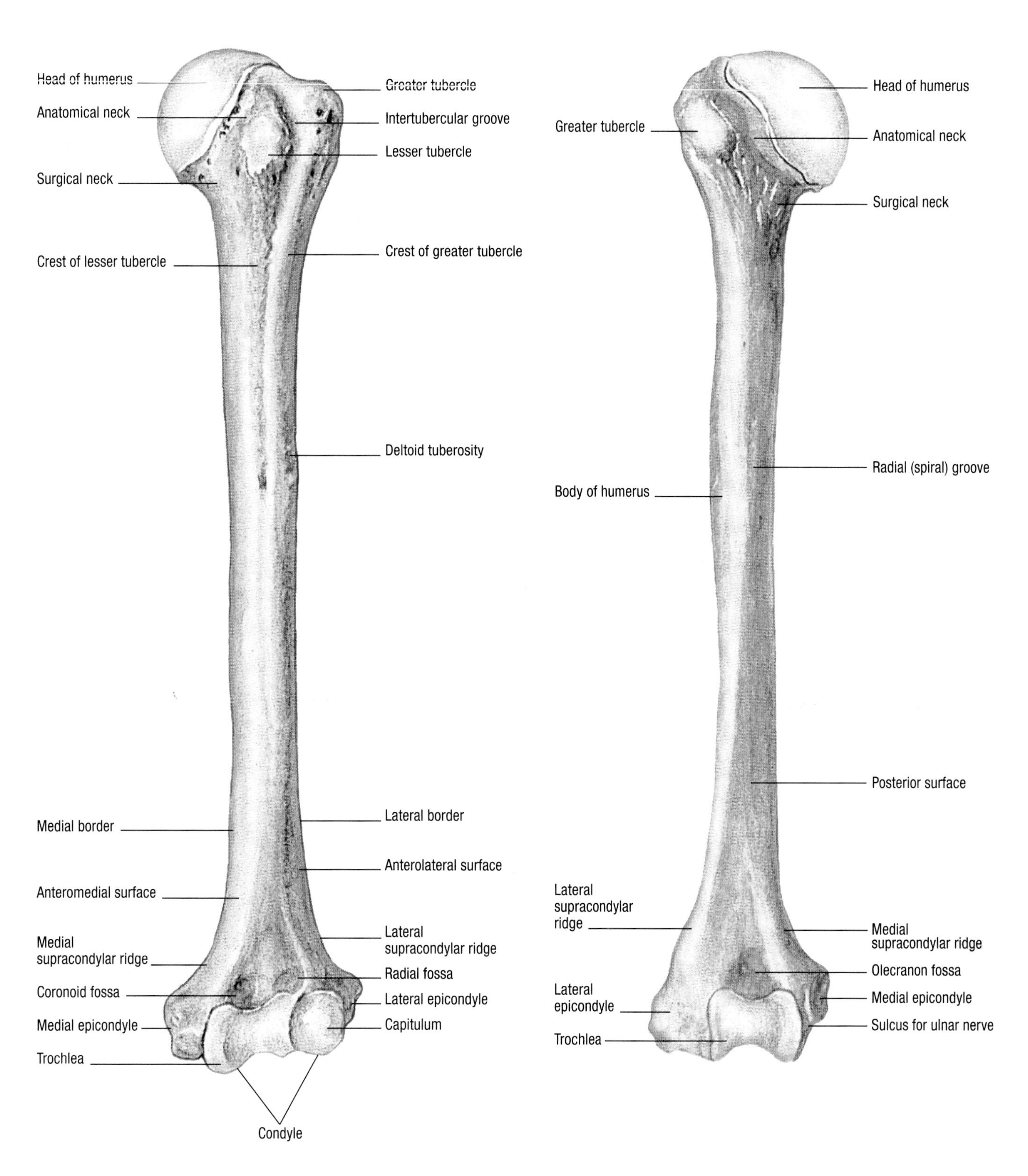

Anterior view
Head of humerus
Anatomical neck
Surgical neck
Crest of lesser tubercle
Greater tubercle
Intertubercular groove
Lesser tubercle
Crest of greater tubercle
Deltoid tuberosity
Medial border
Anteromedial surface
Medial supracondylar ridge
Coronoid fossa
Medial epicondyle
Trochlea
Lateral border
Anterolateral surface
Lateral supracondylar ridge
Radial fossa
Lateral epicondyle
Capitulum
Condyle
Posterior view
Greater tubercle
Body of humerus
Lateral supracondylar ridge
Lateral epicondyle
Trochlea
Head of humerus
Anatomical neck
Surgical neck
Radial (spiral) groove
Posterior surface
Medial supracondylar ridge
Olecranon fossa
Medial epicondyle
Sulcus for ulnar nerve

## SHOULDER JOINT SOCKET

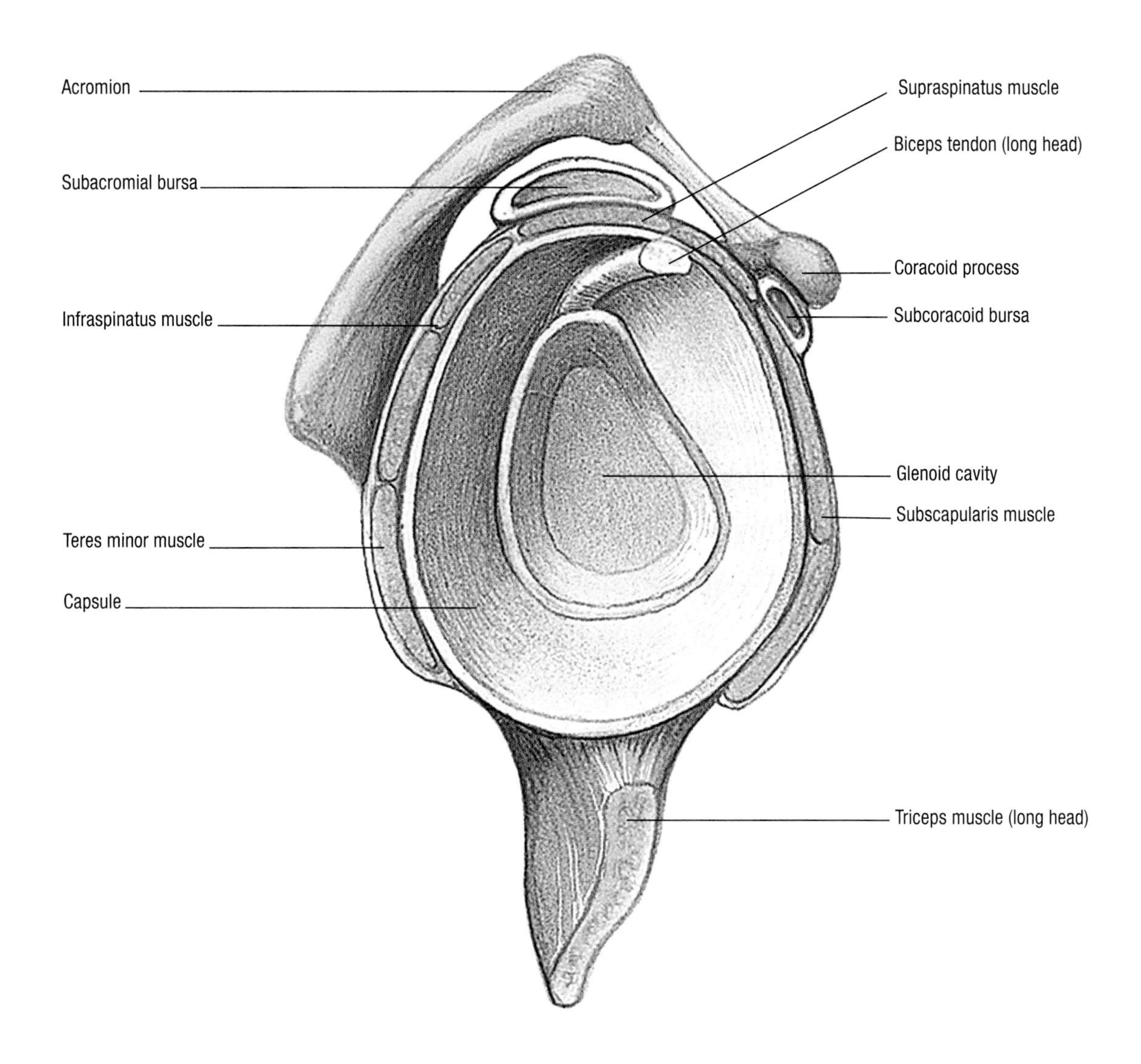
Acromion
Subacromial bursa
Infraspinatus muscle
Teres minor muscle
Capsule
Supraspinatus muscle
Biceps tendon (long head)
Coracoid process
Subcoracoid bursa
Glenoid cavity
Subscapularis muscle
Triceps muscle (long head)

## SHOULDER JOINT LIGAMENTS
### Anterior view

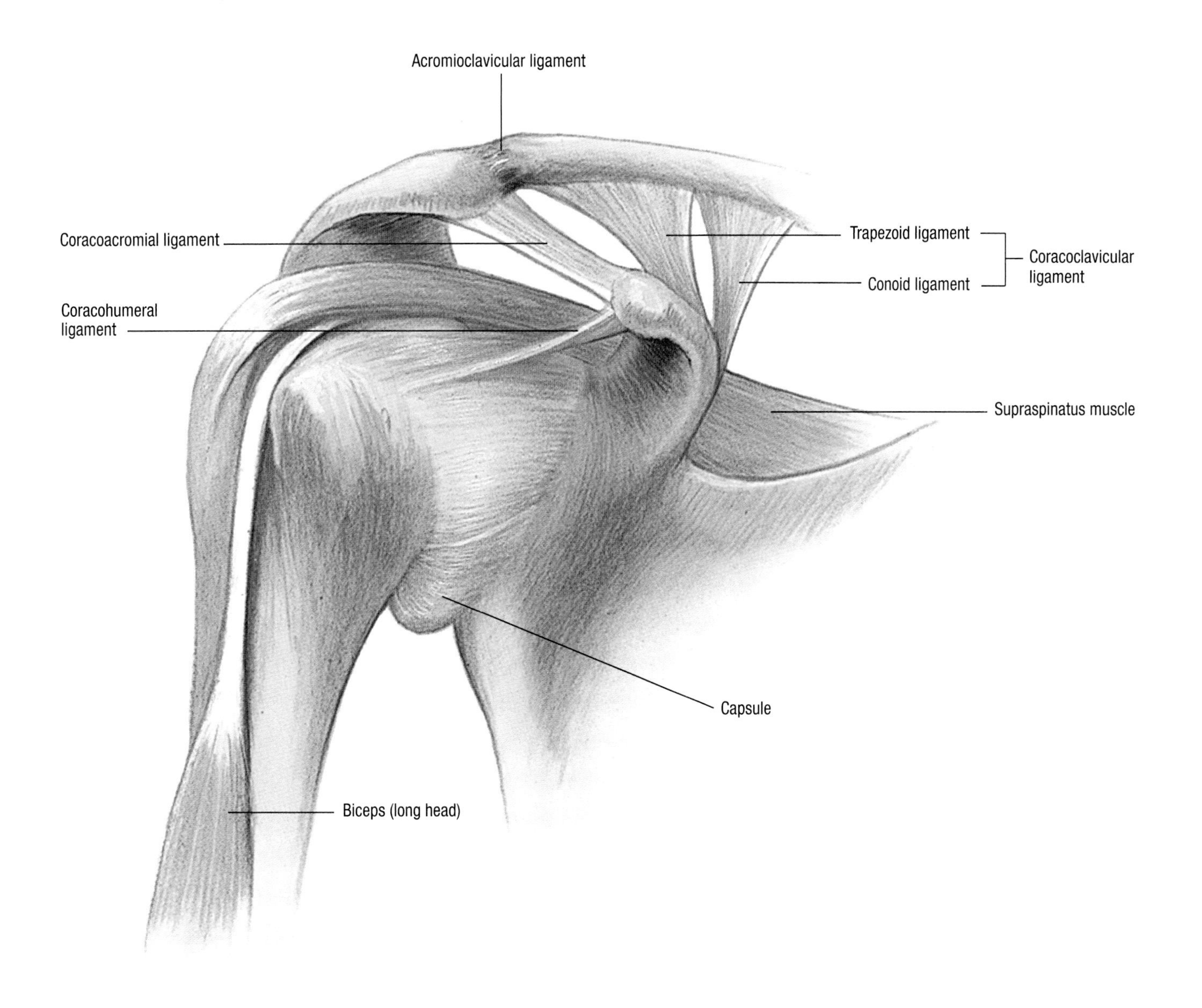
Acromioclavicular ligament
Coracoacromial ligament
Trapezoid ligament
Conoid ligament
Coracoclavicular ligament
Coracohumeral ligament
Supraspinatus muscle
Capsule
Biceps (long head)

# Shoulder and upper limb muscles

This group of muscles includes muscles with origins in the axial skeleton as well as those with origins on the appendicular skeleton. For an extensive list of origin and insertion sites, see *Guide to skeletal muscles,* Appendix C.

## Pectoral girdle muscles

Seven muscles on each side move the scapulae (shoulder blades):

- trapezius – elevates, depresses, rotates, adducts, and stabilizes the scapula
- rhomboideus major and rhomboideus minor – adduct, stabilize, and rotate the scapula
- levator scapulae – elevates and adducts the scapula and bends the neck laterally
- pectoralis minor – draws the scapula forward and downward
- serratus anterior – stabilizes, abducts, and rotates the scapula upward and helps to abduct and raise the arm
- subclavius – stabilizes and depresses the shoulder.

## Muscles of the humerus

Muscles that move the humerus are divided into those with origins on the axial skeleton and those with origins on the scapulae.

- Axial skeleton origins:
  - pectoralis major flexes, adducts, and medially rotates the arm
  - latissimus dorsi extends, adducts, and medially rotates the arm; it also pulls the shoulder downward.
- Scapular origins:
  - deltoideus abducts, rotates, and extends the arm
  - supraspinatus abducts and causes slight lateral rotation of the arm
  - infraspinatus laterally rotates the arm
  - subscapularis medially rotates the arm
  - teres major adducts, extends, and medially rotates the arm
  - teres minor laterally rotates the arm; helps to hold humeral head in glenoid cavity
  - coracobrachialis flexes and abducts the arm; helps to resist downward dislocation.

### Rotator cuff

Collectively, the subscapularis, supraspinatus, infraspinatus, and teres minor, which surround the glenohumeral joint, are called the rotator cuff. They help keep the joint stable and prevent dislocation by resisting the natural push of the humeral head against the tendon surface. A common injury is a tear or rupture of the tendons of these muscles.

**Clinical tip**

### Rotator cuff injuries

Rotator cuff injuries can vary in degree from inflammation of a tendon to complete tear from the bone. When evaluating a patient for suspected rotator cuff injury, look for a history of any of the following:

- Repetitive overhead arm motion, such as from throwing a ball in sports
- Repetitive occupational heavy lifting or frequent reaching, such as in furniture moving
- Sudden trauma or injury, such as a fall onto an outstretched hand or a collision
- Degenerative changes in the shoulder over many years from effects of the aging process involving shoulder motion
- General overuse of the shoulder, resulting in inflammation and impingement syndrome (squeezing of the rotator cuff between the head of the humerus and the acromion, which traps the inflamed rotator cuff tendon under the acromion). Signs and symptoms that can indicate a rotator cuff injury include:
  - Pain over the deltoid muscle at the top and outer side of the shoulder, especially when the arm is raised or extended laterally from the body
  - Pain or weakness on outward or inward rotation of the arm
  - Weakness in the shoulder, especially when lifting
  - "Clicking" or "popping" sensation when moving the shoulder.

## UPPER LIMB MUSCLES
### Anterior view

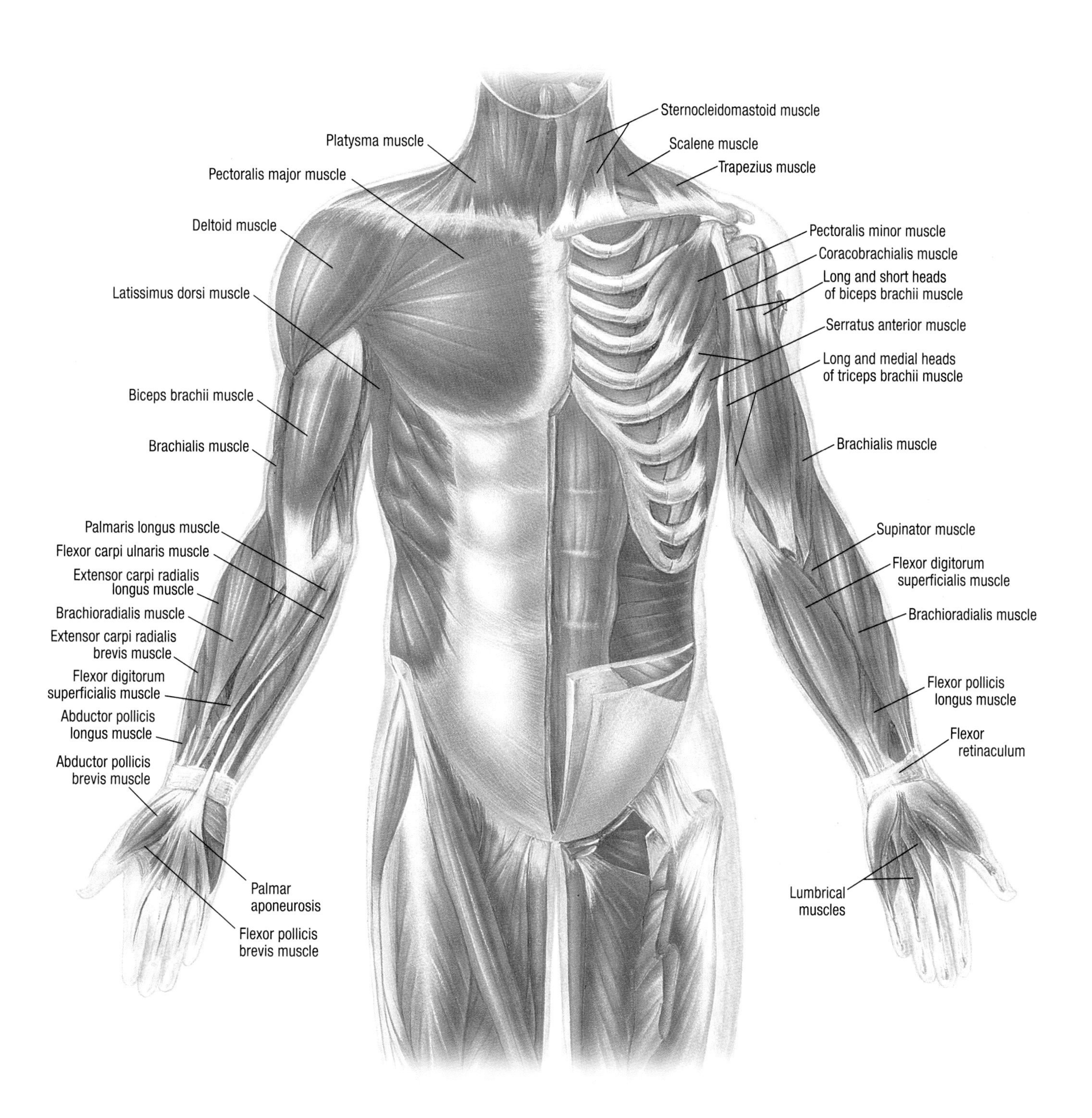
Sternocleidomastoid muscle
Platysma muscle
Scalene muscle
Trapezius muscle
Pectoralis major muscle
Deltoid muscle
Pectoralis minor muscle
Coracobrachialis muscle
Long and short heads of biceps brachii muscle
Latissimus dorsi muscle
Serratus anterior muscle
Long and medial heads of triceps brachii muscle
Biceps brachii muscle
Brachialis muscle
Brachialis muscle
Palmaris longus muscle
Supinator muscle
Flexor carpi ulnaris muscle
Flexor digitorum superficialis muscle
Extensor carpi radialis longus muscle
Brachioradialis muscle
Brachioradialis muscle
Extensor carpi radialis brevis muscle
Flexor digitorum superficialis muscle
Flexor pollicis longus muscle
Abductor pollicis longus muscle
Flexor retinaculum
Abductor pollicis brevis muscle
Palmar aponeurosis
Lumbrical muscles
Flexor pollicis brevis muscle

## UPPER LIMB MUSCLES
### Anterior view (dissection)

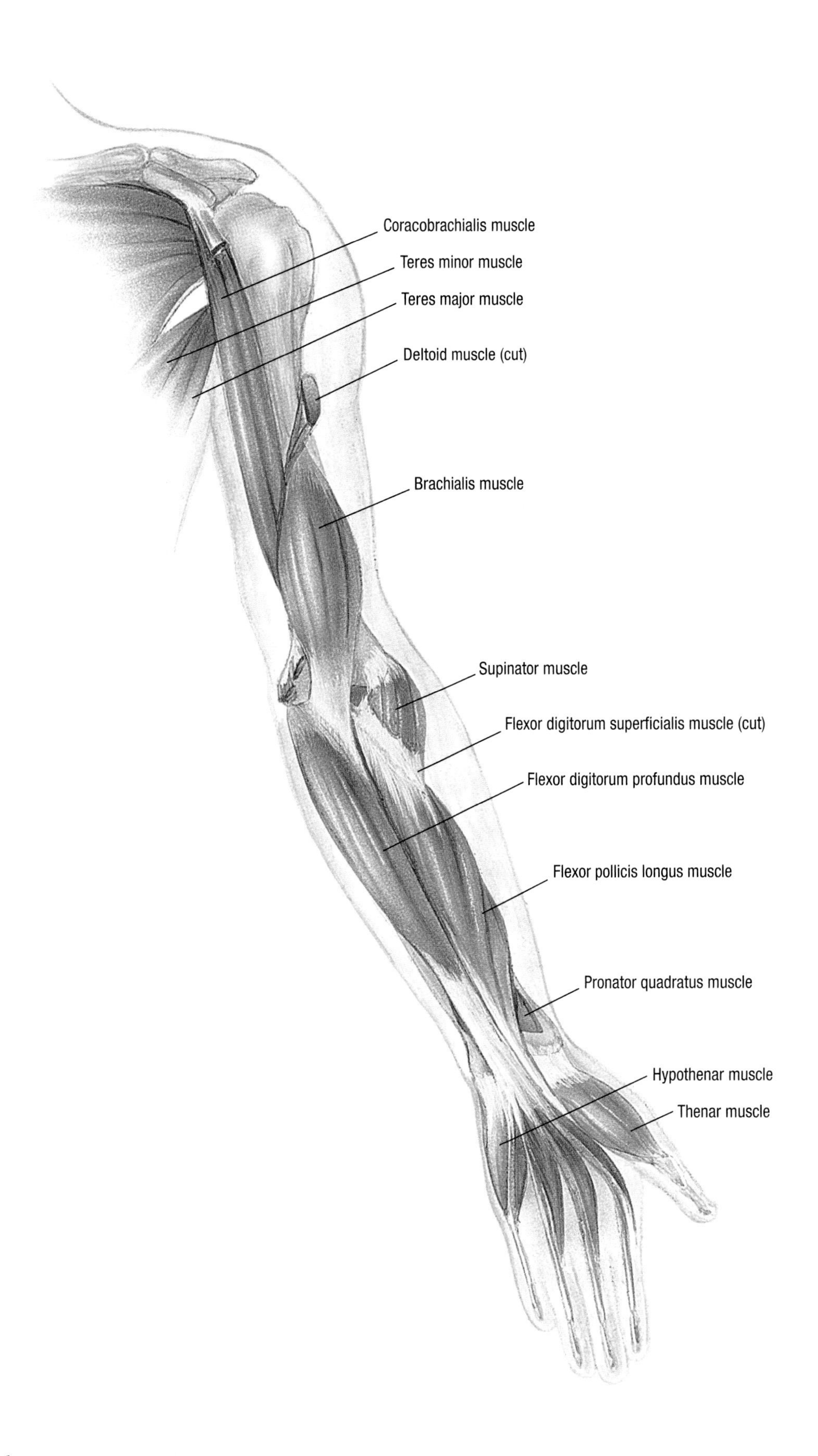
Coracobrachialis muscle
Teres minor muscle
Teres major muscle
Deltoid muscle (cut)
Brachialis muscle
Supinator muscle
Flexor digitorum superficialis muscle (cut)
Flexor digitorum profundus muscle
Flexor pollicis longus muscle
Pronator quadratus muscle
Hypothenar muscle
Thenar muscle

## UPPER LIMB MUSCLES
### Posterior view

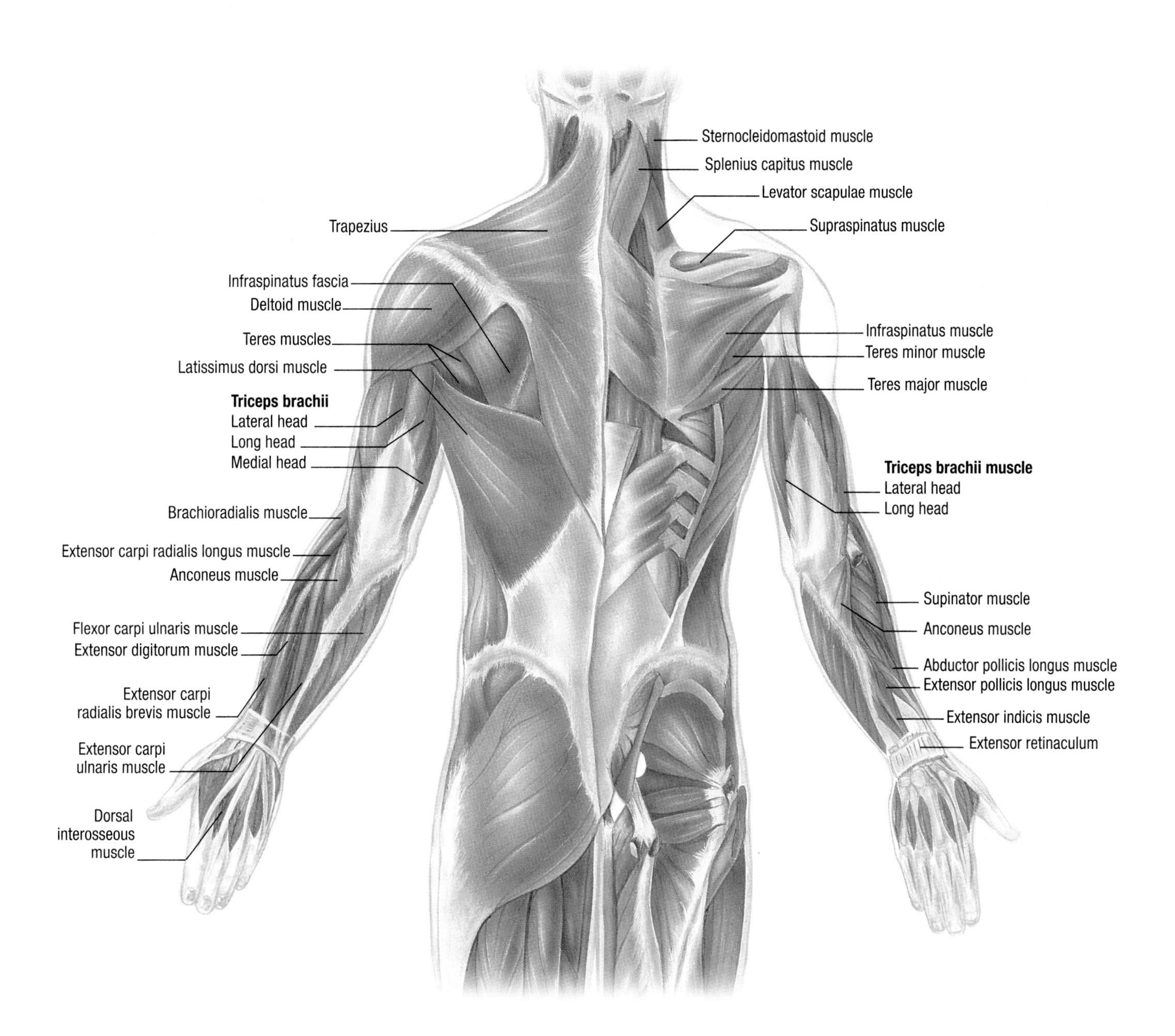
Sternocleidomastoid muscle
Splenius capitus muscle
Levator scapulae muscle
Trapezius
Supraspinatus muscle
Infraspinatus fascia
Deltoid muscle
Teres muscles
Latissimus dorsi muscle
Infraspinatus muscle
Teres minor muscle
Teres major muscle
Triceps brachii
Lateral head
Long head
Medial head
Triceps brachii muscle
Lateral head
Long head
Brachioradialis muscle
Extensor carpi radialis longus muscle
Anconeus muscle
Supinator muscle
Anconeus muscle
Flexor carpi ulnaris muscle
Extensor digitorum muscle
Abductor pollicis longus muscle
Extensor pollicis longus muscle
Extensor carpi radialis brevis muscle
Extensor indicis muscle
Extensor retinaculum
Extensor carpi ulnaris muscle
Dorsal interosseous muscle

## UPPER LIMB MUSCLES
### Posterior view (dissection)

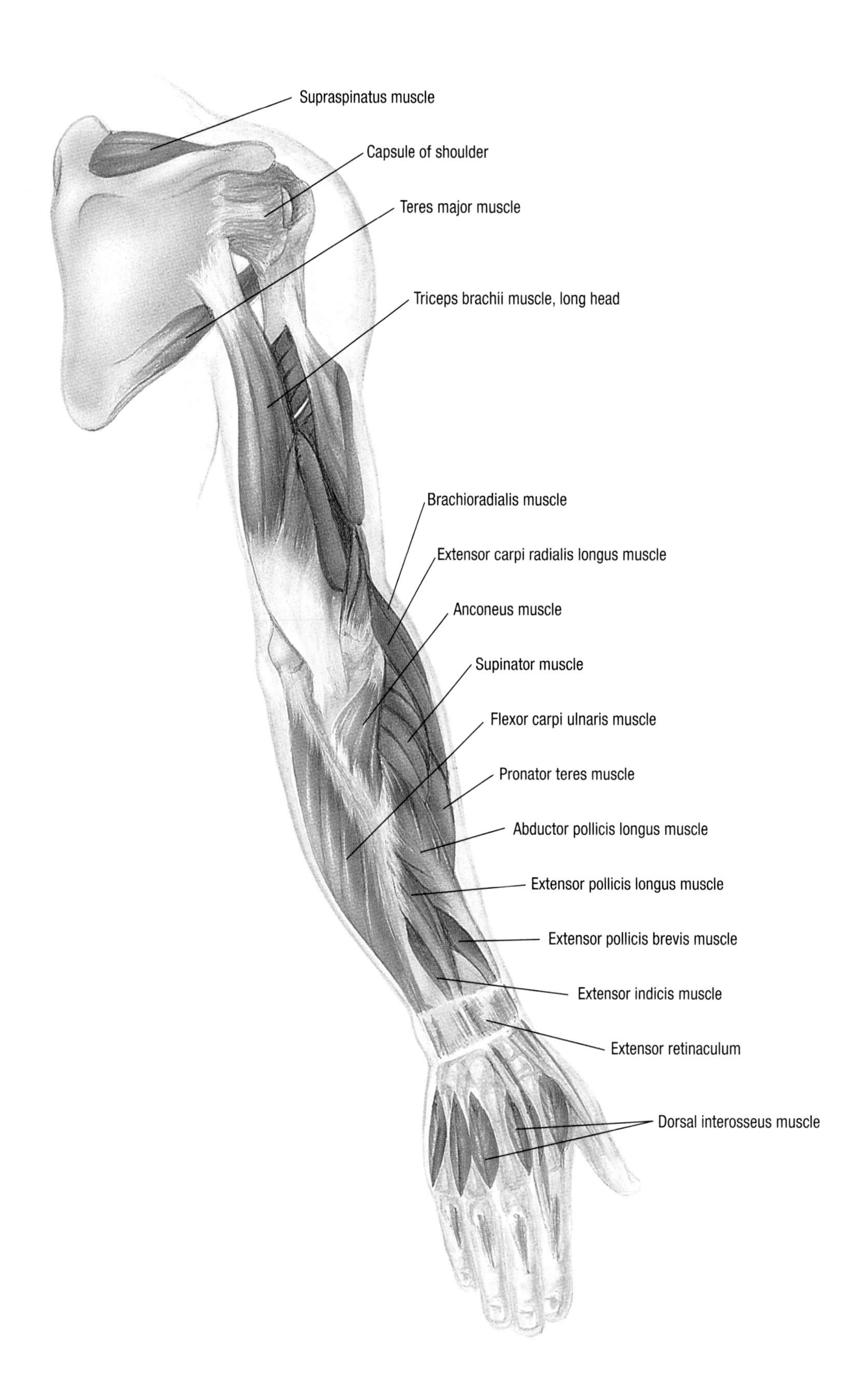
Supraspinatus muscle
Capsule of shoulder
Teres major muscle
Triceps brachii muscle, long head
Brachioradialis muscle
Extensor carpi radialis longus muscle
Anconeus muscle
Supinator muscle
Flexor carpi ulnaris muscle
Pronator teres muscle
Abductor pollicis longus muscle
Extensor pollicis longus muscle
Extensor pollicis brevis muscle
Extensor indicis muscle
Extensor retinaculum
Dorsal interosseus muscle

## UPPER LIMB MUSCLES

**Lateral view**

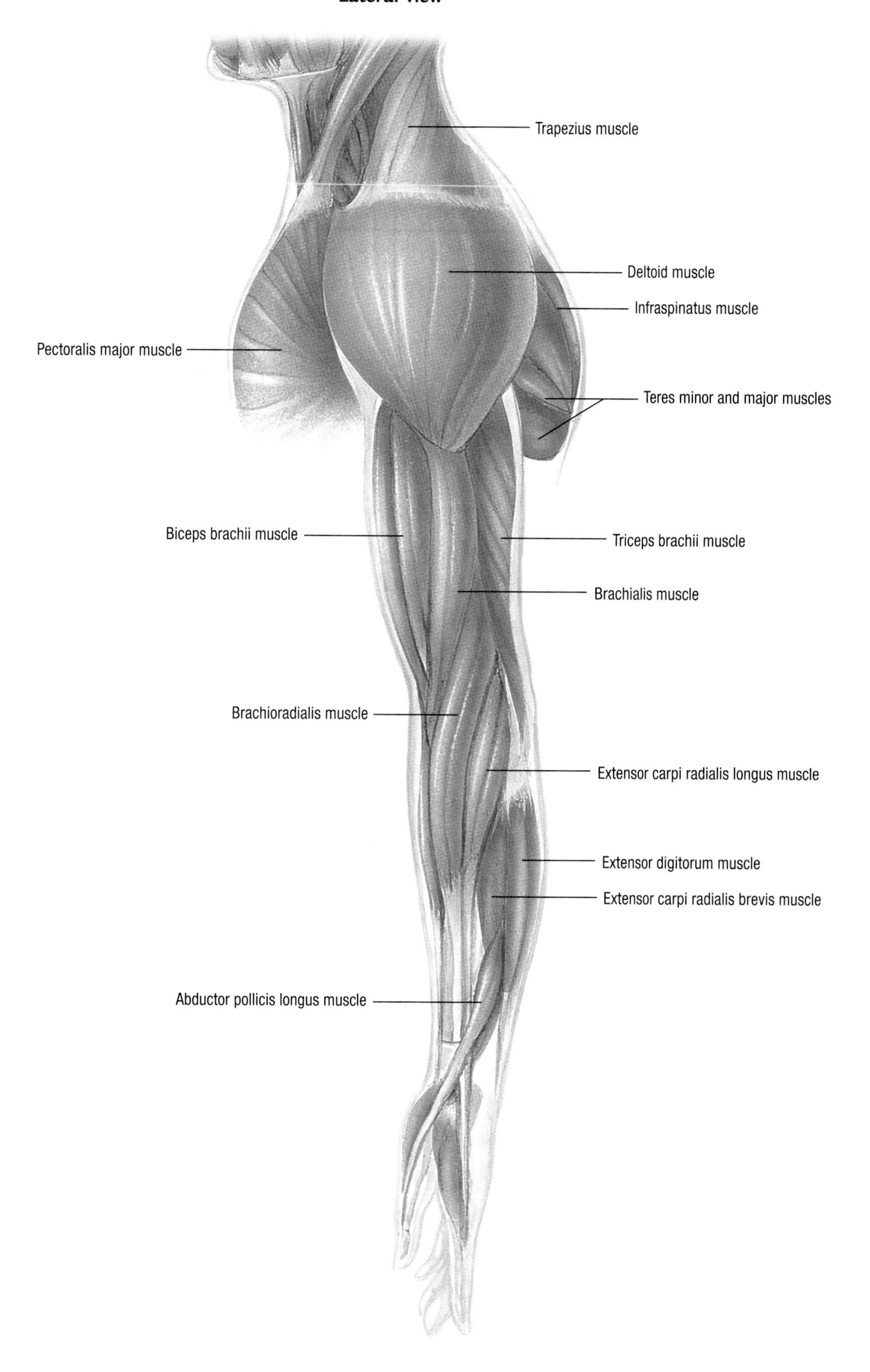
Trapezius muscle
Deltoid muscle
Infraspinatus muscle
Pectoralis major muscle
Teres minor and major muscles
Biceps brachii muscle
Triceps brachii muscle
Brachialis muscle
Brachioradialis muscle
Extensor carpi radialis longus muscle
Extensor digitorum muscle
Extensor carpi radialis brevis muscle
Abductor pollicis longus muscle

# Elbow and forearm

The elbow connects the arm and the forearm. It accommodates the radioulnar articulation and permits flexion and extension of the forearm.

## Elbow joints

The humerus articulates with the forearm bones, the radius, and the ulna, at the elbow. The elbow has the capability for two types of movement:

- extension and flexion (humeroulnar and humeroradial joints) (See *Elbow bones,* page 133.)
- pronation and supination (radioulnar joint) (See *Forearm bones,* page 132.)

The two main ligaments of the elbow are the ulnar (medial) collateral and radial (lateral) collateral ligaments. The olecranon bursa is posterior to the elbow.

## Forearm

The forearm consists of two bones, the ulna and the radius. Pivot joints formed by the two bones at the proximal and distal articulations permit pronation and supination.

**Clinical tip**

Muscle weakness in the forearm can be a result of an acute radial nerve injury. Possible causes include a fracture that compresses or lacerates the radial nerve fibers, or repeated compression of the nerve by some other means.

### Ulna

The ulna is the longer of the two forearm bones. It articulates with the humerus to form the elbow, and with the radius at its proximal and distal ends. The olecranon and coronoid processes are at the proximal end of the ulna. The styloid process is on the medial side of its distal head. The ulnar nerve, which originates in the medial and lateral cords of the brachial plexus, innervates the muscles and skin on the medial part of the hand and on the ulnar side of the forearm.

### Radius

The radius is the shorter of the two forearm bones. Proximally, the upper concavity of the head of the radius articulates with the capitulum of the humerus. Medially, the head articulates with the radial fossa of the ulna. The annular ligament attaches the head of the radius to the humerus. The radial tuberosity (protuberance) is immediately below the proximal head. At the distal end are the medial ulnar notch and the lateral styloid process. The radial nerve, the largest branch of the brachial plexus, begins on each side as a continuation of the posterior trunk.

### Muscles

Five muscles move the forearm. Three are flexors and two are extensors:

- flexors:
  - biceps brachii (also supinates forearm)
  - brachialis
  - brachioradialis.
- extensors:
  - triceps brachii (also helps stabilize humerus during adduction)
  - anconeus.

**Clinical tip**

A common elbow condition is lateral epicondylitis, sometimes referred to as tennis elbow. This condition typically results from indirect trauma to the tendon of the forearm extensor muscles. Evaluate for these signs:

- complaint of a sharp pain during activity, subsiding to a dull ache at rest
- localized tenderness and mild swelling over the epicondyle — tenderness is typically found distal to the bony prominence
- atrophy of forearm muscles (on occasion).

## Blood supply

Branches of the brachial artery supply the elbow and forearm. The ulnar artery arises near the elbow, distributing blood to the forearm, wrist, and hand. This artery passes obliquely in a distal direction to become the superficial palmar arch.

The radial artery begins at the bifurcation of the brachial artery, passing in 12 branches to the forearm, wrist, and hand. The radial recurrent artery, part of the radial artery, arises distal to the elbow and ascends between the branches of the radial nerve, supplying several muscles of the arm and elbow. The superficial cephalic, basilic, and median veins along with the deep radial and ulnar veins provide drainage for the area.

## ELBOW AND FOREARM BONES

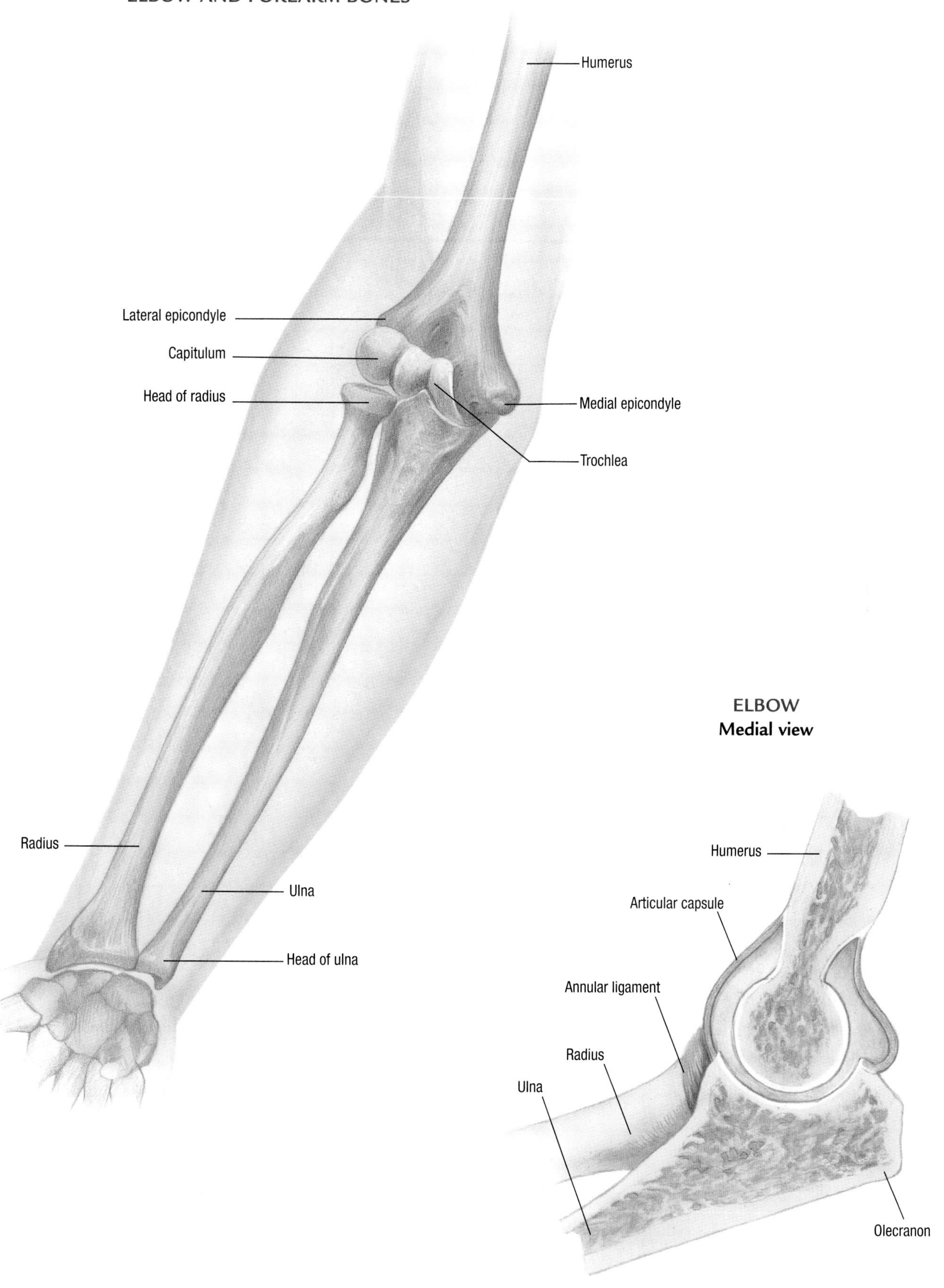
Humerus
Lateral epicondyle
Capitulum
Head of radius
Medial epicondyle
Trochlea
Radius
Ulna
Head of ulna
ELBOW
Medial view
Humerus
Articular capsule
Annular ligament
Radius
Ulna
Olecranon

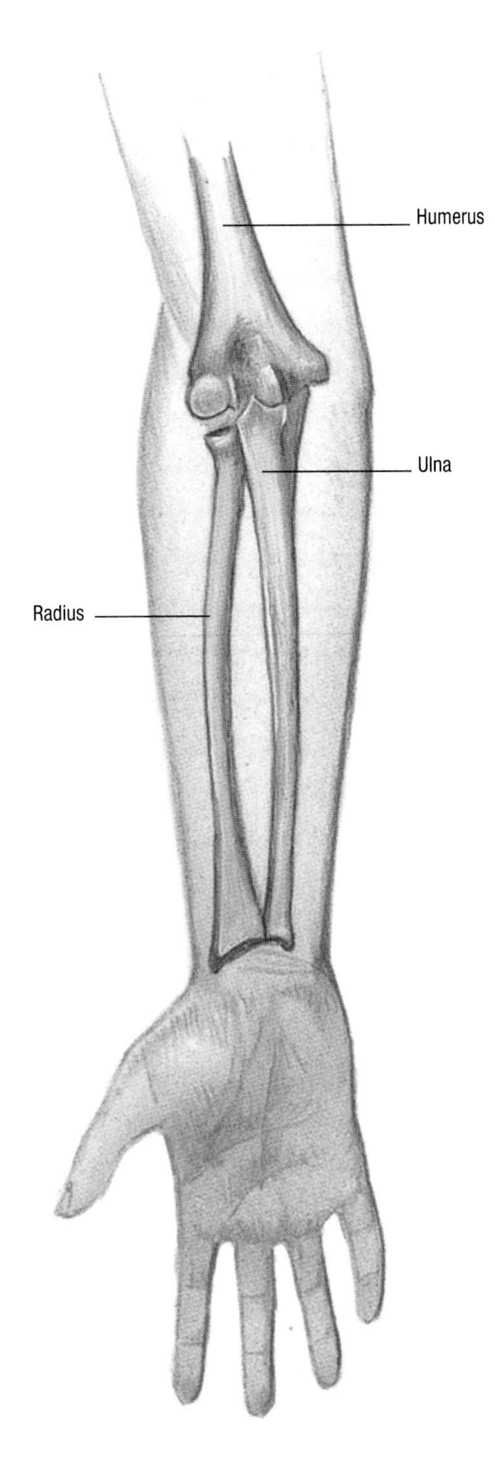
Supination
Humerus
Ulna
Radius

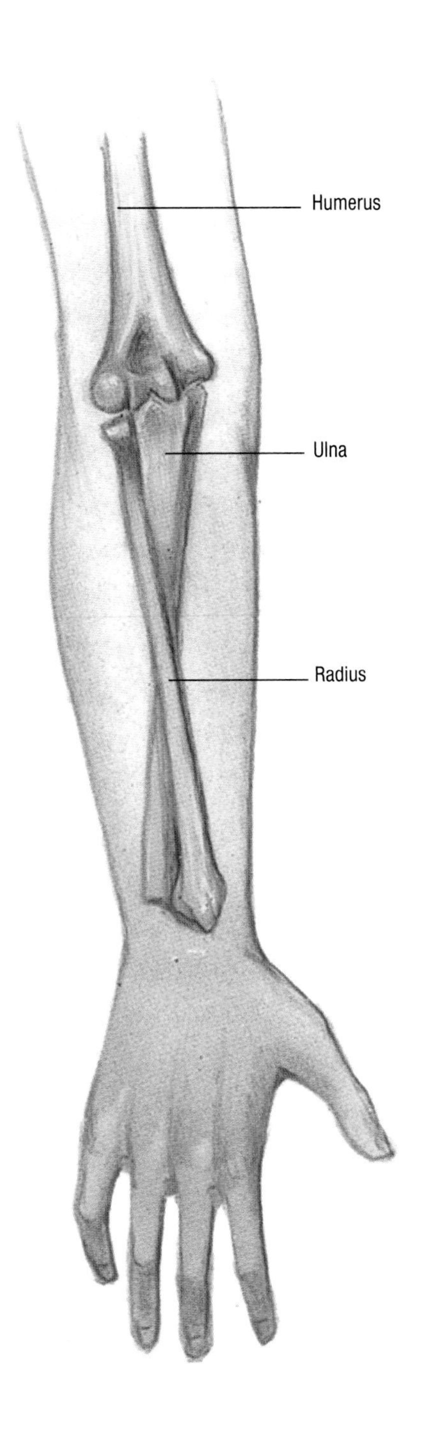
Pronation
Humerus
Ulna
Radius

## ELBOW BONES
### Posterior view

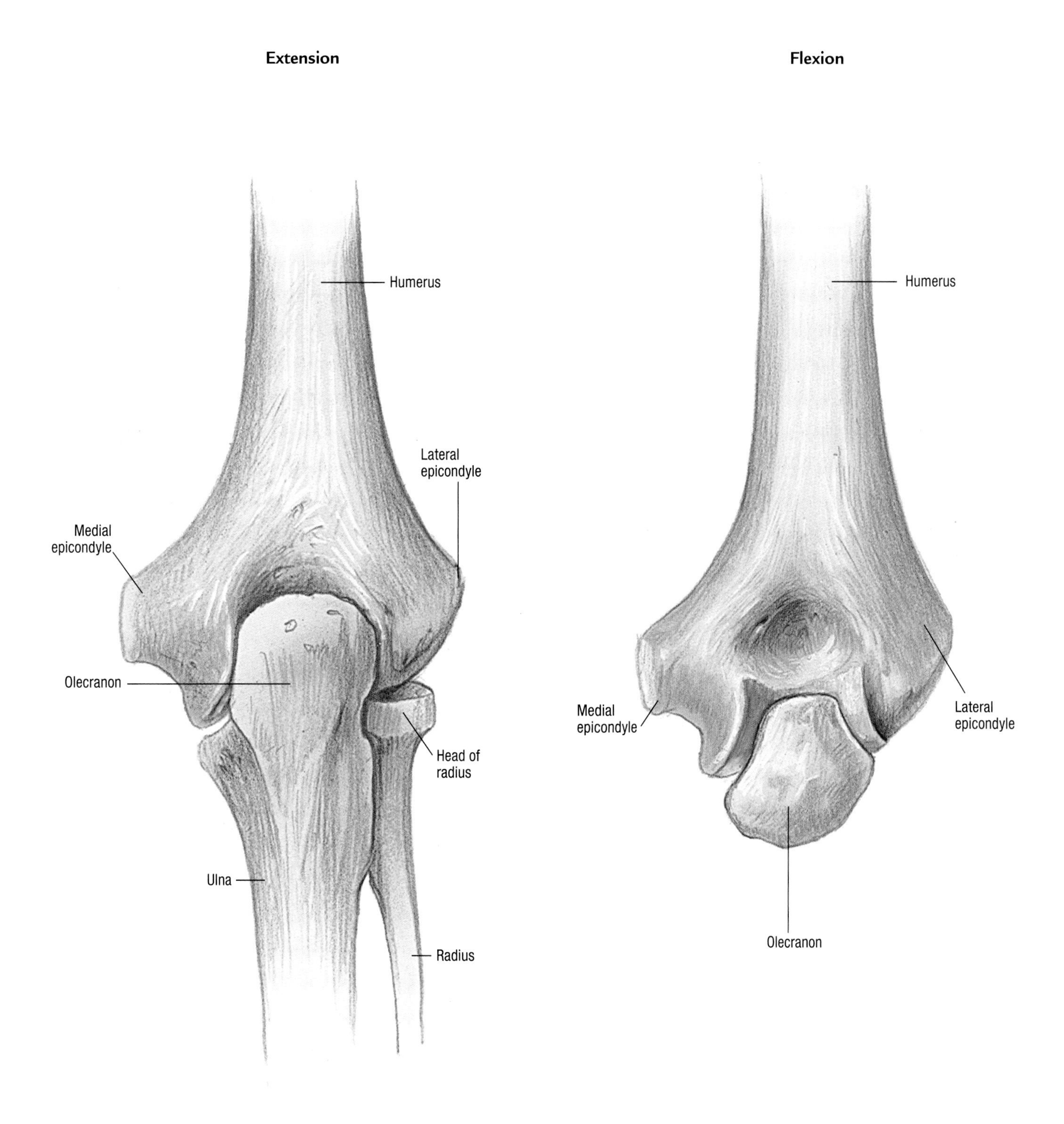
Extension
Flexion
Humerus
Lateral epicondyle
Medial epicondyle
Olecranon
Head of radius
Ulna
Radius
Humerus
Medial epicondyle
Lateral epicondyle
Olecranon

# Wrist and hand

The hand and wrist have a total of 27 bones, including carpal bones (wrist), metacarpals (palm), and phalanges (digits or fingers). Numerous muscles and nerves are responsible for the gross and fine motor movements of the wrists and hands. (See *Wrist and hand nerves*, page 139.) The muscles are discussed in detail on page 142.

### Wrist

The eight carpals (short bones) of the wrist are joined tightly by ligaments. (See *Wrist and hand bones,* pages 135 and 136, *Carpal bones* and *Cross section of wrist*, page 137.) The bones are arranged in two rows, the proximal and distal rows:

- proximal row:
  - scaphoid
  - lunate
  - triquetral
  - pisiform.
- distal row:
  - trapezium
  - trapezoid
  - capitate
  - hamate.

The carpals in the proximal row articulate with each other and with the bones in the distal row. The distal row bones also articulate with each other. Together with the distal end of the radius and the inferior surface of the triangular fibrocartilage, the proximal row of carpal bones forms the radiocarpal joint. This joint has its own synovial cavity. The carpal bones are limited to gliding movements. The flexor retinaculum, which forms the carpal tunnel for passage of the flexor tendons, maintains wrist concavity.

## Carpal tunnel

The carpal tunnel is a passage created by the carpal bones and the flexor retinaculum through which the median nerve and the flexor tendons pass. Compression of the median nerve inside the tunnel from injury or repetitive movement results in carpal tunnel syndrome. (See *Median nerve path to wrist*, page 138.)

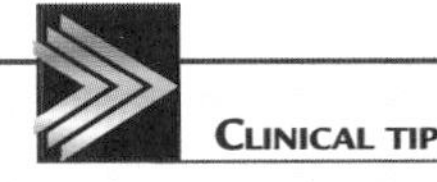

CLINICAL TIP

#### CARPAL TUNNEL SYNDROME

Numbness and tingling in the hand can be secondary to compression of the median nerve as it passes through the bones and ligaments of the carpal tunnel. Mild cases can be treated with wrist splints. More severe cases may require division of the transverse carpal ligament, which will relieve the pressure on the nerve.

## Anatomical snuffbox

This triangular hollow formed by the tendons on the back of the hand is apparent near the radial aspect of the wrist when the thumb is abducted and extended. Radial arteries and veins pass through this space, heading toward the anterior radial surface and the dorsal aspect of the first intermetacarpal space.

### Palm

The five metacarpals, the longest of the hand bones, radiate from the carpus to form the framework of the palm. Beginning with the thumb (pollex), they are numbered 1 through 5 (instead of named).

The base of each metacarpal articulates with a carpal bone. They also articulate with each other. The palmar aponeurosis is a ribbonlike fibrous tissue that runs from the tendon in the palm to the bases of the fingers. It's also called the volar fascia. The outer surface of the palm has several grooves that facilitate flexibility.

### Fingers

Each hand has 14 phalanges, or tapering bones. The thumb has only proximal and distal phalanges. The other four fingers have proximal, middle, and distal phalanges. The joints between the metacarpals and phalanges permit flexion and extension. (See *Thumb movement*, page 142.)

### Innervation

The median nerve, which passes through the carpal tunnel, innervates the intrinsic muscles and the skin on the lateral part of the hand. The ulnar nerve enters distally to the wrist along the border of the pisiform. Branches of the nerve are distributed through the hand.

### Blood supply

Two major arteries for the hand are the radial and ulnar arteries. Branches of these two arteries join to form the deep and superficial palmar arches. One of the numerous arteries supplying the fingers is the palmar metacarpal artery. It originates in the deep palmar arch. Corresponding superficial and deep palmar vein arches provide drainage for the area. (See *Hand ligaments*, pages 140 and 141.)

## WRIST AND HAND BONES
**Palmar view**

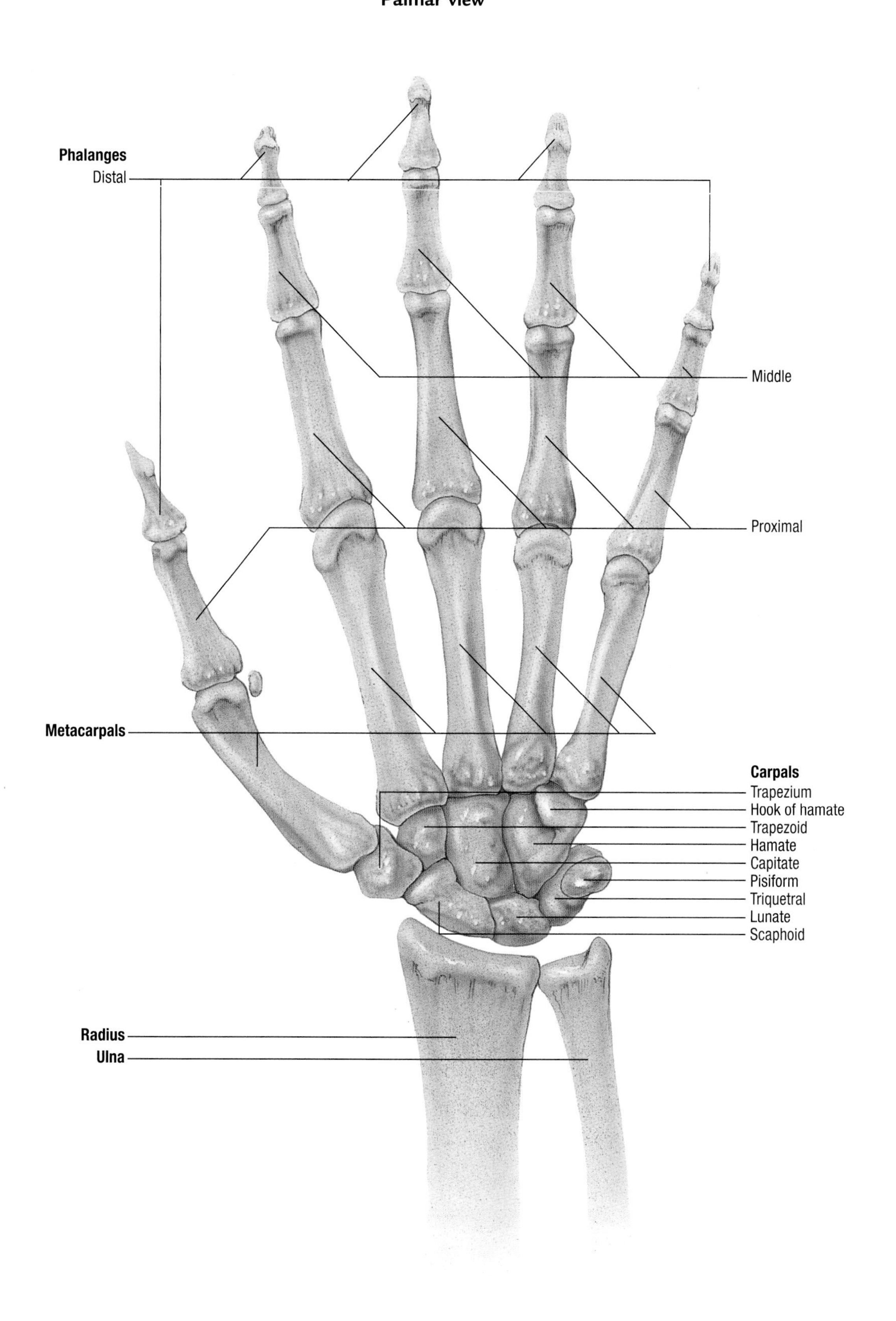

Phalanges
Distal
Middle
Proximal
Metacarpals
Carpals
Trapezium
Hook of hamate
Trapezoid
Hamate
Capitate
Pisiform
Triquetral
Lunate
Scaphoid
Radius
Ulna

## WRIST AND HAND BONES
### Dorsal view

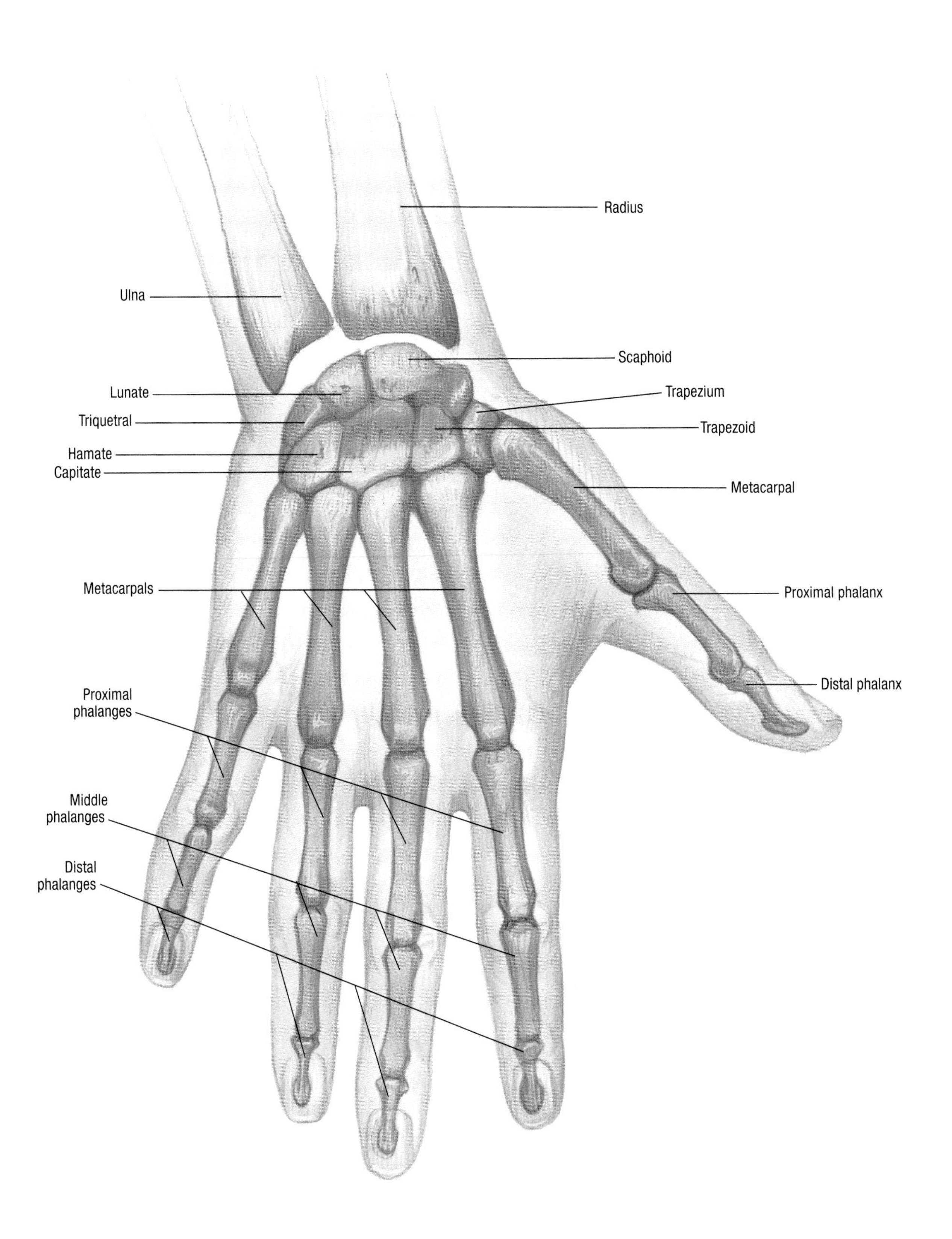
Radius
Ulna
Scaphoid
Lunate
Trapezium
Triquetral
Trapezoid
Hamate
Capitate
Metacarpal
Metacarpals
Proximal phalanx
Distal phalanx
Proximal phalanges
Middle phalanges
Distal phalanges

## CARPAL BONES
**Palmar view**

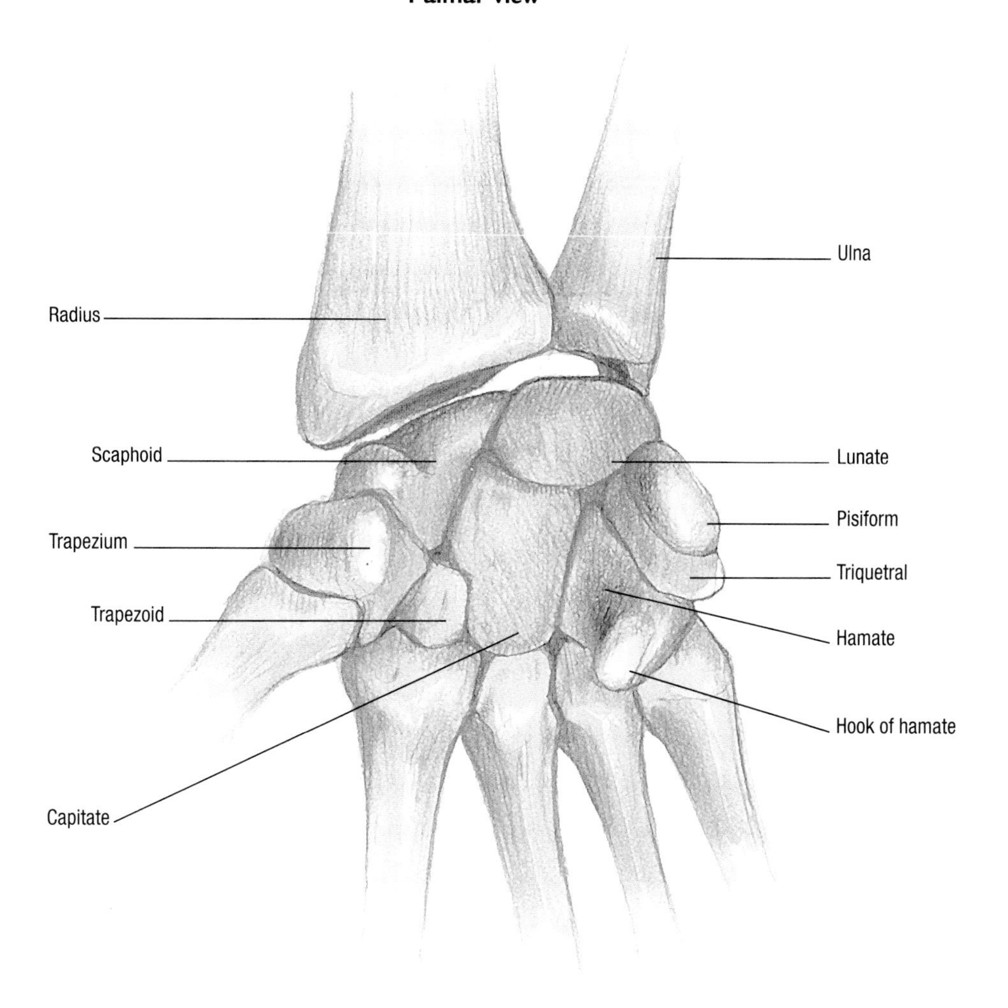

Ulna
Radius
Scaphoid
Lunate
Pisiform
Trapezium
Triquetral
Trapezoid
Hamate
Hook of hamate
Capitate


## CROSS SECTION OF WRIST

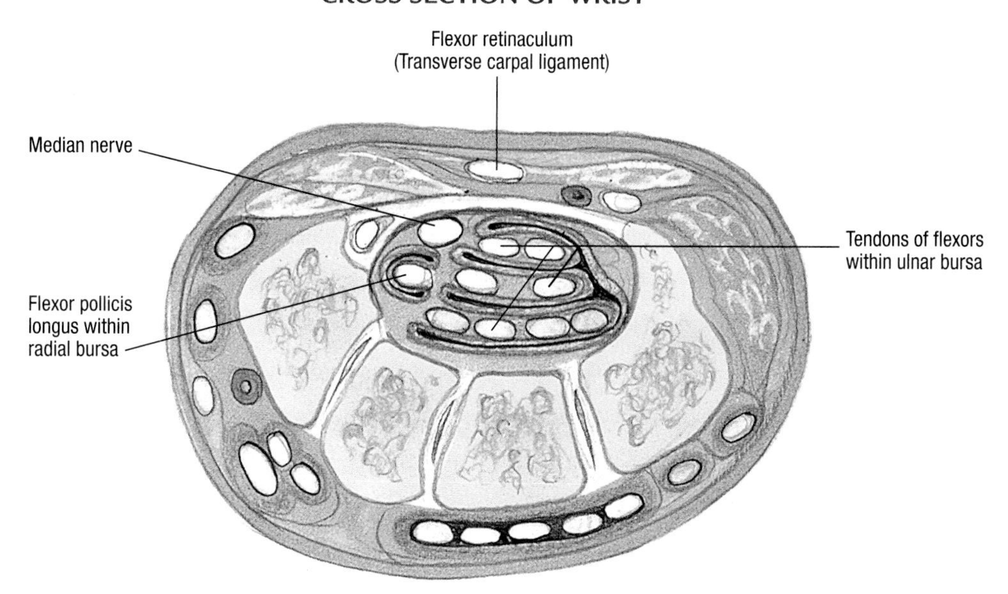

Flexor retinaculum
(Transverse carpal ligament)
Median nerve
Tendons of flexors
within ulnar bursa
Flexor pollicis
longus within
radial bursa

## MEDIAN NERVE PATH TO WRIST

**Palmar view**

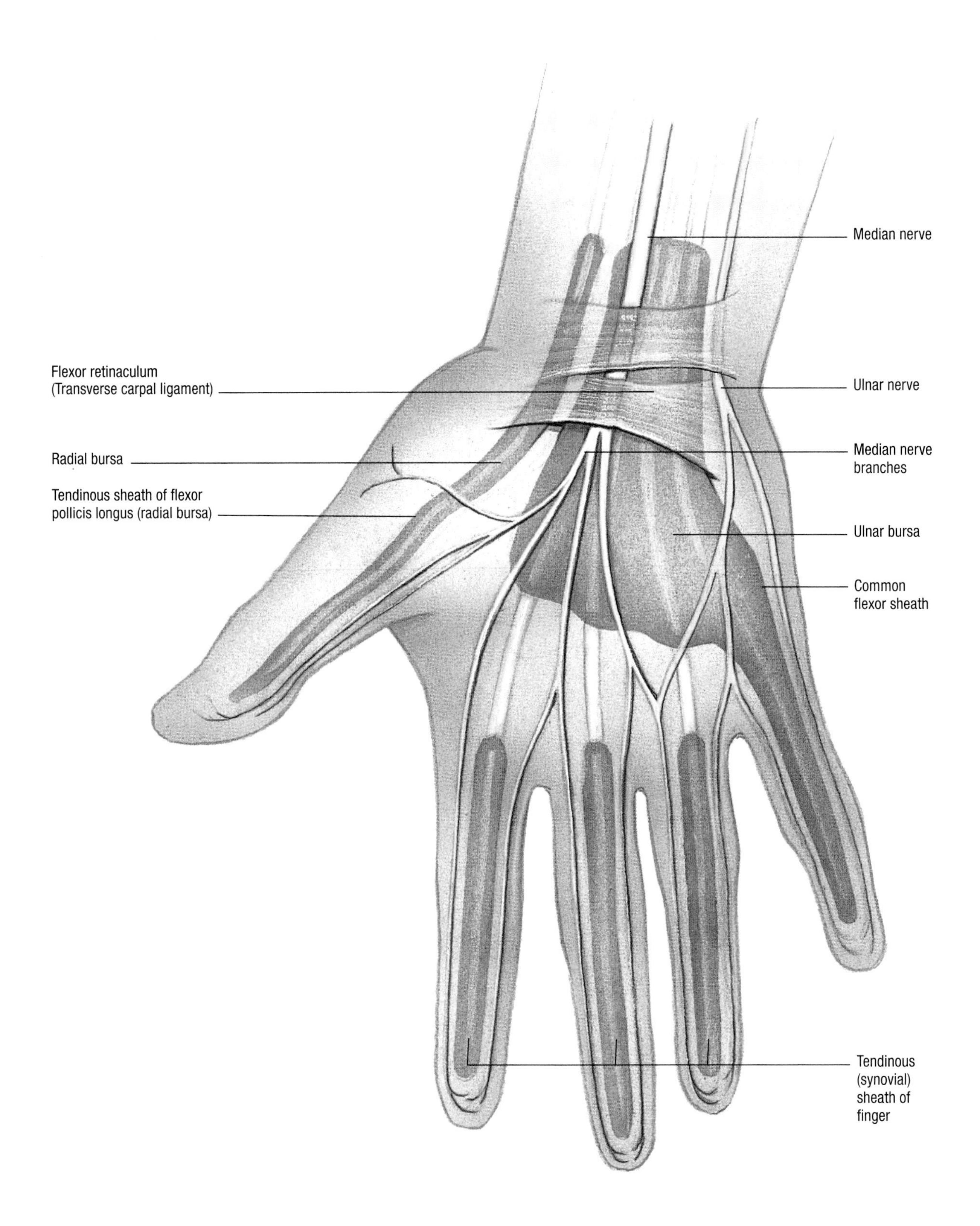

Median nerve
Flexor retinaculum
(Transverse carpal ligament)
Ulnar nerve
Radial bursa
Median nerve
branches
Tendinous sheath of flexor
pollicis longus (radial bursa)
Ulnar bursa
Common
flexor sheath
Tendinous
(synovial)
sheath of
finger

# WRIST AND HAND NERVES

## Palmar view

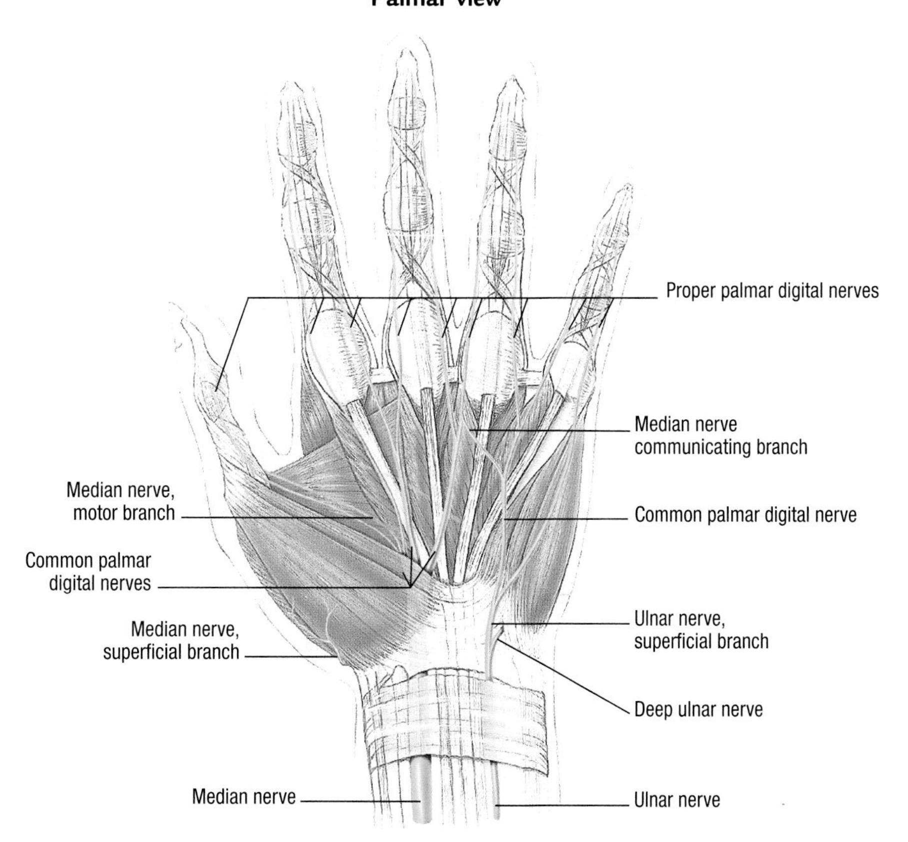
Proper palmar digital nerves
Median nerve
communicating branch
Median nerve,
motor branch
Common palmar digital nerve
Common palmar
digital nerves
Ulnar nerve,
superficial branch
Median nerve,
superficial branch
Deep ulnar nerve
Median nerve
Ulnar nerve

## Dorsal view

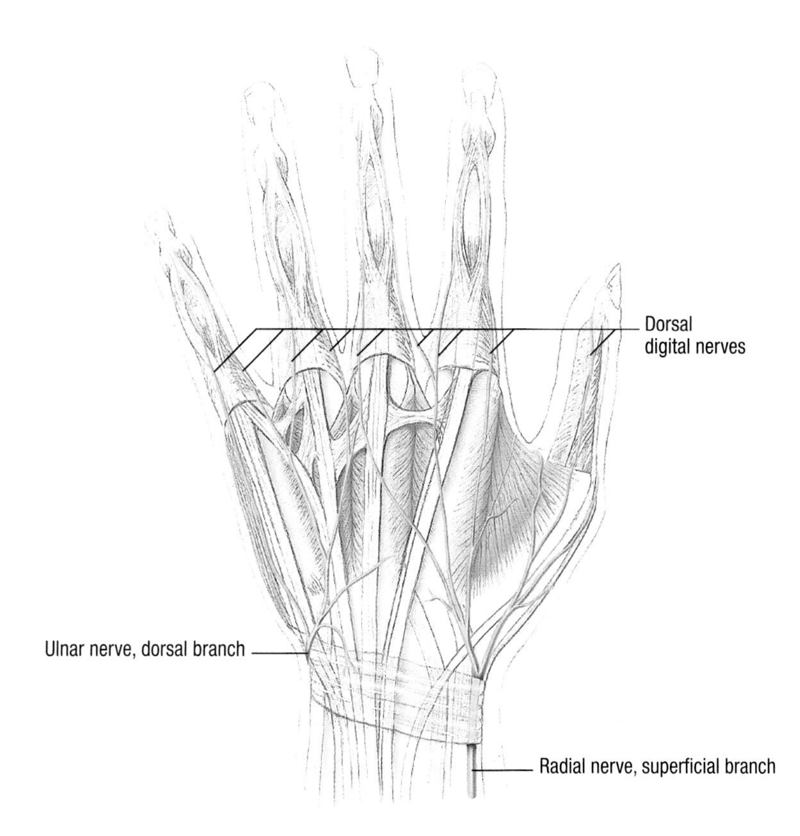
Dorsal
digital nerves
Ulnar nerve, dorsal branch
Radial nerve, superficial branch

## HAND LIGAMENTS
### Dorsal view

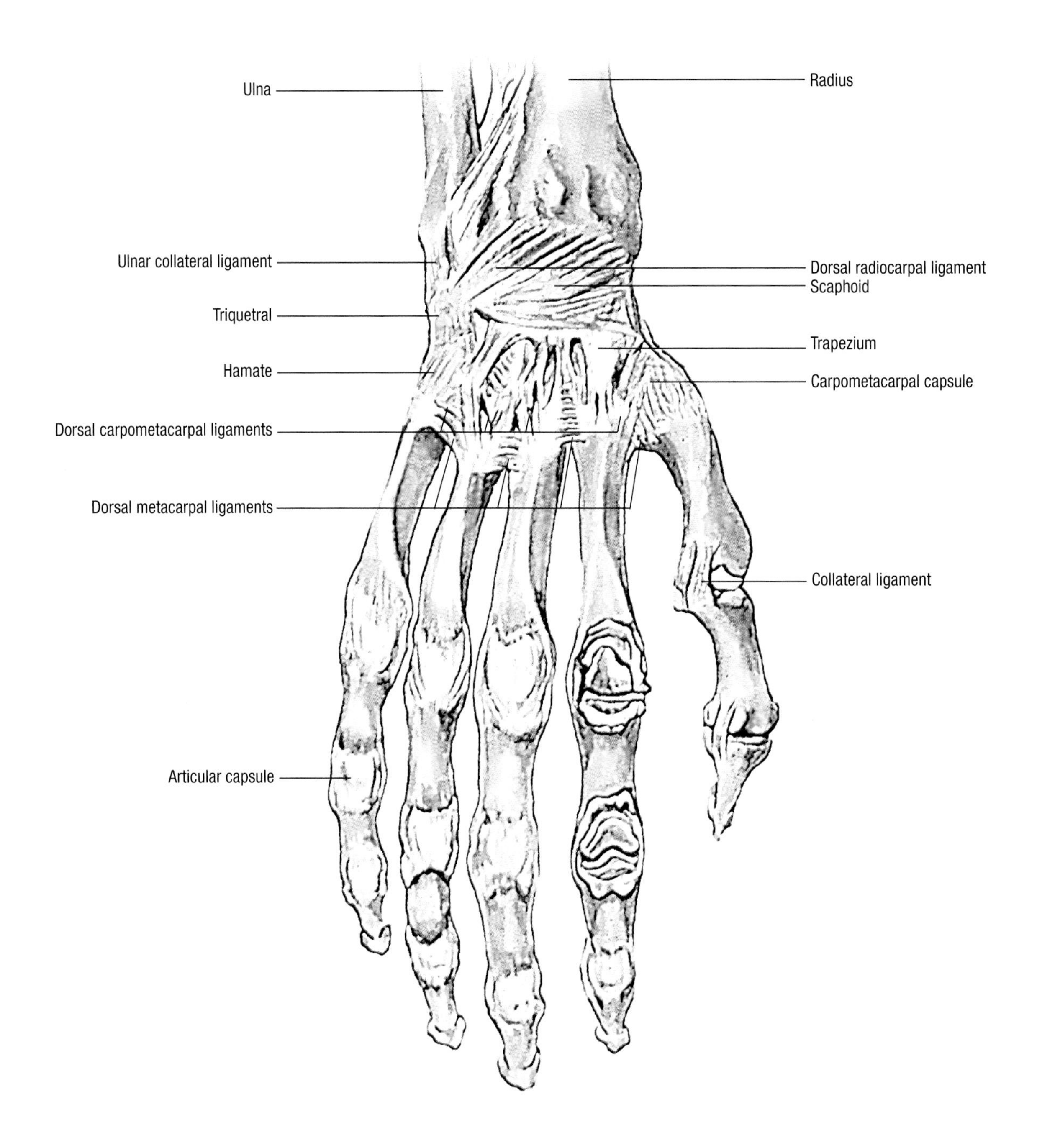
Ulna
Radius
Ulnar collateral ligament
Dorsal radiocarpal ligament
Scaphoid
Triquetral
Trapezium
Hamate
Carpometacarpal capsule
Dorsal carpometacarpal ligaments
Dorsal metacarpal ligaments
Collateral ligament
Articular capsule

## HAND LIGAMENTS
### Palmar view

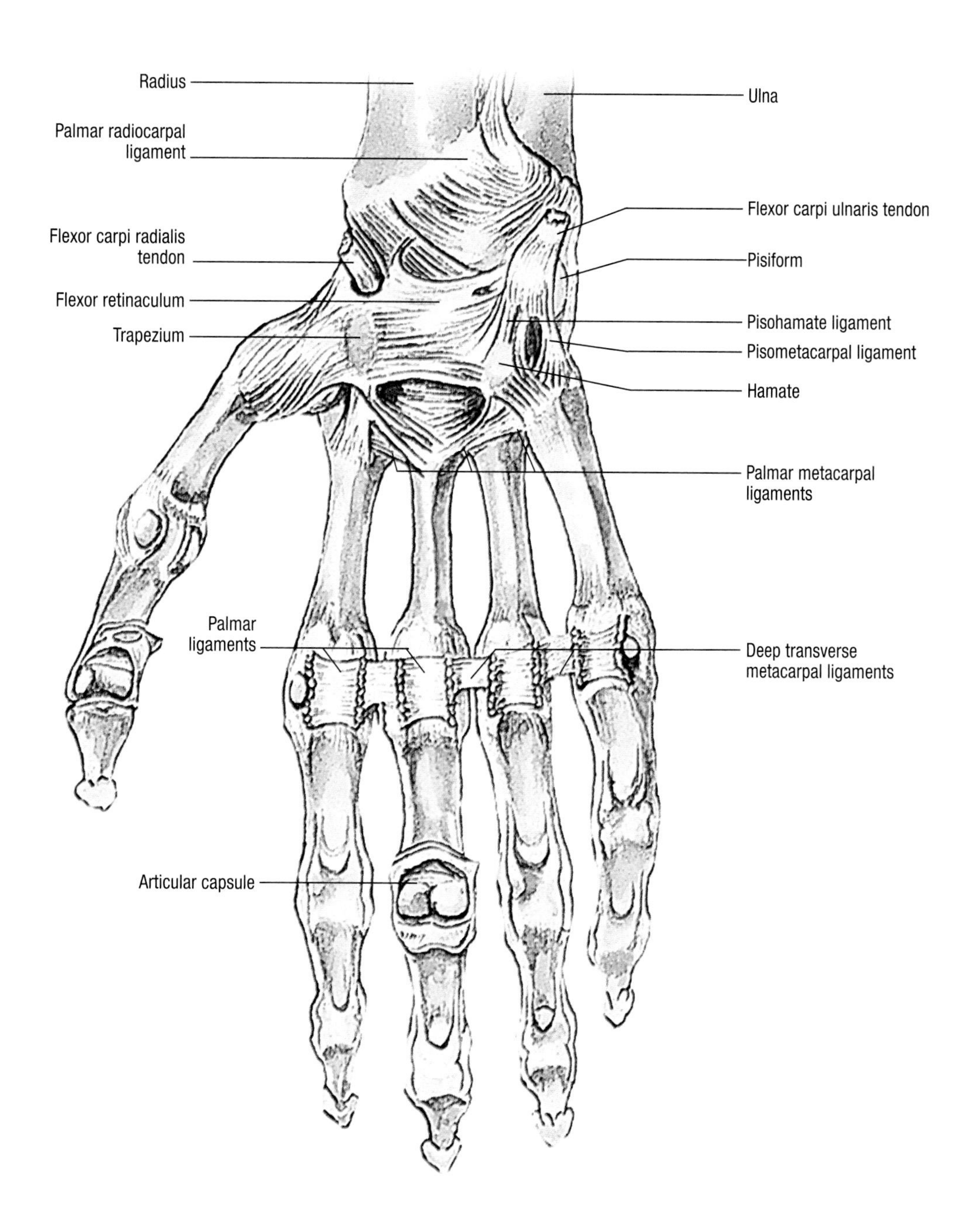
Radius
Palmar radiocarpal ligament
Flexor carpi radialis tendon
Flexor retinaculum
Trapezium
Ulna
Flexor carpi ulnaris tendon
Pisiform
Pisohamate ligament
Pisometacarpal ligament
Hamate
Palmar metacarpal ligaments
Palmar ligaments
Deep transverse metacarpal ligaments
Articular capsule

# Wrist and hand muscles

Muscles that move the wrist, hand, and fingers include the anterior superficial muscles, anterior deep muscles, posterior superficial muscles, and posterior deep muscles; the hand also has several intrinsic muscles.

## Anterior muscles

Five muscles are classified as anterior superficial muscles:
- Pronator teres pronates the forearm (rotates it forward).
- Flexor carpi radialis flexes the wrist and abducts the hand.
- Palmaris longus flexes the wrist.
- Flexor carpi ulnaris flexes the wrist and adducts the hand.
- Flexor digitorum superficialis flexes the wrist and fingers.

Three muscles are classified as anterior deep muscles:
- Flexor digitorum profundus flexes the wrist and fingers.
- Flexor pollicis longus flexes the thumb and helps flex the wrist.
- Pronator quadratus pronates the forearm.

## Posterior muscles

Five muscles are classified as posterior superficial muscles:
- Extensor carpi radialis longus extends the wrist and abducts the wrist.
- Extensor carpi radialis brevis extends the wrist and abducts the hand at the wrist.
- Extensor digitorum communis extends the fingers and wrist.
- Extensor digiti minimi extends the little finger.
- Extensor carpi ulnaris extends the wrist and adducts the hand.

Five muscles are classified as posterior deep muscles:
- Supinator supinates the forearm (rotates it backward).
- Abductor pollicis longus abducts the thumb and hand and extends the thumb.
- Extensor pollicis brevis extends the thumb.
- Extensor pollicis longus extends and abducts the thumb.
- Extensor indicis extends the index finger.

## Intrinsic muscles

The intrinsic muscles of the hand include the thenar, hypothenar, and midpalmar muscle groups. (See *Finger muscles and tendons*, page 145.)

The thenar region has four muscles:
- Abductor pollicis brevis abducts and extends the thumb.
- Opponens pollicis pulls the thumb in front of the palm.
- Flexor pollicis brevis flexes and adducts the thumb.
- Adductor pollicis adducts the thumb.

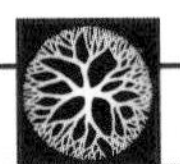

**Physiology**

### Thumb movement

The carpometacarpal joint at the base of thumb is the body's only saddle joint, and it has multiple functions:
- flexion and extension
- abduction and adduction
- opposition and repositioning.

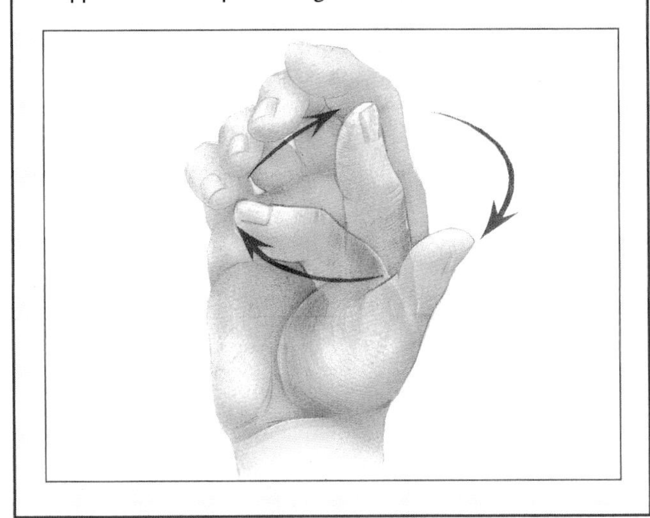

The hypothenar region has four muscles:
- Palmaris brevis pulls the skin toward the middle of the palm.
- Abductor digiti minimi manus abducts the little finger.
- Flexor digiti minimi brevis manus flexes the little finger.
- Opponens digiti minimi rotates and moves the little finger in front of the palm into opposition with the thumb.

The midpalmar region has three groups of muscles:
- Lumbricales manus extend the interphalangeal joints and flex the metacarpophalangeal joints.
- Interossei dorsales manus abduct the fingers.
- Interossei palmares adduct the fingers.

## WRIST AND HAND MUSCLES, LIGAMENTS, AND TENDONS

**Palmar view**

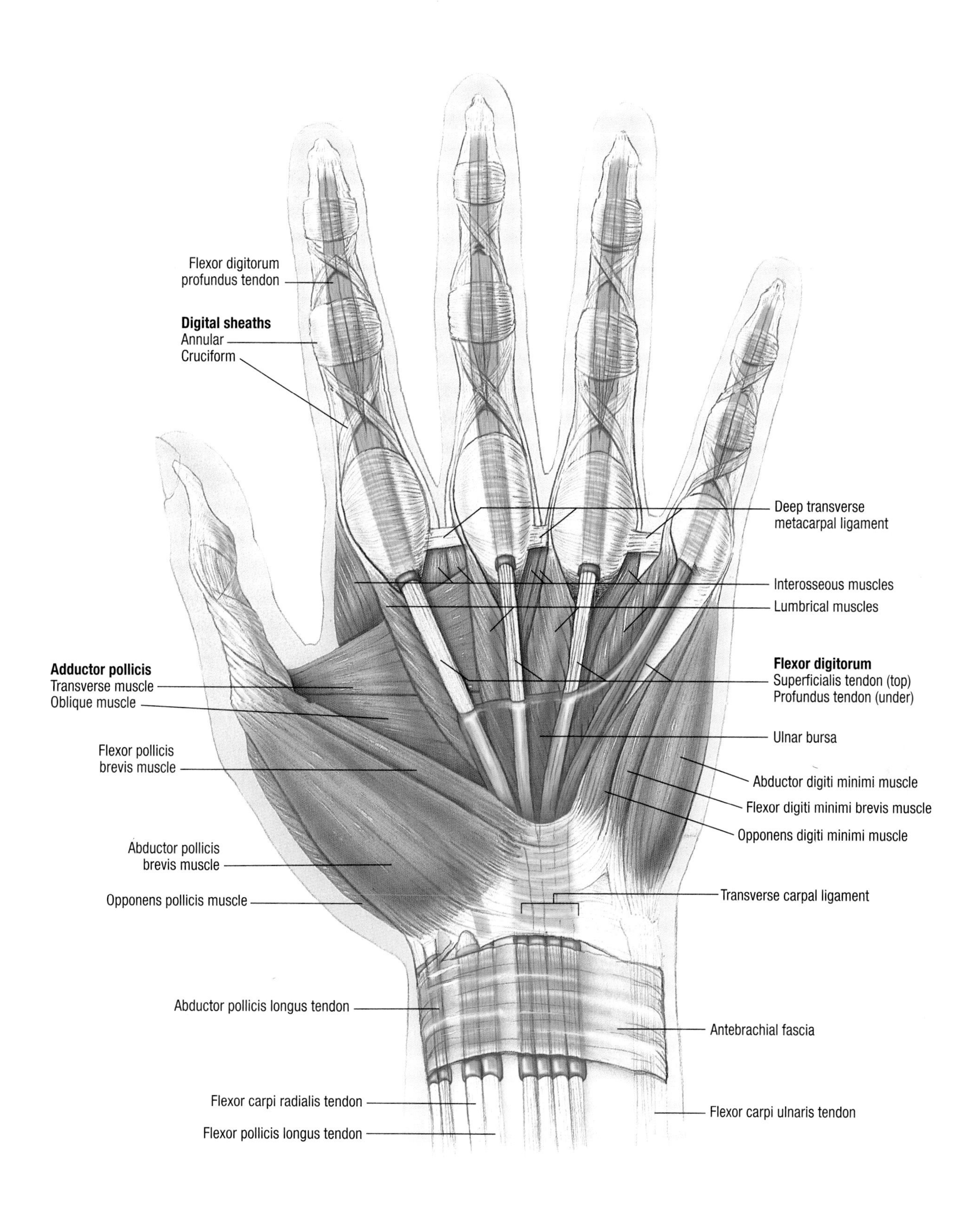

Flexor digitorum profundus tendon
Digital sheaths
Annular
Cruciform
Deep transverse metacarpal ligament
Interosseous muscles
Lumbrical muscles
Adductor pollicis
Transverse muscle
Oblique muscle
Flexor digitorum
Superficialis tendon (top)
Profundus tendon (under)
Ulnar bursa
Flexor pollicis brevis muscle
Abductor digiti minimi muscle
Flexor digiti minimi brevis muscle
Opponens digiti minimi muscle
Abductor pollicis brevis muscle
Opponens pollicis muscle
Transverse carpal ligament
Abductor pollicis longus tendon
Antebrachial fascia
Flexor carpi radialis tendon
Flexor carpi ulnaris tendon
Flexor pollicis longus tendon

## WRIST AND HAND MUSCLES, LIGAMENTS, AND TENDONS

### Dorsal view

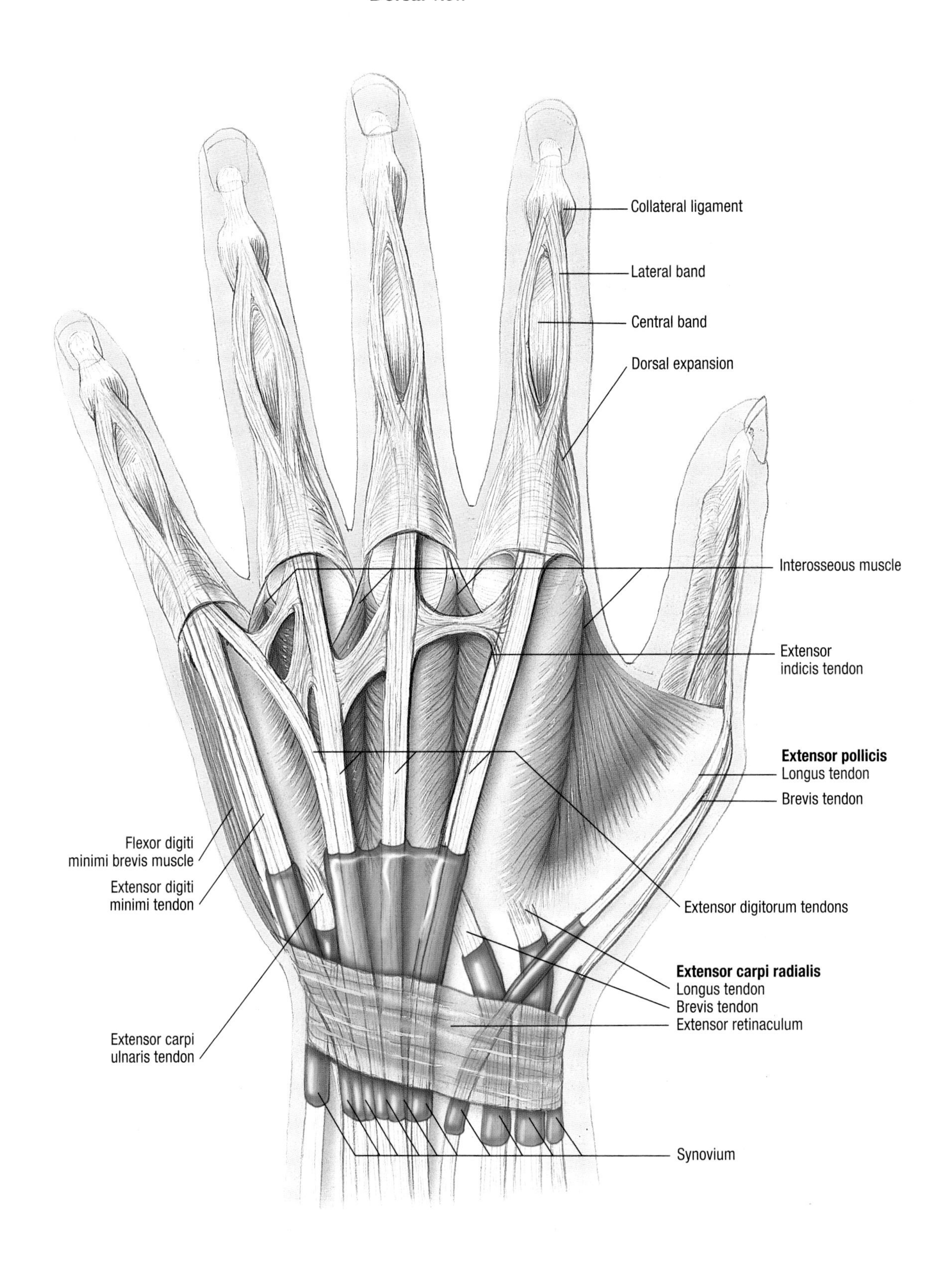

Collateral ligament
Lateral band
Central band
Dorsal expansion
Interosseous muscle
Extensor
indicis tendon
Extensor pollicis
Longus tendon
Brevis tendon
Flexor digiti
minimi brevis muscle
Extensor digiti
minimi tendon
Extensor digitorum tendons
Extensor carpi radialis
Longus tendon
Brevis tendon
Extensor retinaculum
Extensor carpi
ulnaris tendon
Synovium

## FINGER MUSCLES AND TENDONS

### Extension

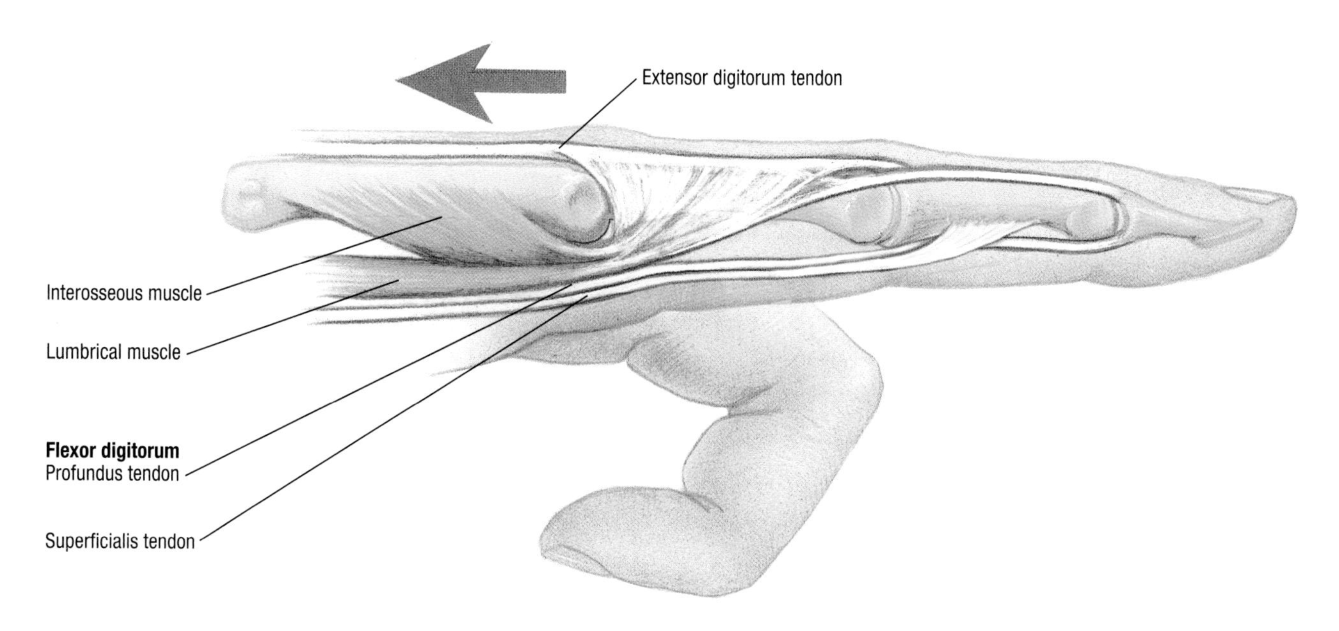

Extensor digitorum tendon
Interosseous muscle
Lumbrical muscle
**Flexor digitorum**
Profundus tendon
Superficialis tendon


### Flexion

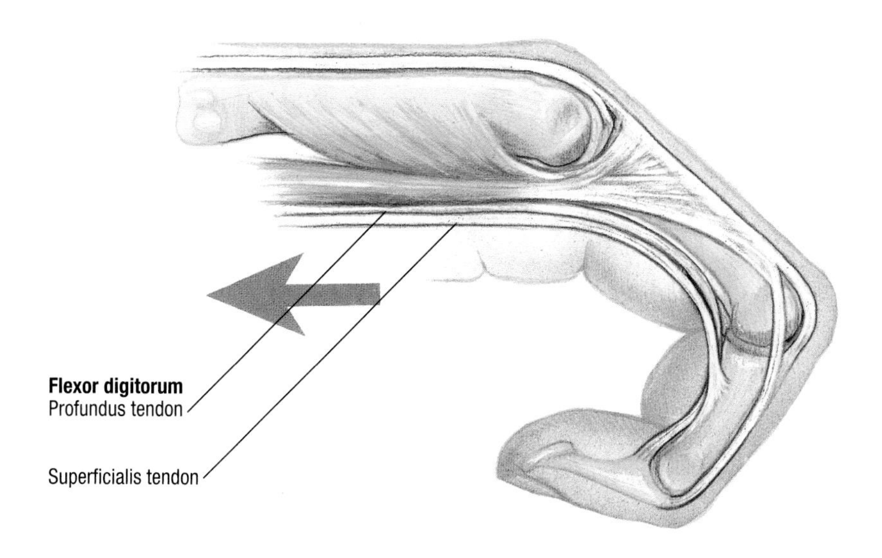

**Flexor digitorum**
Profundus tendon
Superficialis tendon

# PART V

# THORAX

# Overview of thorax

The thorax is between the neck and the thoracic diaphragm. The chest wall of the thorax – the thoracic cage, muscles, and surrounding skin – forms a protective enclosure for the heart, lungs, and great vessels of the thoracic cavity.

## Thoracic cavity

The thoracic cavity lies within the chest wall surrounded by the diaphragm, the scalene muscles and fascia of the neck, and the ribs, intercostal muscles, vertebrae, sternum, and ligaments. The superior thoracic aperture, also called the thoracic inlet, provides communication between the thoracic cavity and the neck. The inferior thoracic aperture, also called the thoracic outlet, provides communication between the thoracic and abdominal cavities.

### Mediastinum

The mediastinum, the space between the lungs, contains the following organs and structures:

- heart and pericardium
- thoracic aorta
- pulmonary artery and veins
- venae cavae and azygos veins
- thymus, lymph nodes, and vessels
- trachea, esophagus, and thoracic duct
- vagus, cardiac, and phrenic nerves.

#### Thymus

Located superior to the trachea and distal to the thyroid gland, the thymus reaches maximal size at puberty, and then starts to atrophy. The thymus consists primarily of lymphoid tissue that produces T cells, which are important in cell-mediated immunity. The thymus also produces the peptide hormones thymosin and thymopoietin, which promote growth of peripheral lymphoid tissue. (For more information on the thymus, see Part IX, Body systems.)

### Pleural cavities

Each of the pleural cavities contains a lung and is lined with serosal membranes called the pleura. The lungs are discussed in greater detail on page 176.

## Thoracic cage

The thoracic cage, also called the bony thorax, is composed of bone and cartilage. It protects the thoracic contents, supports the lungs, and provides space for them to expand and contract. The thoracic cage also supports the shoulder (pectoral) girdle. (For additional information and illustrations of the thoracic cage, see Part III, Back and spine.)

The vertebral column and 12 pairs of ribs form the posterior portion of the thoracic cage. (See *Thoracic cage, posterior view,* page 151.) The ribs extend from the thoracic vertebrae toward the anterior thorax. Like the vertebrae, they're numbered from top to bottom.

The anterior thoracic cage consists of the manubrium, sternum, xiphoid process, and ribs. (See *Thoracic cage, anterior view,* page 150.) It protects the lungs and mediastinal organs that lie between the right and left pleural cavities.

### Sternum

The sternum is the flat bone that joins the ribs anteriorly to close the wall of the bony thorax. It consists of the manubrium (triangular superior portion), body (middle and largest portion), and xiphoid process (inferior smaller portion). The connection between the manubrium and the sternum is called the manubriosternal articulation. The junction between the manubrium and the main body of the sternum can be palpated and is referred to as the angle of Louis or the sternal angle.

The manubrium lies at the level of vertebrae T3 and T4 and articulates with the clavicles and the first two pairs of ribs. The manubrium has a superior depression called the suprasternal or jugular notch.

**Clinical tip**

Because the suprasternal notch isn't covered by the rib cage, as is the rest of the thorax, the trachea and aortic pulsations can be palpated here.

The main body of the sternum is at the level of vertebrae T5 to T9, and the xiphoid process, at the level of T10 and the upper part of the liver. Fracture of the xiphoid process can result in injury to the upper liver. Cartilages of the first seven pairs of ribs articulate with the sternum at the costal notches.

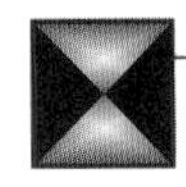

**Age-related changes**

The sternum has four segments, called sternabrae, which typically fuse by age 25. Three transverse ridges on the sternum denote these fused joints.

### Ribs

The first seven pairs of ribs connect directly to the sternum by costal cartilages. They are called true ribs and are numbered 1 through 7.

The remaining five pairs are called false ribs. Pairs 8 through 10 attach to the costal cartilages immediately superior to them. Pairs 11 and 12 are "free-floating"; that is, they aren't attached to any part of the anterior thoracic cage.

The lower parts of the rib cage (costal margins) near the xiphoid process form the borders of the costal angle, normally an angle of about 90 degrees. Posteriorly, each rib articulates with a vertebra at a vertebrocostal joint, which allows a restricted gliding movement.

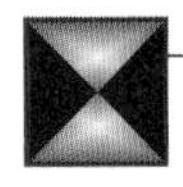

**Age-related changes**

With age, changes in calcium metabolism may increase the anteroposterior chest diameter and cause calcification of costal cartilages, which reduces mobility of the chest wall.

## THORACIC CONTENTS

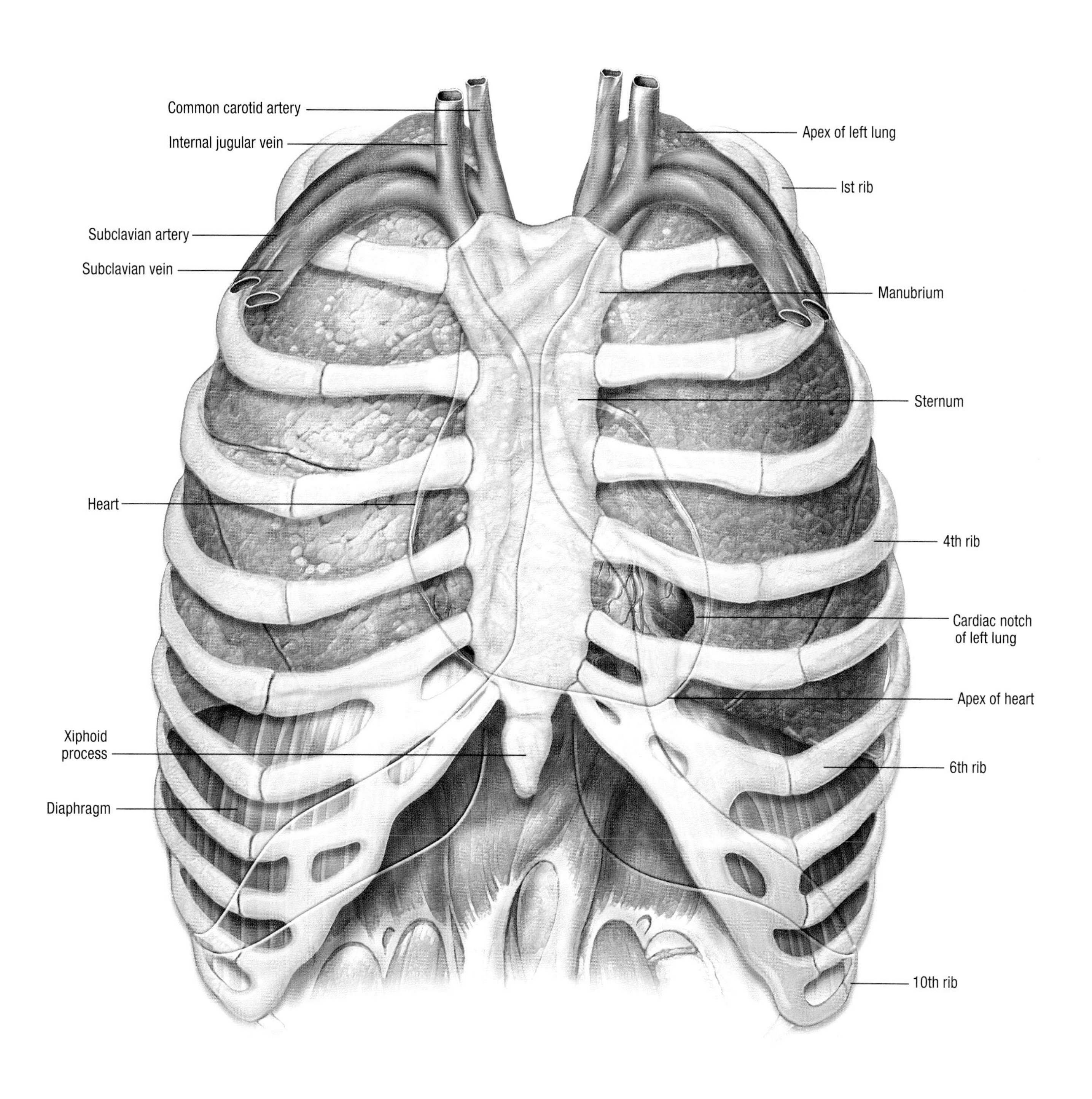

Common carotid artery
Internal jugular vein
Subclavian artery
Subclavian vein
Heart
Xiphoid process
Diaphragm
Apex of left lung
Ist rib
Manubrium
Sternum
4th rib
Cardiac notch of left lung
Apex of heart
6th rib
10th rib

# THORACIC CAGE
## Anterior view

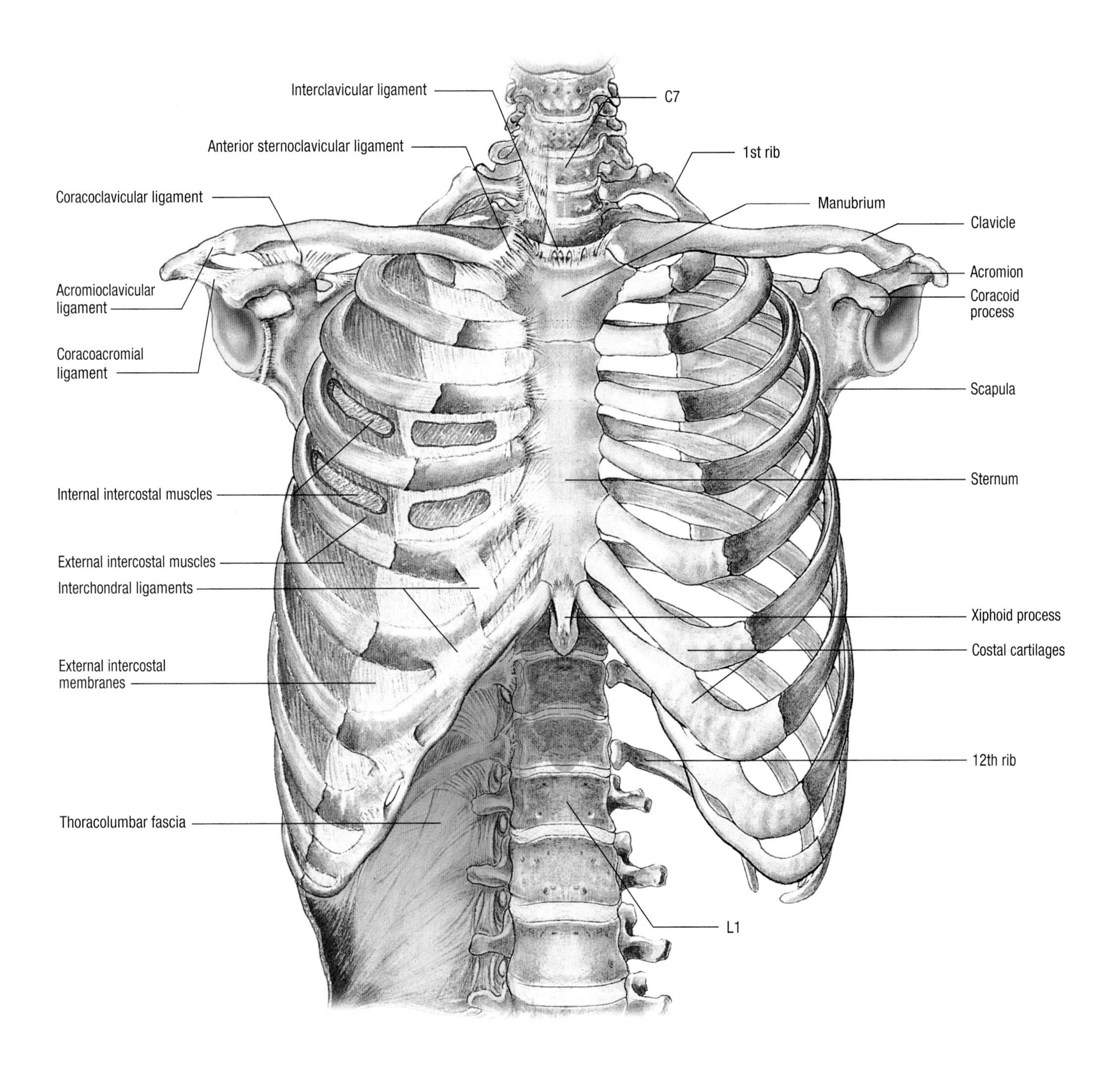
Interclavicular ligament
C7
Anterior sternoclavicular ligament
1st rib
Coracoclavicular ligament
Manubrium
Clavicle
Acromion
Acromioclavicular ligament
Coracoid process
Coracoacromial ligament
Scapula
Sternum
Internal intercostal muscles
External intercostal muscles
Interchondral ligaments
Xiphoid process
Costal cartilages
External intercostal membranes
12th rib
Thoracolumbar fascia
L1

## THORACIC CAGE
**Posterior view**

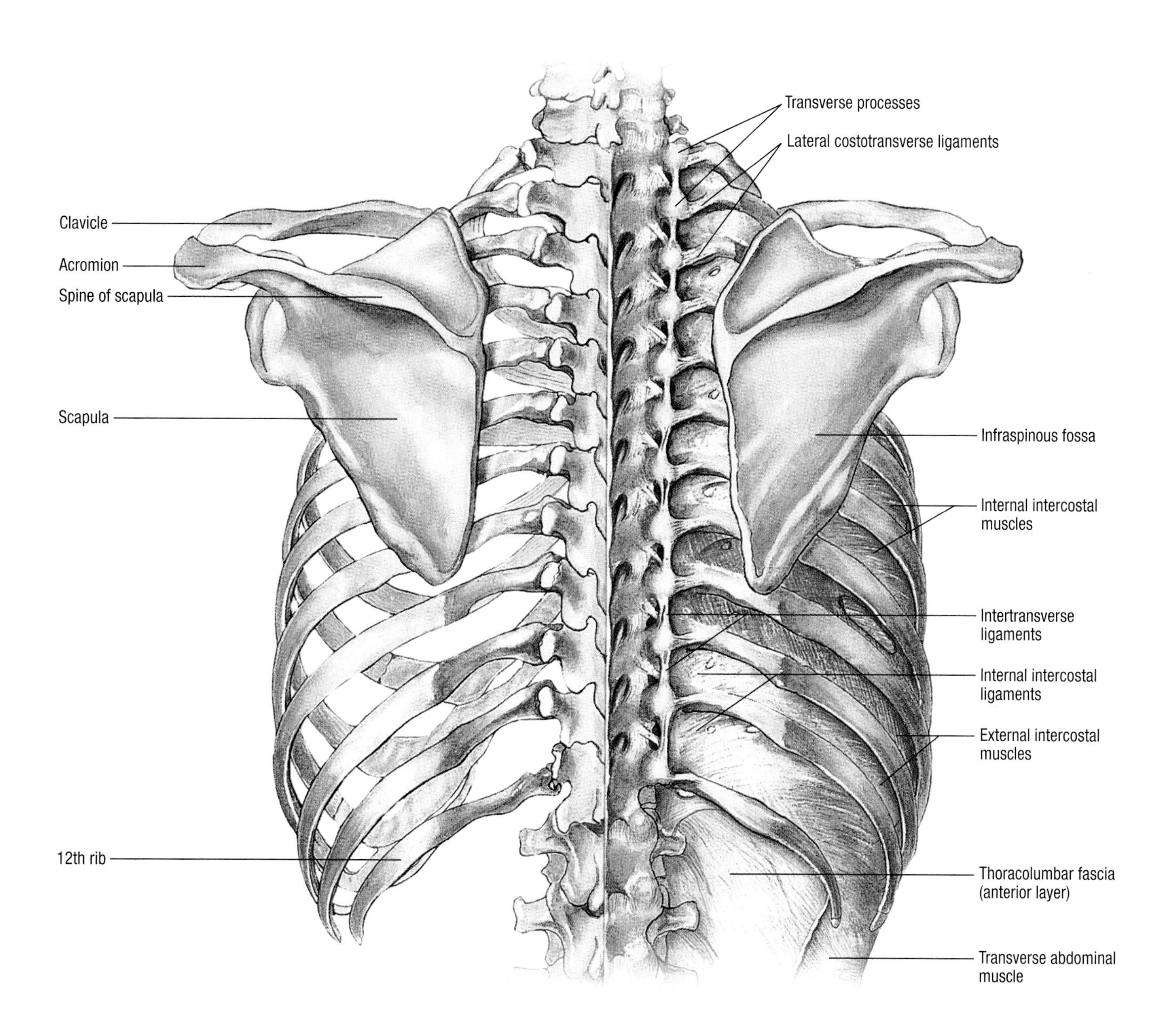

Transverse processes
Lateral costotransverse ligaments
Clavicle
Acromion
Spine of scapula
Scapula
Infraspinous fossa
Internal intercostal muscles
Intertransverse ligaments
Internal intercostal ligaments
External intercostal muscles
12th rib
Thoracolumbar fascia (anterior layer)
Transverse abdominal muscle

# Breast

The breasts are located on either side of the anterior chest wall over the greater pectoral and the anterior serratus muscles: vertically between the second or third and the sixth or seventh ribs, and horizontally between the sternal border and the midaxillary line. Proportions of glandular, adipose, and fibrous tissue vary with age, weight, sex, and other factors such as pregnancy. A small triangle of tissue, called the tail of Spence, projects into the axilla. Cooper's ligaments – fibrous bands attached to the chest wall musculature – support each breast.

## Nipples

Each breast has a centrally located pigmented area of erectile tissue ringed by an areola that's darker than the adjacent tissue. The tissue in the nipple responds to cold, friction, and sexual stimulation. Sebaceous glands, also called Montgomery's tubercles, and hair follicles are scattered on the areolar surface.

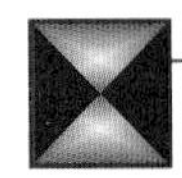

**Age-related changes**

Older men may have gynecomastia, or breast enlargement. Possible causes include age-related hormonal alterations, adverse effects of certain medications, and cirrhosis, leukemia, or thyrotoxicosis.

## Mammary glands

The mammary glands are specialized accessory glands that secrete milk. Although present in both sexes, they normally function only in the female of childbearing age. (See *Breast changes through life,* page 155.)

The mammary gland is formed from many small tubules grouped into a lobule. Several lobules constitute a lobe, each of which has an interlobular duct. Many of these ducts combine to form a lactiferous duct, which terminates at the nipple.

### Milk production, secretion, and drainage

Each mammary gland contains 15 to 25 lobes, separated by fibrous connective tissue and fat. Within the lobes are clustered acini – tiny, saclike secretory elements that produce the ingredients of milk and secrete it during lactation.

The ducts draining the lobules converge to form excretory (lactiferous) ducts and sinuses (ampullae), which store milk during lactation. They convey milk to and through the nipples. These ducts drain onto the nipple surface through 15 to 20 openings. Montgomery's tubercles in the areola produce sebum, which lubricates the areola and nipple during breastfeeding.

## Lymph nodes

Several chains of lymph nodes drain different areas of the breast and axilla. (See *Lymph nodes of breast and axillary region,* page 154.) The node chains and the areas they drain are as follows:

- pectoral – most of the breast and anterior chest
- brachial – most of the arm
- subscapular – posterior chest wall and part of the arm
- midaxillary – pectoral, brachial, and subscapular nodes
- internal mammary nodes – mammary lobes.

**Clinical tip**

Each group of axillary lymph nodes is palpated in a different way, as follows:

- Central axillary nodes – press your fingers downward and in toward the chest wall.
- Pectoral and anterior nodes – grasp the anterior axillary fold between your thumb and fingers, and palpate inside the borders of the pectoral muscles.
- Lateral nodes – press your fingers along the upper inner arm, and try to compress these nodes against the humerus.
- Subscapular or posterior nodes – stand behind your patient and press your fingers to feel the inside of the muscle of the posterior axillary fold.

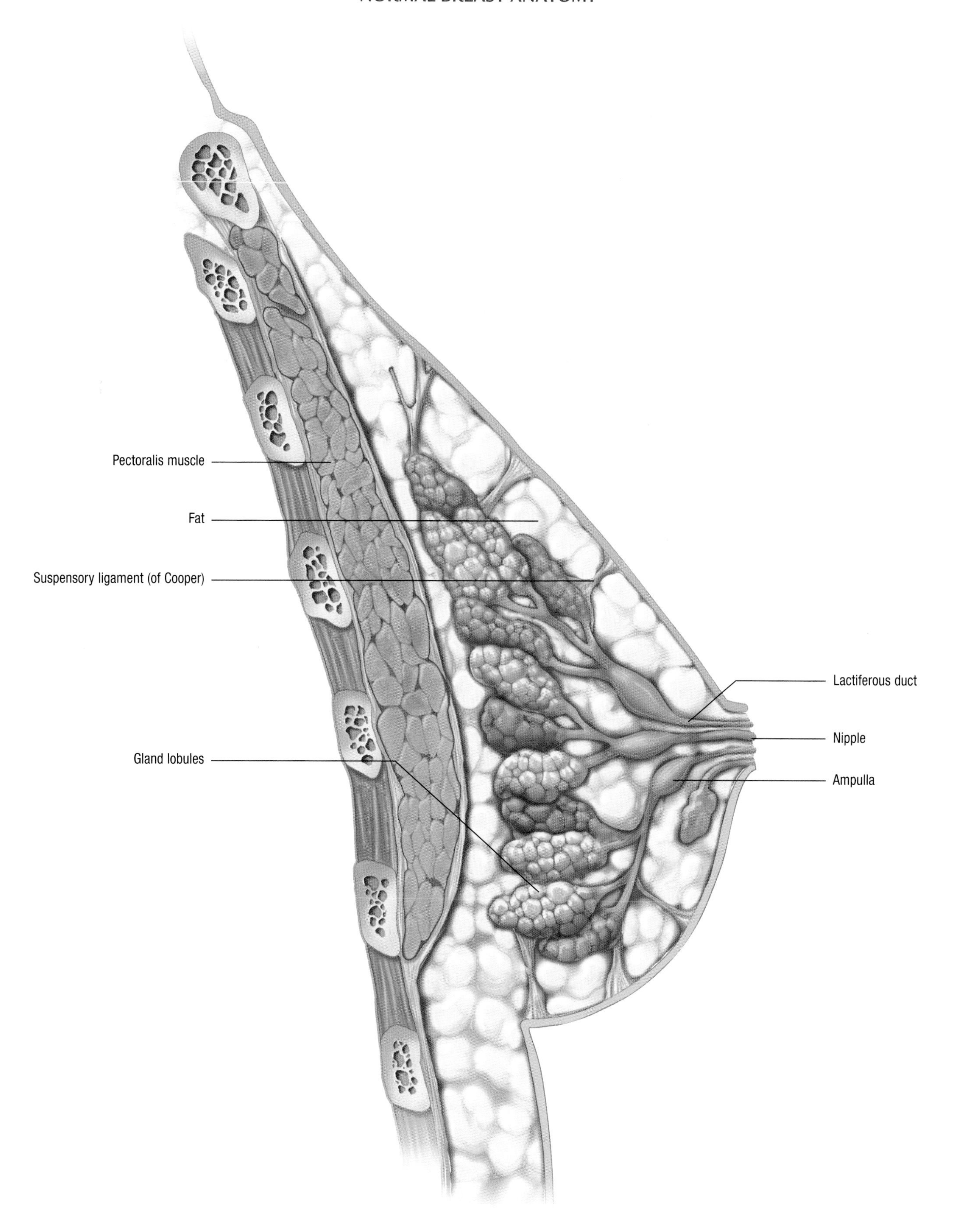
Pectoralis muscle
Fat
Suspensory ligament (of Cooper)
Gland lobules
Lactiferous duct
Nipple
Ampulla

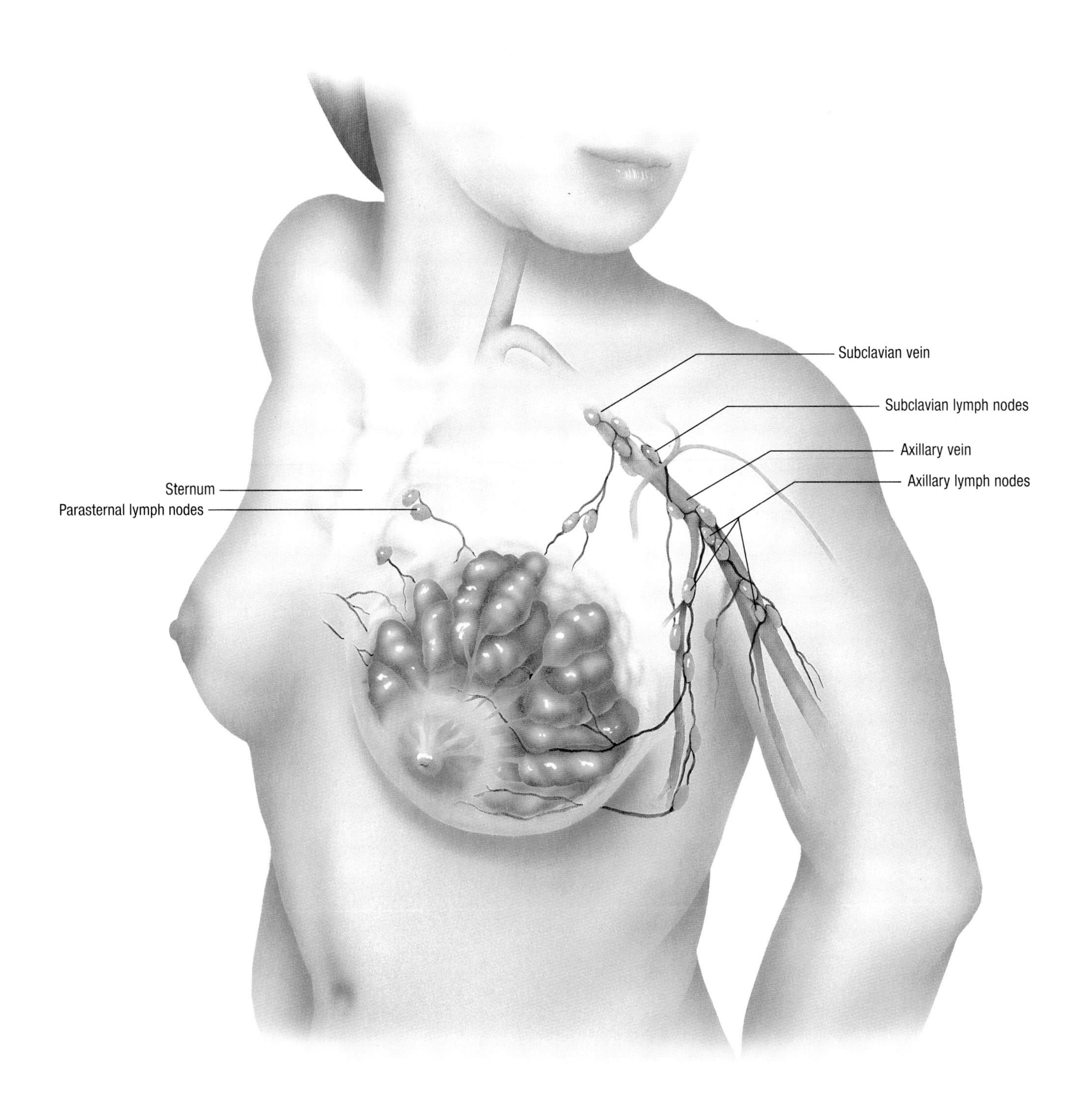
Subclavian vein
Subclavian lymph nodes
Axillary vein
Axillary lymph nodes
Sternum
Parasternal lymph nodes

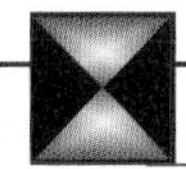

AGE-RELATED CHANGES

## BREAST CHANGES THROUGH LIFE

Throughout life, a woman's breasts typically change – The illustrations below depict the changes from after puberty to the postmenopausal years.

### Post puberty

Terminal tubules are surrounded by dense connective tissue. The secretory epithelial cells lining the tubule lumen are underdeveloped at this stage.

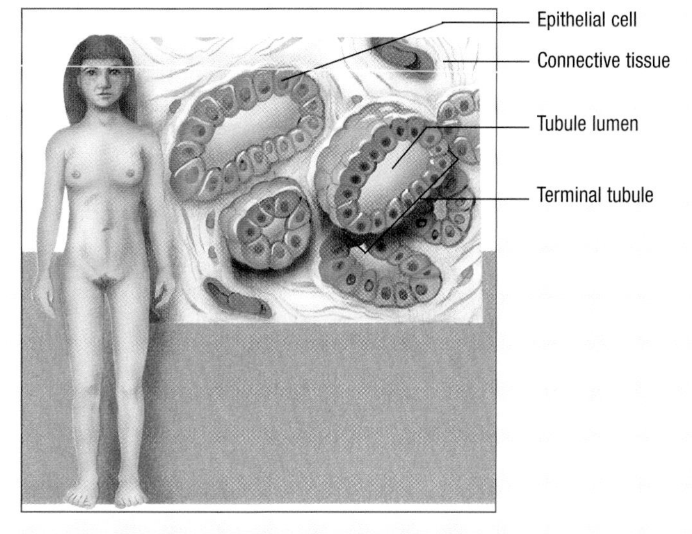
Epithelial cell
Connective tissue
Tubule lumen
Terminal tubule

### Pregnancy

Terminal tubules are converted to acinar (grapelike) structures. The lumina are enlarged and connective tissue is compressed. The secretory epithelial cells are now mature.

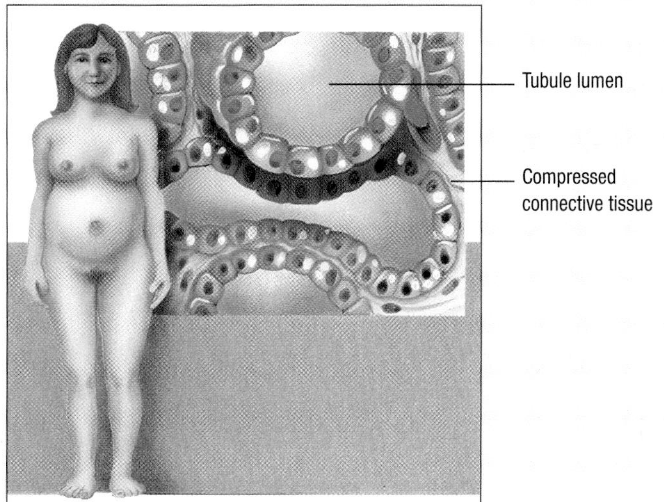
Tubule lumen
Compressed connective tissue

### Post menopause

Shrinkage and collapse of lobules and disappearance of terminal tubules occur as connective tissue increases and hardens

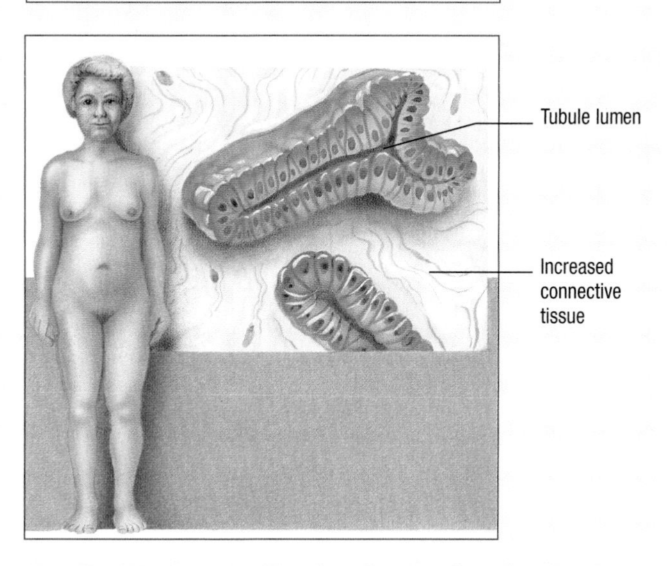
Tubule lumen
Increased connective tissue

# Muscles, nerves, and blood supply of the thorax

The thorax contains vital structures that enable such functions as breathing to occur. Its major muscles are the thoracic wall and upper limb muscles and diaphragm. The intercostal nerves supply innervation, and various blood vessels lie within and supply the thorax.

## Muscles

Muscles of the thorax include thoracic wall muscles, the diaphragm, and some muscles of the upper limb.

### Thoracic wall

Anterior thoracic muscles are reviewed below. Posterior thoracic muscles can be seen in the illustration, *Back muscles*, page 86.

- Internal and external intercostal muscles (11 pairs of each) connect adjoining ribs.
  - Internal intercostals depress the ribs during expiration.
  - External intercostal muscles raise the ribs during inspiration.
- Subcostal muscles originate on the inner surface of each rib near the costal angle and insert on the inner surface of the first, second, or third rib below. They raise the ribs during inspiration.
- Transverse thoracic muscles attach the posterior surface of the lower sternum to the internal surface of costal cartilages 2 through 6. They pull the ribs downward during expiration.

### Diaphragm

The diaphragm, a dome-shaped sheet of muscle, separates the thorax from the abdomen. (See *Diaphragm, right half*, page 158.) The diaphragm has three openings, one each for the aorta, inferior vena cava, and esophagus. The superior surface of the diaphragm is covered by the pleura.

The diaphragm, assisted by the intercostal muscles, plays a vital role in breathing. During inspiration, the diaphragm contracts and flattens, drawing air into the lungs. When inspiration is complete, the diaphragm relaxes and resumes its domed shape.

**Clinical tip**

An elevated diaphragm may be a sign of atelectasis or paralysis of the diaphragm.

### Upper limb muscles

The following muscles of the upper limb attach to the ribs.

- pectoralis major – flexes, adducts, and rotates the arm; may act as an accessory muscle in respiration
- pectoralis minor – draws the scapula forward and downward; helps to elevate ribs during forced inspiration
- subclavius – stabilizes and depresses the shoulder
- serratus anterior – stabilizes, abducts, and rotates the scapula upward, and helps to abduct and raise the arm.

## Intercostal nerves

The intercostal nerves are the primary nerves of the thorax. They pass through the thorax's intercostal spaces, innervating the intercostal and abdominal muscles. (See *Intercostal nerves*, page 159.)

## Blood supply

The thoracic wall receives blood from the thoracic aorta and two of its branches, the subclavian and axillary arteries. (See *Thoracic wall arteries, anterior view*, page 160.)

### Thoracic aorta

The thoracic portion of the aorta extends from the aortic arch inferiorly through the mediastinal cavity to the opening in the diaphragm. Branches of the thoracic aorta provide the blood supply for the viscera and walls of the thoracic cavity, as follows:

- Visceral arteries
  - pericardial – posterior pericardium
  - bronchial – primarily the walls of the bronchi
  - esophageal – thoracic portion of the esophagus
- Arteries to the thoracic cavity walls
  - intercostals – intercostal muscles, pleurae, muscles, and skin
  - superior phrenic – upper surface of the diaphragm.

### Thoracic veins

The internal jugular and subclavian veins of each side join at the first right costal cartilage to form the superior vena cava. An intercostal vein runs alongside each intercostal artery. Each side has 11 posterior intercostal veins and one subcostal vein. Most posterior intercostal veins empty into the azygos venous system, which in turn empties into the superior vena cava at the fourth thoracic vertebra. (See *Thoracic wall veins, anterior view*, page 161.)

The following veins provide the azygous drainage:

- intercostals
- hemiazygos
- bronchials
- several esophageal, mediastinal, and pericardial veins.

Other veins of the thoracic region include the internal thoracic veins, which accompany the internal thoracic arteries, and the brachiocephalic vein, which receives blood from some of the intercostal spaces.

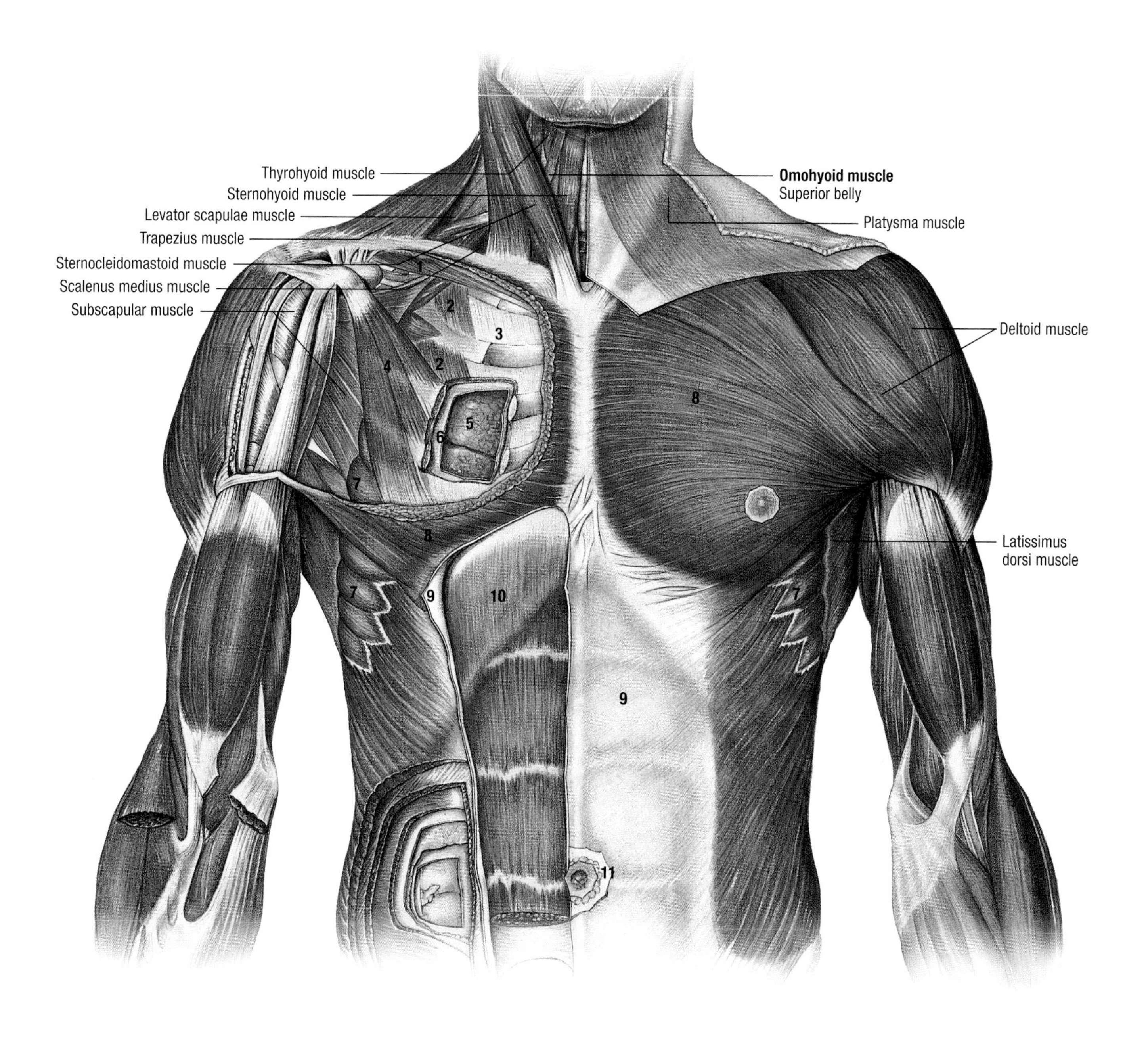
Thyrohyoid muscle
Sternohyoid muscle
Levator scapulae muscle
Trapezius muscle
Sternocleidomastoid muscle
Scalenus medius muscle
Subscapular muscle
Omohyoid muscle
Superior belly
Platysma muscle
Deltoid muscle
Latissimus dorsi muscle
1
2
3
4
2
5
6
7
8
8
7
9
10
7
9
11

**Key**

1 Subclavius muscle
2 External intercostal muscles
3 External intercostal membranes
4 Pectoralis minor muscle
5 Lung
6 Pleura
7 Serratus anterior muscle
8 Pectoralis major muscle
9 Rectus sheath (anterior layer)
10 Rectus abdominis muscle
11 Superficial fascia

## DIAPHRAGM
### Right half

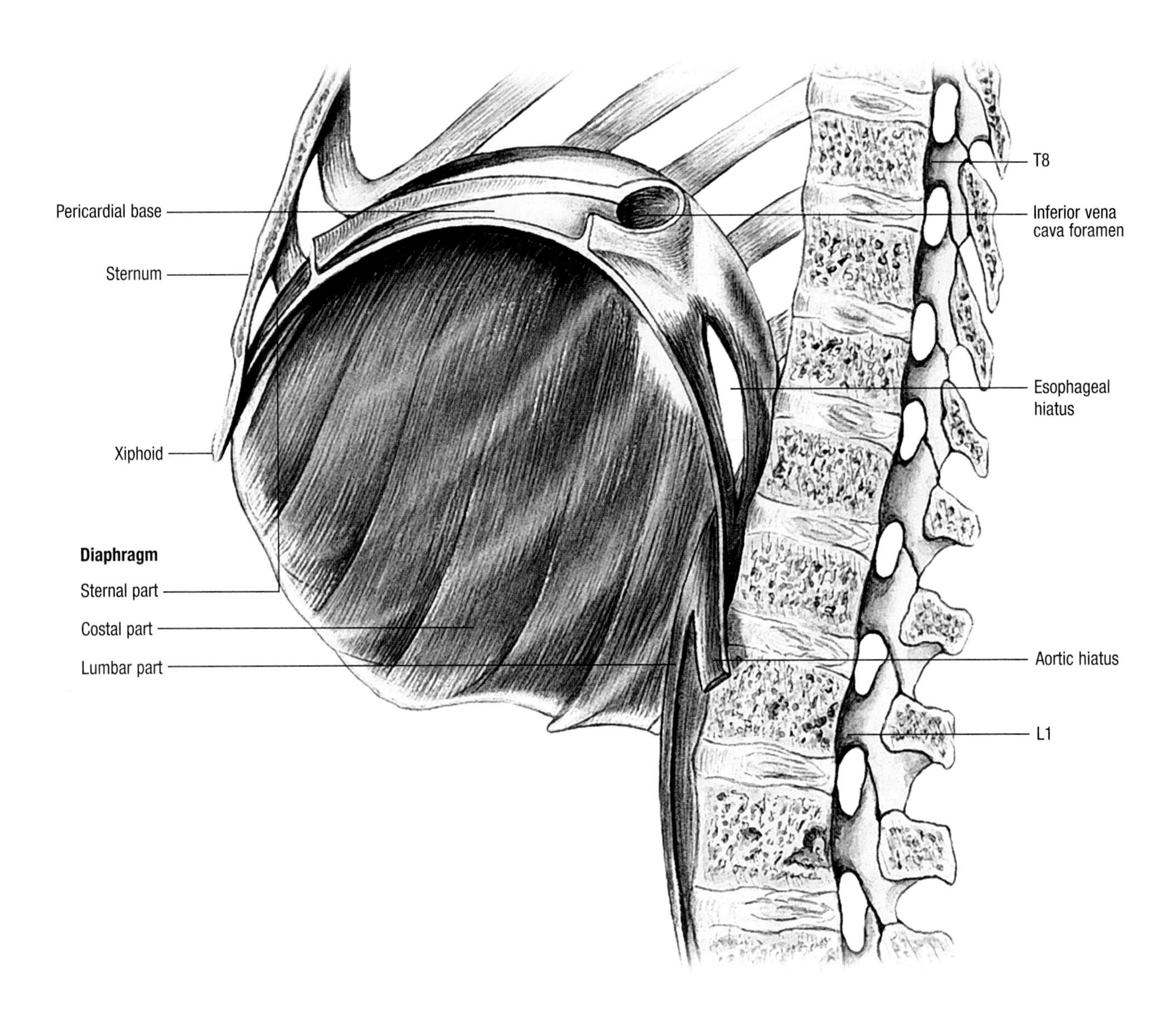

Pericardial base
Sternum
Xiphoid
Diaphragm
Sternal part
Costal part
Lumbar part
T8
Inferior vena cava foramen
Esophageal hiatus
Aortic hiatus
L1

## INTERCOSTAL NERVES

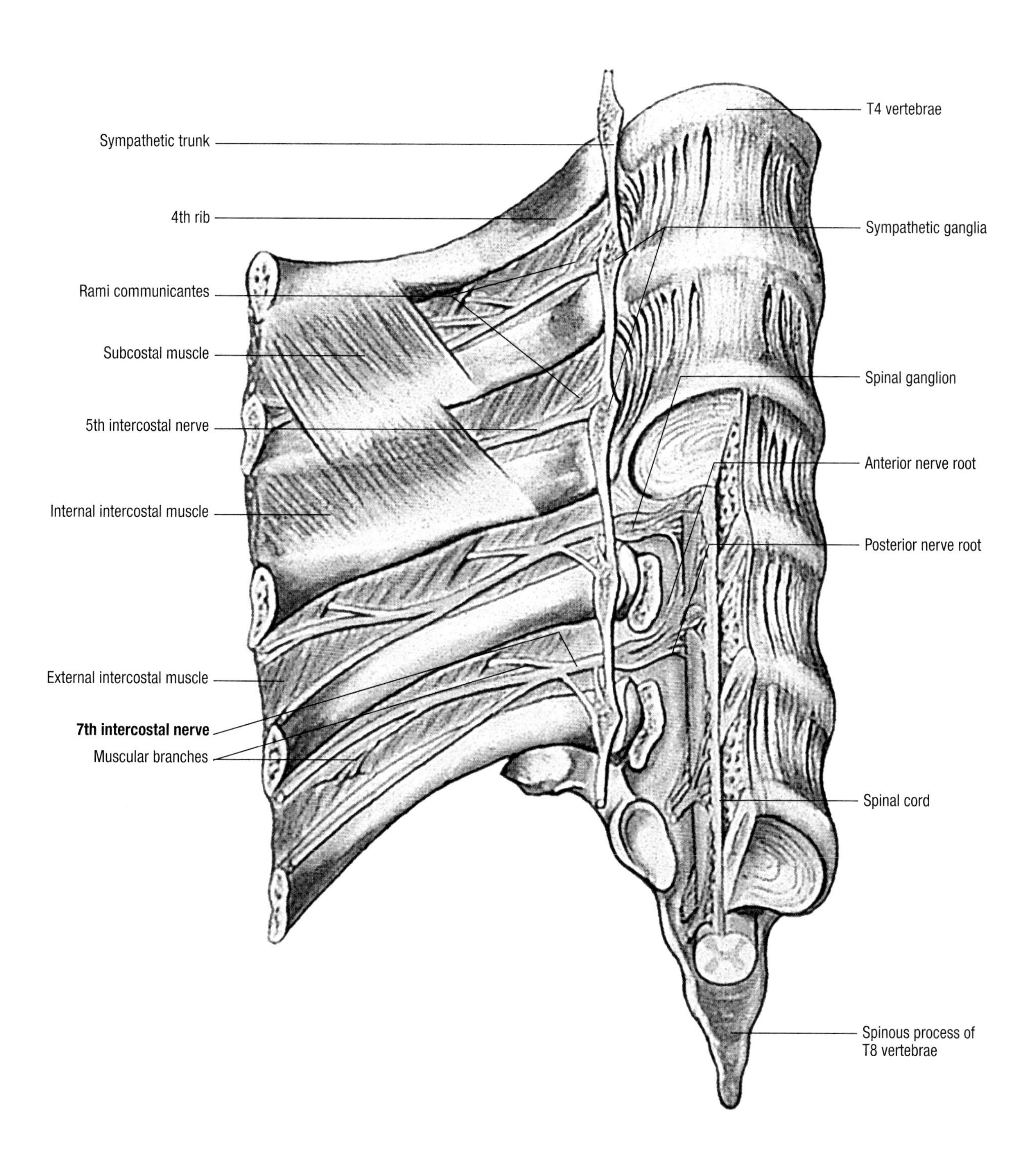
Sympathetic trunk
4th rib
Rami communicantes
Subcostal muscle
5th intercostal nerve
Internal intercostal muscle
External intercostal muscle
7th intercostal nerve
Muscular branches
T4 vertebrae
Sympathetic ganglia
Spinal ganglion
Anterior nerve root
Posterior nerve root
Spinal cord
Spinous process of
T8 vertebrae

## THORACIC WALL ARTERIES
### Anterior view

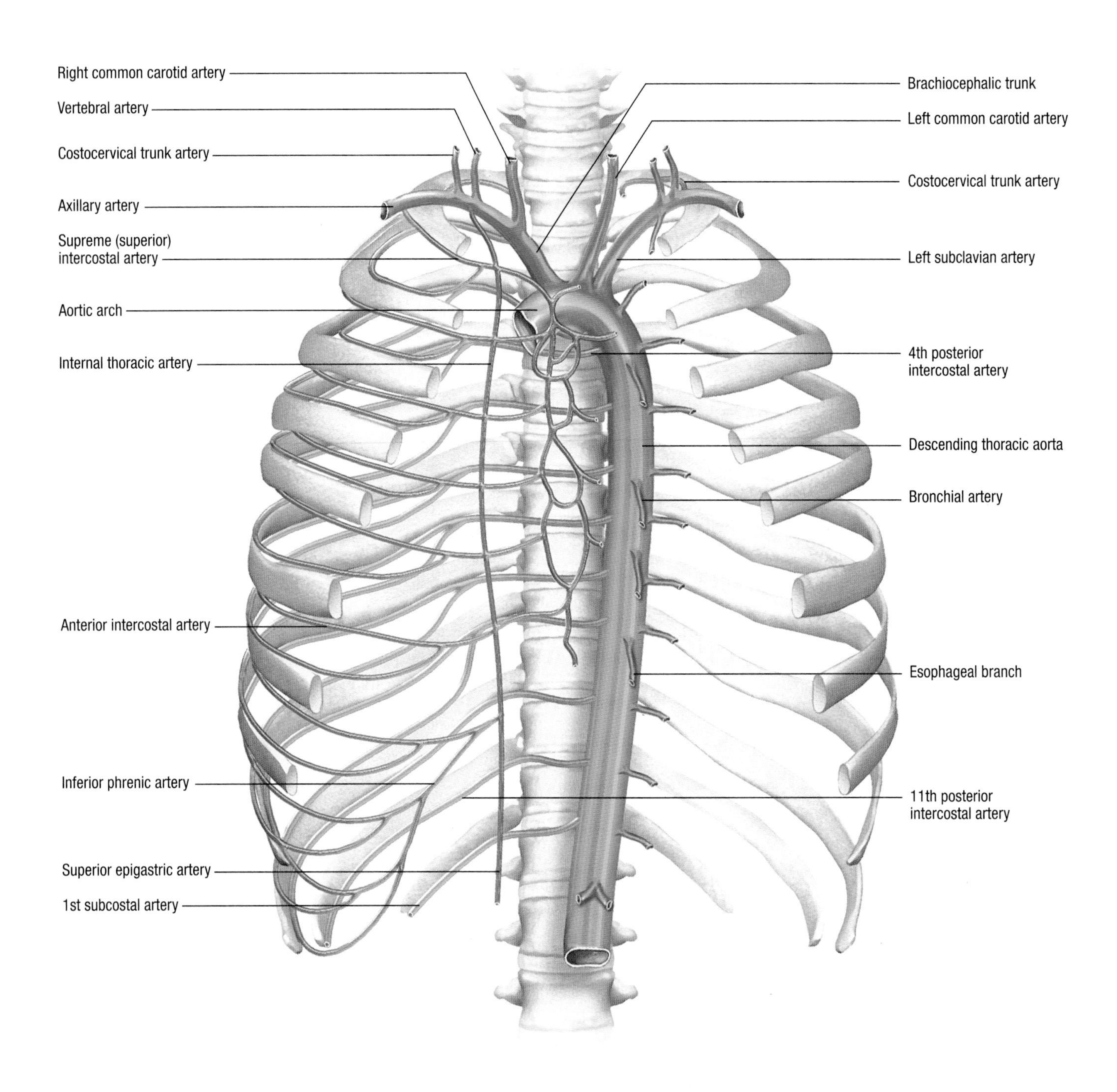

Right common carotid artery
Vertebral artery
Costocervical trunk artery
Axillary artery
Supreme (superior) intercostal artery
Aortic arch
Internal thoracic artery
Anterior intercostal artery
Inferior phrenic artery
Superior epigastric artery
1st subcostal artery
Brachiocephalic trunk
Left common carotid artery
Costocervical trunk artery
Left subclavian artery
4th posterior intercostal artery
Descending thoracic aorta
Bronchial artery
Esophageal branch
11th posterior intercostal artery

## THORACIC WALL VEINS
**Anterior view**

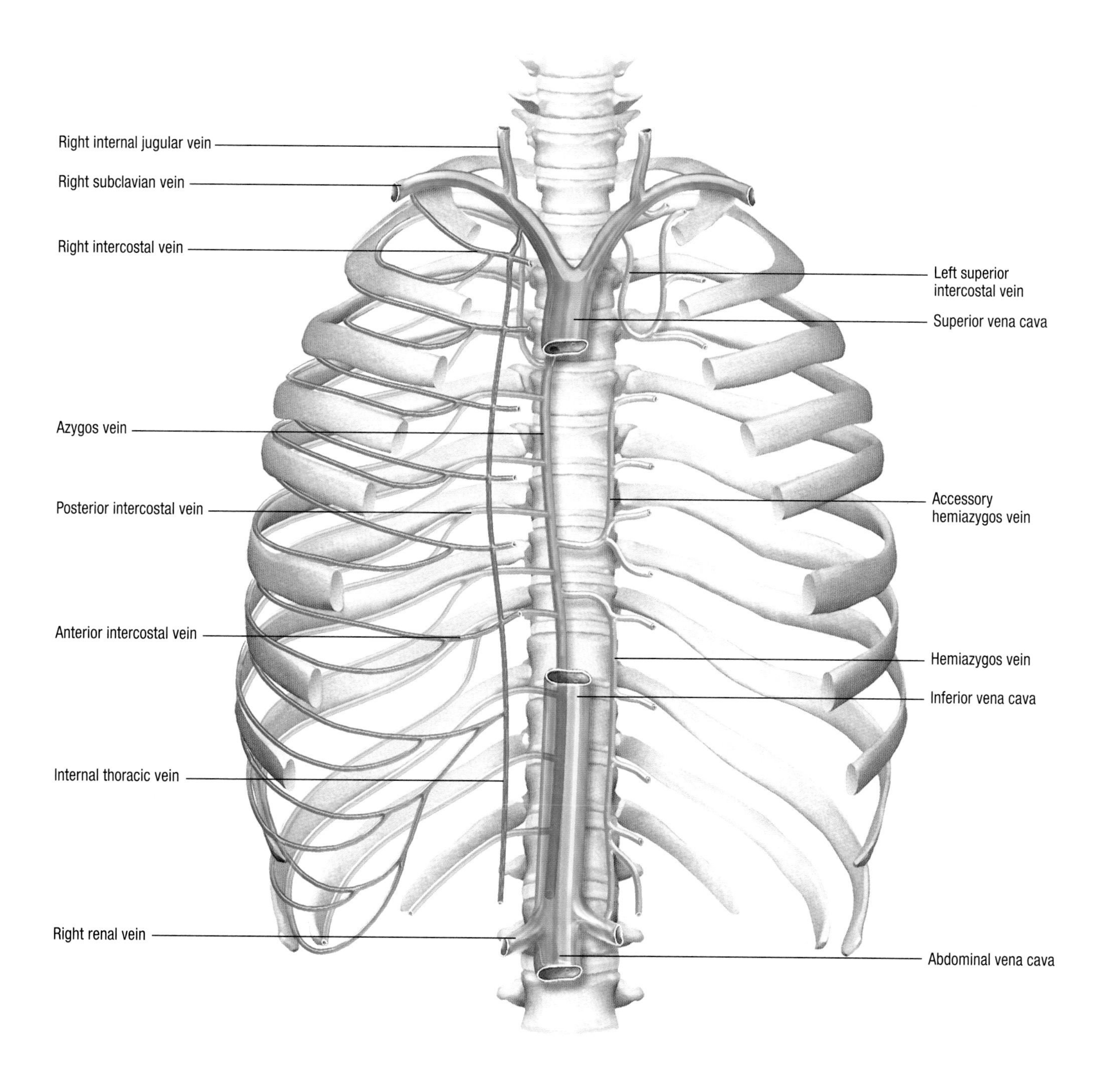

Right internal jugular vein
Right subclavian vein
Right intercostal vein
Left superior intercostal vein
Superior vena cava
Azygos vein
Accessory hemiazygos vein
Posterior intercostal vein
Anterior intercostal vein
Hemiazygos vein
Inferior vena cava
Internal thoracic vein
Right renal vein
Abdominal vena cava

# Heart

The heart is roughly a cone-shaped mass of specialized muscles and nerves that pumps the body's entire volume of blood to and from the lungs (right ventricle and left atrium) and to and from all the other organs (left ventricle and right atrium).

About the size of a closed fist, the heart weighs approximately:

- 10½ to 12½ oz (300 to 355 g) in an adult male
- 9 to 10½ oz (255 to 300 g) in an adult female.

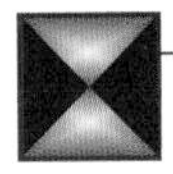

**Age-related changes**

With age, the heart becomes slightly smaller and loses contractile strength, although ventricular wall thickness increases.

## Location

The heart lies beneath the sternum in the mediastinum, between the 2nd and 6th ribs. About one-third of the organ lies to the right of the midsternal line; the remainder, to the left. In most people, the heart rests obliquely, its right side almost in front of the left, the broad part at the top, and the pointed end (apex) at the bottom.

**Clinical tip**

The position of the heart varies with body build. In a tall, thin person, it's more vertical than in a short stocky person. Although not clinically significant, assessment may change slightly. For example, the apical pulse, normally noted in the 4th to 5th intercostal space at the midclavicular line, may shift slightly to the right in a taller, thinner person. Although palpation of apical pulsation may be difficult in a stocky individual, placement is usually unchanged.

## Heart structure

The heart consists of the pericardium, the heart wall, four chambers, and four valves. (See *Heart, right interior view*, page 165, and *Heart, left interior view*, page 166.) For more information on heart valves, see Heart valves, page 168.

### Pericardium

The pericardium is a fibroserous sac that surrounds the heart and the roots of the great vessels.

The serous pericardium, the thin, smooth inner portion, consists of two layers:

- parietal – lines the inside of the fibrous pericardium
- visceral – adheres to the surface of the heart.

The fibrous pericardium, composed of tough, white fibrous tissue, fits loosely around the heart, protecting the heart and serous membrane.

The pericardial space between the two layers contains a few drops of pericardial fluid, which lubricates the surfaces of the space and allows the heart to move easily during contraction.

### Heart wall

The wall of the heart consists of three layers – epicardium, myocardium, and endocardium. (See *Layers of heart muscle*, page 167.)

- The epicardium is the outer layer of the heart wall and the visceral layer of the pericardium.
- The myocardium is the thickest layer of the heart wall. It is composed of interlacing bundles of cardiac muscle fibers that cause the heart to contract.
- The endocardium is the innermost lining of the heart, composed of endothelial tissue with small blood vessels and bundles of smooth muscle. It lines the myocardium and covers the valves of the heart.

Irreglar bundles of muscle that project on the inner surface are called trabeculae carneae. There are three types: ridges along the wall, extensions across the ventricular cavity, and papillary muscles.

### Heart chambers

The heart contains four hollow chambers: two atria and two ventricles.

#### Atria

The atria, the upper chambers, are separated by the interatrial septum. They receive blood returning to the heart and pump blood only to the ventricles.

- Left atrium: forms the uppermost part of the heart's left border, extending to the left of and behind the right atrium. It's smaller but has thicker walls than the right atrium
- Right atrium: in front and to the right of the left atrium. It receives blood from the superior and inferior venae cavae.

#### Ventricles

The right and left ventricles are composed of highly developed musculature. They are larger and thicker-walled than the atria. (See *Cardiac muscle*, page 167.) The ventricles receive blood from the atria.

- Right ventricle, pumps blood to the lungs.
- Left ventricle, larger than the right ventricle, pumps blood through the systemic circulation.

## HEART
### Anterior view

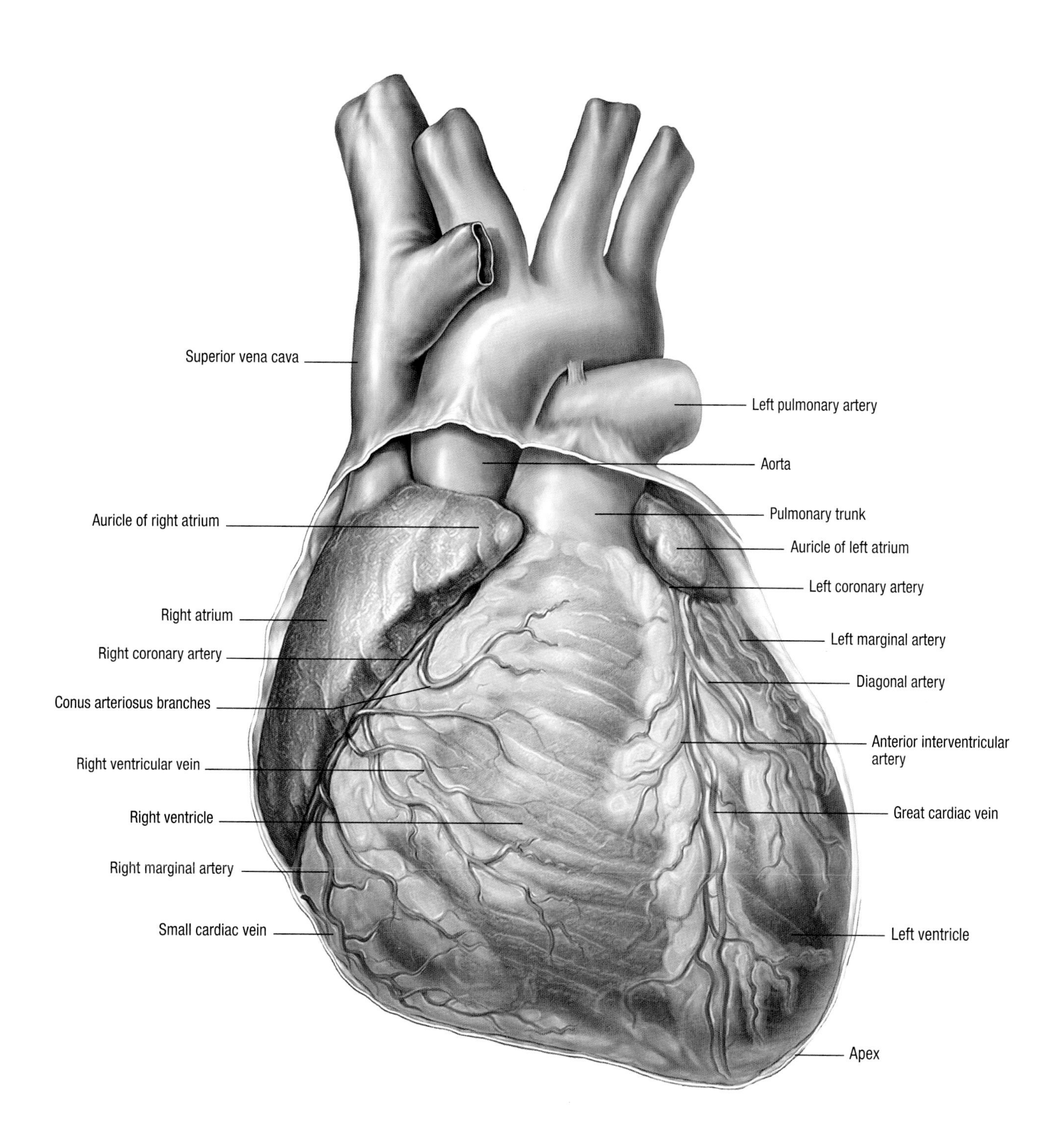

Superior vena cava
Left pulmonary artery
Aorta
Pulmonary trunk
Auricle of right atrium
Auricle of left atrium
Left coronary artery
Right atrium
Left marginal artery
Right coronary artery
Diagonal artery
Conus arteriosus branches
Anterior interventricular artery
Right ventricular vein
Great cardiac vein
Right ventricle
Right marginal artery
Small cardiac vein
Left ventricle
Apex

## HEART
**Posterior view**

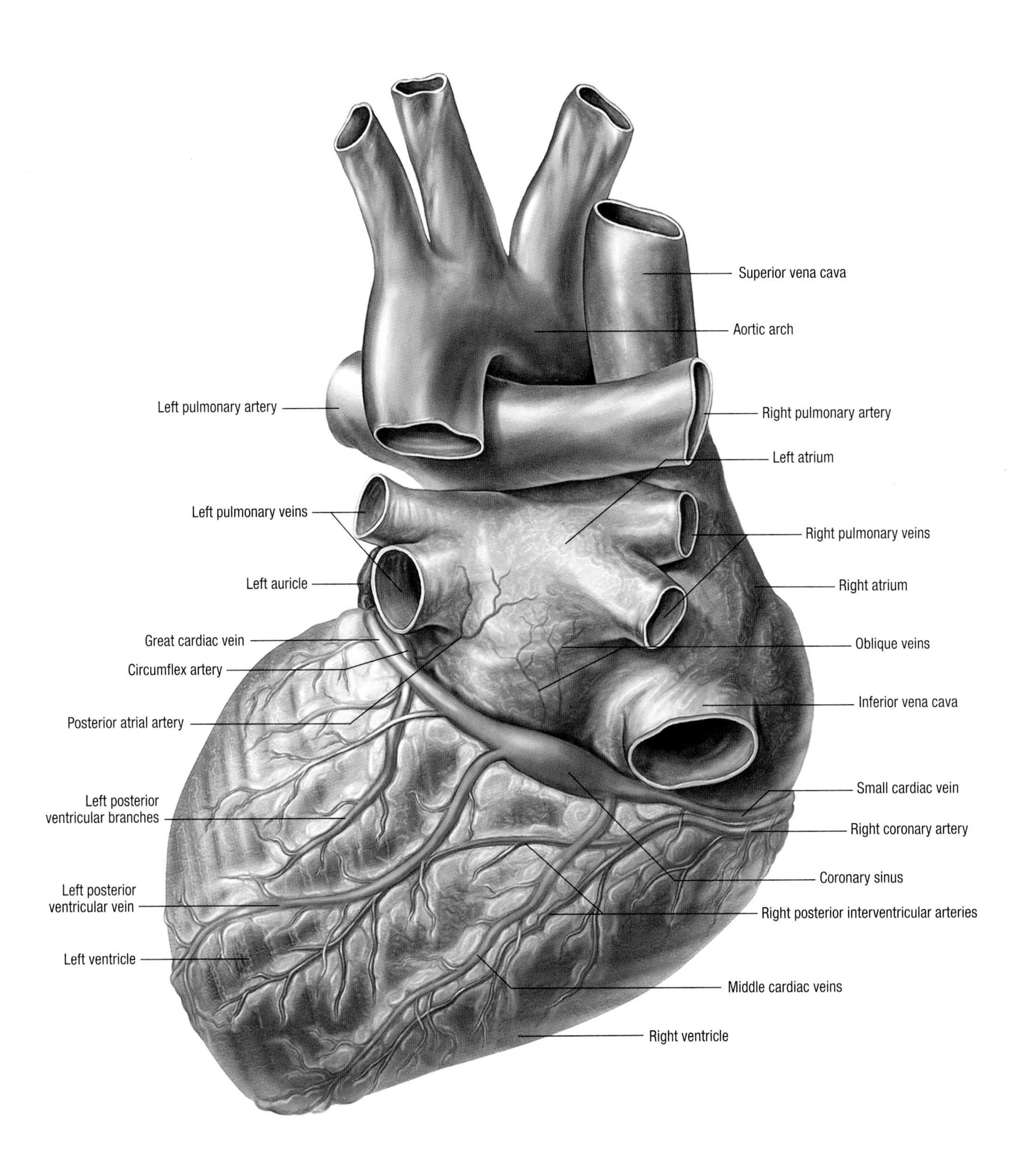

Superior vena cava
Aortic arch
Left pulmonary artery
Right pulmonary artery
Left atrium
Left pulmonary veins
Right pulmonary veins
Left auricle
Right atrium
Great cardiac vein
Oblique veins
Circumflex artery
Inferior vena cava
Posterior atrial artery
Small cardiac vein
Left posterior
ventricular branches
Right coronary artery
Coronary sinus
Left posterior
ventricular vein
Right posterior interventricular arteries
Left ventricle
Middle cardiac veins
Right ventricle

## HEART
### Right interior view

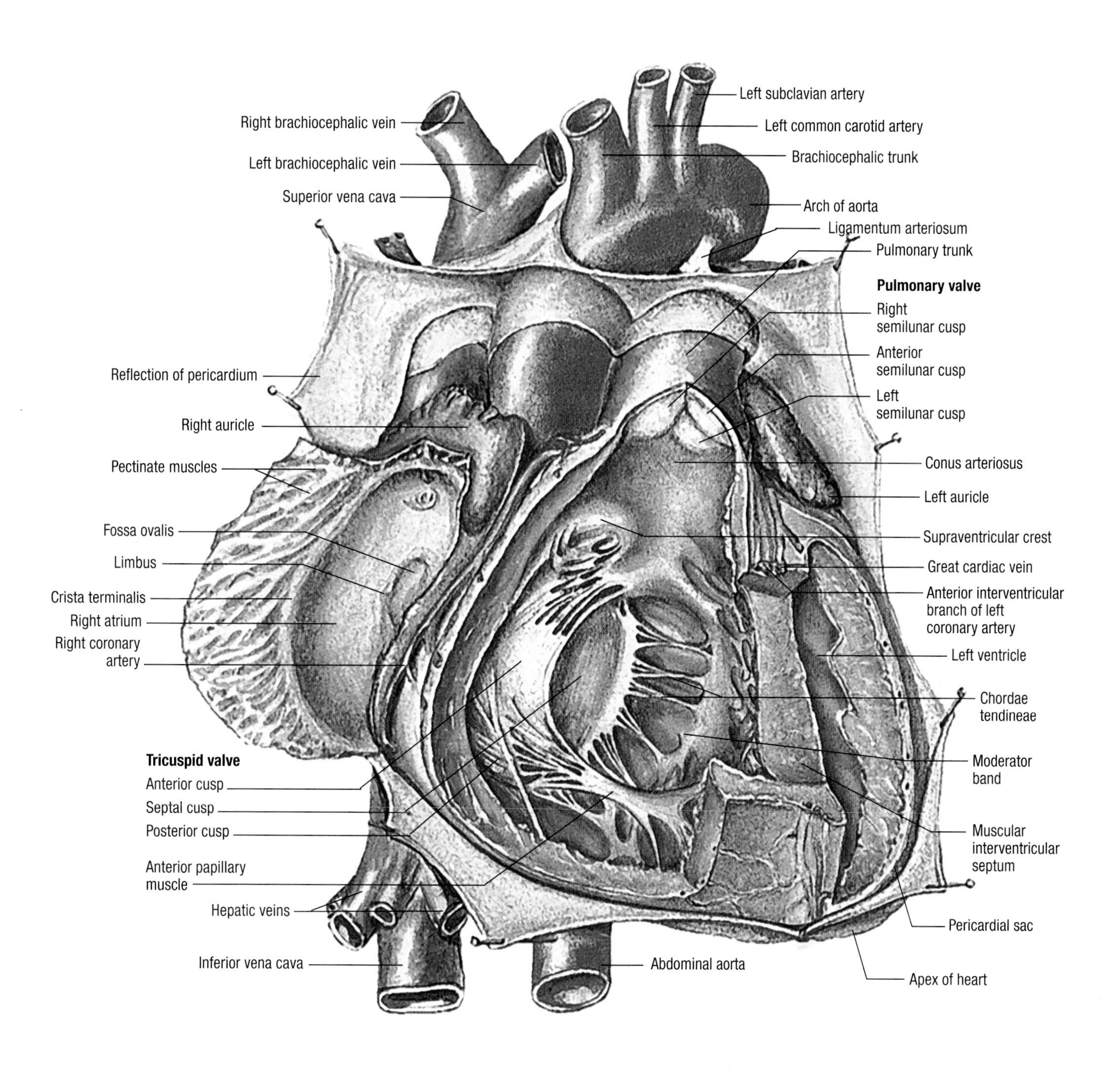

Left subclavian artery
Right brachiocephalic vein
Left common carotid artery
Brachiocephalic trunk
Left brachiocephalic vein
Superior vena cava
Arch of aorta
Ligamentum arteriosum
Pulmonary trunk
Pulmonary valve
Right semilunar cusp
Anterior semilunar cusp
Left semilunar cusp
Reflection of pericardium
Right auricle
Pectinate muscles
Conus arteriosus
Left auricle
Fossa ovalis
Supraventricular crest
Limbus
Great cardiac vein
Crista terminalis
Anterior interventricular branch of left coronary artery
Right atrium
Right coronary artery
Left ventricle
Chordae tendineae
Tricuspid valve
Anterior cusp
Septal cusp
Posterior cusp
Moderator band
Muscular interventricular septum
Anterior papillary muscle
Hepatic veins
Pericardial sac
Inferior vena cava
Abdominal aorta
Apex of heart

# HEART
## Left interior view

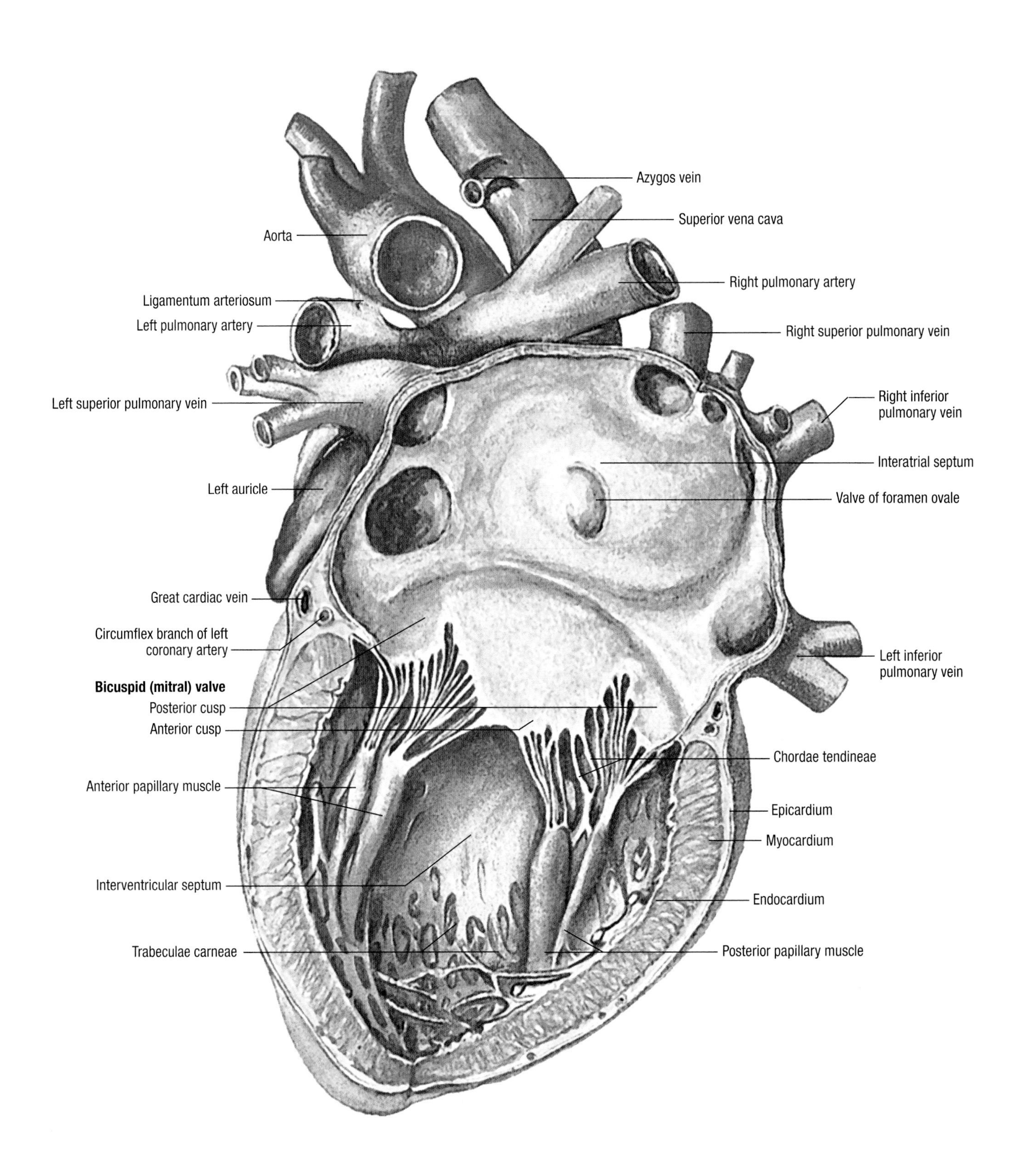

Azygos vein
Superior vena cava
Aorta
Right pulmonary artery
Ligamentum arteriosum
Left pulmonary artery
Right superior pulmonary vein
Left superior pulmonary vein
Right inferior pulmonary vein
Interatrial septum
Left auricle
Valve of foramen ovale
Great cardiac vein
Circumflex branch of left coronary artery
Left inferior pulmonary vein
Bicuspid (mitral) valve
Posterior cusp
Anterior cusp
Chordae tendineae
Anterior papillary muscle
Epicardium
Myocardium
Interventricular septum
Endocardium
Trabeculae carneae
Posterior papillary muscle

## LAYERS OF HEART MUSCLE

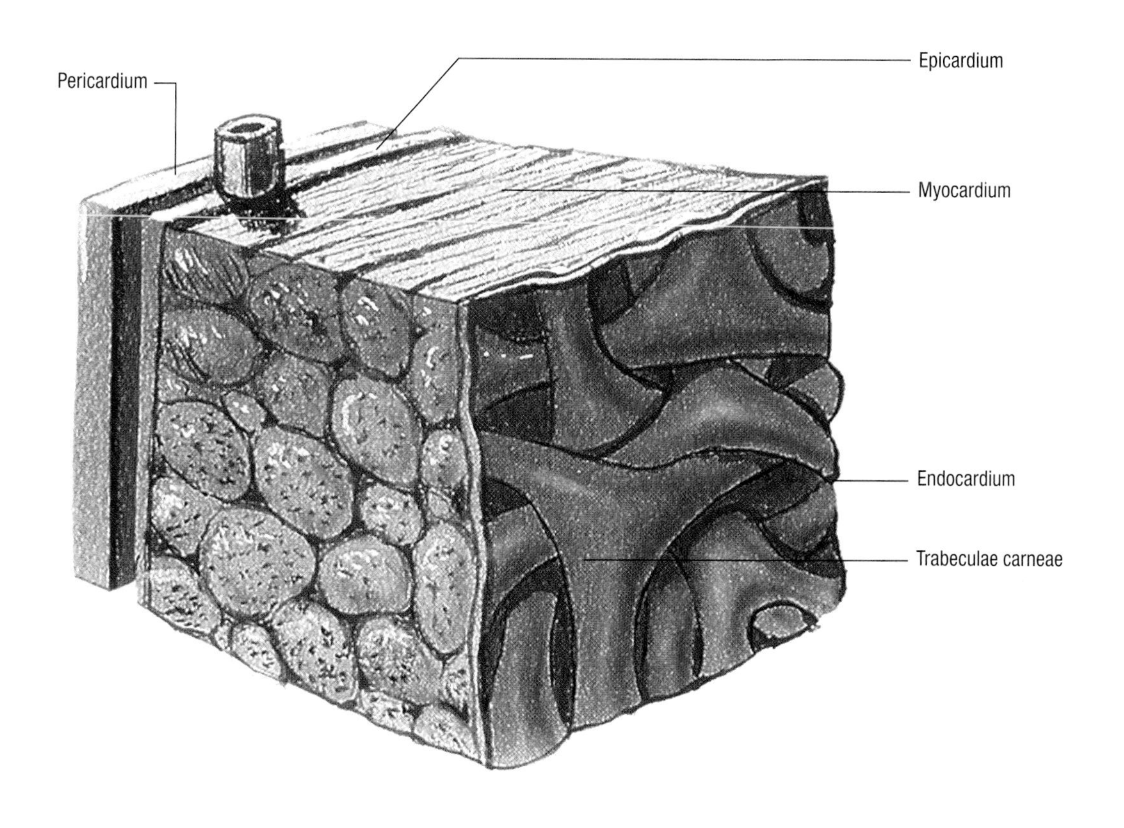
Pericardium
Epicardium
Myocardium
Endocardium
Trabeculae carneae

## CARDIAC MUSCLE

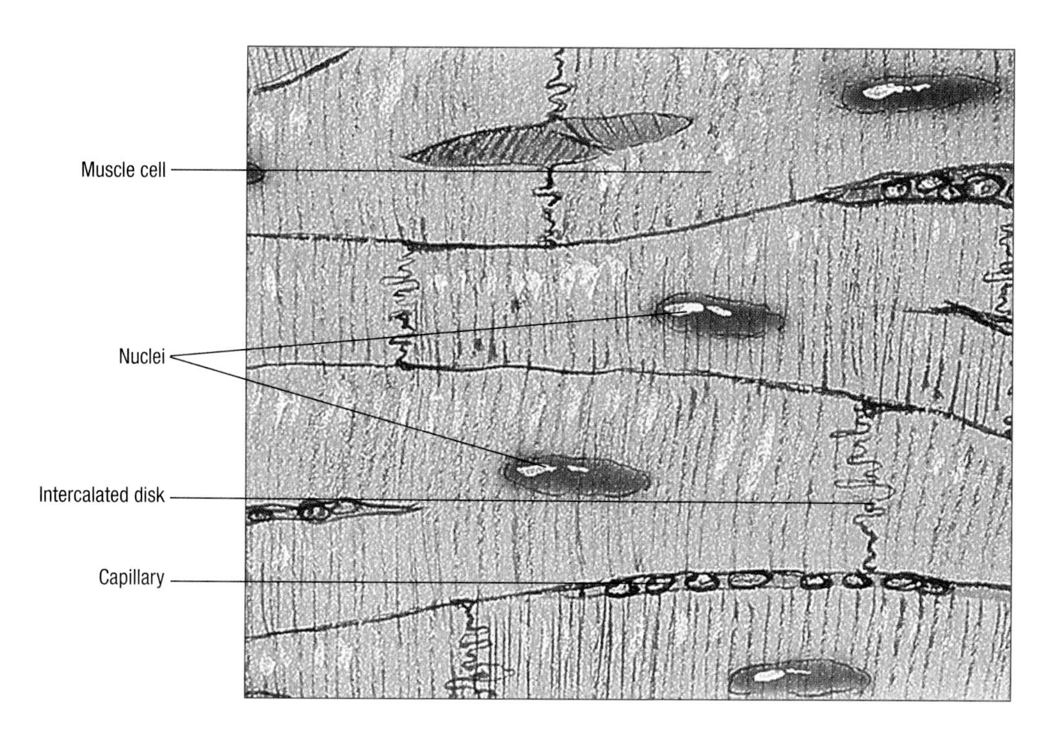
Muscle cell
Nuclei
Intercalated disk
Capillary

# Heart valves

Four valves keep blood flowing in one direction through the heart: two atrioventricular (AV) valves – bicuspid (mitral) and tricuspid – and two semilunar valves – aortic and pulmonary.

## Atrioventricular valves

The AV valves separate the atria from the ventricles. Each AV valve allows blood to flow only from an atrium into a ventricle. Chordae tendineae, thin tendinous strings that extend from the cusps of the atrioventricular valves to the papillary muscles of the heart, keep the valve from inverting and allowing retrograde blood flow.

The right AV valve is also called the tricuspid valve because it has three triangular cusps, or leaflets. The left AV valve, which has two cusps, is called the bicuspid or mitral valve.

## Semilunar valves

Each semilunar valve – the pulmonary valve and the aortic valve – has three cusps shaped like half-moons. Both valves open and close in response to pressure changes brought on by ventricular contraction and blood ejection. They permit one-way blood flow from the heart into the artery.

The pulmonary valve guards the opening between the right ventricle and the pulmonary artery. The aortic valve guards the opening between the left ventricle and the aorta.

## Cardiac circulation

Blood from systemic circulation enters the right atrium via the superior and inferior venae cavae and coronary sinus. Passing through the right AV (tricuspid) valve, it flows from the right atrium into the right ventricle. It exits through the pulmonary (semilunar) valve and enters into pulmonary circulation.

Blood returns from the pulmonary circuit to the left atrium. As it passes through the left AV (bicuspid or mitral) valve, it flows from the left atrium into the left ventricle. Blood exits the left ventricle via the aortic (semilunar) valve and enters systemic circulation. (See *Cardiac circulation*, page 173.)

## Valvular dysfunction

Common valvular dysfunctions include:

- Regurgitation (insufficiency): permits backward flow of blood; may be caused by any disease of the heart or weakening of the heart muscles.
- Stenosis: narrowing or constriction of the valve; usually related to atherosclerosis or calcification.
- Prolapse: drooping of the cusps of the valve (most common in the mitral valve); may be congenital, idiopathic, or secondary to some other condition.
- Endocarditis: inflammation of the valve lining; usually caused by a microorganism or an autoimmune reaction.

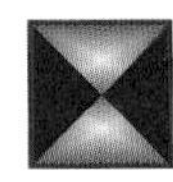

**Age-related changes**

**Fibrotic and sclerotic changes thicken and stiffen heart valves. Consequent rigidity and incomplete closure may cause systolic murmurs.**

**Clinical Tip**

### Assessing heart murmurs

A murmur is a heart sound created by vibrations from turbulent blood flow. Turbulence can occur when blood flows through a partially obstructed opening or from a smaller chamber to a large one.

| MURMUR | LOCATION | PITCH/QUALITY | ASSESSMENT TIPS |
|---|---|---|---|
| Systolic ejection | Left sternal border | High/harsh | • Use diaphragm.<br>• Have patient cough or exercise briefly.<br>• Murmur is normal in children and in thin and hypertensive patients. |
| Aortic valvular stenosis | Base, apex, Erb's point | Medium/harsh | • Use diaphragm or bell.<br>• Thrill may be palpable. |
| Tricuspid regurgitation | Tricuspid area | High/scratchy or blowing | • Use diaphragm.<br>• Have patient breathe deeply and slowly. |
| Mitral stenosis | Apex | Low/rumbling | • Use bell.<br>• Have patient lie in a partially recumbent position. |
| Mitral regurgitation | Apex | Medium high/blowing | • Use diaphragm.<br>• Auscultate after exercise. |
| Aortic regurgitation | Base or apex | High/blowing, musical | • Use diaphragm.<br>• Have patient lean forward and hold breath after exhaling. |

## HEART VALVES

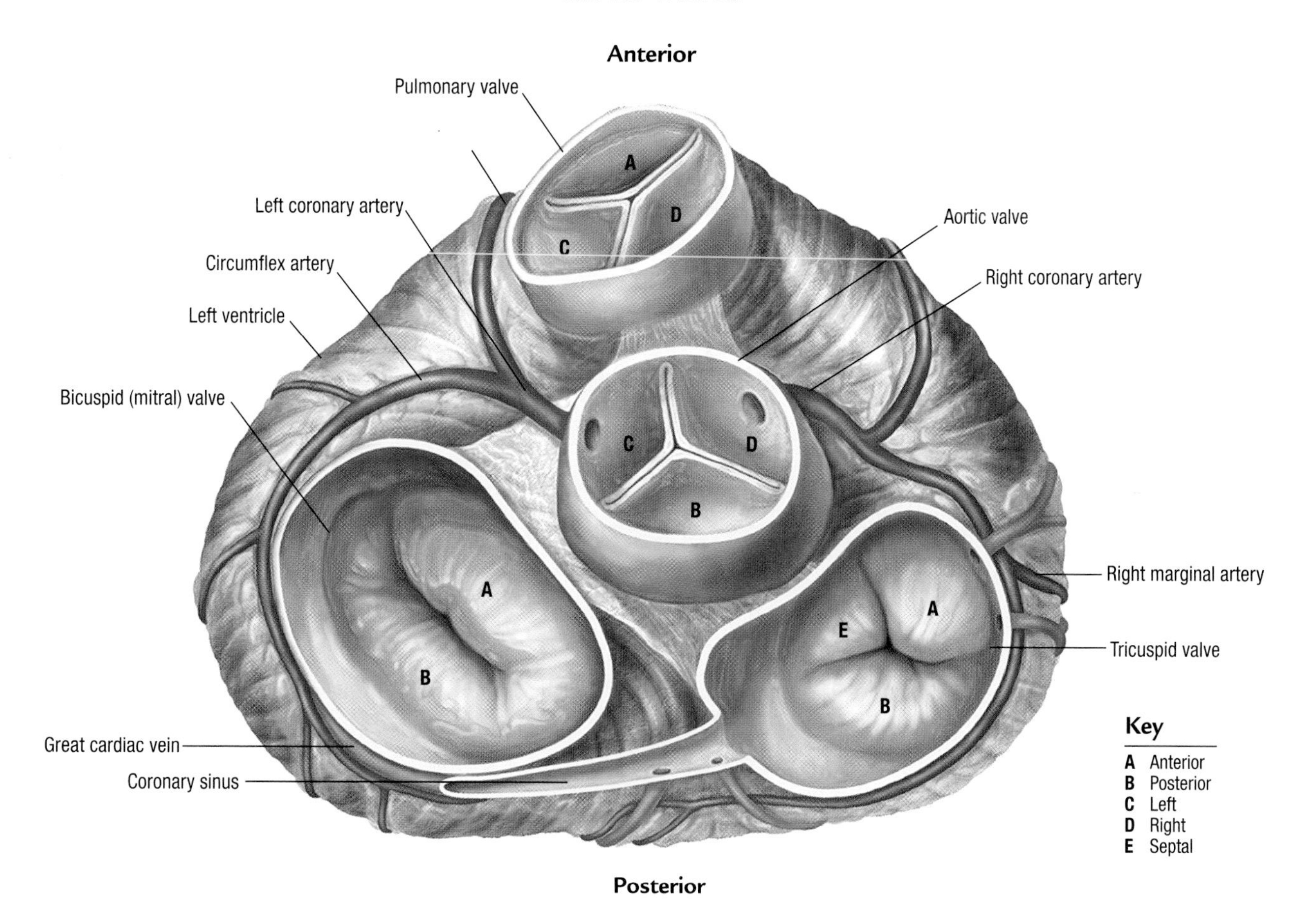
Anterior
Pulmonary valve
Left coronary artery
Circumflex artery
Left ventricle
Bicuspid (mitral) valve
Aortic valve
Right coronary artery
Right marginal artery
Tricuspid valve
Great cardiac vein
Coronary sinus
A
B
C
D
E
Key
A Anterior
B Posterior
C Left
D Right
E Septal
Posterior

## ATRIOVENTRICULAR VALVES

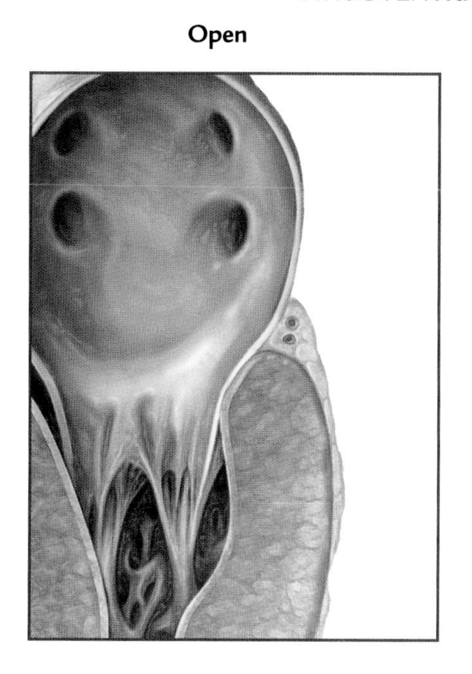
Open

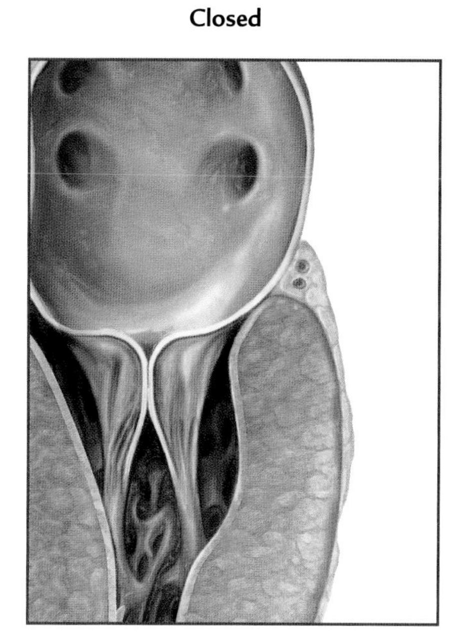
Closed

# Coronary circulation

The heart has its own blood supply, the coronary arteries, which deliver oxygen and nutrients to the heart tissues.

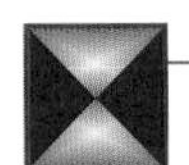

**AGE-RELATED CHANGES**

Coronary artery blood flow diminishes by 35% between ages 20 and 60.

## ARTERIAL BLOOD SUPPLY

The major blood vessels of the heart are the left and right coronary arteries, which branch from the base of the aorta.

The wall of each coronary artery has three layers:

- tunica adventitia – the outermost supportive layer
- tunica media – a thin, fibrous layer
- tunica intima – endothelial lining. The tunica intima secretes chemical mediators that regulate the activity and integrity of the coronary vasculature. Injury to this layer can initiate atherosclerotic plaque formation.

### Left coronary artery

The left coronary artery branches to form the anterior interventricular artery (left anterior descending branch) and the circumflex artery. Blockage or other damage to either of these vessels often carries a poor prognosis.

Each coronary artery supplies a specific area of cardiac muscle, as follows:

- Anterior interventricular artery – the anterior wall of the left ventricle, the anterior interventricular septum, the bundle branches, and portions of the right ventricle. In some persons, a lateral diagonal branch supplies the anterior surface of the heart.
- Circumflex artery – lateral and posterior portions of the left ventricle and the left atrium. One branch, the left marginal artery, also supplies the left ventricle.

### Right coronary artery

The right coronary artery branches into the posterior interventricular artery (also called the posterior descending artery) and the marginal artery. Each artery supplies a specific area of cardiac muscle, as follows:

- Right coronary artery – sinoatrial (SA) and atrioventricular nodes, the right atrium and ventricle, and the inferior wall of the left ventricle. (*Note*: In some persons, the left coronary artery supplies the SA node. Less often, a branch of the left coronary artery supplies the atrioventricular node.)
- Posterior interventricular artery – both ventricles and the interventricular septum. (*Note*: In 20% of the population, this vessel arises from the circumflex artery.)
- Marginal artery – the right ventricle and the apex.

## VENOUS DRAINAGE

Extensive capillary beds join coronary arteries with cardiac veins, which empty into the coronary sinus. Most of the deoxygenated blood empties from the coronary sinus into the right atrium.

A vascular sinus is a small, thin-walled vein that lacks smooth muscle and is therefore unable to change its diameter. Two veins drain into the coronary sinus:

- great cardiac vein – drains anterior portion of the heart
- middle cardiac vein – drains posterior portion of the heart.

**CLINICAL TIP**

## CORONARY ARTERY DISEASE

In coronary artery disease, fatty, fibrous plaques, including calcium deposits, progressively narrow the coronary artery lumina, which reduces the volume of blood that can flow through them. Atherosclerosis is the most common cause of coronary artery disease (CAD). This can lead to myocardial ischemia.

Review controllable risk factors (below) with your patient and discuss possible options to decrease individual risks.

- Hypertension
- High low-density and low high-density lipoproteins
- Smoking
- Stress
- Obesity
- Inactivity

The illustrations below depict the buildup of atherosclerosis.

**Normal coronary artery**

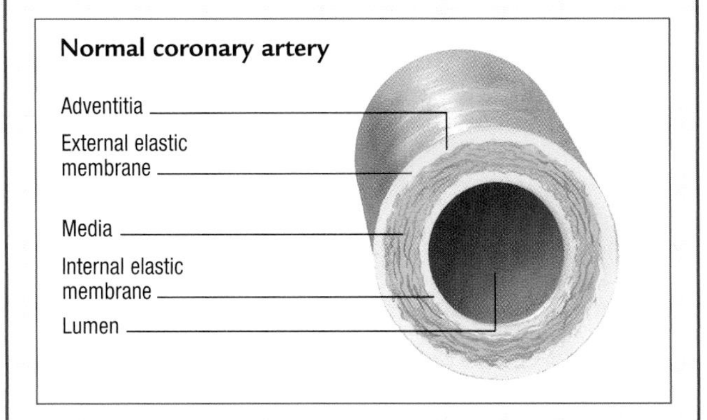
Adventitia
External elastic membrane
Media
Internal elastic membrane
Lumen

**Fibrous plaque**

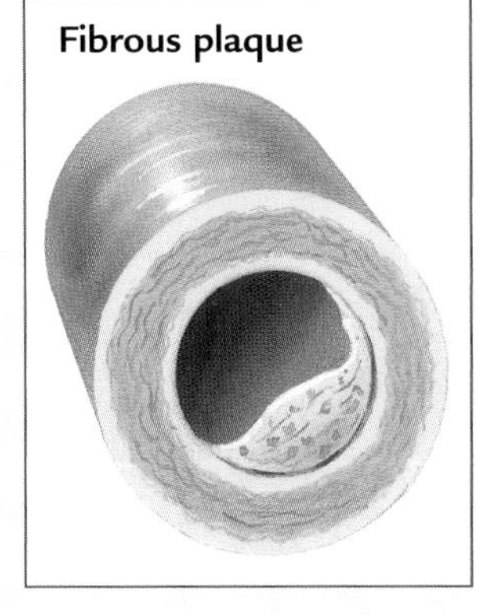

**Complicated plaque**

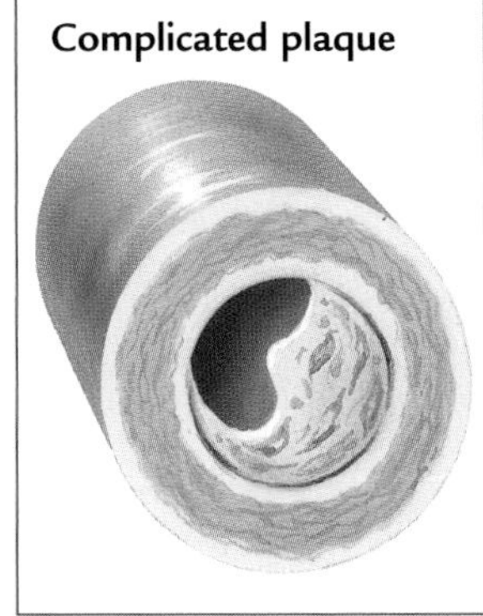

## CORONARY ARTERIES
### Anterior view

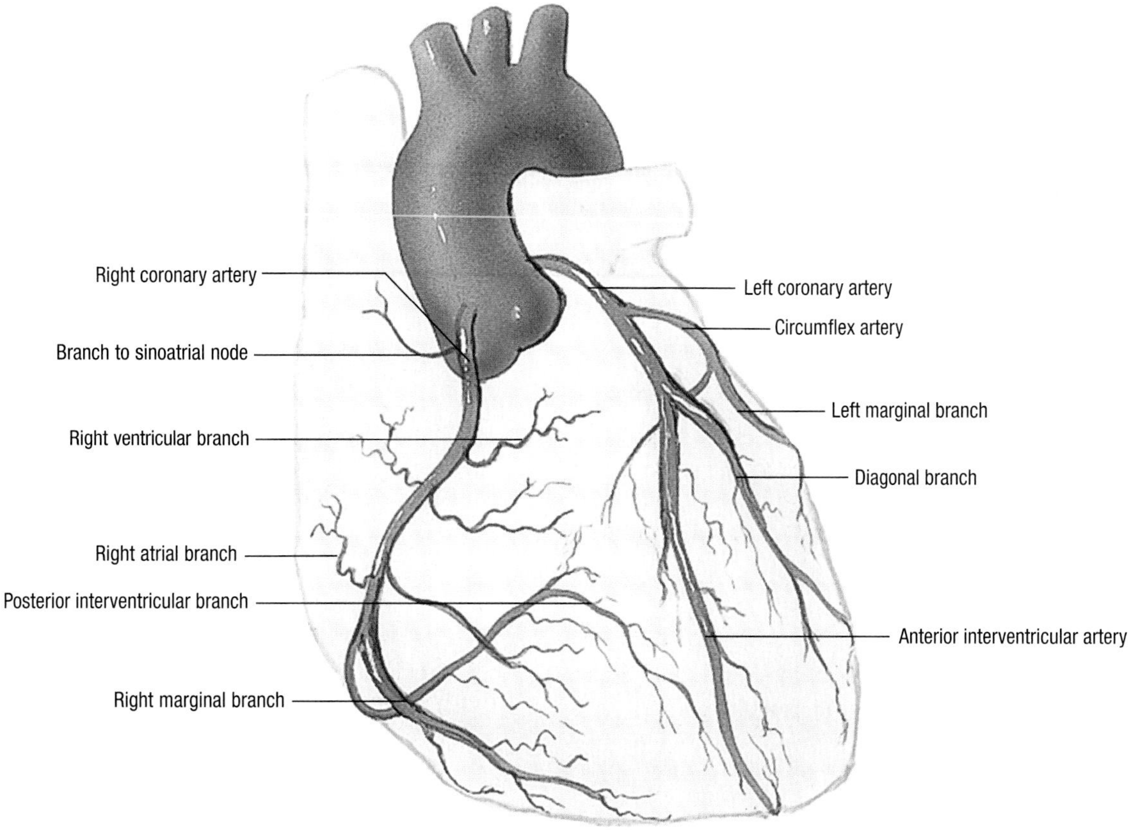

Right coronary artery
Branch to sinoatrial node
Right ventricular branch
Right atrial branch
Posterior interventricular branch
Right marginal branch
Left coronary artery
Circumflex artery
Left marginal branch
Diagonal branch
Anterior interventricular artery

## CORONARY VEINS
### Posterior view

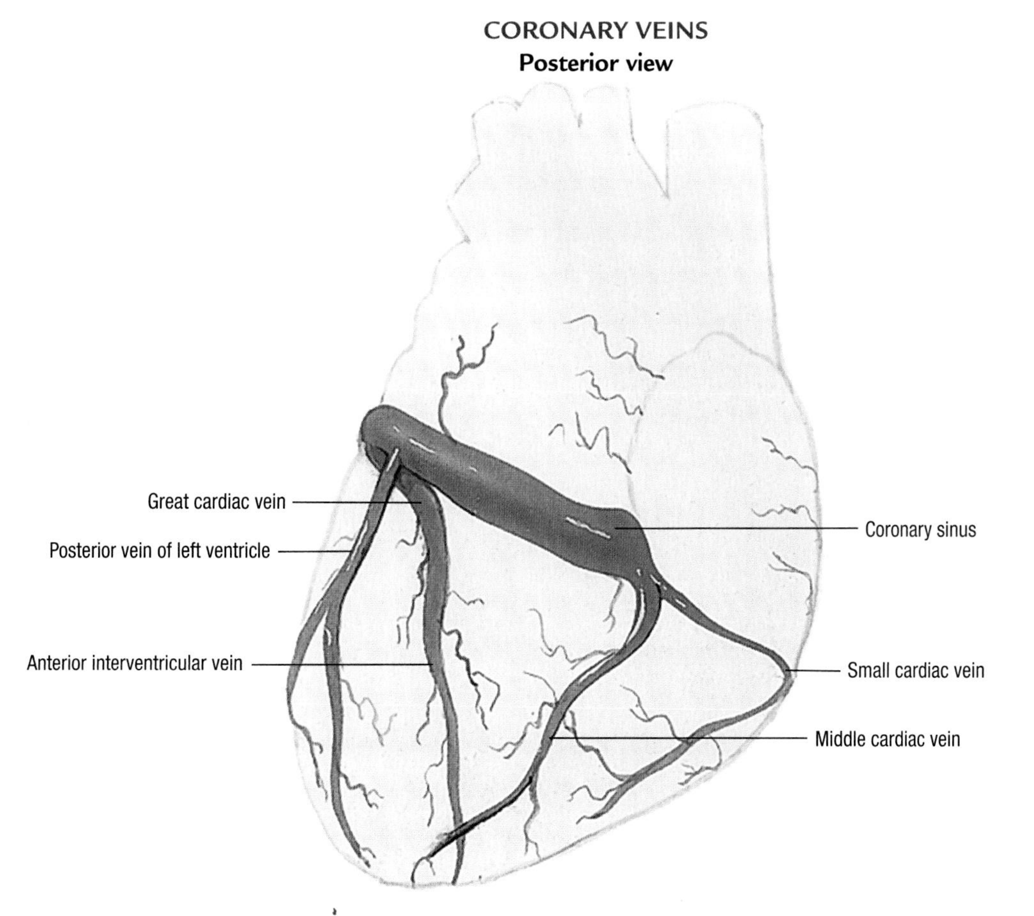
Great cardiac vein
Posterior vein of left ventricle
Anterior interventricular vein
Coronary sinus
Small cardiac vein
Middle cardiac vein

# Cardiac cycle

The sequence of electrical and mechanical events leading to the contraction is known as the *cardiac* cycle. The heart is the only organ that causes itself to contract through intrinsic mechanisms. These mechanisms are driven by the sinoatrial node, the atrioventricular node, the bundle of His, and Purkinje fibers.

Mechanical events correlate with auscultation of heart sounds. The terms systole and diastole describe the filling and emptying phases of the heart cycle.

## Electrical events

An electrical conduction system regulates myocardial contractions. This system includes the nerve fibers of the autonomic nervous system and specialized nerves and fibers in the heart.

### Autonomic innervation

The autonomic nervous system increases or decreases heart activity to meet the body's metabolic needs. The autonomic system does not initiate cardiac contraction; rather, it regulates the length of the cardiac cycle. See Part IX, Body systems, for more about the autonomic nervous system.

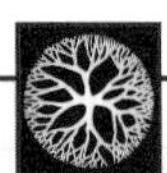

**Physiology**

#### Systole and diastole

**Systole**

At the beginning of systole, the ventricles contract. The rising pressure in the ventricles closes the mitral and tricuspid valves to keep blood from flowing backward into the atria.

When ventricular pressure exceeds the pressure in the pulmonary artery and the aorta, the aortic and pulmonary semilunar valves open, and blood flows into the aorta and the pulmonary artery. Systole is auscultated after the first heart sound ($S_1$), which correlates with closure of the mitral and tricuspid valves.

**Diastole**

When the ventricles empty and relax, ventricular pressure falls below that in the pulmonary artery and the aorta. At the beginning of diastole, the semilunar valves close to prevent the backflow of blood into the ventricles. The mitral and tricuspid valves open, and blood starts to flow into the ventricles from the atria. When atrial and ventricular pressures become nearly equal, the atria contract to send the remaining blood to the ventricles. Diastole is auscultated after the second heart sound ($S_2$), which correlates with closure of the aortic and pulmonary valves.

Both sympathetic and parasympathetic nerves participate in the control of cardiac function. With the body at rest, the parasympathetic nervous system controls the heart through branches of the vagus nerve (cranial nerve X); heart rate and electrical impulse propagation are very slow. In times of activity or stress, the sympathetic nervous system takes control. It increases heart rate and force of contraction.

### Pacemakers

Specialized pacemaker cells in the myocardium control the heart rate and rhythm (a property known as automaticity). However, under certain circumstances, any myocardial muscle cell can control the rate and rhythm of contractions. The sinoatrial and atrioventricular nodes are clusters of pacemaker cells that serve different complementary functions.

#### Sinoatrial node

Located on the endocardial surface of the right atrium, near the superior vena cava, the SA node is the dominant pacemaker. Rhythmic impulses originating in the SA node spread through the right and left atria, initiating atrial contraction.

#### Atrioventricular node

This cluster of cells in the septal wall of the right atrium provides the only electrical connection between the atria and the ventricles. Between contractions, the AV node slows conduction, delaying ventricular activity while blood flows from the atria. Impulses travel through the AV node and a network of fibers called the bundle of His (atrioventricular bundle) to the Purkinje fibers.

The bundle of His arises in the AV node and continues along the right interventricular septum, where it divides to form the right and left bundle branches. Purkinje fibers from the distal portions of the left and right bundle branches fan across the subendocardial surface of the ventricles, from the endocardium through the myocardium, causing the ventricles to contract.

## Mechanical events

The cardiac cycle is the period from the beginning of one heartbeat to the beginning of the next. During this cycle, electrical and mechanical events must occur in the proper sequence and to the proper degree to provide adequate blood flow to all body parts. The cardiac cycle has two phases: systole (contraction and emptying) and diastole (relaxation and filling).

## CARDIAC CONDUCTION

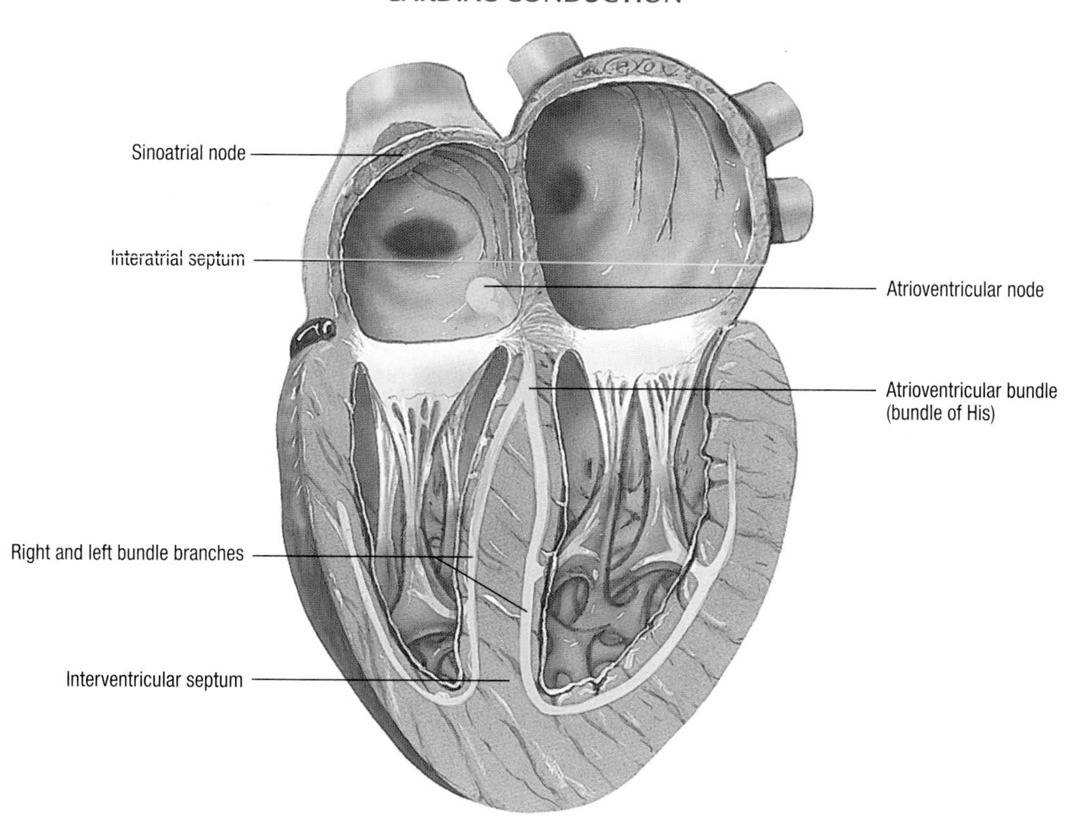

Sinoatrial node
Interatrial septum
Atrioventricular node
Atrioventricular bundle (bundle of His)
Right and left bundle branches
Interventricular septum


## CARDIAC CIRCULATION

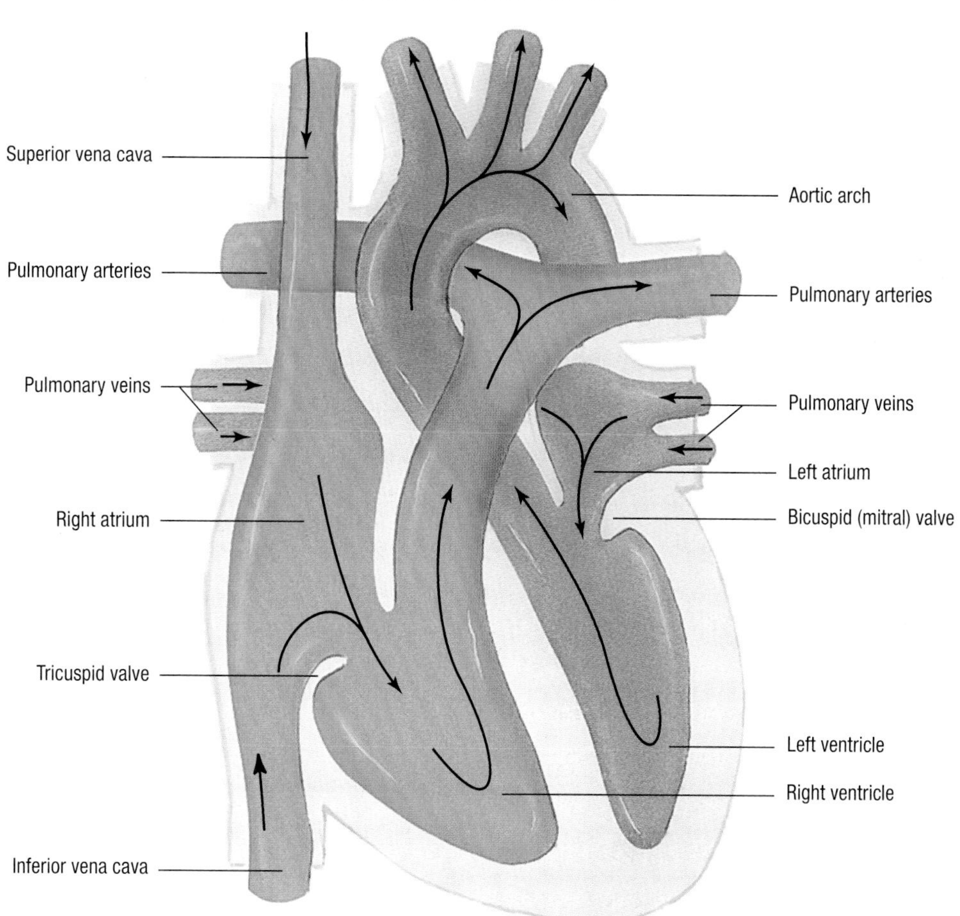

Superior vena cava
Aortic arch
Pulmonary arteries
Pulmonary arteries
Pulmonary veins
Pulmonary veins
Left atrium
Right atrium
Bicuspid (mitral) valve
Tricuspid valve
Left ventricle
Right ventricle
Inferior vena cava

# Trachea and bronchi

The trachea and bronchi permit air movement to the lungs for the purpose of gas exchange. The trachea begins at the larynx, the transition point between the upper and lower airways. During swallowing, the larynx rises to meet the epiglottis, closing the airway and helping prevent the accidental aspiration of food or fluid into the lower airways of the respiratory tract. The trachea and bronchi are lined with respiratory mucosa.

## Trachea

The trachea (windpipe) is a membranous tube that measures 4″ to 5″(10 to 12.5 cm) long, extending from the level of the 6th cervical vertebra to the 5th thoracic vertebra, where it branches into the right and left main bronchi. Dorsally, the trachea contacts the esophagus.

A series of C-shaped rings of hyaline cartilage strengthen the trachea and prevent it from collapsing during inspiration. The trachea is lined with ciliated pseudostratified columnar epithelium and mucus-secreting goblet cells, which trap inhaled debris. Ciliary action moves it toward the pharynx for removal by coughing.

## Bronchi

The main function of the bronchi is to distribute air to the lungs. The bronchi begin at the sternal angle, behind the manubrium at about T5, where the trachea bifurcates into two primary bronchi outside the lungs. At the point of bifurcation, a cartilaginous wedge – the carina – points upward into the trachea. The lumen of the right primary bronchus is larger than that of the left because the carina does not sit symmetrically between the two branches.

The right primary bronchus divides into three secondary bronchi; the left primary bronchus divides into two secondary bronchi. The secondary bronchi branch into tertiary bronchi, which, in turn, branch into increasingly smaller bronchioles.

**Clinical tip**

Breath sounds differ according to the site of auscultation:

- Tracheal – stethoscope over the trachea. Normal sounds are harsh, high-pitched, and discontinuous; occurring on inspiration and expiration.
- Bronchial – stethoscope next to the trachea. Normal sounds are loud, high-pitched, and discontinuous; longer on inspiration than expiration.
- Bronchovesicular – stethoscope next to the upper third of the sternum and between the scapula. Normal sounds are medium-pitched and continuous; heard on inspiration and expiration.
- Vesicular – stethoscope over the remaining areas of the lungs. Normal sounds are soft and low-pitched; prolonged on inspiration and shortened on expiration.

## TRACHEOBRONCHIAL TREE

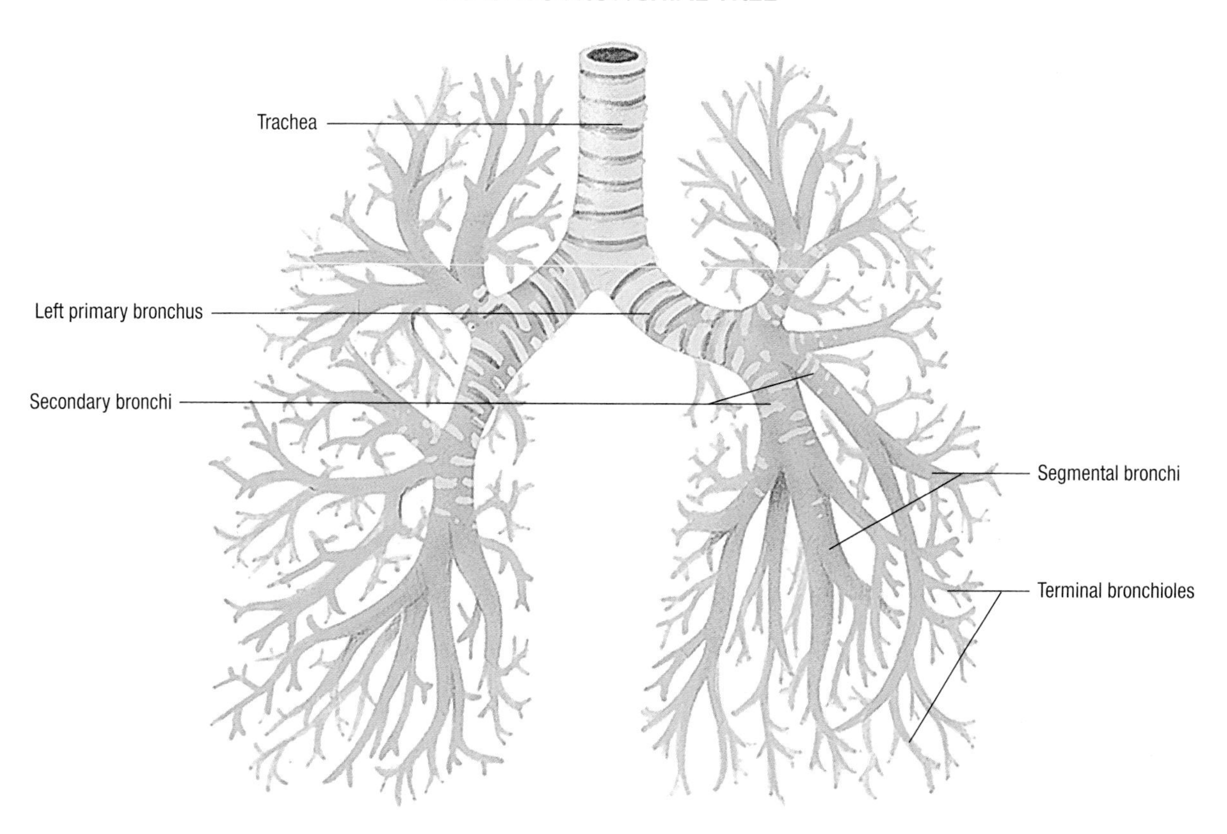
Trachea
Left primary bronchus
Secondary bronchi
Segmental bronchi
Terminal bronchioles

## RESPIRATORY MUCOSA

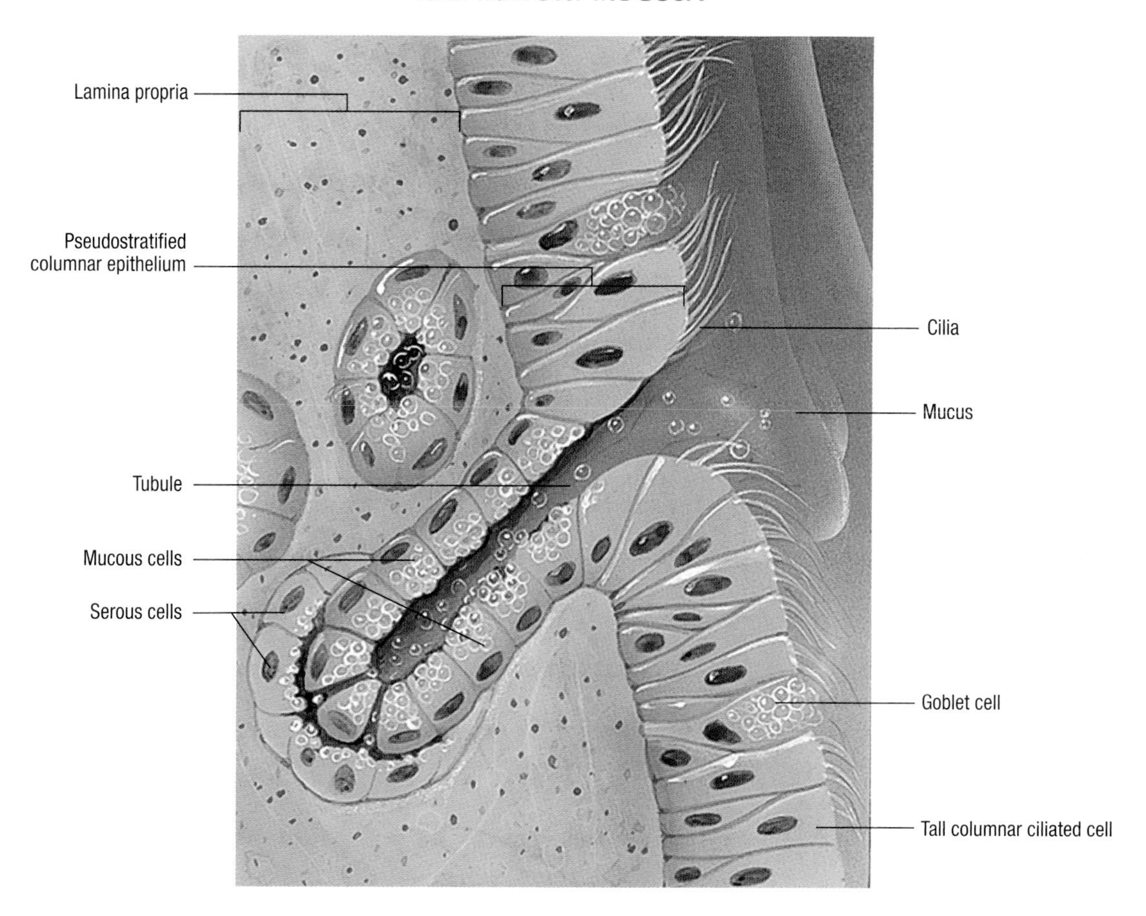
Lamina propria
Pseudostratified columnar epithelium
Cilia
Mucus
Tubule
Mucous cells
Serous cells
Goblet cell
Tall columnar ciliated cell

# Lung

The lungs fill the pleural divisions of the thoracic cavity; they extend from the root of the neck to the diaphragm. The lungs are the main component of the respiratory system; they distribute air and exchange gases.

The right and left lungs are separated by the mediastinum, which contains the heart, blood vessels, and other midline structures; fissures divide each lung into lobes. Each primary bronchus enters its respective lung at the hilus, an indentation on the mediastinal surface. The bronchi and pulmonary blood vessels are bound together by connective tissue to form the root of the lung. The base, the inferior surface of the lung, rests on the diaphragm. The apex, the most superior portion of the lung, projects above the clavicle.

## Pleura

Each lung is enclosed in a pleura, a protective, double-layered serous membrane. The visceral pleura covers the lungs and dips into the fissures. At the hilus, the visceral pleura folds back to form the parietal pleura, which lines the chest wall and covers the diaphragm.

## Pleural cavity

The space between the lungs and chest wall is called the pleural cavity. This space usually contains a small amount of fluid, which lubricates the lungs as they expand and contract. Pleurisy (inflammation of the parietal and visceral pleura) may result from pneumonia or another disorder and may lead to respiratory failure. When fully expanded, the lungs completely fill the pleural cavity, and the parietal and visceral pleurae come in contact.

**Clinical tip**

Suspect a pleural friction rub if you hear a low-pitched, grating, or rubbing sound during inhalation and exhalation when auscultating lung sounds. The rub may be caused by pleural inflammation, which causes two layers of pleura to rub together. Your patient may complain of pain in the area of the rub.

## Lobes of the lung

The right lung is separated into three lobes (superior, middle, and inferior) by the horizontal and oblique fissures. The left lung, smaller than the right, is separated into two lobes (upper and lower) by the oblique fissure; a concavity (cardiac notch) accommodates the heart. (See *Lung lobes*, page 179.)

### Lobe function

Functionally, each lobe is divided into bronchopulmonary segments. The right lung has 10 bronchopulmonary segments; the left lung has 8.

The bronchioles are flexible branches of the bronchi; air travels through them to gas exchange sites in the lungs. (*See Bronchial branches,* page 178.) The diameter of the bronchioles varies with the respiratory phase; it increases slightly during inspiration and decreases slightly during expiration. Unlike bronchi, bronchioles have no cartilage in their walls. Bronchioles branch repeatedly to form respiratory bronchioles, which conduct air and participate in gas exchange, and become progressively smaller to form terminal bronchioles, which conduct air but do not exchange gases. Eventually, terminal bronchioles branch into alveolar ducts, which terminate in clusters called alveoli. (See page 182 for more information on the alveoli.)

## Blood supply

Blood circulates through the lungs via the pulmonary and systemic circulatory systems. In the pulmonary circulation, pulmonary arteries branch profusely into the pulmonary capillaries, which surround the alveoli. Pulmonary arteries carry deoxygenated blood to the lungs, and pulmonary veins return oxygenated blood to the heart. (See *Pulmonary arteries*, page 180, and *Pulmonary veins*, page 181.) The systemic circulation supplies blood directly to lung tissues. This blood travels through the bronchial arteries (one artery on the right and two on the left). Oxygenated blood returns via the bronchial veins.

## Lymphatic system

The lungs contain a network of lymphatic vessels. Some lymphatics course along the interpulmonary bronchi through the pulmonary lymph nodes; others from the visceral pleura drain lymph from the lung surface to the hilum. Lymph from the lung passes through the tracheobronchial and superior tracheobronchial lymph nodes.

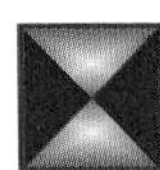

**Age-related changes**

The following changes occur in the lung as a person ages:

- Diffusing capacity declines.
- Loss of inspiratory and expiratory muscle strength diminishes vital capacity.
- Decreasing elastic recoil capability results in an elevated residual volume; thus, aging alone can cause emphysema.
- Closing of some airways produces poor ventilation of the basal areas, resulting in both a decreased surface area for gas exchange and a reduced partial pressure of oxygen. The normal partial pressure of oxygen in arterial blood (80 to 105 mm Hg) falls to 70 to 85 mm Hg; oxygen saturation (normal, 91% to 100%) falls by 5%.
- The lungs become more rigid, and the number and size of alveoli decline.
- Thickened mucus leads to formation of mucus plugs and heightens the risk of pulmonary infection.

## LOCATION OF LUNGS AND PLEURAE

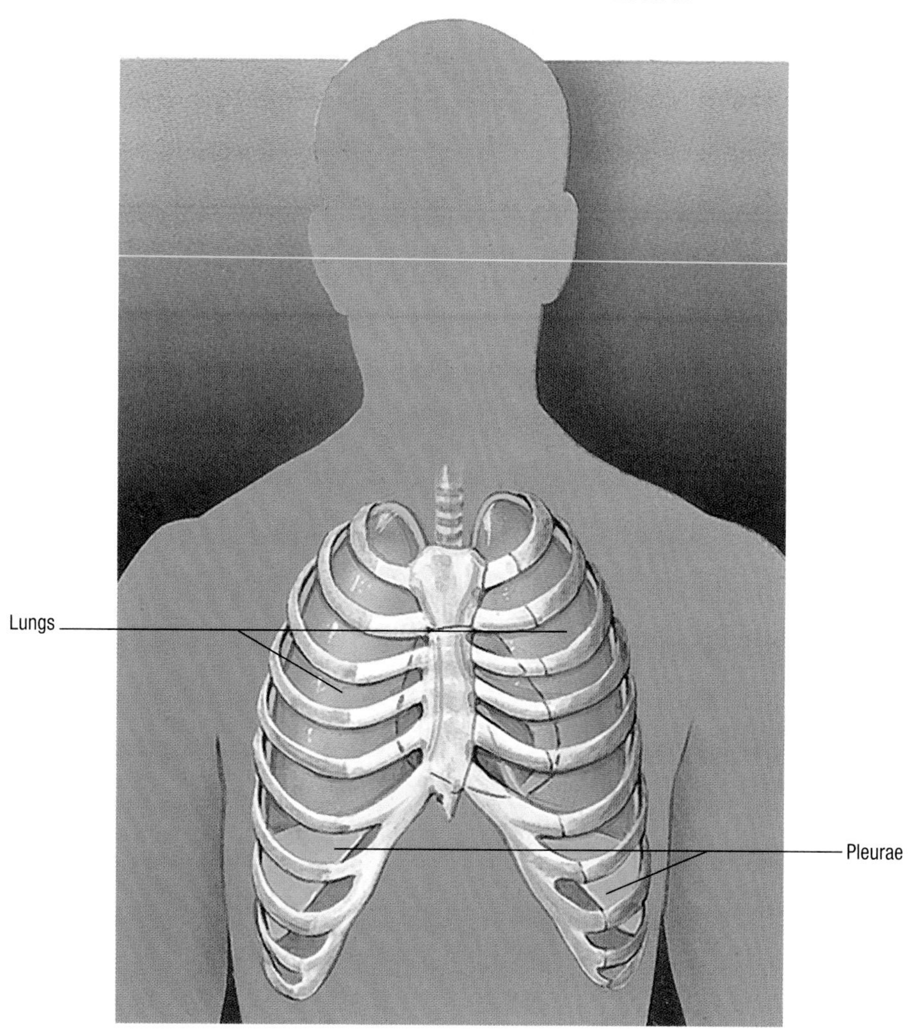

Lungs
Pleurae


## PLEURAE AND CAVITIES

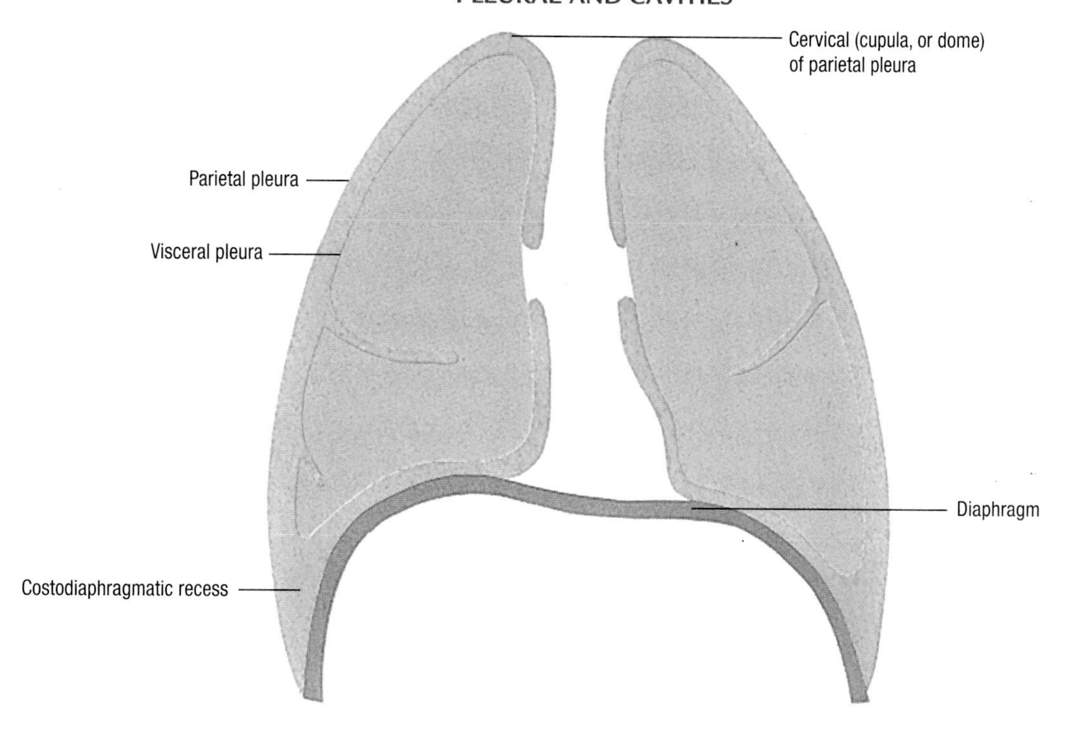

Cervical (cupula, or dome) of parietal pleura
Parietal pleura
Visceral pleura
Diaphragm
Costodiaphragmatic recess

## BRONCHIAL BRANCHES

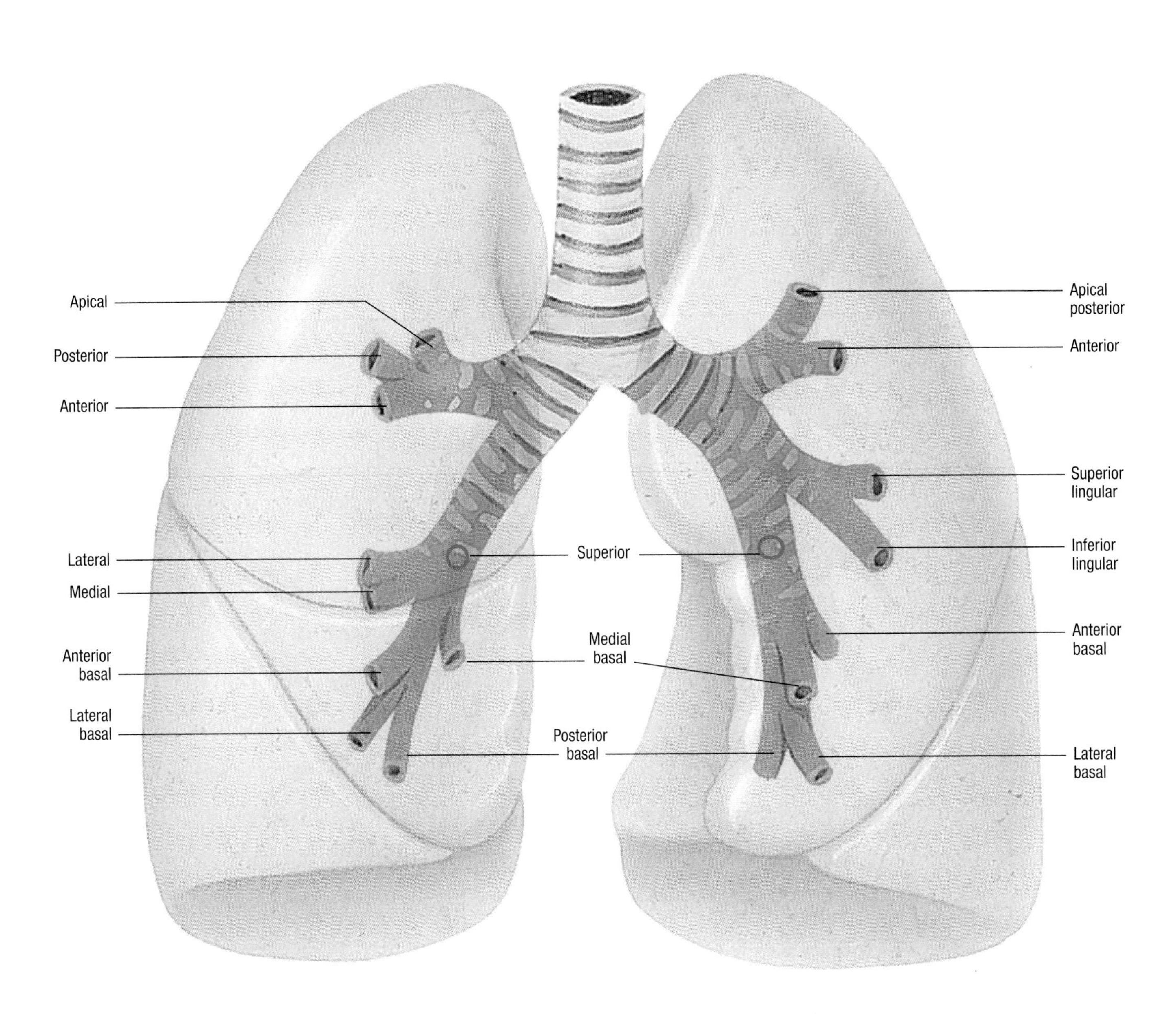

Apical
Posterior
Anterior
Lateral
Medial
Anterior basal
Lateral basal
Superior
Medial basal
Posterior basal
Apical posterior
Anterior
Superior lingular
Inferior lingular
Anterior basal
Lateral basal

## LUNG LOBES

### Anterior view

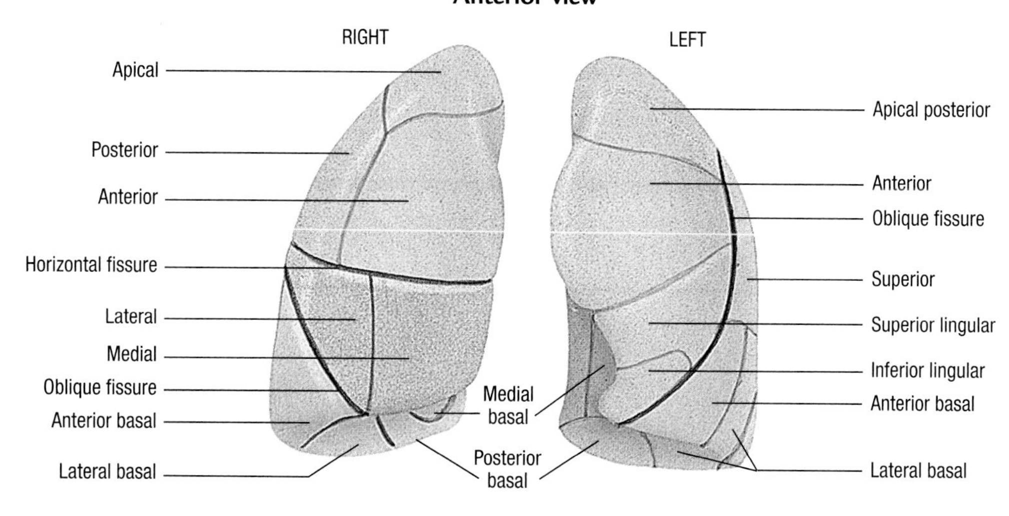

RIGHT
LEFT
Apical
Posterior
Anterior
Horizontal fissure
Lateral
Medial
Oblique fissure
Anterior basal
Lateral basal
Medial basal
Posterior basal
Apical posterior
Anterior
Oblique fissure
Superior
Superior lingular
Inferior lingular
Anterior basal
Lateral basal


### Posterior view

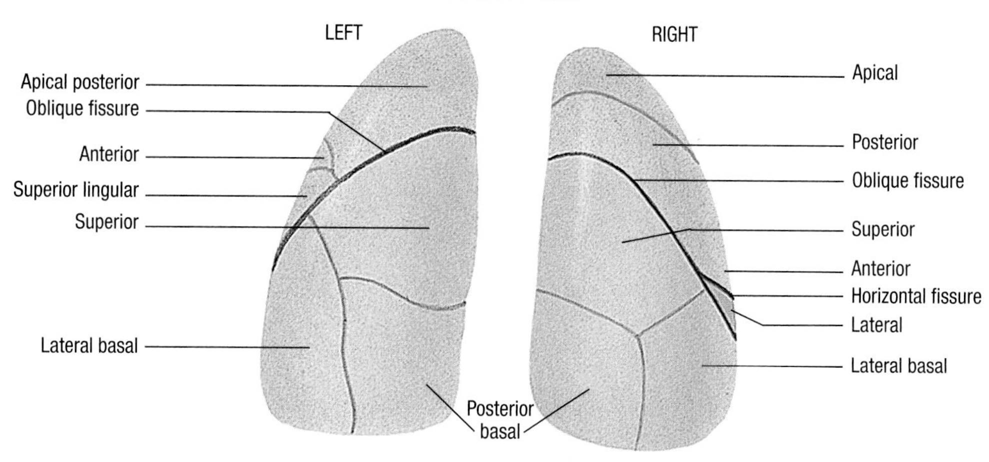

LEFT
RIGHT
Apical posterior
Oblique fissure
Anterior
Superior lingular
Superior
Lateral basal
Posterior basal
Apical
Posterior
Oblique fissure
Superior
Anterior
Horizontal fissure
Lateral
Lateral basal


### Medial view

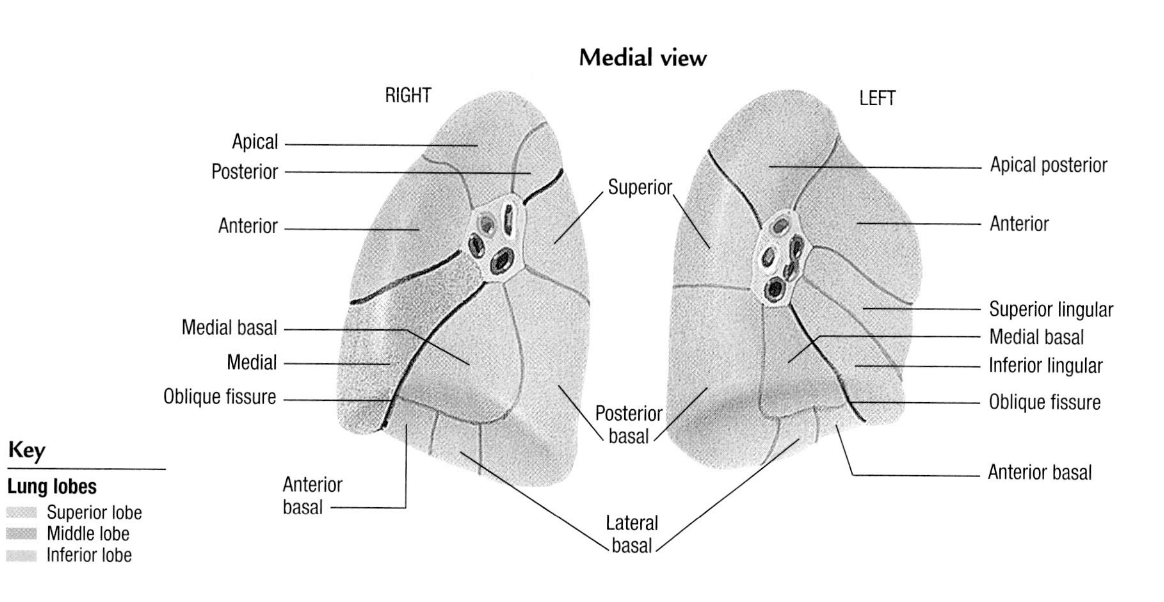

RIGHT
LEFT
Apical
Posterior
Anterior
Medial basal
Medial
Oblique fissure
Anterior basal
Superior
Posterior basal
Lateral basal
Apical posterior
Anterior
Superior lingular
Medial basal
Inferior lingular
Oblique fissure
Anterior basal
Key
Lung lobes
Superior lobe
Middle lobe
Inferior lobe

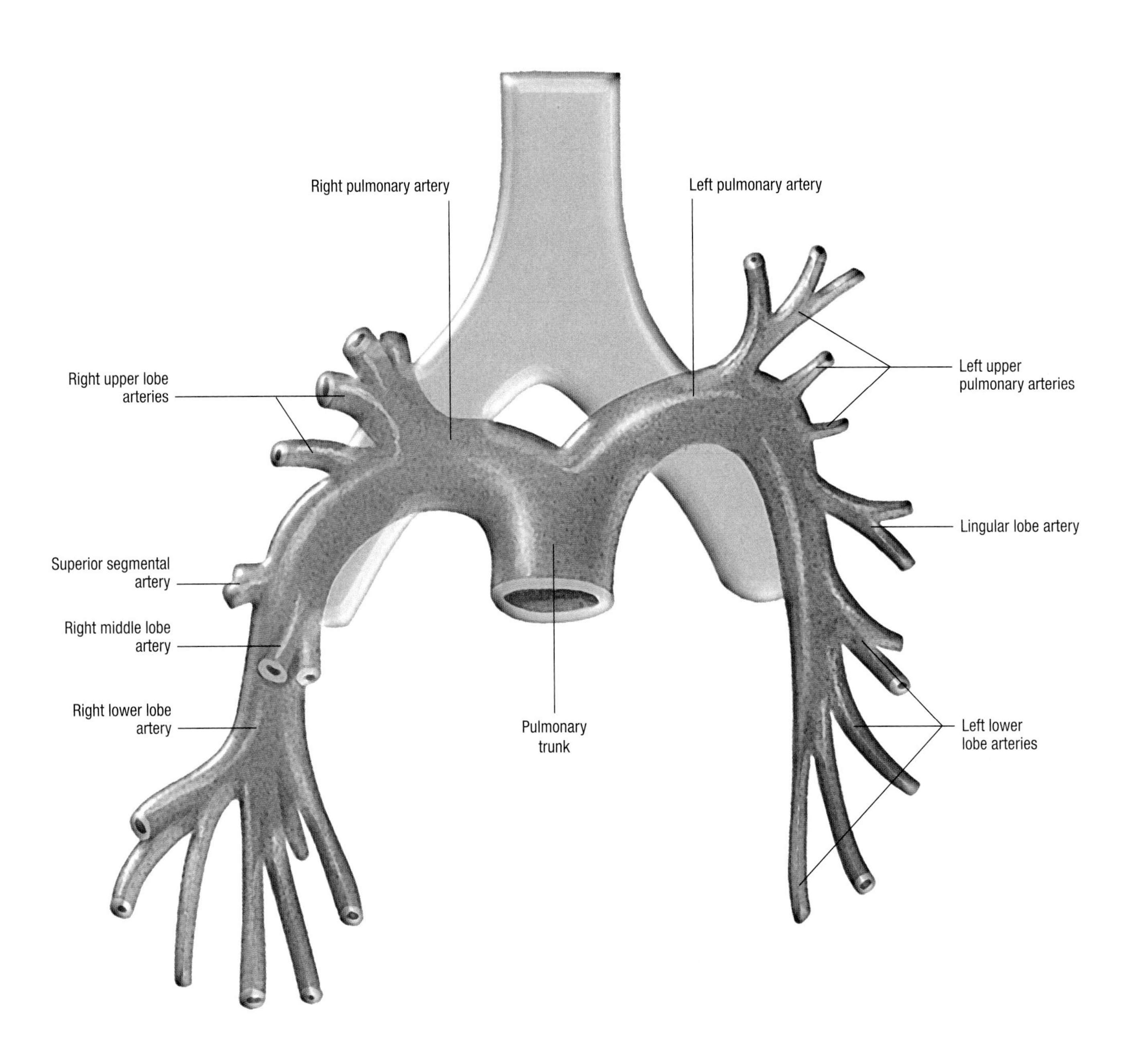
Right pulmonary artery
Left pulmonary artery
Right upper lobe arteries
Left upper pulmonary arteries
Lingular lobe artery
Superior segmental artery
Right middle lobe artery
Right lower lobe artery
Pulmonary trunk
Left lower lobe arteries

## PULMONARY VEINS

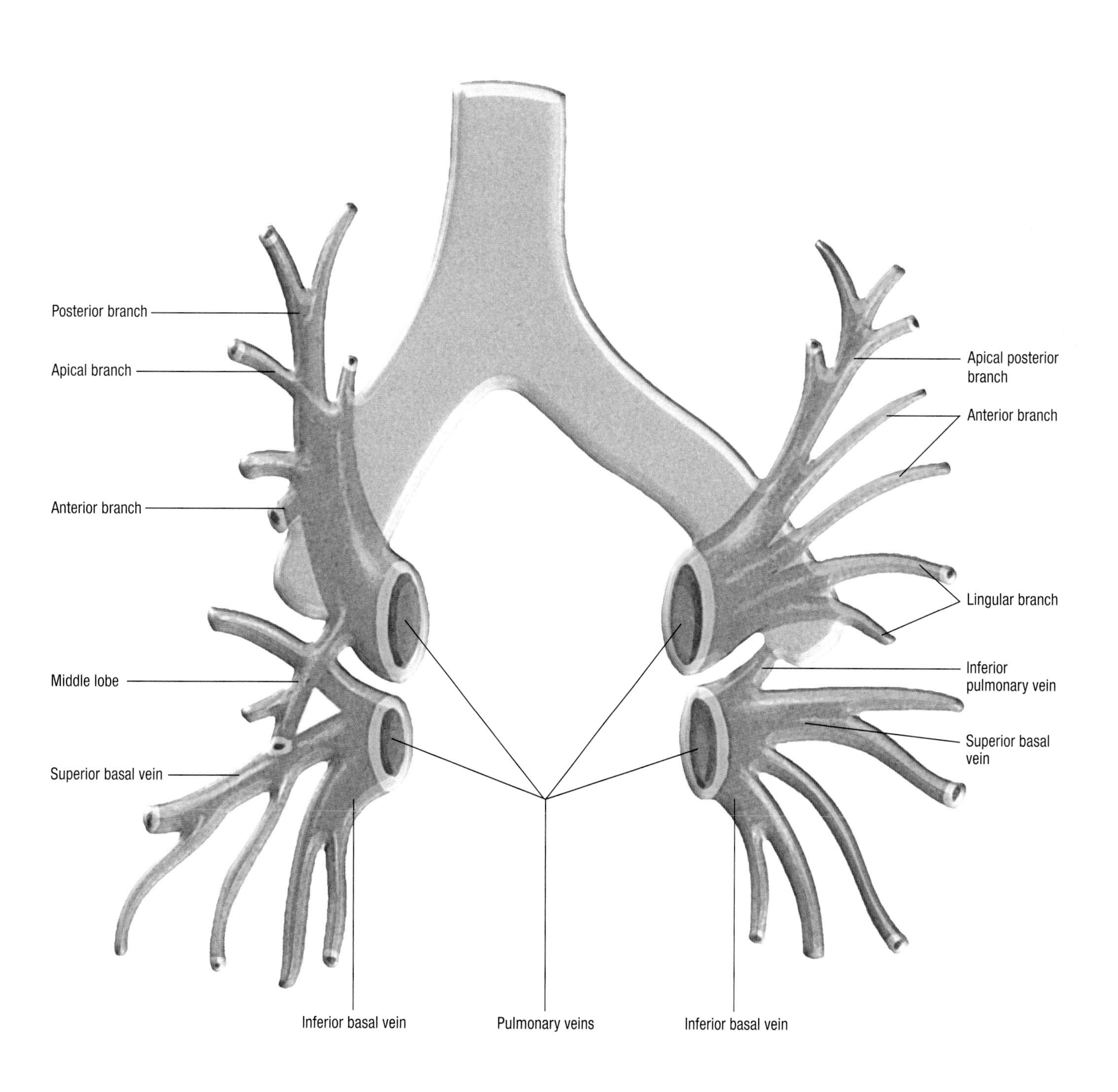

Posterior branch
Apical branch
Anterior branch
Middle lobe
Superior basal vein
Inferior basal vein
Pulmonary veins
Inferior basal vein
Apical posterior branch
Anterior branch
Lingular branch
Inferior pulmonary vein
Superior basal vein

# Alveoli

Alveoli are tiny, grapelike clusters of air sacs at the ends of bronchioles. (See *Alveolus, cross section,* pages 184 and 185.) The lungs in a typical adult contain about 300 million alveoli, which are surrounded by an extensive network of capillaries. The alveoli exchange oxygen and carbon dioxide by diffusion with these capillaries. Alveoli are separated by thin, vascular partitions called alveolar septa, which are lined by flat squamous cells (Type I) and small numbers of secretory cells (Type II). The secretory cells produce a lipoprotein surfactant. The surfactant reduces the cohesive surface tension of water molecules in the alveoli, so that the fluid in the alveoli spreads as a thin film, rather than coalescing into droplets. This allows the alveoli to expand uniformly during inspiration and prevents alveolar collapse during expiration.

## Surfactant

Without surfactant, surface tension could restrict alveolar expansion during inspiration or cause alveolar collapse during expiration. Surfactant is commonly lacking in neonates born before 28 weeks' gestation, causing respiratory distress syndrome of the newborn. Surfactant deficiency in adults also causes alveolar collapse and adult respiratory distress syndrome.

## Physiology

Oxygen and carbon dioxide diffusion between the alveoli, blood, and tissues depends on the concentrations and pressures of these gases. In a mixture of gases, the pressure exerted by each gas (partial pressure) is independent of that of the other gases and directly corresponds to its concentration in the total mixture.

Diffusion refers to the exchange of gases — particularly oxygen and carbon dioxide — between the alveoli and capillaries, and between body cells and RBCs. Diffusing molecules move from an area of higher concentration to one of lower concentration; a gas diffuses from an area with a high partial pressure of the gas to one with a low partial pressure.

Oxygen diffuses into the tissues from the arterial blood, and carbon dioxide diffuses in the opposite direction, from the tissues into the blood. Inspired air has a $Po_2$ of 158 mm Hg and a $Pco_2$ of 0.3 mm Hg. When blood returns to the heart from the systemic circulation, the right ventricle pumps it to the lungs, where it passes through the pulmonary capillaries. The $Po_2$ of blood in the venae cavae is 40 mm Hg; the $Pco_2$ is 47 mm Hg. Because alveolar air has a higher $Po_2$ (100 mm Hg) and a lower $Pco_2$ (40 mm Hg) than the blood in the pulmonary capillaries, oxygen diffuses from alveolar air into the pulmonary capillaries, and carbon dioxide diffuses in the opposite direction. As a result, arterial blood has a $Po_2$ of 97 mm Hg, which is almost the same as the alveolar $Po_2$. Its $Pco_2$ is 40 mm Hg, the same as the alveolar $Pco_2$. In the tissues, the $Po_2$ is lower (40 mm Hg) and the $Pco_2$ is higher (about 60 mm Hg) than in arterial blood.

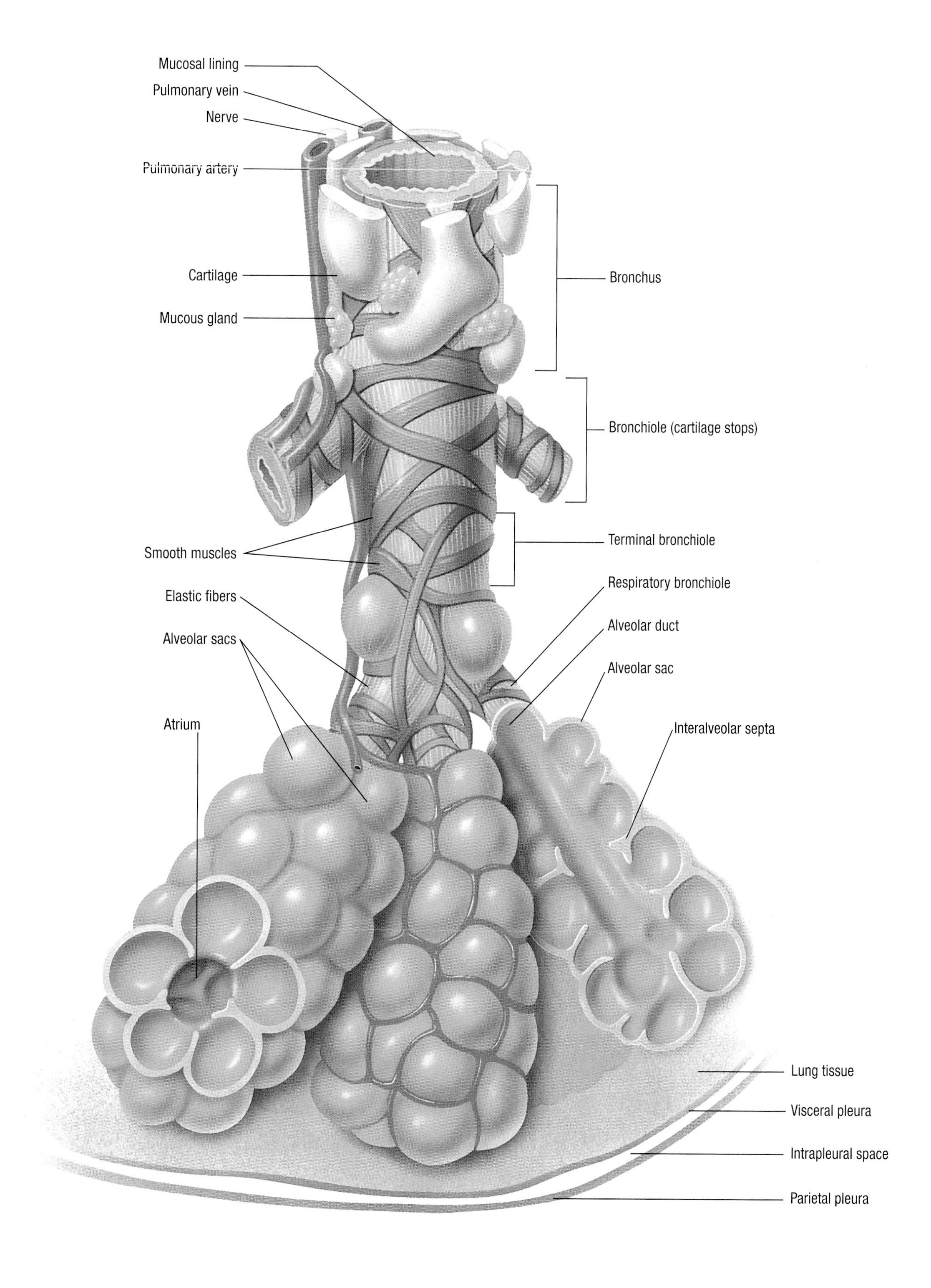
Mucosal lining
Pulmonary vein
Nerve
Pulmonary artery
Cartilage
Mucous gland
Bronchus
Bronchiole (cartilage stops)
Smooth muscles
Terminal bronchiole
Respiratory bronchiole
Elastic fibers
Alveolar duct
Alveolar sacs
Alveolar sac
Atrium
Interalveolar septa
Lung tissue
Visceral pleura
Intrapleural space
Parietal pleura

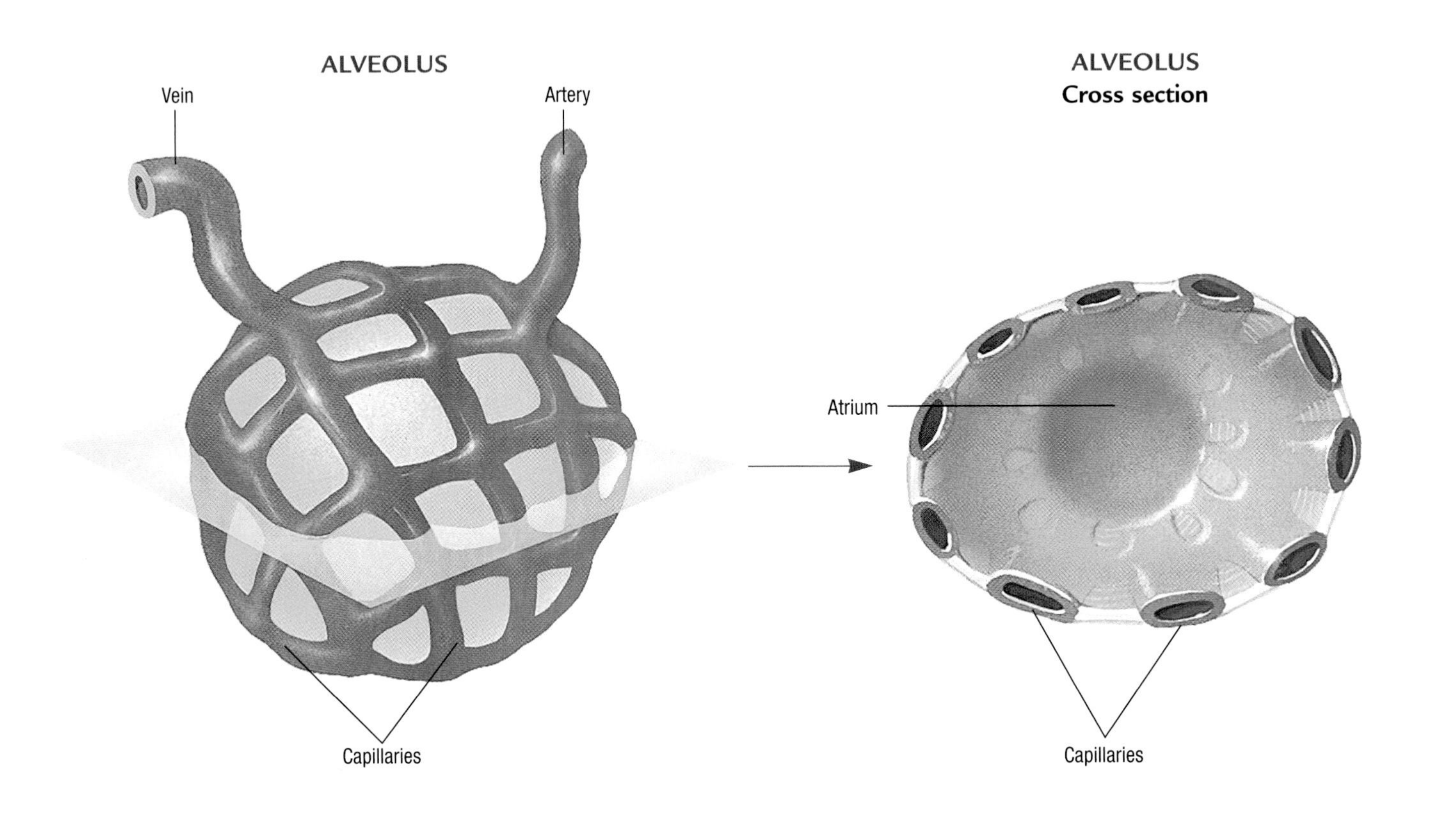
ALVEOLUS
Vein
Artery
Capillaries
ALVEOLUS
Cross section
Atrium
Capillaries

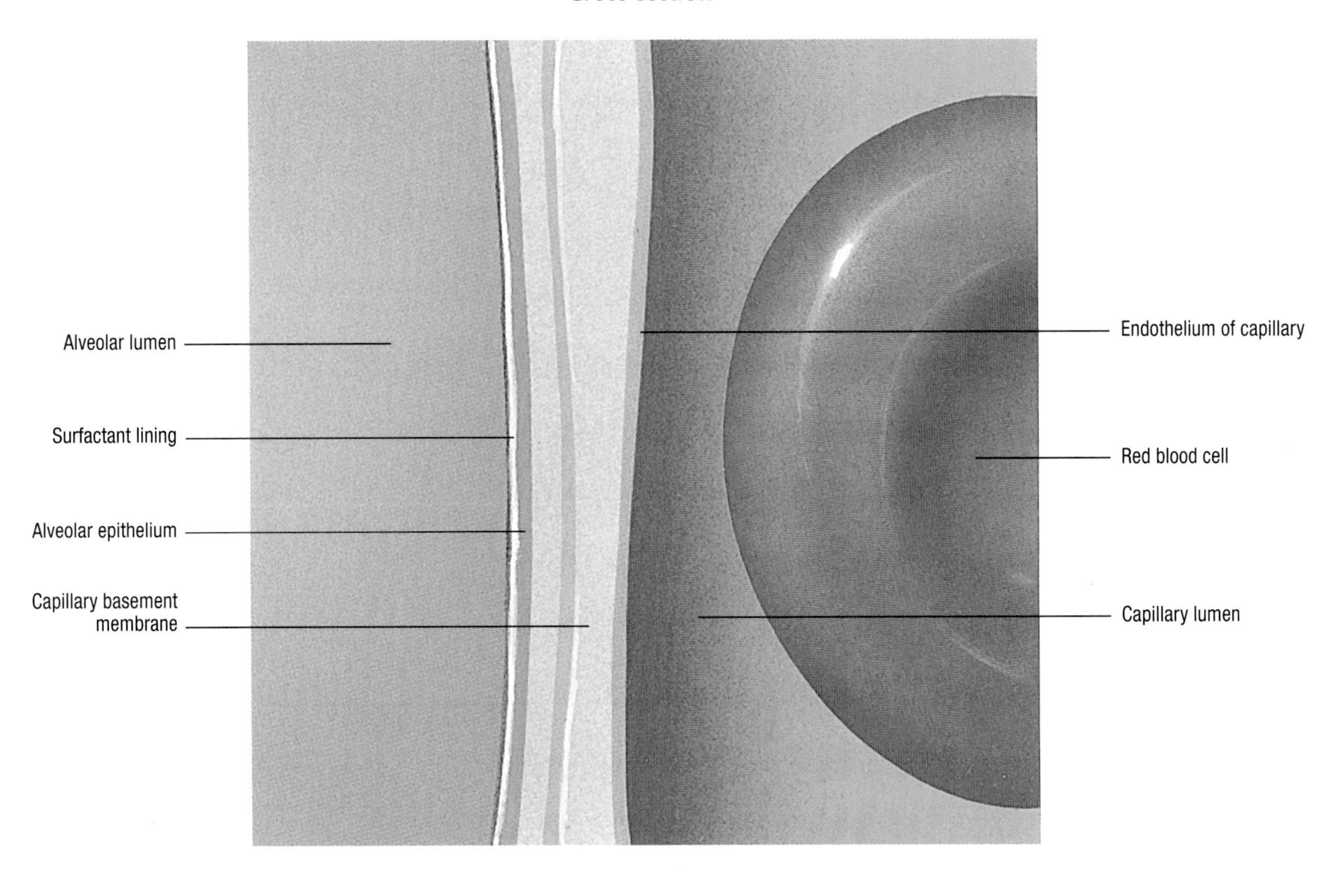
ALVEOLUS AND CAPILLARY
Cross section
Alveolar lumen
Surfactant lining
Alveolar epithelium
Capillary basement membrane
Endothelium of capillary
Red blood cell
Capillary lumen

ALVEOLUS
**Cross section**

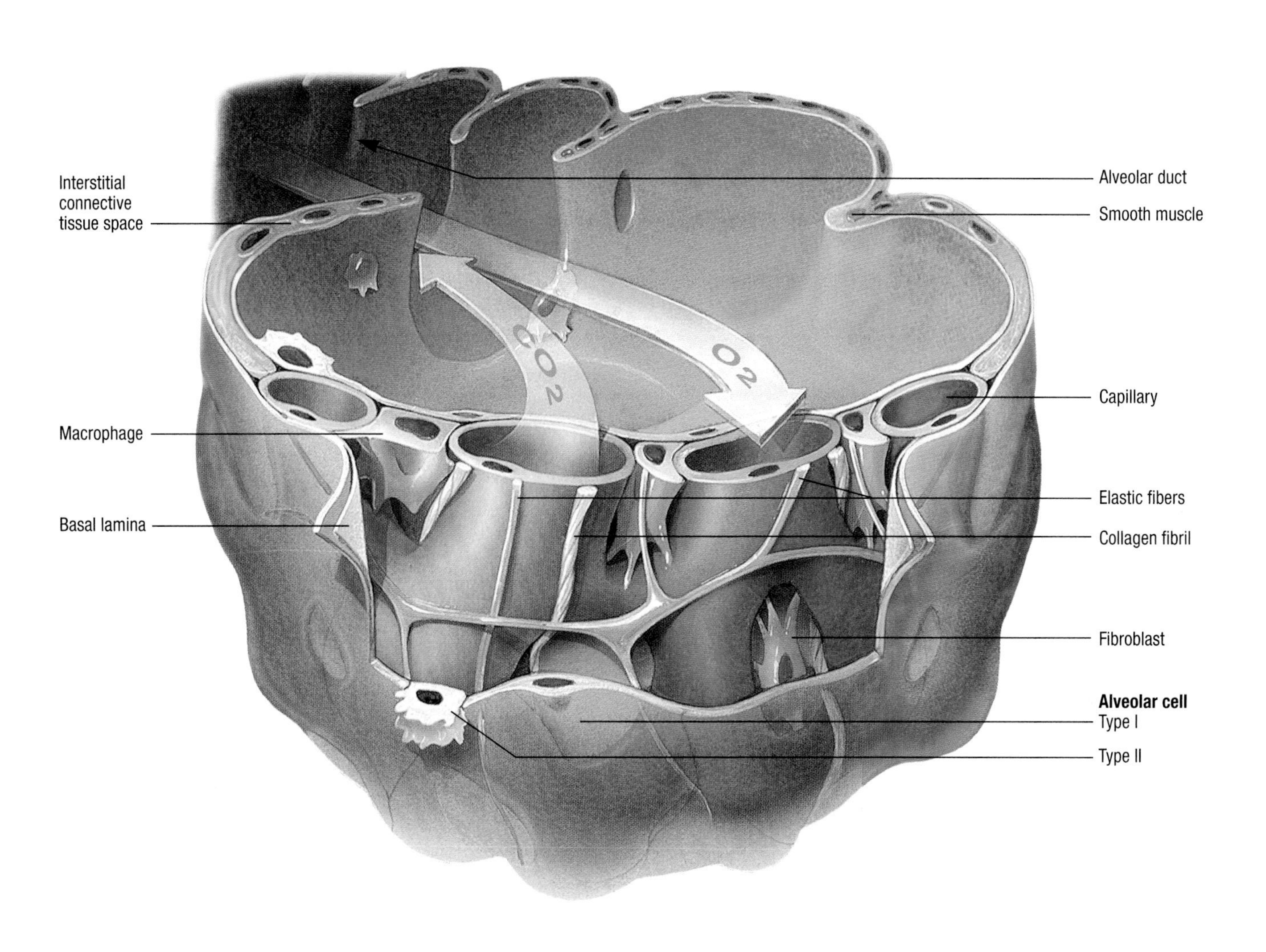
Interstitial connective tissue space
Macrophage
Basal lamina
Alveolar duct
Smooth muscle
Capillary
Elastic fibers
Collagen fibril
Fibroblast
Alveolar cell
Type I
Type II
$CO_2$
$O_2$

# PART VI

# ABDOMEN

# Overview of abdomen

The abdomen is the portion of the trunk between the thorax and the pelvis. It consists of the abdominal cavity, the peritoneum, the abdominal viscera, and the abdominal wall. Subcutaneous tissue and varying quantities of fat cover the abdominal wall.

## Abdominal cavity

The abdominal cavity is a subdivision of the abdominopelvic cavity. No muscles or membranes separate these two cavities. (The pelvic cavity is described in Part VII, Pelvis.) The abdominal cavity extends into the thoracic cage via the space under the dome of the diaphragm. This provides protection for some of the abdominal organs.

## Abdominal divisions

For purposes of reference, the abdomen is typically divided into four quadrants or nine regions.

### Quadrants

The right and left upper quadrants and the right and left lower quadrants are defined by the following:

- vertical median plane
- horizontal transumbilical plane through the umbilicus and the L3-L4 intervertebral disk.

### Regions

Abdominal regions are determined by the vertical midclavicular planes through the midpoint of the clavicles – the subcostal plane, at the inferior border of the 10th costal cartilage; and the transtubercular plane, between the iliac tubercles and L5. (See illustration depicting these regions in Part I, General overview.)

## Peritoneum

The serosa lining the abdominal cavity is called the peritoneum. It covers the abdominal wall and the viscera and is divided into parietal and visceral layers; the potential space between the layers is called the peritoneal cavity.

Folds of the peritoneum called mesenteries hold the intestines and other gastrointestinal organs in place; blood vessels, lymphatic vessels, and nerves run through the mesenteries. Other folds, called omenta, attach to the stomach.

### Visceral peritoneum

The visceral peritoneum is the outer covering of the gastrointestinal organs. In the esophagus and rectum, it's also called the *tunica adventitia*; elsewhere in the GI tract, it's called the *tunica serosa*.

The visceral peritoneum covers most of the abdominal organs. It lies next to the parietal peritoneum. Blood vessels, nerves, and lymphatics travel within the two layers of the visceral peritoneum that make up the mesenteries. These mesenteries attach the jejunum and ileum to the posterior abdominal wall to prevent twisting.

### Parietal peritoneum

The parietal peritoneum is the serous membrane that lines the abdominal and pelvic walls. The space between the parietal peritoneum and the transversalis fascia contains loose, fatty extraperitoneal tissue. Terms used to describe the abdominal areas use this sac as a point of reference, for example:

- retroperitoneal – behind the peritoneum
- preperitoneal – in front of (ventral to) the peritoneum
- subperitoneal – between the peritoneum and the pelvic floor.

## Blood vessels

Arteries of the abdominal surface include the superior epigastric, descending from the internal mammary artery; the inferior epigastric and deep circumflex iliac arteries from the external iliac arteries; and the superficial circumflex iliac and superficial epigastric arteries from the femoral artery.

The thoracic aorta is continuous with the abdominal aorta. The abdominal aorta has several visceral and parietal branches. (See *Branches of abdominal aorta and portal vein*, page 191.) Ultimately, it bifurcates to form the right and left iliac arteries. (See *Visceral arteries,* page 192.)

Small end arteries join arterioles and capillary beds, which in turn join venules and progressively larger veins that carry venous blood to the vena cava. The junction of the common iliac veins forms the inferior vena cava, which passes behind the liver and through the diaphragm.

### Portal system

The network of veins returning from the abdominal organs to the liver is called the portal system. (See *Portal system*, page 193.) The portal vein, formed by the superior mesenteric and splenic veins behind the pancreas, ascends anterior to the inferior vena cava. In the liver, it forms right and left branches, which branch further to form the hepatic veins, which, in turn, are branches of the inferior vena cava.

## Innervation

The surface of the abdomen is innervated by intercostal nerves T7 through T11, subcostal nerve T12, and the iliohypogastric and ilioinguinal nerves (L1). These nerves supply the anterolateral muscles and carry sensory impulses from the skin. (See *Abdominal nerves*, page 194, and *Abdominal cutaneous nerves*, page 195.)

## Lymphatic drainage

Superficial abdominal lymph nodes above the umbilicus drain to the axillary lymph nodes or the parasternal lymph nodes. Lymph nodes below the umbilicus drain to the superficial inguinal lymph nodes. Other lymph nodes in the abdominal region include the lumbar, iliac, and deep inguinal nodes. (See *Lymphatic vessels and nodes*, pages 196 and 197.)

The lumbar trunk drains the deep lymphatics from the abdominal wall and delivers lymph to the cisterna chyli. The intestinal trunk receives lymph form the stomach, intestines, pancreas, spleen, and liver.

## ABDOMINAL CONTENTS
### Anterior view

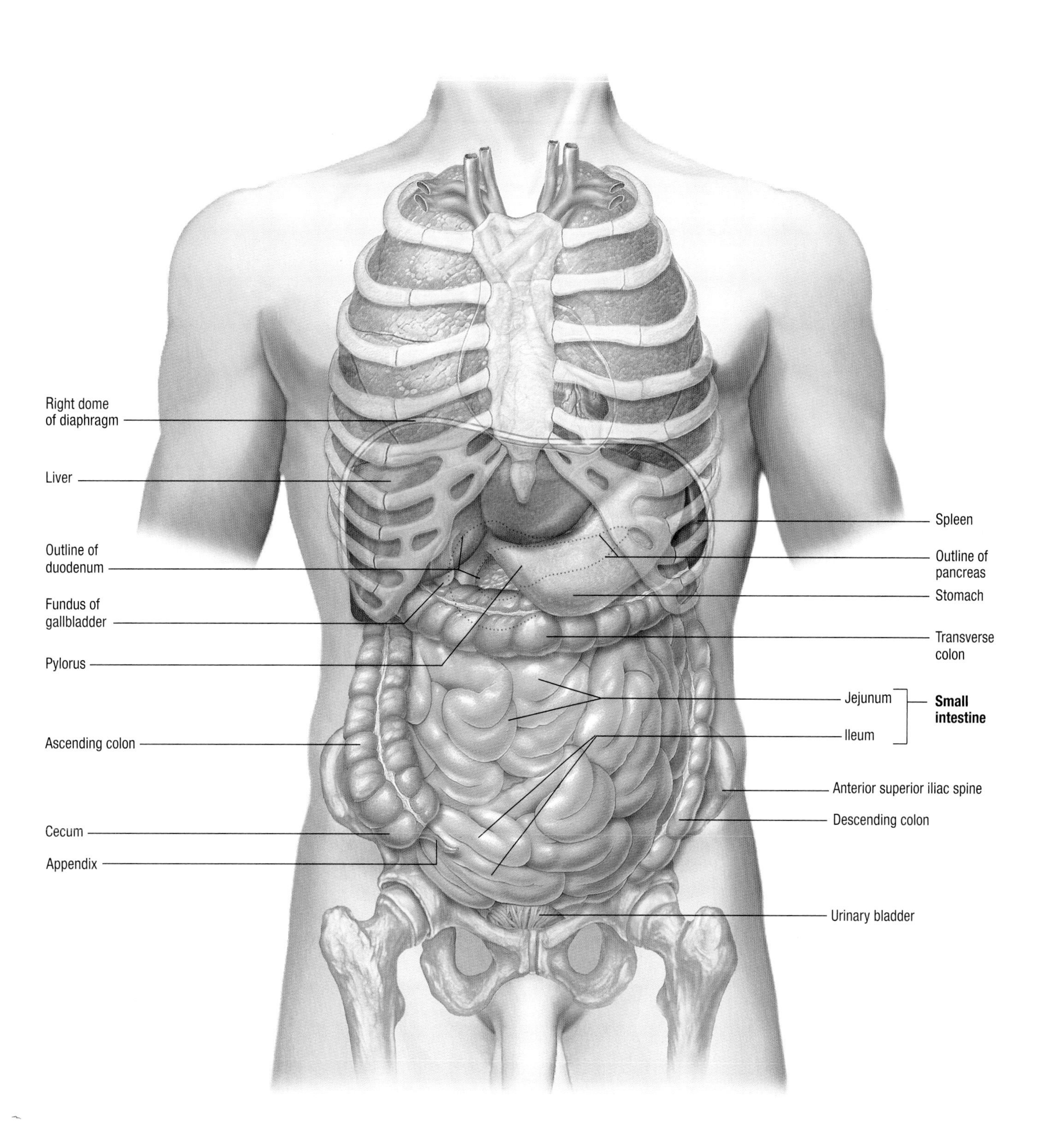

Right dome
of diaphragm
Liver
Outline of
duodenum
Fundus of
gallbladder
Pylorus
Ascending colon
Cecum
Appendix
Spleen
Outline of
pancreas
Stomach
Transverse
colon
Jejunum
Ileum
Small
intestine
Anterior superior iliac spine
Descending colon
Urinary bladder

## ABDOMINAL CONTENTS
### Posterior view

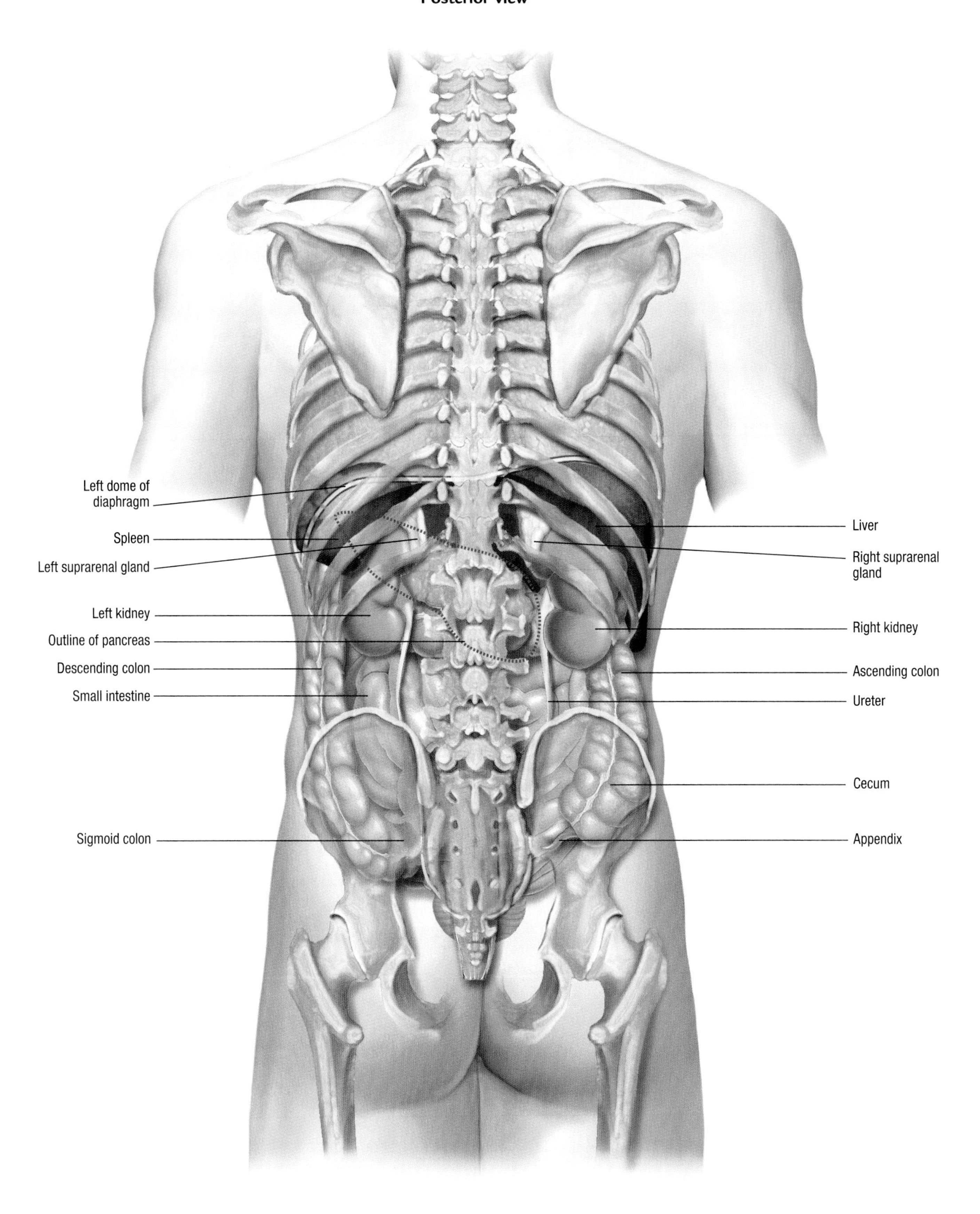

Left dome of diaphragm
Spleen
Left suprarenal gland
Left kidney
Outline of pancreas
Descending colon
Small intestine
Sigmoid colon
Liver
Right suprarenal gland
Right kidney
Ascending colon
Ureter
Cecum
Appendix

# BRANCHES OF ABDOMINAL AORTA AND PORTAL VEIN

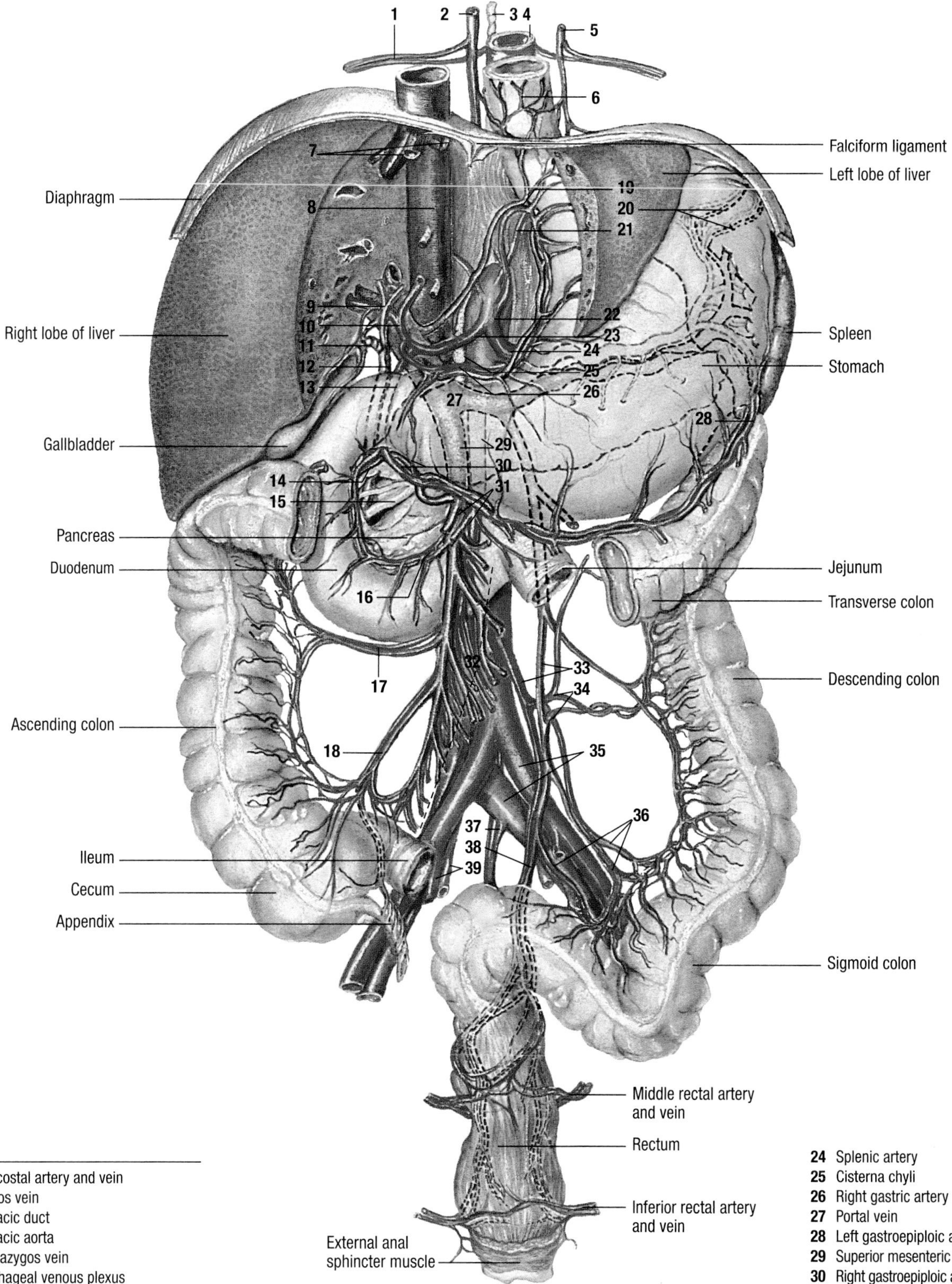

1
2
3
4
5
6
7
8
9
10
11
12
13
14
15
16
17
18
19
20
21
22
23
24
25
26
27
28
29
30
31
32
33
34
35
36
37
38
39
Diaphragm
Right lobe of liver
Gallbladder
Pancreas
Duodenum
Ascending colon
Ileum
Cecum
Appendix
Falciform ligament
Left lobe of liver
Spleen
Stomach
Jejunum
Transverse colon
Descending colon
Sigmoid colon
Middle rectal artery and vein
Rectum
Inferior rectal artery and vein
External anal sphincter muscle


**Key**

**1** Intercostal artery and vein
**2** Azygos vein
**3** Thoracic duct
**4** Thoracic aorta
**5** Hemiazygos vein
**6** Esophageal venous plexus
**7** Hepatic veins
**8** Inferior vena cava
**9** Common hepatic duct
**10** Proper hepatic artery
**11** Cystic duct
**12** Common bile duct
**13** Gastroduodenal artery
**14** Anterior superior pancreaticoduodenal artery and vein
**15** Pancreatic duct
**16** Anterior inferior pancreaticoduodenal artery and vein
**17** Right colic artery and vein
**18** Ileocolic artery and vein
**19** Left gastric artery and vein
**20** Short gastric arteries and veins
**21** Inferior phrenic artery
**22** Celiac trunk
**23** Common hepatic artery
**24** Splenic artery
**25** Cisterna chyli
**26** Right gastric artery
**27** Portal vein
**28** Left gastroepiploic artery and vein
**29** Superior mesenteric artery and vein
**30** Right gastroepiploic artery and vein
**31** Middle colic artery and vein
**32** Jejunal and ileal arteries and veins
**33** Inferior mesenteric artery and vein
**34** Left colic artery and vein
**35** Common iliac arteries and veins
**36** Sigmoid arteries and veins
**37** Middle sacral artery and vein
**38** Superior rectal artery and vein
**39** Right internal iliac artery and vein

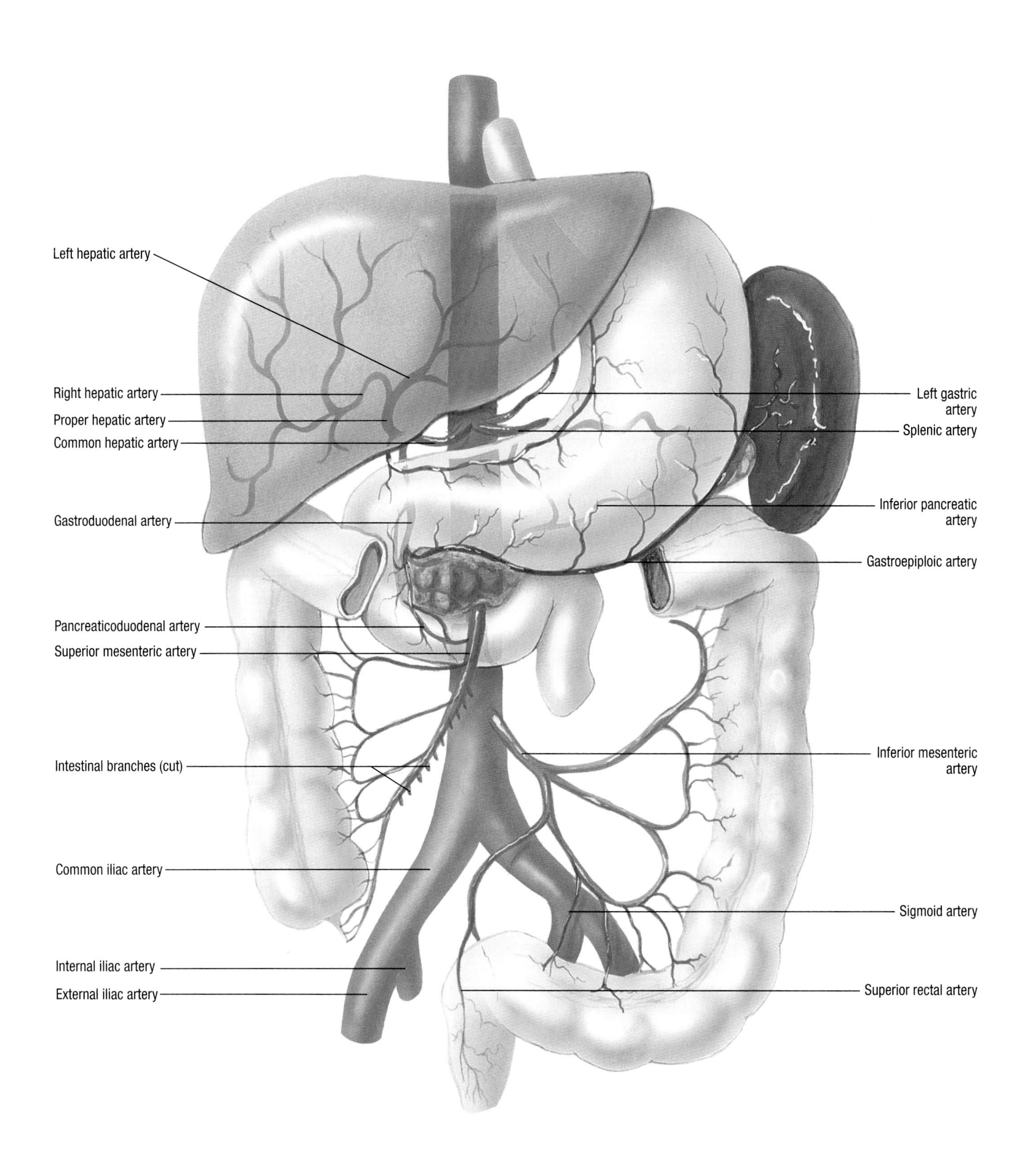
Left hepatic artery
Right hepatic artery
Proper hepatic artery
Common hepatic artery
Gastroduodenal artery
Pancreaticoduodenal artery
Superior mesenteric artery
Intestinal branches (cut)
Common iliac artery
Internal iliac artery
External iliac artery
Left gastric artery
Splenic artery
Inferior pancreatic artery
Gastroepiploic artery
Inferior mesenteric artery
Sigmoid artery
Superior rectal artery

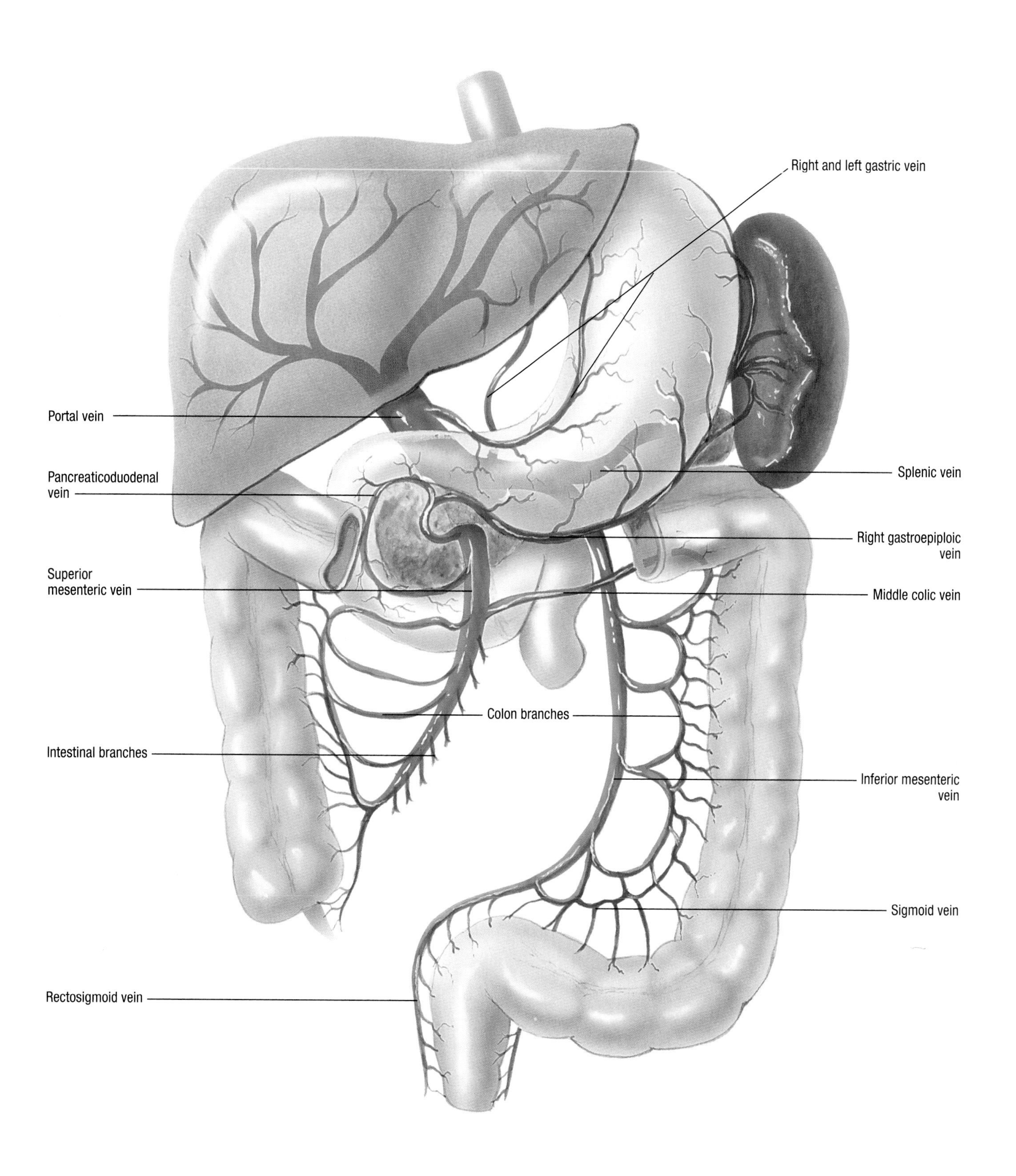
Right and left gastric vein
Portal vein
Splenic vein
Pancreaticoduodenal vein
Right gastroepiploic vein
Superior mesenteric vein
Middle colic vein
Colon branches
Intestinal branches
Inferior mesenteric vein
Sigmoid vein
Rectosigmoid vein

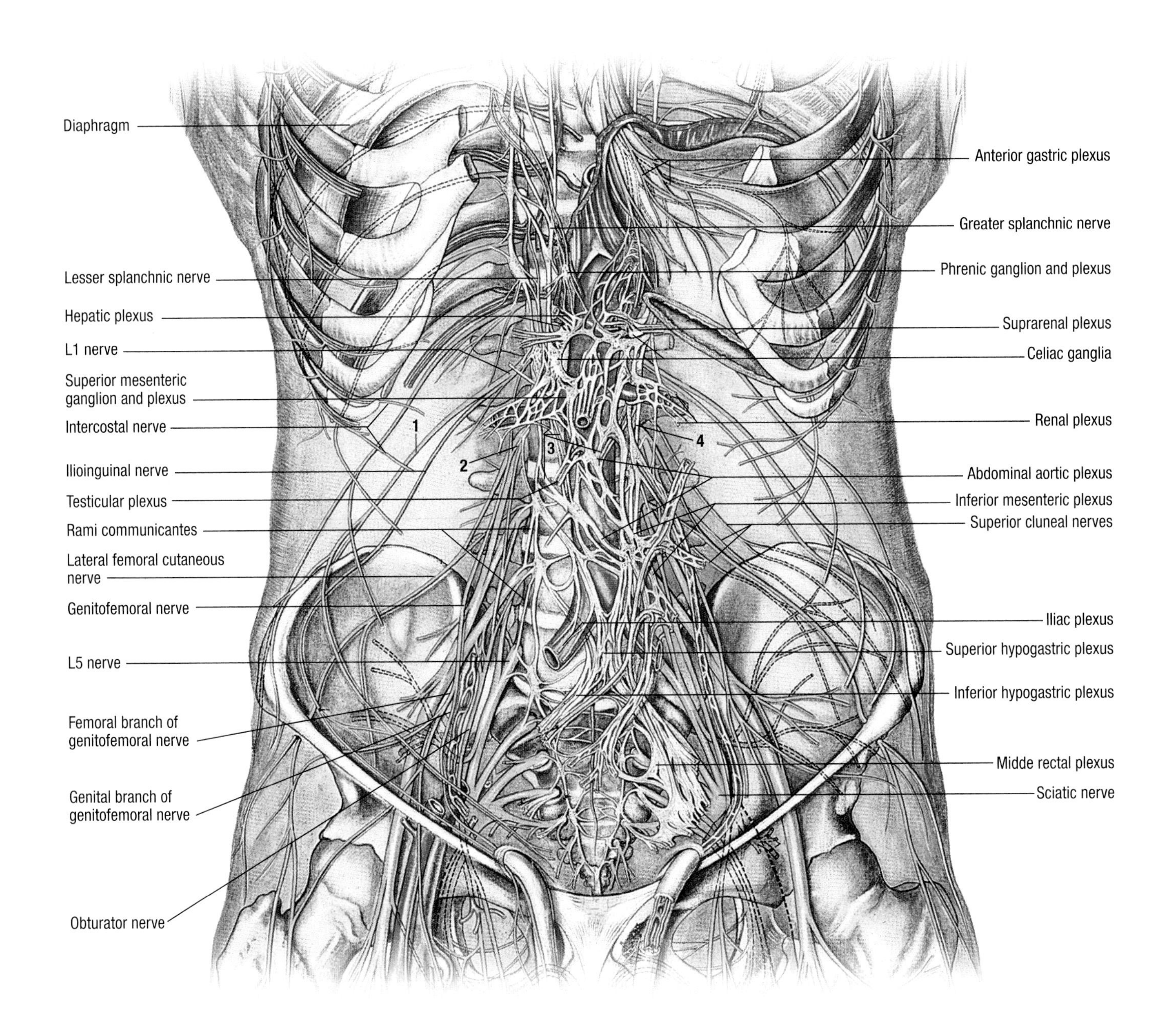
Diaphragm
Anterior gastric plexus
Greater splanchnic nerve
Phrenic ganglion and plexus
Lesser splanchnic nerve
Hepatic plexus
Suprarenal plexus
L1 nerve
Celiac ganglia
Superior mesenteric ganglion and plexus
Renal plexus
Intercostal nerve
1
4
3
2
Ilioinguinal nerve
Abdominal aortic plexus
Testicular plexus
Inferior mesenteric plexus
Superior cluneal nerves
Rami communicantes
Lateral femoral cutaneous nerve
Genitofemoral nerve
Iliac plexus
Superior hypogastric plexus
L5 nerve
Inferior hypogastric plexus
Femoral branch of genitofemoral nerve
Midde rectal plexus
Sciatic nerve
Genital branch of genitofemoral nerve
Obturator nerve

**Key**

1 Iliohypogastric nerve
2 L2 nerve
3 Lumbar splanchnic nerves
4 Sympathetic trunk

# ABDOMINAL CUTANEOUS NERVES
## Lateral view

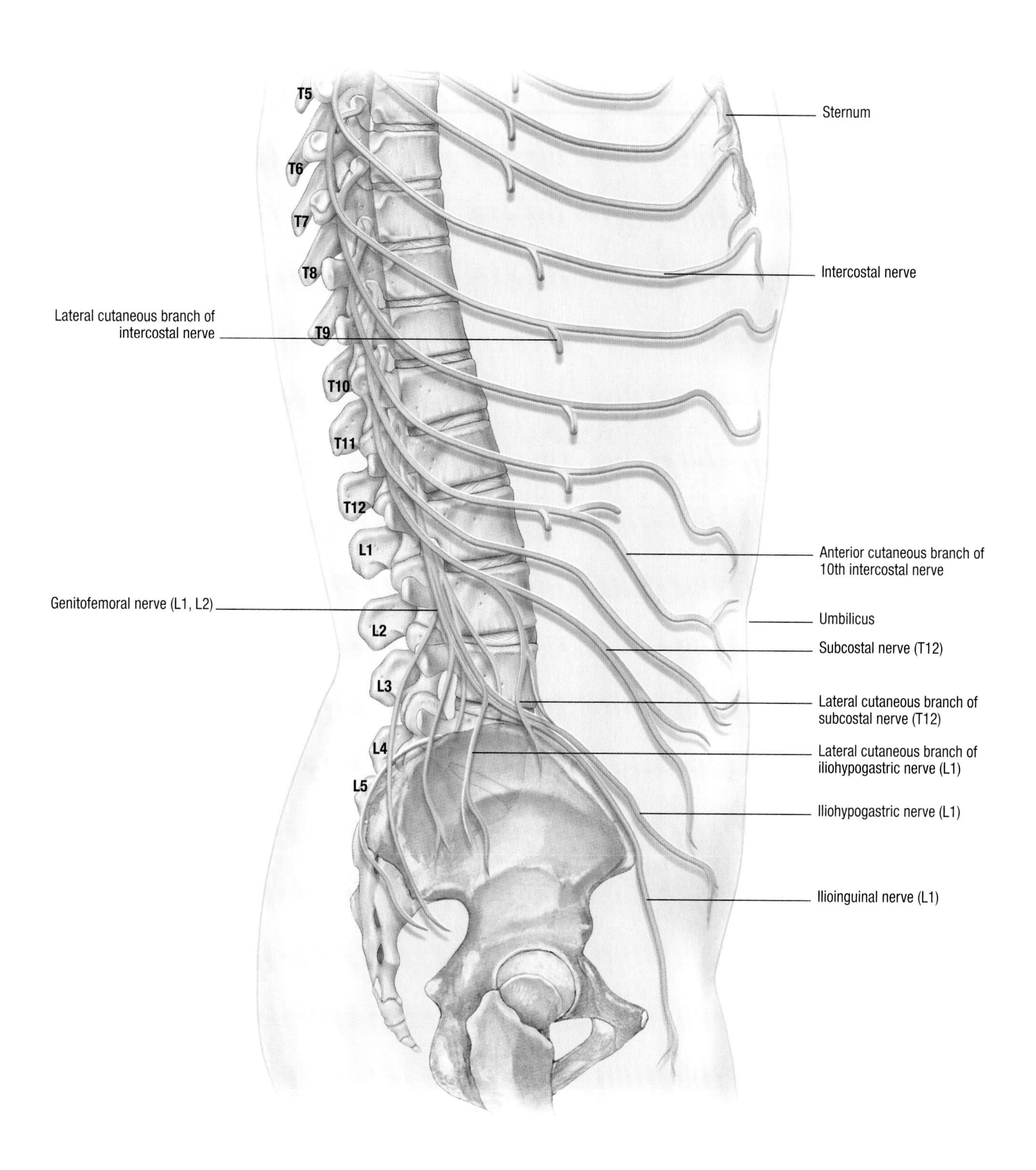
T5
T6
T7
T8
T9
T10
T11
T12
L1
L2
L3
L4
L5
Sternum
Intercostal nerve
Lateral cutaneous branch of intercostal nerve
Anterior cutaneous branch of 10th intercostal nerve
Genitofemoral nerve (L1, L2)
Umbilicus
Subcostal nerve (T12)
Lateral cutaneous branch of subcostal nerve (T12)
Lateral cutaneous branch of iliohypogastric nerve (L1)
Iliohypogastric nerve (L1)
Ilioinguinal nerve (L1)

## LYMPHATIC VESSELS AND NODES
### Posterior wall of abdomen and inguinal region

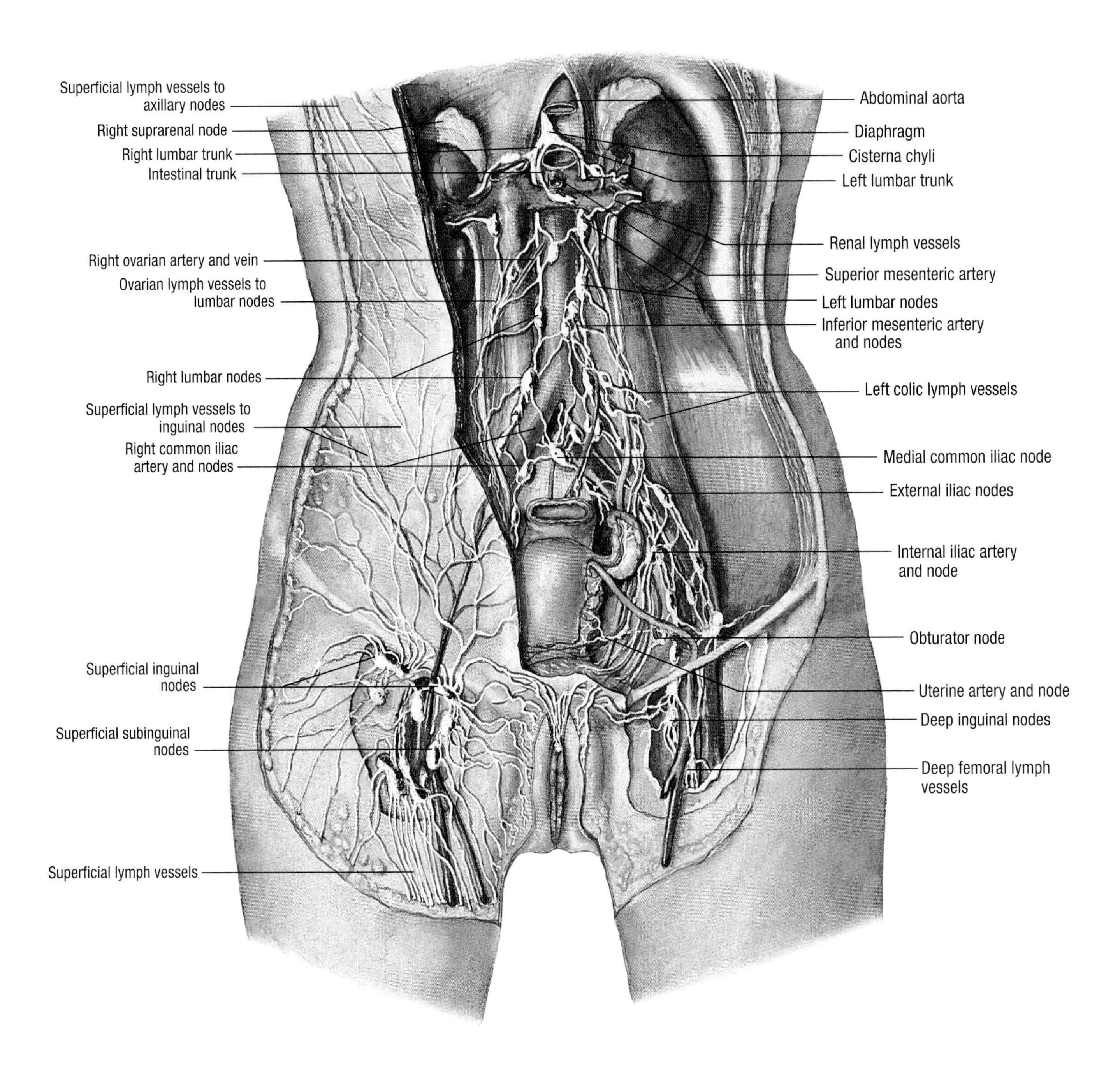

Superficial lymph vessels to axillary nodes
Right suprarenal node
Right lumbar trunk
Intestinal trunk
Right ovarian artery and vein
Ovarian lymph vessels to lumbar nodes
Right lumbar nodes
Superficial lymph vessels to inguinal nodes
Right common iliac artery and nodes
Superficial inguinal nodes
Superficial subinguinal nodes
Superficial lymph vessels
Abdominal aorta
Diaphragm
Cisterna chyli
Left lumbar trunk
Renal lymph vessels
Superior mesenteric artery
Left lumbar nodes
Inferior mesenteric artery and nodes
Left colic lymph vessels
Medial common iliac node
External iliac nodes
Internal iliac artery and node
Obturator node
Uterine artery and node
Deep inguinal nodes
Deep femoral lymph vessels

## LYMPHATIC VESSELS AND NODES
### Stomach, pancreas, spleen, and biliary tract

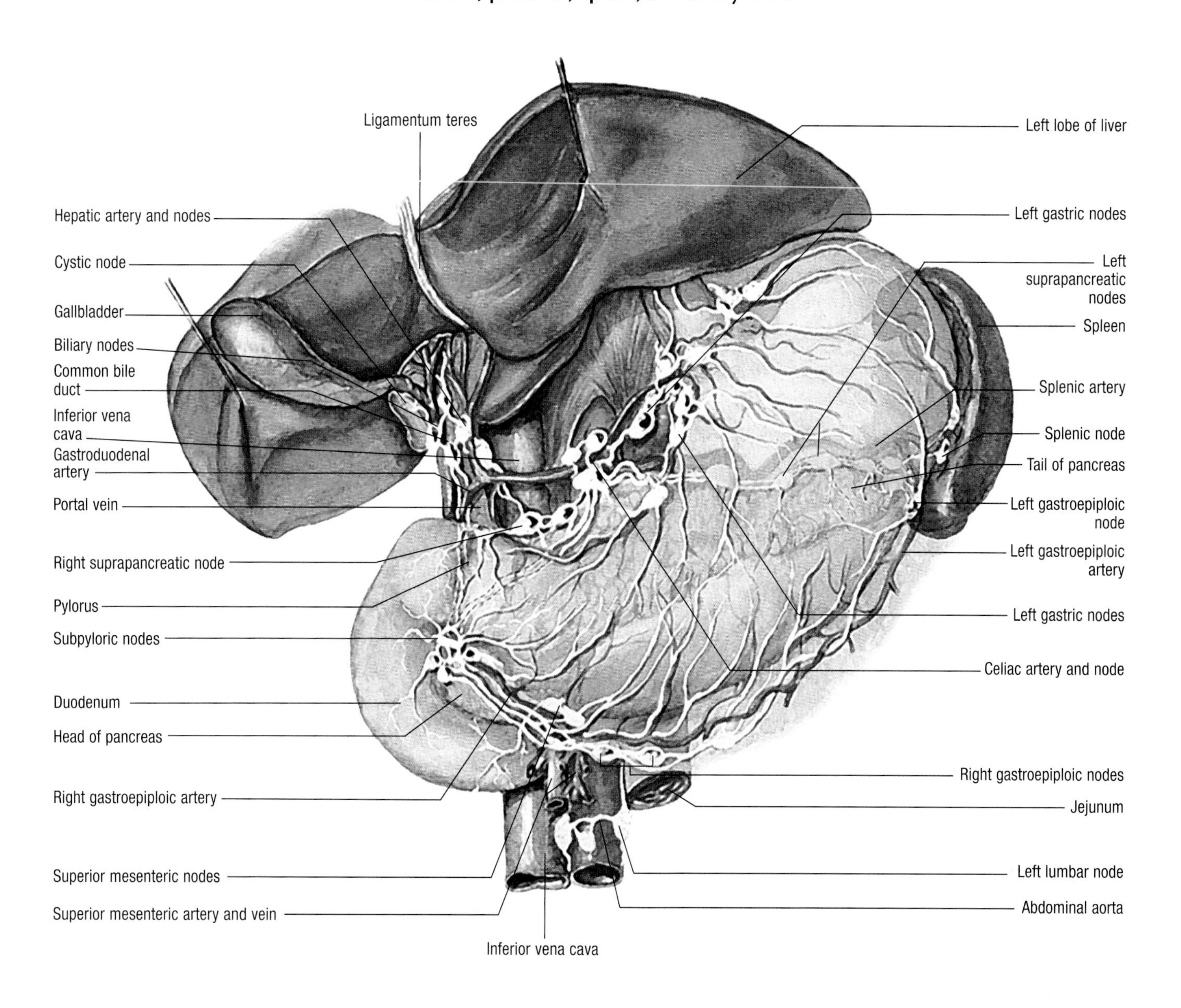

Ligamentum teres
Left lobe of liver
Hepatic artery and nodes
Left gastric nodes
Cystic node
Left suprapancreatic nodes
Gallbladder
Spleen
Biliary nodes
Common bile duct
Splenic artery
Inferior vena cava
Splenic node
Gastroduodenal artery
Tail of pancreas
Portal vein
Left gastroepiploic node
Left gastroepiploic artery
Right suprapancreatic node
Pylorus
Left gastric nodes
Subpyloric nodes
Celiac artery and node
Duodenum
Head of pancreas
Right gastroepiploic nodes
Right gastroepiploic artery
Jejunum
Superior mesenteric nodes
Left lumbar node
Superior mesenteric artery and vein
Abdominal aorta
Inferior vena cava


## INTERNAL ILIAC LYMPH NODE
### Transverse section

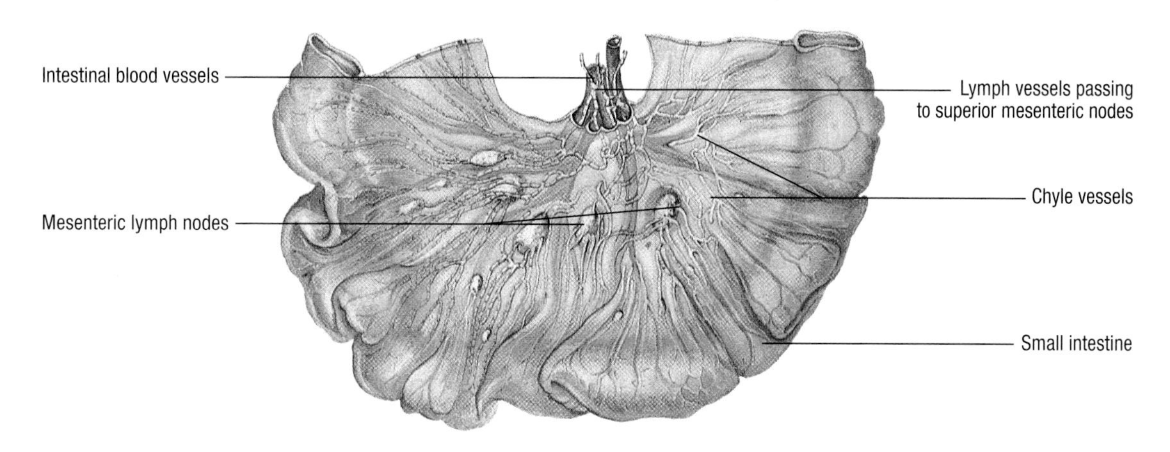

Intestinal blood vessels
Lymph vessels passing to superior mesenteric nodes
Chyle vessels
Mesenteric lymph nodes
Small intestine

# Abdominal muscles

Abdominal muscles can be classified by area – anterior, lateral, and posterior. Anterior and lateral (anterolateral) muscles are discussed together because the boundaries between the two are not clearly defined.

## Anterolateral muscles

Five pairs of muscles, three flat and two vertical, lie in the anterolateral area.

### Flat muscles

The three flat muscles are the internal and external abdominal oblique muscles and the transversus abdominis muscle. The union of the aponeuroses of the three flat abdominal muscles forms the linea alba, a thin, whitish median line in the anterior abdominal wall. The outer layer of the linea alba forms the tendinous rectus sheath, which has anterior and posterior layers and extends from the symphysis pubis to the xiphoid process.

- Internal and external abdominal obliques – compress and support the abdominopelvic cavity and viscera, and help flex and rotate the vertebral column
- Transversus abdominis – compresses and supports the abdominopelvic cavity and viscera.

### Vertical muscles

Within the rectus sheath lie blood vessels, nerves, parts of the lymphatic system, and two vertical muscles – the rectus abdominis and the pyramidalis.

- Rectus abdominis muscles
  - extend the length of the ventral aspect of the abdomen and are separated by the linea alba; continuity disrupted by tendinous intersections running transversely across the abdomen
  - compress the abdominopelvic cavity and flex the vertebral column
  - act as accessory muscles of respiration during active expiration by pulling the lower chest downward, depressing the lower ribs.
- Pyramidalis – a small, triangular muscle inferior to the rectus abdominis muscle, which tenses the linea alba.

### Inguinal region

At the lower end of the abdomen, the aponeuroses of the external oblique muscles, which extend from the anterior superior iliac spine to the pubic tubercle, split lateral to the pubis to leave a gap – the superficial inguinal ring. The lower border of the aponeurosis as it passes from spine to pubis is the inguinal ligament. At a deeper level, a gap called the deep inguinal ring forms in the transversalis fascia. The oblique pathway between the superficial and deep inguinal rings, parallel to the inguinal ligament, is the inguinal canal, which is about 4 cm in length. The canal is a passageway for blood vessels and nerves and for the spermatic cord in males and the round ligament in females. (See *Inguinal region*, page 201.) Three cartilaginous inguinal rings keep the canal open:

- internal, or deep – at the interior opening
- external, or superficial – at the distal end
- middle – between the two.

**Clinical tip**

Remember these tips when assessing for an inguinal hernia:

- Inspect the area for visible bulges.
- Place the index and middle fingers of each hand over each external inguinal ring, and ask the patient to bear down or cough to briefly increase intra-abdominal pressure.
- Then, with the patient relaxed, gently insert the middle or index finger (adult male), or the little finger (young child), into the scrotal sac; follow the spermatic cord upward to the external inguinal ring, to an opening just above and lateral to the pubic tubercle, known as Hesselbach's triangle. Holding the finger at this spot, ask the patient to bear down or cough again. A hernia will feel like a soft mass or bulge.

## Posterior muscles

The posterior abdominal wall muscles are the psoas major, the iliacus, and the quadratus lumborum. (See *Posterior abdominal wall muscles, anterior view*, page 200.)

- Psoas major
  - originates from the transverse processes and bodies of the lumbar vertebrae and the fibrocartilages; inserts at the lesser trochanter of the femur
  - flexes the thigh and rotates the thigh laterally; flexes the vertebral column.
- Iliacus
  - originates at the iliac fossa and joins with the psoas major to form the iliopsoas
  - assists with thigh flexion.
- Quadratus lumborum
  - originates at the iliac crest and inserts on the 12th rib and upper four lumbar vertebrae; located in the posterior abdominal wall
  - pulls the thoracic cage toward the pelvis.

**Clinical tip**

The iliopsoas sign can suggest such conditions as appendicitis, peritonitis, or disease of an adjoining structure, such as the pancreas or a kidney.

- Position the patient supine with legs straight.
- Instruct patient to raise his right leg upward as you exert slight pressure with your hand; repeat the maneuver with the left leg.
- Increased abdominal pain is a positive result, reflecting inflammation of the psoas major muscle.

## ABDOMINAL WALL MUSCLES
### Anterior view

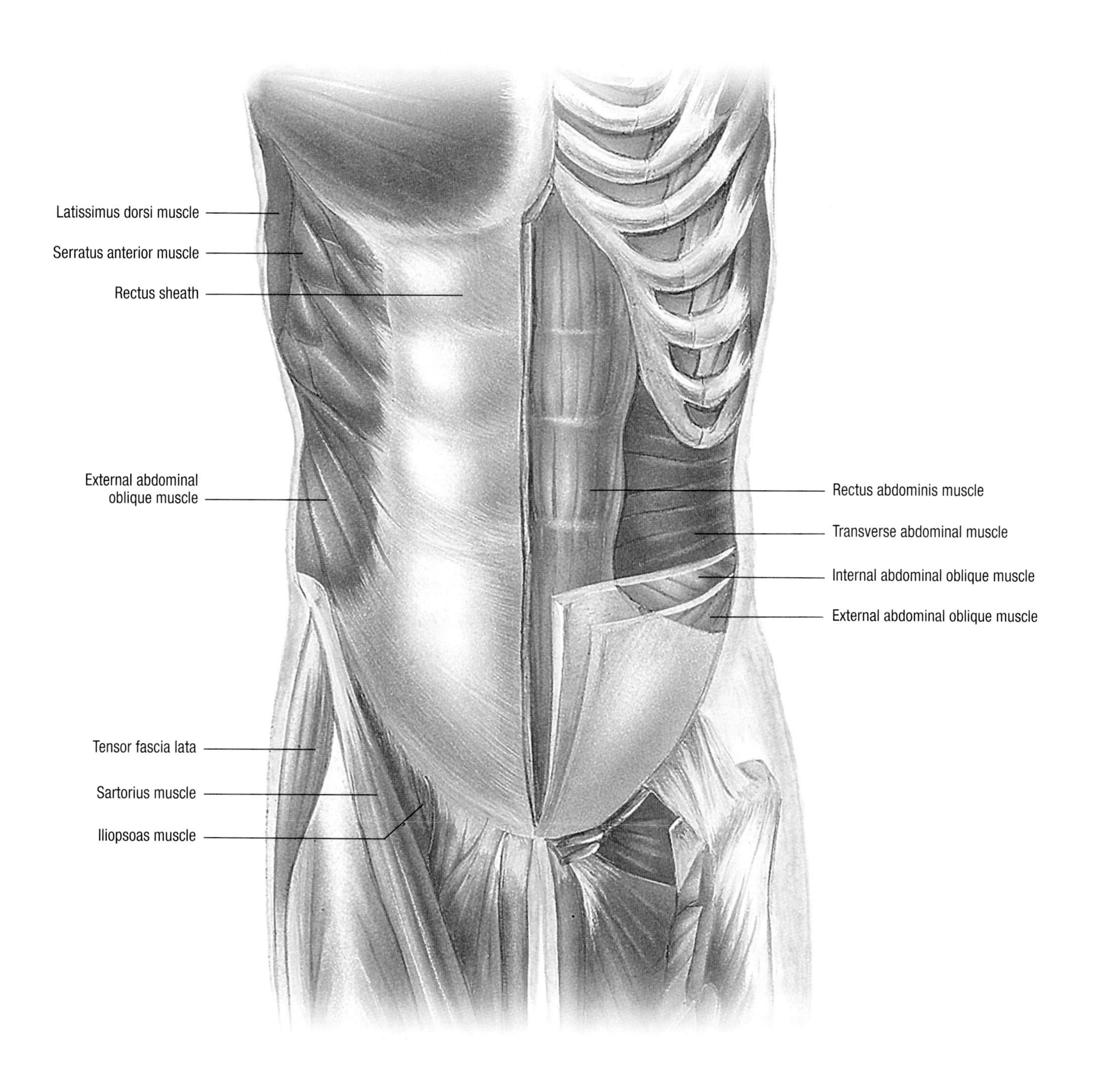

Latissimus dorsi muscle
Serratus anterior muscle
Rectus sheath
External abdominal oblique muscle
Tensor fascia lata
Sartorius muscle
Iliopsoas muscle
Rectus abdominis muscle
Transverse abdominal muscle
Internal abdominal oblique muscle
External abdominal oblique muscle

# POSTERIOR ABDOMINAL WALL MUSCLES
## Anterior view

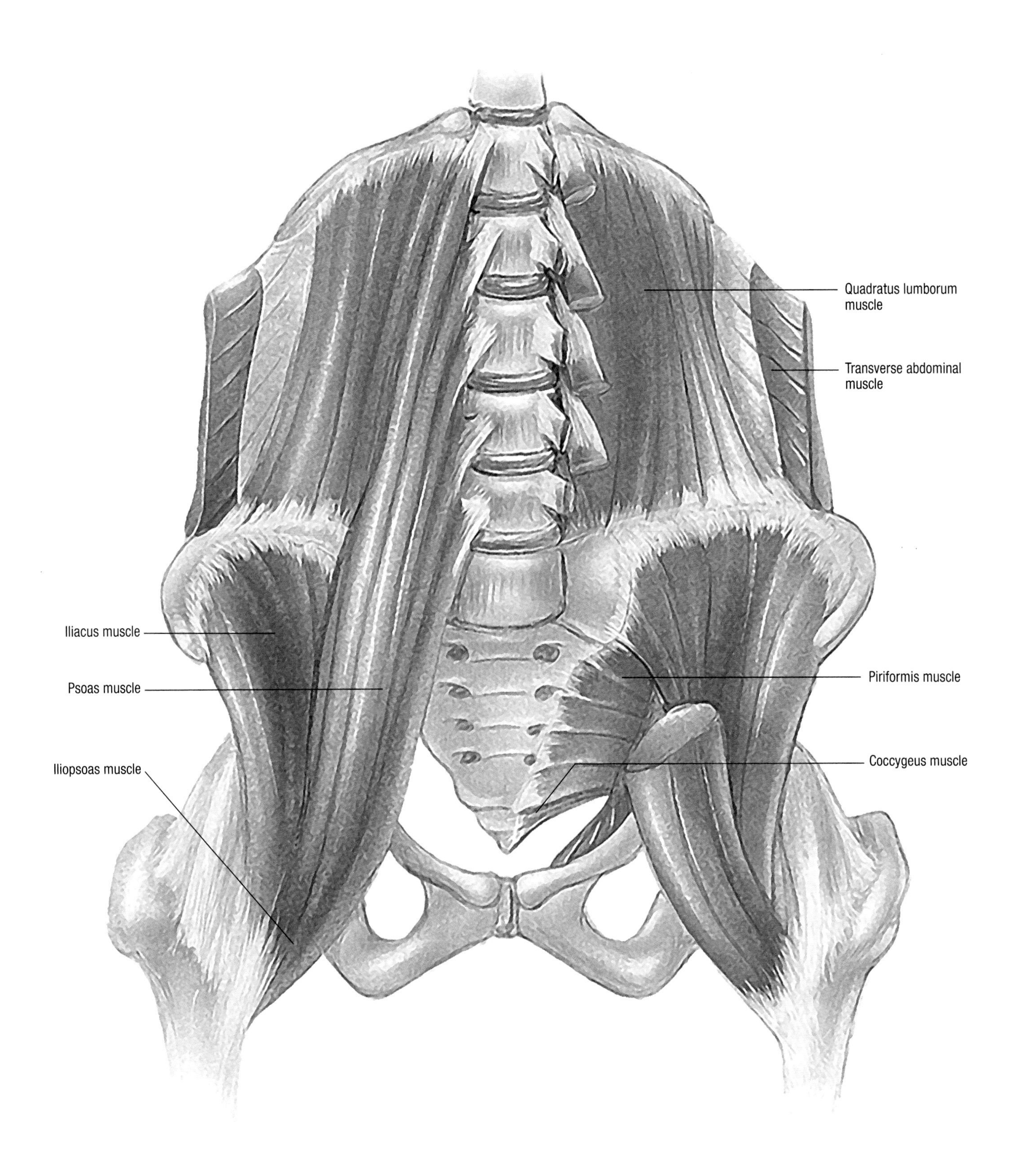

Quadratus lumborum muscle
Transverse abdominal muscle
Iliacus muscle
Psoas muscle
Iliopsoas muscle
Piriformis muscle
Coccygeus muscle

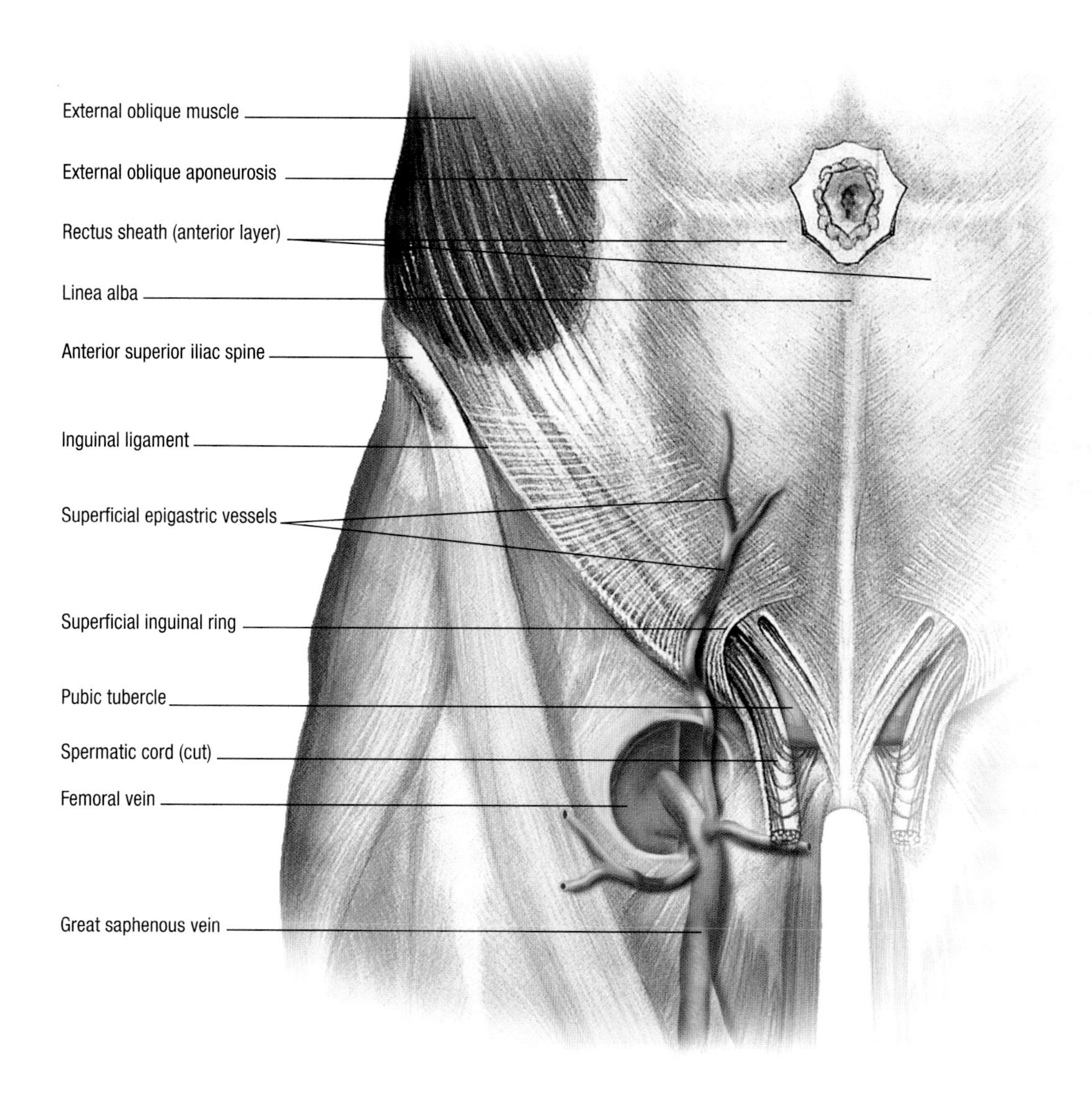
External oblique muscle
External oblique aponeurosis
Rectus sheath (anterior layer)
Linea alba
Anterior superior iliac spine
Inguinal ligament
Superficial epigastric vessels
Superficial inguinal ring
Pubic tubercle
Spermatic cord (cut)
Femoral vein
Great saphenous vein

# Esophagus

The esophagus is a collapsible muscular tube – about 10 inches (25 cm) long – that lies dorsal to the trachea and ventral to the vertebral column. It extends from the oropharynx anterior to the vertebral column, enters the mediastinum, leaves the thorax via the esophageal hiatus, and joins the stomach. The point where the esophagus ends and the stomach begins is the esophagogastric junction.

## Esophageal wall

The esophageal wall contains four layers, as follows from the lumen outward:

- mucosa – epithelium, lamina propria, and glands
- submucosa – connective tissue, blood vessels, and glands
- muscularis (middle layer) – upper third, striated muscle; middle third, striated and smooth; lower third, smooth muscle
- adventitia – connective tissue that merges with connective tissue of surrounding structures.

**Clinical tip**

Esophageal cancer is most common in men over age 60. It is also more common in African Americans and alcoholics. Esophageal tumors are usually fungating and infiltrating. In most cases, the tumor partially constricts the lumen of the esophagus. It metastasizes early by way of the submucosal lymphatic system.

## Innervation

The muscularis layer contains the nerve supply for the esophagus. Similar to the rest of the GI tract, nervous innervation is supplied by the myenteric plexus. The myenteric plexus contains fibers from both divisions of the autonomic nervous system. It controls GI motility.

## Circulation and lymphatic drainage

The submucosa contains the blood supply and lymphatic drainage system of the esophagus. The lymph tissue is especially important because the GI tract is continually exposed to bacteria and other deleterious agents from the environment. Bronchial and esophageal arteries supply blood and nutrients. Esophageal veins that empty into the hemiazygos vein in front of the vertebral column provide venous drainage.

## Function

The esophagus does not produce enzymes or absorb nutrients. Its main functions are secretion of mucus and movement of food.

### Food transport

Swallowing triggers movement of oral contents from the oropharynx into the esophagus through the cricopharyngeal sphincter, a thick ring of muscle. Passage of food and liquids from the pharynx to the stomach usually takes about 4 to 8 seconds. However, very soft foods or liquids may take as little as 1 second.

Peristalsis, the rhythmic contraction and relaxation of smooth muscle, propels liquids and solids through the esophagus into the stomach. It is controlled by the medulla oblongata.

The esophagus ends at the lower esophageal sphincter (LES). When the sphincter relaxes, the food bolus enters the stomach. Normally, the lower esophageal sphincter closes after food enters the stomach and opens only during swallowing, belching, or vomiting.

**Clinical tip**

If the lower esophageal sphincter fails to close or remain closed, backflow of gastric juices, or esophageal reflux, occurs. Evaluate for symptoms of reflux such as:

- heartburn, relieved with antacids or sitting upright
- sour or bitter taste in the throat
- regurgitation.

If the LES fails to relax as food approaches, the food cannot move into the stomach. This is known as achalasia. Key symptoms to be monitored include excruciating abdominal and chest pain mimicking a myocardial infarction. Achalasia is caused by malfunction of the myenteric plexus.

## ESOPHAGUS

Nasal cavity

Tongue

Oropharynx

Esophagus

Aorta

Liver

Gallbladder

Portal vein

Celiac trunk

Stomach

## ESOPHAGOGASTRIC JUNCTION

Esophagus

Diaphragm

Stomach

# Stomach

The stomach temporarily stores food and liquids while digestion begins. It is a collapsible, pouchlike structure about 10 inches (25 cm) long and capable of holding 2 to 4 qt. Attached to the lower end of the esophagus, it lies immediately inferior to the diaphragm and extends to the duodenal portion of the small intestine. It lies in the left upper quadrant of the abdominal cavity.

The lateral surface of the stomach is called the greater curvature; the medial surface, the lesser curvature. The lesser omentum layer of the peritoneum extends around the stomach, and the greater omentum is found along the greater curvature of the stomach. The interior of the stomach is lined with rows of folds or wrinkles, called rugae.

**CLINICAL TIP**

The size and position of the stomach vary among individuals. One contributing factor is the degree of distention. The markedly distended stomach – for example, after a very large meal – impedes downward movement of the diaphragm during inspiration and may cause shortness of breath.

## REGIONS

The stomach has four main regions – the cardia, fundus, body, and pylorus.

- cardia – immediately distal to gastroesophageal junction of the stomach and esophagus
- fundus – the enlarged portion distal to the cardia, lying above and to the left of the gastroesophageal opening
- body – the middle portion of the stomach, distal to the fundus and tapering in size
- pylorus – the lower portion, between the body and the gastroduodenal junction.

## MUSCLE LAYERS

The stomach has three layers of smooth muscle – the outer longitudinal, the middle circular, and the inner oblique muscles. Contraction of these muscles helps mix and break the contents into a suspension of nutrients called chyme and propels it into the duodenum.

## BLOOD SUPPLY AND LYMPHATIC DRAINAGE

The celiac trunk and its branches provide the arterial blood supply to the stomach. The left gastric artery supplies the lesser curvature of the fundus and the body of the stomach. The right gastric artery is a loop that supplies the lesser curvature and then forms an anastomosis with the left gastric artery. The left and right gastro-omental arteries supply the greater curvature. Many other small arterial branches provide an extensive collateral circulation to the stomach.

Gastric veins run parallel to the arteries of the stomach. The following are the major abdominal veins and their flow patterns:

- left and right gastric veins – directly into the portal vein
- left gastro-omental vein – into the splenic vein
- right gastro-omental vein – into the superior mesenteric vein
- splenic and superior mesenteric veins – form the portal vein.

Lymph is drained by gastric lymphatic vessels to the gastric and gastro-omental lymph nodes, near the curvatures. Lymphatic drainage from the stomach may also pass through the pancreaticosplenic, pyloric, or pancreaticoduodenal nodes.

## INNERVATION

The stomach has a parasympathetic and a sympathetic nerve supply. Parasympathetic nerves include anterior and posterior branches of the vagus nerves. Sympathetic innervation is received from the celiac plexus via the greater splanchnic nerve and its branches.

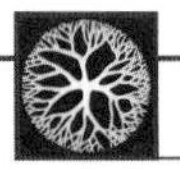

**PHYSIOLOGY**

### SITES AND MECHANISMS OF GASTRIC SECRETION

The body of the stomach lies between the lower esophageal, or cardiac, sphincter (LES) and the pyloric sphincter. Between these sphincters lie the cardia, fundus, body, antrum, and pylorus. These areas have a rich variety of mucosal cells that help the stomach carry out its tasks.

#### Glands and secretions

Three types of glands secrete 2 to 3 liters of gastric juice daily through the stomach's gastric pits:

- cardiac glands near the LES – thin mucus
- pyloric glands in the pylorus – thin mucus
- gastric glands in the body and fundus – hydrochloric acid (HCl), pepsinogen, intrinsic factor, and mucus.

Specialized cells line the gastric glands, gastric pits, and surface epithelium. Mucous cells in the necks of the gastric glands produce a thin mucus; those in the surface epithelium, a protective alkaline mucus. Both substances lubricate gastric contents and protect the stomach from self-digestion by corrosive enzymes and acids.

***Gastrin and pepsinogen***

Argentaffin cells in gastric glands produce the hormone gastrin. Chief cells, primarily in the fundus, produce pepsinogen – the inactive precursor of the proteolytic enzyme pepsin, which breaks proteins into polypeptides.

***Hydrochloric acid and intrinsic factor***

Large parietal cells scattered throughout the fundus secrete HCl and intrinsic factor. HCl degrades pepsinogen into pepsin and maintains the acid environment favorable for pepsin activity. It also helps disintegrate nucleoproteins and collagens, hydrolyzes sucrose, and inhibits excess growth of bacteria. Intrinsic factor promotes vitamin $B_{12}$ absorption in the small intestine.

## STOMACH REGIONS

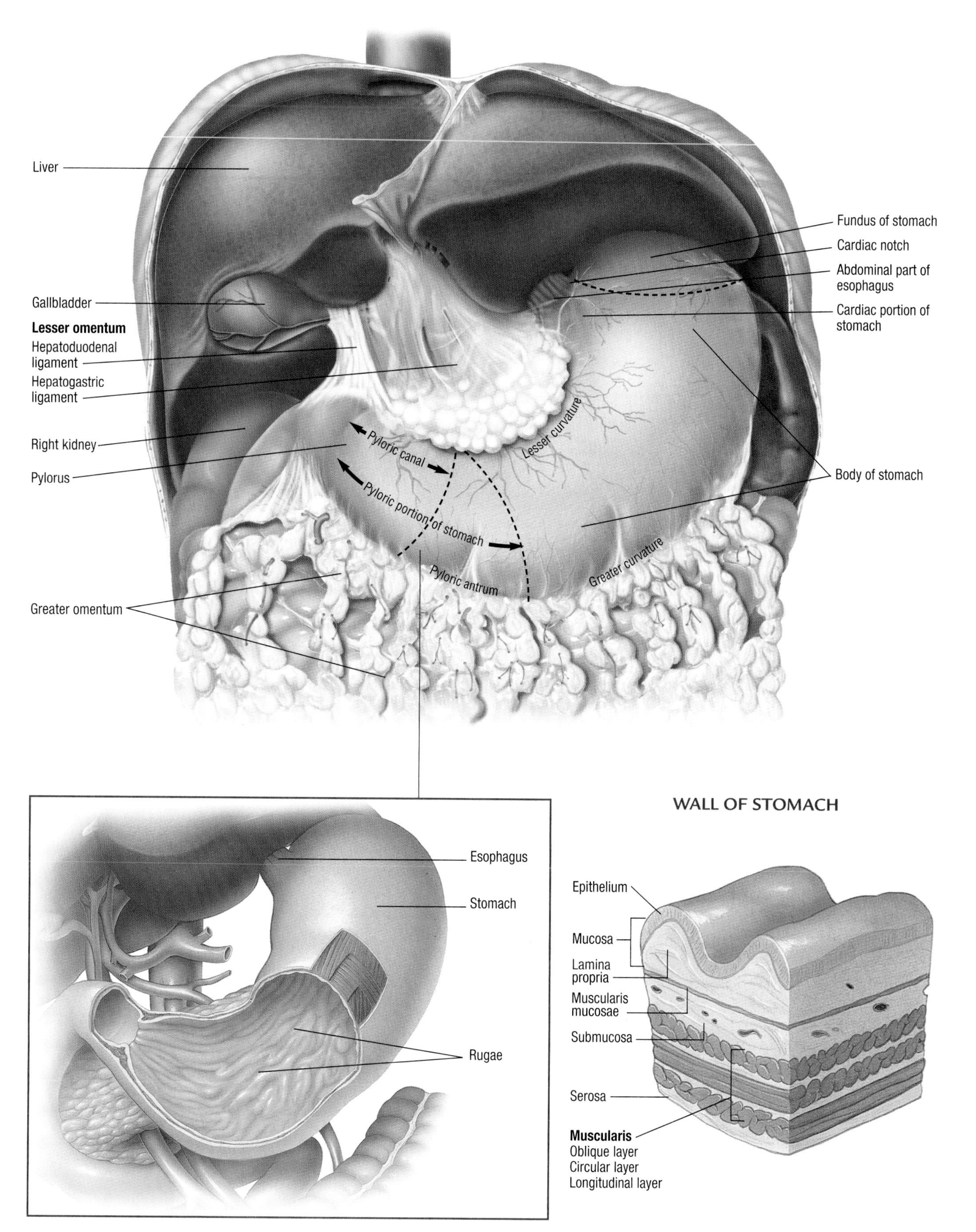

Liver
Gallbladder
Lesser omentum
Hepatoduodenal ligament
Hepatogastric ligament
Right kidney
Pylorus
Greater omentum
Fundus of stomach
Cardiac notch
Abdominal part of esophagus
Cardiac portion of stomach
Body of stomach
Lesser curvature
Pyloric canal
Pyloric portion of stomach
Pyloric antrum
Greater curvature
Esophagus
Stomach
Rugae
WALL OF STOMACH
Epithelium
Mucosa
Lamina propria
Muscularis mucosae
Submucosa
Serosa
Muscularis
Oblique layer
Circular layer
Longitudinal layer

# Small intestine

The longest organ of the GI tract, the small intestine is a tube about 21′ (6.35 m) long and 1″ (2.5 cm) in diameter. The small intestine extends from the pyloric sphincter to the ileocecal valve, where it joins the large intestine. The small intestine has three major divisions:
• duodenum, the first division – roughly 10″ (25 cm) long
• jejunum, the middle portion – about 8′ (2.5 m) long
• ileum, the third portion – approximately 12′ (3.7 m) long.

## Intestinal wall

The intestinal wall consists of several layers: mucosa, submucosa, muscularis (consisting of circular and longitudinal smooth muscles ), and serosa (visceral peritoneum). Peyer's patches, located in the ileum, are collections of lymphatic tissue in the submucosa. Structural features of the wall of the small intestine significantly increase its absorptive surface area. These features include:
• plicae circulares – circular folds of the intestinal mucosa
• villi – fingerlike projections of the mucosa
• microvilli – tiny cytoplasmic projections on the surfaces of epithelial cells.

### Intestinal juice

Intestinal crypts, simple glands in the grooves separating villi, secrete digestive juices and provide replacement epithelial cells for the villi. The small intestine secretes 2 to 3 liters per day of intestinal juice. This slightly alkaline liquid (pH about 7.6) contains water, mucus, and the following enzymes:
• carbohydrate-digesting – maltase, sucrase, and lactase
• protein-digesting – peptidase
• nucleic acid-digesting – ribonuclease and deoxyribonuclease.

Brunner's submucosal glands in the duodenum secrete an alkaline mucus to protect the walls of the mucosa from the enzymes.

## Functions

The main function of the small intestine is to complete digestion. Segmented contractions mix its contents, and peristalsis propels the contents toward the large intestine.

**Clinical tip**

Abnormal peristalsis can cause constipation, diarrhea, or both, as occurs in irritable bowel syndrome. If your patient exhibits signs of abnormal peristalsis, expect treatment to include stress reduction, smoking cessation, high-fiber diet, antispasmodics, laxatives, antidiarrheal agents, or tranquilizers, depending on the diagnosis.

Most of the nutrients, water, and electrolytes in foods are digested and absorbed during the 6- to 8-hour passage through the small intestine. The intestinal hormones cholecystokinin (CCK) and secretin regulate gallbladder function, as well as pancreatic fluid and bile secretion. (See *Gallbladder*, page 216.)

As digestion reaches completion, nutrient molecules are absorbed through the wall of the small intestine into the circulatory system, which transports them throughout the body.

## Blood supply and lymphatic drainage

The superior mesenteric artery, which branches off from the aorta, supplies the small intestine. The superior mesenteric vein and its branches return blood from the small intestine via the hepatic portal system. Lymphatic drainage of the intestine occurs via intestinal nodes and the intestinal trunk, which is part of the thoracic duct system.

## Innervation

The sympathetic and parasympathetic divisions of the autonomic nervous system control contraction of smooth muscles in the intestinal wall.

### Sympathetic

The splanchnic nerve passes through the celiac or solar plexus. Distally, postganglionic fibers innervate the small intestine. Sympathetic stimulation slows motility of the small intestine.

### Parasympathetic

The vagus (X) nerve supplies a vast distribution of parasympathetic fibers. Postganglionic fibers from the plexus associated with the vagus nerve innervate the small intestine. Parasympathetic stimulation of the small intestine causes increased motility.

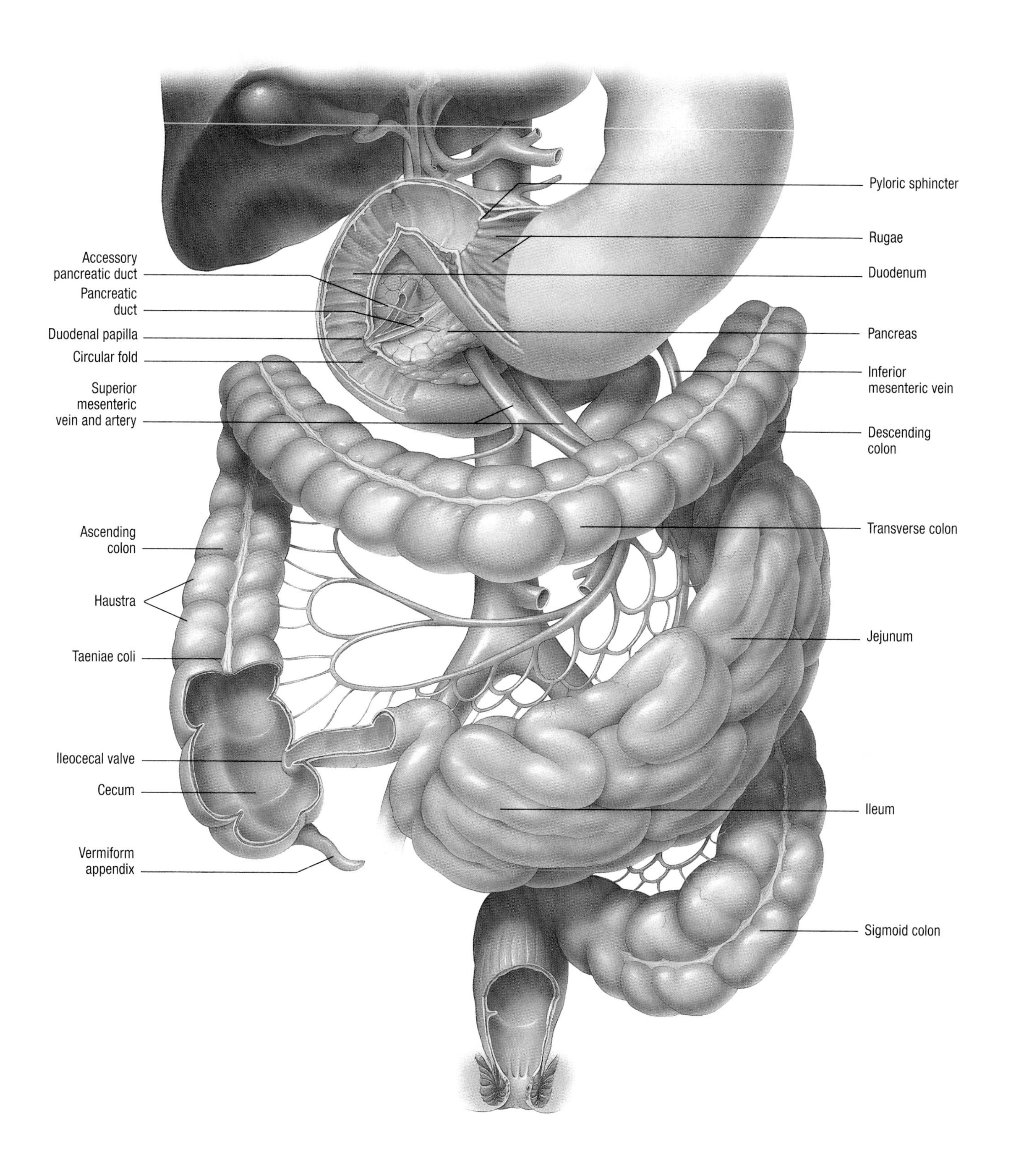

Pyloric sphincter
Rugae
Accessory pancreatic duct
Duodenum
Pancreatic duct
Duodenal papilla
Pancreas
Circular fold
Inferior mesenteric vein
Superior mesenteric vein and artery
Descending colon
Ascending colon
Transverse colon
Haustra
Jejunum
Taeniae coli
Ileocecal valve
Cecum
Ileum
Vermiform appendix
Sigmoid colon

# Large intestine

The large intestine extends from the ileocecal valve (between the ileum and the first segment of the large intestine) to the anus and is about 59 inches (1.5 m) long.

## Segments

The large intestine has five segments:

- cecum – a saclike structure beginning just below the ileocecal valve and extending a few inches; the appendix is a blind pouch attached to the cecum
- ascending colon – runs upward along the right posterior abdominal wall, then turns sharply under the liver at the hepatic flexure
- transverse colon – extends horizontally across the abdomen above the small intestine and below the liver, stomach, and spleen; turns downward at the left colic flexure, near the spleen
- descending colon – extends down the left side of the abdomen into the pelvic cavity
- sigmoid colon – descends through the pelvic cavity, where it becomes the rectum. The rectum, the last few inches (about 20 cm) of the large intestine, terminates at the anus.

Unlike those of the rest of the GI tract, longitudinal muscles do not form a continuous layer around the large intestine. Instead, three bands of longitudinal muscle, called taeniae coli, run the length of the colon. Contractions gather the colon into bands (haustra), giving the colon its "puckered" appearance.

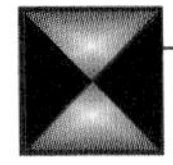

**Age-related changes**

Structural changes of the large intestine that occur with age may have functional effects:

- decreased motility
- decreased bowel wall tone
- decreased anal sphincter tone.

## Functions

The large intestine absorbs water, secretes mucus, and eliminates digestive wastes.

Peristalsis and segmenting contractions move the intestinal contents slowly; water and minerals are absorbed from the contents, leaving a residue of fecal material. Intestinal bacteria act on the residue, releasing decomposition products and intestinal gases. These bacteria also synthesize vitamin K and some B vitamins, which are absorbed into the colon.

Peristalsis propels fecal material into the rectum. Its arrival triggers reflex contraction of rectal smooth muscle and relaxation of the internal anal sphincter (defecation reflex). Evacuation of fecal contents is the result of voluntary relaxation of the external anal sphincter and bearing-down efforts.

**Clinical tip**

Normal bowel sounds are high-pitched, gurgling noises caused by air mixing with fluid during peristalsis. The noises vary in pitch, frequency, and intensity, and occur regularly from 5 to 34 times per minute.

## Blood supply and lymphatic drainage

The superior and inferior mesenteric and middle sacral arteries are branches of the aorta. The superior mesenteric artery supplies the large intestine; the inferior mesenteric, the large intestine and rectum; and the middle sacral, the rectum.

Venous drainage of the large intestine occurs via the hepatic portal system. The inferior mesenteric vein drains the rectum, sigmoid, and descending portions of the colon.

Lymphatic drainage of the intestine occurs via the intestinal trunk, which is part of the thoracic duct system. Lymph nodes lie along the intestinal trunk.

## Innervation

The sympathetic and parasympathetic divisions of the autonomic nervous system control contraction of both layers of smooth muscle in the intestinal wall.

### Sympathetic

The splanchnic nerve terminates in the celiac or solar plexus. Distally, postganglionic fibers innervate the small intestine. Sympathetic stimulation leads to decreased motility in the small intestine.

### Parasympathetic

The vagus (X) nerve supplies a vast distribution of parasympathetic fibers. Preganglionic fibers form the sacral parasympathetic outflow. These fibers form the pelvic splanchnic nerves. They pass through the celiac plexus and form a network of postganglionic fibers, which innervate the large intestine. Parasympathetic stimulation of the large intestine increases motility.

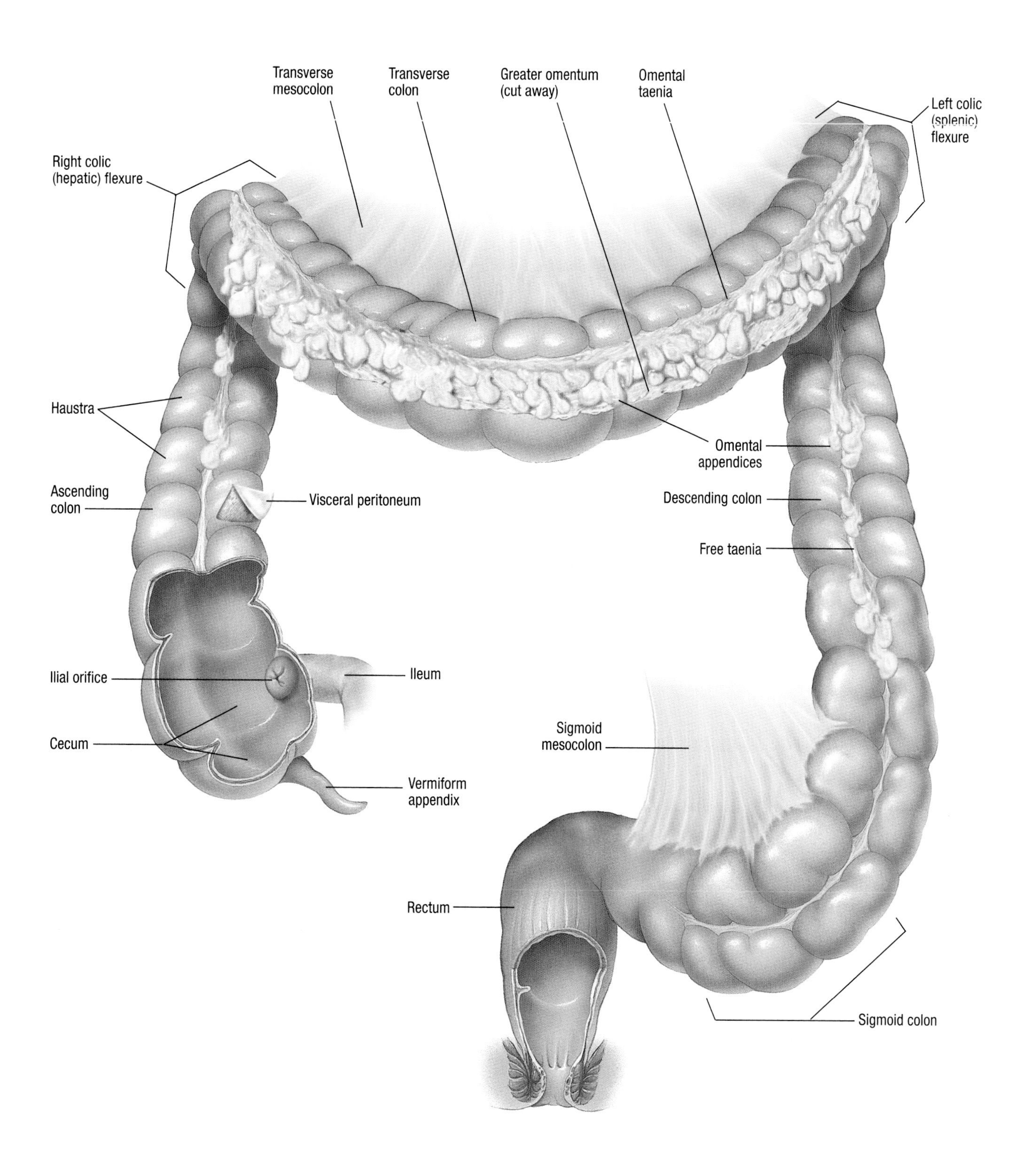
Transverse mesocolon
Transverse colon
Greater omentum (cut away)
Omental taenia
Left colic (splenic) flexure
Right colic (hepatic) flexure
Haustra
Omental appendices
Ascending colon
Visceral peritoneum
Descending colon
Free taenia
Ilial orifice
Ileum
Cecum
Sigmoid mesocolon
Vermiform appendix
Rectum
Sigmoid colon

# Liver

The liver is the largest organ in the body. It is immediately inferior to the diaphragm and partially anterior to the stomach. The liver consists of a right lobe, left lobe, caudate lobe (behind the right lobe), and quadrate lobe (below the left lobe). (See *Liver segments*, page 212, and *Liver segments with biliary draining areas*, page 213.)

The liver's functional unit, the lobule, is a plate of hepatic cells, or hepatocytes, that encircle a central vein and radiate outward. (See *Liver lobule*, page 215.) Separating the hepatocyte plates from each other are sinusoids, which make up the liver's capillary system. The sinusoids carry oxygenated blood from the hepatic artery, and nutrient-rich blood from the portal vein. Unoxygenated blood leaves through the central vein and flows through hepatic veins to the inferior vena cava.

## Blood and lymph

A fold of visceral peritoneum called the lesser omentum covers most of the liver; the hepatic artery and hepatic portal vein (both of which enter the liver) and the common bile duct and hepatic veins (both of which leave the liver) all pass through the lesser omentum. (See *Distribution of vessels and ducts,* page 214.) The intestinal trunk provides lymphatic drainage.

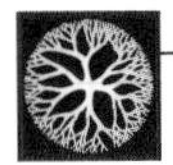

**Physiology**

Reticuloendothelial macrophages (Kupffer's cells) lining the sinusoids remove bacteria and toxins that have entered the blood through the intestinal capillaries.

## Bile

The liver recycles bile salts in the blood to form bile, which drains into bile ducts (canaliculi). The canaliculi merge into the right and left hepatic ducts to form the common hepatic duct. (See *Duct system*, page 214.) This duct joins the cystic duct from the gallbladder to form the common bile duct, which empties into the duodenum.

**Physiology**

The liver secretes 500 to 1,000 ml of bile daily. Bile is a complex secretion composed of cholesterol, lecithin (a phospholipid), bile salts (composed of cholesterol and amino acid derivatives), minerals, bile pigments (derived from hemoglobin breakdown), and water. Bile emulsifies fats into small globules for more efficient digestion; bile salts promote the absorption of fats and fat-soluble vitamins.

## Functions

A specialized accessory organ, the liver has digestive, metabolic, and regulatory functions; its chief digestive function is production of bile, which acts as a fat emulsifier in the small intestine.

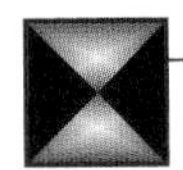

**Age-related changes**

Hepatic function decreases with age. Because many drugs are metabolized in the liver, the elderly are more at risk for drug toxicity.

The liver receives the products of digestion through the portal vein and processes them into physiologically essential substances. Its actions include:

- conversion of absorbed hexoses, amino acids, and lipid digestion products into whatever nutrient mixture the body needs for metabolic processes
- formation of ketone bodies from products of lipid metabolism
- production of blood proteins (such as albumin and globulin), lipoproteins, and proteins involved with blood coagulation
- storage of a small reserve of fat and glycogen, iron, and vitamins A, $B_{12}$, D, E, and K
- detoxification or excretion of many wastes, toxins, and drugs, such as antibiotics and steroid hormones.

**Clinical tip**

Hepatomegaly is often associated with hepatitis and other liver diseases. Liver borders may be difficult to assess. Here's one way to estimate the span of the liver.

- Identify the upper border of liver dullness. Start in the right midclavicular line in the area of lung resonance, and percuss downward toward the liver. Use a pen to mark the spot where the sound changes to dullness.
- Start in the right midclavicular line at a level below the umbilicus, and lightly percuss upward toward the liver. Mark the spot where the sound changes from tympany to dullness.
- Use a ruler to measure the vertical span between the two marks. Normal liver span in an adult is 2 ½" to 4 ¾" (6.5 to 12 cm).

## LIVER
### Anterior view

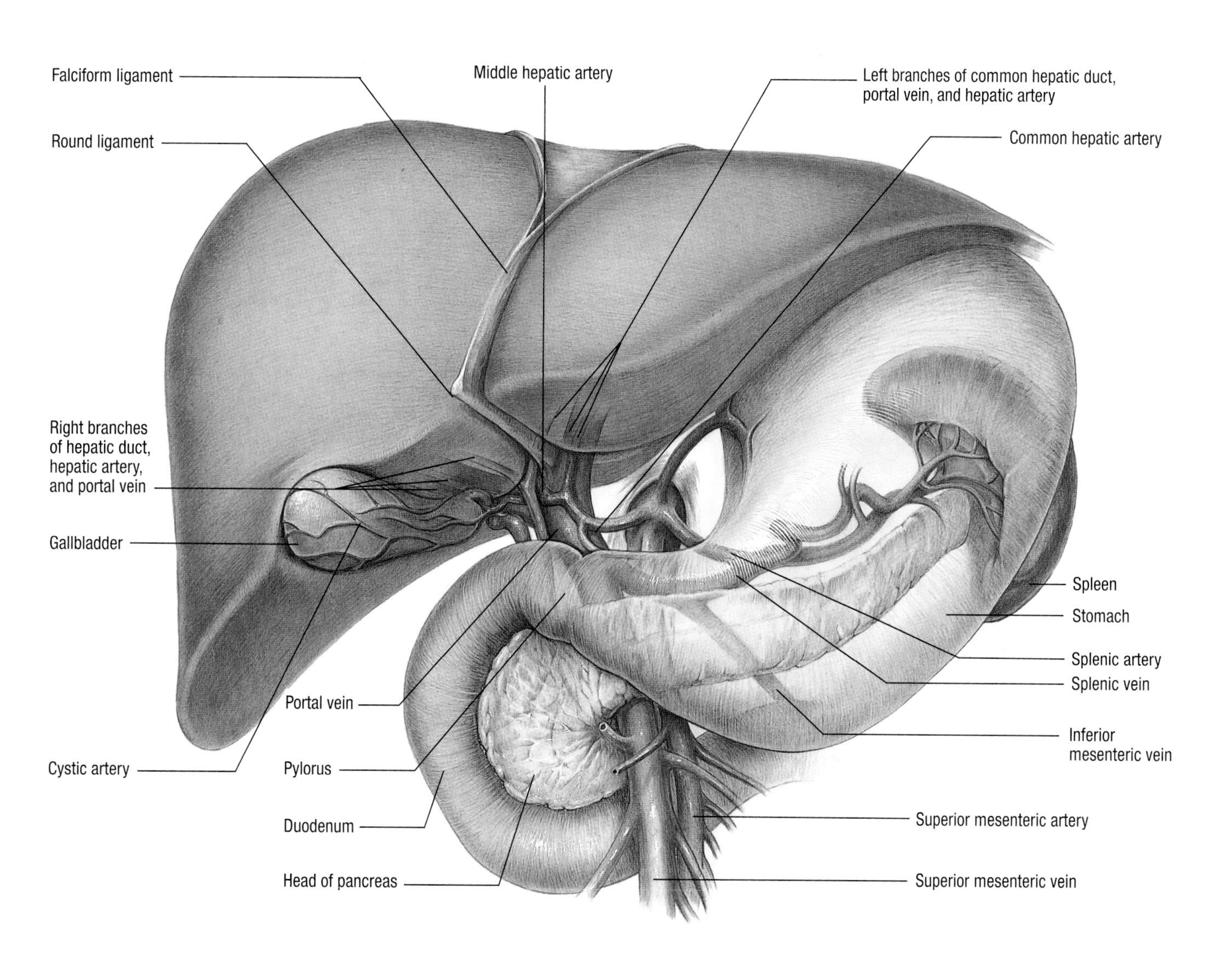

Falciform ligament
Middle hepatic artery
Left branches of common hepatic duct, portal vein, and hepatic artery
Round ligament
Common hepatic artery
Right branches of hepatic duct, hepatic artery, and portal vein
Gallbladder
Spleen
Stomach
Splenic artery
Splenic vein
Portal vein
Inferior mesenteric vein
Cystic artery
Pylorus
Duodenum
Superior mesenteric artery
Head of pancreas
Superior mesenteric vein

## LIVER SEGMENTS
**Posterior view**

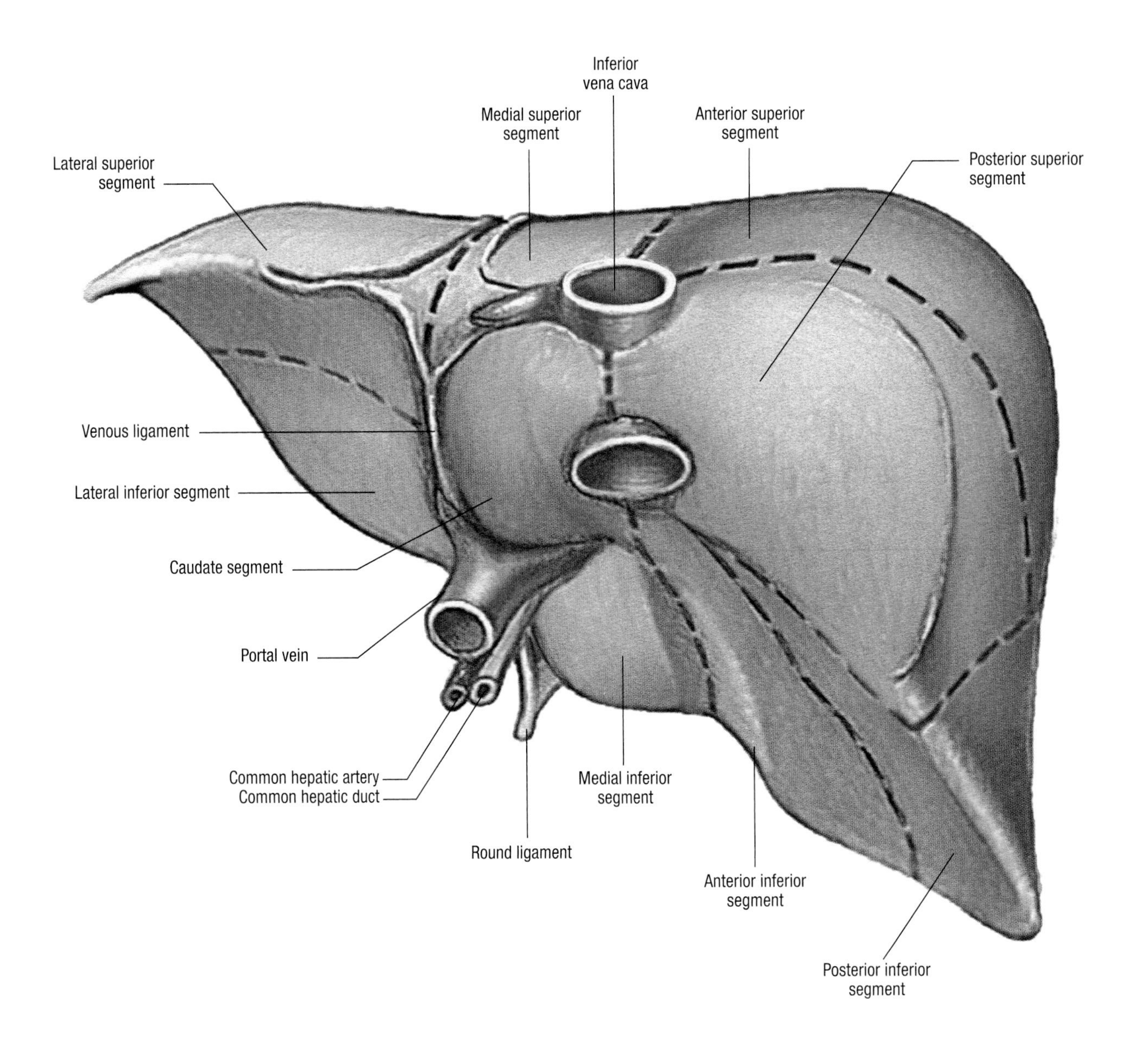

Inferior vena cava
Medial superior segment
Anterior superior segment
Lateral superior segment
Posterior superior segment
Venous ligament
Lateral inferior segment
Caudate segment
Portal vein
Common hepatic artery
Common hepatic duct
Medial inferior segment
Round ligament
Anterior inferior segment
Posterior inferior segment

## LIVER SEGMENTS WITH BILIARY DRAINING AREAS

**Visceral view**

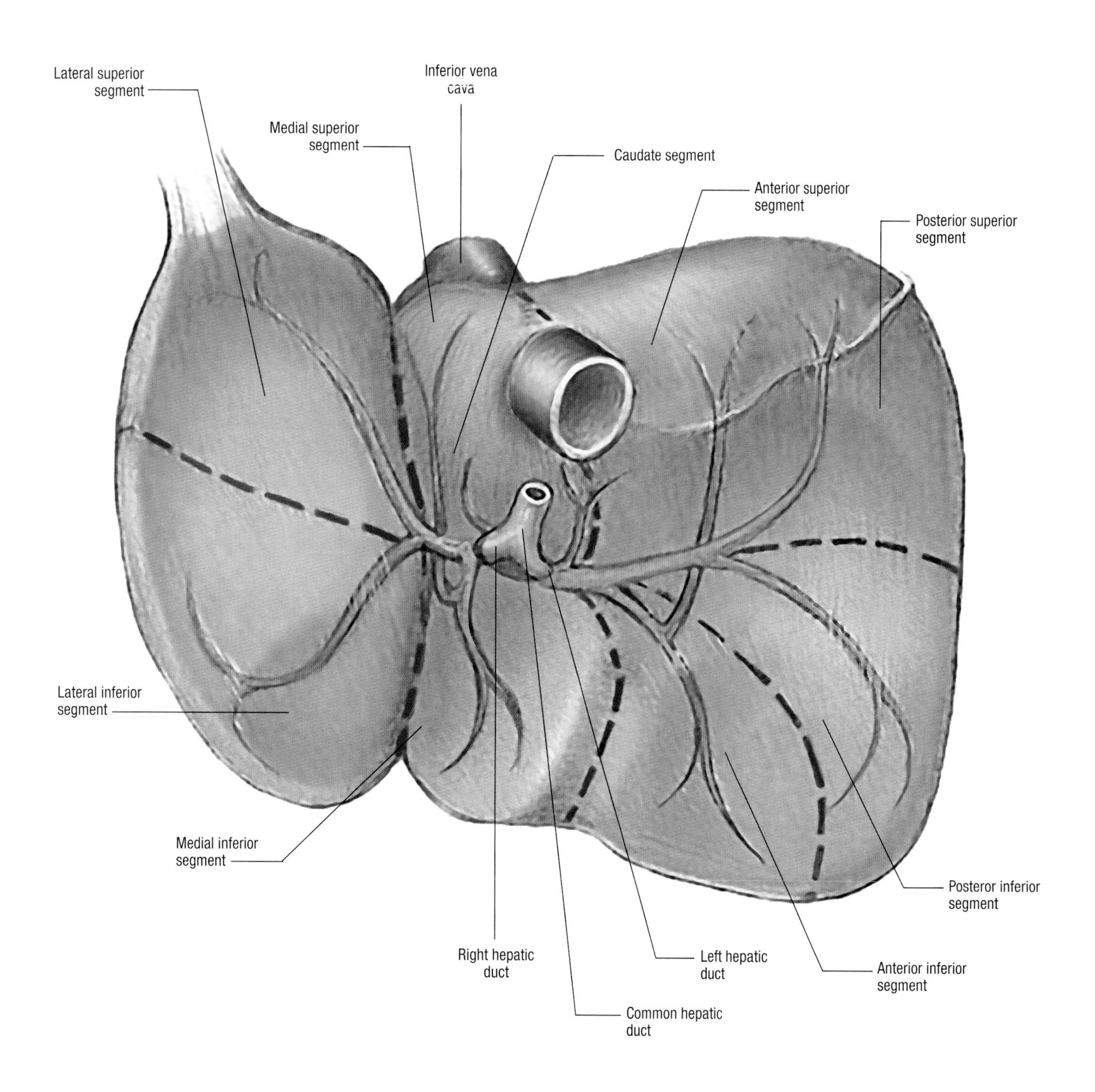
Lateral superior segment
Inferior vena cava
Medial superior segment
Caudate segment
Anterior superior segment
Posterior superior segment
Lateral inferior segment
Medial inferior segment
Posteror inferior segment
Right hepatic duct
Left hepatic duct
Anterior inferior segment
Common hepatic duct

## DISTRIBUTION OF VESSELS AND DUCTS

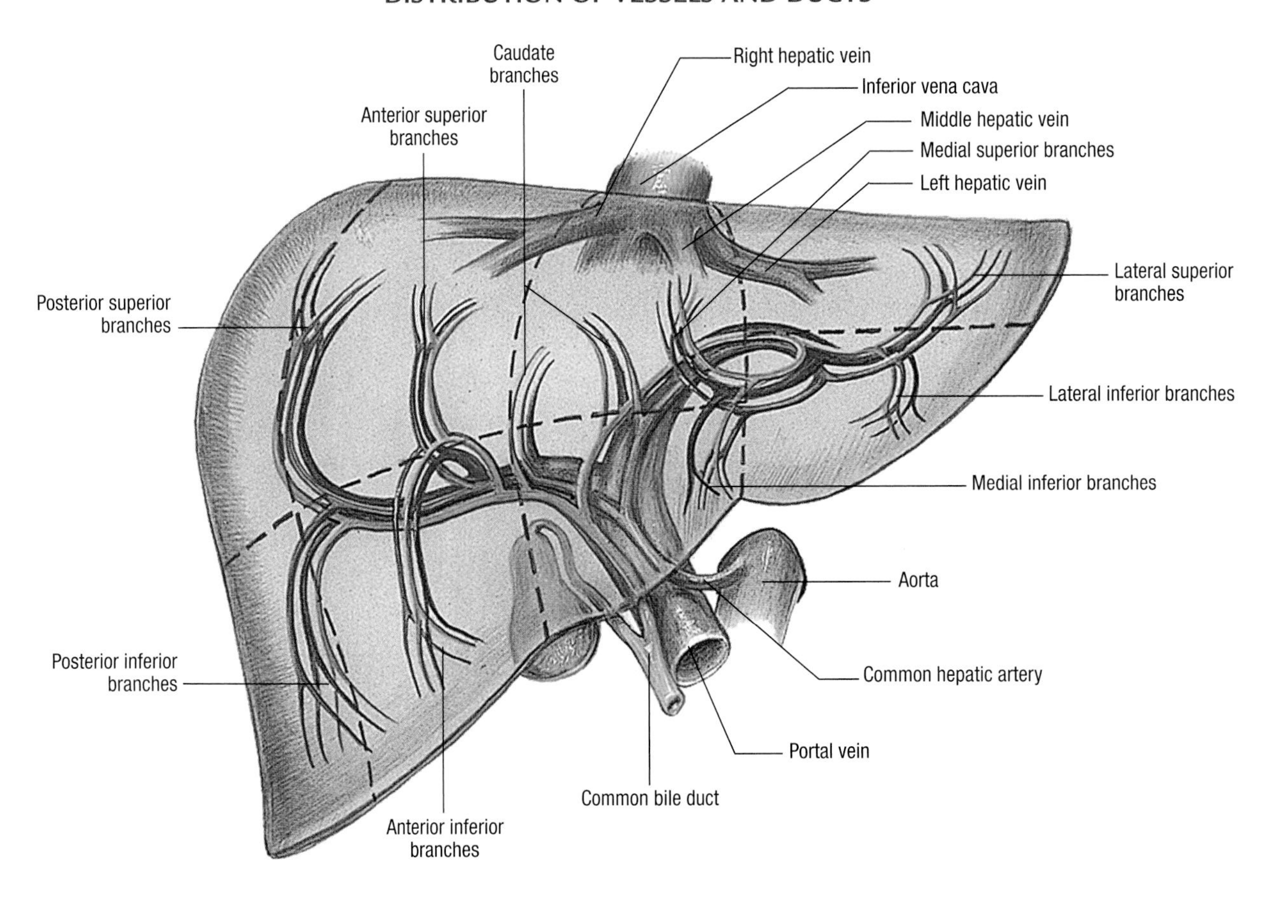

Caudate branches
Right hepatic vein
Inferior vena cava
Anterior superior branches
Middle hepatic vein
Medial superior branches
Left hepatic vein
Lateral superior branches
Posterior superior branches
Lateral inferior branches
Medial inferior branches
Aorta
Common hepatic artery
Posterior inferior branches
Portal vein
Common bile duct
Anterior inferior branches


## DUCT SYSTEM

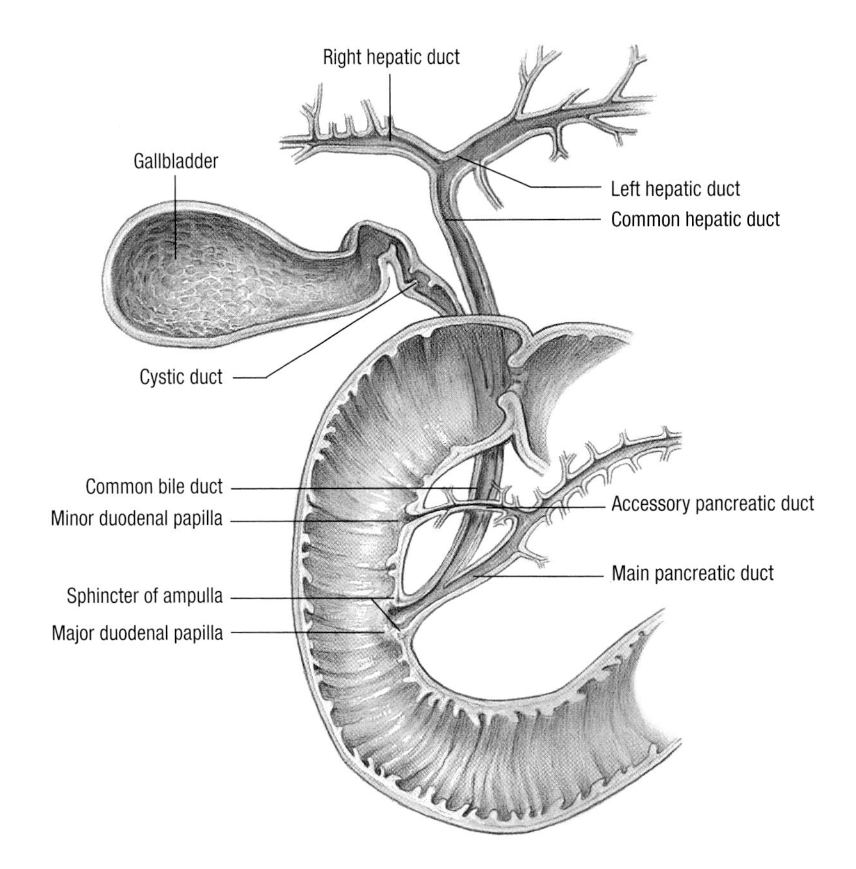

Right hepatic duct
Gallbladder
Left hepatic duct
Common hepatic duct
Cystic duct
Common bile duct
Accessory pancreatic duct
Minor duodenal papilla
Main pancreatic duct
Sphincter of ampulla
Major duodenal papilla

## LIVER LOBULE

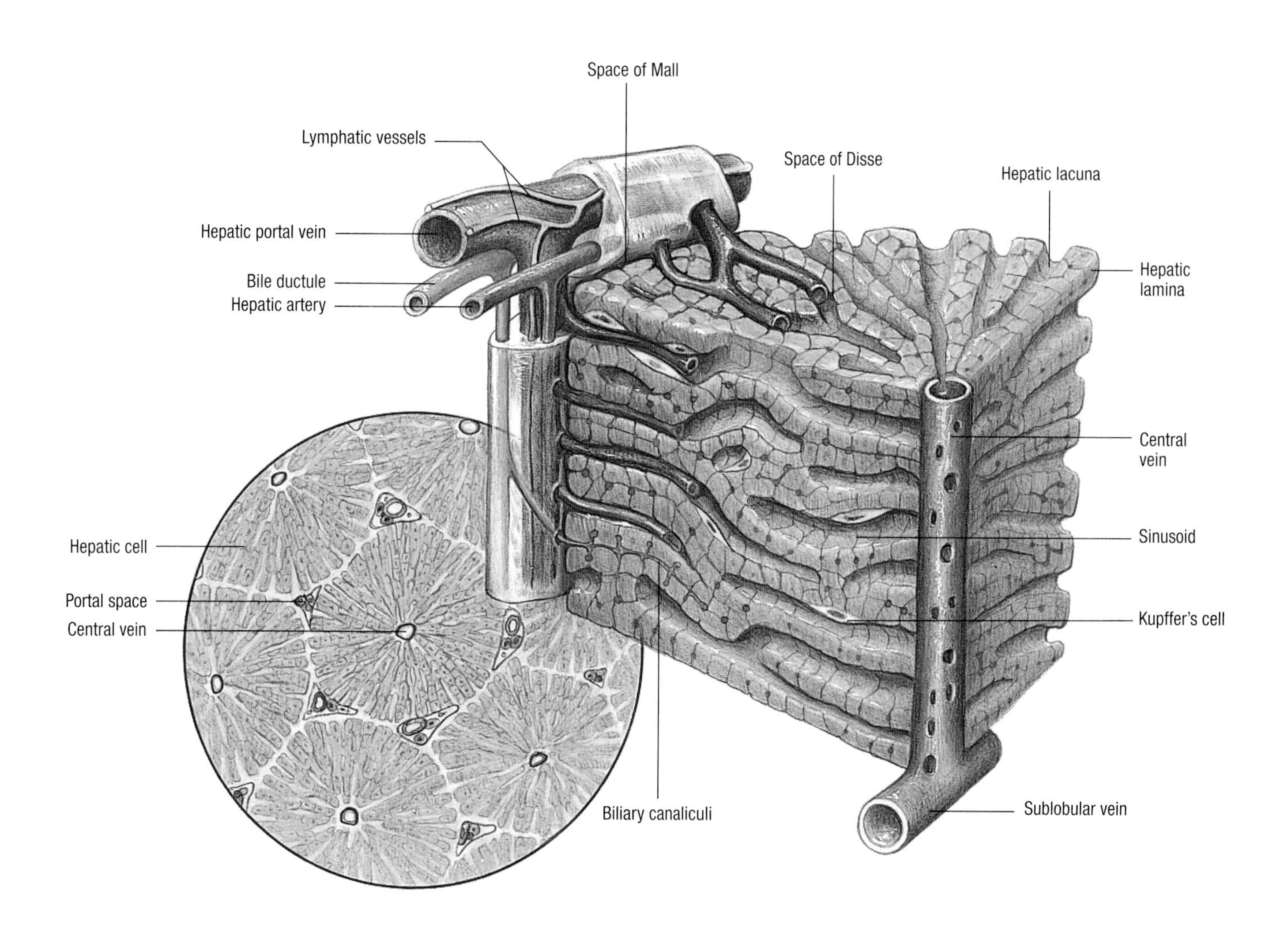

Space of Mall
Lymphatic vessels
Space of Disse
Hepatic lacuna
Hepatic portal vein
Bile ductule
Hepatic artery
Hepatic lamina
Central vein
Sinusoid
Hepatic cell
Portal space
Central vein
Kupffer's cell
Biliary canaliculi
Sublobular vein

# Gallbladder

The gallbladder is a muscular, pear-shaped sac, roughly 3 to 4 inches (7.5 to 10 cm) long. It is attached to the ventral surface of the liver between the right and quadrate lobes. The cystic duct connects the gallbladder to the common bile duct.

The gallbladder's lining is folded into rugae (similar to those in the stomach). Its middle layer consists of smooth-muscle fibers that contract to eject bile. Its outer surface is covered with visceral peritoneum.

## Functions

The gallbladder stores and concentrates bile produced by the liver. Bile can enter the duodenum only when the ampulla of Oddi's sphincter is open, which occurs during digestion. Secretion of the hormone cholecystokinin after a fatty meal stimulates gallbladder contraction and relaxation of Oddi's sphincter. When the sphincter closes, bile is shunted to the gallbladder for storage.

### Regulation of cholesterol

Bile salts form micelles, which have a lipophilic (lipid-attracting) center and a hydrophilic (water-attracting) periphery; these structures regulate the solubility of cholesterol in bile. Cholesterol dissolves in the centers of the micelles. Cholesterol may precipitate from bile and form gallstones if its concentration in bile exceeds the capacity of the micelles to hold cholesterol in solution.

**Clinical tip**

Signs and symptoms of acute cholecystitis often strike after a person eats a meal that is rich in fats. They may occur at night and suddenly awaken a patient. These signs and symptoms include:

- acute abdominal pain in the right upper quadrant that may radiate to the back, between the shoulders, or to the front of the chest
- colic
- belching, flatulence, indigestion
- light-headedness
- chills
- nausea, vomiting
- low-grade fever
- jaundice.

Untreated cholecystitis can progress to chronic cholecystitis, empyema, hydrops, gangrene, peritonitis, or pancreatitis.

**Age-related changes**

The acute form of cholecystitis is most commonly found in young adults. The chronic form of cholecystitis is most often found in the elderly.

### Circulation

The cystic vein drains the gallbladder and empties into the hepatic portal vein. Branches of the celiac artery supply oxygenated blood and nutrients to the gallbladder.

## GALLBLADDER

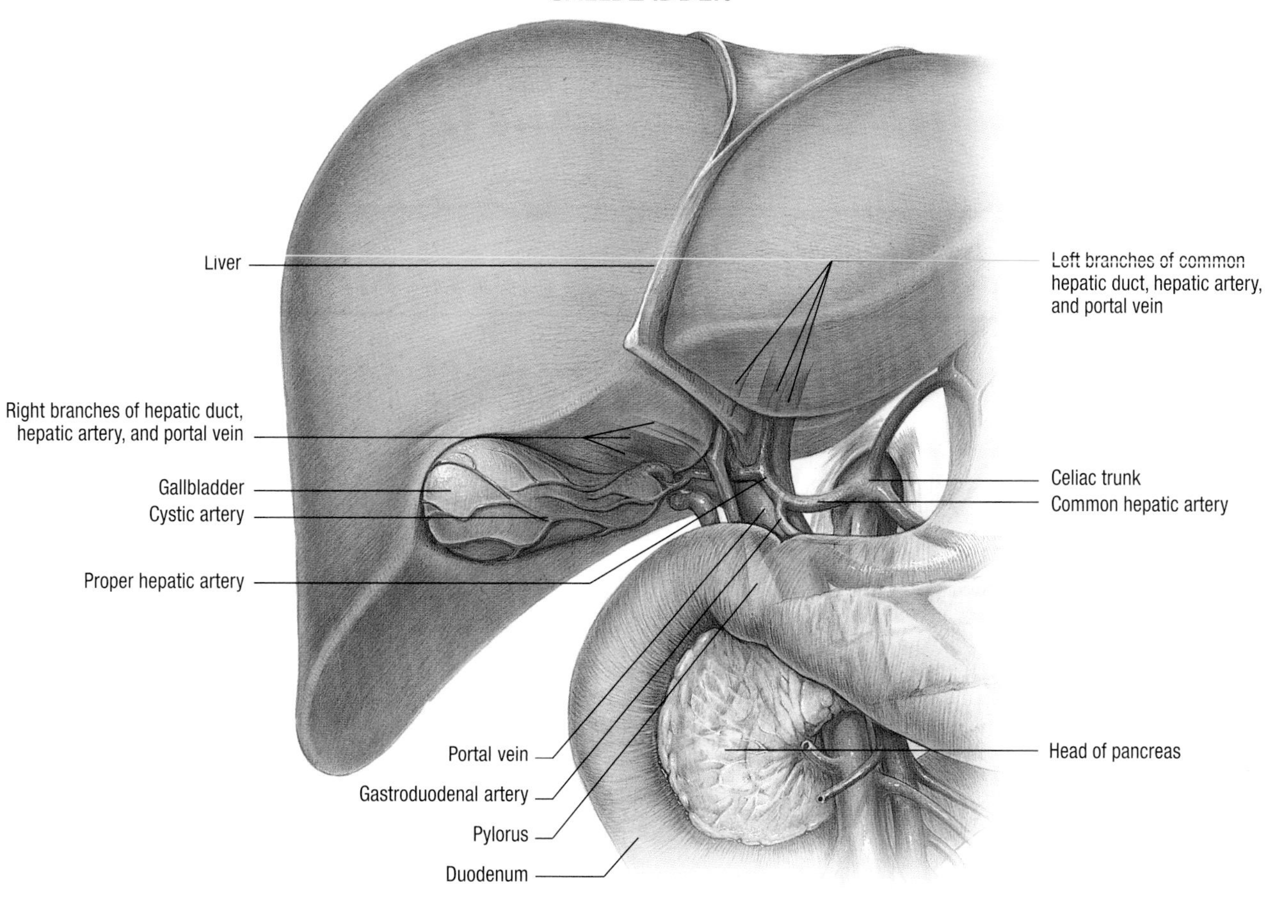
Liver
Left branches of common hepatic duct, hepatic artery, and portal vein
Right branches of hepatic duct, hepatic artery, and portal vein
Gallbladder
Cystic artery
Proper hepatic artery
Celiac trunk
Common hepatic artery
Head of pancreas
Portal vein
Gastroduodenal artery
Pylorus
Duodenum

## GALLBLADDER DUCT SYSTEM

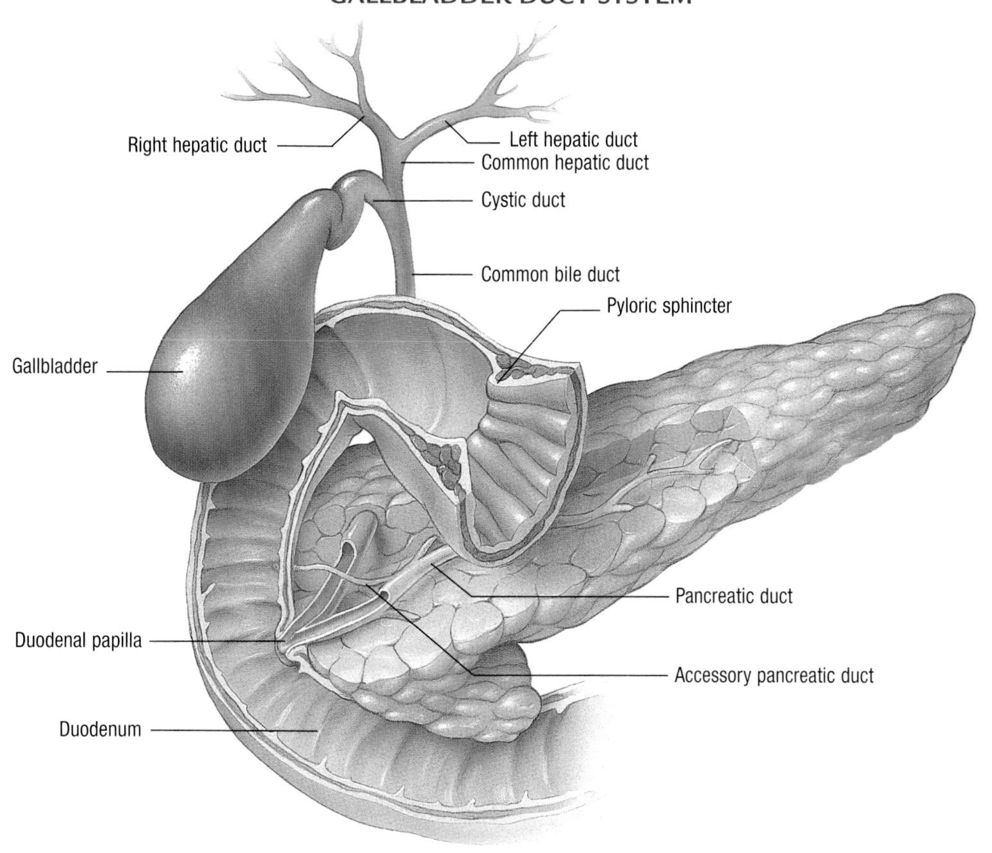
Right hepatic duct
Left hepatic duct
Common hepatic duct
Cystic duct
Common bile duct
Pyloric sphincter
Gallbladder
Pancreatic duct
Duodenal papilla
Accessory pancreatic duct
Duodenum

# Pancreas

The pancreas is an elongated, triangular gland. It is a somewhat flat organ, about 6 to 9 inches (15 to 23 cm) long. The pancreas lies transversely dorsal to the stomach, between the spleen and the duodenum.

## Structure and location

The pancreas has four main parts: heads, neck, body, and tail. The head and neck extend into the curve of the duodenum. A portion of the head – the uncinate process – hooks to the left. The body of the pancreas runs leftward, posterior to the omental bursa, the part of the peritoneal cavity dorsal to the stomach. The tail lies anterior to the left kidney and extends to the spleen.

### Pancreatic ducts

The pancreas has two ducts, which drain the pancreatic secretions. The main pancreatic duct runs from the tail to the head and joins with the bile duct to form the hepatopancreatic ampulla, which empties into the duodenum at the major duodenal papilla. The accessory pancreatic duct drains part of the head of the pancreas. Usually, it unites with the main pancreatic duct, but in some cases it may empty directly into the duodenum at the minor duodenal papilla.

## Functions

The pancreas performs both exocrine and endocrine functions.

### Exocrine

Scattered exocrine cells secrete about 1,500 ml of alkaline pancreatic fluid into the duodenum daily. Pancreatic fluid contains potent digestive enzymes. The pancreas also secretes inactive precursors of proteolytic (protein-digesting) enzymes.

- The intestinal enzyme enterokinase activates trypsinogen to trypsin. In turn, trypsin activates chymotrypsinogen and procarboxypeptidase to form chymotrypsin and carboxypeptidase.
- The small intestine releases the hormone secretin when acidic chyme is expelled into the duodenum. This hormone stimulates the pancreas to secrete large volumes of pancreatic fluid, which is slightly alkaline and raises the pH of intestinal contents.
- The small intestine also releases cholecystokinin (CCK) when lipids and proteins enter the duodenum. This stimulation causes secretion of pancreatic fluid, which is rich in digestive enzymes, and stimulates gallbladder contraction.

### Pancreatic secretions

Secretions from the pancreas assist with the digestion of many substances. The chart below summarizes the functions of enzymes contained in the pancreatic secretions.

| ENZYME | FUNCTION |
|---|---|
| Trypsin, chymotrypsin, and carboxypeptidase | Digest protein |
| Ribonuclease and deoxyribonuclease | Digest nucleic acids |
| Amylase | Digests starch |
| Lipase | Digests fats and other lipids |
| Cholesterol esterase | Splits cholesterol esters into cholesterol and fatty acids |

### Endocrine

The endocrine function of the pancreas involves about 1 million small cell clusters – the islets of Langerhans – each composed of three types of cells:

- alpha – secrete glucagon to increase blood glucose in response to decreased blood glucose levels
- beta – secrete insulin to lower high blood glucose; absent or defective in people with type II diabetes mellitus
- delta – secrete somatostatin, which inhibits secretion of glucagon and insulin.

Other cells within islet of Langerhans cells secrete pancreatic polypeptide, a hormone whose function is not fully understood.

**Clinical tip**

When assessing the patient with suspected pancreatitis, look for two classic signs:

- Cullen's sign: bluish periumbilical discoloration
- Turner's sign: bluish flank discoloration.

## PANCREAS

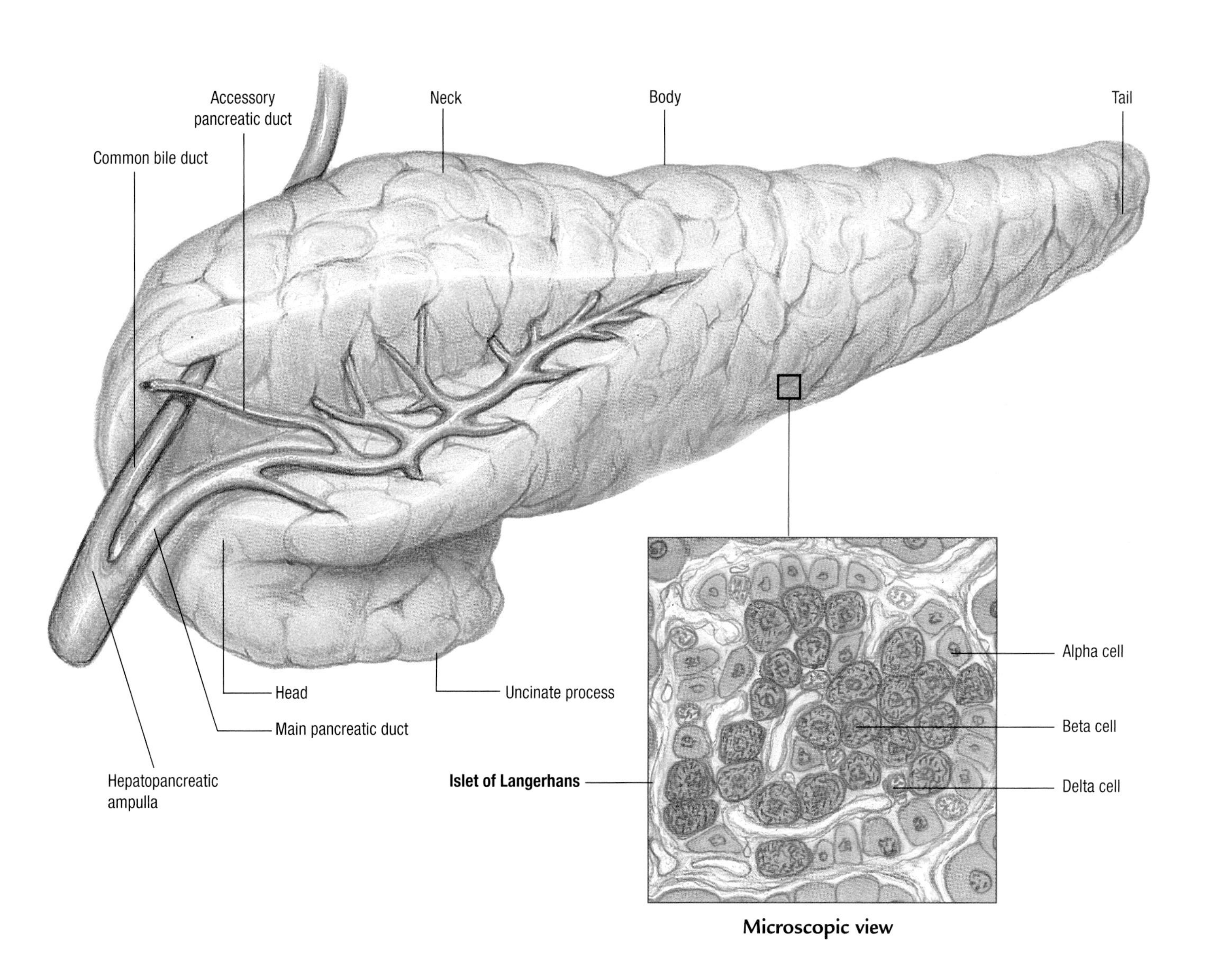

Accessory pancreatic duct
Neck
Body
Tail
Common bile duct
Head
Uncinate process
Main pancreatic duct
Hepatopancreatic ampulla
Islet of Langerhans
Alpha cell
Beta cell
Delta cell


Microscopic view

# Spleen

The spleen is a dark red, ovoid, fist-sized structure that is the largest lymphatic organ. The spleen stores blood and 20% to 30% of platelets. It's surrounded by a dense fibrous capsule, called the splenic pulp, from which bands of connective tissue extend into its interior.

The spleen lies at about the level of the 10th rib in the left midaxillary line of the left upper abdominal quadrant, dorsal to the stomach and below the diaphragm. The diaphragmatic surface of the spleen is curved to fit the shape of the diaphragm.

## Splenic pulp

Splenic pulp is characterized as white or red:
- white pulp – compact masses of lymphocytes surrounding branches of the splenic artery
- red pulp – blood-filled sinusoids, supported by a framework of reticular fibers containing mononuclear phagocytes, lymphocytes, plasma cells, and monocytes; cords of splenic tissue surround and separate the sinusoids. Long, narrow endothelial cells line the splenic sinusoids; they lie parallel to the long axis of the sinusoids and are supported by fenestrated basement membrane.

## Blood supply

Blood flows into the spleen from branches of the splenic arteries; most enters the pulp cords, and some flows directly into the splenic sinusoids.

Blood cells pass from the meshwork of cells and fibers between the splenic cords into the sinusoid lumina through long, slitlike openings between adjacent endothelial cells. This forces the blood to flow through the framework of reticular fibers, macrophages, and other cells before entering the sinusoids.

## Function

The spleen filters out bacteria and other foreign substances that enter the bloodstream. Splenic phagocytes then perform several functions:
- remove filtered bacteria and foreign substances
- engulf and break down worn-out red blood cells (RBCs); this action releases hemoglobin, which is broken down into its components by splenic phagocytes.
- selectively retain and destroy damaged or abnormal RBCs, such as those containing a large amount of abnormal hemoglobin
- interact with lymphocytes to initiate an immune response.

## Splenectomy

Injury or disease may require splenectomy, which affects the body's defense mechanisms, especially elimination of bacteria and production of antibodies. Consequently, the individual becomes susceptible to serious blood infections caused by various pathogenic organisms.

**Clinical tip**

A normal spleen isn't palpable; an enlarged spleen is. Conditions that cause splenomegaly include mononucleosis, trauma, and illnesses that destroy red blood cells, such as sickle cell anemia and some cancers.

To assess the spleen:
- Ask the patient to breathe deeply.
- Percuss along the 9th to 11th intercostal spaces on the left. Listen for a change from tympany to dullness. Measure the area of dullness.
- Then with your right hand on the patient's abdomen, press upward and inward toward the spleen. If you do feel the spleen, stop palpating immediately because compression can cause rupture.

## Visceral surface

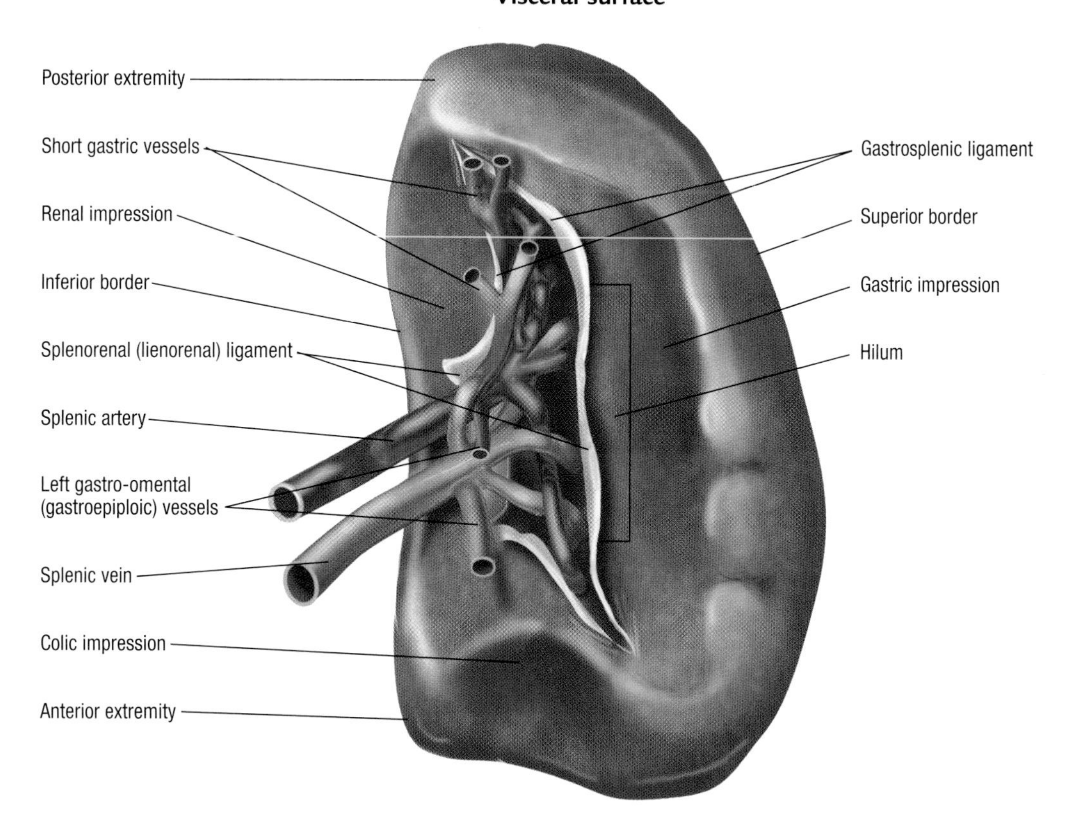
Posterior extremity
Short gastric vessels
Renal impression
Inferior border
Splenorenal (lienorenal) ligament
Splenic artery
Left gastro-omental
(gastroepiploic) vessels
Splenic vein
Colic impression
Anterior extremity
Gastrosplenic ligament
Superior border
Gastric impression
Hilum

## Diaphragmatic surface

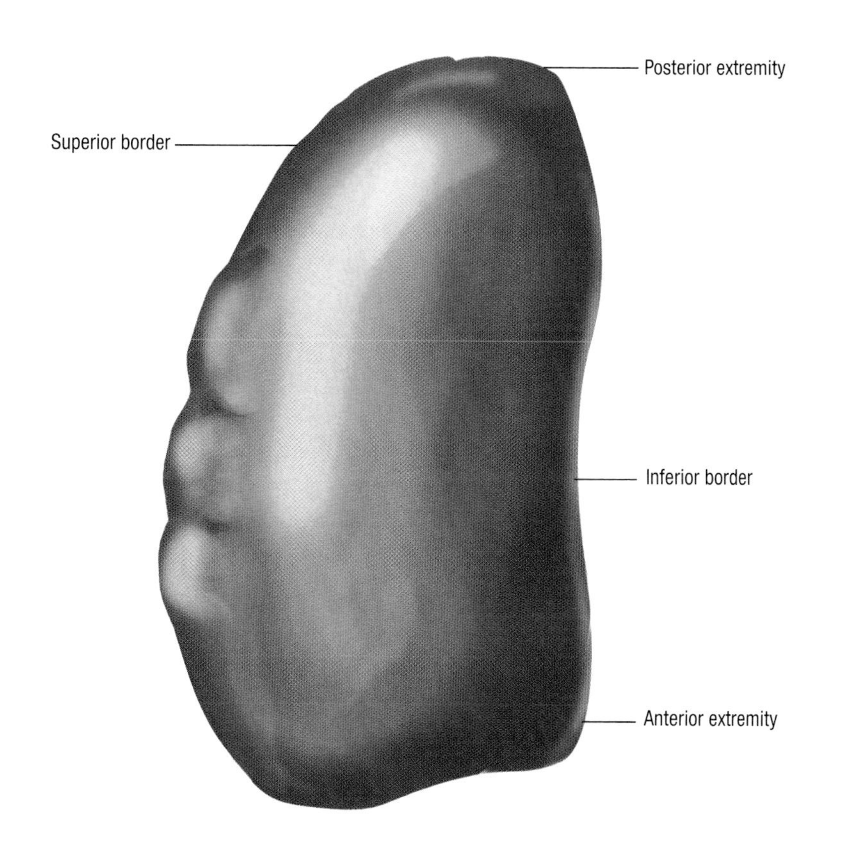
Posterior extremity
Superior border
Inferior border
Anterior extremity

# Kidney and ureter

Two kidneys and two ureters are the parts of the urinary system that are located in the abdomen. See page 260 for a complete discussion of the remainder of the urinary system — the bladder and urethra.

## Kidneys

The kidneys are a pair of reddish, bean-shaped organs, 4.5 to 5 inches (11.5 to 12.5 cm) long and about 2.5 inches (6.5 cm) wide. These highly vascular organs perform two essential functions of the urinary system: forming urine and maintaining homeostasis.

The kidneys are located on either side of the lumbar spine. They lie retroperitoneally (external to the peritoneal lining of the abdominal cavity) in front of the muscles attached to the vertebral column. The kidneys are partially protected by the 11th and 12th ribs, the abdominal contents, the vertebral muscles, and a layer of fat surrounding each kidney. The right kidney, which is found below the liver mass, is a bit lower than the left. The kidneys shift with changes in body position.

**Clinical tip**

The following procedure tells how to palpate the right kidneys; to palpate the left kidney, reverse the positions:

- Have the patient lie supine. Stand on the patient's right side, and place your left hand under the back and your right hand on the abdomen.
- Instruct the patient to inhale deeply, which moves the kidney downward. During inhalation, press up with your left hand and down with your right. You should be able to "capture" the kidney between your hands.

### Outer surface

The following structures cover the kidneys:

- true capsule (fibrous capsule, renal capsule) — thin, transparent fibrous membrane
- perirenal fat (adipose capsule) — protects kidneys from trauma
- renal fascia — anchors the kidneys to the abdominal wall and surrounding organs.

### Structures

The major parts of the kidney include the following structures: (See *Pattern of renal parenchyma*, page 227.)

- cortex — site of blood filtration
- renal columns — extensions of renal cortex
- medulla — inner portion of the kidney, composed of renal pyramids and tubular structures
- renal artery — supplies blood to the kidneys
- renal pyramids — conelike structures that make up the renal medulla
- renal calyx — extension of the renal pelves that encloses the papilla of the renal pyramid; urine is emptied into it.
- renal vein — carries filtered blood back to the systemic circulation
- renal pelvis — receives urine through the calyces; it is the expanded proximal end of the ureter
- renal plexus — a network of autonomic nerve fibers and ganglia that innervates kidneys and ureters
- hilus — notch where ureters, renal blood vessels, lymphatic vessels, and nerves enter and leave the kidney
- renal sinus — cavity composed of renal pelvis, calyces, vessels, nerves, and fatty tissue.

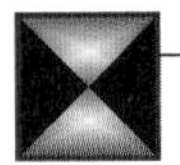

**Age-related changes**

After age 40, a person's renal function may diminish. By age 90, it may decrease by as much as 50%.

### Functions

The major functions of the kidneys are:

- maintaining fluid and acid-base balance
- detoxifying the blood and eliminating waste
- regulating blood pressure
- producing the hormone erythropoietin, which regulates production of RBCs
- participating in activation of vitamin D.

## Ureters

The ureters are fibromuscular tubes, about 10 to 12 inches (25 to 30 cm) long and 2 to 8 mm in diameter in adults. Because the left kidney is higher than the right, the left ureter is usually slightly longer than the right.

The ureters are narrowest where they originate, at the renal pelvis (ureteropelvic junction). They then run obliquely downward to the urinary bladder.

### Structure

The ureteral walls have three layers:

- mucosa — innermost layer, transitional epithelium
- muscularis — middle layer, smooth muscle
- fibrous coat — outer layer; extensions of the fibrous coat hold the ureter in place.

### Function

Urine flows through the ureters from the kidneys to the bladder. Filling of the bladder constricts the ureters at the ureterovesicular junction, where they enter the bladder. Peristaltic waves, occurring about one to five times each minute, move urine through the ureters.

## KIDNEYS AND URETERS
**Anterior view**

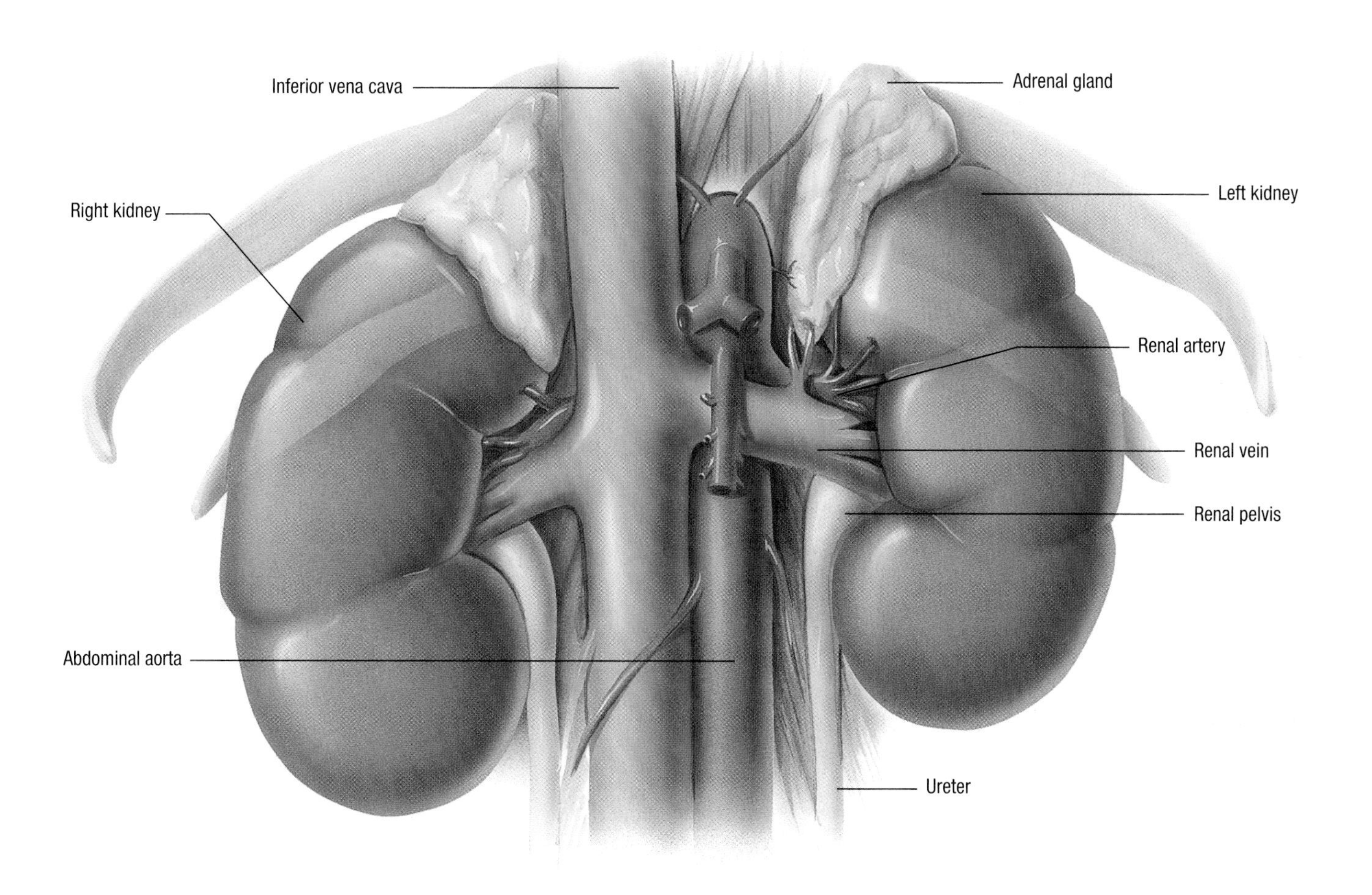
Inferior vena cava
Adrenal gland
Left kidney
Right kidney
Renal artery
Renal vein
Renal pelvis
Abdominal aorta
Ureter

## KIDNEYS
### Anterior relations

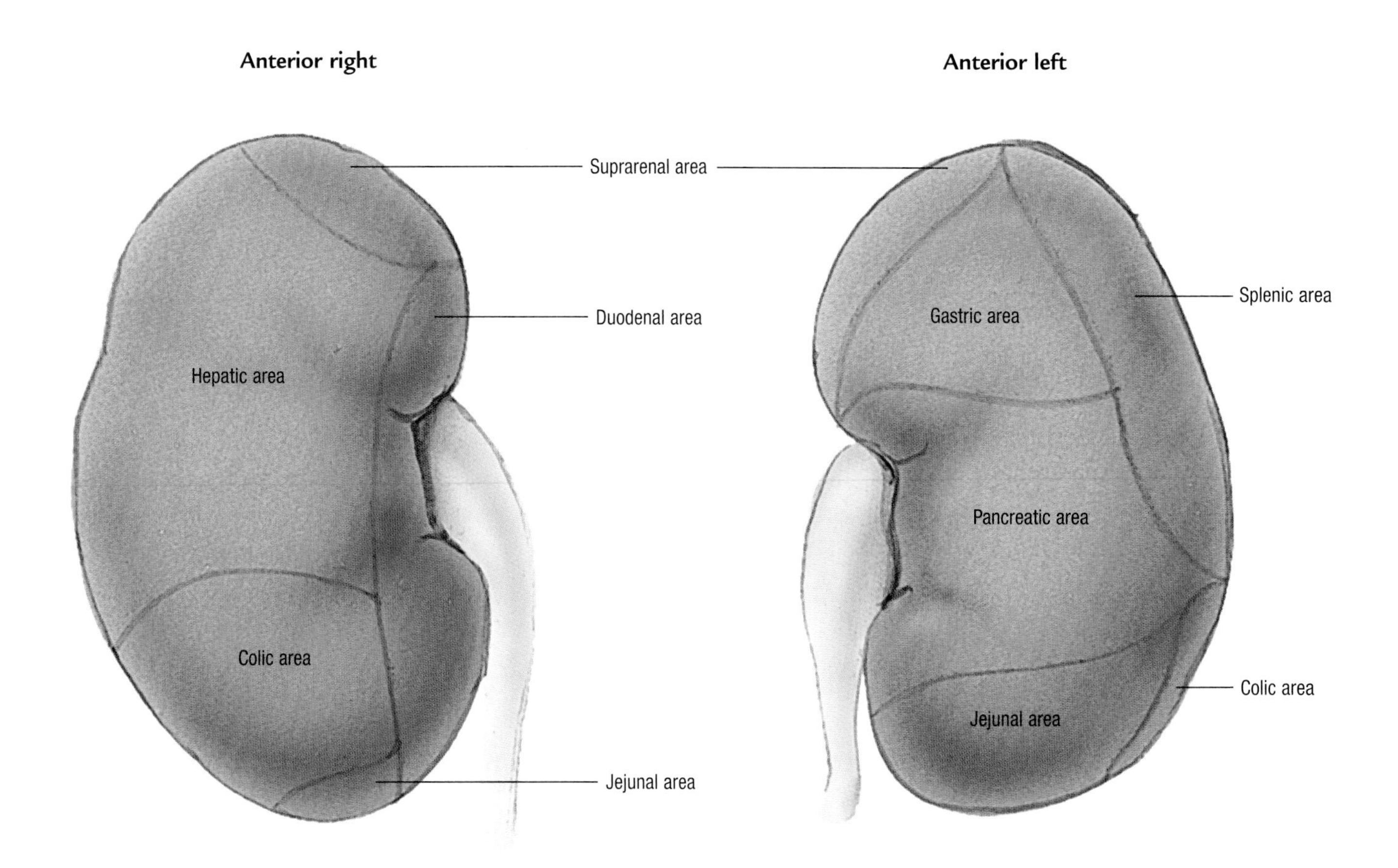
Anterior right
Anterior left
Suprarenal area
Duodenal area
Hepatic area
Colic area
Jejunal area
Gastric area
Splenic area
Pancreatic area
Colic area
Jejunal area

## KIDNEYS
### Posterior relations

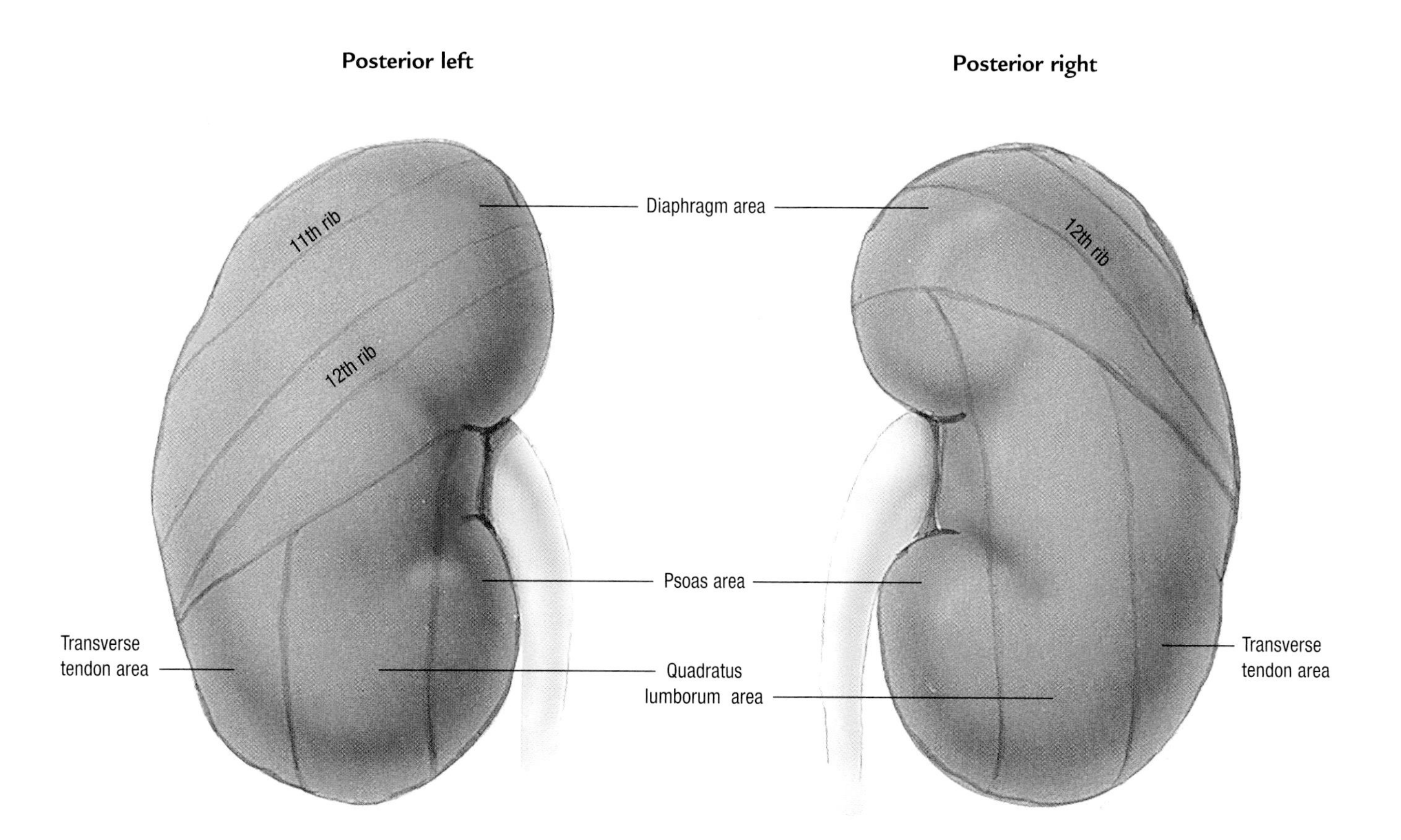

Posterior left
Posterior right
11th rib
12th rib
12th rib
Diaphragm area
Psoas area
Transverse tendon area
Quadratus lumborum area
Transverse tendon area

## LEFT KIDNEY AND ADRENAL GLAND
### Cross section

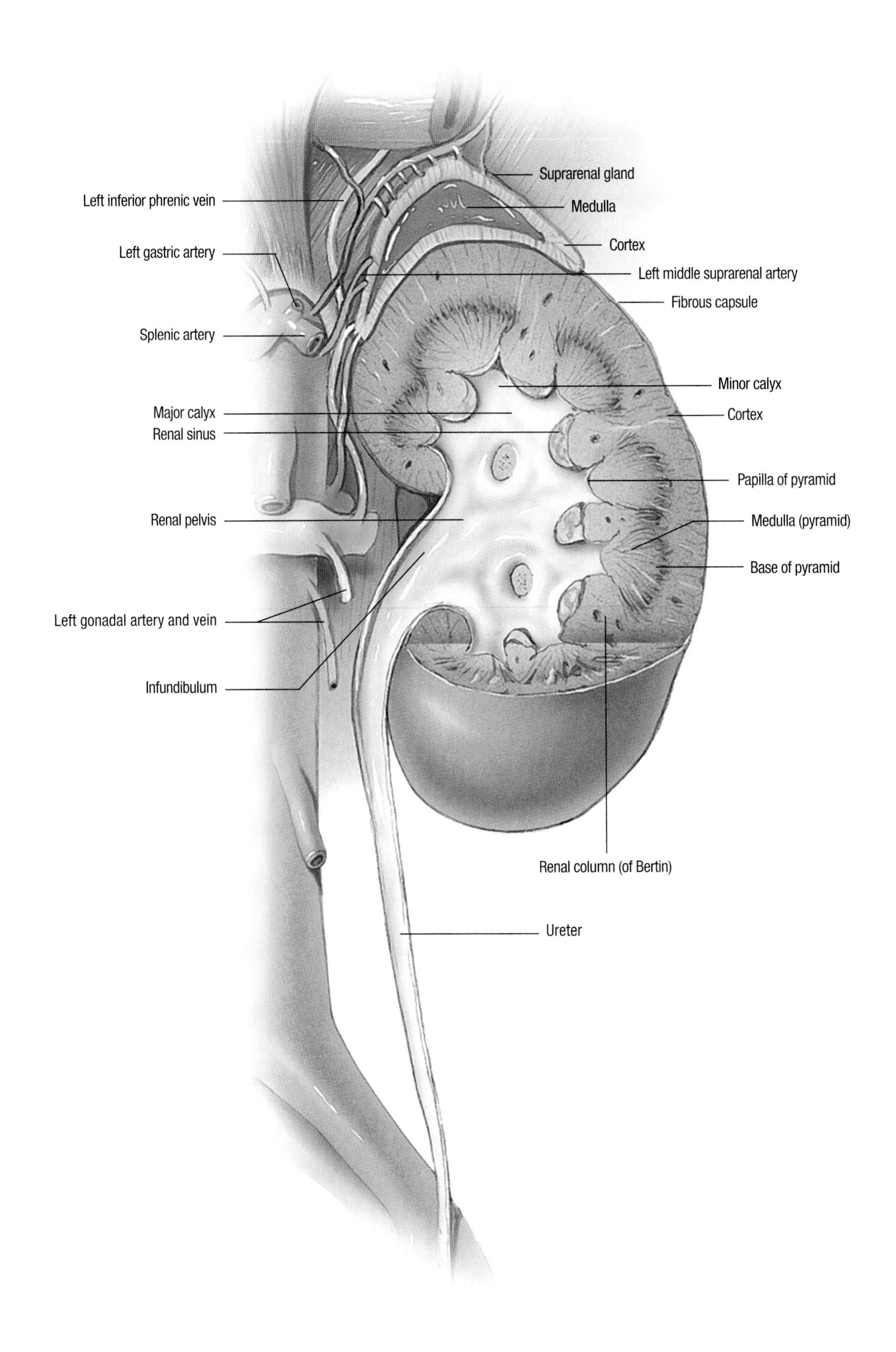

Suprarenal gland
Medulla
Cortex
Left middle suprarenal artery
Fibrous capsule
Left inferior phrenic vein
Left gastric artery
Splenic artery
Minor calyx
Cortex
Major calyx
Renal sinus
Papilla of pyramid
Renal pelvis
Medulla (pyramid)
Base of pyramid
Left gonadal artery and vein
Infundibulum
Renal column (of Bertin)
Ureter

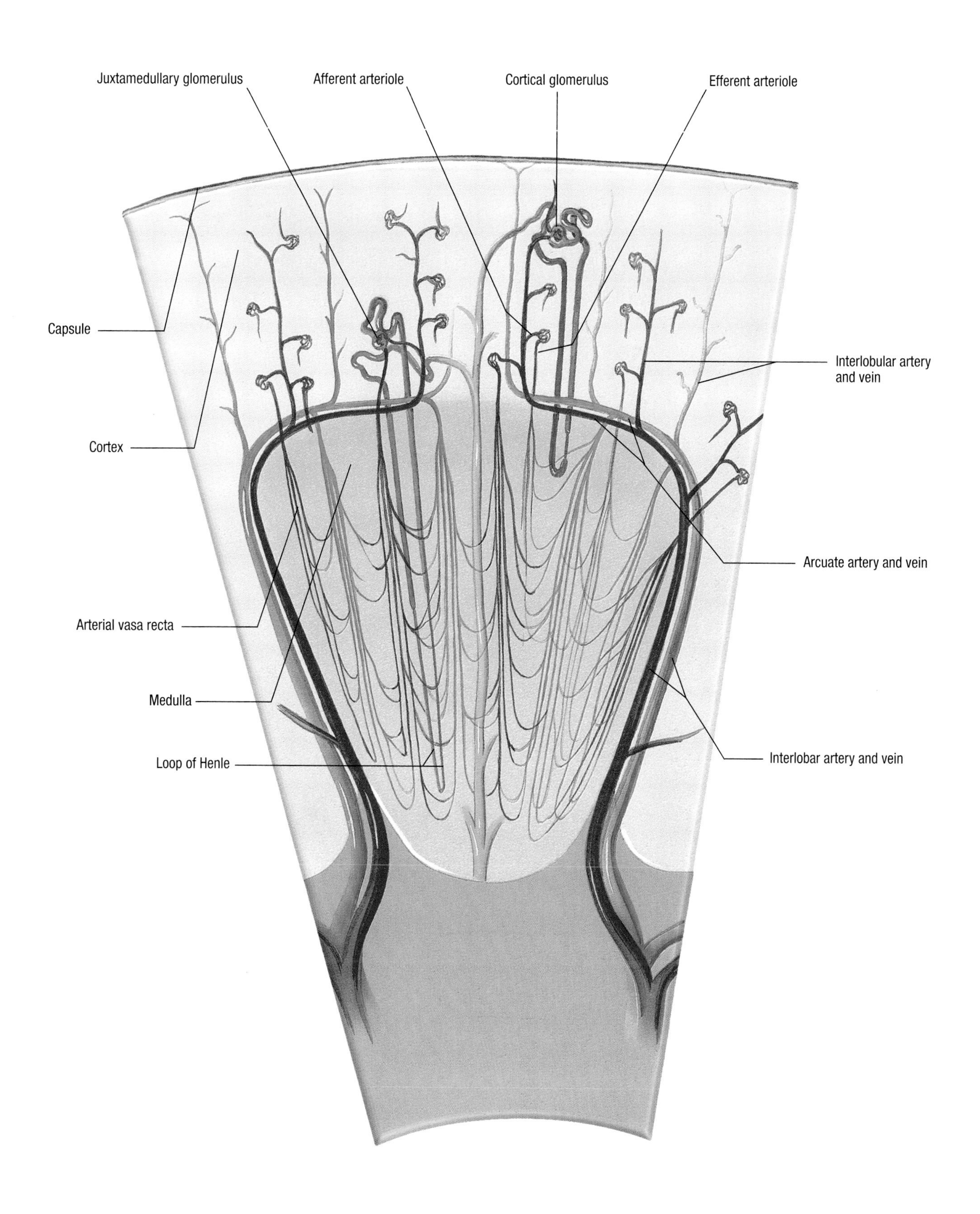
Juxtamedullary glomerulus
Afferent arteriole
Cortical glomerulus
Efferent arteriole
Capsule
Interlobular artery and vein
Cortex
Arcuate artery and vein
Arterial vasa recta
Medulla
Interlobar artery and vein
Loop of Henle

# Nephron

The nephron is the functional and structural unit of the kidney. Each kidney contains more than 1 million nephrons, which do the following:

- selectively reabsorb and secrete fluids and electrolytes
- mechanically filter wastes.

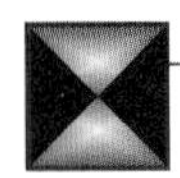

**Age-related changes**

As a person ages, tubular reabsorption and renal concentrating ability decline because the size and number of functioning nephrons decrease.

Each nephron is a long tubule with a closed end, called Bowman's capsule. Components of the nephron include:

- renal corpuscle
  - glomerulus – a network of twisted capillaries; normally produces protein-free and cell-free filtrate, which passes into the proximal convoluted tubules
  - Bowman's capsule – encloses the glomerulus; collects filtrate
- proximal convoluted tubule – reabsorbs glucose, amino acids, metabolites, and electrolytes from filtrate; reabsorbed substances return to circulation
- loop of Henle – a U-shaped segment between the proximal and distal convoluted tubules; extends into the medulla, where it concentrates filtrate through electrolyte exchange and reabsorption to produce hyperosmolar fluid
- distal convoluted tubule – reabsorbs sodium under influence of aldosterone; site from which filtrate enters the collecting tubule
- collecting tubule – distal end of nephron; site of final concentration; empties into papillary ducts.

**Clinical tip**

The glomerular filtration rate (GFR) is the rate at which the glomeruli filter blood, normally about 120 ml/minute. GFR depends on:

- permeability of capillary walls
- vascular pressure
- filtration pressure
- clearance.

Clearance is the complete removal of substance from the blood. The most accurate measure of glomerular filtration is creatinine clearance – the glomeruli filter creatinine, but the tubules don't reabsorb it.

Here's more about how the GFR affects clearance measurements for a substance in the blood:

- If the tubules neither reabsorb nor secrete the substance – as happens with creatinine – clearance equals the GFR.
- If the tubules reabsorb the substance, clearance is less than the GFR.
- If the tubules secrete the substance, clearance exceeds the GFR.
- If the tubules reabsorb and secrete the substance, clearance may be less than, equal to, or greater than the GFR.

## Hormonal effects on the nephron

Two extrarenal hormones regulate fluid balance in the kidney:

- antidiuretic hormone (ADH), produced by the pituitary gland
- aldosterone, produced by the adrenal cortex.

### ADH

ADH alters the collecting tubules' permeability to water. High ADH concentration in plasma increases permeability to water and leads to excretion of a highly concentrated but small volume of urine. If ADH concentration is low, the tubules are less permeable to water, and the kidney excretes a larger volume of dilute urine.

### Aldosterone

Aldosterone regulates water reabsorption by the distal tubules and changes urine concentration by increasing sodium reabsorption. A high plasma aldosterone concentration increases sodium and water reabsorption by the tubules and decreases sodium and water excretion in the urine. A low plasma aldosterone concentration promotes sodium and water excretion.

Aldosterone also helps control the secretion of potassium by the distal tubules. A high aldosterone concentration increases the excretion of potassium. Other factors that affect potassium secretion include:

- the amount of potassium ingested
- the number of hydrogen ions secreted
- potassium levels in the cells
- the concentration of sodium in the distal tubule
- the glomerular filtration rate.

## NEPHRON AND COLLECTING TUBULE

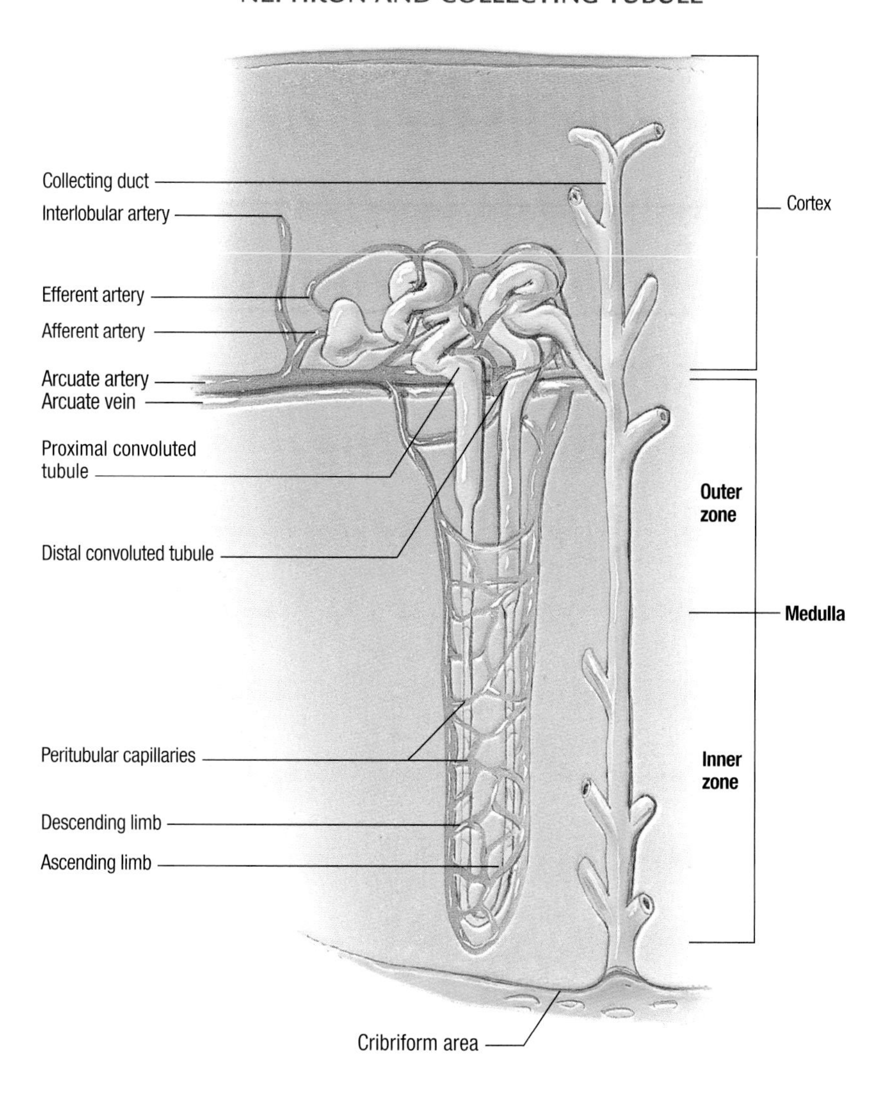
Collecting duct
Interlobular artery
Efferent artery
Afferent artery
Arcuate artery
Arcuate vein
Proximal convoluted tubule
Distal convoluted tubule
Peritubular capillaries
Descending limb
Ascending limb
Cribriform area
Cortex
Outer zone
Medulla
Inner zone

## RENAL CORPUSCLE

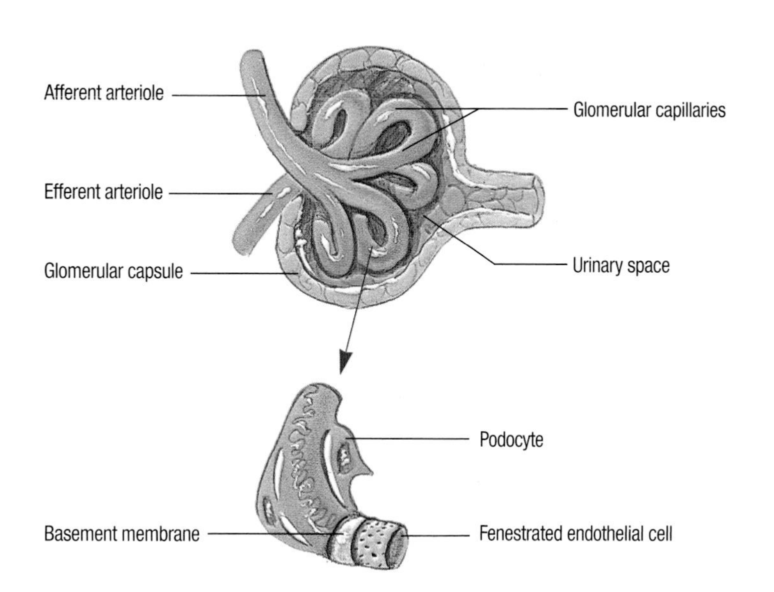
Afferent arteriole
Efferent arteriole
Glomerular capsule
Glomerular capillaries
Urinary space
Podocyte
Basement membrane
Fenestrated endothelial cell

# Blood flow, innervation, and lymphatic drainage

Blood enters the kidneys via large renal arteries, which branch from the abdominal aorta. The right and left renal arteries provide 25% of the total cardiac output to the kidneys; they deliver blood to the kidneys at a rate of about 1,200 ml/minute.

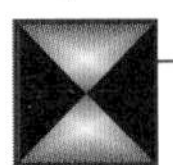

**AGE-RELATED CHANGES**

The most common causes of hypertension before adolescence are
- renal disease (78%)
- renal artery disease (12%).

## RENAL ARTERIES

The renal arteries enter the kidney through the hilus.

Each renal artery branches into five segmental (lobar) arteries, which correspond to the areas of the kidney they supply, as follows:
- superior (apical)
- anterosuperior segmental
- anteroinferior segmental
- inferior segmental
- posterior segmental.

Each lobar artery, in turn, branches into several interlobar arteries, which pass between the renal pyramids and columns. Each interlobar artery branches into arcuate arteries at the junction of the renal medulla and renal cortex. Each arcuate artery divides to form interlobular arteries, from which arise the afferent arterioles, which deliver blood to the glomerular capillaries.

### Renal capillaries

The glomerular capillaries regroup to form efferent arterioles, which are smaller in diameter than afferent arterioles. This size difference helps maintain glomerular pressure. Efferent arterioles form:
- peritubular capillary networks that surround the proximal convoluted tubule and collect reabsorbed electrolytes
- vasa recta – long loops that run along the loop of Henle.

**PHYSIOLOGY**

The kidneys help regulate blood pressure by producing and secreting the enzyme renin in response to an actual or perceived decline in extracellular fluid volume.

The juxtaglomerular apparatus, which regulates blood flow through the glomerulus and regulates blood pressure by producing renin, consists of the macula densa and juxtaglomerular cells.
- The macula densa is an area of compact, heavily nucleated cells in the distal convoluted tubule, where the tubule makes contact with the juxtaglomerular cells. It is sensitive to changes in salt content of the filtrate in the tubule.
- The juxtaglomerular cells, which contain renin granules, are found in the wall of the afferent arteriole at the point where it is in contact with the distal convoluted tubule.

Changes in salt concentration and therefore in perceived fluid volume trigger release of renin and formation of angiotensinogen, which is converted to angiotensin II in the lungs. Angiotensin II raises arterial blood pressure by increasing peripheral vasoconstriction and stimulating aldosterone secretion. This increase in aldosterone promotes the reabsorption of sodium and water to correct fluid deficits and inadequate blood flow (renal ischemia).

Causes of hypertension include fluid and electrolyte imbalance and renin-angiotensin hyperactivity. Hypertension can damage blood vessels and cause hardening of the kidneys (nephrosclerosis), one of the major causes of renal failure.

### Renal veins

The peritubular capillaries merge to form the interlobular veins, which in turn merge to form the arcuate, interlobar, and renal veins. The renal vein leaves the kidney through the hilus and flows into the inferior vena cava.

## INNERVATION

The renal plexus, a network of autonomic nerve fibers and ganglia, innervates the kidneys and ureters. Nerves distributed alongside the blood vessels regulate the circulation of blood through the kidney by altering the diameter of the vessels.

## LYMPHATICS

Lymph from the kidneys and adrenal glands drains into the cisterna chyli. The cisterna chyli is a portion of the thoracic (left lymphatic) duct, which is the main collecting duct of the lymphatic system.

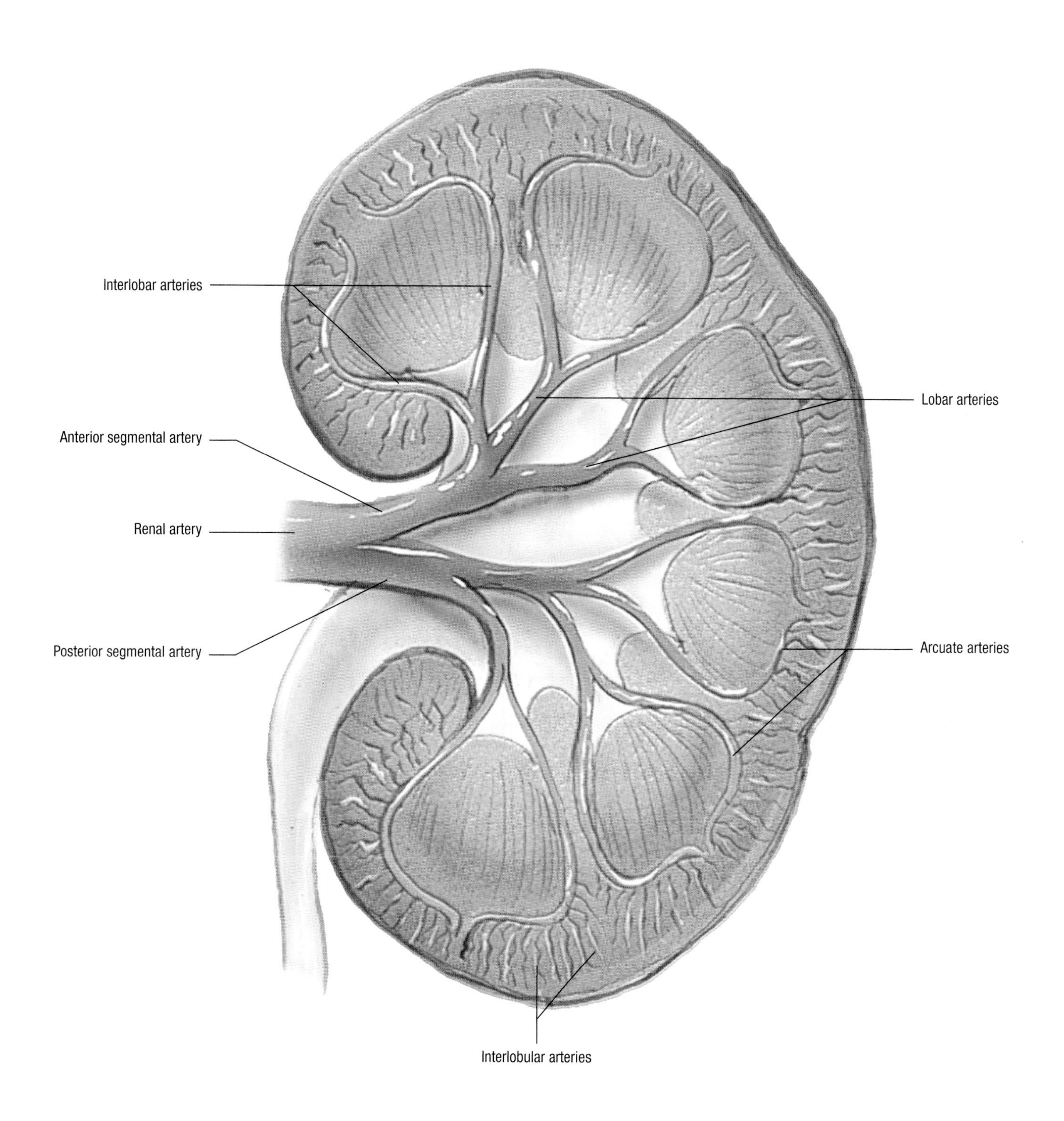
Interlobar arteries
Lobar arteries
Anterior segmental artery
Renal artery
Posterior segmental artery
Arcuate arteries
Interlobular arteries

# PART VII

# PELVIS

# Overview of pelvis

The lowest area of the trunk, the pelvis is bounded by hip bones laterally and anteriorly, and by sacrum and coccyx posteriorly. It consists of *bony pelvis*, *pelvic cavity*, and *perineum*.

## Function

Functions of the pelvis include:
- protecting the bladder, a portion of the reproductive organs, and part of the large intestine
- supporting the contents of the abdominopelvic cavity
- forming the fixed axis of the birth canal (in females).

## Borders

The upper border *(pelvic inlet)* is bounded by the *sacral promontory*, *arcuate lines* of the ilia, *symphysis pubis*, and *pubic crest*.

The lower border (*pelvic outlet*) is bounded by the *pubic arch*, *sacrotuberal ligaments*, part of the *ischium*, and *coccyx*. It's closed by the *pelvic diaphragm* (muscles and fasciae of the pelvic floor).

**Clinical tip**

The pelvis is separated into two major areas: the greater, or false, pelvis and the lesser, or true, pelvis. The greater pelvis — above the pelvic inlet — contains portions of the ileum and the sigmoid colon. The lesser pelvis — between the pelvic inlet and pelvic outlet — contains the bladder and female reproductive organs.

## Pelvic girdle

Formed by the hip (*innominate*) bones, the pelvic girdle connects the lower limbs to the axial skeleton. Ventrally, the *innominate bones* articulate with each other. Dorsally, they articulate with the sacrum. The joint where the sacrum and ilium meet, called the *sacroiliac* articulation, has limited movement. Together with the sacrum and coccyx, the two bones of the pelvic girdle form the bony pelvis.

## Bony pelvis

The bony pelvis is the pelvic skeletal structure. It's formed by four bones — two innominates, the sacrum, and the coccyx. A deep, cup-shaped socket (acetabulum) on the lateral surface of the innominate at the juncture of the ilium, ischium, and pubis articulates with the head of the femur, the thigh bone.

### Ilium

The large, flaring ilium forms the superior and largest portion of each innominate bone. The curved upper rim of each ilium — a bony projection known as the hip — forms the iliac crest.

### Ischium

The ischium forms the posteroinferior part of the innominate bone. Its inferior surface thickens to form the ischial tuberosity. The sacrospinal ligament attaches to the ischial spine, a pointed projection on the posterior surface of the ischium.

### Pubis

The pubis, also called pubic bone, forms the anterior part of the innominate bone. It joins the ischium posterolaterally to form the obturator foramen, an opening in the pelvis. The two pubic bones articulate anteriorly at the *symphysis pubis*.

**Pelvic differences**

The female pelvis is lighter than the male pelvis. Other common differences are outlined below.
- The acetabulum is smaller in a female.
- The obturator foramen is oval in a female, round in a male.
- The pubic arch and subpubic angle are wider in a female.
- The lesser pelvis is wider and shallower in a female.
- The greater pelvis is shallower in a female.
- The pelvic outlet is larger in a female.
- The pelvic inlet is an oval or rounded shape in a female, heart-shaped in a male.

The illustration on page 235 shows the differences in shape of the female pelvis superimposed over the male pelvis.

## Pelvic openings

The bony pelvis contains several pairs of openings that serve as passageways for organs, vessels, and nerves:
- greater sciatic foramen, in front of and above the ileum and the rim of the great sciatic notch — greater sciatic nerve, the largest nerve in the body
- lesser sciatic foramen, in front of the ischial tuberosity — lesser sciatic nerve
- obturator foramen, a large opening in the lower portion of each innominate bone covered by the tough, fibrous obturator membrane — obturator internus and externus muscles originate on the inner and outer surfaces, respectively.

## Pelvic cavity

The pelvic cavity is the lower portion of the abdominopelvic cavity, which extends into the pelvis. No structure separates the pelvic and abdominal cavities. The pelvic cavity contains:
- bladder, parts of the ureters, and rectum
- blood vessels, lymphatics, and nerves
- internal genital organs.

## Perineum

A complex structure of muscles, blood vessels, fasciae, nerves, and lymphatics, the perineum lies between the symphysis pubis and the coccyx, bordered on the sides by the ischial tuberosities. In males, it's bounded by the anus and the scrotum; in females, it's bounded by the anus and the vulva.

### Muscles

Muscles of the perineum include the ischiocavernous, the superficial transverse perineus, and the bulbocavernous muscles.

### Blood supply

The major arteries supplying the perineum are the middle sacral artery and the internal and external iliac arteries. Venous drainage is provided by the internal and external iliac veins.

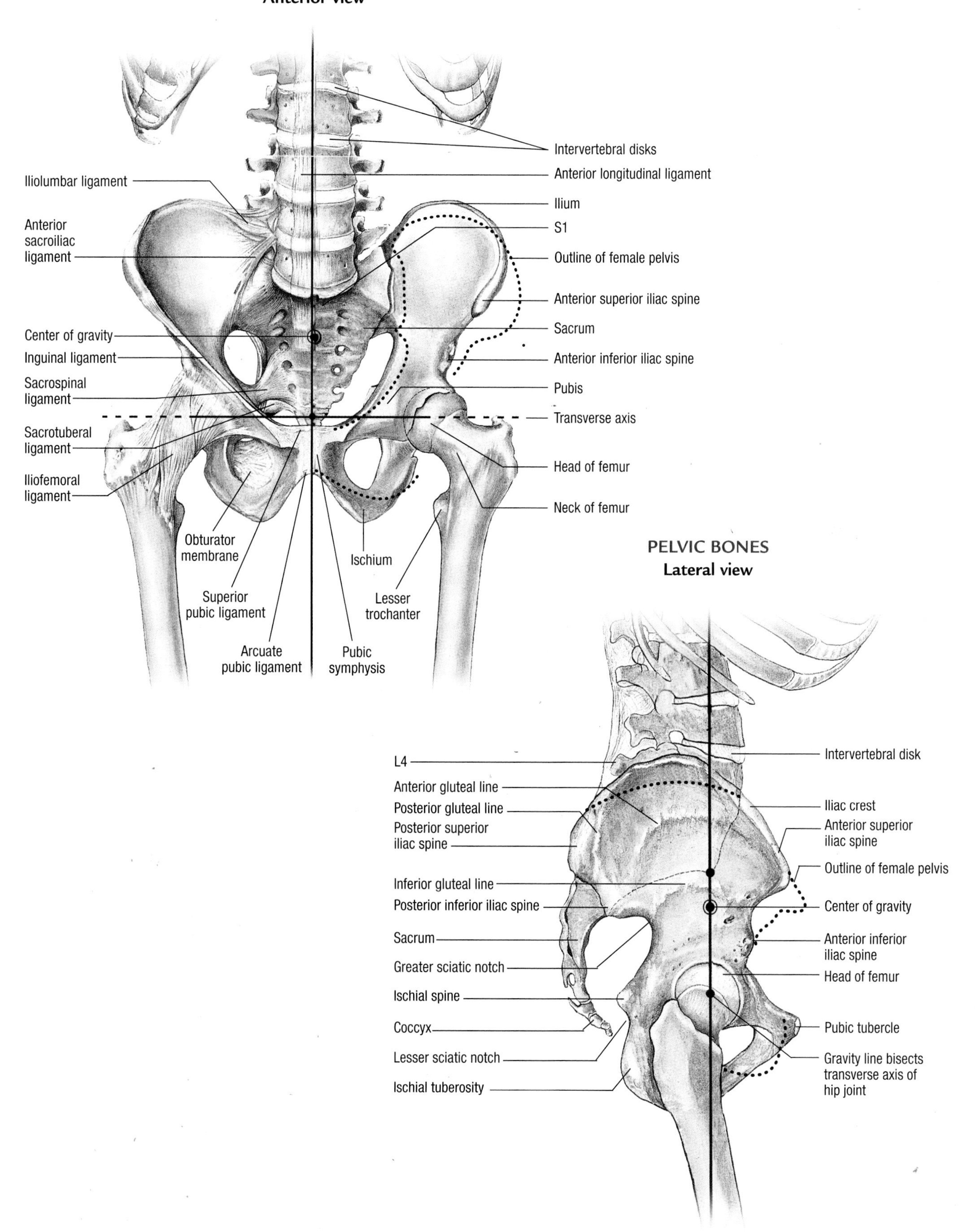
PELVIC BONES
Anterior view
Intervertebral disks
Anterior longitudinal ligament
Iliolumbar ligament
Ilium
S1
Anterior sacroiliac ligament
Outline of female pelvis
Anterior superior iliac spine
Sacrum
Center of gravity
Anterior inferior iliac spine
Inguinal ligament
Pubis
Sacrospinal ligament
Transverse axis
Sacrotuberal ligament
Head of femur
Iliofemoral ligament
Neck of femur
Obturator membrane
Ischium
Superior pubic ligament
Lesser trochanter
Arcuate pubic ligament
Pubic symphysis
PELVIC BONES
Lateral view
L4
Intervertebral disk
Anterior gluteal line
Posterior gluteal line
Iliac crest
Posterior superior iliac spine
Anterior superior iliac spine
Outline of female pelvis
Inferior gluteal line
Posterior inferior iliac spine
Center of gravity
Sacrum
Anterior inferior iliac spine
Greater sciatic notch
Head of femur
Ischial spine
Pubic tubercle
Coccyx
Lesser sciatic notch
Gravity line bisects transverse axis of hip joint
Ischial tuberosity

# Ligaments, muscles, nerves, and vessels

Ligaments and muscles provide support and, in some cases, attachments for pelvic contents. Muscle pairs and deep fasciae in the pelvic floor are accessory structures to the bony pelvis.

The area is rich with nerves, blood vessels, and lymphatic vessels and nodes that serve the pelvic area and pass through the pelvis en route to the lower extremity.

## Ligaments

Ligaments support the joints of the pelvis and some of the structures that lie within it. Ligaments include:

- iliolumbar (2) – extend from the ilium to the 4th and 5th lumbar vertebrae
- pubic – connects the pubic bones at the pubic symphysis
- sacrococcygeal – connects the sacrum to the base of the coccyx
- sacroiliac (4) – connect the sacrum and ilium (two pairs of ligaments, anterior and posterior)
- sacrospinal (2) – connect the ischial spine to the sacrum and coccyx
- sacrotuberal (2) – extend from each ischial tuberosity to the inferior and posterior superior iliac spines; also extend to the lower sacrum and coccyx.

## Muscles

Muscles constitute a portion of the pelvic floor. (See *Pelvic wall and floor muscles,* page 239.) Other muscles, originating in the pelvis, move the lower limb, particularly the hip joint and femur.

### Pelvic diaphragm

The downwardly convex pelvic diaphragm, located at the caudal aspect of the body wall, forms the pelvic floor and supports the pelvic organs. It is made up of the levator ani and coccygeus muscles along with the fasciae above and below them. The pelvic diaphragm ligaments, fasciae, and muscles are anchored to the perineal body. The ano-urogenital hiatus is a passageway through the pelvic diaphragm for the rectum, urethra, and vagina (in females).

### Other pelvic muscles

- Obturator externus – a flat triangular muscle on the external surface of the pelvic wall; provides support
- Obturator internus – in the pelvis and upper leg; rotates the thigh laterally and abducts it
- Piriformis – a flat, pyramidal muscle almost parallel with the gluteus medius; rotates the thigh. The coccygeus and piriformis muscles form a bed for the coccygeal and sacral nerve plexuses.

## Nerves

The *sacral plexus* is a network of motor and sensory nerves formed by the lumbosacral trunk from the 4th and 5th lumbar and the 1st, 2nd, and 3rd sacral nerves (L4 through S3). (See *Pelvic nerves,* pages 240 and 241.)

The sciatic nerve, also called the great sciatic nerve, arises from the sacral plexus and innervates portions of the lower extremity. (See Part VIII, Lower limb.)

The pudendal nerve, a branch of the sacral plexus, innervates most of the skin and muscles of the perineum. After arising from the 2nd, 3rd, and 4th sacral nerves, it passes between the piriformis and coccygeus muscles, and leaves the pelvis through the greater sciatic foramen.

Other key nerves in the pelvic area include:

- inferior and superior gluteal nerves
- obturator nerve.

## Vessels

Major arteries of the pelvis include the internal pudendal branch of the internal iliac artery and the external pudendal artery, a branch of the common femoral artery. (See *Pelvic blood vessels,* pages 242 and 243.) The superior vesical artery, a branch of the umbilical artery, supplies the bladder. The following branches of the inferior epigastric arteries supply pelvic structures:

- median sacral
- ovarian
- pubic
- superior rectal.

Venous drainage of the area is provided by branches of the internal pudendal vein, which empties into the internal iliac vein. There are two exceptions.

- The right ovarian vein drains into the inferior vena cava.
- The superior rectal vein drains into the inferior mesenteric vein.

## Lymphatic drainage

A number of lymph nodes lie within the abdominopelvic area. They include:

- lateral aortic
- inferior mesenteric
- common, external, and internal iliac
- superficial inguinal
- deep inguinal
- sacral
- pararectal.

(See Part VI, Abdomen, for a detailed illustration of the pelvic and abdominal lymphatic systems.)

## PELVIC LIGAMENTS
### Anterior view

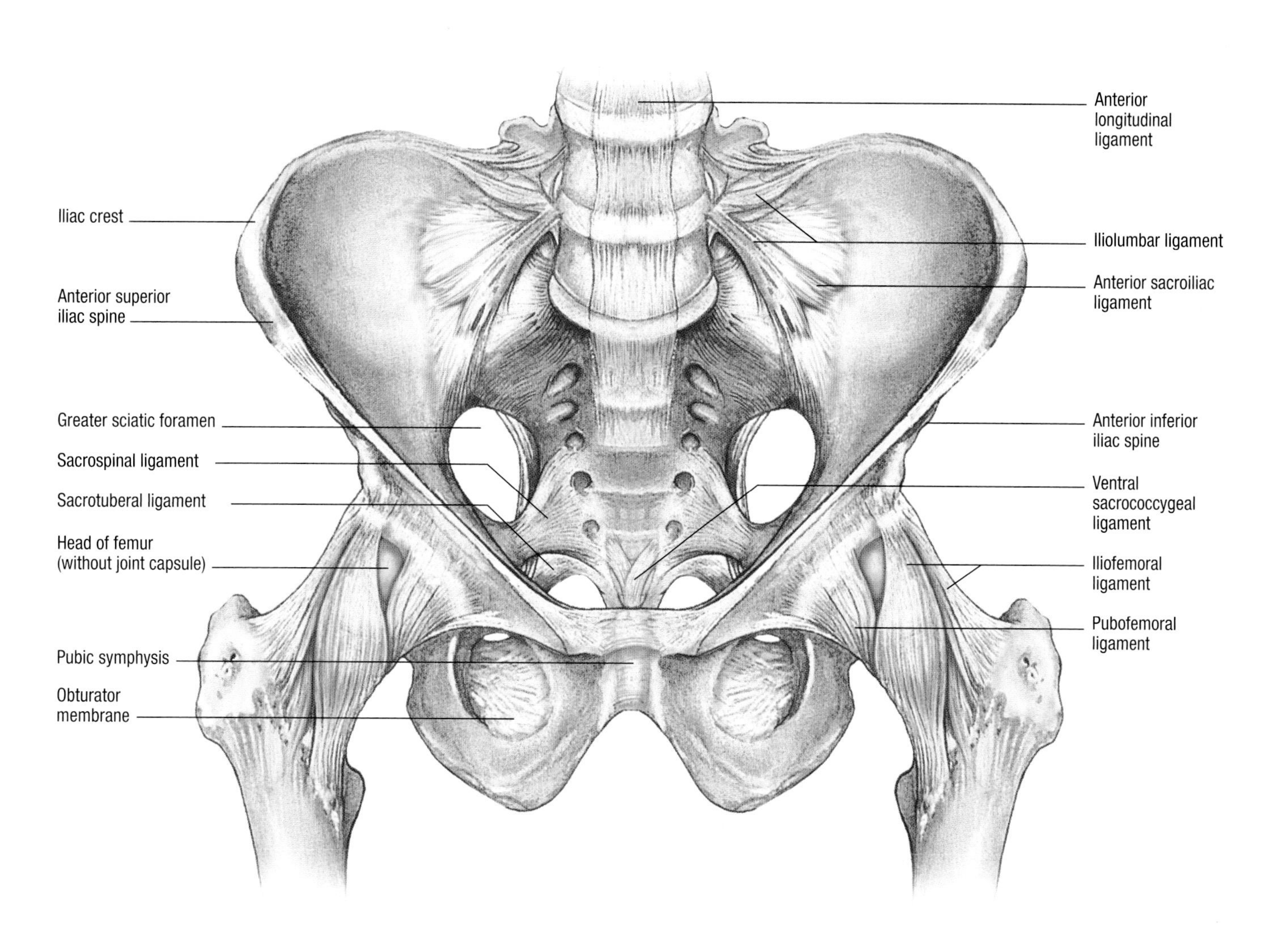

Anterior longitudinal ligament
Iliac crest
Iliolumbar ligament
Anterior superior iliac spine
Anterior sacroiliac ligament
Greater sciatic foramen
Anterior inferior iliac spine
Sacrospinal ligament
Ventral sacrococcygeal ligament
Sacrotuberal ligament
Head of femur (without joint capsule)
Iliofemoral ligament
Pubofemoral ligament
Pubic symphysis
Obturator membrane

## PELVIC LIGAMENTS
### Posterior view

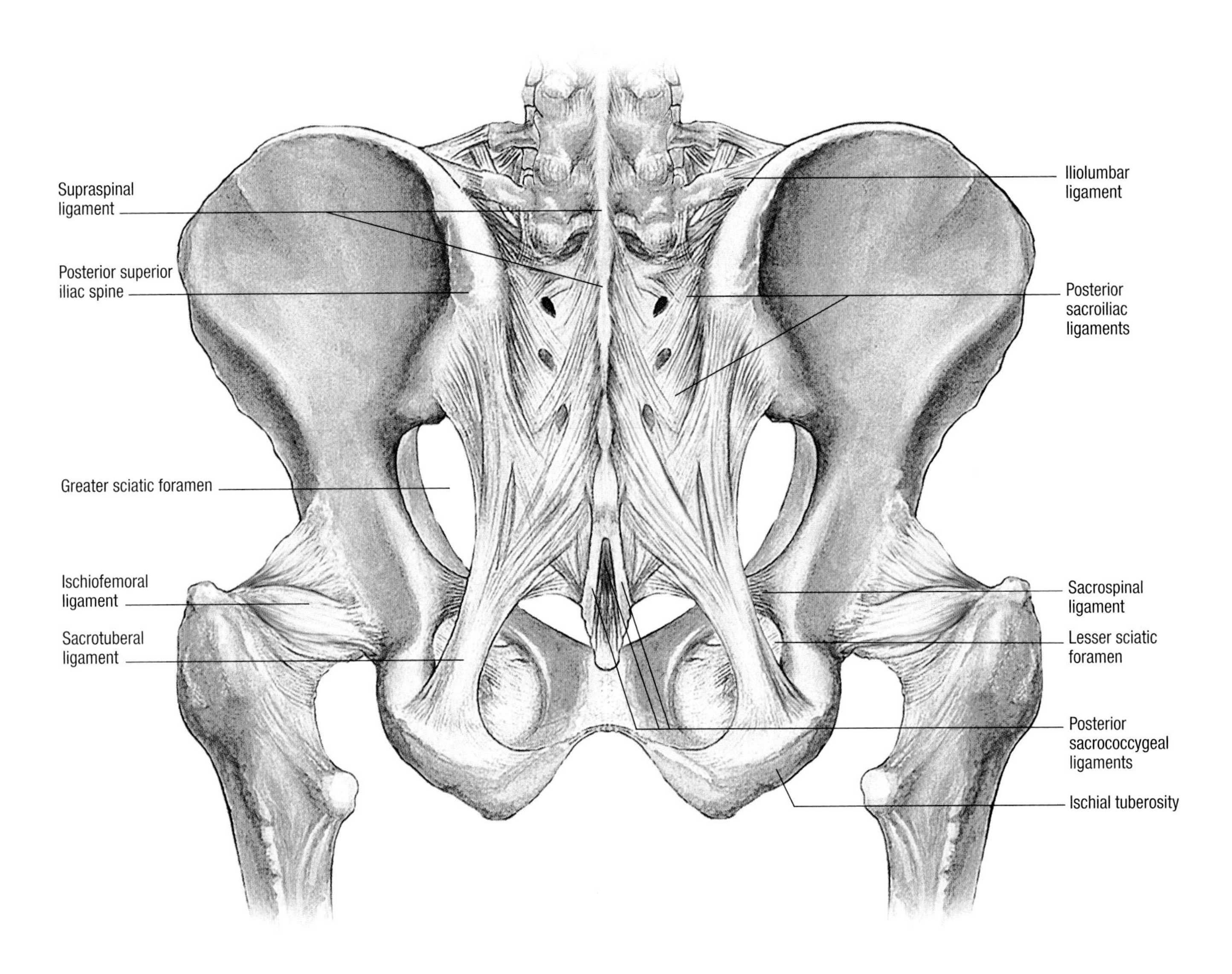
Supraspinal ligament
Posterior superior iliac spine
Greater sciatic foramen
Ischiofemoral ligament
Sacrotuberal ligament
Iliolumbar ligament
Posterior sacroiliac ligaments
Sacrospinal ligament
Lesser sciatic foramen
Posterior sacrococcygeal ligaments
Ischial tuberosity

## PELVIC WALL AND FLOOR MUSCLES
### Superior view

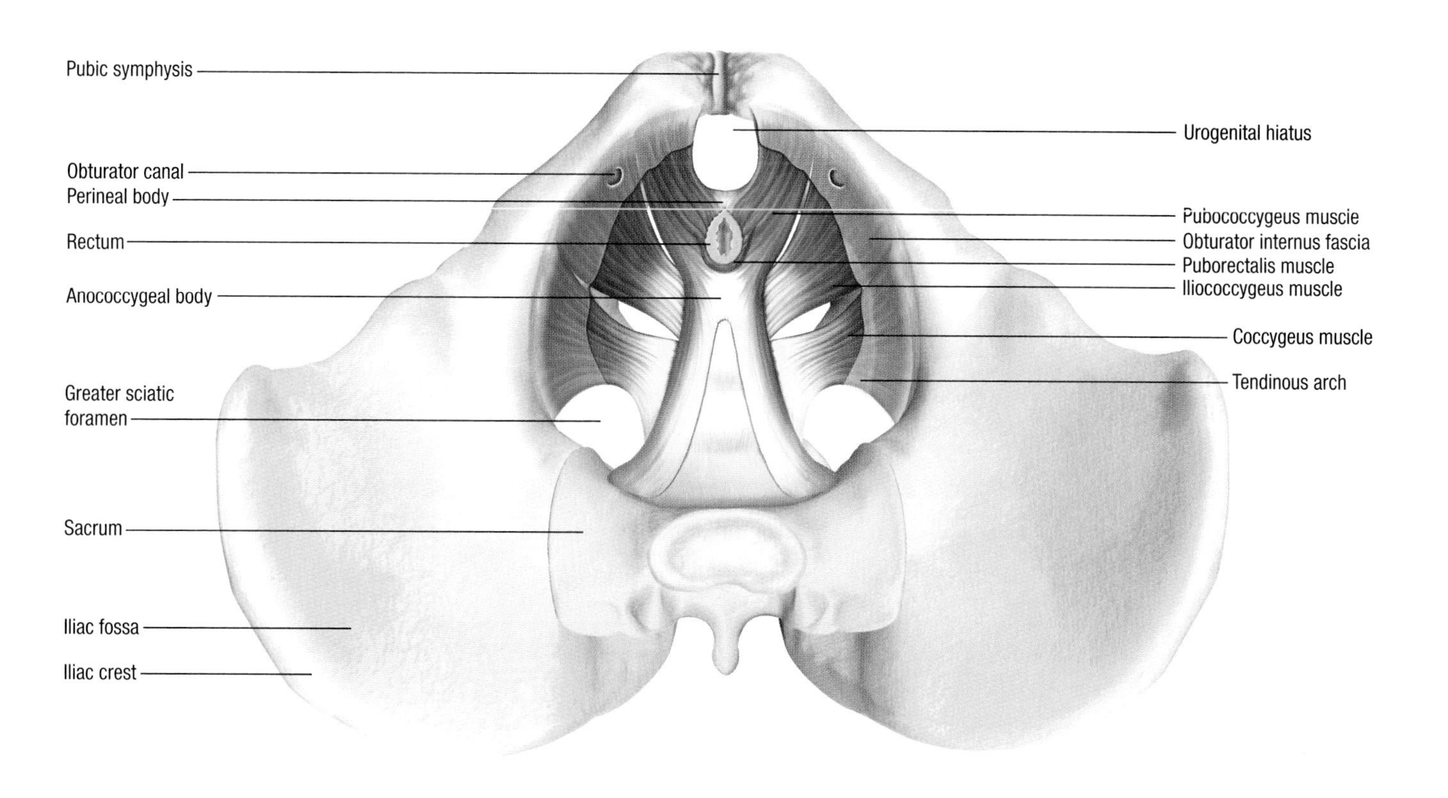
Pubic symphysis
Obturator canal
Perineal body
Rectum
Anococcygeal body
Greater sciatic foramen
Sacrum
Iliac fossa
Iliac crest
Urogenital hiatus
Pubococcygeus muscle
Obturator internus fascia
Puborectalis muscle
Iliococcygeus muscle
Coccygeus muscle
Tendinous arch

## PELVIC WALL AND FLOOR MUSCLES
### Inferior view

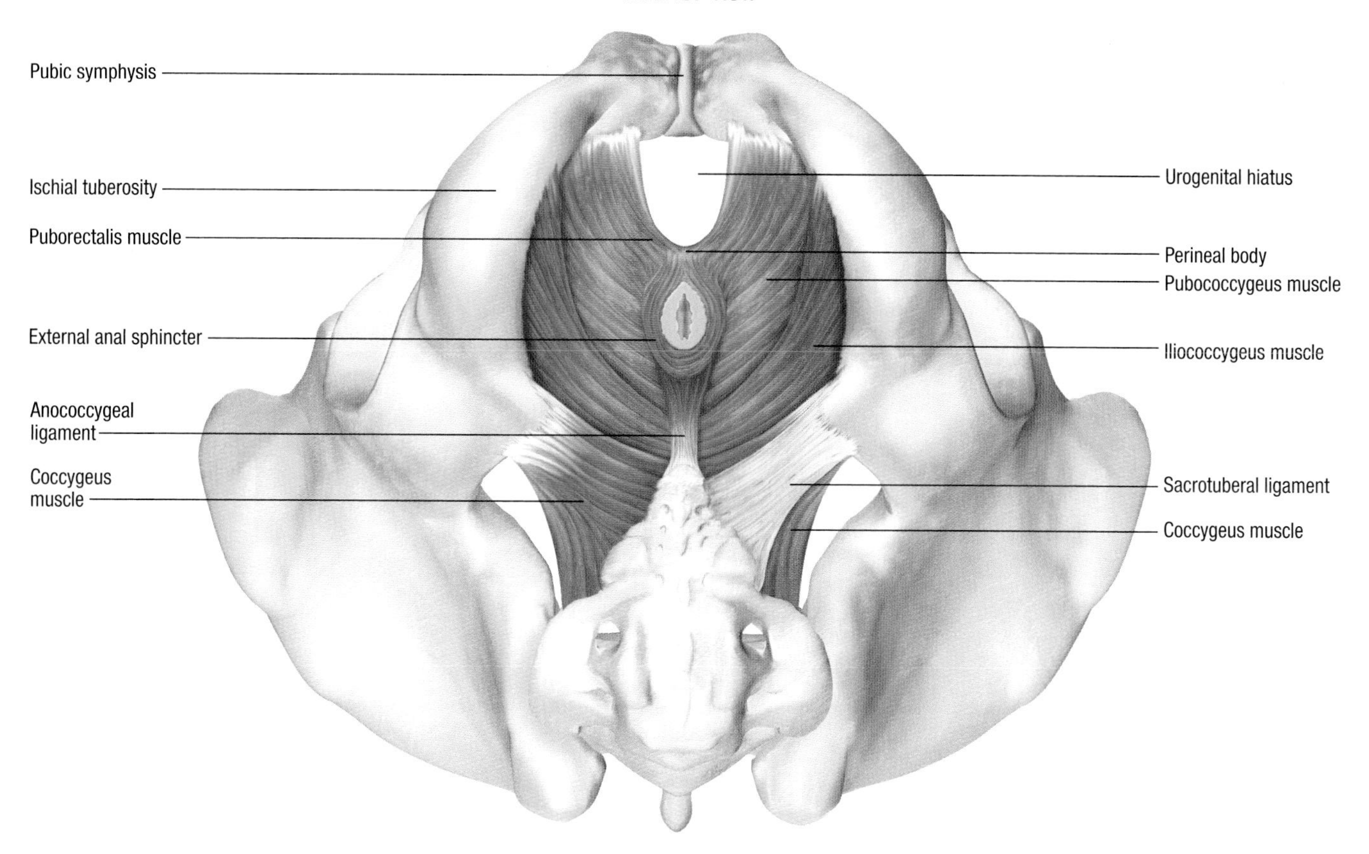
Pubic symphysis
Ischial tuberosity
Puborectalis muscle
External anal sphincter
Anococcygeal ligament
Coccygeus muscle
Urogenital hiatus
Perineal body
Pubococcygeus muscle
Iliococcygeus muscle
Sacrotuberal ligament
Coccygeus muscle

## PELVIC NERVES
### Male

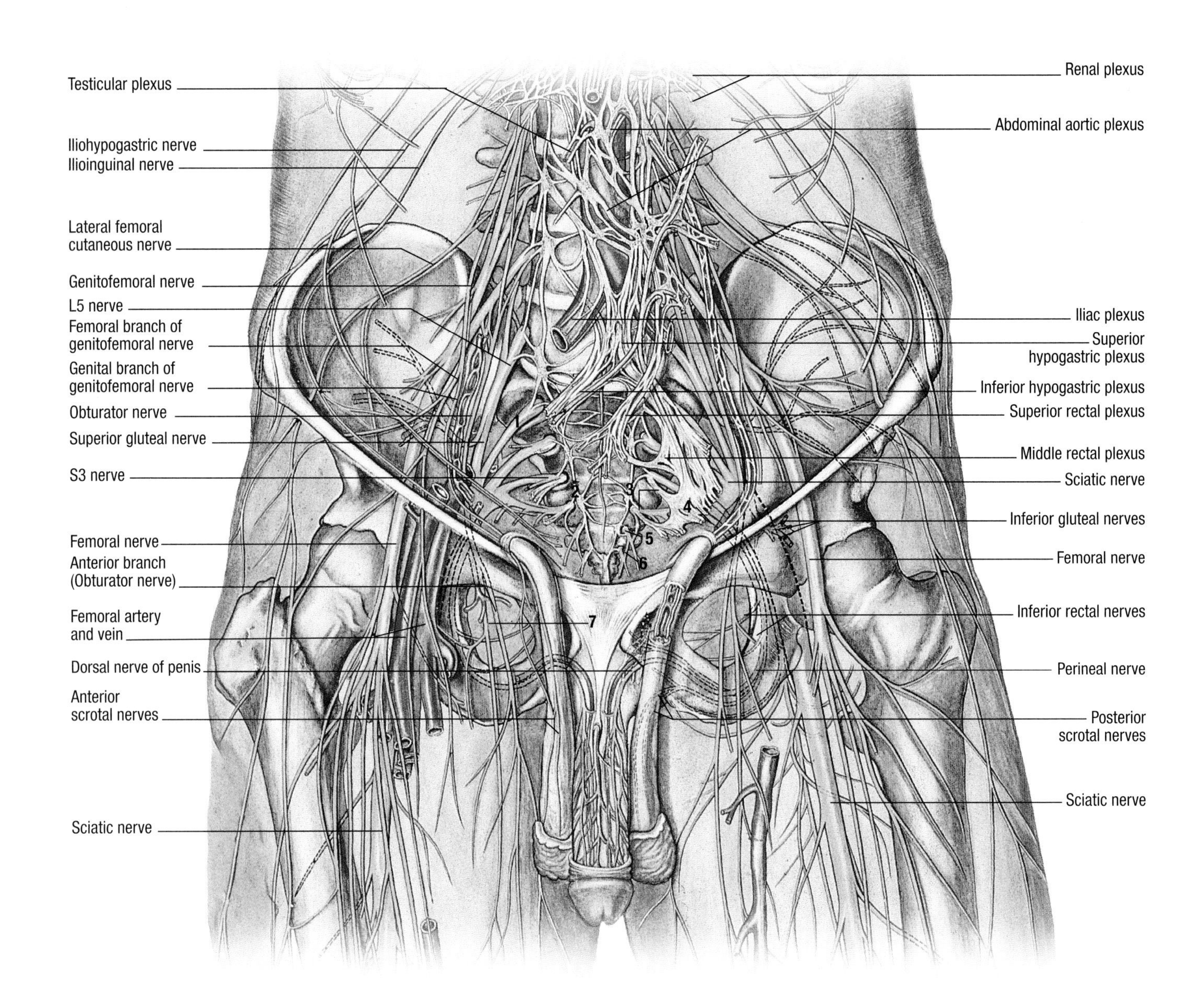

Testicular plexus
Iliohypogastric nerve
Ilioinguinal nerve
Lateral femoral cutaneous nerve
Genitofemoral nerve
L5 nerve
Femoral branch of genitofemoral nerve
Genital branch of genitofemoral nerve
Obturator nerve
Superior gluteal nerve
S3 nerve
Femoral nerve
Anterior branch (Obturator nerve)
Femoral artery and vein
Dorsal nerve of penis
Anterior scrotal nerves
Sciatic nerve
Renal plexus
Abdominal aortic plexus
Iliac plexus
Superior hypogastric plexus
Inferior hypogastric plexus
Superior rectal plexus
Middle rectal plexus
Sciatic nerve
Inferior gluteal nerves
Femoral nerve
Inferior rectal nerves
Perineal nerve
Posterior scrotal nerves
Sciatic nerve
1
3
4
5
6
7


### Key

1 S1 nerve
2 Sympathetic trunk
3 Pelvic splanchnic nerves
4 Pudendal nerve
5 S5 nerve
6 Coccygeal nerve
7 Posterior branch (Obturator nerve)

## PELVIC NERVES
**Female — Sagittal section**

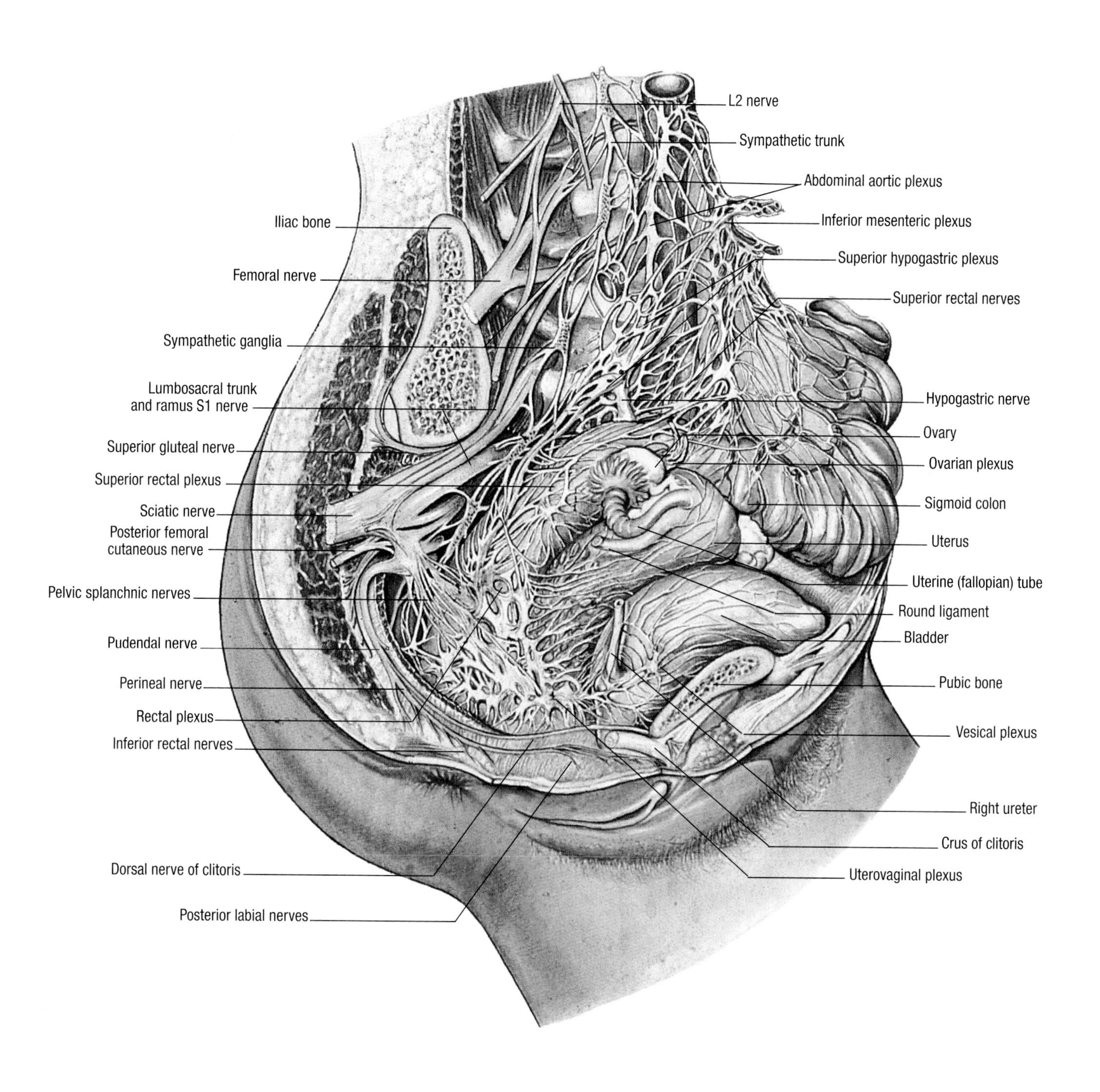

L2 nerve
Sympathetic trunk
Abdominal aortic plexus
Inferior mesenteric plexus
Superior hypogastric plexus
Superior rectal nerves
Hypogastric nerve
Ovary
Ovarian plexus
Sigmoid colon
Uterus
Uterine (fallopian) tube
Round ligament
Bladder
Pubic bone
Vesical plexus
Right ureter
Crus of clitoris
Uterovaginal plexus
Iliac bone
Femoral nerve
Sympathetic ganglia
Lumbosacral trunk
and ramus S1 nerve
Superior gluteal nerve
Superior rectal plexus
Sciatic nerve
Posterior femoral
cutaneous nerve
Pelvic splanchnic nerves
Pudendal nerve
Perineal nerve
Rectal plexus
Inferior rectal nerves
Dorsal nerve of clitoris
Posterior labial nerves

## PELVIC BLOOD VESSELS
### Male

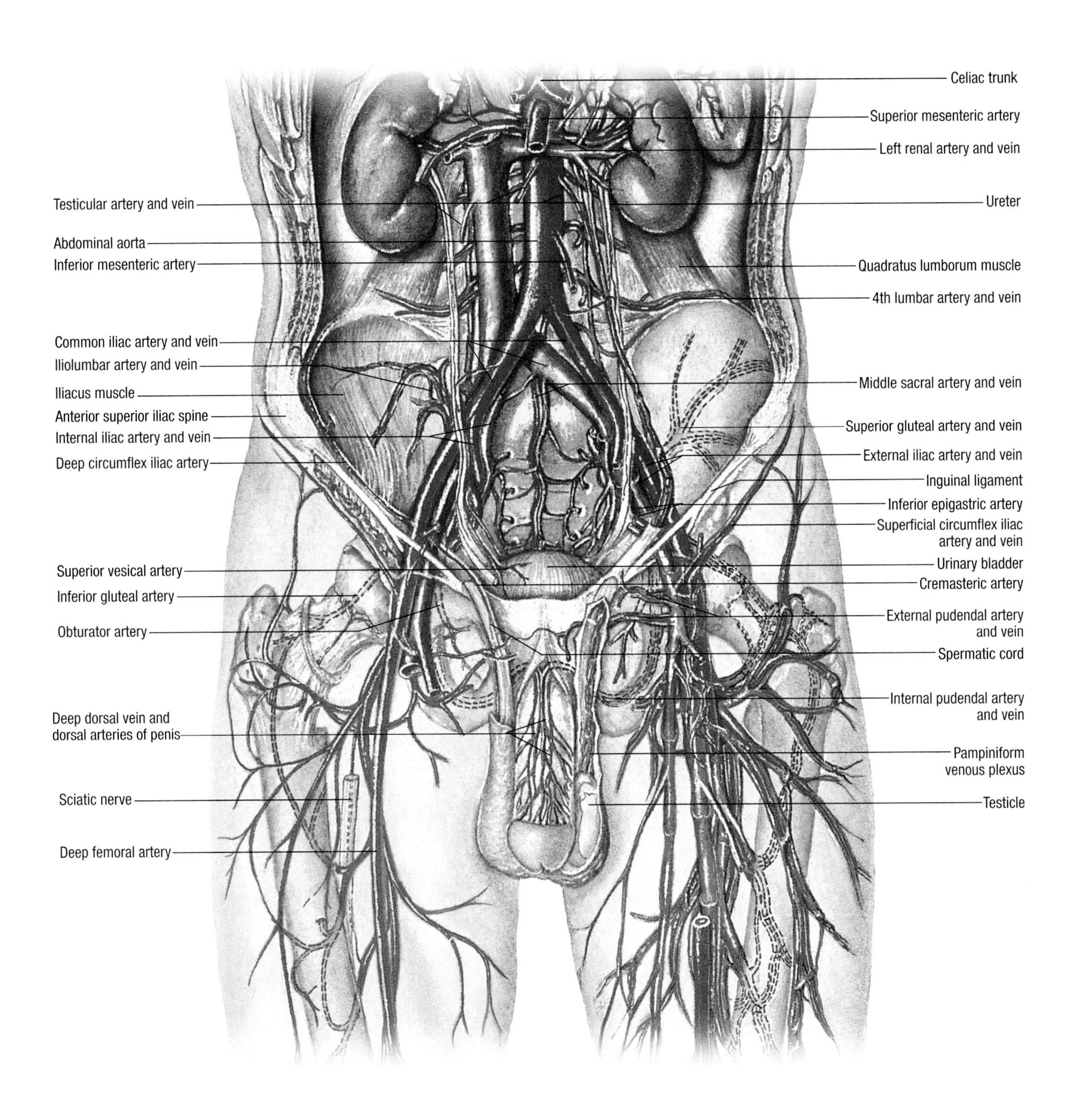

Celiac trunk
Superior mesenteric artery
Left renal artery and vein
Testicular artery and vein
Ureter
Abdominal aorta
Inferior mesenteric artery
Quadratus lumborum muscle
4th lumbar artery and vein
Common iliac artery and vein
Iliolumbar artery and vein
Middle sacral artery and vein
Iliacus muscle
Anterior superior iliac spine
Superior gluteal artery and vein
Internal iliac artery and vein
External iliac artery and vein
Deep circumflex iliac artery
Inguinal ligament
Inferior epigastric artery
Superficial circumflex iliac artery and vein
Superior vesical artery
Urinary bladder
Inferior gluteal artery
Cremasteric artery
External pudendal artery and vein
Obturator artery
Spermatic cord
Internal pudendal artery and vein
Deep dorsal vein and dorsal arteries of penis
Pampiniform venous plexus
Sciatic nerve
Testicle
Deep femoral artery

## PELVIC BLOOD VESSELS
### Female — Posterior view

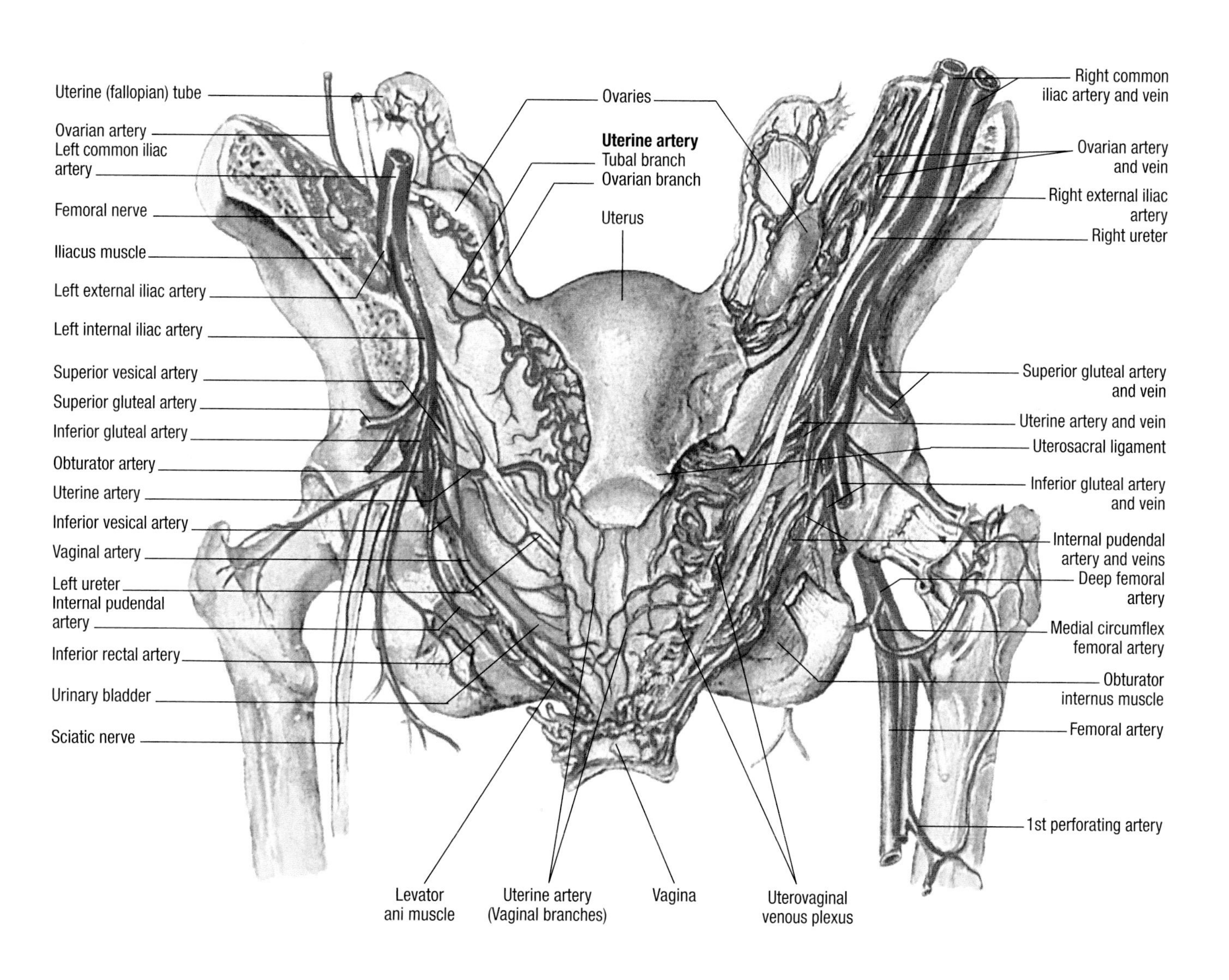

Uterine (fallopian) tube
Ovarian artery
Left common iliac artery
Femoral nerve
Iliacus muscle
Left external iliac artery
Left internal iliac artery
Superior vesical artery
Superior gluteal artery
Inferior gluteal artery
Obturator artery
Uterine artery
Inferior vesical artery
Vaginal artery
Left ureter
Internal pudendal artery
Inferior rectal artery
Urinary bladder
Sciatic nerve
Ovaries
**Uterine artery**
Tubal branch
Ovarian branch
Uterus
Right common iliac artery and vein
Ovarian artery and vein
Right external iliac artery
Right ureter
Superior gluteal artery and vein
Uterine artery and vein
Uterosacral ligament
Inferior gluteal artery and vein
Internal pudendal artery and veins
Deep femoral artery
Medial circumflex femoral artery
Obturator internus muscle
Femoral artery
1st perforating artery
Levator ani muscle
Uterine artery (Vaginal branches)
Vagina
Uterovaginal venous plexus

# Male pelvic structures

In the male, pelvic structures include gastrointestinal, reproductive, and urinary organs. The scrotum, the root of the penis, and a portion of the spongy parts of the urethra make up part of the male perineum. (See *Male perineum,* page 246.) The bladder and the rectum are similar in both sexes. (See Bladder, page 260, and Rectum and anal canal, page 262.)

## Penis

The penis is the organ of copulation and the terminal duct of the urinary tract. (See *Penis* and *Male bladder and urethra,* page 247, and *Vasculature and innervation*, page 246.)

The penis consists of an attached root; a free shaft or body covered with thin, loose skin; and an enlarged tip, the glans penis. Internally, the cylinder-shaped penile shaft consists of three columns of erectile tissue bound together by heavy fibrous tissue:

- two corpora cavernosa
- corpus spongiosum, on the underside, encases the urethra; its enlarged proximal end forms the bulb of the penis.

The glans penis, at the distal end of the shaft, is a cone-shaped structure formed from the corpus spongiosum. Its lateral margin forms a ridge of tissue known as the corona. In an uncircumcised male, a skin flap, the foreskin or prepuce, covers the corona and much of the glans. The urethral meatus opens through the glans to allow urination and ejaculation.

Blood enters the penis through the internal pudendal artery and flows through the penile artery into the corpora cavernosa. Venous blood returns through the internal iliac vein to the vena cava.

## Scrotum

The penis meets the scrotum, or scrotal sac, at the penoscrotal junction. Posterior to the penis and anterior to the anus, the scrotum is an extra-abdominal pouch consisting of a thin layer of skin overlying a tighter, musclelike layer. This musclelike layer overlies the tunica vaginalis, a serous membrane lining the internal scrotal cavity. Externally, the median raphe separates right and left halves of the scrotal skin. Internally, a septum divides the scrotum into two sacs, each containing a testis, an epididymis, and a spermatic cord.

### Function

The apparent function of the scrotum is to keep the testes cool. Spermatogenesis requires a temperature below that of the body. The dartos muscle, a smooth muscle in the superficial fascia, causes scrotal skin to wrinkle, which helps regulate temperature. The cremaster muscle, rising from the internal oblique muscle, helps to govern temperature by elevating the testis.

**Age-related changes**

The reproductive system in the aging male undergoes the following changes:

- slowing of testosterone production, which results in decreased libido as well as atrophy and softening of the testes
- approximately 50% to 70% decrease in sperm production between ages 60 and 80
- prostate gland enlargement with decreasing secretions
- decreased volume and viscosity of seminal fluid
- slower and weaker physiologic reaction during intercourse; longer refractory period.

### Lymphatic system

Lymph nodes from the penis, scrotal surface, and anus drain into the superficial inguinal lymph nodes. Lymph nodes from the testes drain into the lateral aortic and preaortic lymph nodes in the abdomen.

# MALE PELVIC STRUCTURES
## Sagittal section

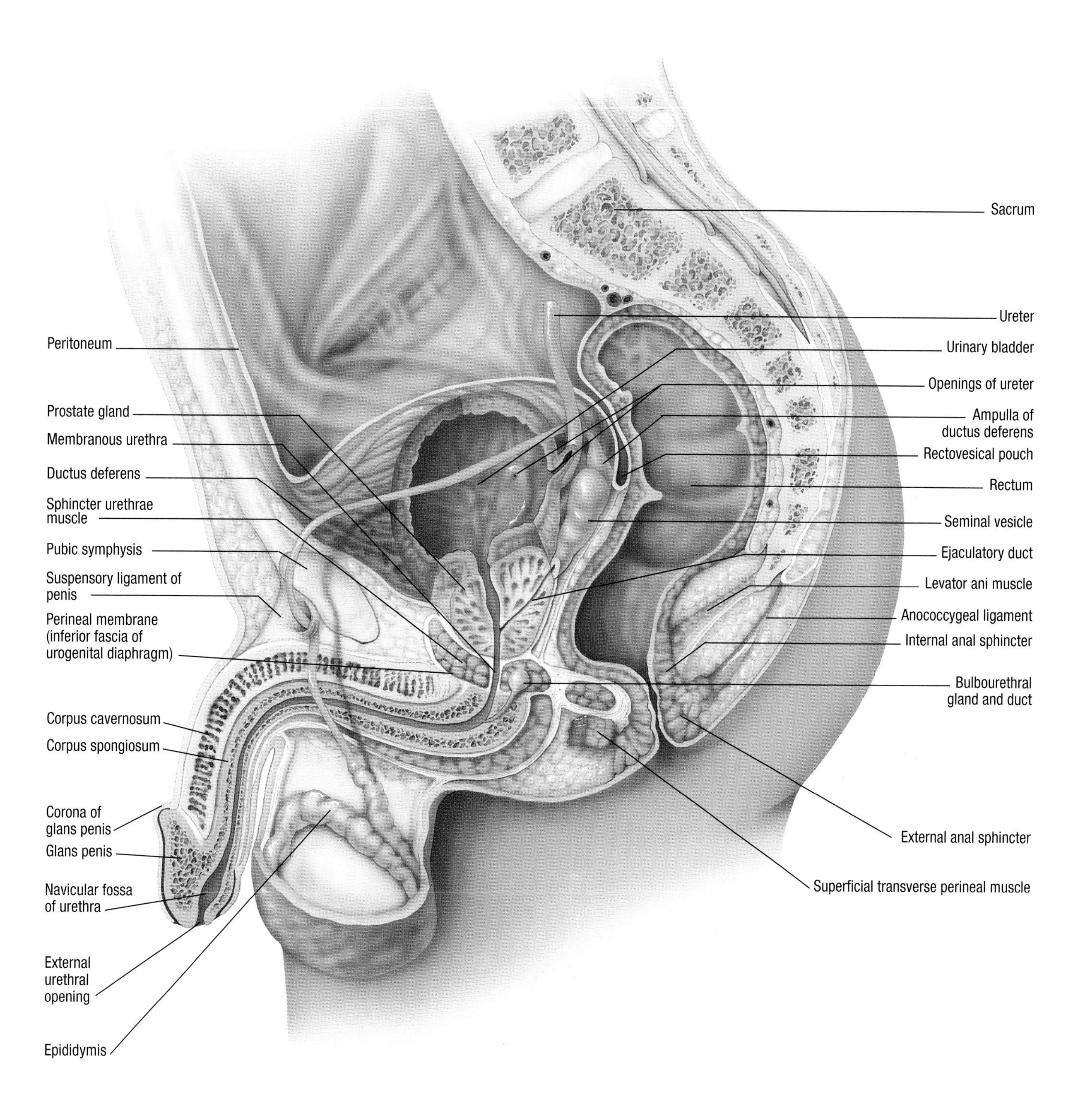

Sacrum
Ureter
Urinary bladder
Openings of ureter
Ampulla of ductus deferens
Rectovesical pouch
Rectum
Seminal vesicle
Ejaculatory duct
Levator ani muscle
Anococcygeal ligament
Internal anal sphincter
Bulbourethral gland and duct
External anal sphincter
Superficial transverse perineal muscle
Peritoneum
Prostate gland
Membranous urethra
Ductus deferens
Sphincter urethrae muscle
Pubic symphysis
Suspensory ligament of penis
Perineal membrane (inferior fascia of urogenital diaphragm)
Corpus cavernosum
Corpus spongiosum
Corona of glans penis
Glans penis
Navicular fossa of urethra
External urethral opening
Epididymis

## VASCULATURE AND INNERVATION

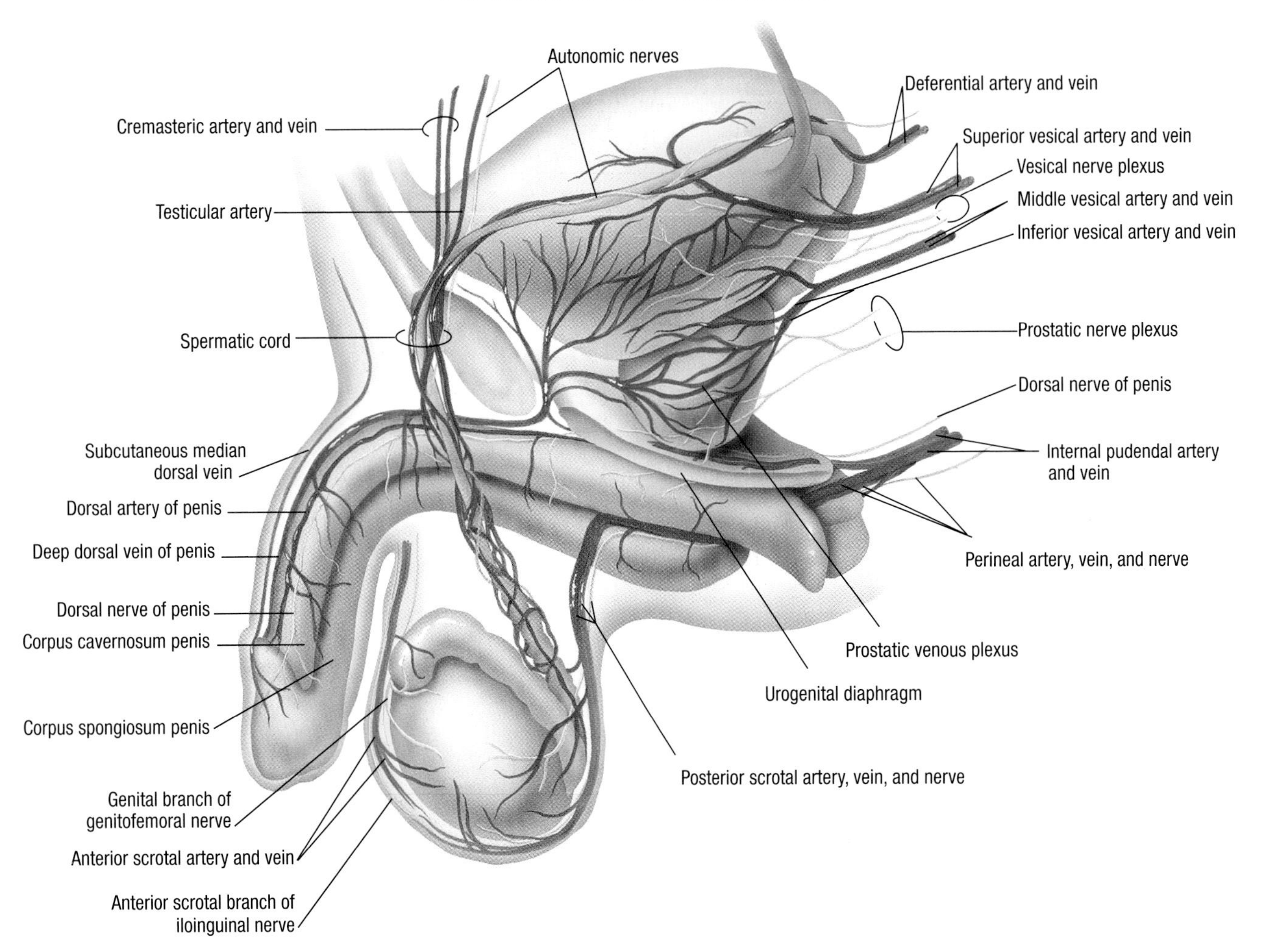
Autonomic nerves
Deferential artery and vein
Cremasteric artery and vein
Superior vesical artery and vein
Vesical nerve plexus
Middle vesical artery and vein
Testicular artery
Inferior vesical artery and vein
Spermatic cord
Prostatic nerve plexus
Dorsal nerve of penis
Subcutaneous median dorsal vein
Internal pudendal artery and vein
Dorsal artery of penis
Deep dorsal vein of penis
Perineal artery, vein, and nerve
Dorsal nerve of penis
Corpus cavernosum penis
Prostatic venous plexus
Urogenital diaphragm
Corpus spongiosum penis
Posterior scrotal artery, vein, and nerve
Genital branch of genitofemoral nerve
Anterior scrotal artery and vein
Anterior scrotal branch of iloinguinal nerve

## MALE PERINEUM

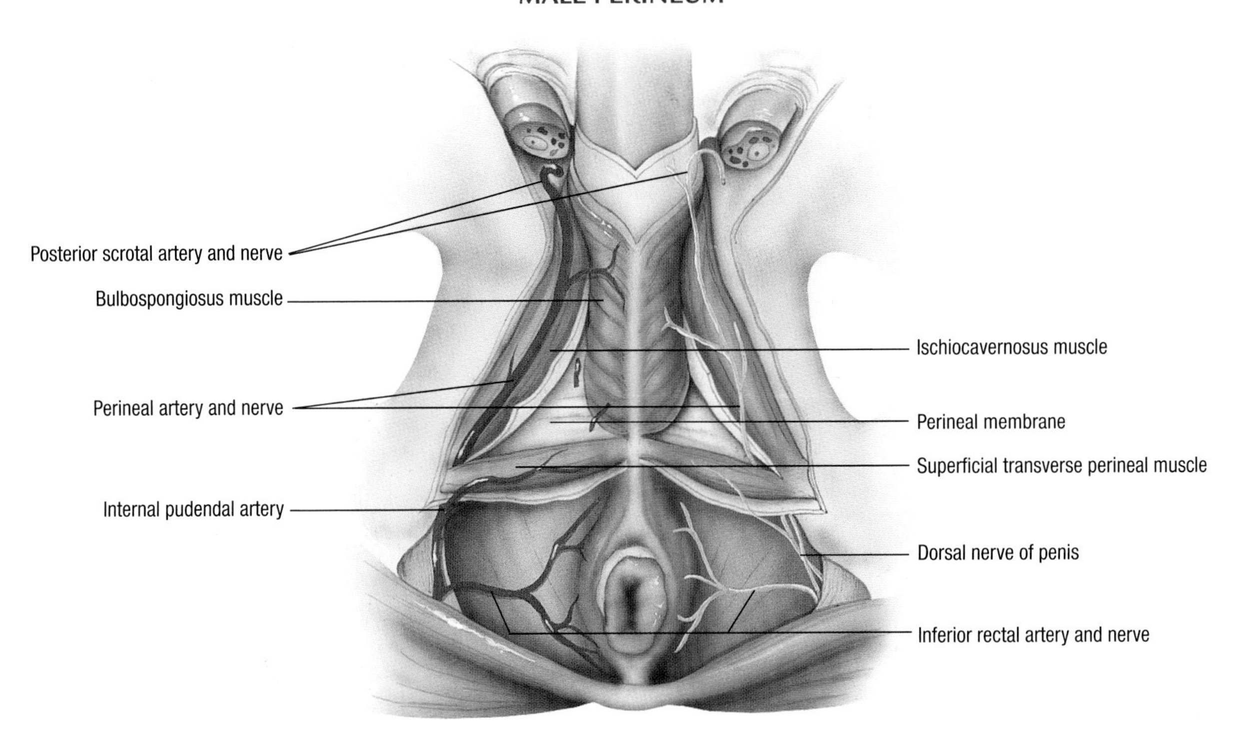
Posterior scrotal artery and nerve
Bulbospongiosus muscle
Ischiocavernosus muscle
Perineal artery and nerve
Perineal membrane
Superficial transverse perineal muscle
Internal pudendal artery
Dorsal nerve of penis
Inferior rectal artery and nerve

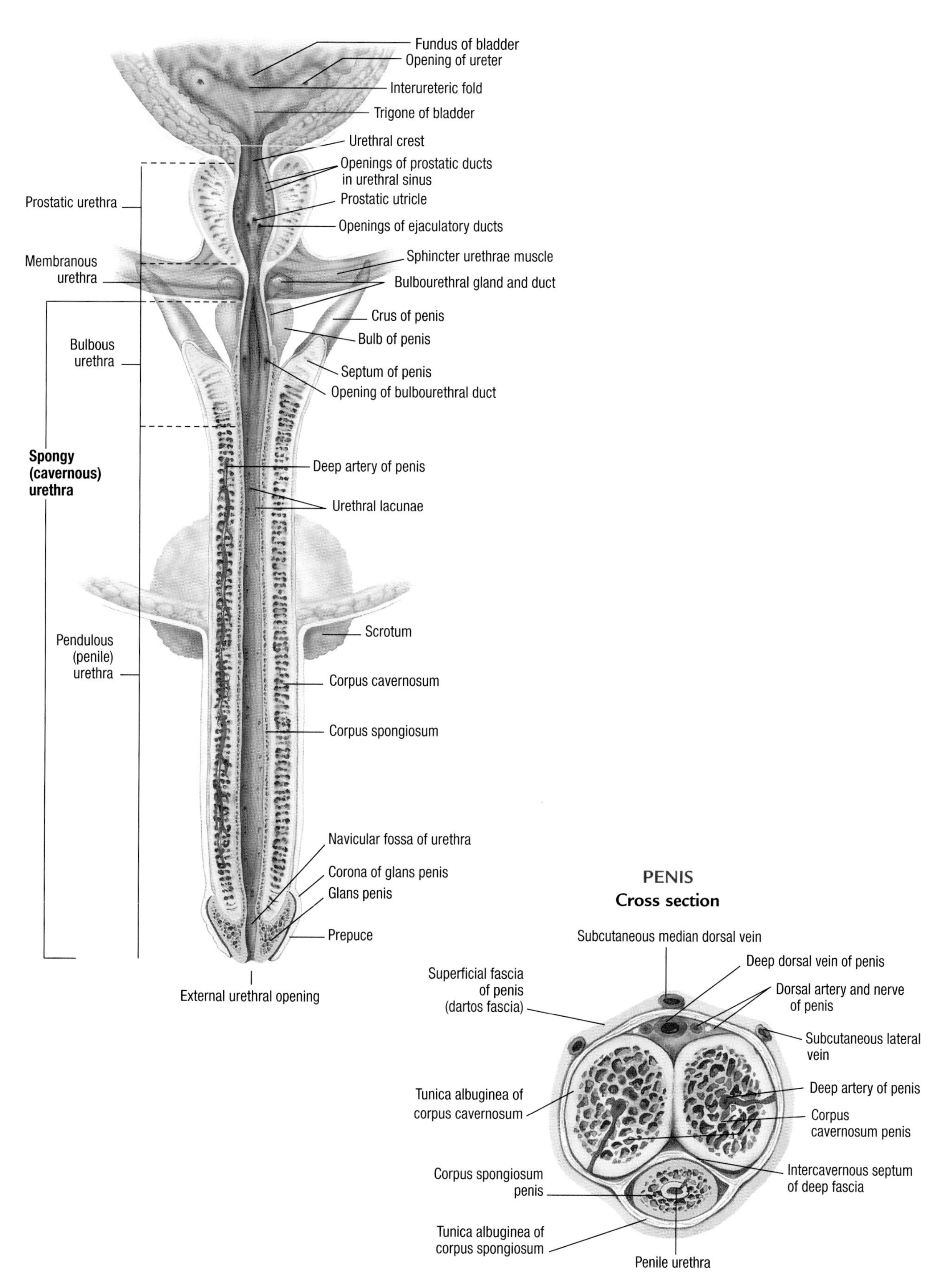
Fundus of bladder
Opening of ureter
Interureteric fold
Trigone of bladder
Urethral crest
Openings of prostatic ducts in urethral sinus
Prostatic utricle
Openings of ejaculatory ducts
Sphincter urethrae muscle
Bulbourethral gland and duct
Crus of penis
Bulb of penis
Septum of penis
Opening of bulbourethral duct
Deep artery of penis
Urethral lacunae
Scrotum
Corpus cavernosum
Corpus spongiosum
Navicular fossa of urethra
Corona of glans penis
Glans penis
Prepuce
External urethral opening
Prostatic urethra
Membranous urethra
Bulbous urethra
Spongy (cavernous) urethra
Pendulous (penile) urethra
PENIS
Cross section
Subcutaneous median dorsal vein
Deep dorsal vein of penis
Dorsal artery and nerve of penis
Superficial fascia of penis (dartos fascia)
Subcutaneous lateral vein
Deep artery of penis
Tunica albuginea of corpus cavernosum
Corpus cavernosum penis
Intercavernous septum of deep fascia
Corpus spongiosum penis
Tunica albuginea of corpus spongiosum
Penile urethra

# Testes, duct system, and accessory reproductive glands

The testes (testicles) – paired oval structures in the scrotum – are the male gonads. They also function as part of the endocrine system. Each testis measures about 2″ (5 cm) long by 1″ (2.5 cm) wide and weighs about 1/2 oz (14 g). The testes are enveloped in two connective tissue layers – the outer *tunica vaginalis* and the deeper *tunica albuginea*. Extensions of the tunica albuginea separate each testis into several hundred *lobules*. Each lobule contains one to four seminiferous tubules, small tubes where spermatogenesis takes place. (See *Testis*, page 250.)

The spermatic cord is a connective tissue sheath encasing autonomic nerve fibers, blood vessels, lymph vessels, and the vas deferens (also called ductus deferens). The spermatic cord travels from the testis through the inguinal canal, leaving the scrotum through the external inguinal ring and ending at the internal inguinal ring. The inguinal canal lies between the two rings. (To assess for inguinal hernia, see Part VI, Abdomen.)

## Duct system

The duct system – epididymis, vas deferens, and urethra – conveys sperm from the testes to the ejaculatory ducts.

### Epididymis

A coiled tube about 20 feet (6.1 m) long, the epididymis is superior to the testis, along its posterior border. At the lower border of the testis, the epididymis turns upward to join the vas deferens. During *ejaculation*, smooth muscle in the epididymis contracts, ejecting spermatozoa into the vas deferens.

### Vas deferens

The vas deferens leads from the testes to the abdominal cavity. About 18 inches (46 cm) long, it extends upward through the inguinal canal, arches over the urethra, and descends behind the bladder. Its enlarged portion, called the ampulla, merges with the duct of the seminal vesicle to form the short ejaculatory duct. After passing through the prostate gland, the vas deferens empties into the urethra.

### Urethra

The urethra conveys urine and semen to the tip of the penis. This small tube, leading from the floor of the bladder to the exterior, consists of three parts:

- prostatic – drains the bladder; surrounded by the prostate gland
- membranous – passes through the urogenital diaphragm
- spongy – about 75% of the entire tube.

## Accessory reproductive glands

Accessory glands produce most of the semen. They include:

- seminal vesicles – paired sacs at the base of the bladder
- bulbourethral (Cowper's) glands – paired glands inferior to the prostate
- prostate gland.

## Spermatogenesis

Sperm formation, or spermatogenesis, begins when a male reaches puberty and normally continues throughout life. The process is stimulated by male sex hormones. Sperm formation occurs in four stages.

### First stage

The primary germinal epithelial cells, called spermatogonia, grow and develop into primary spermatocytes. Both spermatogonia and primary spermatocytes contain 46 chromosomes – 44 autosomes and the two sex chromosomes, X and Y.

### Second stage

Primary spermatocytes divide to form secondary spermatocytes. No new chromosomes are formed in this stage; the pairs only divide. Each secondary spermatocyte contains half the number of autosomes, 22; one secondary spermatocyte contains an X chromosome, the other, a Y chromosome.

### Third stage

Each secondary spermatocyte divides to form spermatids.

### Fourth stage

The spermatids undergo a series of structural changes that transform them into mature spermatozoa, or sperm. Each spermatozoon has a head, neck, body, and tail. The head contains the nucleus; the tail, a large amount of adenosine triphosphate (ATP), which provides energy for sperm motility. Newly mature sperm pass from the seminiferous tubules through the vasa recta into the epididymis, where they mature. Only a small number of sperm can be stored in the epididymis. Most of them move into the vas deferens, where they're stored until sexual stimulation triggers emission.

## Prostate gland

Lying under the bladder and surrounding the urethra, the prostate gland is approximately 1½″ (4 cm) in diameter. It consists of three lobes – the left and right lateral lobes and the median lobe. (See *Prostate*, page 250.) The gland continuously secretes prostatic fluid, a thin, milky, alkaline fluid that adds volume to semen during sexual activity. This fluid enhances sperm motility and may improve the chance for conception by neutralizing the acidity of both the urethra and the vagina.

## Semen

Consisting of spermatozoa and accessory gland secretions, semen is a viscous, white secretion with a slightly alkaline pH (7.8 to 8). The seminal vesicles produce roughly 60% of the fluid portion of the semen; the prostate gland produces about 30%. A viscid fluid secreted by the bulbourethral glands also becomes part of the semen.

## BLADDER, TESTES, AND PROSTATE
### Posterior view

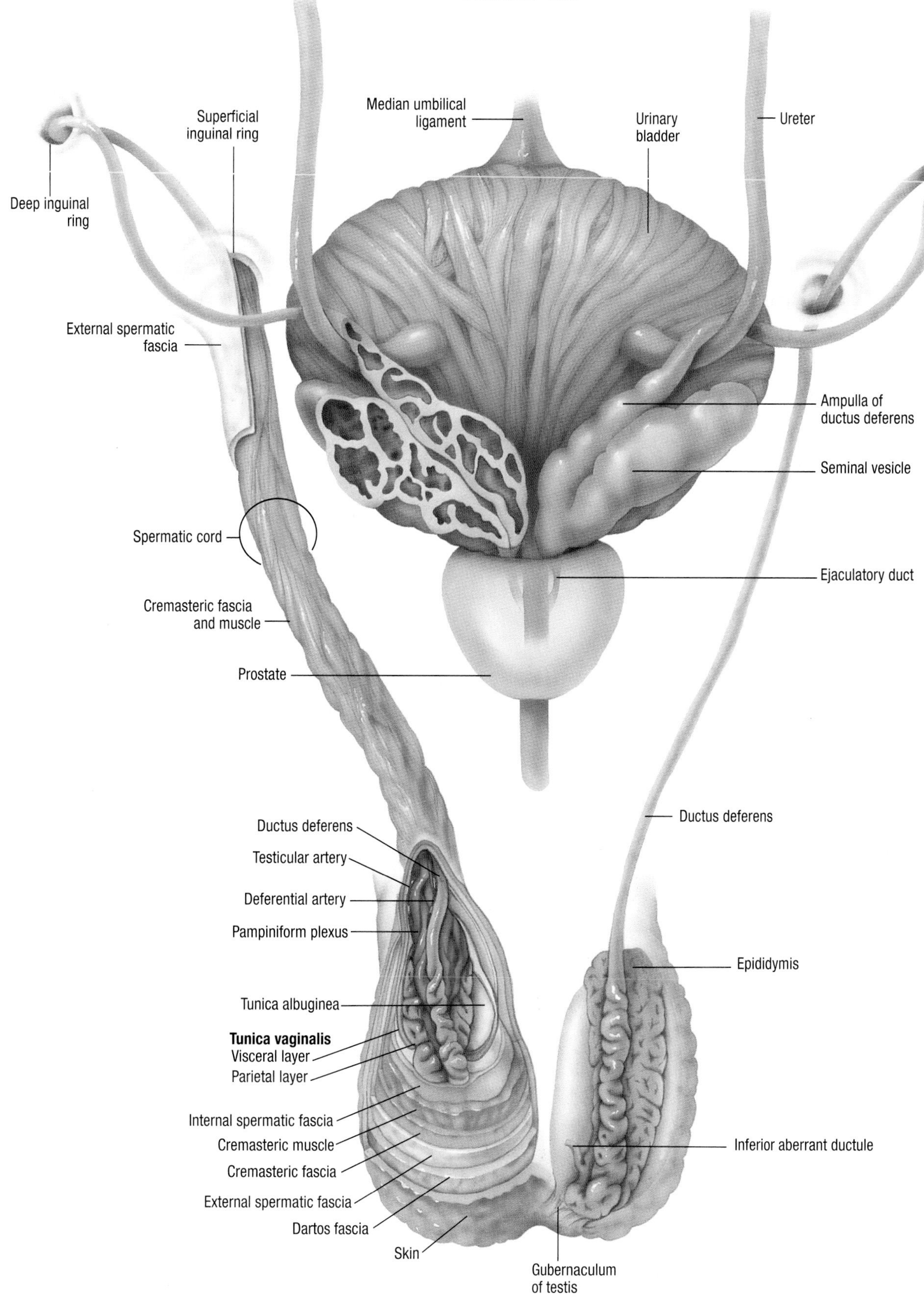

Superficial inguinal ring
Median umbilical ligament
Urinary bladder
Ureter
Deep inguinal ring
External spermatic fascia
Ampulla of ductus deferens
Seminal vesicle
Spermatic cord
Ejaculatory duct
Cremasteric fascia and muscle
Prostate
Ductus deferens
Ductus deferens
Testicular artery
Deferential artery
Pampiniform plexus
Epididymis
Tunica albuginea
**Tunica vaginalis**
Visceral layer
Parietal layer
Internal spermatic fascia
Cremasteric muscle
Inferior aberrant ductule
Cremasteric fascia
External spermatic fascia
Dartos fascia
Skin
Gubernaculum of testis

## TESTIS
### Sagittal section

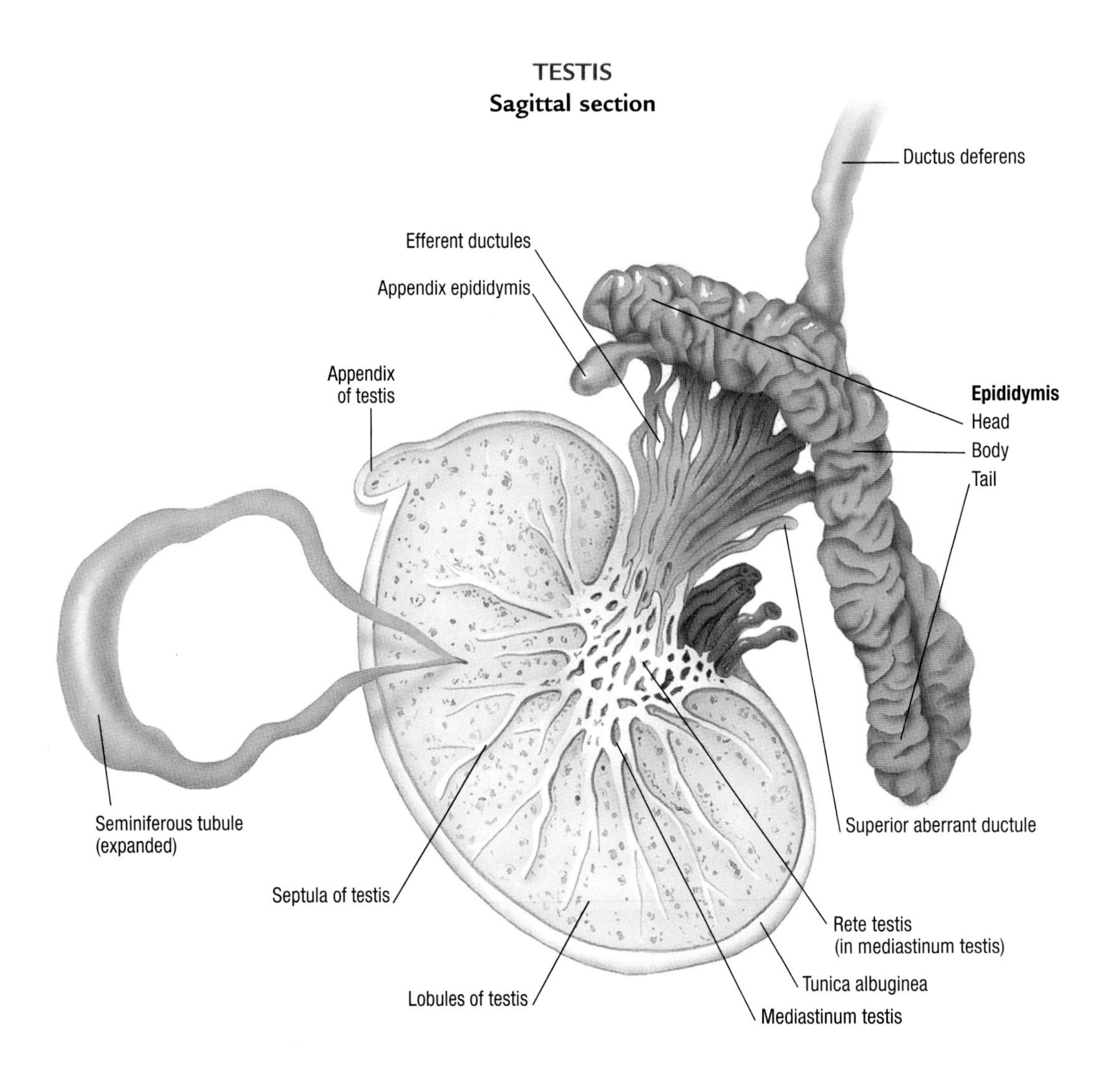

Ductus deferens
Efferent ductules
Appendix epididymis
Appendix of testis
Epididymis
Head
Body
Tail
Seminiferous tubule (expanded)
Superior aberrant ductule
Septula of testis
Rete testis (in mediastinum testis)
Tunica albuginea
Lobules of testis
Mediastinum testis


## PROSTATE

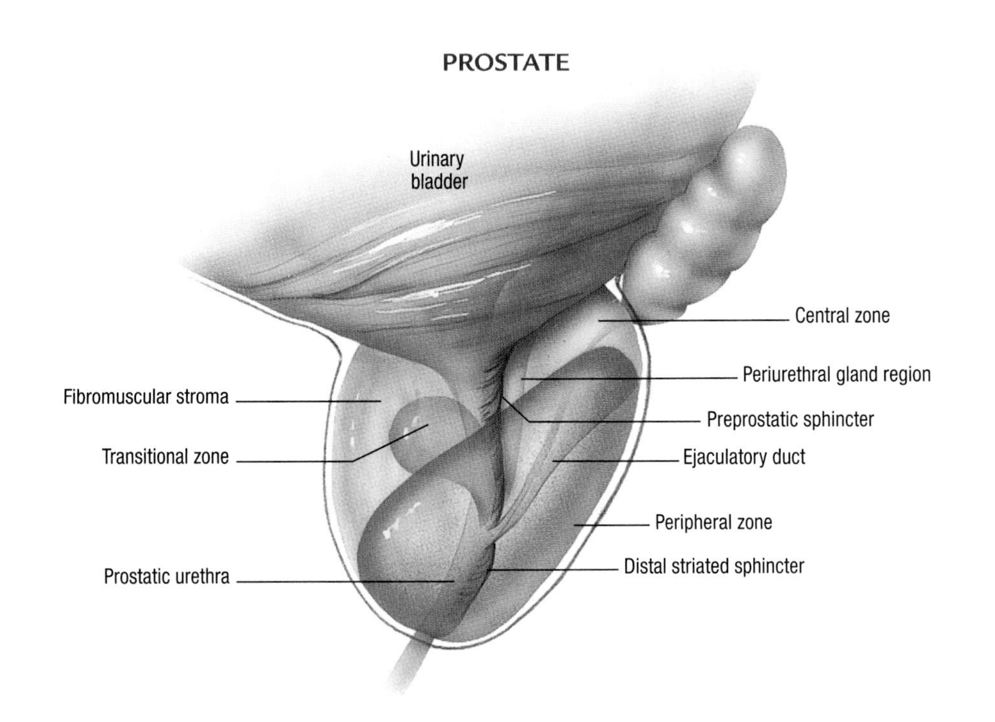

Urinary bladder
Central zone
Periurethral gland region
Fibromuscular stroma
Preprostatic sphincter
Transitional zone
Ejaculatory duct
Peripheral zone
Distal striated sphincter
Prostatic urethra

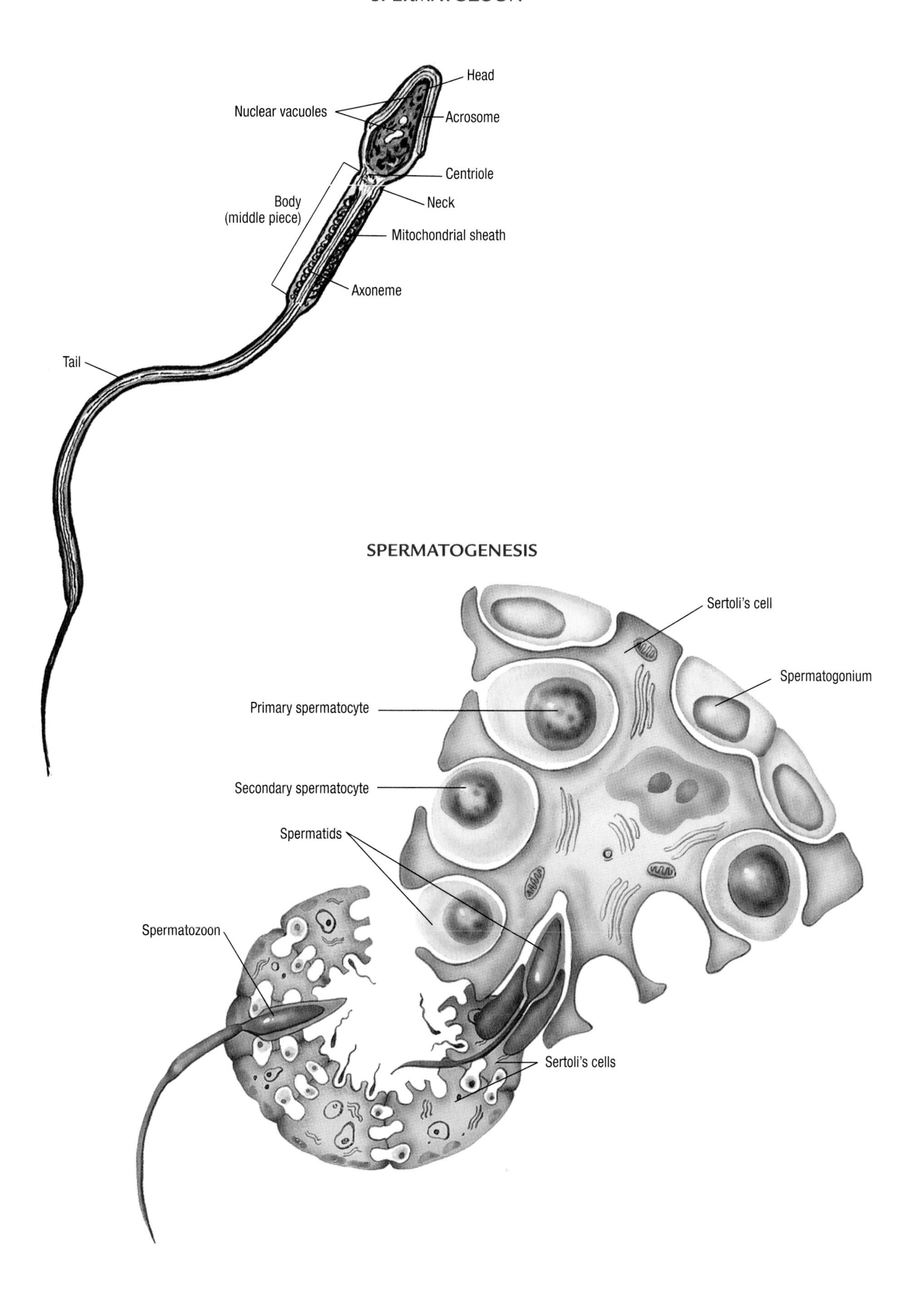
SPERMATOZOON
Head
Nuclear vacuoles
Acrosome
Centriole
Body
(middle piece)
Neck
Mitochondrial sheath
Axoneme
Tail
SPERMATOGENESIS
Sertoli's cell
Spermatogonium
Primary spermatocyte
Secondary spermatocyte
Spermatids
Spermatozoon
Sertoli's cells

# Female pelvic structures

The female pelvis includes reproductive, urinary, and gastrointestinal structures. Reproductive structures include the external and internal genitalia. Hormonal influences determine the development and function of these structures and affect fertility, childbearing, and the ability to experience sexual pleasure. The bladder, anus, and rectum, also located in the pelvis, are discussed on pages 260 and 262, respectively.

## Internal genitalia

The female internal genitalia are specialized organs whose main function is reproduction. They include the vagina, uterus, fallopian tubes, and ovaries. (See *Ovary, fallopian tube, uterus, and vagina*, page 254.)

### Vagina

The vagina is an elastic muscular tube that lies between the urethra and the rectum at a 45-degree angle to the long axis of the body. It is approximately 2½" to 2¾" (6 to 7 cm) long anteriorly and 3½" (9 cm) long posteriorly. The vaginal wall has three tissue layers – epithelium, loose connective tissue, and muscle. The uterine cervix projects into the upper portion of the vagina. Four *fornices* – recesses in the vaginal wall – surround the cervix. The cervix projects into the upper portion of the vagina. The lower cervical opening is the external os; the upper opening is the internal os.

#### Function

The vagina has three main functions:
- accommodating the penis during coitus
- channeling menstrual blood from the uterus
- serving as the birth canal during childbirth.

#### Vasculature

The upper, middle, and lower sections of the vagina have separate blood supplies. Branches of the uterine arteries supply blood to the upper vagina; the inferior vesical arteries supply blood to the middle vagina; and the hemorrhoidal and internal pudendal arteries feed into the lower vagina. Blood returns through the vaginal plexus, which drains through the vaginal vein to the internal iliac vein.

### Uterus

The uterus is a small, firm, pear-shaped, muscular organ that lies between the bladder and rectum, usually at almost a 90-degree angle to the vagina (although other locations may be normal). The mucous membrane lining the uterus is called the endometrium; the muscular layer, the myometrium. The uterus has three functions:
- menstruation and rejuvenation of the endometrium
- holding the product of conception (pregnancy)
- contracting during labor and assisting with delivery of the fetus and placenta.

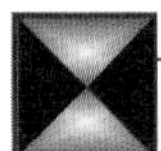

**Age-related changes**

Over a woman's lifetime, the uterine corpus and the cervix change in size, as do the percentages of space these parts occupy. For example, before the first menstrual period (menarche), one-third of the uterus may be uterine corpus and two-thirds may be cervix. After childbirth, proportions may be reversed.

After menopause, the cervix and vagina atrophy. The cervical endometrium and myometrium become thinner, and the vaginal mucosa becomes thin and dry.

### Fallopian tubes

Fertilization occurs in one of two fallopian tubes attached to the uterus at the upper angles of the uterine *fundus* (body of the uterus). These narrow cylinders of muscle fibers are about 2¾ to 5½" (7 to 14 cm) long. The curved, distal portion of the fallopian tube, called the ampulla, ends in the funnel-shaped infundibulum, where fingerlike projections, called fimbriae, move in waves that sweep the mature ovum released from the ovary into the fallopian tube.

### Ovaries

The ovaries are almond-shaped organs located on either side of the uterus at the distal ends of the fallopian tubes. They measure approximately 1¼" to 1½" (3 to 4 cm) long, ¾" (2 cm) wide, and ¼" to ½" (0.6 to 1.25 cm) thick. (See *Ovary* and *Unfertilized ovum*, page 255.)

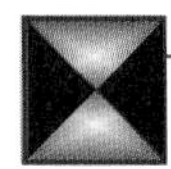

**Age-related changes**

The size, shape, and position of the ovaries vary with age. Round, smooth, and pink at birth, they enlarge, flatten, and turn grayish by puberty. During childbearing years, they take on an almond shape and a rough, pitted surface; after menopause, they shrink and turn white.

#### Function

The ovaries' main function is to produce mature *ova*. At birth, each ovary contains approximately 50,000 *graafian follicles*. During the childbearing years, one graafian follicle produces a mature ovum during the first half of each menstrual cycle. As the ovum matures, the follicle ruptures and the ovum is swept into the fallopian tube. The ovaries also produce estrogen, progesterone, and small quantities of androgens.

### Pelvic ligaments

Several ligaments in the pelvis support reproductive structures including:
- round ligament – supports the uterus
- broad ligaments – attaches to the uterus; surrounds ovaries and fallopian tubes
- uterosacral ligaments – attaches to cervix and passes to sacrum
- cardinal (lateral cervical) ligaments – attaches to uterine cervix and a portion of the vagina.

## FEMALE PELVIC ORGANS
### Medial section

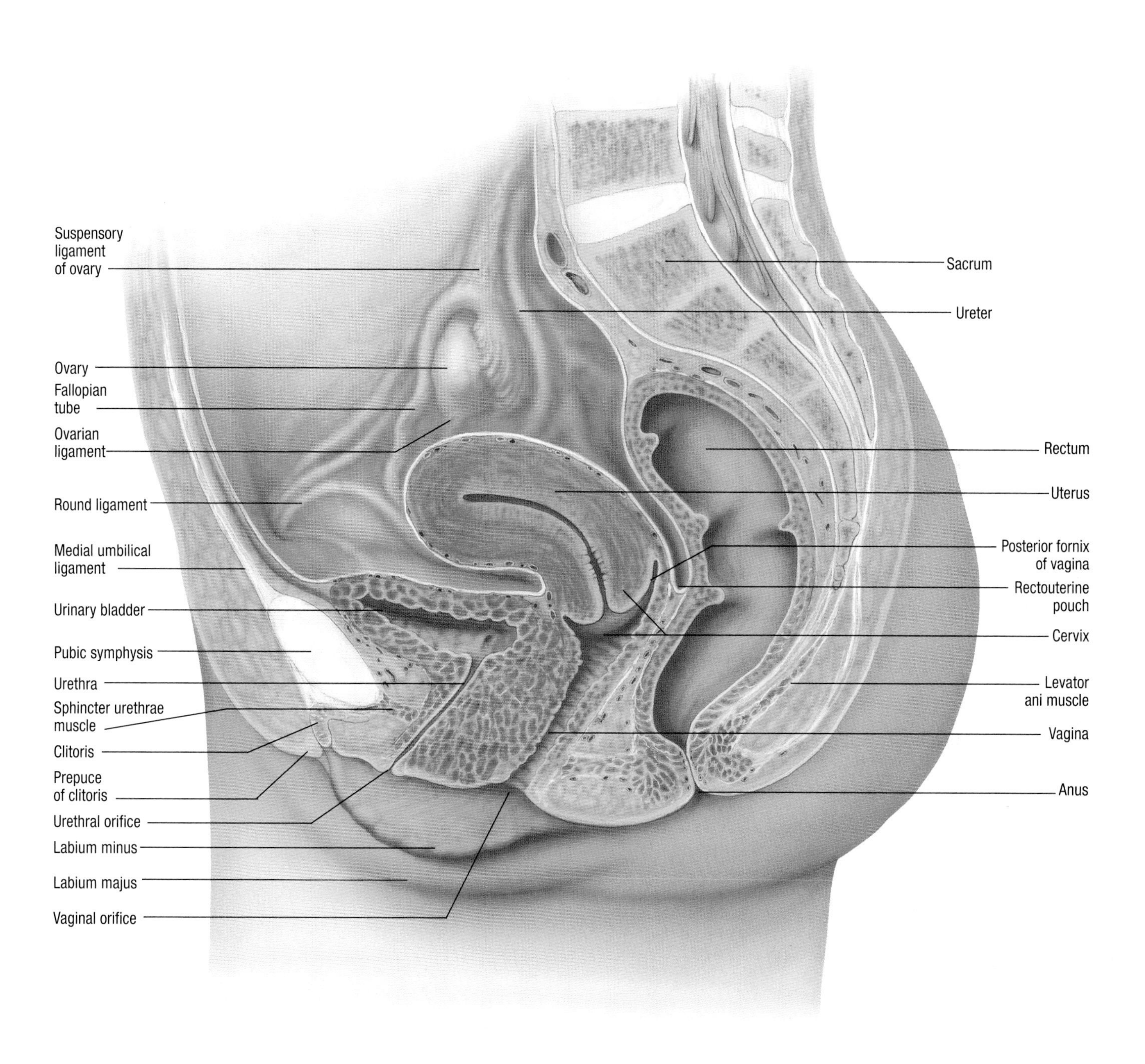

Suspensory ligament of ovary
Ovary
Fallopian tube
Ovarian ligament
Round ligament
Medial umbilical ligament
Urinary bladder
Pubic symphysis
Urethra
Sphincter urethrae muscle
Clitoris
Prepuce of clitoris
Urethral orifice
Labium minus
Labium majus
Vaginal orifice
Sacrum
Ureter
Rectum
Uterus
Posterior fornix of vagina
Rectouterine pouch
Cervix
Levator ani muscle
Vagina
Anus

## OVARY, FALLOPIAN TUBE, UTERUS, AND VAGINA

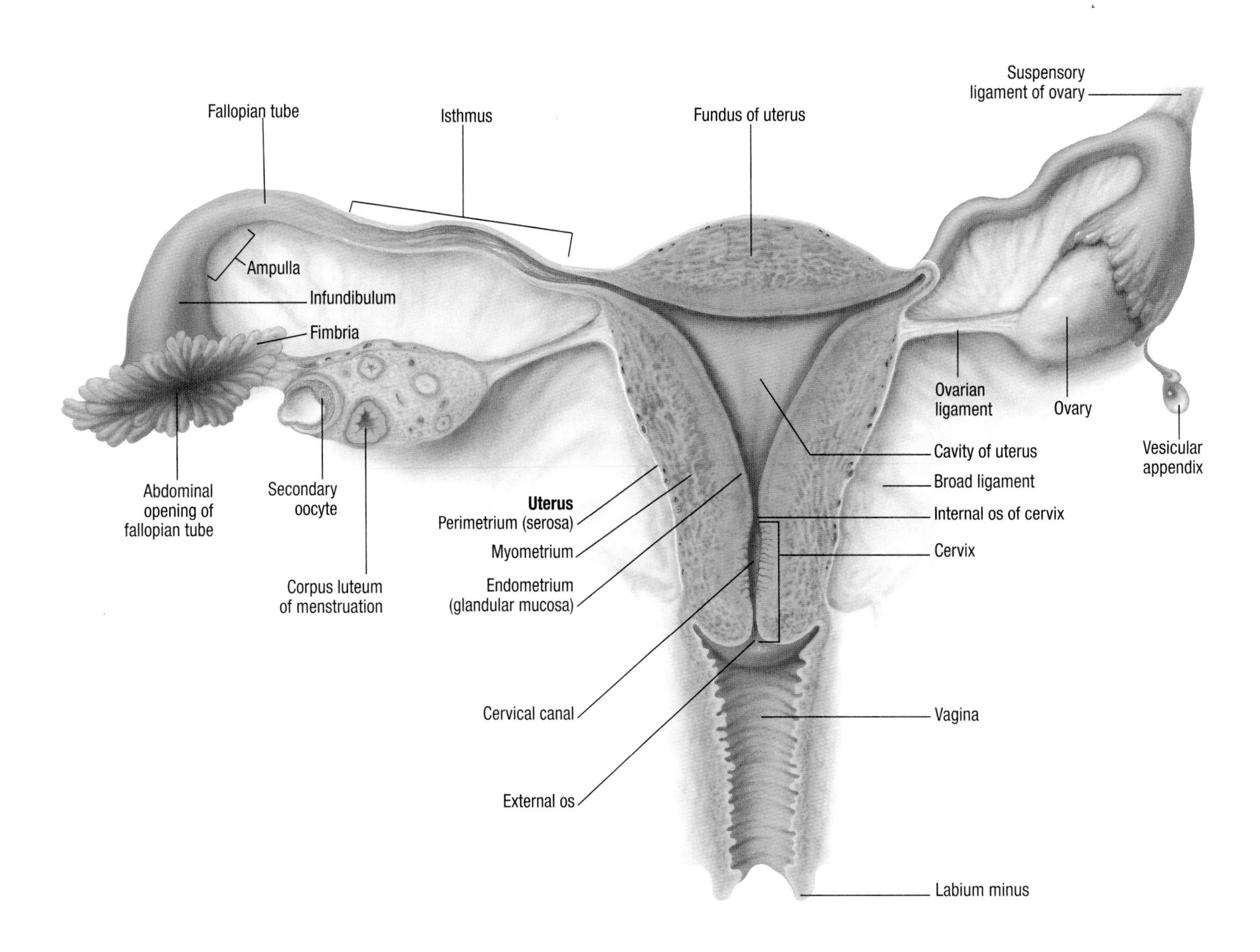

Suspensory ligament of ovary
Fallopian tube
Isthmus
Fundus of uterus
Ampulla
Infundibulum
Fimbria
Ovarian ligament
Ovary
Vesicular appendix
Cavity of uterus
Broad ligament
Abdominal opening of fallopian tube
Secondary oocyte
Uterus
Perimetrium (serosa)
Myometrium
Endometrium (glandular mucosa)
Internal os of cervix
Cervix
Corpus luteum of menstruation
Cervical canal
Vagina
External os
Labium minus

## OVARY

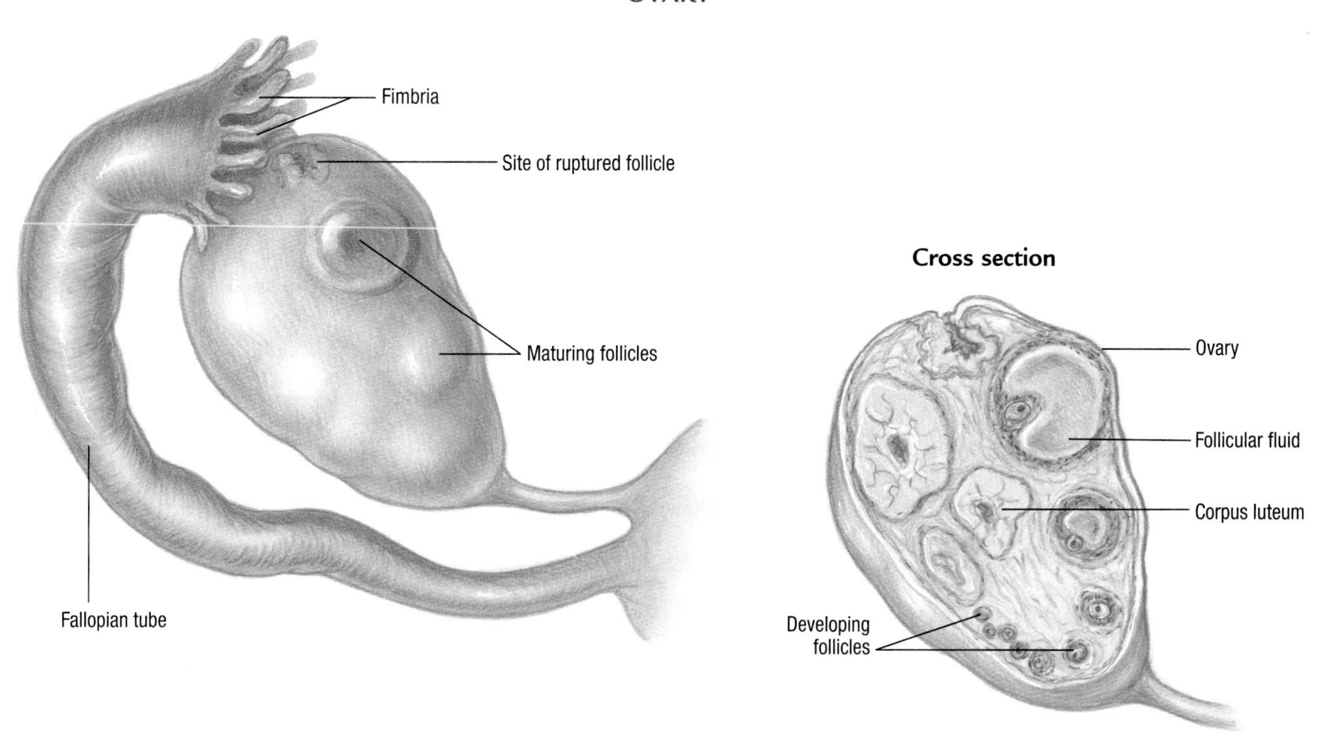
Fimbria
Site of ruptured follicle
Maturing follicles
Fallopian tube
Cross section
Ovary
Follicular fluid
Corpus luteum
Developing
follicles

## UNFERTILIZED OVUM

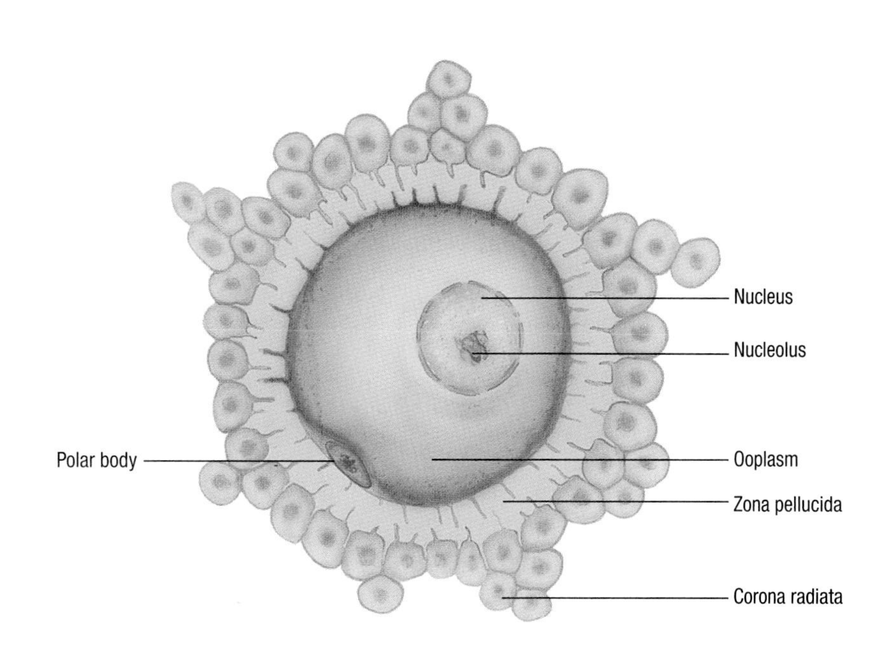
Nucleus
Nucleolus
Polar body
Ooplasm
Zona pellucida
Corona radiata

# Female external genitalia

The vulva, also called external genitalia, consists of external female genitalia visible on inspection. It includes the mons pubis, labia majora, labia minora, clitoris, and vestibule.

## Mons pubis

The mons pubis is a rounded cushion of fatty and connective tissue covered by skin and coarse, curly hair in a triangular pattern over the symphysis pubis (the joint formed by the union of the pubic bones anteriorly).

## Labia majora

The labia majora are two raised folds of adipose and connective tissue that border the vulva on either side, extending from the mons pubis to the perineum. The labia are highly vascular and contain many sensory nerve endings, which makes them sensitive to pain, pressure, touch, sexual stimulation, and temperature extremes.

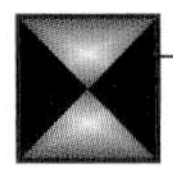

**Age-related changes**

After menarche, the outer surfaces of the labia are covered with pubic hair. The inner surface is pink and moist.

## Labia minora

The labia minora are two moist, dark pink to red folds of mucosal tissue that lie medial to the labia majora. Each labium minorum has an upper and a lower section; each upper section divides into an upper and a lower lamella. The two upper lamellae join to form the prepuce, a hoodlike covering over the clitoris. The two lower lamellae form the frenulum, the posterior portion of the clitoris.

The labia taper down and back from the clitoris to the perineum, where they join to form the fourchette, a thin tissue fold along the anterior edge of the perineum.

The labia minora contain sebaceous glands, which secrete a lubricant that also acts as a bactericide. Like the labia majora, they are rich in blood vessels and nerve endings, making them highly responsive to stimulation. Swelling in response to sexual stimulation triggers other changes that prepare the genitalia for *coitus*.

## Clitoris

The clitoris is the small, protuberant organ just beneath the arch of the mons pubis. It contains erectile tissue, venous cavernous spaces, and specialized sensory corpuscles that are stimulated during sexual activity.

## Vestibule

The vestibule is an oval area bounded anteriorly by the clitoris, laterally by the labia minora, and posteriorly by the *fourchette*. The urethral meatus is the slitlike opening below the clitoris.

In the center of the vestibule is the vaginal orifice, which may be completely or partially covered by the hymen, a tissue membrane. Presence or absence of a hymen is no indication of whether a female has been sexually active. The hymen may rupture for any number of reasons – including tampon use, sexual activity, and other physical causes.

The vestibule contains two pairs of mucus-producing glands:

- Skene's glands – on either side of the urethral meatus
- Bartholin's glands – on either side of the inner vaginal orifice.

**Age-related changes**

With advancing age and declining estrogen levels, the following changes occur:

- loss of pubic hair
- flattening of the labia majora
- shrinking of vulval tissue
- loss of tissue elasticity.

## FEMALE PERINEUM

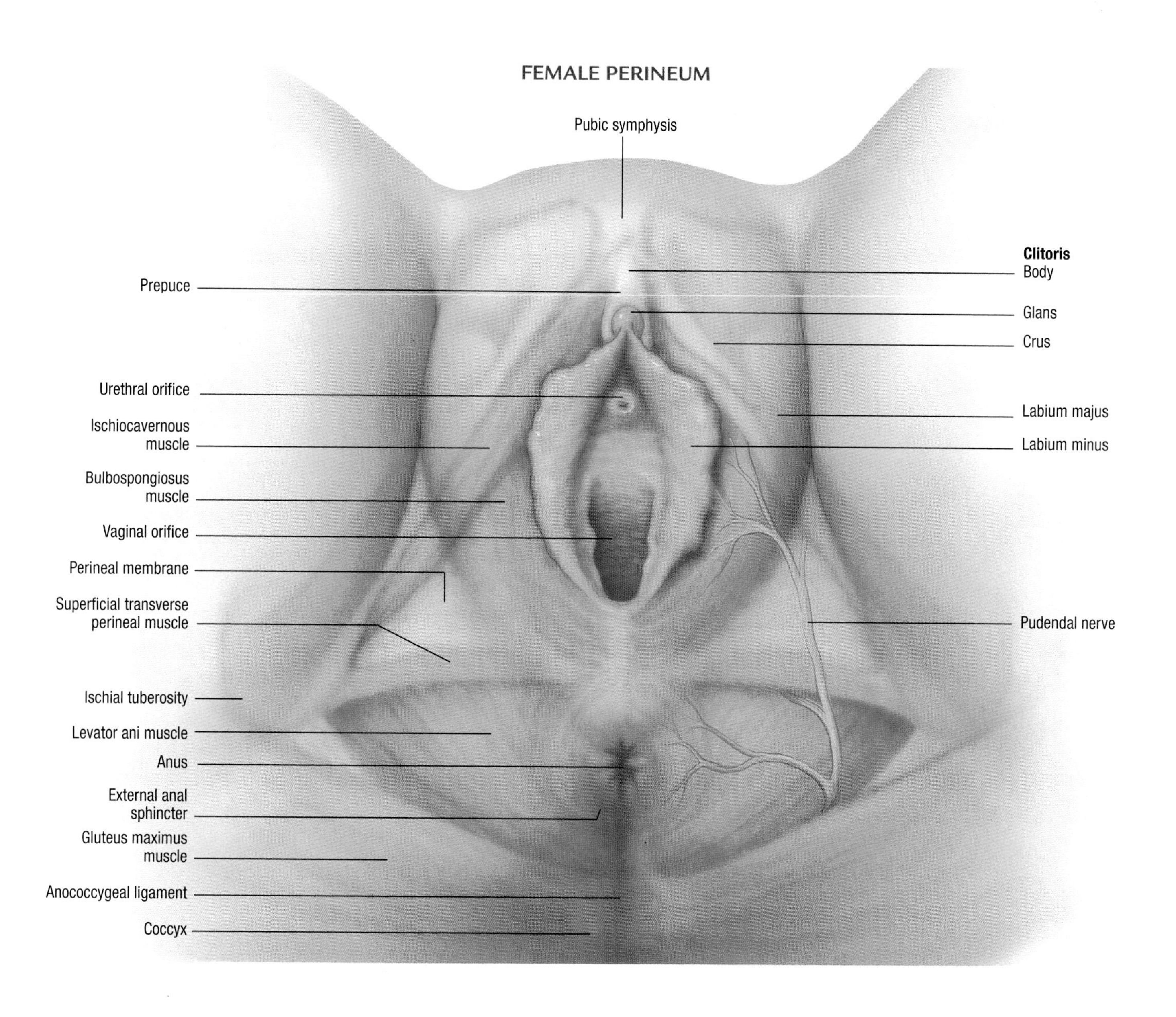

Pubic symphysis
Clitoris
Body
Glans
Crus
Prepuce
Urethral orifice
Ischiocavernous muscle
Bulbospongiosus muscle
Vaginal orifice
Perineal membrane
Superficial transverse perineal muscle
Labium majus
Labium minus
Pudendal nerve
Ischial tuberosity
Levator ani muscle
Anus
External anal sphincter
Gluteus maximus muscle
Anococcygeal ligament
Coccyx


## PERINEAL VESSELS AND NERVES

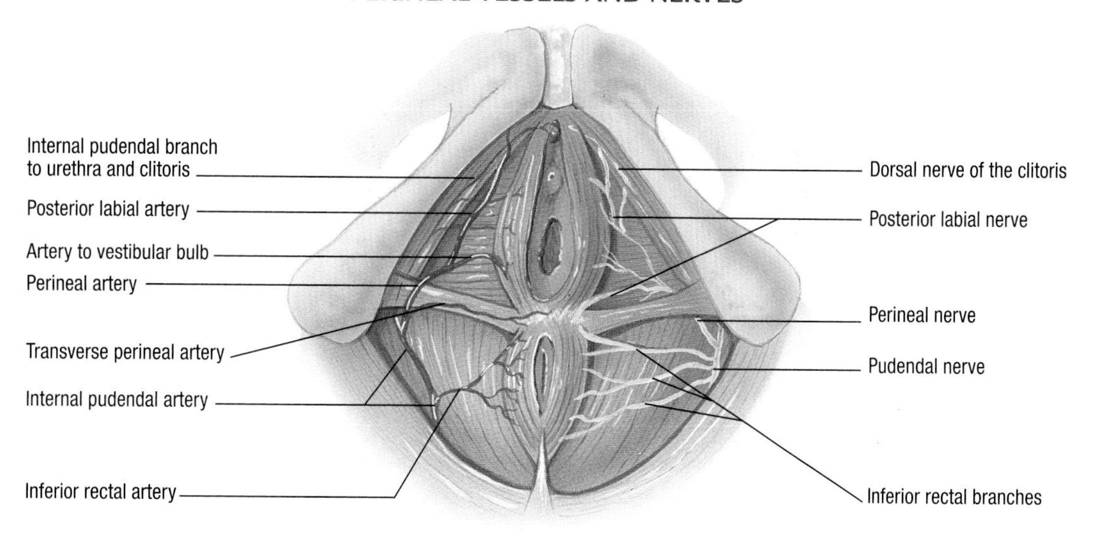

Internal pudendal branch to urethra and clitoris
Posterior labial artery
Artery to vestibular bulb
Perineal artery
Transverse perineal artery
Internal pudendal artery
Inferior rectal artery
Dorsal nerve of the clitoris
Posterior labial nerve
Perineal nerve
Pudendal nerve
Inferior rectal branches

# Menstruation and pregnancy

The female body undergoes a series of changes throughout life as a result of menstruation or pregnancy.

## Menstruation

Menstruation usually begins around age 9 to 12 years. During the menstrual cycle, a series of anatomic changes occur to prepare the endometrium to receive a fertilized ovum. If the ovum is successfully fertilized, pregnancy occurs.

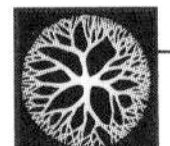

**Physiology**

The hypothalamus, ovaries, and pituitary gland secrete hormones that affect the buildup and shedding of the endometrium during the menstrual cycle. The menstrual cycle usually occurs over 28 days, although the normal cycle may range from 22 to 34 days. The cycle is regulated by fluctuating hormone levels that, in turn, are regulated by negative and positive feedback mechanisms involving the hypothalamus, pituitary gland, and ovaries.

### Endometrial changes

The hormonal changes of the menstrual cycle trigger a series of changes in the uterine endometrium, as follows:

- menstrual (preovulatory) phase – endometrium exfoliates and sheds
- proliferative (follicular) phase and ovulation – endometrium proliferates
- luteal (secretory) phase – endometrium becomes thick and secretory to prepare for implantation of a fertilized ovum
- premenstrual phase – in absence of fertilization, estrogen and progesterone levels drop and the endometrium shrinks.

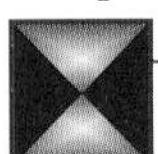

**Age-related changes**

Cessation of menses usually occurs between ages 40 and 55 as ovarian follicles stop responding to follicle-stimulating hormone and luteinizing hormone, both released by the pituitary gland. The term *menopause* applies after one year without menses. *Climacteric* refers to the transitional years from reproductive fertility to infertility during which several physiologic changes, including menopause, occur.

## Fertilization

Fertilization is the union of a spermatozoon (sperm) and an ovum. Sperm move through the female reproductive tract by two mechanisms:

- flagellar propulsion
- rhythmic contractions of the uterine muscles.

Although a single ejaculation deposits several hundred million sperm, the acidity of vaginal secretions destroys most of them. Only the sperm that enter the cervical canal, where they are protected by cervical mucus, survive.

Fertilization occurs in the distal portion of the fallopian tubes. Enzymes in the sperm's acrosome allow the sperm to penetrate the zona pellucida on the outer surface of the ovum. Fusion of the nuclei of sperm and ovum forms a cell nucleus with 46 chromosomes. The fertilized ovum is a *zygote*.

## Pregnancy

Pregnancy causes a number of changes, including the following:

- endocrine
  - enlarged thyroid gland
- breast
  - enlarged and nodular
  - nipple enlarged and more erectile
  - thick, yellowish colostrum secreted
  - darkened areolae
  - prominent Montgomery glands
  - more apparent venous pattern
- abdominal
  - enlarged abdomen
  - purplish striae
  - linea nigra (brownish pigmented line) on midline
  - decreased muscle tone
  - displaced abdominal organs
  - loss of lower esophageal sphincter tone
- genitourinary
  - thickened vaginal wall
  - bluish vagina
  - increased, more acidic, thicker vaginal secretions
  - softened isthmus (Hegar's sign)
  - straightened uterus rising in the pelvis
  - softened cervix
  - pressure on the bladder, less capacity
- musculoskeletal
  - stretched pelvic ligaments
  - relaxed joints.

## FERTILIZATION AND IMPLANTATION

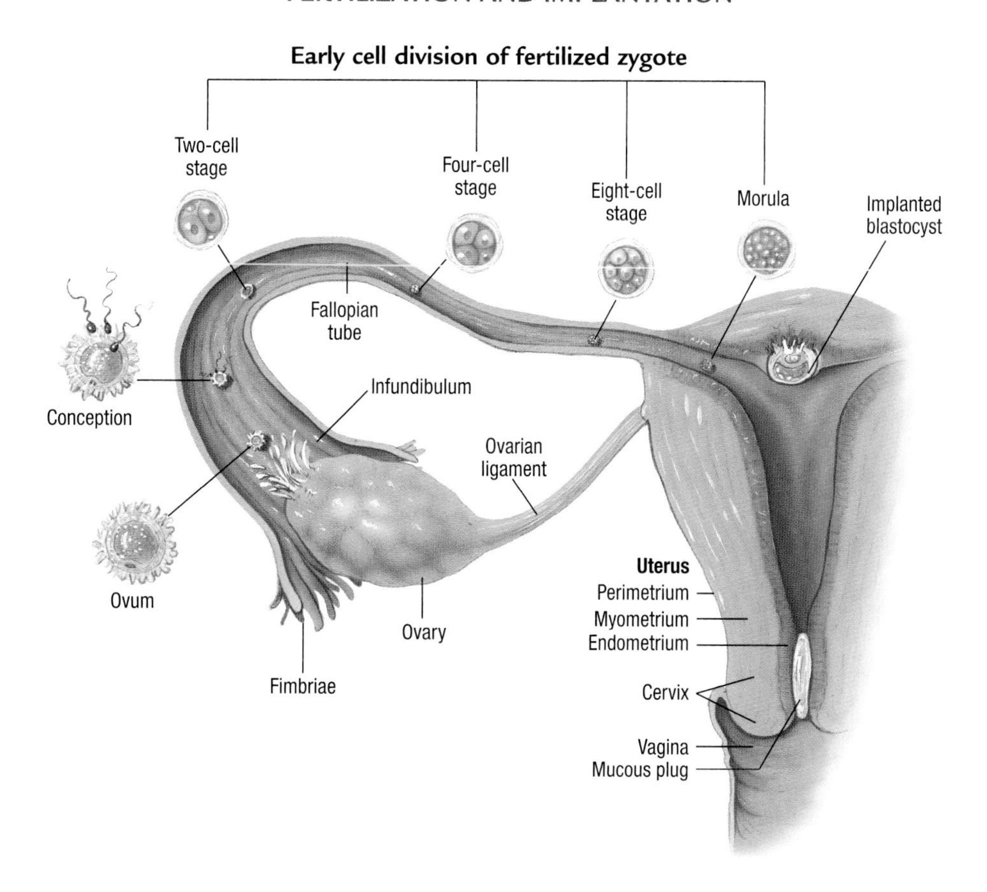

Early cell division of fertilized zygote
Two-cell stage
Four-cell stage
Eight-cell stage
Morula
Implanted blastocyst
Fallopian tube
Conception
Infundibulum
Ovarian ligament
Ovum
Ovary
Fimbriae
Uterus
Perimetrium
Myometrium
Endometrium
Cervix
Vagina
Mucous plug


## OVARIAN AND UTERINE CHANGES DURING THE MENSTRUAL CYCLE

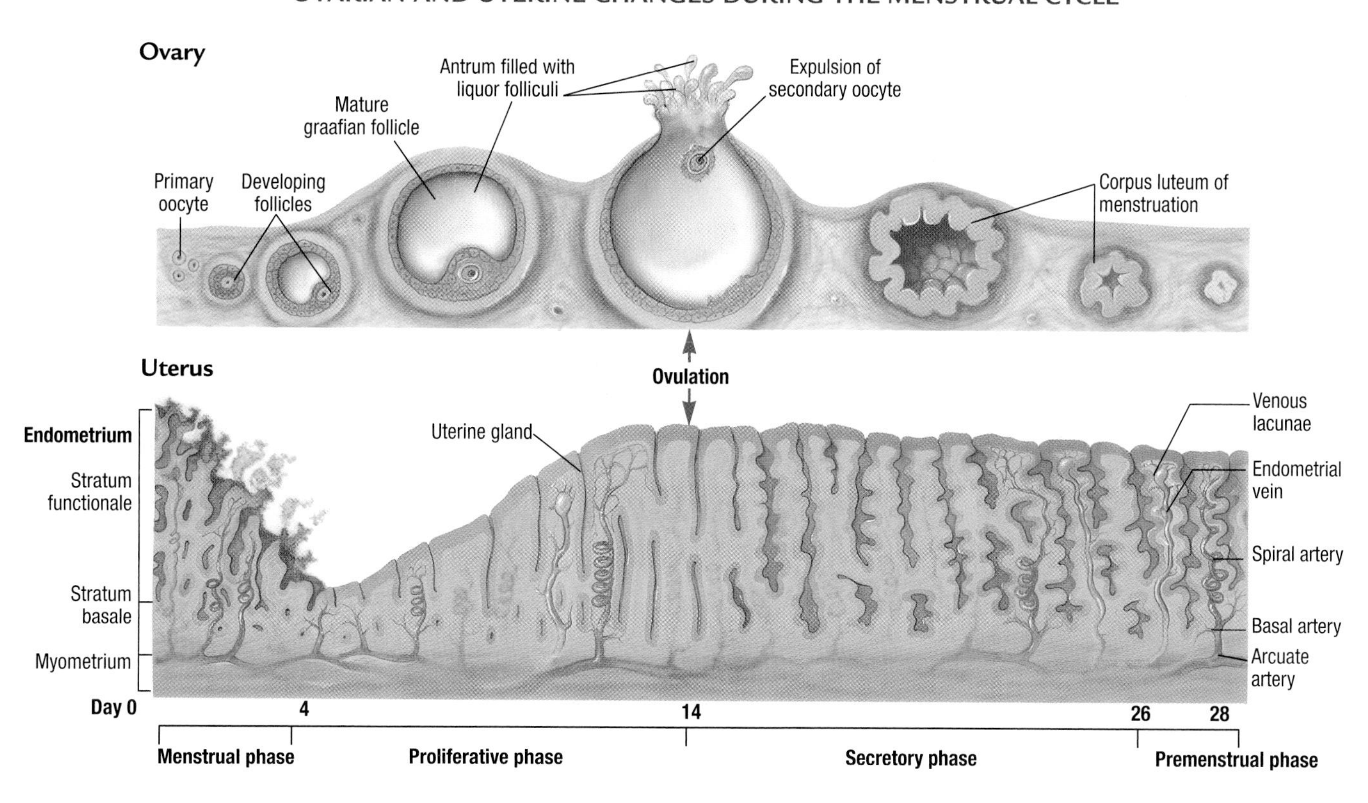

Ovary
Antrum filled with liquor folliculi
Expulsion of secondary oocyte
Mature graafian follicle
Primary oocyte
Developing follicles
Corpus luteum of menstruation
Uterus
Ovulation
Endometrium
Uterine gland
Stratum functionale
Stratum basale
Myometrium
Venous lacunae
Endometrial vein
Spiral artery
Basal artery
Arcuate artery
Day 0
4
14
26
28
Menstrual phase
Proliferative phase
Secretory phase
Premenstrual phase

# Bladder

The bladder is a hollow, sphere-shaped, muscular organ. The ureters carry urine from the kidneys to the bladder by *peristaltic* contractions. Bladder capacity is 500 to 1,000 ml in normal adults, less in children and elderly people.

## Location and position

The bladder lies anterior and inferior to the pelvic cavity and posterior to the symphysis pubis. In males, the bladder is above the prostate gland and anterior to the seminal vesicals. In females, the bladder is anterior to the top of the vagina and uterine fundus. Loose connective tissue between the symphysis pubis and the bladder permits it to stretch as it fills.

Position can vary with the fullness of the bladder:

- empty – behind the pubic bone
- full – moves superiorly above the pubic bone.

## Characteristics

The neck of the bladder is continuous with the urethra. The apex is connected to the umbilicus by the median umbilical ligament.

The *trigone,* a triangular area at the base of the bladder, contains three openings:

- posterior boundary – two ureteral orifices
- anterior boundary – urethral orifice.

## Bladder wall

The wall of the bladder consists of several layers:

- mucosa – internal lining, made of transitional epithelium cells
- detrusor – three layers of muscle tissue
  - inner and outer detrusor layers – longitudinal muscle fibers
  - middle layer – fibers in a circular arrangement.

### Contraction and relaxation

Urination is a sequence of involuntary (reflex) and voluntary (learned or intentional) processes. When urine fills the bladder, parasympathetic nerve fibers in the bladder wall cause the bladder to contract and the internal sphincter (at the internal urethral orifice) to relax; this is called the *micturition reflex.* Then the cerebrum stimulates voluntary relaxation and contraction of the external sphincter (about ¾″ [2 cm] distal to the internal sphincter).

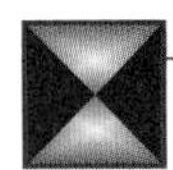

**Age-related changes**

Incontinence is the lack of voluntary control over urination. It's normal in infants and children under age 2. As long as the neurons of the external sphincter muscle remain immature, an infant voids whenever the bladder is full enough to initiate a reflex stimulus.

## Innervation

Parasympathetic innervation is provided by the pelvic splanchnic nerves, which cause the evacuation of urine by inhibiting the internal sphincter and stimulating detrusor muscle contractions. Sympathetic stimulation originates from spinal nerves L1 and L2 and proceeds through the pelvic plexus.

## Blood supply

Branches of the internal iliac arteries – including the umbilical artery and the inferior and superior vesicals – supply the bladder. In females, the vaginal arteries also supply the bladder. Branches of the internal iliac veins provide venous drainage.

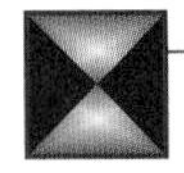

**Age-related changes**

Bladder changes that occur with advancing age include:

- reduction in size and capacity
- weakening of muscles, causing incomplete emptying and chronic urine retention.

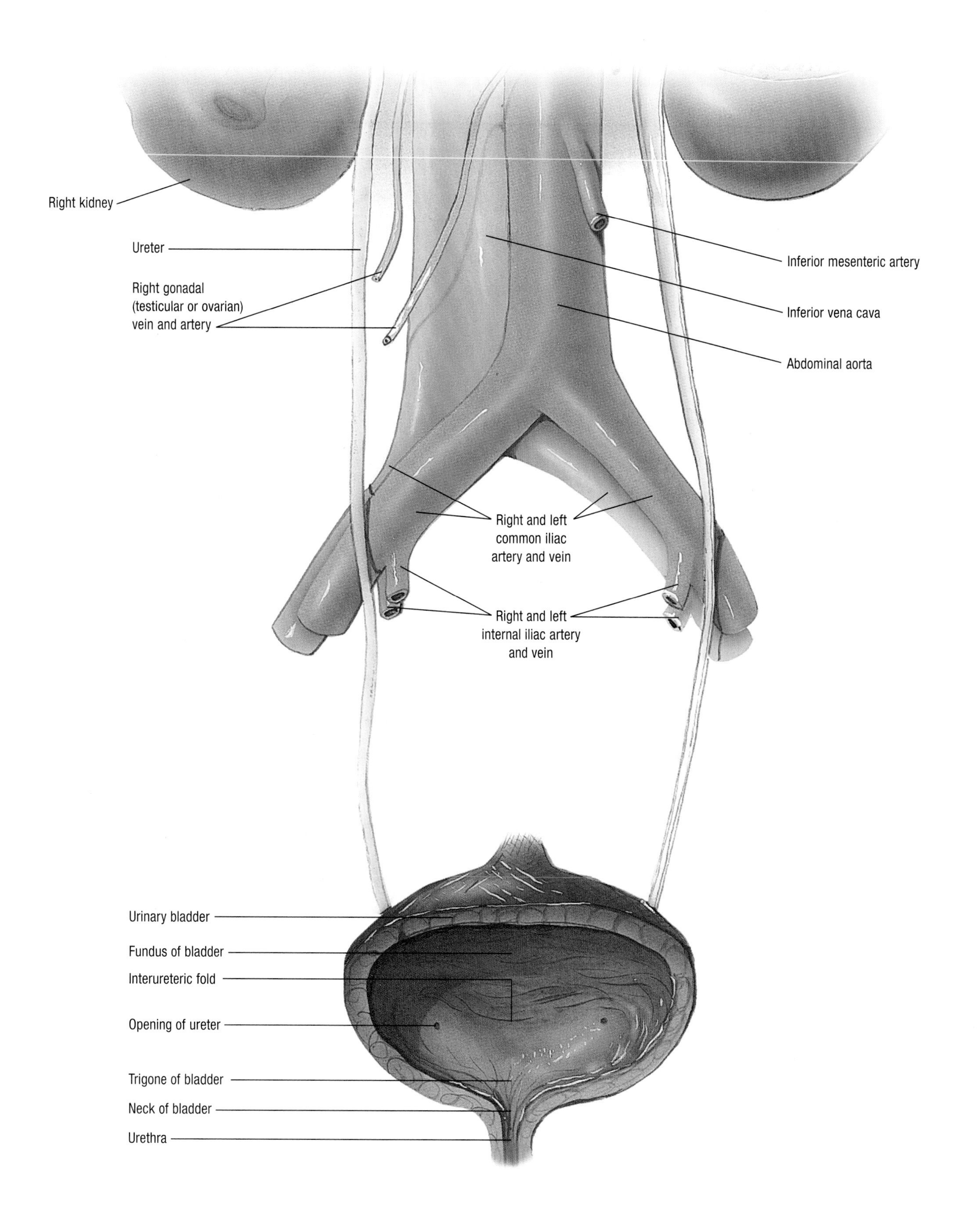
Right kidney
Ureter
Right gonadal
(testicular or ovarian)
vein and artery
Inferior mesenteric artery
Inferior vena cava
Abdominal aorta
Right and left
common iliac
artery and vein
Right and left
internal iliac artery
and vein
Urinary bladder
Fundus of bladder
Interureteric fold
Opening of ureter
Trigone of bladder
Neck of bladder
Urethra

# Rectum and anal canal

The rectum and the anal canal are the terminal structures in the lower digestive system. The rectum is the last 8 inches (20 cm) of the GI tract. The last inch (2 to 3 cm) of the rectum is the anal canal.

## Structure

The rectum lies anterior to the sacrum and coccyx, beginning in front of the S3 vertebra. At the anorectal flexure, the rectum passes through the pelvic diaphragm and is thereafter known as the anal canal.

The anal canal extends to the anus. At this point, the rectal walls form longitudinal folds – the anal or rectal columns – that surround the rectal veins and arteries. At the distal end of these columns, at the pectinate line, the tissue changes to mucosa.

**Clinical tip**

Hemorrhoids are varicosities of the superior rectal veins, which may protrude through the external anal sphincter or into the rectum. Internal hemorrhoids are those above the pectinate line; external hemorrhoids, below.

Except during defecation, the anal canal remains closed by the action of two muscles:
- internal sphincter – smooth muscle regulated by the autonomic nervous system
- external sphincter – skeletal muscle under voluntary control.

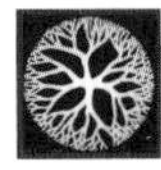

**Physiology**

In the lower colon, long and relatively sluggish contractions cause propulsive waves, or mass movements. Normally occurring several times a day, these movements propel intestinal contents into the rectum and produce the urge to defecate. The dilated area of the rectum where the mass accumulates is known as the ampulla of the rectum. Defecation normally results from a combination of the defecation reflex, a sensory and parasympathetic nerve-mediated response, and the voluntary relaxation of the external anal sphincter.

## Innervation

Parasympathetic stimulation from the sacral spinal nerves to the rectum increases:
- gut and sphincter tone
- frequency, strength, and velocity of smooth-muscle contractions. Vagal stimulation also increases motor and secretory activities.

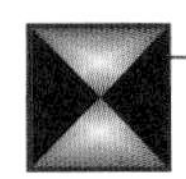

**Age-related changes**

Changes that occur in the rectum and anal canal with advancing age include diminution of:
- motility
- bowel wall tone
- anal sphincter tone.

## Blood supply and drainage

Arteries of the rectal and anal area include the superior, middle, and inferior rectal arteries and the middle sacral artery. Venous drainage is provided by:
- superior rectal vein via the portal venous system
- inferior and middle rectal veins via the internal iliac veins.

The principal lymphatic drainage for the upper rectal area are the aortic nodes. The lower rectal area drains into the iliac nodes.

**Clinical tip**

Examination of the rectum and anus can reveal a number of medical conditions, including:
- fissures
- lesions
- inflammation
- discharge
- rectal prolapse
- external hemorrhoids.

To prepare for a rectal examination:
- Put on gloves, and spread the buttocks to expose the anus and surrounding tissue. On initial inspection, expect the perianal skin to be somewhat darker than that of the surrounding area.
- Next, apply a water-soluble lubricant to your gloved index finger and prepare to palpate the rectum.
- Tell the patient to relax. Alert the patient to the possibility of feeling pressure during the examination.
- Insert your finger into the patient's rectum. Rotate your finger clockwise and then counterclockwise to assess the rectal wall. It should feel soft, without masses, fecal impaction, or tenderness.

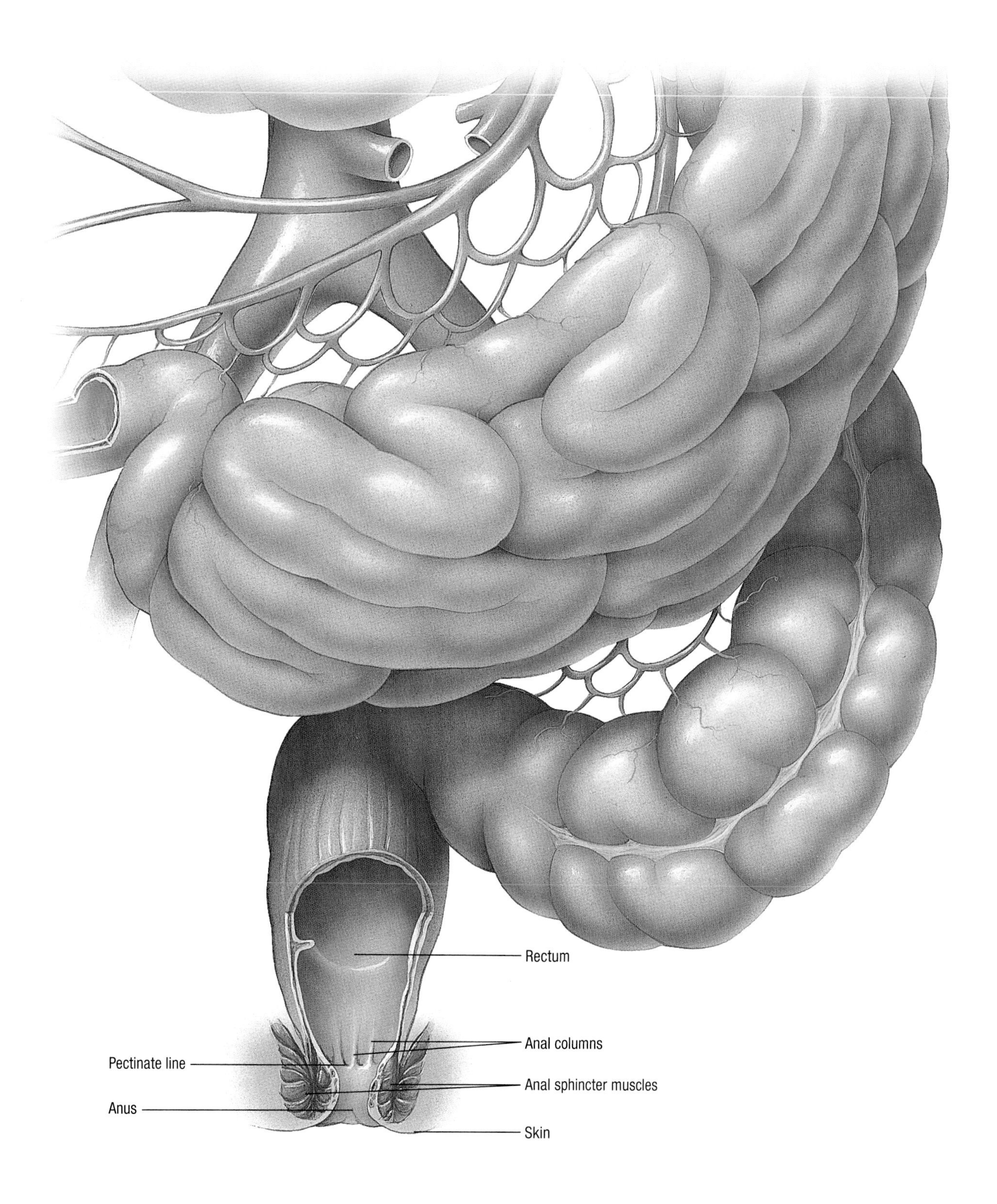
Rectum
Anal columns
Pectinate line
Anal sphincter muscles
Anus
Skin

# PART VIII

# LOWER LIMB

# Overview of lower limb

The lower limb, including the pelvic girdle, is part of the appendicular skeleton. The lower limb performs the following functions:
- supporting weight
- providing a means for movement
- maintaining balance.

Commonly, the lower limb is referred to as the leg. Anatomically, however, the lower limb is divided into three main areas:
- thigh – from the hip to the knee
- leg – from the knee to the ankle
- foot.

The initial pages in this section provide a brief summary of lower limb structures; they're followed by detailed descriptions of structures in each region of the lower limb: hip; femur, tibia, and fibula; knee; and ankle and foot.

## Bones

The pelvic girdle, formed by the hip bones and sacrum, which attaches the lower limbs to the axial skeleton, is described in Part VII, Pelvis.

Each lower limb consists of 30 bones:
- femur (1)
- patella (1)
- tibia (1)
- fibula (1)
- tarsals (7)
- metatarsals (5)
- phalanges (14).

The strong, thick bones of the lower limbs support greater weight and greater forces than do those of the upper limbs.

## Joints and ligaments

The joints of the lower limbs (except for the symphysis pubis) are diarthrotic (freely movable). Reinforcing ligaments, consisting of fibrous connective tissue, connect bones in the joint and strengthen the joint capsule.

Joints of the lower extremity (including the pelvic girdle) and their joint classifications include:
- sacroiliac – synovial, gliding
- symphysis pubis – cartilaginous, symphysis
- hip – synovial, ball-and-socket
- knee (tibiofemoral) – synovial, modified hinge
- proximal ends of tibia and fibula – synovial, gliding
- distal ends of tibia and fibula – fibrous (syndesmosis)
- ankle – synovial, hinge
- intertarsal – synovial, gliding.

Ligaments of the lower extremity include:
- iliofemoral – iliac region of innominate (hip) bone and femur
- patellar – patella and tibia
- arcuate popliteal – lateral condyle of femur and styloid process of fibula
- oblique popliteal – femur and head of tibia
- tibial collateral – medial condyles of femur and tibia
- fibular collateral – lateral condyles of femur and fibula
- anterior and posterior cruciates – tibia and femur
- talofibular – talus and fibula.

**Clinical tip**

**A cracking sound when a joint moves is a normal sound that occurs when a tendon or ligament slips over a bone. Crepitus – crunching or grating you can hear and feel when a joint with roughened articular surfaces moves – is not normal. It occurs in patients with rheumatoid arthritis or osteoarthritis or when broken pieces of bone rub together.**

## Bursae

Lined with synovial membrane and filled with synovial fluid, bursae are flattened fibrous sacs that decrease stress on nearby tissues by acting as a cushion. They are present in areas where ligaments, muscles, skin, or tendons rub against each other or against bone, such as the following:
- hip – greater trochanteric and lesser trochanteric bursae
- gluteal region – trochanteric, ischial, and gluteofemoral bursae
- knee – anterior, medial, and lateral bursae
- foot – retrocalcaneal and intercalcaneal bursae.

## Innervation

Nerves from the lumbar and sacral plexuses innervate the lower extremities. (See *Lower limb nerves, anterior view,* page 269.) Several branches of these nerves provide innervation throughout the extremity. Major nerves and branches include:
- femoral – saphenous, popliteal, and anterior femoral cutaneous
- obturator – cutaneous branch
- sciatic
  - common fibular (peroneal) – superficial and deep fibular, lateral sural cutaneous
  - tibial – medial sural cutaneous, medial and lateral plantar
- superior and inferior gluteal
- posterior femoral cutaneous – cluneal
- lateral femoral cutaneous.

# LOWER LIMB BONES
## Anterior view

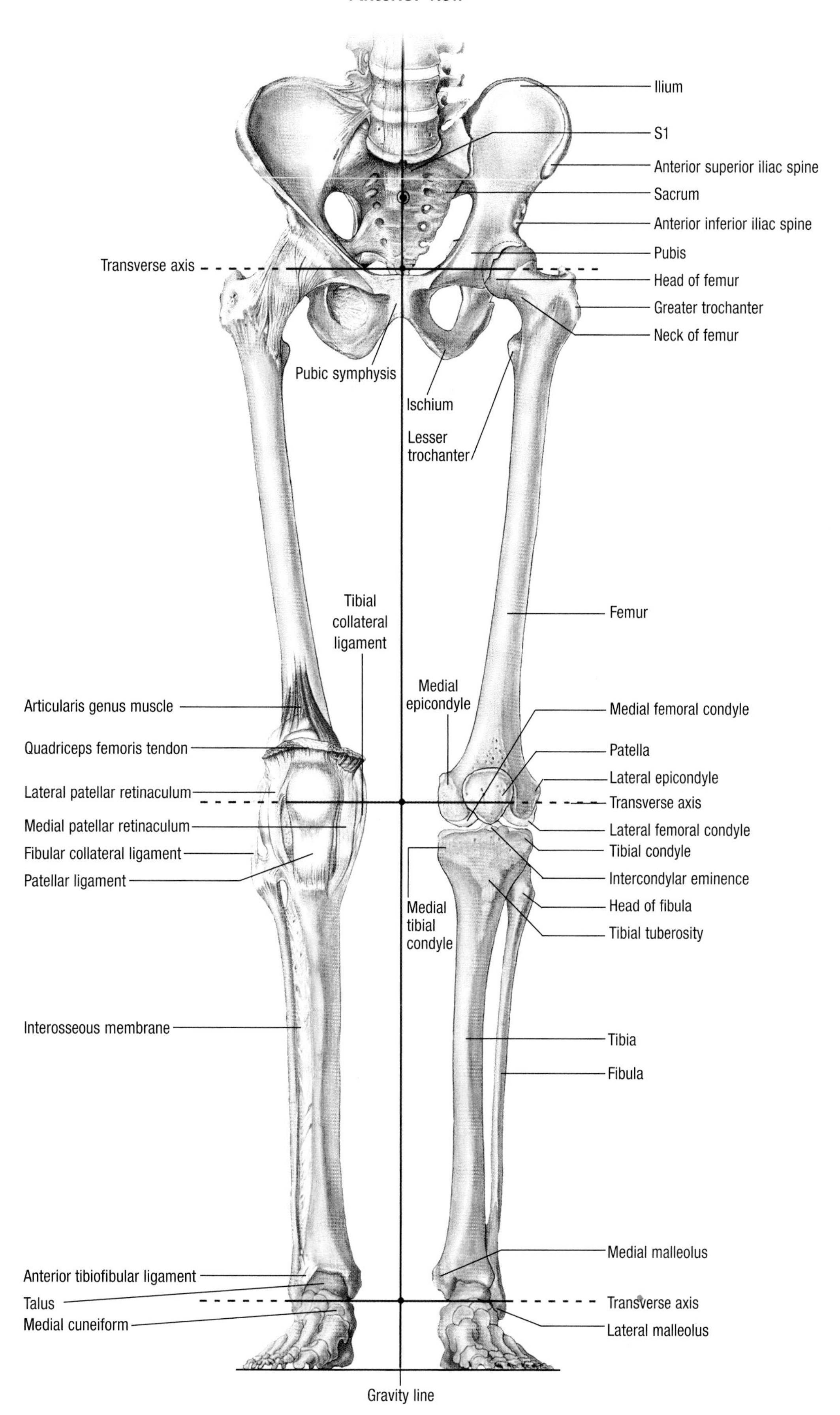

Ilium
S1
Anterior superior iliac spine
Sacrum
Anterior inferior iliac spine
Transverse axis
Pubis
Head of femur
Greater trochanter
Neck of femur
Pubic symphysis
Ischium
Lesser trochanter
Tibial collateral ligament
Femur
Medial epicondyle
Articularis genus muscle
Medial femoral condyle
Quadriceps femoris tendon
Patella
Lateral patellar retinaculum
Lateral epicondyle
Transverse axis
Medial patellar retinaculum
Lateral femoral condyle
Fibular collateral ligament
Tibial condyle
Patellar ligament
Intercondylar eminence
Medial tibial condyle
Head of fibula
Tibial tuberosity
Interosseous membrane
Tibia
Fibula
Medial malleolus
Anterior tibiofibular ligament
Talus
Transverse axis
Medial cuneiform
Lateral malleolus
Gravity line

## LOWER LIMB BONES
### Posterior view

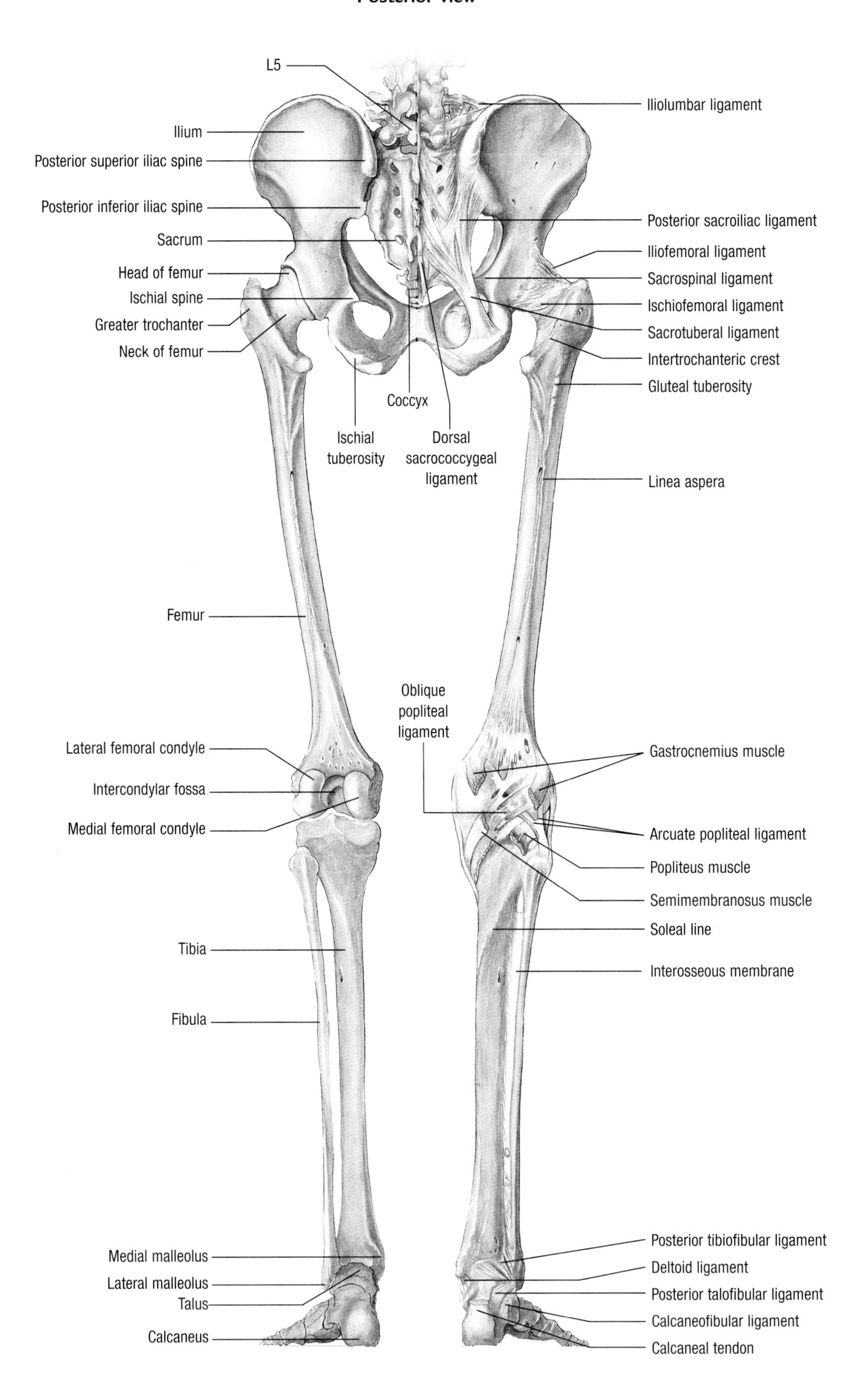

L5
Ilium
Posterior superior iliac spine
Posterior inferior iliac spine
Sacrum
Head of femur
Ischial spine
Greater trochanter
Neck of femur
Iliolumbar ligament
Posterior sacroiliac ligament
Iliofemoral ligament
Sacrospinal ligament
Ischiofemoral ligament
Sacrotuberal ligament
Intertrochanteric crest
Gluteal tuberosity
Coccyx
Ischial tuberosity
Dorsal sacrococcygeal ligament
Linea aspera
Femur
Oblique popliteal ligament
Lateral femoral condyle
Intercondylar fossa
Medial femoral condyle
Gastrocnemius muscle
Arcuate popliteal ligament
Popliteus muscle
Semimembranosus muscle
Soleal line
Interosseous membrane
Tibia
Fibula
Medial malleolus
Lateral malleolus
Talus
Calcaneus
Posterior tibiofibular ligament
Deltoid ligament
Posterior talofibular ligament
Calcaneofibular ligament
Calcaneal tendon

# LOWER LIMB NERVES

## Anterior view

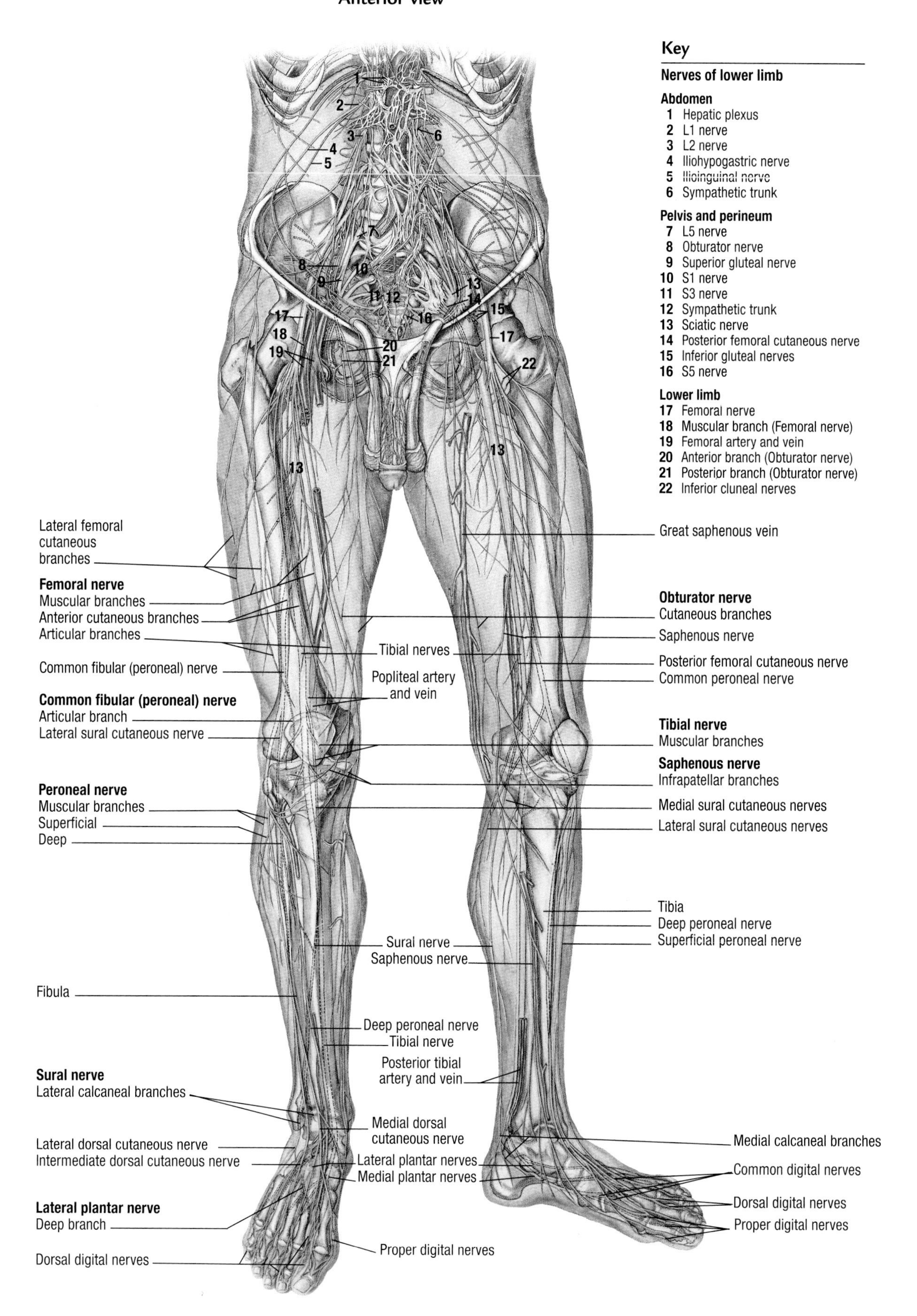

Key
Nerves of lower limb
Abdomen
1 Hepatic plexus
2 L1 nerve
3 L2 nerve
4 Iliohypogastric nerve
5 Ilioinguinal nerve
6 Sympathetic trunk
Pelvis and perineum
7 L5 nerve
8 Obturator nerve
9 Superior gluteal nerve
10 S1 nerve
11 S3 nerve
12 Sympathetic trunk
13 Sciatic nerve
14 Posterior femoral cutaneous nerve
15 Inferior gluteal nerves
16 S5 nerve
Lower limb
17 Femoral nerve
18 Muscular branch (Femoral nerve)
19 Femoral artery and vein
20 Anterior branch (Obturator nerve)
21 Posterior branch (Obturator nerve)
22 Inferior cluneal nerves
Lateral femoral cutaneous branches
Femoral nerve
Muscular branches
Anterior cutaneous branches
Articular branches
Common fibular (peroneal) nerve
Common fibular (peroneal) nerve
Articular branch
Lateral sural cutaneous nerve
Peroneal nerve
Muscular branches
Superficial
Deep
Fibula
Sural nerve
Lateral calcaneal branches
Lateral dorsal cutaneous nerve
Intermediate dorsal cutaneous nerve
Lateral plantar nerve
Deep branch
Dorsal digital nerves
Tibial nerves
Popliteal artery and vein
Sural nerve
Saphenous nerve
Deep peroneal nerve
Tibial nerve
Posterior tibial artery and vein
Medial dorsal cutaneous nerve
Lateral plantar nerves
Medial plantar nerves
Proper digital nerves
Great saphenous vein
Obturator nerve
Cutaneous branches
Saphenous nerve
Posterior femoral cutaneous nerve
Common peroneal nerve
Tibial nerve
Muscular branches
Saphenous nerve
Infrapatellar branches
Medial sural cutaneous nerves
Lateral sural cutaneous nerves
Tibia
Deep peroneal nerve
Superficial peroneal nerve
Medial calcaneal branches
Common digital nerves
Dorsal digital nerves
Proper digital nerves

# Blood supply and lymphatics

Arteries, veins, and lymphatic vessels and nodes form a network throughout the lower limb.

## Arteries

The main source of the blood supply to the lower extremities is the abdominal aorta and the major arteries that branch from it. Major arteries and their branches include the following:

- common iliac
- branches of aorta
  - external iliac
  - femoral
  - popliteal
- distal arteries
  - peroneal
  - anterior tibial
  - posterior tibial
  - dorsalis pedis
  - dorsal arch
  - lateral plantar
  - medial plantar.

## Veins

The venous system of the lower extremities begins at the bifurcation of the inferior vena cava, forming the common iliac veins. (See *Major lower limb veins, anterior view,* page 272.) Names of veins parallel those of nearby arteries. The veins of the lower limb have the body's greatest number of valves, which help prevent backflow. Major vessels and their branches include:

- femoral veins
- superficial veins (located in subcutaneous tissue)
  - great saphenous
  - small saphenous
- deep veins (travel with arteries and nerves)
  - popliteal
  - peroneal
  - anterior tibial
  - posterior tibial
  - dorsal venous arch
  - dorsalis pedis
  - lateral plantar
  - medial plantar.

**Age-related changes**

Peripheral vascular disease is any disorder of the blood vessels of the extremities, especially one that interferes with blood flow. It can assume many forms in the elderly. Include assessment for the following when evaluating elderly patients for suspected peripheral vascular disease:

- aneurysms
- arteriosclerosis
- varicose veins
- deep vein thrombosis
- thrombophlebitis
- stasis ulcers
- chronic venous insufficiency.

## Lymphatic drainage

The thoracic duct (also known as the left lymphatic duct) provides lymphatic drainage from lymph nodes in the lower extremities. (See *Lymphatic drainage of lower limb, anterior view,* page 273.) Major regions and their drainage include:

- deep gluteal region – gluteal lymph nodes to iliac lymph nodes to lateral aortic lymph nodes
- thigh and superficial gluteal tissue – superficial inguinal lymph nodes
- knee – popliteal lymph nodes to deep inguinal lymph nodes.

**Clinical tip**

To grade edema, press your fingertip over a bony surface. Hold for a few seconds. Note the indentation. Grades of edema are as follows:

+1 – slight indentation
+2 – moderate pitting, lasting for a few seconds
+3 – deep indentation, returning slowly
+4 – deeper indentation, returning very slowly.

# MAJOR LOWER LIMB ARTERIES
## Anterior view

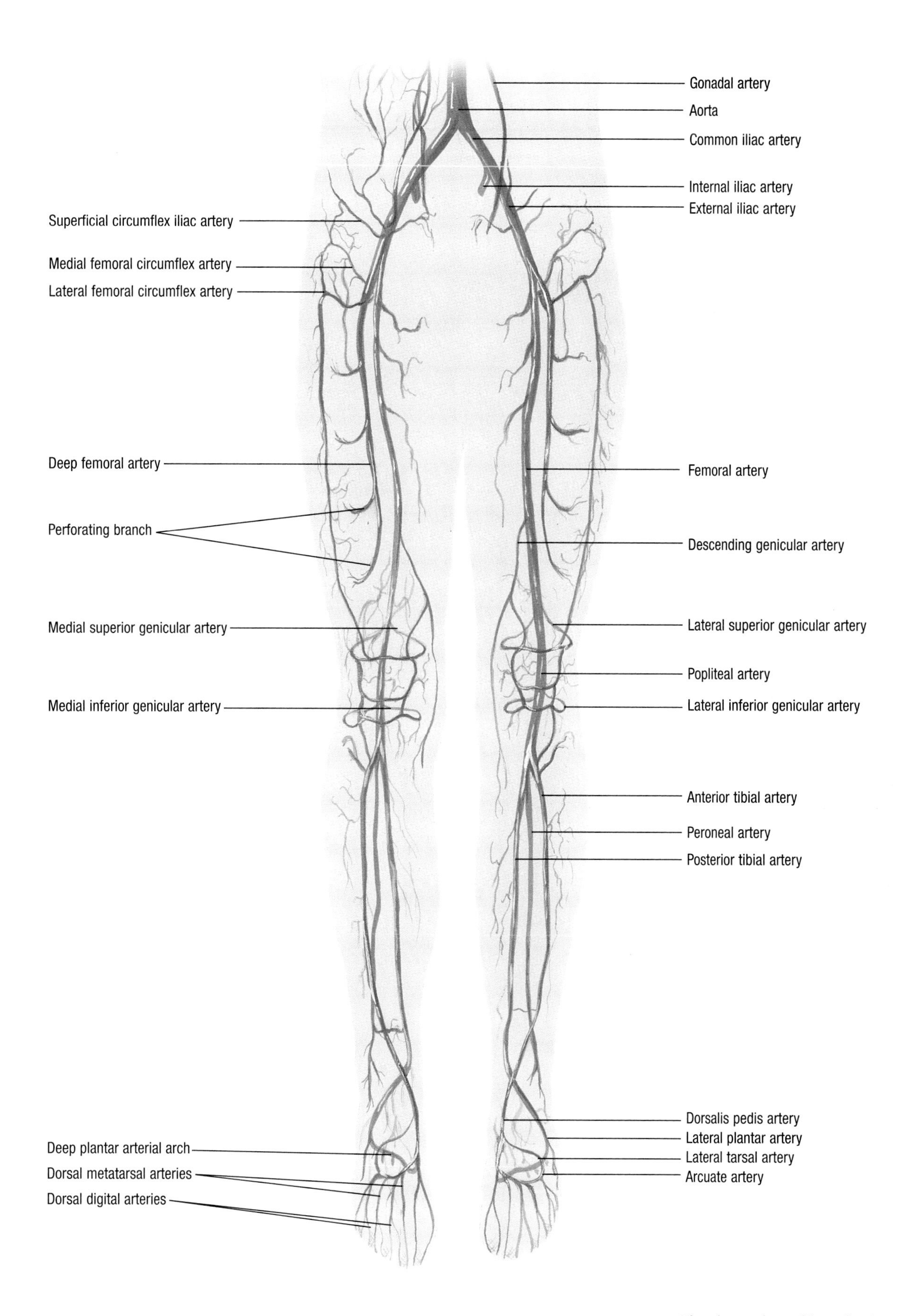

Gonadal artery
Aorta
Common iliac artery
Internal iliac artery
External iliac artery
Superficial circumflex iliac artery
Medial femoral circumflex artery
Lateral femoral circumflex artery
Deep femoral artery
Femoral artery
Perforating branch
Descending genicular artery
Medial superior genicular artery
Lateral superior genicular artery
Popliteal artery
Medial inferior genicular artery
Lateral inferior genicular artery
Anterior tibial artery
Peroneal artery
Posterior tibial artery
Dorsalis pedis artery
Lateral plantar artery
Deep plantar arterial arch
Lateral tarsal artery
Dorsal metatarsal arteries
Arcuate artery
Dorsal digital arteries

# MAJOR LOWER LIMB VEINS
## Anterior view

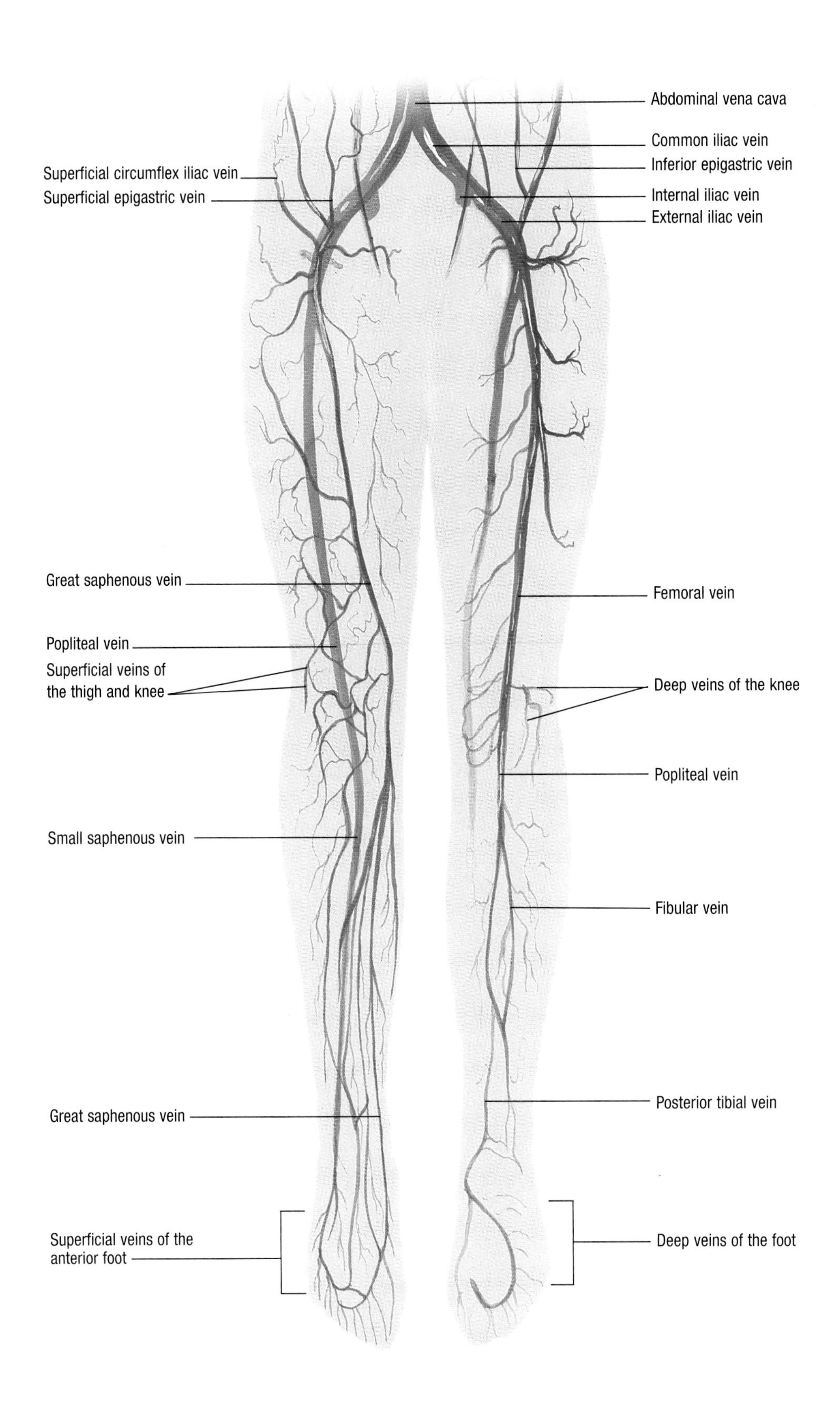

Abdominal vena cava
Common iliac vein
Inferior epigastric vein
Superficial circumflex iliac vein
Superficial epigastric vein
Internal iliac vein
External iliac vein
Great saphenous vein
Femoral vein
Popliteal vein
Superficial veins of the thigh and knee
Deep veins of the knee
Popliteal vein
Small saphenous vein
Fibular vein
Great saphenous vein
Posterior tibial vein
Superficial veins of the anterior foot
Deep veins of the foot

# LYMPHATIC DRAINAGE OF LOWER LIMB
## Anterior view

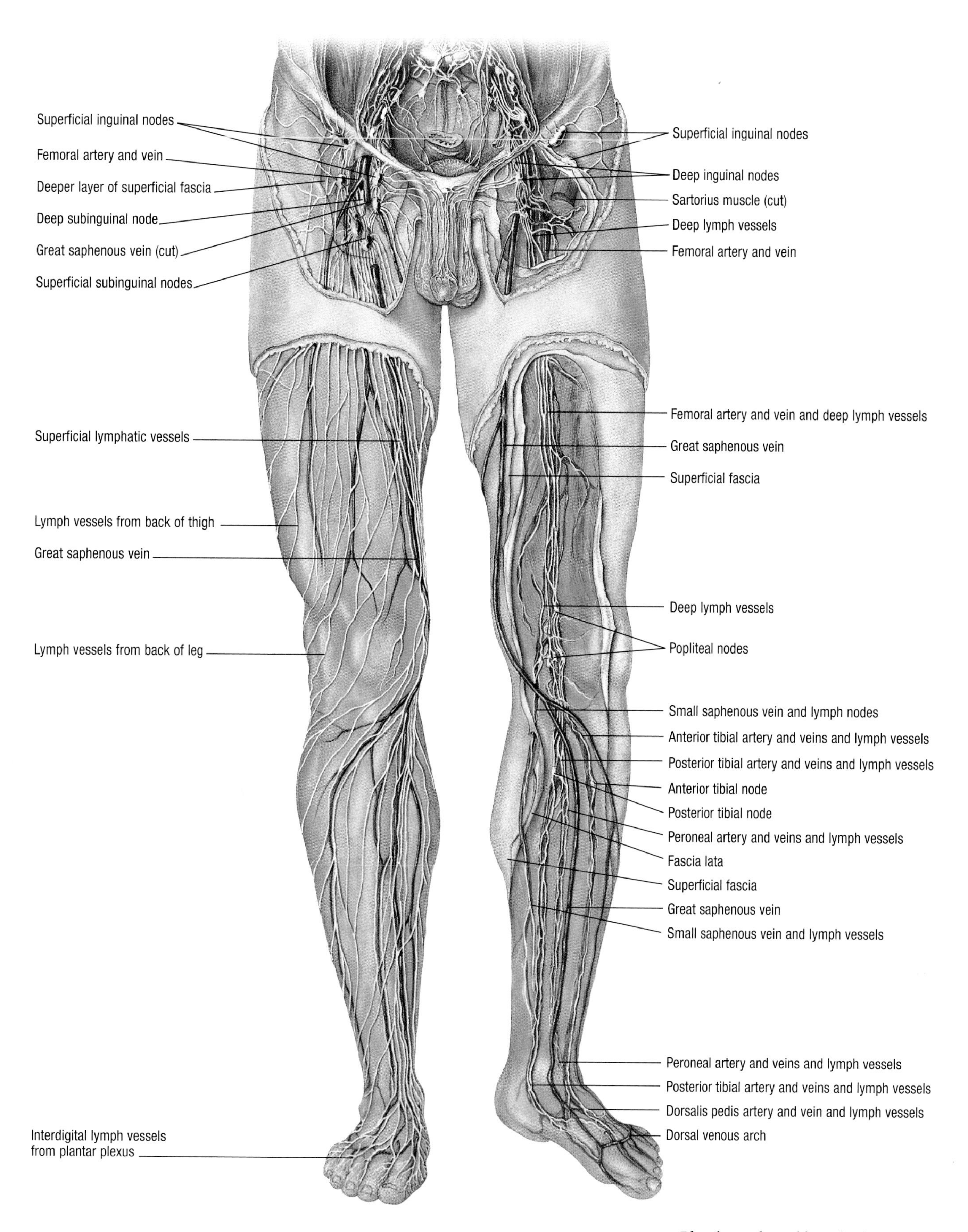

Superficial inguinal nodes
Femoral artery and vein
Deeper layer of superficial fascia
Deep subinguinal node
Great saphenous vein (cut)
Superficial subinguinal nodes
Superficial inguinal nodes
Deep inguinal nodes
Sartorius muscle (cut)
Deep lymph vessels
Femoral artery and vein
Superficial lymphatic vessels
Lymph vessels from back of thigh
Great saphenous vein
Lymph vessels from back of leg
Femoral artery and vein and deep lymph vessels
Great saphenous vein
Superficial fascia
Deep lymph vessels
Popliteal nodes
Small saphenous vein and lymph nodes
Anterior tibial artery and veins and lymph vessels
Posterior tibial artery and veins and lymph vessels
Anterior tibial node
Posterior tibial node
Peroneal artery and veins and lymph vessels
Fascia lata
Superficial fascia
Great saphenous vein
Small saphenous vein and lymph vessels
Peroneal artery and veins and lymph vessels
Posterior tibial artery and veins and lymph vessels
Dorsalis pedis artery and vein and lymph vessels
Dorsal venous arch
Interdigital lymph vessels
from plantar plexus

# Muscles

In the following lists, muscles of the lower extremities are organized by divisions of the lower limb – the thigh, leg, and ankle and foot.

## Thigh

The following muscles move the thigh:
- psoas major – flexes and rotates laterally
- iliacus – flexes and rotates laterally
- gluteus maximus – extends and rotates laterally
- gluteus medius – abducts
- gluteus minimus – abducts and rotates medially
- tensor fasciae latae – flexes and abducts
- adductor longus – flexes, rotates, and adducts
- adductor brevis – flexes, rotates, and adducts
- adductor magnus – flexes, extends, and adducts
- piriformis – laterally rotates and abducts
- obturator externus – rotates laterally
- obturator internus – rotates laterally
- pectineus – flexes, rotates laterally, adducts.

The three hamstring muscles in the posterior aspect of the thigh are sometimes classified as leg muscles. They extend the thigh and flex the leg. The three hamstring muscles are:
- biceps femoris
- semitendinosus
- semimembranosus.

**Clinical tip**

Diagnostic terms:
- *Pulled groin* – usually refers to a strain, stretched or torn tendons around adductor muscles of the thigh
- *Pulled hamstring* – various tears in or around the hamstring muscles.

## Leg

The three hamstring muscles are active in leg flexion. The quadriceps femoris – the major muscle group of the leg – has four distinct parts, each usually described as a separate muscle. All four of the following extend the leg; the rectus femoris also flexes the thigh:
- vastus lateralis
- vastus medialis
- vastus intermedius
- rectus femoris.

Two other significant muscles are the gracilis and the sartorius:
- gracilis – flexes knee and adducts thigh
- sartorius – flexes leg and thigh; abducts and rotates leg and thigh laterally. (See *Knee muscles, lateral view,* page 277.)

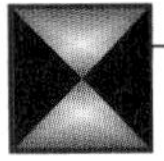

**Age-related changes**

Changes with age include:
- decreased muscle mass
- weakness
- difficulty in tandem (heel-to-toe) walking.

Many elderly people walk with shorter steps and a wider leg stance to achieve better balance and stable weight distribution.

## Ankle and foot

The following muscles move the foot and toes:
- gastrocnemius – plantarflexes foot
- soleus – plantarflexes ankle
- peroneus longus and peroneus brevis – plantarflex and evert foot
- peroneus tertius – dorsiflexes and everts foot
- tibialis anterior – dorsiflexes and inverts foot
- tibialis posterior – plantarflexes and inverts foot
- flexor digitorum longus – flexes toes; plantarflexes and inverts foot
- extensor digitorum longus – extends toes; dorsiflexes and everts foot.

(See *Foot muscles, right foot – plantar view,* page 277.)

## LOWER LIMB MUSCLES
### Anterior view

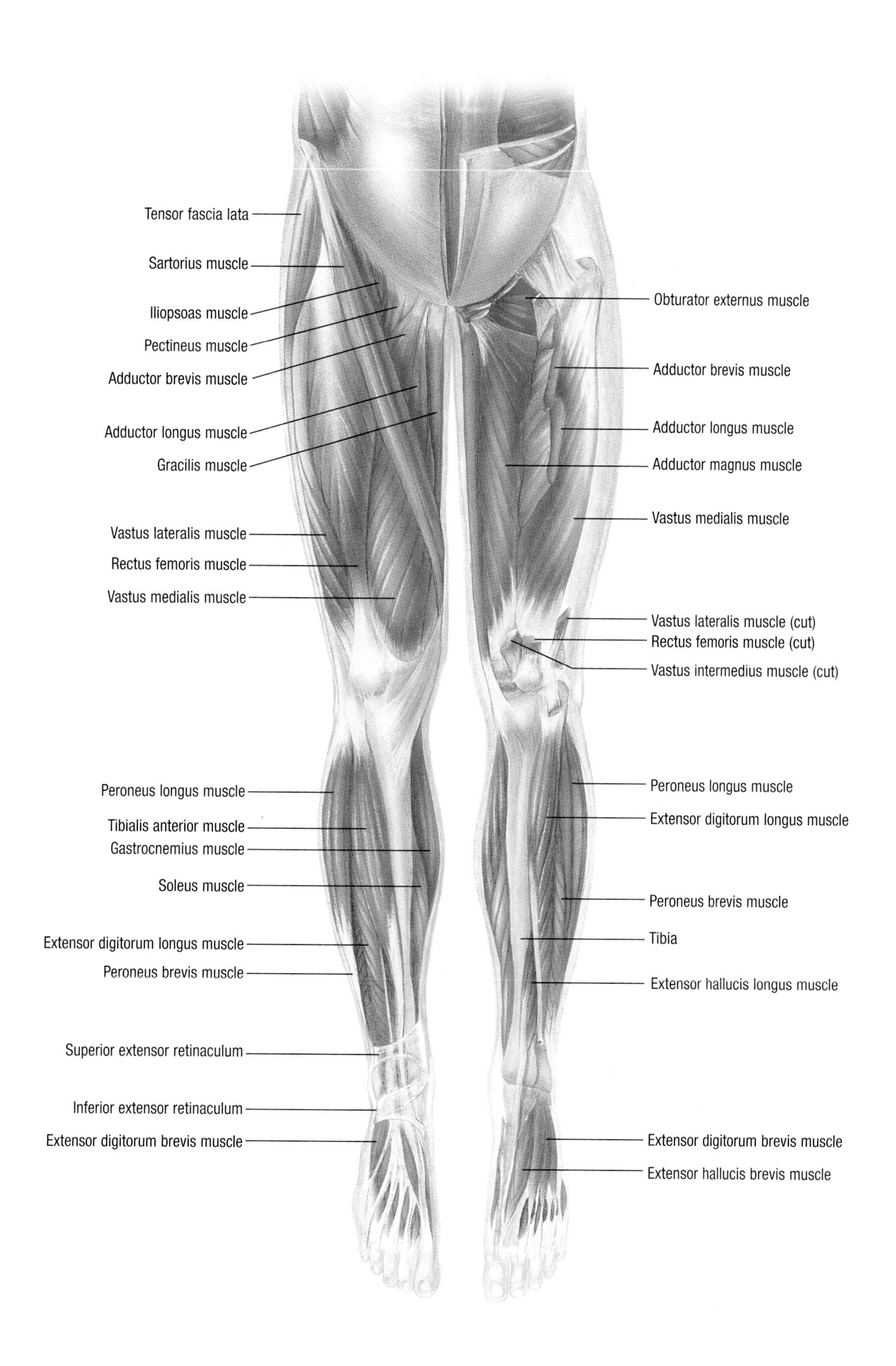
Tensor fascia lata
Sartorius muscle
Iliopsoas muscle
Pectineus muscle
Adductor brevis muscle
Adductor longus muscle
Gracilis muscle
Vastus lateralis muscle
Rectus femoris muscle
Vastus medialis muscle
Peroneus longus muscle
Tibialis anterior muscle
Gastrocnemius muscle
Soleus muscle
Extensor digitorum longus muscle
Peroneus brevis muscle
Superior extensor retinaculum
Inferior extensor retinaculum
Extensor digitorum brevis muscle
Obturator externus muscle
Adductor brevis muscle
Adductor longus muscle
Adductor magnus muscle
Vastus medialis muscle
Vastus lateralis muscle (cut)
Rectus femoris muscle (cut)
Vastus intermedius muscle (cut)
Peroneus longus muscle
Extensor digitorum longus muscle
Peroneus brevis muscle
Tibia
Extensor hallucis longus muscle
Extensor digitorum brevis muscle
Extensor hallucis brevis muscle

# LOWER LIMB MUSCLES
## Posterior view

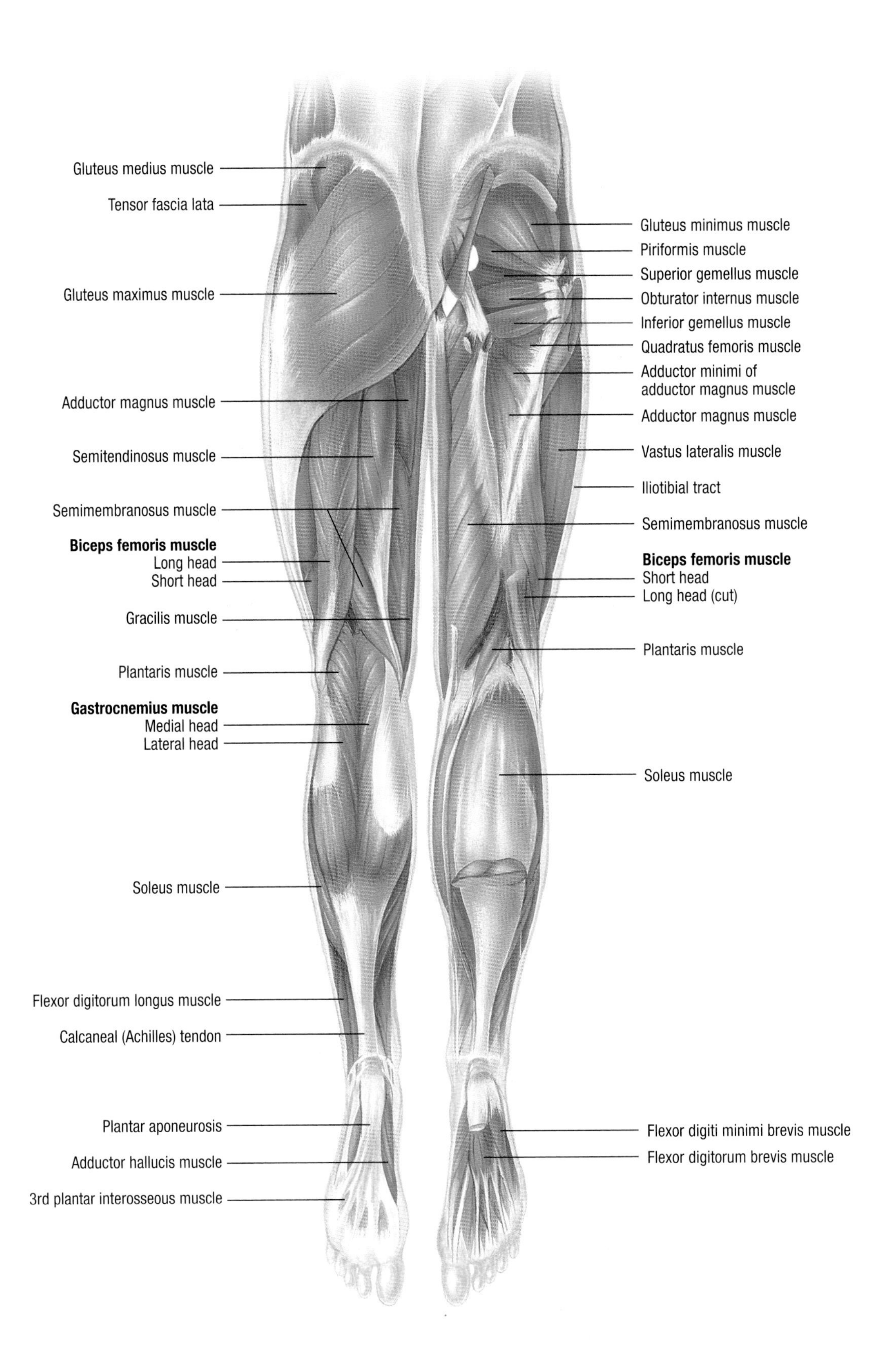

Gluteus medius muscle
Tensor fascia lata
Gluteus maximus muscle
Adductor magnus muscle
Semitendinosus muscle
Semimembranosus muscle
Biceps femoris muscle
Long head
Short head
Gracilis muscle
Plantaris muscle
Gastrocnemius muscle
Medial head
Lateral head
Soleus muscle
Flexor digitorum longus muscle
Calcaneal (Achilles) tendon
Plantar aponeurosis
Adductor hallucis muscle
3rd plantar interosseous muscle
Gluteus minimus muscle
Piriformis muscle
Superior gemellus muscle
Obturator internus muscle
Inferior gemellus muscle
Quadratus femoris muscle
Adductor minimi of
adductor magnus muscle
Adductor magnus muscle
Vastus lateralis muscle
Iliotibial tract
Semimembranosus muscle
Biceps femoris muscle
Short head
Long head (cut)
Plantaris muscle
Soleus muscle
Flexor digiti minimi brevis muscle
Flexor digitorum brevis muscle

## KNEE MUSCLES
### Lateral view

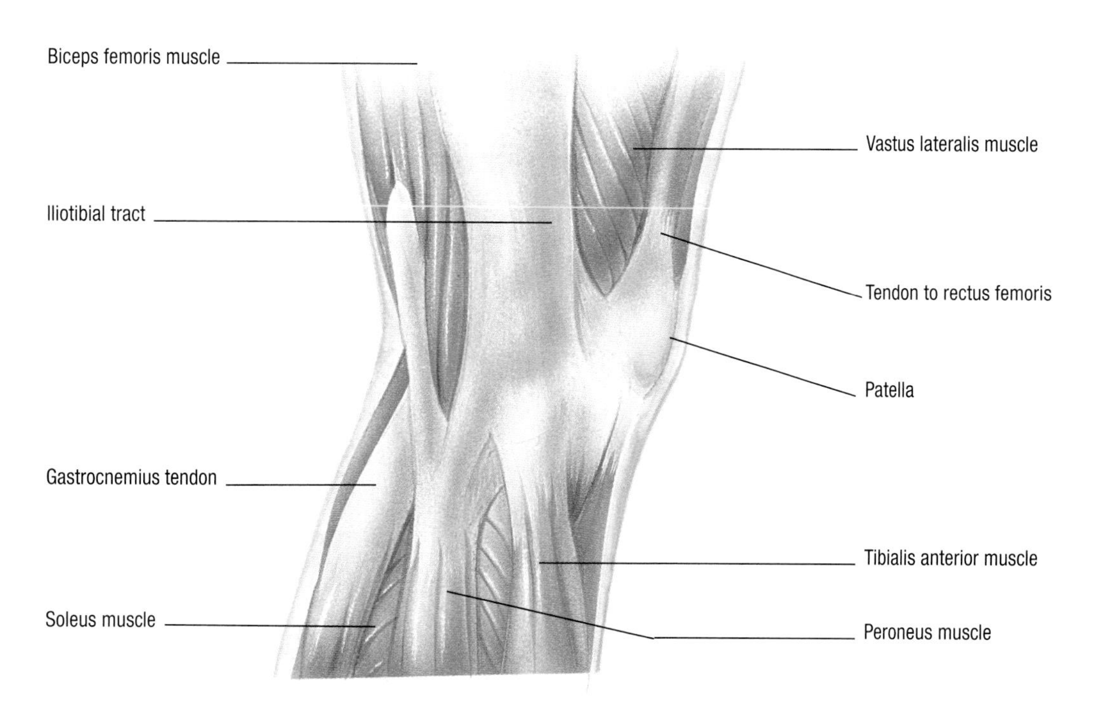
Biceps femoris muscle
Vastus lateralis muscle
Iliotibial tract
Tendon to rectus femoris
Patella
Gastrocnemius tendon
Tibialis anterior muscle
Soleus muscle
Peroneus muscle

## FOOT MUSCLES
### Right foot — Plantar view

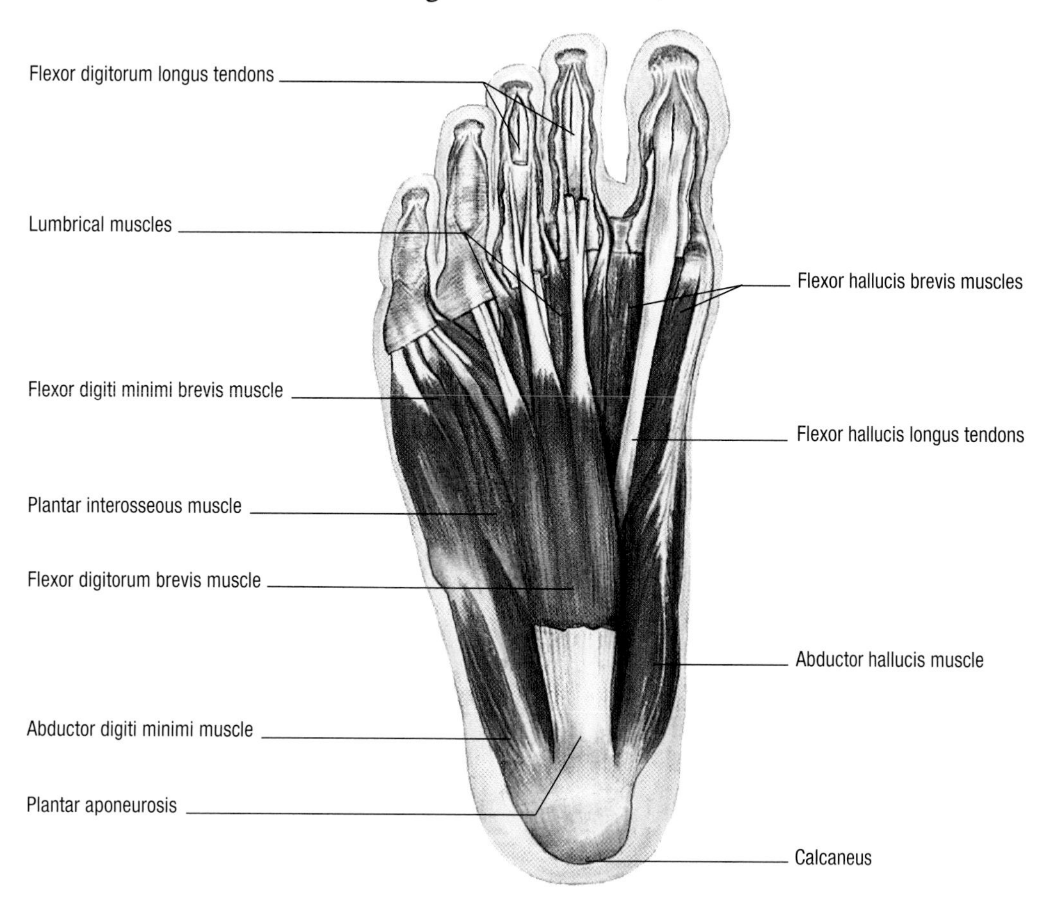
Flexor digitorum longus tendons
Lumbrical muscles
Flexor hallucis brevis muscles
Flexor digiti minimi brevis muscle
Flexor hallucis longus tendons
Plantar interosseous muscle
Flexor digitorum brevis muscle
Abductor hallucis muscle
Abductor digiti minimi muscle
Plantar aponeurosis
Calcaneus

# Long bones

The lower extremity contains three long bones – femur, tibia, and fibula.

## Femur

The femur, or thigh bone, is the longest, thickest, and strongest bone in the body. The ball-like proximal head of the femur articulates with the acetabulum of the pelvis. The greater trochanter is a bony projection located superolateral to the neck of the femur; the trochanteric bursa lies nearby. The lesser trochanter is inferior and medial to the greater.

The posterior femoral shaft has a rough vertical ridge known as the *linea aspera*. The distal features of the femur are the lateral and medial epicondyles and the lateral and medial condyles, all of which articulate with the tibia.

## Tibia

The tibia is the weight-bearing bone of the leg. (See *Tibia and fibula, anterior and posterior views,* pages 280 and 281.) The following features are significant:

- proximal
  - medial and lateral condyles articulate with femur anteriorly and posteriorly
  - tibial tuberosity attaches the patellar ligament.
- distal
  - medial malleolus articulates with talus of the ankle
  - medial malleolus (a downward projection) forms the medial bulge of the ankle
  - fibular notch articulates with the fibula.

**Clinical tip**

Take these steps to assess the tibiofemoral joint:

- Flex the patient's knee to 90 degrees.
- Place his foot on the examining table.
- Using your thumbs, press the tibiofemoral joint.
- Palpate along the tibial margins.
- Palpate the collateral ligament.

Tenderness or bony ridges around the joint margins suggest osteoarthritis. Tender ligaments may signify injury.

### Learning about compartments

Intermuscular septa, bones, and fasciae – sheets or bands of fibrous tissue – divide tissues of the thigh and leg into sections called compartments. The thigh has anterior, medial, and posterior compartments. Compartments of the leg are termed anterior, lateral, and posterior.

During infection or after trauma, compartment boundaries can confine accumulated fluid, inflammatory products, or even circulating blood and lymph in a limited space. The consequent increased pressure within the compartment is called *compartment syndrome*.

Compartment syndrome can be limb threatening and may require surgical intervention in the form of a fasciotomy – an incision in the deep fascial boundary of the compartment – to release the pressure and restore normal flow of blood, lymph, and extracellular fluid.

#### Signs of compartment syndrome

- progressive, intense, lower leg pain that increases with passive muscle stretching
- muscle weakness and paresthesias
- absent pulse and paralysis – suggest irreversible muscle ischemia.

## Fibula

The fibula, a non-weight-bearing bone, parallels the tibia, but is much smaller. It articulates with the tibia proximally at the lateral tibial condyle and distally at the talus, the tallest of the tarsal bones. The lateral malleolus, an inferior projection, lies at the distal end of the fibula.

## FEMUR

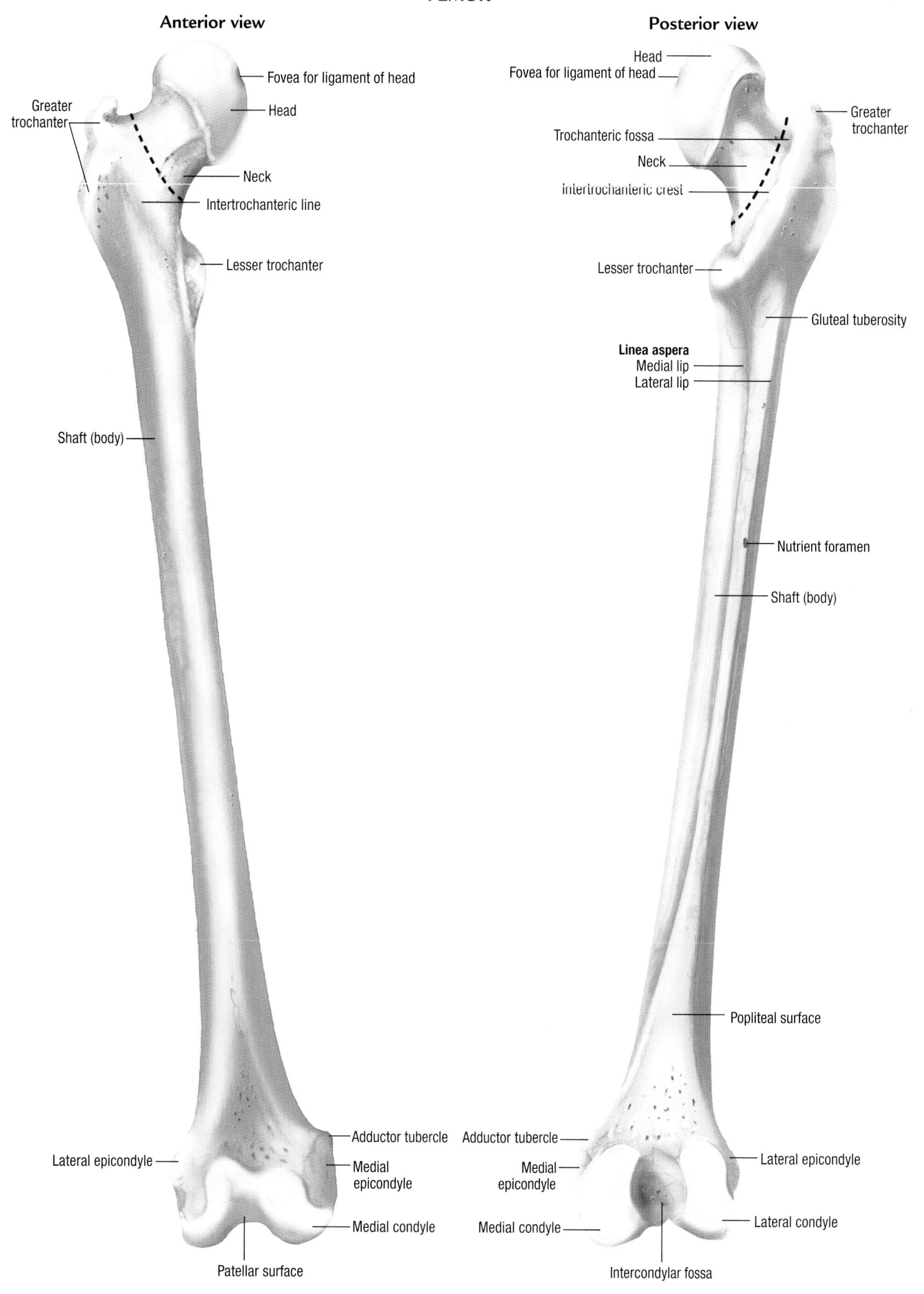

Anterior view
Fovea for ligament of head
Greater trochanter
Head
Neck
Intertrochanteric line
Lesser trochanter
Shaft (body)
Adductor tubercle
Lateral epicondyle
Medial epicondyle
Medial condyle
Patellar surface
Posterior view
Head
Fovea for ligament of head
Greater trochanter
Trochanteric fossa
Neck
Intertrochanteric crest
Lesser trochanter
Gluteal tuberosity
Linea aspera
Medial lip
Lateral lip
Nutrient foramen
Shaft (body)
Popliteal surface
Adductor tubercle
Lateral epicondyle
Medial epicondyle
Medial condyle
Lateral condyle
Intercondylar fossa

## TIBIA AND FIBULA
### Anterior view

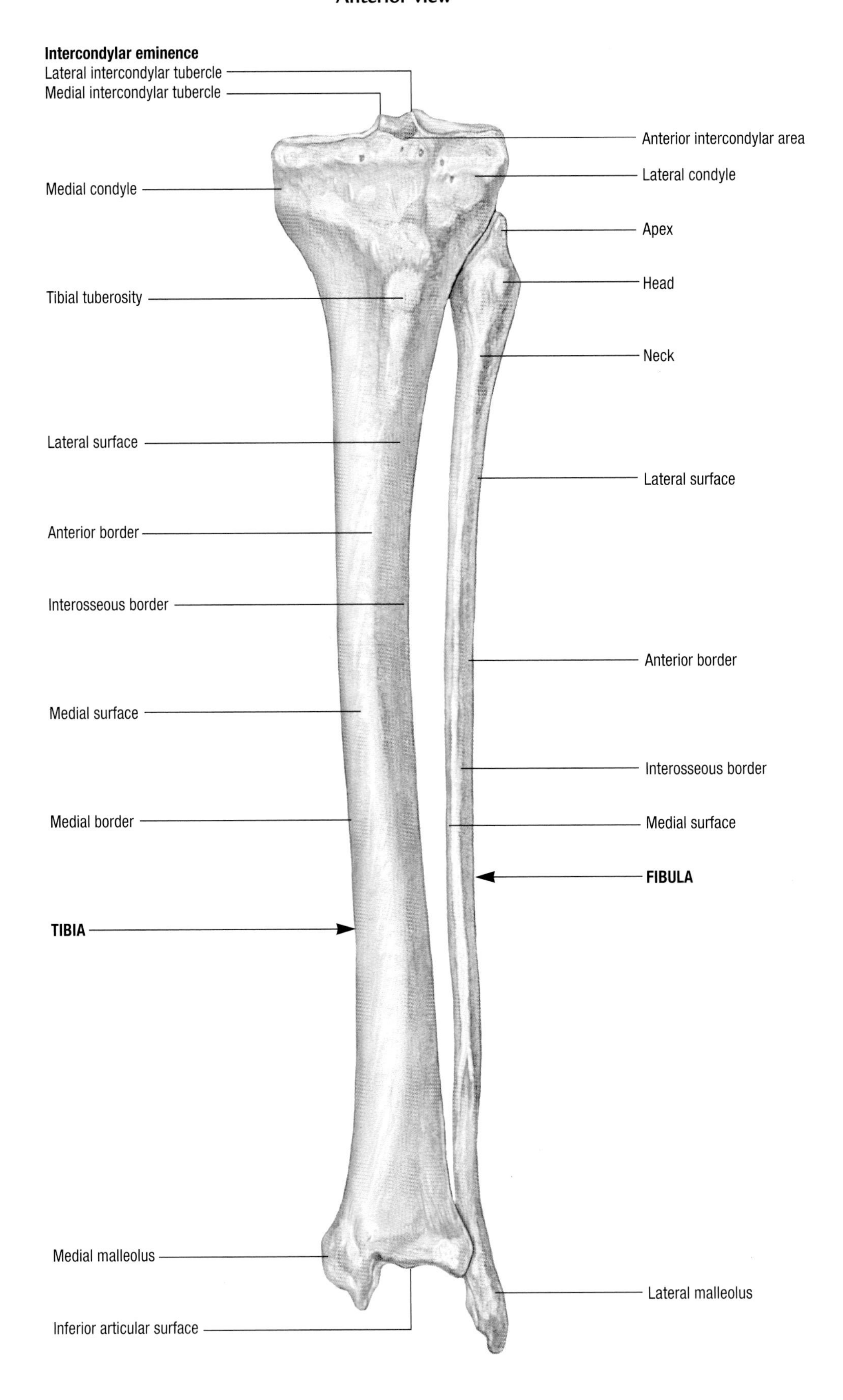

Intercondylar eminence
Lateral intercondylar tubercle
Medial intercondylar tubercle
Anterior intercondylar area
Lateral condyle
Medial condyle
Apex
Head
Tibial tuberosity
Neck
Lateral surface
Lateral surface
Anterior border
Interosseous border
Anterior border
Medial surface
Interosseous border
Medial border
Medial surface
FIBULA
TIBIA
Medial malleolus
Lateral malleolus
Inferior articular surface

## TIBIA AND FIBULA
### Posterior view

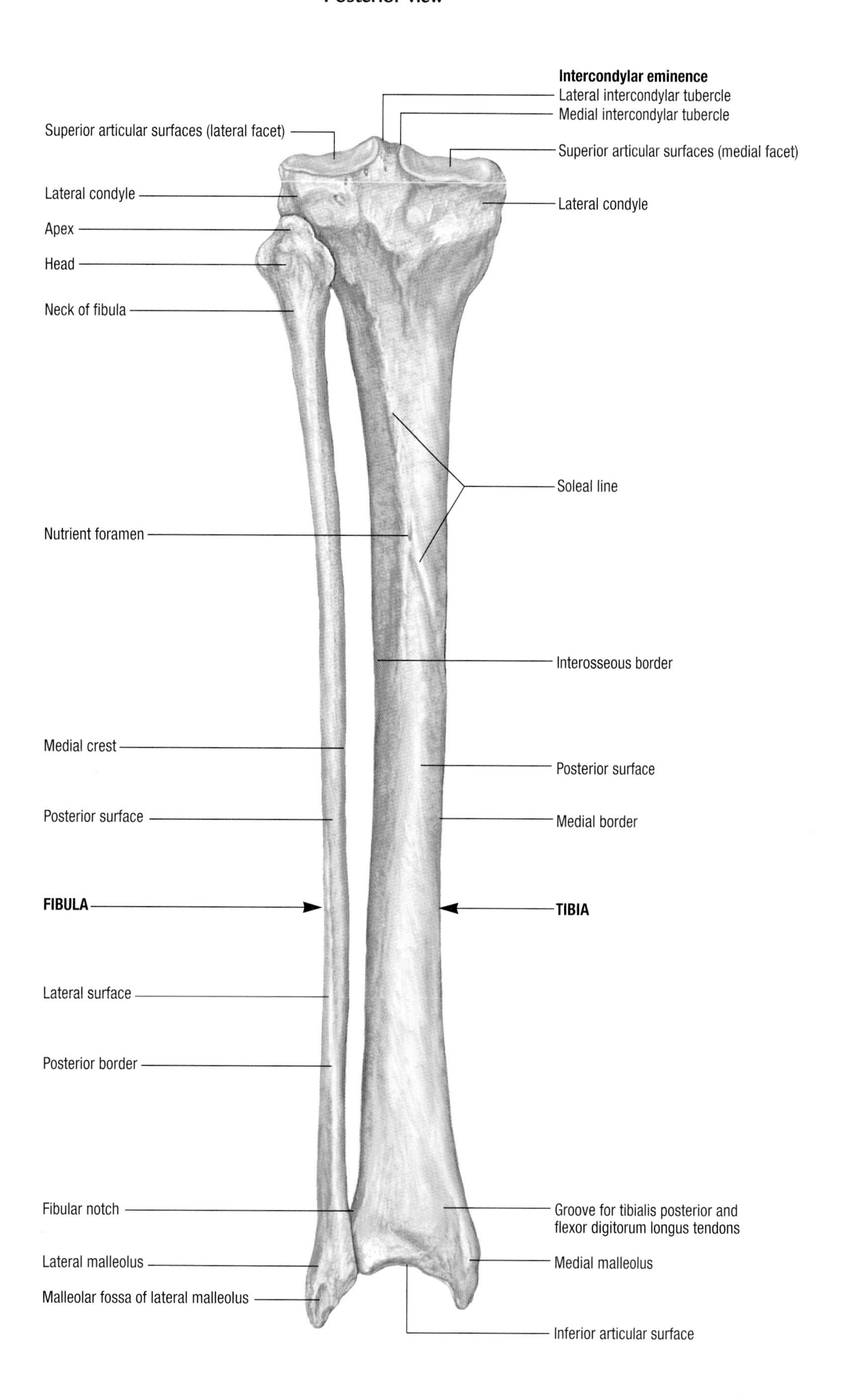

Intercondylar eminence
Lateral intercondylar tubercle
Medial intercondylar tubercle
Superior articular surfaces (lateral facet)
Superior articular surfaces (medial facet)
Lateral condyle
Lateral condyle
Apex
Head
Neck of fibula
Soleal line
Nutrient foramen
Interosseous border
Medial crest
Posterior surface
Posterior surface
Medial border
FIBULA
TIBIA
Lateral surface
Posterior border
Fibular notch
Groove for tibialis posterior and flexor digitorum longus tendons
Lateral malleolus
Medial malleolus
Malleolar fossa of lateral malleolus
Inferior articular surface

# Hip

The lower limbs are connected to the axial skeleton by the pelvic girdle. Together with the sacrum, the bones of the hip form the pelvic girdle. Three bones – the ilium, ischium, and pubis – which fuse by adulthood to form the innominate bone, constitute the hip bone. A detailed discussion of the pelvic girdle and illustrations can be found in Part VII, Pelvis.

## Hip joint

A deep, round socket (acetabulum) on the lateral surface of the hip bone articulates with the head of the femur. (See *Hip joint capsule, anterior and posterior views,* page 284, and *Acetabulum,* page 285.) This joint is so stable that dislocation of the hip joint is a rare event.

Additional structural features of the hip joint include:

- articular capsule – surrounds joint where the femur articulates with the acetabulum; consists of outer capsule and inner synovial membrane
- iliofemoral ligament – one of the body's strongest; extends from ilium to anterior intertrochanteric line (between the greater and lesser trochanters) of the femur
- ischiofemoral ligament – spirals superolaterally to the neck of the femur from the ischium
- pubofemoral ligament – extends from the superior ramus of the pubis to the intertrochanteric line of the femur
- zona orbicularis – ring around the femoral neck; made of circular fibers of the articular capsule.

**Clinical tip**

A prosthesis can restore hip mobility and stability and relieve pain. Common causes of hip degeneration include primary degenerative arthritis, severe chronic arthritis, joint trauma, and avascular necrosis, commonly caused by corticosteroid therapy.

### Bursae

Each hip joint has five bursae: deep trochanteric, iliac, iliopectineal, obturator, and trochanteric bursae.

### Blood supply

The femoral head receives blood from two arteries, the medial and lateral femoral circumflex arteries. The placement of these arteries makes them subject to injury when the hip undergoes trauma, such as fracture or dislocation.

Other arteries that supply the hip and lower extremity include:

- popliteal
- anterior and posterior tibial
- peroneal
- dorsalis pedis
- lateral and medial plantar
- common, internal, and external iliac.

The principal veins of the hip and lower extremity include:

- great saphenous
- femoral
- popliteal
- peroneal
- posterior and anterior tibial
- dorsal venous arch
- common, internal, and external iliac.

**Clinical tip**

To assess internal and external rotation of the hip:

- Have the patient lie down and lift one thigh and leg while keeping the knee straight.
- Turn the patient's thigh, leg, and foot medially and laterally. The normal ranges of motion are 40 degrees internal rotation and 45 degrees external.

## HIP JOINT
**Anterior view**

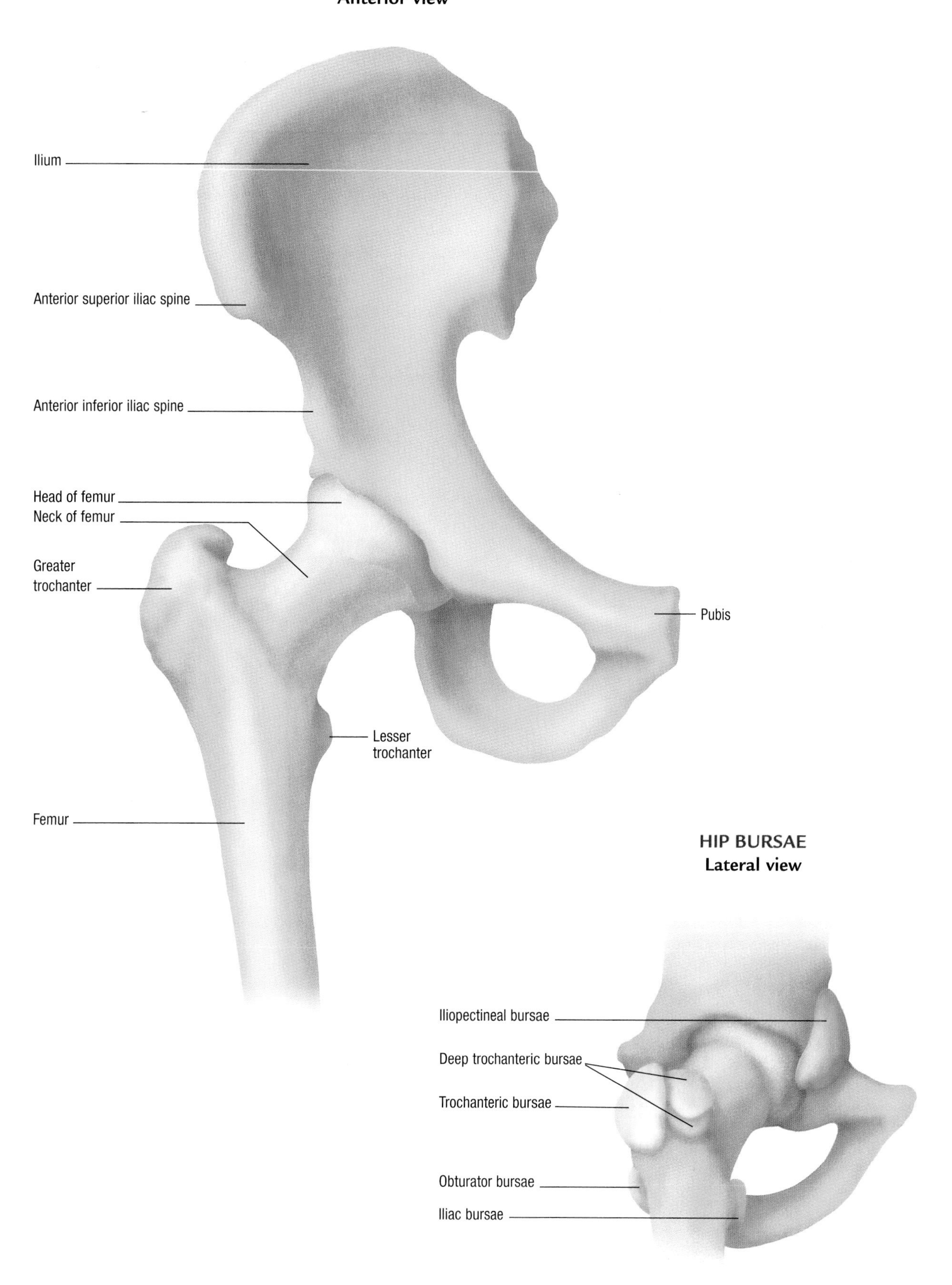

Ilium
Anterior superior iliac spine
Anterior inferior iliac spine
Head of femur
Neck of femur
Greater trochanter
Pubis
Lesser trochanter
Femur
HIP BURSAE
Lateral view
Iliopectineal bursae
Deep trochanteric bursae
Trochanteric bursae
Obturator bursae
Iliac bursae

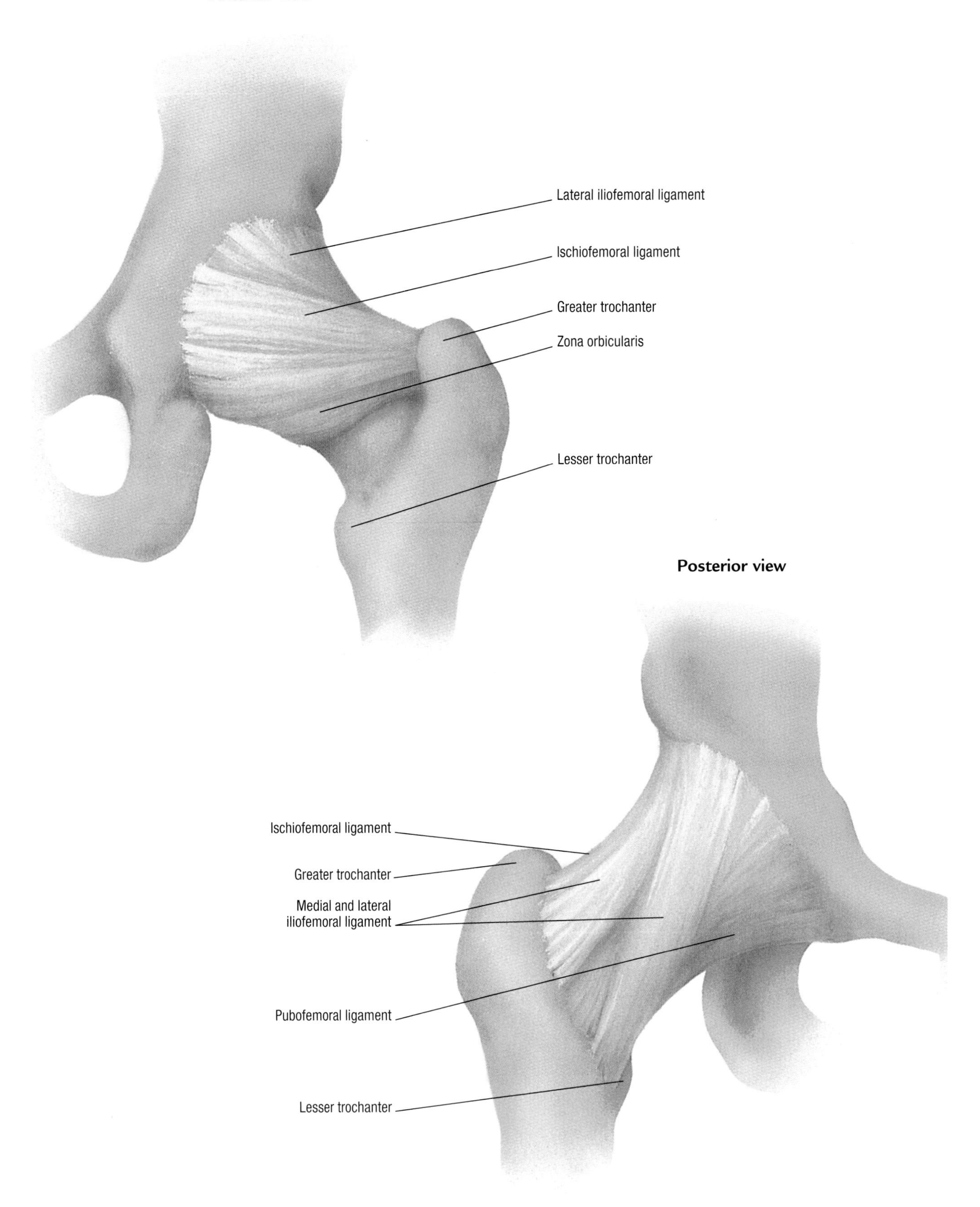
Anterior view
Lateral iliofemoral ligament
Ischiofemoral ligament
Greater trochanter
Zona orbicularis
Lesser trochanter
Posterior view
Ischiofemoral ligament
Greater trochanter
Medial and lateral iliofemoral ligament
Pubofemoral ligament
Lesser trochanter

## ACETABULUM

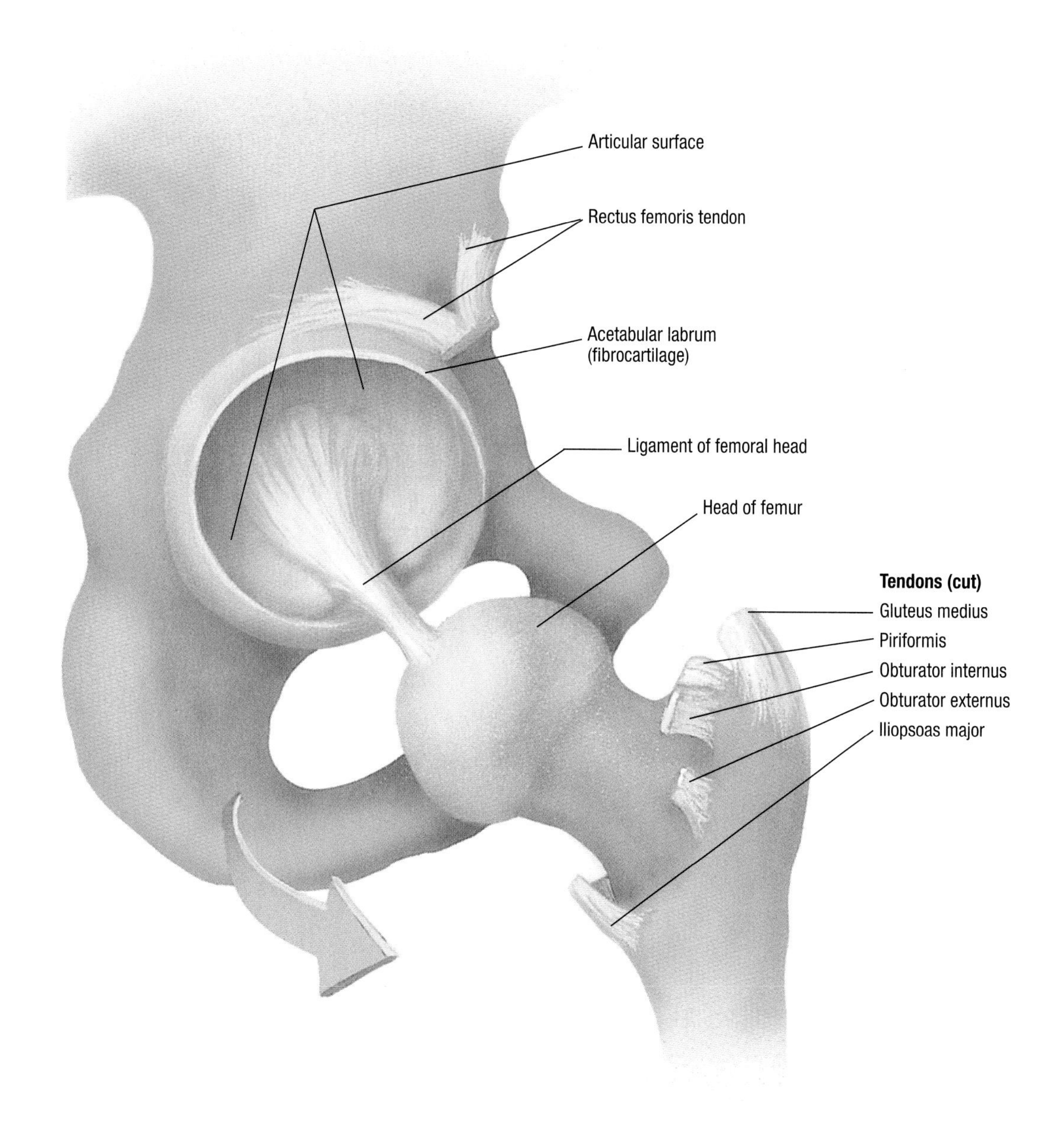

Articular surface
Rectus femoris tendon
Acetabular labrum (fibrocartilage)
Ligament of femoral head
Head of femur
Tendons (cut)
Gluteus medius
Piriformis
Obturator internus
Obturator externus
Iliopsoas major

# Knee

The knee is the largest joint of the body. This synovial hinge joint includes the femur superiorly and the tibia distally. It's covered anteriorly by the patella or kneecap. (See *Knee joint, right knee – anterior and posterior views,* pages 288 and 289.) Cartilage cushions the end of each bone, and synovial fluid fills the joint space. See page 274 for a description of the quadriceps and hamstring muscles, which flex the leg.

## Patella

The patella articulates with the anterior surfaces of the femur and tibia between the femoral condyles. It is a triangular, sesamoid bone (fibrocartilage in a tendon over a bone) attached to the quadriceps muscle via the patellar ligament. The proximal end of the triangle is the base, and the pointed distal end is the apex. Two facets on the dorsal surface articulate with the medial and lateral femoral condyles.

## Menisci and ligaments

Menisci are crescent-shaped interarticular fibrocartilages in certain joints, including the knee. Ligaments are bands of connective tissue between the articular ends of bones.

### Menisci

The medial and lateral menisci are connected by the transverse ligament:

- medial meniscus
  - semicircular
  - anterior end attached to anterior intercondylar fossa of tibia
  - posterior end attached to posterior intercondylar fossa of tibia between posterior cruciate ligament and lateral meniscus.
- lateral meniscus
  - nearly circular
  - attached to anterior intercondylar eminence of tibia, lateral and posterior to anterior cruciate ligament
  - attached to posterior intercondylar eminence, anterior to posterior end of medial meniscus.

(See *Tibial plateau, superior and lateral views,* page 302.)

### Ligaments

The transverse ligaments lie between the two menisci. Many other ligaments are active in knee function, including the following:

- posterior cruciate – between posterior intercondylar fossa of tibia and lateral aspect of medial condyle of femur
- anterior cruciate – between posterior part of medial aspect of lateral condyle of femur and the area anterior to the intercondylar eminence of tibia
- tibial (medial) collateral – between medial epicondyle of femur and medial condyle of tibia
- fibular (lateral) collateral – between lateral epicondyle of femur and head of fibula
- patellar – between apex of patella and tibial tuberosity; considered part of quadriceps tendon.

(See *Knee ligaments, anterolateral, posteromedial, and posterior views,* page 291.)

**Clinical tip**

The most common sports injury to the knee is rupture of the tibial collateral ligament with tearing of the anterior cruciate ligament and the medial meniscus. It is caused by trauma to the lateral portion of the knee.

## Bursae and other structures

The principal bursae of the knee are described according to location:

- anterior (prepatellar, infrapatellar)
- medial
- lateral.

(See *Knee bursae,* page 289.)

Other structures of the knee include:

- infrapatellar fat pads – soft tissue on either side of the patellar tendon
- suprapatellar pouch – an extension of the synovial cavity that includes part of the quadriceps
- synovial cavity – surrounds the joint, allowing free movement.

**Clinical tip**

The bulge sign indicates excess fluid in the knee. To assess for this sign:

- Ask the patient to lie down.
- Palpate the knee.
- After palpating, firmly stroke the medial side of the knee two to four times to displace excess fluid.
- Tap the lateral aspect of the knee just behind the lateral margin of the patella while observing for a fluid wave.

## RIGHT KNEE

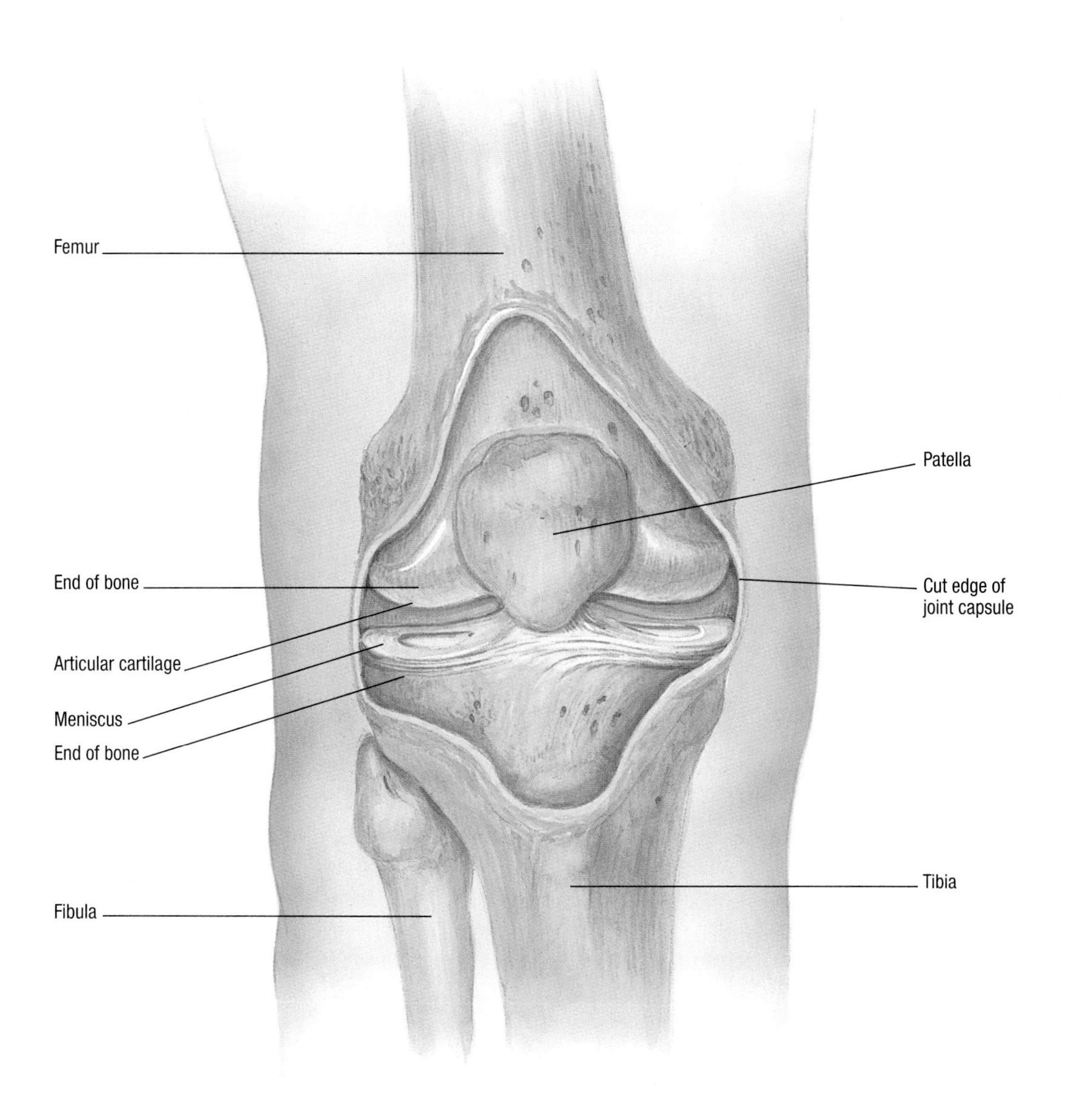

Femur
Patella
End of bone
Cut edge of
joint capsule
Articular cartilage
Meniscus
End of bone
Tibia
Fibula

## KNEE JOINT
**Right knee (flexed) — Anterior view**

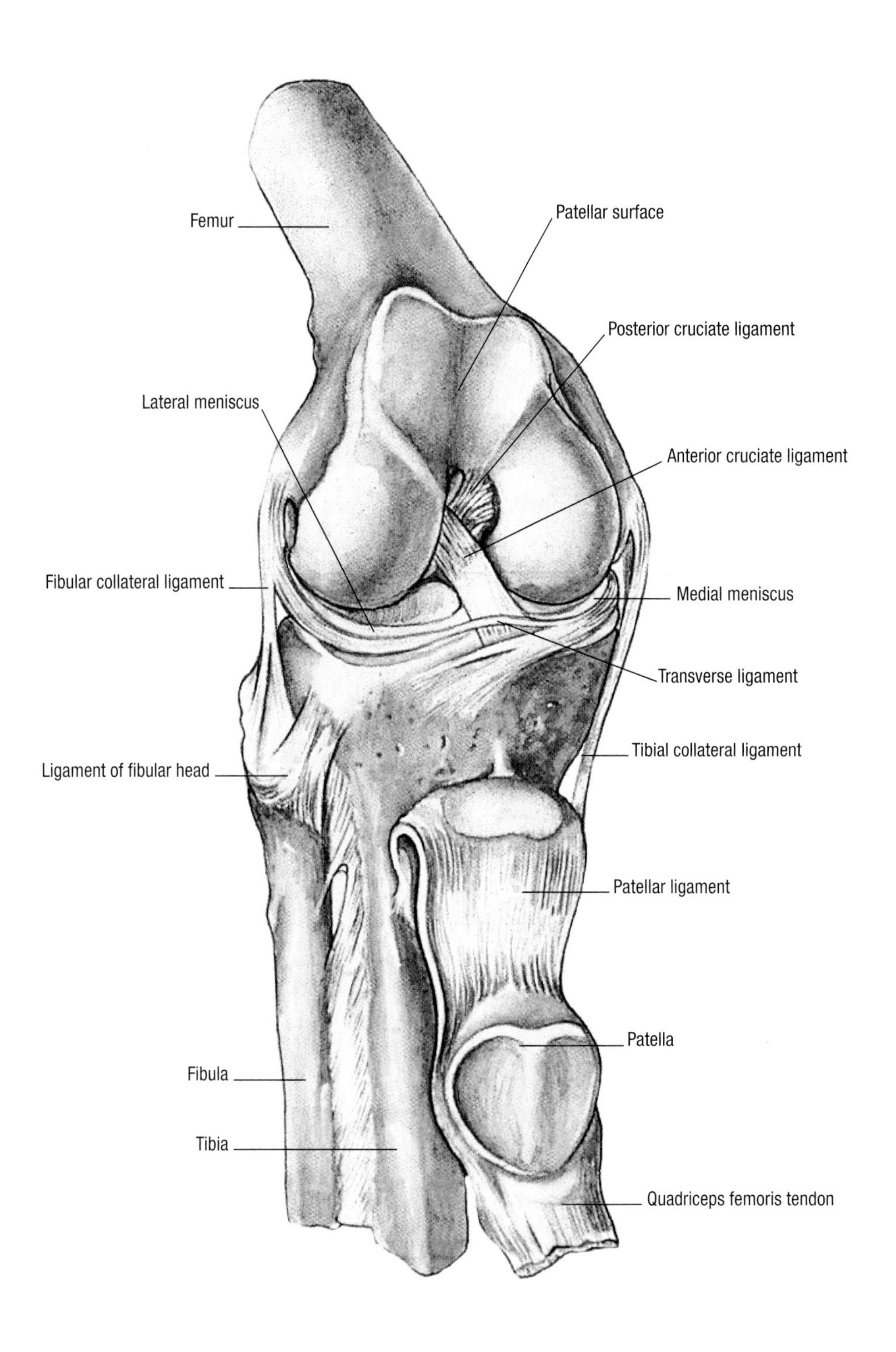
Femur
Patellar surface
Posterior cruciate ligament
Lateral meniscus
Anterior cruciate ligament
Fibular collateral ligament
Medial meniscus
Transverse ligament
Tibial collateral ligament
Ligament of fibular head
Patellar ligament
Patella
Fibula
Tibia
Quadriceps femoris tendon

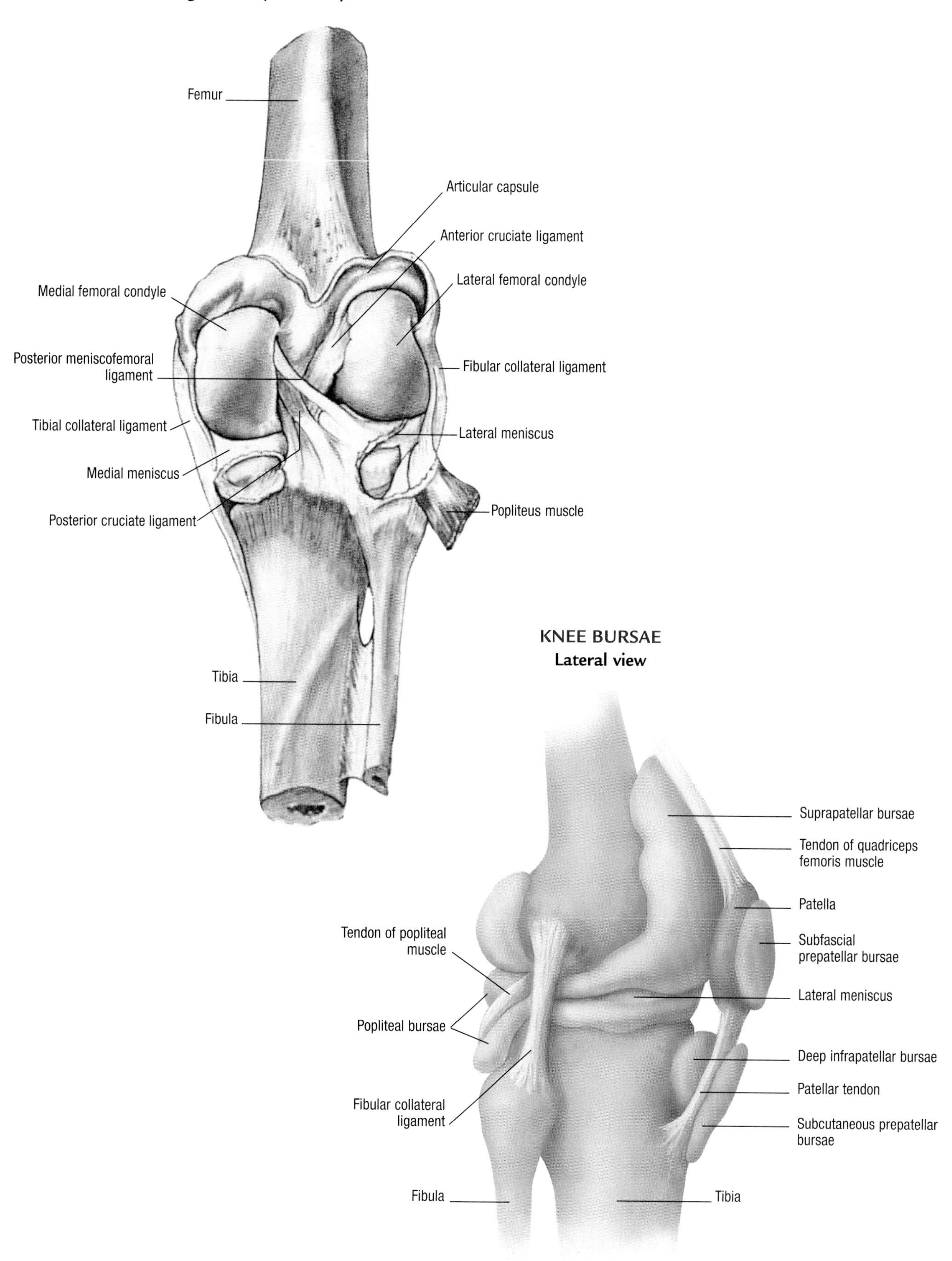
KNEE JOINT
Right knee (extended) — Posterior view
Femur
Articular capsule
Anterior cruciate ligament
Lateral femoral condyle
Medial femoral condyle
Posterior meniscofemoral ligament
Fibular collateral ligament
Tibial collateral ligament
Lateral meniscus
Medial meniscus
Popliteus muscle
Posterior cruciate ligament
Tibia
Fibula
KNEE BURSAE
Lateral view
Suprapatellar bursae
Tendon of quadriceps femoris muscle
Patella
Tendon of popliteal muscle
Subfascial prepatellar bursae
Lateral meniscus
Popliteal bursae
Deep infrapatellar bursae
Patellar tendon
Fibular collateral ligament
Subcutaneous prepatellar bursae
Fibula
Tibia

## TIBIAL PLATEAU
### Superior view

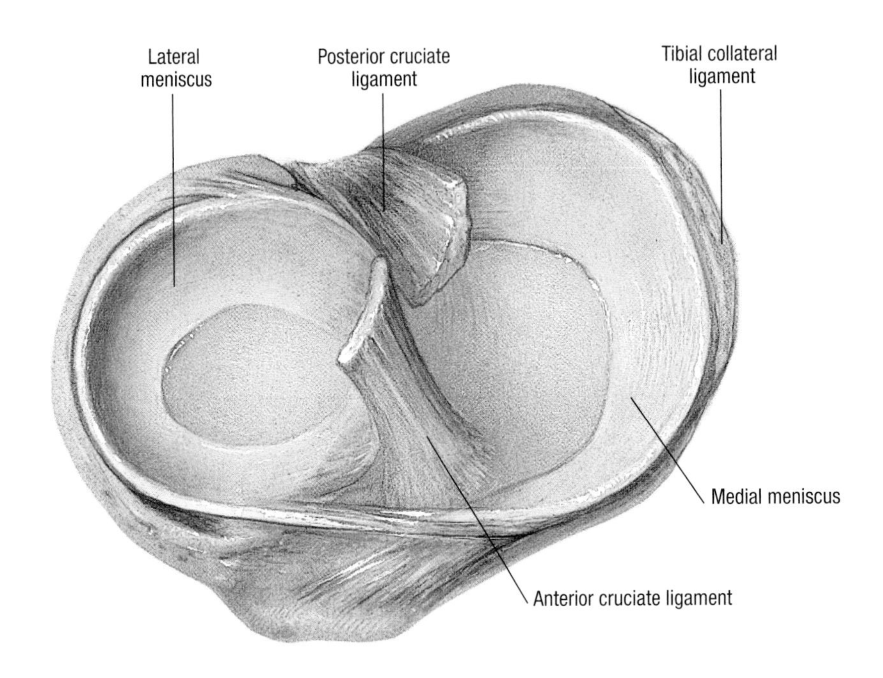

Lateral meniscus
Posterior cruciate ligament
Tibial collateral ligament
Medial meniscus
Anterior cruciate ligament


## TIBIAL PLATEAU
### Lateral view

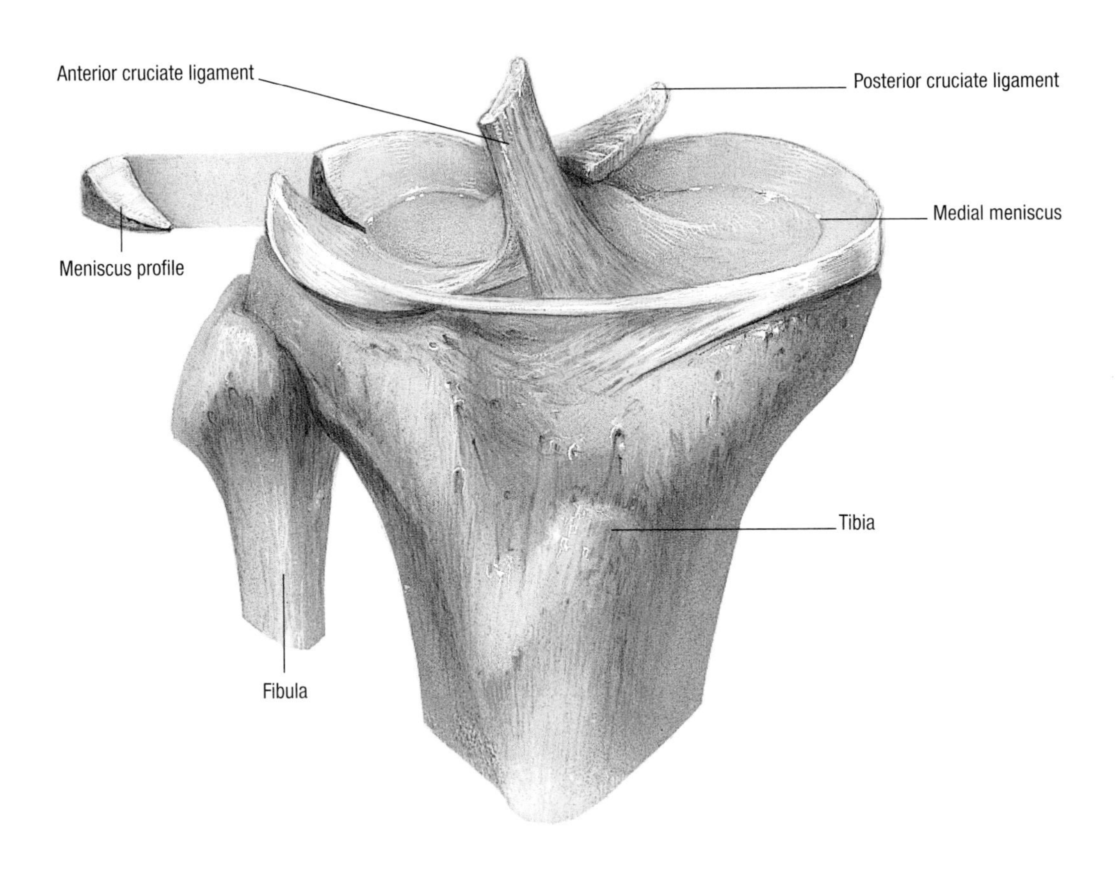

Anterior cruciate ligament
Posterior cruciate ligament
Medial meniscus
Meniscus profile
Tibia
Fibula

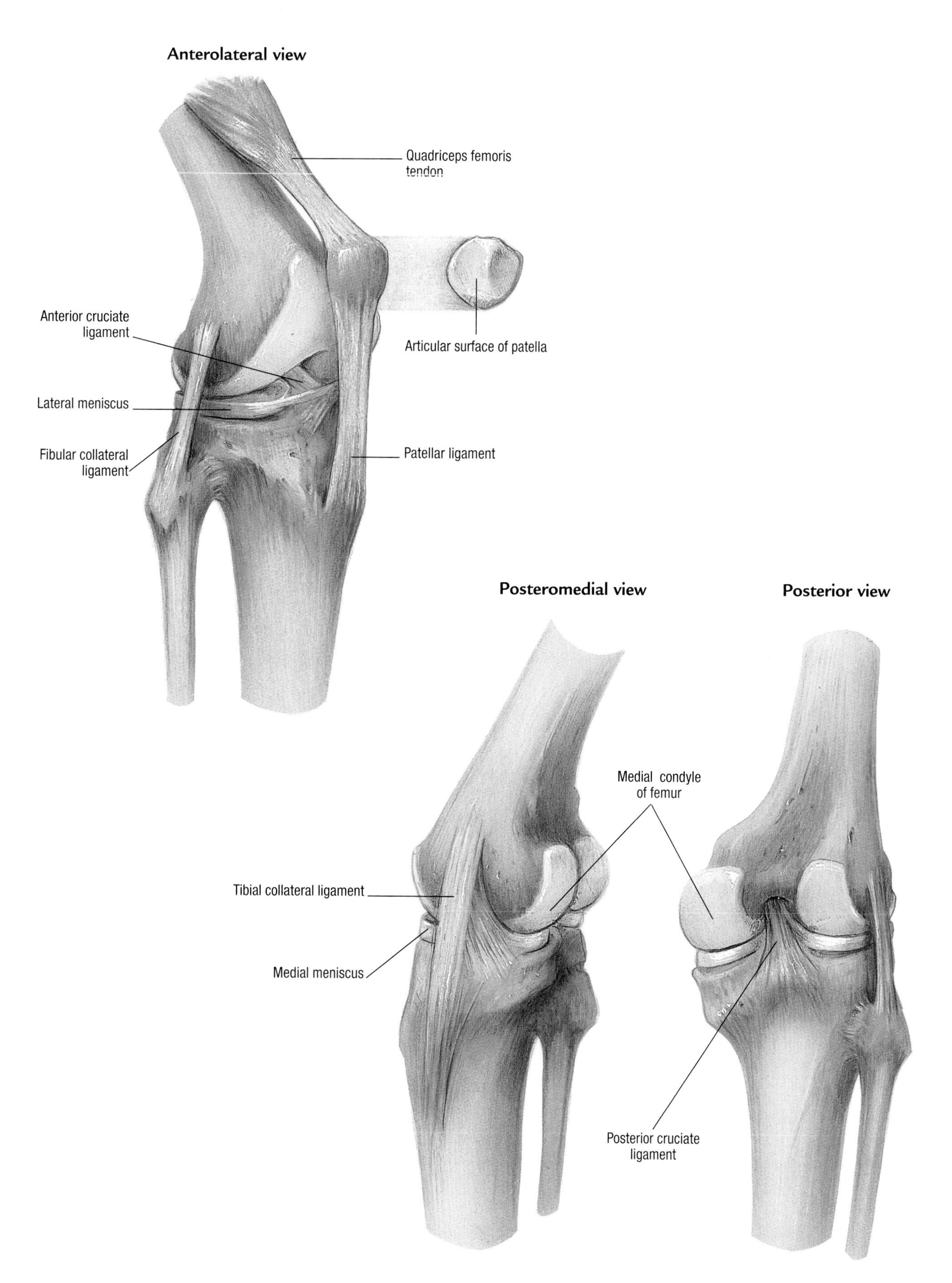
Anterolateral view
Quadriceps femoris tendon
Articular surface of patella
Anterior cruciate ligament
Lateral meniscus
Fibular collateral ligament
Patellar ligament
Posteromedial view
Posterior view
Medial condyle of femur
Tibial collateral ligament
Medial meniscus
Posterior cruciate ligament

# Ankle and foot

The ankle is the joint between the leg and the foot. The foot supports the body and acts as a lever during locomotion. The nine muscles that move the bones of the feet are described in detail on page 274.

## Bones

The foot bones include the tarsals (ankle), metatarsals (instep), and phalanges (toes).

### Tarsals

The seven tarsals correspond to the carpals of the wrist:

- talus (ankle bone)
- calcaneus (heel bone)
- cuboid
- navicular
- medial, intermediate, and lateral cuneiforms (wedge shape).

The talus bears the entire weight of the lower extremity. It transmits part of the weight to the calcaneus, which is the largest and strongest tarsal bone.

### Metatarsals

The metatarsals are numbered 1 through 5, beginning medially. The 1st metatarsal bears more weight than the other metatarsals. Each metatarsal has a proximal base, a shaft, and a distal head.

The metatarsals articulate proximally with the cuneiform bones and the cuboid. Distally, they articulate with the proximal phalanges.

### Phalanges

The 14 phalanges in the toes correspond to those in the fingers. The great toe has only two phalanges, one proximal and one distal; each of the other toes has a proximal, a middle, and a distal phalanx.

**Clinical tip**

When evaluating for signs of advanced gout, look for bumps protruding from the great toe. These bumps signify the presence of urate crystals, which develop into hard, irregular, yellow-white nodules called tophi.

## Arches of the foot

The bones of the foot form two arches. The tarsal and metatarsal bones form the longitudinal arch, which begins at the calcaneus and has medial and lateral portions:

- medial (inner) – calcaneus, talus, navicular, and cuneiforms
- lateral (outer) – calcaneus, cuboid, and two most lateral metatarsals.

The transverse arch is formed by the second and third cuneiforms and second through fourth metatarsals.

## Joints

The bones of the foot and ankle form six groups of joints, named for adjoining bones, as follows:

- ankle – between the tibia and fibula and talus
- intertarsal
- tarsometatarsal
- intermetatarsal
- metatarsophalangeal
- interphalangeal.

### Differentiating arterial and venous insufficiency

Arterial insufficiency produces distinctively different assessment findings from those produced by venous insufficiency. Some common differences are outlined below.

| CHARACTERISTIC | ARTERIAL INSUFFICIENCY | VENOUS INSUFFICIENCY |
|---|---|---|
| Pulses | • Decreased or absent | • Present<br>• Possible assessment difficulty due to edema |
| Temperature of skin | • Cool | • Normal |
| Appearance of foot | • Pale and shiny<br>• Deep red when dependent | • Edematous<br>• Cyanotic when dependent |
| Ulcerations | • Around toes | • Around ankle |

## ANKLE AND FOOT
### Frontal view

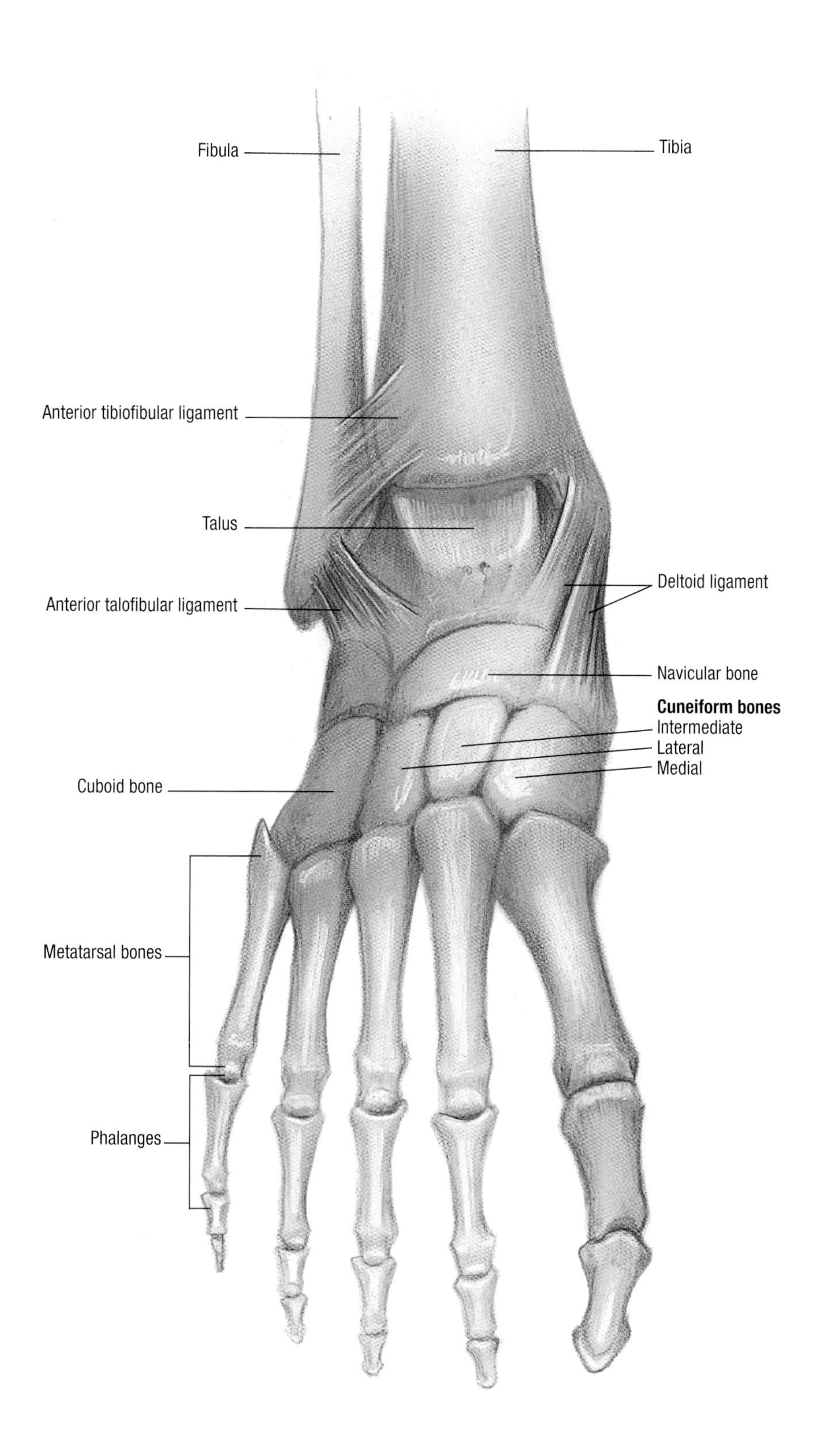

Fibula
Tibia
Anterior tibiofibular ligament
Talus
Deltoid ligament
Anterior talofibular ligament
Navicular bone
**Cuneiform bones**
Intermediate
Lateral
Medial
Cuboid bone
Metatarsal bones
Phalanges

# Tendons, ligaments, nerves, and vessels

The ankle and foot contain many tendons and ligaments. The tendons, which are bands of tissue, attach muscles to bones, enabling the bones to move when skeletal muscles contract. Ligaments, which are bands of strong, flexible, connective tissue, connect the articular ends of bones and provide stability to strengthen and stabilize the joint.

## Tendons

Key tendons include the Achilles, peroneus longus tendon, extensor tendons, and the flexor digitorum and hallucis longus tendons. The Achilles tendon originates from the gastrocnemius and attaches to the calcaneus.

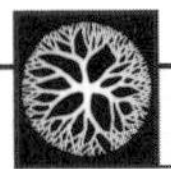

**Physiology**

### Extension and flexion

The muscles and tendons of the ankle joint are responsible for dorsiflexion (upward motion) and plantarflexion (downward motion) of the foot. Inversion (inward motion) and eversion (outward motion) of the foot take place in the joints below the talus.

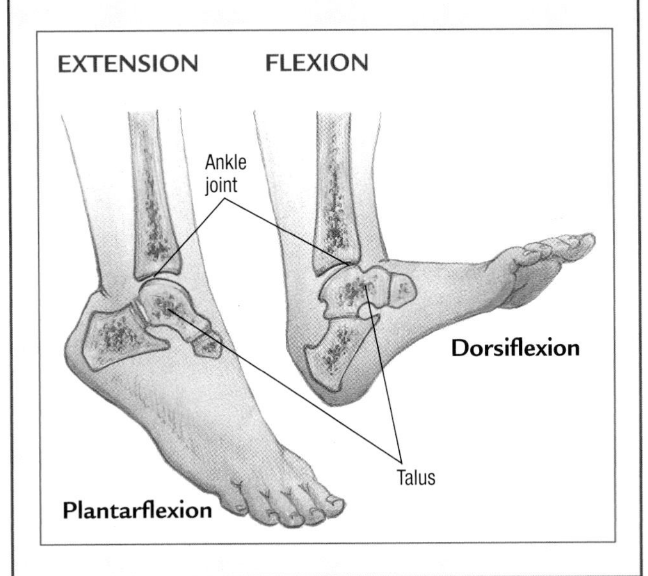
EXTENSION
FLEXION
Ankle joint
Dorsiflexion
Talus
Plantarflexion

## Ligaments

The most significant ligaments in the ankle are medial and lateral ligaments. The lateral ligament consists of three parts: anterior talofibular ligament, posterior talofibular ligament, and calcaneofibular ligament. Ligaments in the foot include the short and long plantar ligaments and the dorsometatarsal ligaments.

## Nerves, blood vessels, and lymphatic drainage

The superficial peroneal nerve supplies the dorsum of the foot. The toes are supplied by the deep peroneal and medial and lateral plantar nerves.

The major arteries of the food include:

- dorsalis pedis
- lateral and medial tarsal
- arcuate
- deep plantar.

Venous drainage of the foot is provided by the dorsal venous arch to the saphenous vein. Lymphatic drainage follows the course of the saphenous vein.

**Clinical tip**

To test the Achilles reflex, have the patient plantarflex or dorsiflex his foot. Then, support the plantar surface and strike the Achilles tendon. The reflex is plantarflexion of the foot and ankle. Absence of this reflex may occur in patients with diabetes mellitus or peripheral neuropathy. An abnormally slow return of the foot to the neutral position after the Achilles tendon reflex may indicate hypothyroidism; an abnormally fast return may occur in hyperthyroidism or pyramidal disease.

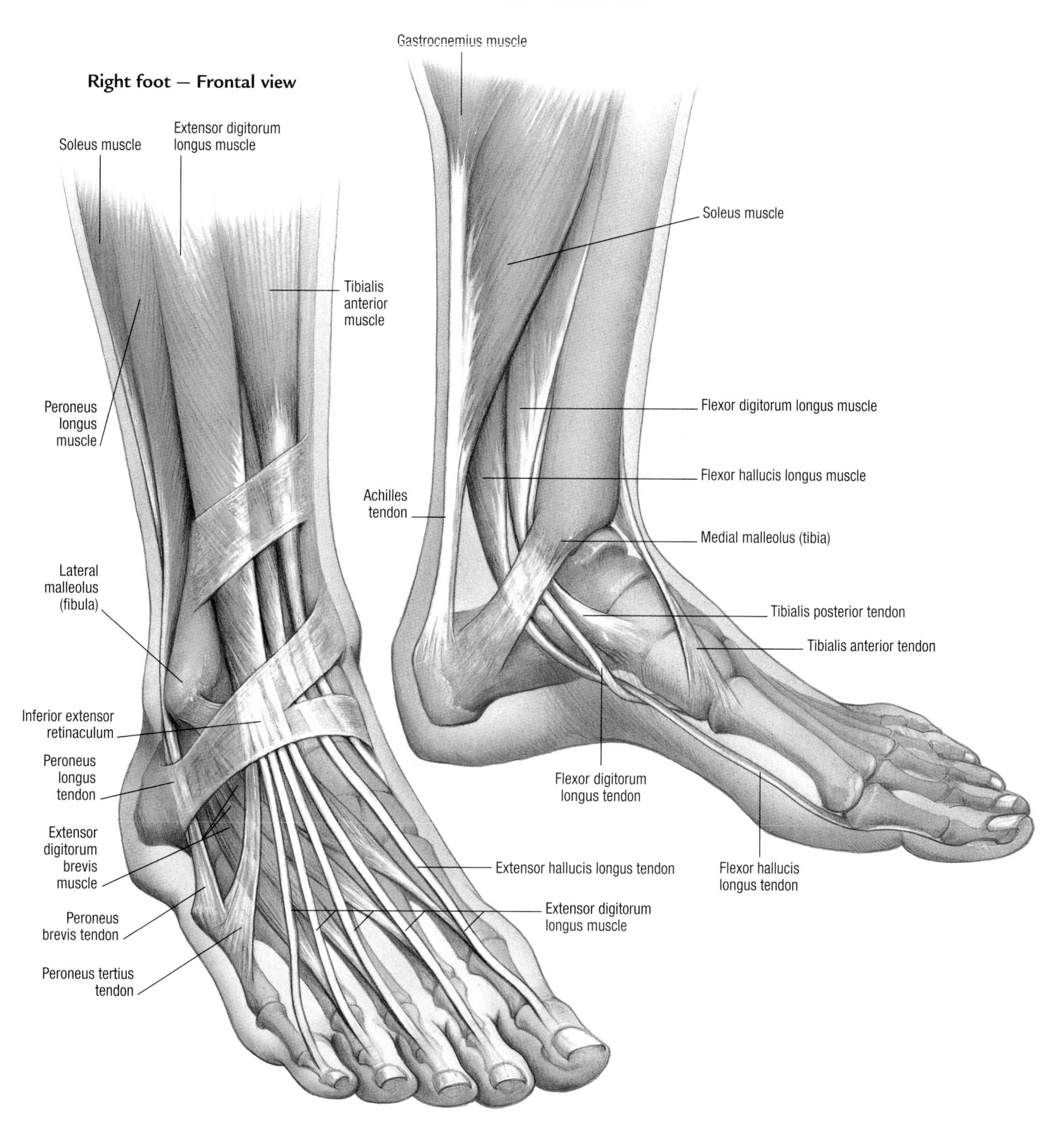
Left foot — Medial view
Gastrocnemius muscle
Soleus muscle
Flexor digitorum longus muscle
Flexor hallucis longus muscle
Achilles tendon
Medial malleolus (tibia)
Tibialis posterior tendon
Tibialis anterior tendon
Flexor digitorum longus tendon
Flexor hallucis longus tendon
Right foot — Frontal view
Soleus muscle
Extensor digitorum longus muscle
Tibialis anterior muscle
Peroneus longus muscle
Lateral malleolus (fibula)
Inferior extensor retinaculum
Peroneus longus tendon
Extensor digitorum brevis muscle
Peroneus brevis tendon
Peroneus tertius tendon
Extensor hallucis longus tendon
Extensor digitorum longus muscle

# ANKLE LIGAMENTS

## Right foot — Medial view

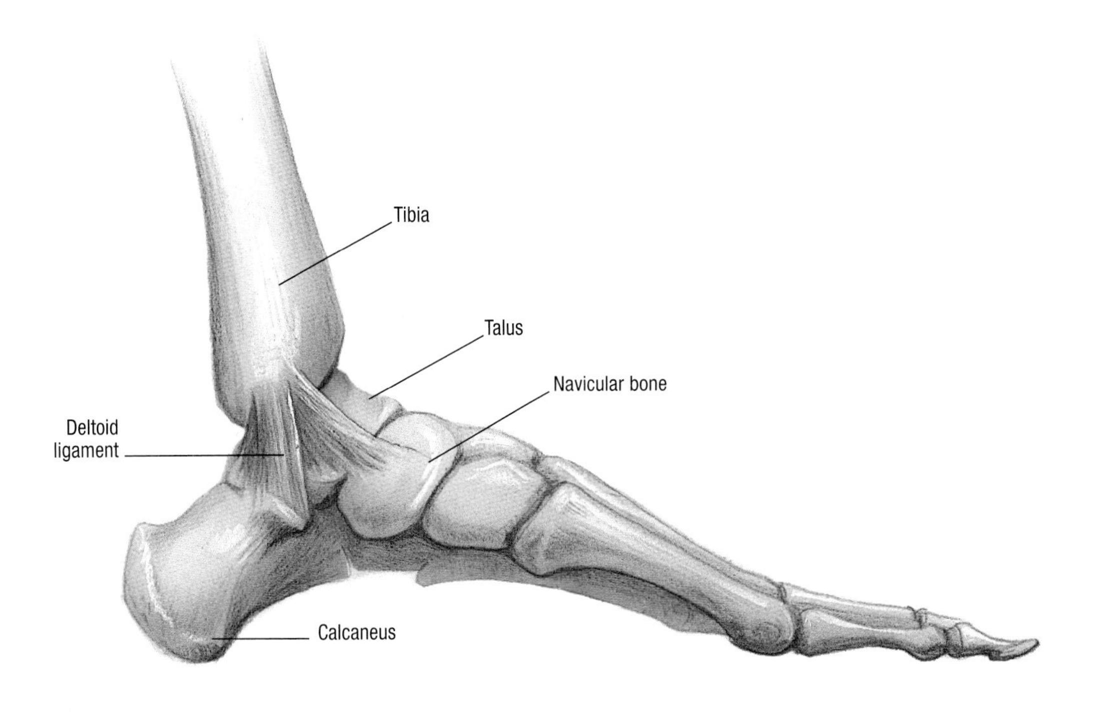

Tibia
Talus
Navicular bone
Deltoid ligament
Calcaneus


## Right foot — Lateral view

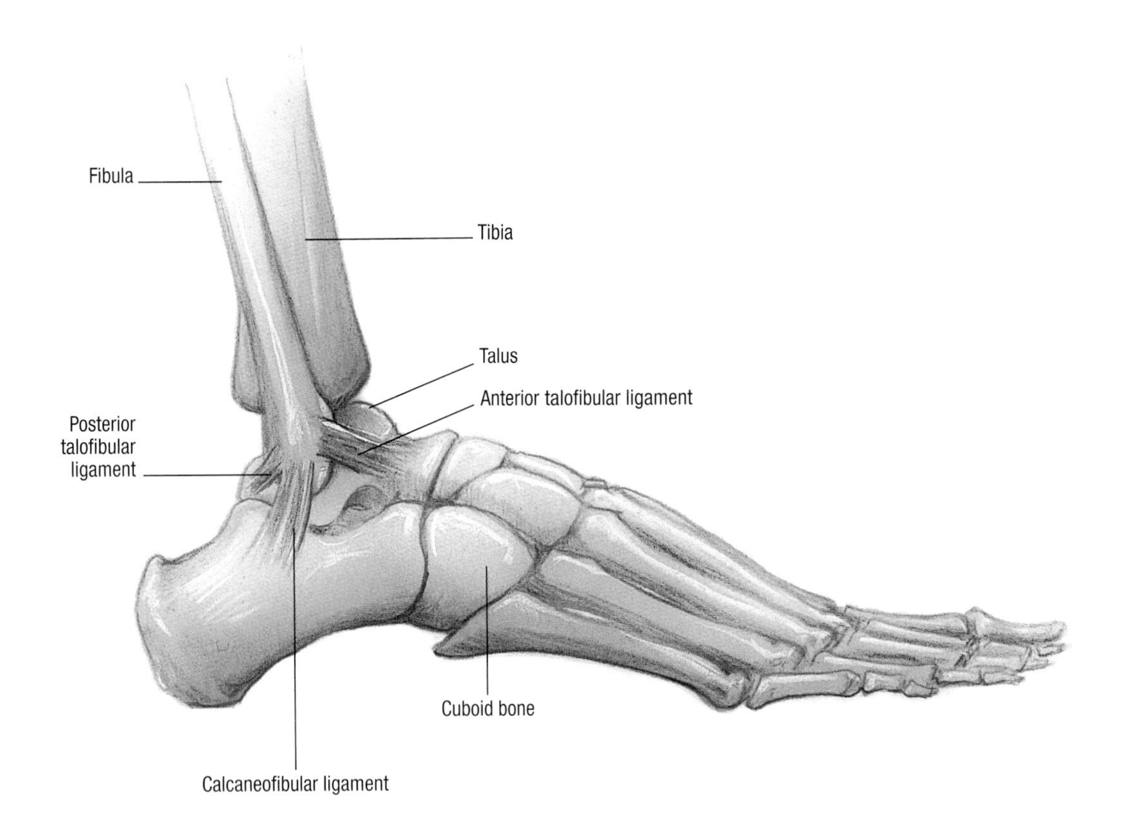

Fibula
Tibia
Talus
Anterior talofibular ligament
Posterior talofibular ligament
Cuboid bone
Calcaneofibular ligament

# FOOT LIGAMENTS

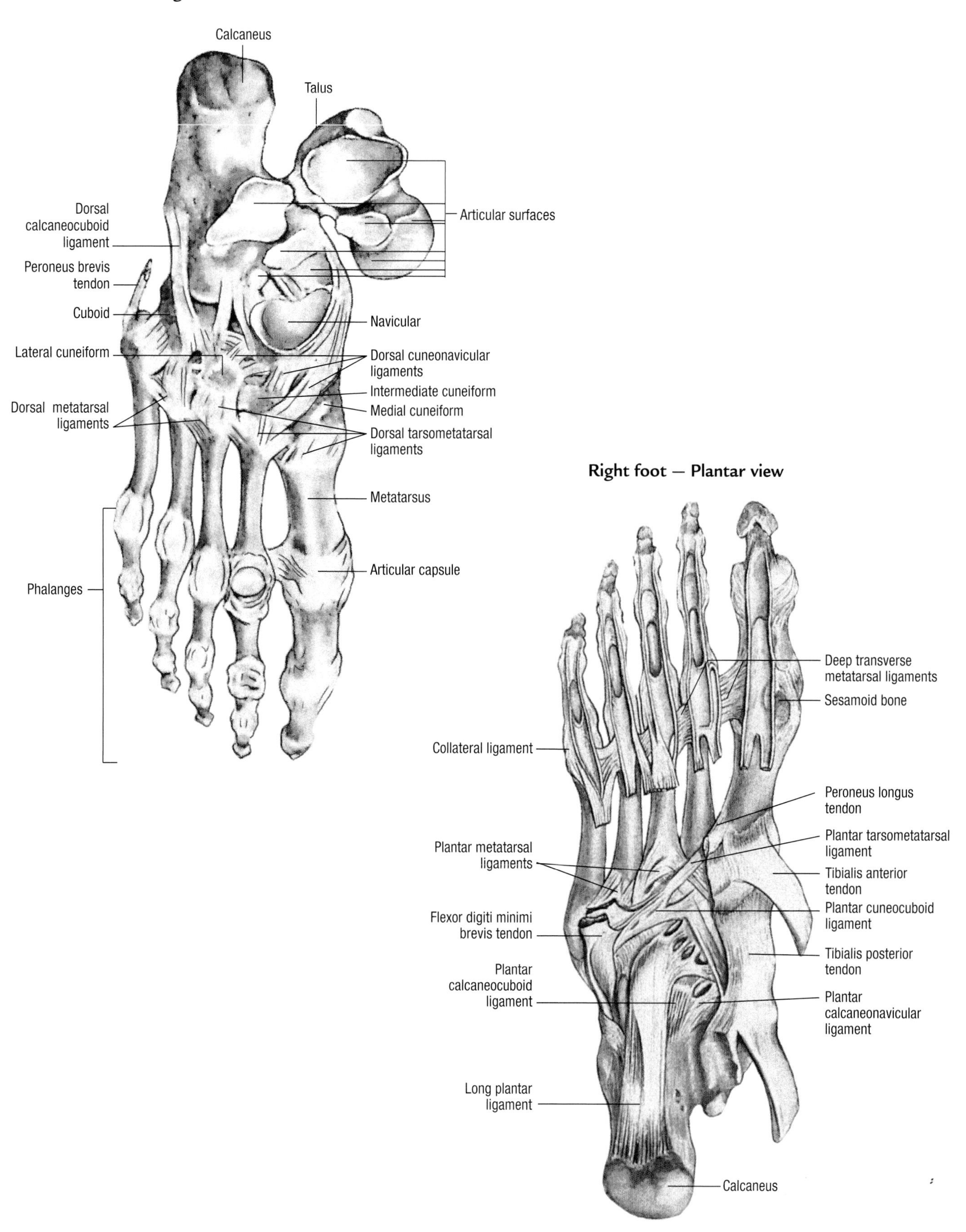
Right foot — Dorsal view
Calcaneus
Talus
Articular surfaces
Dorsal calcaneocuboid ligament
Peroneus brevis tendon
Cuboid
Navicular
Lateral cuneiform
Dorsal cuneonavicular ligaments
Intermediate cuneiform
Dorsal metatarsal ligaments
Medial cuneiform
Dorsal tarsometatarsal ligaments
Metatarsus
Articular capsule
Phalanges
Right foot — Plantar view
Deep transverse metatarsal ligaments
Sesamoid bone
Collateral ligament
Peroneus longus tendon
Plantar tarsometatarsal ligament
Plantar metatarsal ligaments
Tibialis anterior tendon
Plantar cuneocuboid ligament
Flexor digiti minimi brevis tendon
Tibialis posterior tendon
Plantar calcaneocuboid ligament
Plantar calcaneonavicular ligament
Long plantar ligament
Calcaneus

# PART IX

# BODY SYSTEMS

# Skeletal system

The human skeleton, which consists of 206 bones, performs various anatomic (mechanical) and physiologic functions. It forms the body's framework, moving the body and supporting organs and tissues. The bones also serve as storage sites for minerals and produce blood cells.

## Axial skeleton

The axial skeleton consists of 80 bones, which form the body's axis: skull and hyoid (23), auditory ossicles (6), vertebral column (26), and ribs and sternum (25).

## Appendicular skeleton

The appendicular skeleton consists of 126 bones: upper limb (60), lower limb (60), pelvic girdle (2), and pectoral girdle (4).

## Classification

Bones are classified by shape: long (humerus, femur, tibia), short (carpals, tarsals), flat (scapula, skull), irregular (vertebrae, mandible), and sesamoid (patella).

## Structure

Bone consists of layers (*lamellae*) of calcified matrix arranged around central canals (*haversian canals*). *Osteocytes* (bone cells) occupy small cavities (*lacunae*) between the lamellae. Tiny canals (*canaliculi*) between the lacunae radiate in all directions; the canaliculi provide routes for nutrients to reach the osteocytes and for waste to be removed.

A typical long bone has a main shaft (*diaphysis*) and two ends (*epiphyses*). A layer of cartilage separates the epiphyses from the diaphysis at the epiphyseal line. Articular cartilage beneath the articular surface of the epiphysis cushions the joint.

The periosteum surrounds the epiphysis and is thickest around the diaphysis. This tough fibrous sheath consists of an outer fibrous layer and an inner bone-forming layer. The endosteum is a thin membrane lining the medullary cavity, which contains the marrow.

**Clinical tip**

Men's bones may be denser than women's. Bone density and structural integrity decrease after age 30 in women and after age 45 in men.

## Tissue types

Each bone consists of two types of bone tissue: dense, smooth compact bone and spongy cancellous bone.

- Compact bone – diaphyses and outer layers of epiphyses of long bones; outer layers of short, flat, and irregular bones
- Cancellous bone – central regions of epiphyses; inner portions of short, flat, and irregular bones.

## Blood supply

Blood vessels enter compact bone through the nutrient foramen. Blood reaches bone cells by way of arterioles in haversian canals, through blood vessels in Volkmann's canals (which connect the periosteum and the bone matrix), and through vessels in the bone ends. Spongy bone does not have a true haversian system. Instead, a network of thin, latticelike bone, called trabeculae, contains red marrow. Blood vessels from the periosteum penetrate the spongy bone and supply blood directly to the osteocytes.

## Cartilage

Cartilage is a dense fibrous, hyaline, or elastic connective tissue, which supports and shapes various structures. It also cushions and absorbs shock.

## Joints

Points of contact between two bones, joints hold the bones together. Most allow and restrict flexibility and movement.

### Classification

Joints can be classified by extent of movement (function) or by structure. The three major functional types of joints are:

- synarthrosis – immovable, such as between skull bones
- amphiarthrosis – slightly movable, such as between vertebrae
- diarthrosis – freely movable, such as ankle or shoulder joints.

Structural types of joints include fibrous, cartilaginous, and synovial. Fibrous joints do not have a joint cavity and permit little or no movement. The three types of fibrous joints are:

- suture, which connects bones of the skull
- syndesmosis, such as the stylohyoid ligament, which connects the styloid process and the hyoid bone
- gomphosis, which anchors the root of a tooth in its socket.

Cartilaginous joints also have no joint cavity. The articulating bones are connected tightly by cartilage. The two types of cartilaginous joints and examples are: synchondrosis (epiphyseal plate) and symphysis (symphysis pubis).

All synovial joints are diarthrotic. A fibrous capsule that stabilizes the joint structures surrounds each synovial joint and its ligaments. Synovial joints are further classified by the type of movement they allow and by structure, as follows:

- gliding – adjacent bone surfaces move against one another
- hinge – permit movement in only one direction
- pivot – pivot around a stationary bone
- condylar – oval head of one bone fits into depression of a second bone
- saddle – similar to condylar, but allows more movement
- ball-and-socket – spherical head of one bone fits into a socket in the other.

### Bursae

Bursae are small, synovial, fluid-filled sacs located around joints at friction points between tendons, ligaments, and bones. They act as cushions.

### Ligaments and tendons

Ligaments are dense, strong, flexible bands of fibrous connective tissue that tie bones to other bones. Ligaments that connect the joint ends of bones either limit or facilitate movement. They also provide stability.

Tendons are bands of fibrous connective tissue that attach muscles to the fibrous membrane that covers bones (periosteum). Tendons move bones when skeletal muscles contract.

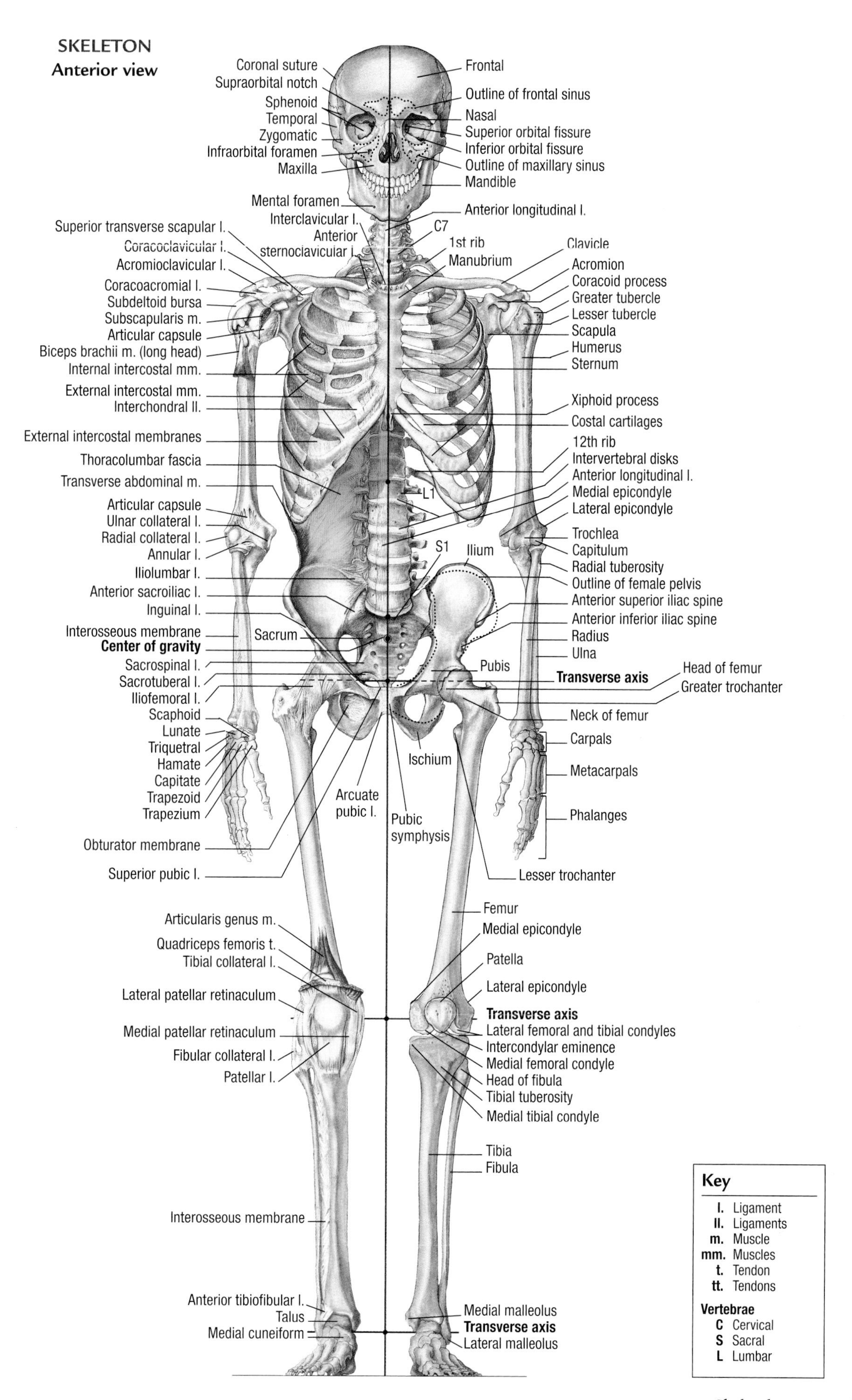
SKELETON
Anterior view
Coronal suture
Supraorbital notch
Sphenoid
Temporal
Zygomatic
Infraorbital foramen
Maxilla
Mental foramen
Interclavicular l.
Anterior sternoclavicular l.
Superior transverse scapular l.
Coracoclavicular l.
Acromioclavicular l.
Coracoacromial l.
Subdeltoid bursa
Subscapularis m.
Articular capsule
Biceps brachii m. (long head)
Internal intercostal mm.
External intercostal mm.
Interchondral ll.
External intercostal membranes
Thoracolumbar fascia
Transverse abdominal m.
Articular capsule
Ulnar collateral l.
Radial collateral l.
Annular l.
Iliolumbar l.
Anterior sacroiliac l.
Inguinal l.
Interosseous membrane
Center of gravity
Sacrospinal l.
Sacrotuberal l.
Iliofemoral l.
Scaphoid
Lunate
Triquetral
Hamate
Capitate
Trapezoid
Trapezium
Obturator membrane
Superior pubic l.
Sacrum
Arcuate pubic l.
Pubic symphysis
Ischium
Articularis genus m.
Quadriceps femoris t.
Tibial collateral l.
Lateral patellar retinaculum
Medial patellar retinaculum
Fibular collateral l.
Patellar l.
Interosseous membrane
Anterior tibiofibular l.
Talus
Medial cuneiform
Frontal
Outline of frontal sinus
Nasal
Superior orbital fissure
Inferior orbital fissure
Outline of maxillary sinus
Mandible
Anterior longitudinal l.
C7
1st rib
Manubrium
Clavicle
Acromion
Coracoid process
Greater tubercle
Lesser tubercle
Scapula
Humerus
Sternum
Xiphoid process
Costal cartilages
12th rib
Intervertebral disks
Anterior longitudinal l.
Medial epicondyle
Lateral epicondyle
L1
S1
Ilium
Trochlea
Capitulum
Radial tuberosity
Outline of female pelvis
Anterior superior iliac spine
Anterior inferior iliac spine
Radius
Ulna
Pubis
Transverse axis
Head of femur
Greater trochanter
Neck of femur
Carpals
Metacarpals
Phalanges
Lesser trochanter
Femur
Medial epicondyle
Patella
Lateral epicondyle
Transverse axis
Lateral femoral and tibial condyles
Intercondylar eminence
Medial femoral condyle
Head of fibula
Tibial tuberosity
Medial tibial condyle
Tibia
Fibula
Medial malleolus
Transverse axis
Lateral malleolus

**Key**

| | |
|---|---|
| **l.** | Ligament |
| **ll.** | Ligaments |
| **m.** | Muscle |
| **mm.** | Muscles |
| **t.** | Tendon |
| **tt.** | Tendons |

**Vertebrae**

| | |
|---|---|
| **C** | Cervical |
| **S** | Sacral |
| **L** | Lumbar |

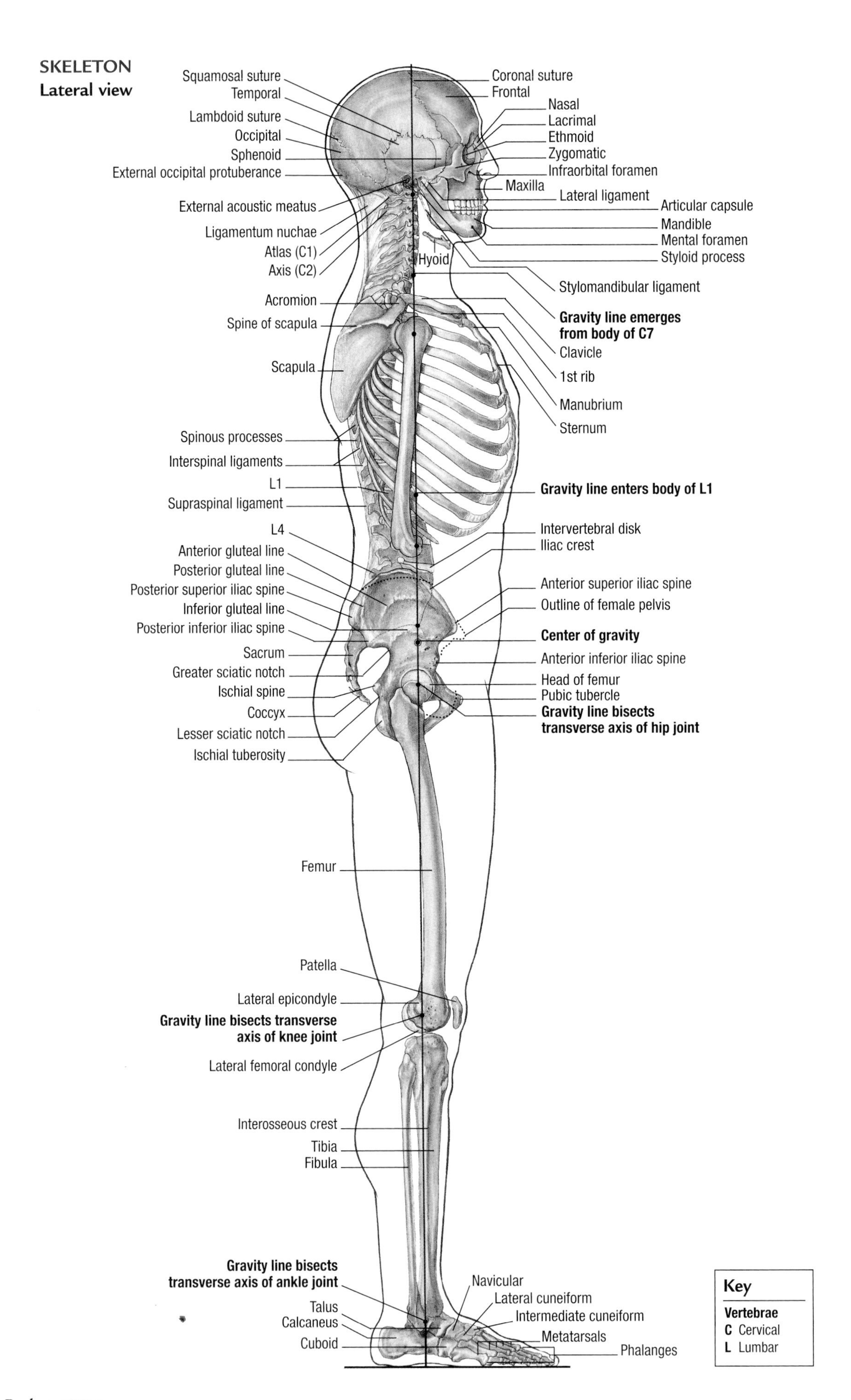

SKELETON
Lateral view
Squamosal suture
Temporal
Lambdoid suture
Occipital
Sphenoid
External occipital protuberance
External acoustic meatus
Ligamentum nuchae
Atlas (C1)
Axis (C2)
Acromion
Spine of scapula
Scapula
Spinous processes
Interspinal ligaments
L1
Supraspinal ligament
L4
Anterior gluteal line
Posterior gluteal line
Posterior superior iliac spine
Inferior gluteal line
Posterior inferior iliac spine
Sacrum
Greater sciatic notch
Ischial spine
Coccyx
Lesser sciatic notch
Ischial tuberosity
Femur
Patella
Lateral epicondyle
Gravity line bisects transverse axis of knee joint
Lateral femoral condyle
Interosseous crest
Tibia
Fibula
Gravity line bisects transverse axis of ankle joint
Talus
Calcaneus
Cuboid
Coronal suture
Frontal
Nasal
Lacrimal
Ethmoid
Zygomatic
Infraorbital foramen
Maxilla
Lateral ligament
Articular capsule
Mandible
Mental foramen
Styloid process
Hyoid
Stylomandibular ligament
Gravity line emerges from body of C7
Clavicle
1st rib
Manubrium
Sternum
Gravity line enters body of L1
Intervertebral disk
Iliac crest
Anterior superior iliac spine
Outline of female pelvis
Center of gravity
Anterior inferior iliac spine
Head of femur
Pubic tubercle
Gravity line bisects transverse axis of hip joint
Navicular
Lateral cuneiform
Intermediate cuneiform
Metatarsals
Phalanges
Key
Vertebrae
C Cervical
L Lumbar

## SKELETON
### Posterior view

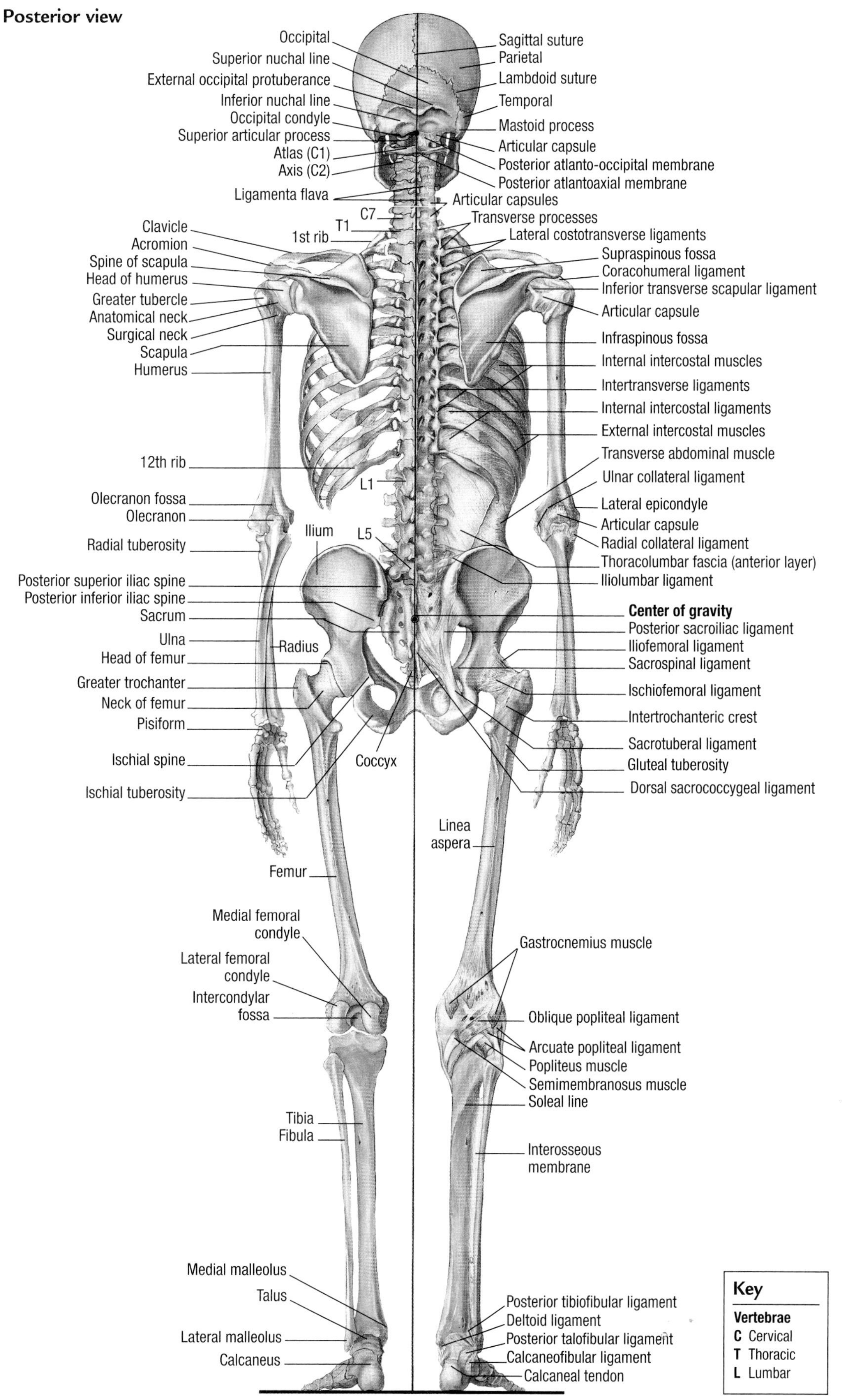
Occipital
Superior nuchal line
External occipital protuberance
Inferior nuchal line
Occipital condyle
Superior articular process
Atlas (C1)
Axis (C2)
Ligamenta flava
C7
T1
1st rib
Clavicle
Acromion
Spine of scapula
Head of humerus
Greater tubercle
Anatomical neck
Surgical neck
Scapula
Humerus
12th rib
L1
Olecranon fossa
Olecranon
Ilium
L5
Radial tuberosity
Posterior superior iliac spine
Posterior inferior iliac spine
Sacrum
Ulna
Radius
Head of femur
Greater trochanter
Neck of femur
Pisiform
Ischial spine
Coccyx
Ischial tuberosity
Sagittal suture
Parietal
Lambdoid suture
Temporal
Mastoid process
Articular capsule
Posterior atlanto-occipital membrane
Posterior atlantoaxial membrane
Articular capsules
Transverse processes
Lateral costotransverse ligaments
Supraspinous fossa
Coracohumeral ligament
Inferior transverse scapular ligament
Articular capsule
Infraspinous fossa
Internal intercostal muscles
Intertransverse ligaments
Internal intercostal ligaments
External intercostal muscles
Transverse abdominal muscle
Ulnar collateral ligament
Lateral epicondyle
Articular capsule
Radial collateral ligament
Thoracolumbar fascia (anterior layer)
Iliolumbar ligament
Center of gravity
Posterior sacroiliac ligament
Iliofemoral ligament
Sacrospinal ligament
Ischiofemoral ligament
Intertrochanteric crest
Sacrotuberal ligament
Gluteal tuberosity
Dorsal sacrococcygeal ligament
Linea aspera
Femur
Medial femoral condyle
Lateral femoral condyle
Intercondylar fossa
Gastrocnemius muscle
Oblique popliteal ligament
Arcuate popliteal ligament
Popliteus muscle
Semimembranosus muscle
Soleal line
Tibia
Fibula
Interosseous membrane
Medial malleolus
Talus
Lateral malleolus
Calcaneus
Posterior tibiofibular ligament
Deltoid ligament
Posterior talofibular ligament
Calcaneofibular ligament
Calcaneal tendon
Key
Vertebrae
C Cervical
T Thoracic
L Lumbar

# Muscular system

The three major types of muscles in the human body, classified according to structure and function, are as follows:
- smooth, nonstriated – involuntary, muscular walls of visceral organs
- striated skeletal – voluntary, skeletal muscles
- striated cardiac – involuntary, heart muscle.

This section deals with skeletal muscle; for information on cardiac and visceral muscle, see Part V, Thorax, and Part VI, Abdomen. The human body has about 600 skeletal muscles.

## Functions

Skeletal muscles move body parts, maintain posture, and generate body heat. Their movement may be voluntary or involuntary.

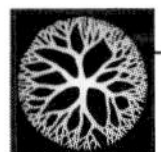

**Physiology**

The skeletal, muscular, and nervous systems work together to produce voluntary movements. Muscles contract when stimulated by impulses from the nervous system. During contraction, the muscle shortens, pulling on the bone to which it's attached by a tendon. Force applied to a tendon moves the bone toward, away from, or around a second bone. Most movement involves groups of muscles rather than one muscle.

## Structure

Skeletal muscle contains cell groups called muscle fibers, which have many nuclei and transverse striations and look like long, striped bands when viewed through a microscope.

Each muscle fiber is surrounded by a membrane, the sarcolemma. In the muscle fiber's cytoplasm (sarcoplasm) are tiny myofibrils, arranged lengthwise. Each myofibril consists of two types of finer fibers called filaments – about 1,500 thick myosin filaments and about 3,000 thin actin filaments. The filaments are stacked in compartments called sarcomeres, the functional units of skeletal muscle. During muscle contraction, the sarcomere shortens when thick and thin filaments slide over each other.

A sheath of connective tissue called the perimysium binds the muscle fibers into a bundle, or fascicle. A stronger sheath, the epimysium, binds fasciculi together to form the fleshy part of the muscle. Fascia connects skin to underlying tissue. It is a fibrous membrane that covers, separates, and supports muscles. The epimysium is the outer layer of connective tissue that surrounds the muscle, and aponeurosis is connective tissue that attaches muscles to bone or other tissues.

Tendons are bands of fibrous connective tissue that attach muscles to the periosteum, a fibrous membrane covering the bone. Tendons enable bones to move when skeletal muscles contract.

## Muscle attachment

Most skeletal muscles are attached to bones, either directly or indirectly. In the human body, indirect attachments outnumber direct attachments.
- Direct attachment – the epimysium of the muscle fuses to the periosteum of the bone, for example, such as the levator scapulae muscle, which connects the cervical vertebrae to the scapula.
- Indirect attachment – the fascia extends past the muscle as a tendon or aponeurosis, which in turn attaches to the bone, such as the pronator teres of the forearm.

### Origin and insertion points

During contraction, one of the bones to which the muscle is attached stays relatively stationary while the other is pulled in the opposite direction. The point where the muscle attaches to the stationary or less movable bone is called the *origin*; it attaches to the more movable bone at the *insertion*. The origin is usually at the proximal end of one bone, and the insertion site at the distal end of the other. (See *Guide to skeletal muscles*, Appendix C, for major origin and insertion sites.)

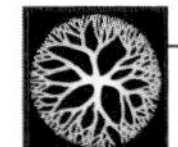

**Physiology**

Muscle develops when existing muscle fibers hypertrophy. Because of such factors as exercise, nutrition, gender, and genetic constitution, muscle strength and size vary among individuals.

## Muscle movements

Types of skeletal muscle movement include:
- flexion – decreases angle between two adjoining bones
- extension – increases angle between two adjoining bones
- abduction – away from the midline of the body
- adduction – toward the midline of the body
- circumduction – circle (a combination of extension, flexion, abduction, and adduction)
- internal (medial) rotation – toward the midline of the body
- external rotation – away from the midline
- supination – upward turning
- pronation – downward turning
- inversion – inward turning
- eversion – outward turning
- retraction and protraction – backward, forward movement.

**Clinical tip**

Use the following 0-to-5 scale to grade muscle strength:
- 5/5 – normal; patient moves joint through full range of motion and against gravity with full resistance
- 4/5 – good; patient completes range of motion against gravity with moderate resistance
- 3/5 – fair; patient completes range of motion against gravity only
- 2/5 – poor; patient completes full range of motion with gravity eliminated (passive motion)
- 1/5 – trace; patient's attempt at muscle contraction is palpable but without joint movement
- 0/5 – zero; no evidence of muscle contraction.

## MUSCULAR SYSTEM
### Anterior view

Skin
Temporalis m.
**Orbicularis oculi muscle**
Orbital part
Palpebral part
Procerus m.
Nasalis m.
Zygomaticus major m.
Masseter m.
Buccinator m.
Depressor anguli oris m.
Depressor labii inferioris m.
Thyrohyoid m.
Sternohyoid m.
Levator scapulae m.
Trapezius m.
Scalenus medius m.
Subscapular m.
**Biceps brachii muscle**
Long head
Short head
Teres major m.
Latissimus dorsi m.
Deltoid m.
**Triceps brachii muscle** — Long head, Lateral head, Medial head
Biceps brachii m.
Brachialis m.
Brachioradialis m.
Bicipital aponeurosis
Flexor carpi radialis
Supinator m.
Extensor carpi radialis longus m.
Flexor digitorum profundus m.
Flexor carpi ulnaris m.
Pronator teres m.
Flexor digitorum superficialis m.
Flexor pollicis longus m.
Flexor retinaculum
Flexor carpi radialis t.
Gluteus medius m.
Tensor fasciae latae m.
Sartorius m.
Gluteus minimus m.
Rectus femoris m.
Iliopsoas m.
Pectineus m.
Vastus intermedius m.
Gracilis m.
Vastus medialis m.
Rectus femoris m.
Iliotibial tract
Biceps femoris m.
Lateral patellar retinaculum
Medial patellar retinaculum
Patellar l.
Peroneus longus m.
Soleus m.
Tibialis anterior m.
Interosseous membrane
Extensor digitorum longus m.
Extensor hallucis longus m.
Peroneus longus t.
Peroneus brevis m.
Tibialis anterior t.
Peroneus tertius m.
Inferior extensor retinaculum
Extensor digitorum brevis m.

Galea aponeurotica
Frontalis m.
Corrugator supercilii m.
Levator labii superioris alaeque nasi m.
**Auricularis muscles**
Superior
Anterior
Levator labii superioris m.
Zygomaticus minor m.
Levator anguli oris m.
Risorius m.
Depressor septi m.
Orbicularis oris m.
Mentalis m.
**Omohyoid muscle** Superior belly
Platysma m.
Sternocleidomastoid m.
Deltoid m.
Coracobrachialis m.
Latissimus dorsi m.
**Triceps brachii muscle**
Long head
Medial head
Lateral head
Biceps brachii m.
Brachialis m.
Brachialis m.
Bicipital aponeurosis
Biceps brachii t.
Supinator m.
Brachioradialis m.
Extensor carpi radialis longus m.
Pronator teres m.
Flexor carpi radialis m.
Palmaris longus m.
Flexor carpi ulnaris m.
Abductor pollicis longus m.
Flexor pollicis longus m.
Pronator quadratus m.
Flexor retinaculum
Palmar aponeurosis
Flexor digitorum superficialis m.
Gluteus medius m.
Tensor fasciae latae m.
Sartorius m.
Pectineus m.
**Adductor muscles**
Brevis
Longus
Magnus
Vastus lateralis m.
Iliotibial tract
Rectus femoris m.
Gastrocnemius m.
Tibialis anterior m.
Extensor digitorum longus m.
Peroneus longus m.
Soleus m.
Peroneus brevis m.
Extensor hallucis longus m.
Superior extensor retinaculum
Extensor digitorum longus tt.
Peroneus tertius t.

**Key**

- **l.** Ligament
- **ll.** Ligaments
- **m.** Muscle
- **mm.** Muscles
- **t.** Tendon
- **tt.** Tendons

**Key**

1. Subclavius m.
2. External intercostal mm.
3. Pectoralis minor m.
4. Serratus anterior m.
5. Pectoralis major m.
6. Rectus sheath (anterior layer)
7. Rectus abdominis m.
8. External abdominal oblique m.
9. Internal abdominal oblique m.
10. Transversus abdominis m.
11. Rectus sheath (posterior layer)
12. Arcuate line
13. Cremaster m.
14. Linea alba
15. Aponeurosis of external abdominal oblique m.

## MUSCULAR SYSTEM
### Posterior view

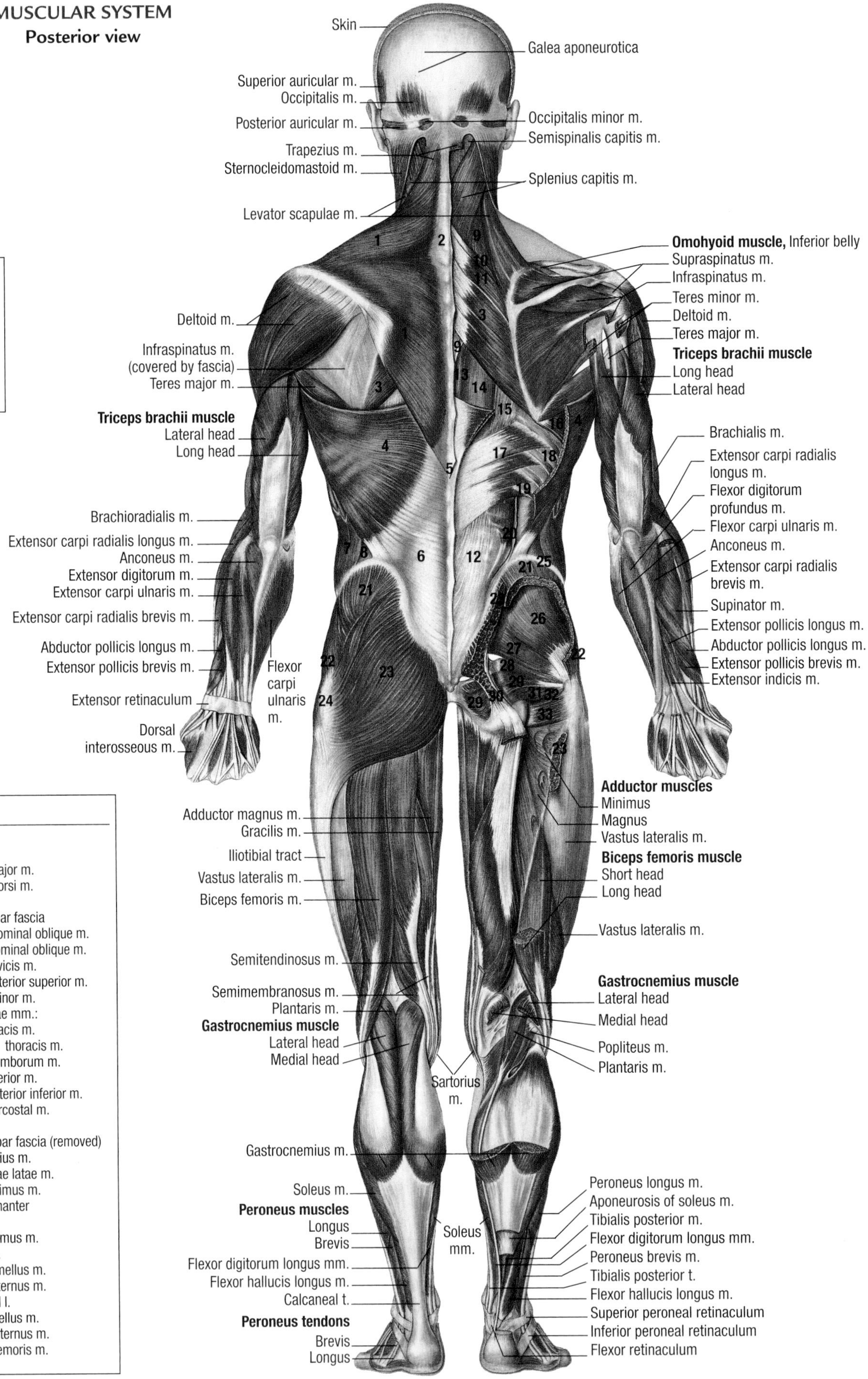

Skin
Galea aponeurotica
Superior auricular m.
Occipitalis m.
Posterior auricular m.
Occipitalis minor m.
Semispinalis capitis m.
Trapezius m.
Sternocleidomastoid m.
Splenius capitis m.
Levator scapulae m.
**Omohyoid muscle,** Inferior belly
Supraspinatus m.
Infraspinatus m.
Teres minor m.
Deltoid m.
Teres major m.
**Triceps brachii muscle**
Long head
Lateral head
Deltoid m.
Infraspinatus m. (covered by fascia)
Teres major m.
**Triceps brachii muscle**
Lateral head
Long head
Brachialis m.
Extensor carpi radialis longus m.
Flexor digitorum profundus m.
Flexor carpi ulnaris m.
Anconeus m.
Extensor carpi radialis brevis m.
Supinator m.
Extensor pollicis longus m.
Abductor pollicis longus m.
Extensor pollicis brevis m.
Extensor indicis m.
Brachioradialis m.
Extensor carpi radialis longus m.
Anconeus m.
Extensor digitorum m.
Extensor carpi ulnaris m.
Extensor carpi radialis brevis m.
Abductor pollicis longus m.
Extensor pollicis brevis m.
Extensor retinaculum
Flexor carpi ulnaris m.
Dorsal interosseous m.
**Adductor muscles**
Minimus
Magnus
Vastus lateralis m.
**Biceps femoris muscle**
Short head
Long head
Vastus lateralis m.
Adductor magnus m.
Gracilis m.
Iliotibial tract
Vastus lateralis m.
Biceps femoris m.
Semitendinosus m.
Semimembranosus m.
Plantaris m.
**Gastrocnemius muscle**
Lateral head
Medial head
Sartorius m.
**Gastrocnemius muscle**
Lateral head
Medial head
Popliteus m.
Plantaris m.
Gastrocnemius m.
Soleus m.
**Peroneus muscles**
Longus
Brevis
Soleus mm.
Flexor digitorum longus mm.
Flexor hallucis longus m.
Calcaneal t.
**Peroneus tendons**
Brevis
Longus
Peroneus longus m.
Aponeurosis of soleus m.
Tibialis posterior m.
Flexor digitorum longus mm.
Peroneus brevis m.
Tibialis posterior t.
Flexor hallucis longus m.
Superior peroneal retinaculum
Inferior peroneal retinaculum
Flexor retinaculum

**Key**

- **l.** Ligament
- **ll.** Ligaments
- **m.** Muscle
- **mm.** Muscles
- **t.** Tendon
- **tt.** Tendons

**Key**

1. Trapezius m.
2. Spine of C7
3. Rhomboid major m.
4. Latissimus dorsi m.
5. Spine of T12
6. Thoracolumbar fascia
7. External abdominal oblique m.
8. Internal abdominal oblique m.
9. Splenius cervicis m.
10. Serratus posterior superior m.
11. Rhomboid minor m.
12. Erector spinae mm.:
13. Spinalis thoracis m.
14. Longissimus thoracis m.
15. Iliocostalis lumborum m.
16. Serratus anterior m.
17. Serratus posterior inferior m.
18. External intercostal m.
19. 12th rib
20. Thoracolumbar fascia (removed)
21. Gluteus medius m.
22. Tensor fasciae latae m.
23. Gluteus maximus m.
24. Greater trochanter
25. Iliac crest
26. Gluteus minimus m.
27. Piriformis m.
28. Superior gemellus m.
29. Obturator internus m.
30. Sacrotuberal l.
31. Inferior gemellus m.
32. Obturator externus m.
33. Quadratus femoris m.

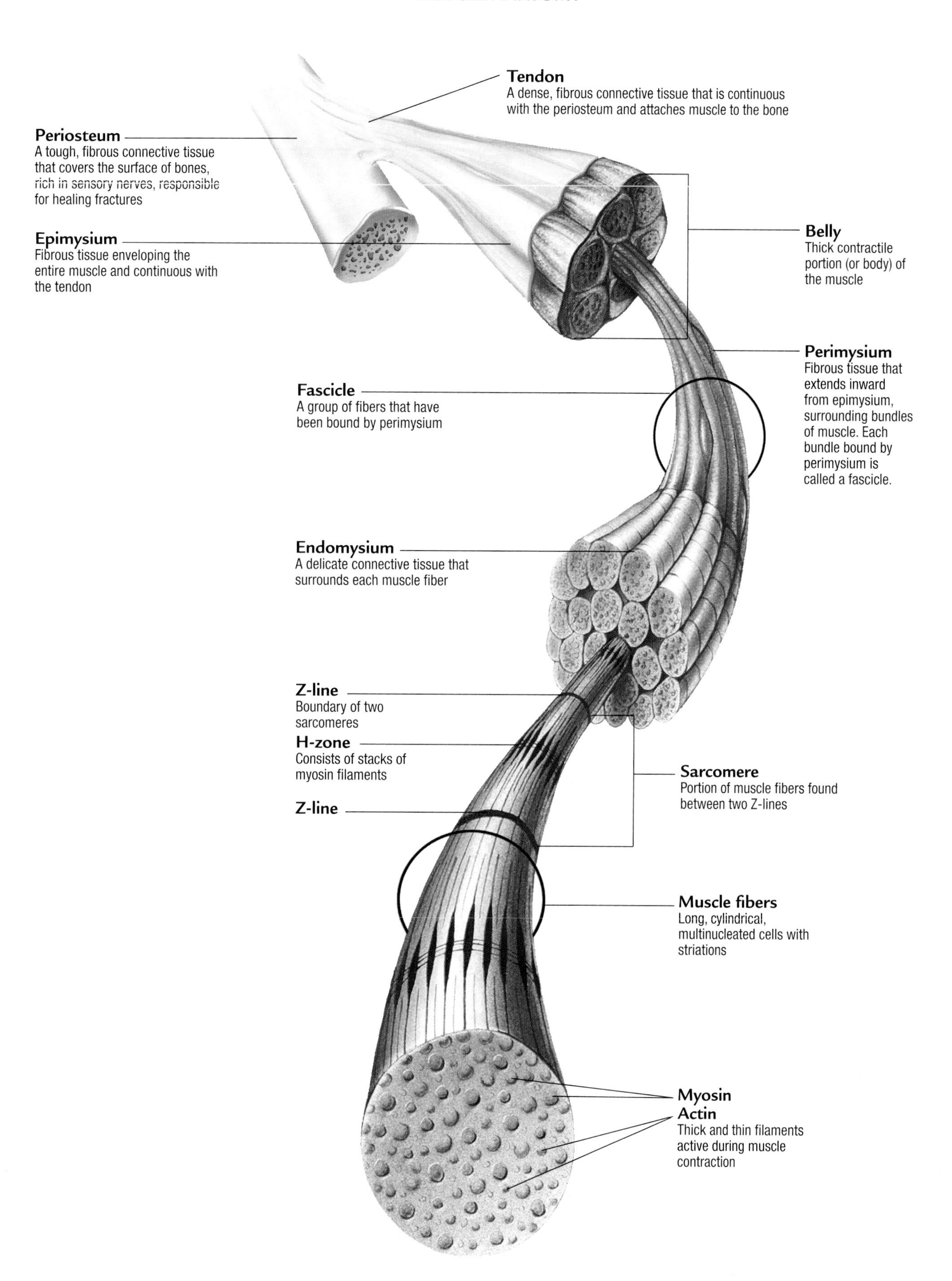
Tendon
A dense, fibrous connective tissue that is continuous with the periosteum and attaches muscle to the bone
Periosteum
A tough, fibrous connective tissue that covers the surface of bones, rich in sensory nerves, responsible for healing fractures
Epimysium
Fibrous tissue enveloping the entire muscle and continuous with the tendon
Belly
Thick contractile portion (or body) of the muscle
Perimysium
Fibrous tissue that extends inward from epimysium, surrounding bundles of muscle. Each bundle bound by perimysium is called a fascicle.
Fascicle
A group of fibers that have been bound by perimysium
Endomysium
A delicate connective tissue that surrounds each muscle fiber
Z-line
Boundary of two sarcomeres
H-zone
Consists of stacks of myosin filaments
Z-line
Sarcomere
Portion of muscle fibers found between two Z-lines
Muscle fibers
Long, cylindrical, multinucleated cells with striations
Myosin
Actin
Thick and thin filaments active during muscle contraction

# Nervous system

The nervous system allows communication among different parts of the body and between the body and the external environment. Nervous tissue is composed of densely packed and intertwined cells. The cells fall into two categories – neurons and neuroglia – which are discussed in Part I, Overview.

## Function

The chief functions of the nervous system are to monitor, integrate, and respond to environmental stimuli. These tasks are carried out in three general steps, summarized as follows:
• Sensory receptors detect changes inside and outside the body
• The central nervous system (CNS) processes (integrates) information from the sensory receptors
• The appropriate muscle or gland responds to information from the CNS.

The nervous and endocrine systems interact to maintain homeostasis. Although certain functions are controlled chiefly by the nervous system and others by the endocrine system, the two systems generally work together; signals from either system affect functions in the other system.

**Clinical tip**

When assessing a patient's neurologic system, always use an orderly approach. Begin with the highest levels of neurologic function and work down to the lowest in each of these five areas:
• mental status and speech
• cranial nerve function
• sensory function
• motor function
• reflexes.

Anatomically, the nervous system is divided into the central nervous system (CNS) and the peripheral nervous system. Functionally, it is divided into the sensory (afferent) division, which conveys impulses from the periphery to the CNS, and the motor (efferent) division, which conveys impulses from the CNS to the periphery. Mixed nerves are both efferent and afferent.

**Physiology**

The action of neurons is responsible for neurotransmission —conduction of electrochemical impulses throughout the nervous system. Neuron activity may be provoked by the following types of stimuli:
• mechanical, such as touch or pressure
• thermal, such as heat or cold
• external chemical, such as environmental toxins
• internal chemical, released by the body (such as histamine).

## Central nervous system

The CNS includes the brain and spinal cord, which together collect and interpret voluntary and involuntary motor and sensory stimuli. The spinal cord provides pathways for nerve impulses to and from the brain:
• *ascending tracts* – bundles of sensory nerve fibers that conduct impulses toward the brain
• *descending tracts* – bundles of motor nerve fibers that conduct impulses down the cord.

## Peripheral nervous system

The peripheral nervous system consists of the cranial nerves, the spinal nerves, and the autonomic nervous system.
• 12 pairs of cranial nerves carry motor or sensory messages or both, primarily between the brain or brain stem and the head and neck. (See Appendix A, *Cranial nerves.* See also Part II, Head and neck, for additional information.)
• 31 pairs of spinal nerves are mixed nerves; each consists of afferent (sensory) and efferent (motor) neurons that carry messages to and from particular body regions, called dermatomes. (See *Spinal nerves*, page 310.)

A dermatome is the cutaneous area (area of skin) supplied by nerve fibers from a single dorsal root and its ganglion. Assessment of a dermatome can help determine which portion of the spinal cord or spinal nerve is not functioning properly. (See *Dermatomes*, page 311.) Actual boundaries of dermatomes are variable, and dermatomes may overlap.

Spinal nerves intermix considerably during the formation of peripheral nerves. The dorsal and ventral rami (branches of spinal nerves) remain somewhat distinct in the thoracic region, but the ventral rami form extensive plexuses in the cervical and lumbosacral regions. Most peripheral nerves contain fibers from two, three, four, or five ventral rami. Consequently, the cutaneous areas supplied by the peripheral nerves do not correspond to the dermatomal distributions of the dorsal roots.

## Autonomic nervous system

The autonomic nervous system contains motor neurons that regulate the activities of the visceral organs and affect the smooth and cardiac muscles and glands. It involves both the central and peripheral nervous systems and consists of two parts:
• sympathetic – controls fight-or-flight reactions
• parasympathetic – maintains baseline body functions, such as heart rate, blood pressure, and temperature.
This system is discussed in detail on page 312.

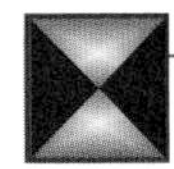

**Age-related changes**

Various degenerative changes in neurons can have such effects as:
• diminished reflexes
• decreased hearing, vision, taste, and smell
• lowed reaction time
• decreased vibratory sense in the ankles.

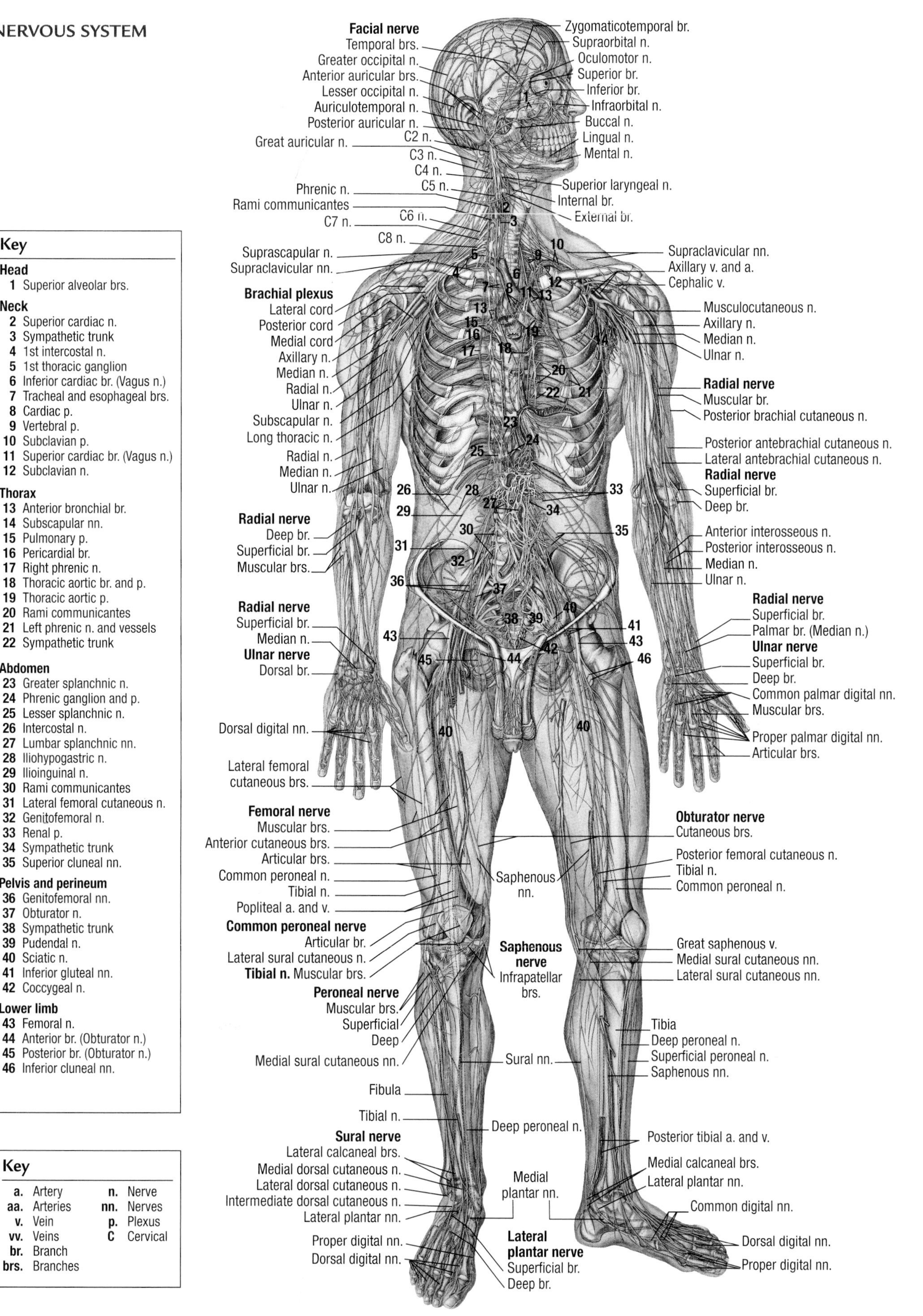
Facial nerve
Temporal brs.
Greater occipital n.
Anterior auricular brs.
Lesser occipital n.
Auriculotemporal n.
Posterior auricular n.
Great auricular n.
C2 n.
C3 n.
C4 n.
C5 n.
Phrenic n.
Rami communicantes
C6 n.
C7 n.
C8 n.
Suprascapular n.
Supraclavicular nn.
Brachial plexus
Lateral cord
Posterior cord
Medial cord
Axillary n.
Median n.
Radial n.
Ulnar n.
Subscapular n.
Long thoracic n.
Radial n.
Median n.
Ulnar n.
Radial nerve
Deep br.
Superficial br.
Muscular brs.
Radial nerve
Superficial br.
Median n.
Ulnar nerve
Dorsal br.
Dorsal digital nn.
Lateral femoral cutaneous brs.
Femoral nerve
Muscular brs.
Anterior cutaneous brs.
Articular brs.
Common peroneal n.
Tibial n.
Popliteal a. and v.
Common peroneal nerve
Articular br.
Lateral sural cutaneous n.
Tibial n. Muscular brs.
Peroneal nerve
Muscular brs.
Superficial
Deep
Medial sural cutaneous nn.
Fibula
Tibial n.
Sural nerve
Lateral calcaneal brs.
Medial dorsal cutaneous n.
Lateral dorsal cutaneous n.
Intermediate dorsal cutaneous n.
Lateral plantar nn.
Proper digital nn.
Dorsal digital nn.
Zygomaticotemporal br.
Supraorbital n.
Oculomotor n.
Superior br.
Inferior br.
Infraorbital n.
Buccal n.
Lingual n.
Mental n.
Superior laryngeal n.
Internal br.
External br.
Supraclavicular nn.
Axillary v. and a.
Cephalic v.
Musculocutaneous n.
Axillary n.
Median n.
Ulnar n.
Radial nerve
Muscular br.
Posterior brachial cutaneous n.
Posterior antebrachial cutaneous n.
Lateral antebrachial cutaneous n.
Radial nerve
Superficial br.
Deep br.
Anterior interosseous n.
Posterior interosseous n.
Median n.
Ulnar n.
Radial nerve
Superficial br.
Palmar br. (Median n.)
Ulnar nerve
Superficial br.
Deep br.
Common palmar digital nn.
Muscular brs.
Proper palmar digital nn.
Articular brs.
Obturator nerve
Cutaneous brs.
Posterior femoral cutaneous n.
Tibial n.
Common peroneal n.
Saphenous nn.
Saphenous nerve
Infrapatellar brs.
Great saphenous v.
Medial sural cutaneous nn.
Lateral sural cutaneous nn.
Tibia
Deep peroneal n.
Superficial peroneal n.
Saphenous nn.
Sural nn.
Deep peroneal n.
Posterior tibial a. and v.
Medial calcaneal brs.
Lateral plantar nn.
Medial plantar nn.
Common digital nn.
Lateral plantar nerve
Superficial br.
Deep br.
Dorsal digital nn.
Proper digital nn.
Key
Head
1 Superior alveolar brs.
Neck
2 Superior cardiac n.
3 Sympathetic trunk
4 1st intercostal n.
5 1st thoracic ganglion
6 Inferior cardiac br. (Vagus n.)
7 Tracheal and esophageal brs.
8 Cardiac p.
9 Vertebral p.
10 Subclavian p.
11 Superior cardiac br. (Vagus n.)
12 Subclavian n.
Thorax
13 Anterior bronchial br.
14 Subscapular nn.
15 Pulmonary p.
16 Pericardial br.
17 Right phrenic n.
18 Thoracic aortic br. and p.
19 Thoracic aortic p.
20 Rami communicantes
21 Left phrenic n. and vessels
22 Sympathetic trunk
Abdomen
23 Greater splanchnic n.
24 Phrenic ganglion and p.
25 Lesser splanchnic n.
26 Intercostal n.
27 Lumbar splanchnic nn.
28 Iliohypogastric n.
29 Ilioinguinal n.
30 Rami communicantes
31 Lateral femoral cutaneous n.
32 Genitofemoral n.
33 Renal p.
34 Sympathetic trunk
35 Superior cluneal nn.
Pelvis and perineum
36 Genitofemoral nn.
37 Obturator n.
38 Sympathetic trunk
39 Pudendal n.
40 Sciatic n.
41 Inferior gluteal nn.
42 Coccygeal n.
Lower limb
43 Femoral n.
44 Anterior br. (Obturator n.)
45 Posterior br. (Obturator n.)
46 Inferior cluneal nn.
Key
a. Artery
aa. Arteries
v. Vein
vv. Veins
br. Branch
brs. Branches
n. Nerve
nn. Nerves
p. Plexus
C Cervical

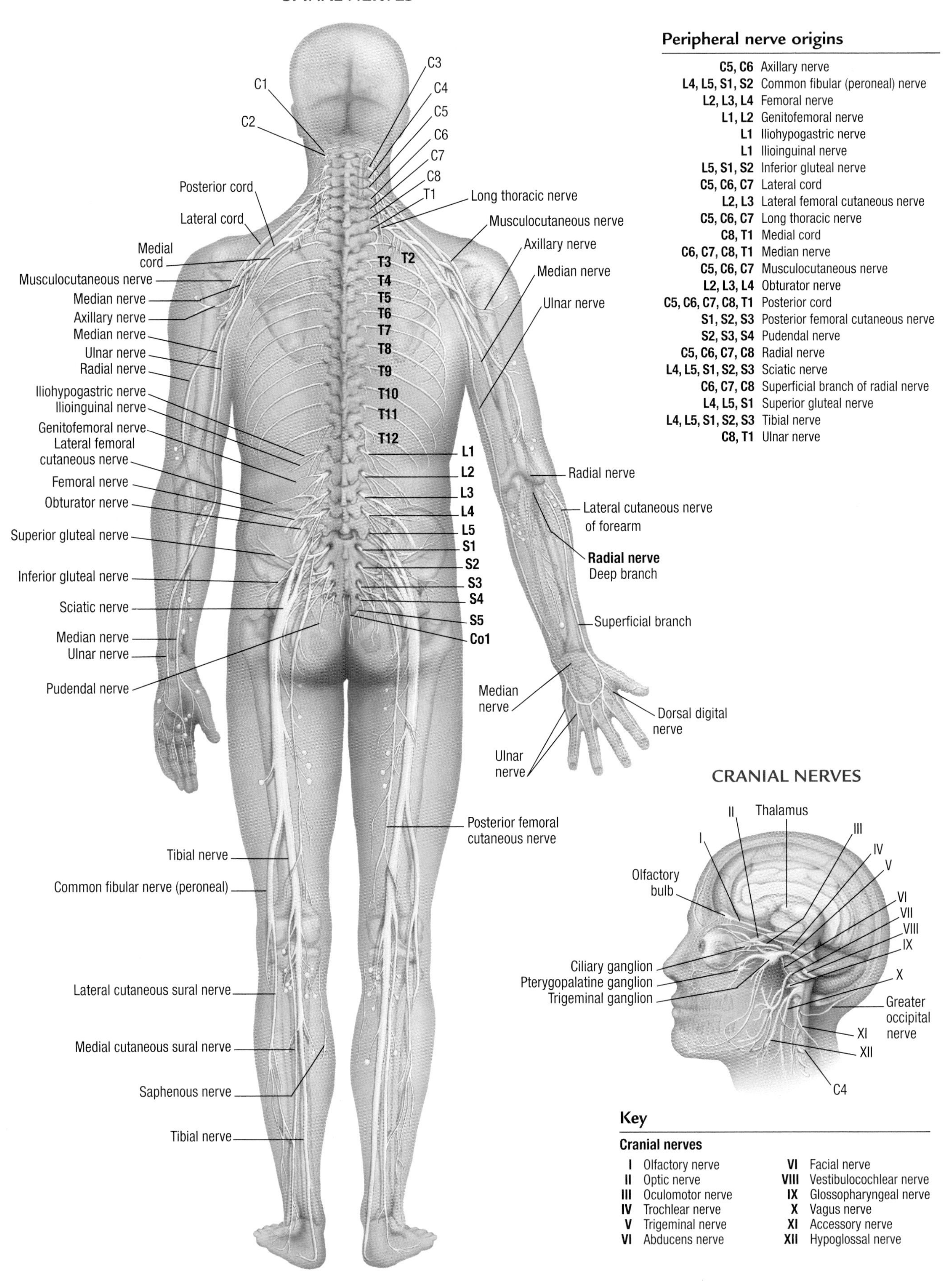
SPINAL NERVES
C1
C2
C3
C4
C5
C6
C7
C8
T1
Posterior cord
Lateral cord
Medial cord
Musculocutaneous nerve
Median nerve
Axillary nerve
Median nerve
Ulnar nerve
Radial nerve
Iliohypogastric nerve
Ilioinguinal nerve
Genitofemoral nerve
Lateral femoral cutaneous nerve
Femoral nerve
Obturator nerve
Superior gluteal nerve
Inferior gluteal nerve
Sciatic nerve
Median nerve
Ulnar nerve
Pudendal nerve
Long thoracic nerve
Musculocutaneous nerve
Axillary nerve
Median nerve
Ulnar nerve
T2
T3
T4
T5
T6
T7
T8
T9
T10
T11
T12
L1
L2
L3
L4
L5
S1
S2
S3
S4
S5
Co1
Radial nerve
Lateral cutaneous nerve of forearm
Radial nerve
Deep branch
Superficial branch
Median nerve
Dorsal digital nerve
Ulnar nerve
Posterior femoral cutaneous nerve
Tibial nerve
Common fibular nerve (peroneal)
Lateral cutaneous sural nerve
Medial cutaneous sural nerve
Saphenous nerve
Tibial nerve
Peripheral nerve origins
C5, C6 Axillary nerve
L4, L5, S1, S2 Common fibular (peroneal) nerve
L2, L3, L4 Femoral nerve
L1, L2 Genitofemoral nerve
L1 Iliohypogastric nerve
L1 Ilioinguinal nerve
L5, S1, S2 Inferior gluteal nerve
C5, C6, C7 Lateral cord
L2, L3 Lateral femoral cutaneous nerve
C5, C6, C7 Long thoracic nerve
C8, T1 Medial cord
C6, C7, C8, T1 Median nerve
C5, C6, C7 Musculocutaneous nerve
L2, L3, L4 Obturator nerve
C5, C6, C7, C8, T1 Posterior cord
S1, S2, S3 Posterior femoral cutaneous nerve
S2, S3, S4 Pudendal nerve
C5, C6, C7, C8 Radial nerve
L4, L5, S1, S2, S3 Sciatic nerve
C6, C7, C8 Superficial branch of radial nerve
L4, L5, S1 Superior gluteal nerve
L4, L5, S1, S2, S3 Tibial nerve
C8, T1 Ulnar nerve
CRANIAL NERVES
II
Thalamus
I
III
IV
V
Olfactory bulb
VI
VII
VIII
IX
Ciliary ganglion
Pterygopalatine ganglion
Trigeminal ganglion
X
Greater occipital nerve
XI
XII
C4
Key
Cranial nerves
I Olfactory nerve
II Optic nerve
III Oculomotor nerve
IV Trochlear nerve
V Trigeminal nerve
VI Abducens nerve
VI Facial nerve
VIII Vestibulocochlear nerve
IX Glossopharyngeal nerve
X Vagus nerve
XI Accessory nerve
XII Hypoglossal nerve

## DERMATOMES
### Cutaneous areas of peripheral nerve innervation

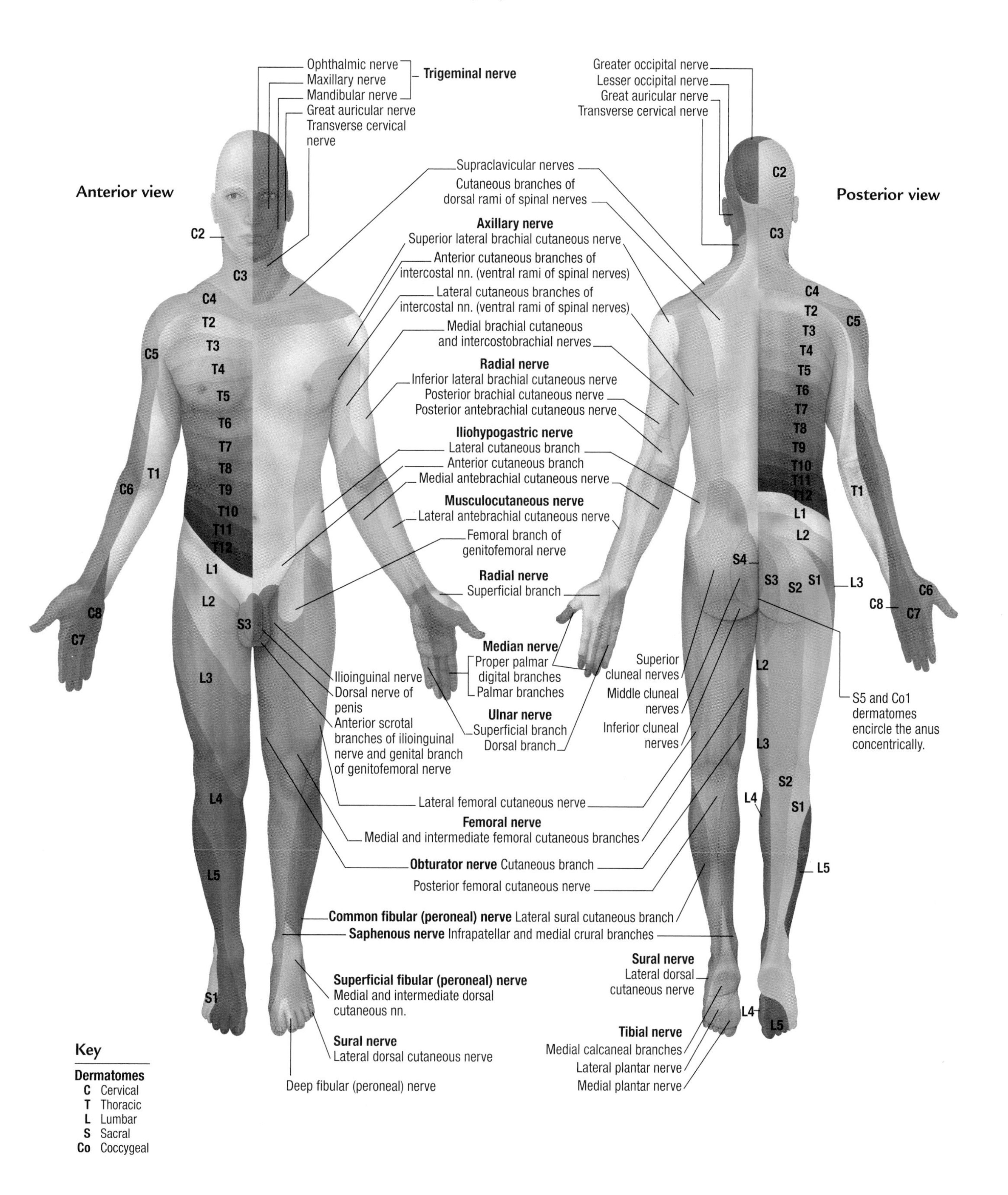

Anterior view
Posterior view
Ophthalmic nerve
Maxillary nerve
Mandibular nerve
Trigeminal nerve
Great auricular nerve
Transverse cervical nerve
Greater occipital nerve
Lesser occipital nerve
Great auricular nerve
Transverse cervical nerve
Supraclavicular nerves
Cutaneous branches of dorsal rami of spinal nerves
Axillary nerve
Superior lateral brachial cutaneous nerve
Anterior cutaneous branches of intercostal nn. (ventral rami of spinal nerves)
Lateral cutaneous branches of intercostal nn. (ventral rami of spinal nerves)
Medial brachial cutaneous and intercostobrachial nerves
Radial nerve
Inferior lateral brachial cutaneous nerve
Posterior brachial cutaneous nerve
Posterior antebrachial cutaneous nerve
Iliohypogastric nerve
Lateral cutaneous branch
Anterior cutaneous branch
Medial antebrachial cutaneous nerve
Musculocutaneous nerve
Lateral antebrachial cutaneous nerve
Femoral branch of genitofemoral nerve
Radial nerve
Superficial branch
Median nerve
Proper palmar digital branches
Palmar branches
Ulnar nerve
Superficial branch
Dorsal branch
Ilioinguinal nerve
Dorsal nerve of penis
Anterior scrotal branches of ilioinguinal nerve and genital branch of genitofemoral nerve
Superior cluneal nerves
Middle cluneal nerves
Inferior cluneal nerves
S5 and Co1 dermatomes encircle the anus concentrically.
Lateral femoral cutaneous nerve
Femoral nerve
Medial and intermediate femoral cutaneous branches
Obturator nerve Cutaneous branch
Posterior femoral cutaneous nerve
Common fibular (peroneal) nerve Lateral sural cutaneous branch
Saphenous nerve Infrapatellar and medial crural branches
Superficial fibular (peroneal) nerve
Medial and intermediate dorsal cutaneous nn.
Sural nerve
Lateral dorsal cutaneous nerve
Deep fibular (peroneal) nerve
Sural nerve
Lateral dorsal cutaneous nerve
Tibial nerve
Medial calcaneal branches
Lateral plantar nerve
Medial plantar nerve
C2
C3
C4
T2
T3
T4
T5
T6
T7
T8
T9
T10
T11
T12
L1
L2
L3
L4
L5
S1
S2
S3
S4
C5
C6
C7
C8
T1
Key
Dermatomes
C Cervical
T Thoracic
L Lumbar
S Sacral
Co Coccygeal

# Autonomic nervous system

The vast autonomic nervous system (ANS) innervates all internal organs, smooth muscles, glands, and end organs. Sometimes known as visceral efferent nerves, the nerves of the ANS carry messages to the viscera from the brain stem and neuroendocrine regulatory centers. The ANS has two major functional and anatomic subdivisions: sympathetic (thoracolumbar) and parasympathetic (craniosacral).

## Sympathetic nervous system

Sympathetic nerves called *preganglionic neurons* originate in the lateral gray horns and leave the spinal cord between the levels of the first thoracic and second lumbar vertebrae, where they enter a chain of small ganglia near the cord. The ganglia spread impulses to *postganglionic* neurons, which reach many organs and glands and can cause widespread, generalized physiologic responses, such as:

- vasoconstriction
- elevated blood pressure
- enhanced blood flow to skeletal muscles
- increased heart rate and contractility
- increased respiratory rate
- smooth-muscle relaxation in bronchioles, GI tract, and urinary tract
- sphincter contraction
- pupillary dilation and ciliary muscle relaxation
- increased sweat gland secretion
- reduced pancreatic secretion.

## Parasympathetic nervous system

Fibers of the parasympathetic nervous system leave the CNS by way of the cranial nerves from the midbrain and medulla and the spinal nerves between the second and fourth sacral vertebrae (S2 to S4). The long preganglionic fiber of each parasympathetic nerve travels to a ganglion near a target organ or gland; the short postganglionic fiber enters the target. The result is a specific response involving only one organ or gland.

The parasympathetic system has little effect on mental or metabolic activity, but its effects on other body activities are myriad and include:

- reduced heart rate, contractility, and conduction velocity
- bronchial smooth-muscle constriction
- increased GI tract tone and peristalsis; sphincter relaxation
- increased bladder tone; sphincter relaxation
- vasodilation of external male genitalia, causing erection
- constricted pupils
- increased pancreatic, salivary, and lacrimal secretions.

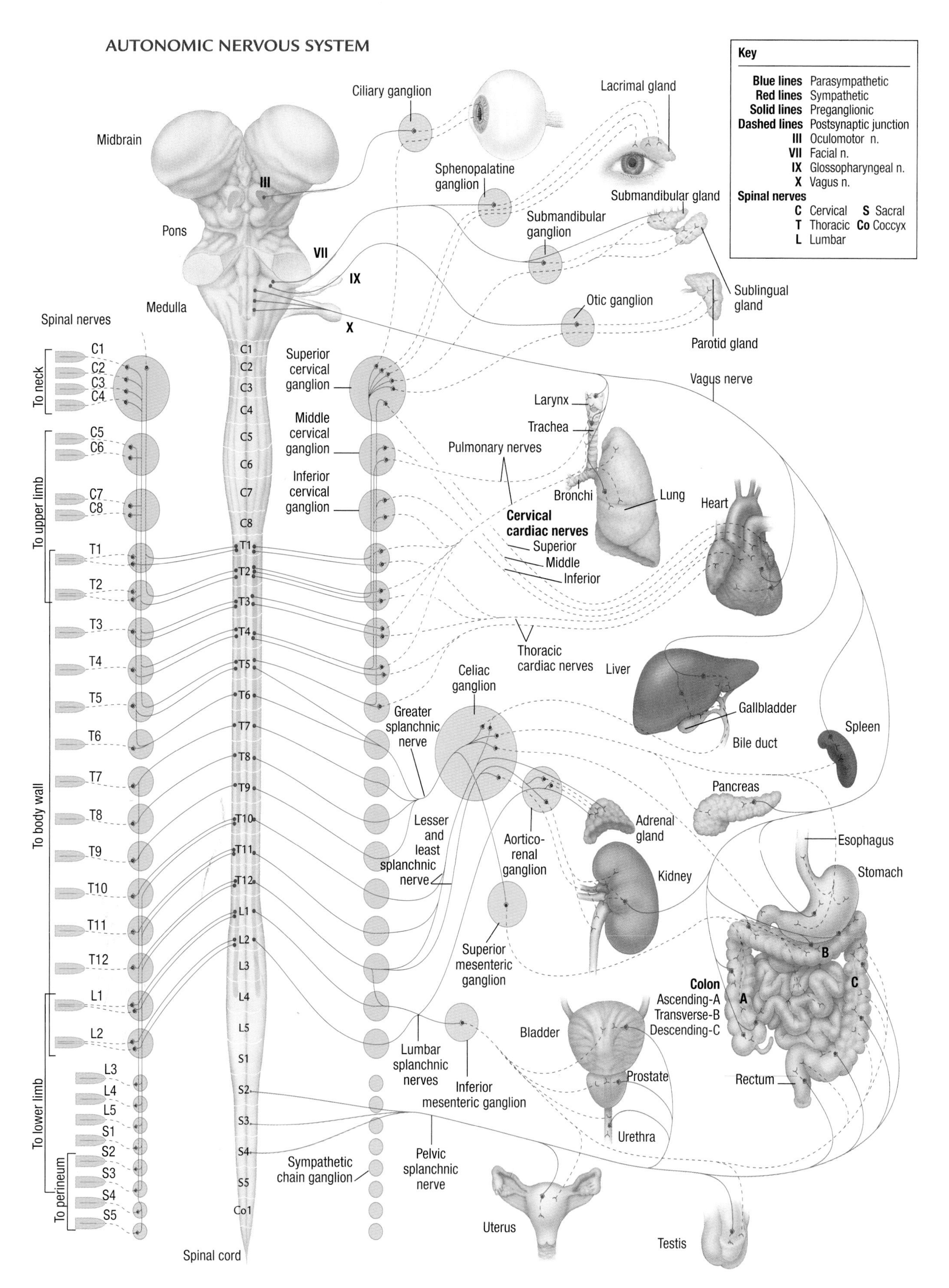

AUTONOMIC NERVOUS SYSTEM
Key
Blue lines Parasympathetic
Red lines Sympathetic
Solid lines Preganglionic
Dashed lines Postsynaptic junction
III Oculomotor n.
VII Facial n.
IX Glossopharyngeal n.
X Vagus n.
Spinal nerves
C Cervical S Sacral
T Thoracic Co Coccyx
L Lumbar
Midbrain
Pons
Medulla
III
VII
IX
X
Ciliary ganglion
Sphenopalatine ganglion
Lacrimal gland
Submandibular gland
Submandibular ganglion
Sublingual gland
Otic ganglion
Parotid gland
Spinal nerves
To neck
To upper limb
To body wall
To lower limb
To perineum
C1
C2
C3
C4
C5
C6
C7
C8
T1
T2
T3
T4
T5
T6
T7
T8
T9
T10
T11
T12
L1
L2
L3
L4
L5
S1
S2
S3
S4
S5
Co1
Spinal cord
Superior cervical ganglion
Middle cervical ganglion
Inferior cervical ganglion
Vagus nerve
Larynx
Trachea
Pulmonary nerves
Bronchi
Lung
Heart
Cervical cardiac nerves
Superior
Middle
Inferior
Thoracic cardiac nerves
Celiac ganglion
Greater splanchnic nerve
Liver
Gallbladder
Bile duct
Spleen
Pancreas
Adrenal gland
Kidney
Lesser and least splanchnic nerve
Aortico-renal ganglion
Superior mesenteric ganglion
Esophagus
Stomach
Colon
Ascending-A
Transverse-B
Descending-C
A
B
C
Bladder
Prostate
Rectum
Urethra
Lumbar splanchnic nerves
Inferior mesenteric ganglion
Sympathetic chain ganglion
Pelvic splanchnic nerve
Uterus
Testis

# Cardiovascular system

The cardiovascular system (sometimes called the circulatory system) consists of the heart, blood vessels, and lymphatics. This network brings life-sustaining oxygen and nutrients to the body's cells, removes metabolic waste products, and carries hormones from one part of the body to another. For information on the lymphatic system, see page 318.

As blood courses through the vascular system, it travels through five distinct types of blood vessels: arteries, arterioles, capillaries, venules, and veins. The structure of each type of vessel differs with its function in the cardiovascular system and the pressure exerted by the volume of blood at various sites in the system.

**Clinical tip**

If you have trouble hearing Korotkoff's sounds, which reflect the systolic and diastolic pressures, try one of these methods:

- Raise the arm.

Palpate for the brachial pulse and mark its location with a pen to avoid losing the pulse spot. Apply the cuff and have the patient raise the arm above his head. Then inflate the cuff to about 30 mm Hg above the patient's systolic pressure. Lower his arm until the cuff reaches heart level, deflate the cuff, and take a reading.

- Make a fist.

Position the patient's arm at heart level. Inflate the cuff to 30 mm Hg above the patient's systolic pressure and ask the patient to make a fist. Have the patient rapidly open and close his hand about 10 times; then deflate the cuff and take the reading.

## Vessel structure

Differences in blood pressure are reflected in vessel structure. (See *Artery wall* and *Vein wall and valve*, page 317.)

- arteries – thick, muscular walls to accommodate the flow of blood at high speeds and pressures
- arterioles – thinner walls that constrict or dilate as needed to control blood flow to the capillaries
- capillaries – walls composed of only a single layer of endothelial cells
- venules – receive blood from capillaries; walls thinner than those of arterioles
- veins – thinner walls but larger diameters than arteries; maintain low blood pressures required for return to heart.

## Blood circulation

About 60,000 miles of arteries, arterioles, capillaries, venules, and veins keep blood circulating to and from every functioning cell in the body. This circulatory network has two loops: pulmonary circulation and systemic circulation. Blood flowing through the chambers doesn't exchange oxygen and other nutrients with myocardial cells. Instead, a specialized part of the systemic circulation, the coronary circulation, supplies blood to the heart.

### Pulmonary circulation

Blood travels to the lungs to pick up oxygen and release carbon dioxide as follows:

- Contraction of the heart forces unoxygenated blood out of the right ventricle through the pulmonary semilunar valve into the pulmonary arteries
- Blood passes through progressively smaller arteries and arterioles into the capillaries of the lungs
- Blood reaches the alveoli and exchanges carbon dioxide for oxygen across the capillary wall
- Oxygenated blood returns via venules and veins to the pulmonary veins, which carry it back to the *left* atrium.

### Systemic circulation

Through the systemic circulation, blood carries oxygen and other nutrients to body cells as it picks up and transports waste products for excretion.

The major artery – the aorta – branches into vessels that supply specific organs and areas of the body.

- The left common carotid artery, left subclavian artery, and innominate artery arise from the arch of the aorta and supply blood to the brain, arms, and upper chest.
- As the aorta descends through the thorax and abdomen, its branches supply the organs of the gastrointestinal and genitourinary systems, the spinal column, and the lower chest and abdominal muscles.
- Then, at the level of the fourth lumbar vertebra, the aorta divides into the iliac arteries, which further divide into the external and internal iliac arteries.

At the end of the arterioles and the beginning of the capillaries, strong sphincters control blood flow into the tissues. These sphincters dilate to permit more flow when needed, close to shunt blood to other areas, or constrict to increase blood pressure.

As the arteries divide into smaller units, the number of vessels increases dramatically, thereby increasing the area of tissue to which blood flows (also called the area of perfusion). Although the capillary bed contains the smallest vessels, it supplies blood to the largest area. The extremely low capillary pressure permits the exchange of nutrients, oxygen, and carbon dioxide with body cells.

From the capillaries, blood flows into venules and, eventually, into veins. Valves in the veins prevent blood backflow, and the pumping action of skeletal muscles aids return of blood to the heart. The veins merge until they form two main branches, the superior vena cava and the inferior vena cava, which return blood to the right atrium.

# MAJOR ARTERIES

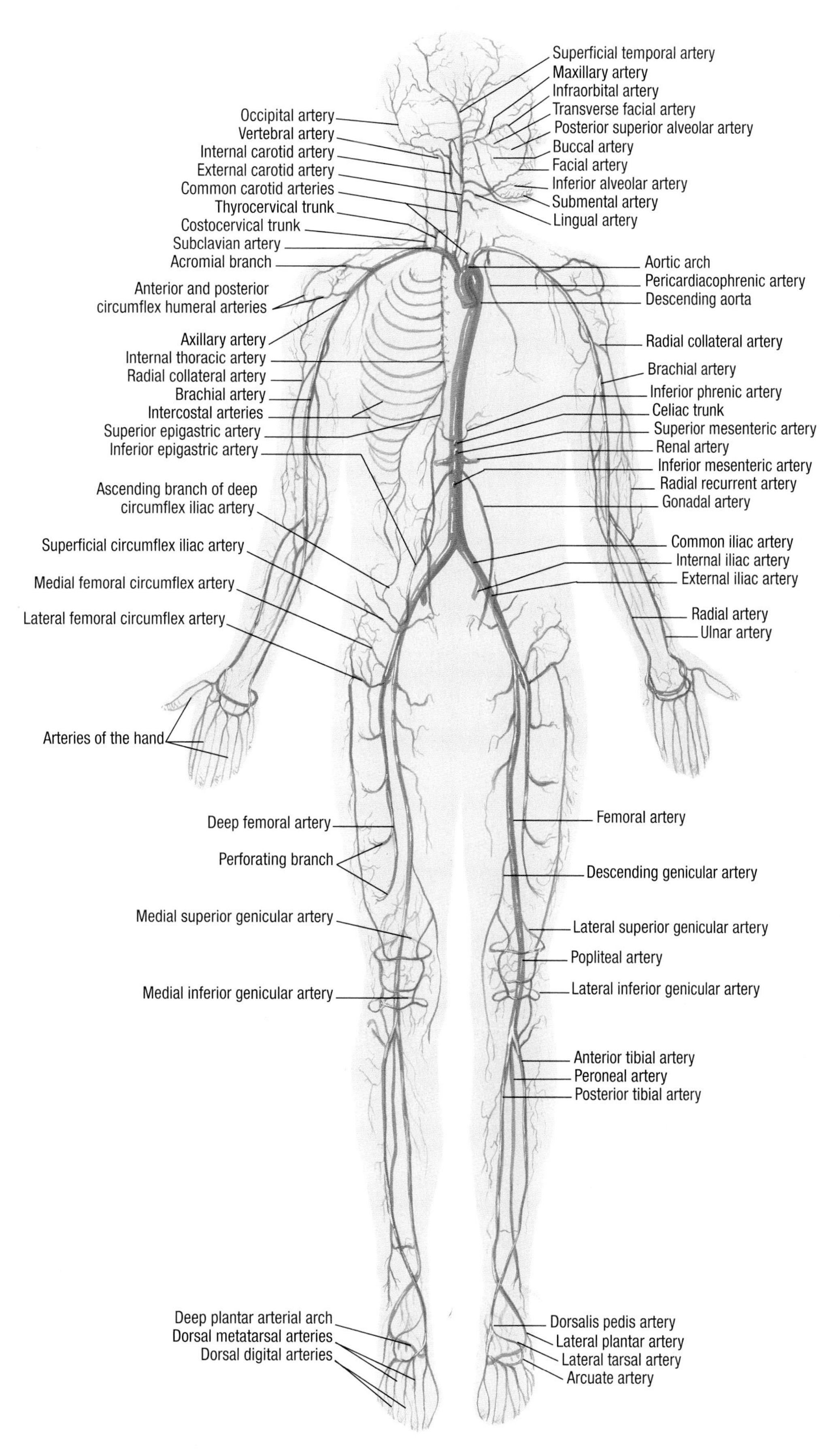
Superficial temporal artery
Maxillary artery
Infraorbital artery
Transverse facial artery
Posterior superior alveolar artery
Buccal artery
Facial artery
Inferior alveolar artery
Submental artery
Lingual artery
Occipital artery
Vertebral artery
Internal carotid artery
External carotid artery
Common carotid arteries
Thyrocervical trunk
Costocervical trunk
Subclavian artery
Acromial branch
Anterior and posterior circumflex humeral arteries
Axillary artery
Internal thoracic artery
Radial collateral artery
Brachial artery
Intercostal arteries
Superior epigastric artery
Inferior epigastric artery
Ascending branch of deep circumflex iliac artery
Superficial circumflex iliac artery
Medial femoral circumflex artery
Lateral femoral circumflex artery
Arteries of the hand
Aortic arch
Pericardiacophrenic artery
Descending aorta
Radial collateral artery
Brachial artery
Inferior phrenic artery
Celiac trunk
Superior mesenteric artery
Renal artery
Inferior mesenteric artery
Radial recurrent artery
Gonadal artery
Common iliac artery
Internal iliac artery
External iliac artery
Radial artery
Ulnar artery
Deep femoral artery
Perforating branch
Medial superior genicular artery
Medial inferior genicular artery
Femoral artery
Descending genicular artery
Lateral superior genicular artery
Popliteal artery
Lateral inferior genicular artery
Anterior tibial artery
Peroneal artery
Posterior tibial artery
Deep plantar arterial arch
Dorsal metatarsal arteries
Dorsal digital arteries
Dorsalis pedis artery
Lateral plantar artery
Lateral tarsal artery
Arcuate artery

## MAJOR VEINS

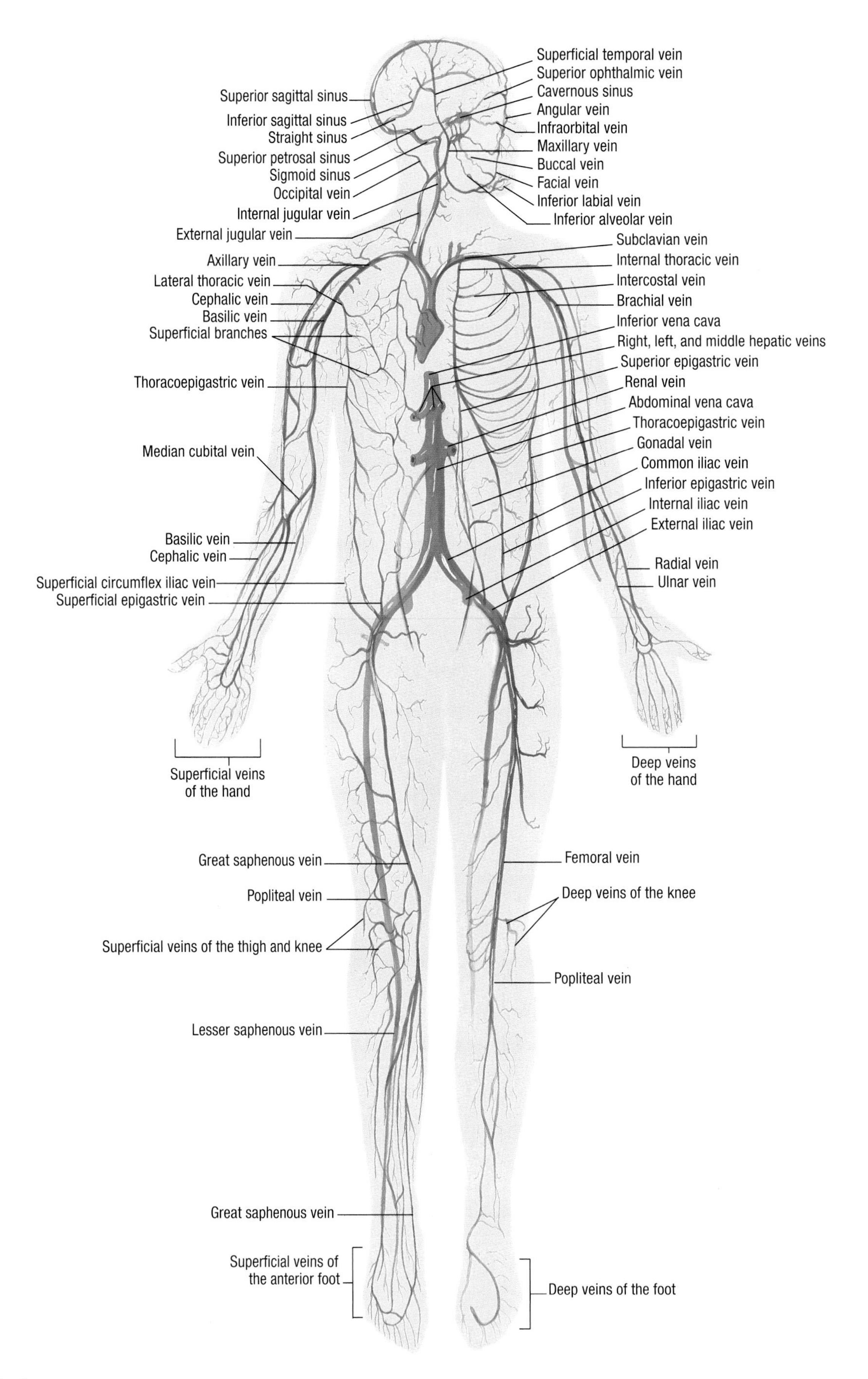

Superficial temporal vein
Superior ophthalmic vein
Cavernous sinus
Angular vein
Infraorbital vein
Maxillary vein
Buccal vein
Facial vein
Inferior labial vein
Inferior alveolar vein
Superior sagittal sinus
Inferior sagittal sinus
Straight sinus
Superior petrosal sinus
Sigmoid sinus
Occipital vein
Internal jugular vein
External jugular vein
Axillary vein
Lateral thoracic vein
Cephalic vein
Basilic vein
Superficial branches
Thoracoepigastric vein
Median cubital vein
Basilic vein
Cephalic vein
Superficial circumflex iliac vein
Superficial epigastric vein
Subclavian vein
Internal thoracic vein
Intercostal vein
Brachial vein
Inferior vena cava
Right, left, and middle hepatic veins
Superior epigastric vein
Renal vein
Abdominal vena cava
Thoracoepigastric vein
Gonadal vein
Common iliac vein
Inferior epigastric vein
Internal iliac vein
External iliac vein
Radial vein
Ulnar vein
Superficial veins of the hand
Deep veins of the hand
Great saphenous vein
Popliteal vein
Superficial veins of the thigh and knee
Lesser saphenous vein
Great saphenous vein
Superficial veins of the anterior foot
Femoral vein
Deep veins of the knee
Popliteal vein
Deep veins of the foot

## ARTERY WALL
**Cross section**

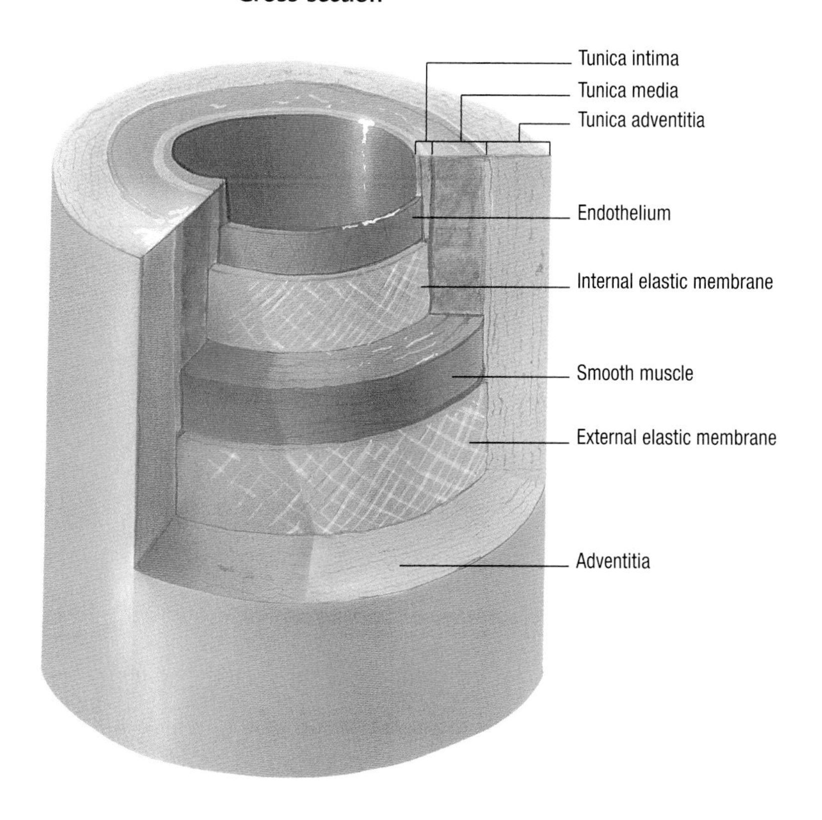
Tunica intima
Tunica media
Tunica adventitia
Endothelium
Internal elastic membrane
Smooth muscle
External elastic membrane
Adventitia

## VEIN WALL AND VALVE
**Cross section**

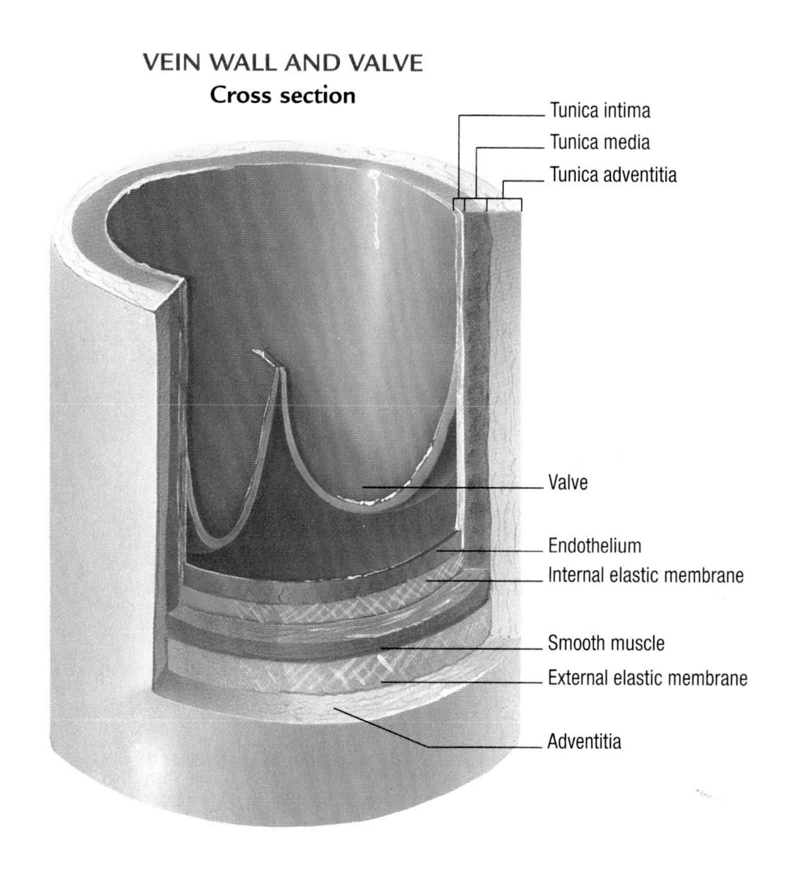
Tunica intima
Tunica media
Tunica adventitia
Valve
Endothelium
Internal elastic membrane
Smooth muscle
External elastic membrane
Adventitia

# Lymphatic system

The lymphatic system, a specialized component of the circulatory system, provides two important functions for the body:

- Collects lymph – excess extracellular fluid from body tissues – and eventually returns it to venous blood
- Defends the body against harmful organisms or chemical toxins.

Key structures of the lymphatic system are lymph nodes; lymphatic vessels; specialized lymphatic organs such as thymus, spleen, and tonsils; and isolated nodules of lymph tissue, such as Peyer's patches in the intestinal wall.

## Lymph nodes

The lymph nodes are small, oval structures that lie along the network of lymphatic vessels. Most abundant in the head, neck, axilla, abdomen, pelvis, and groin, they help remove and destroy antigens (substances capable of triggering an immune response) that circulate in the blood and lymph.

Each lymph node is enclosed by a fibrous capsule whose bands of connective tissue extend into the node and divide it into three compartments: superficial cortex, deep cortex, and medulla.

- The superficial cortex contains follicles made up predominantly of B lymphocytes.
- The deep cortex and interfollicular areas contain mostly T lymphocytes.
- The medulla contains numerous plasma cells, which actively secrete immunoglobulins.

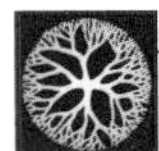

**Physiology**

The bone marrow and thymus play a role in the development of B cells and T cells – the two major types of lymphocytes. Lymphocytes differentiate to become either B cells (which mature in the bone marrow) or T cells (which travel to the thymus and mature there). B cells and T cells are distributed throughout the lymphoid organs, especially the lymph nodes and spleen. They travel through the blood system and the body's network of lymphatic vessels.
In the thymus, some T cells undergo a process called T-cell education, in which the cells are "trained" to recognize other cells from the same body (self cells) and to distinguish them from all other cells (nonself cells).

Usually, lymph travels through more than one lymph node because numerous nodes line the lymphatic vessels that drain a particular region. For example, axillary nodes (under the arms) filter drainage from the arms, and femoral nodes (in the inguinal region) filter drainage from the legs. This arrangement prevents organisms that enter peripheral areas from migrating unchallenged to central areas. Lymph nodes also serve as a primary source of circulating T lymphocytes, which provide specific immune responses.

## Lymph and lymphatic vessels

Lymph is a colorless liquid that flows through the lymphatic system. Lymph originates as fluid that has leaked from blood capillaries. It resembles blood plasma, and consists mainly of water with dissolved salts and protein. It circulates through the lymphatic vessels and is filtered by the lymph nodes.

Lymphatic vessels, typically located alongside blood vessels, form a network that extends throughout most of the body. Lymph seeps into the walls of small blind-ending lymphatic capillaries located throughout most of the body. Wider than blood capillaries, they permit lymph to flow in but not out. A network of lymphatic capillaries collecting lymph from tissues is termed a lymphatic capillary plexus. In most parts of the body, lymphatic vessels and lymphatic capillaries aid the function of veins and blood capillaries by draining many body tissues and enhancing blood return to the heart.

### Lymph drainage

Afferent lymphatic vessels carry lymph into the subcapsular sinus of the lymph node. From there, lymph flows through cortical sinuses and smaller radial medullary sinuses in the node. Phagocytic cells in the deep cortex and medullary sinuses attack antigens that may be carried in lymph. The follicles of the superficial cortex also may trap antigens.

Cleansed lymph leaves the node through efferent lymphatic vessels at the hilum of the node. These vessels drain into lymph node chains, which, in turn, empty into large lymph vessels, or trunks. Trunks empty into two main ducts:

- right lymphatic duct – drains the right side of the head and neck, the right upper limb and pectoral region, and the right half of the thoracic cavity
- thoracic duct – drains all other parts of the body.

Both ducts empty into the subclavian vein.

## Specialized organs and tissues

The tonsils, adenoids, appendix, thymus, and Peyer's patches remove foreign debris in much the same way that lymph nodes do. They are located in food and air passages, where microbial access is greatest.

The spleen is a primary organ for the storage of blood. It filters and removes bacteria and other foreign substances that enter the bloodstream; these substances are promptly removed by splenic phagocytes. Splenic phagocytes interact with lymphocytes to initiate an immune response. For more information about the spleen, see Part VI, Abdomen.

The thymus is a two-lobed mass of lymphoid tissue located over the base of the heart in the mediastinum. It activates T lymphocytes. It gradually atrophies until only a remnant persists in adults.

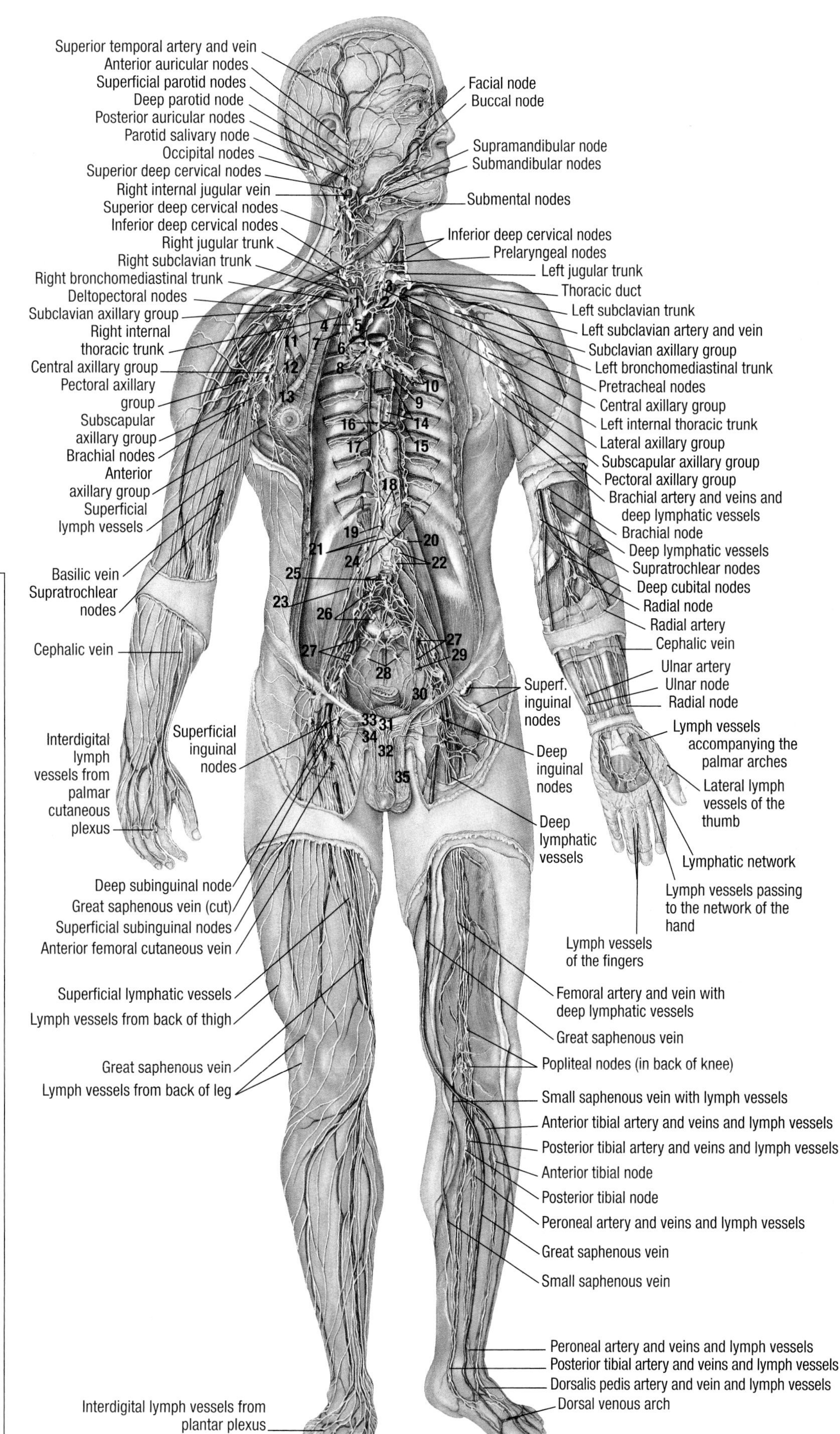

Superior temporal artery and vein
Anterior auricular nodes
Superficial parotid nodes
Deep parotid node
Posterior auricular nodes
Parotid salivary node
Occipital nodes
Superior deep cervical nodes
Right internal jugular vein
Superior deep cervical nodes
Inferior deep cervical nodes
Right jugular trunk
Right subclavian trunk
Right bronchomediastinal trunk
Deltopectoral nodes
Subclavian axillary group
Right internal thoracic trunk
Central axillary group
Pectoral axillary group
Subscapular axillary group
Brachial nodes
Anterior axillary group
Superficial lymph vessels
Basilic vein
Supratrochlear nodes
Cephalic vein
Interdigital lymph vessels from palmar cutaneous plexus
Superficial inguinal nodes
Deep subinguinal node
Great saphenous vein (cut)
Superficial subinguinal nodes
Anterior femoral cutaneous vein
Superficial lymphatic vessels
Lymph vessels from back of thigh
Great saphenous vein
Lymph vessels from back of leg
Interdigital lymph vessels from plantar plexus
Facial node
Buccal node
Supramandibular node
Submandibular nodes
Submental nodes
Inferior deep cervical nodes
Prelaryngeal nodes
Left jugular trunk
Thoracic duct
Left subclavian trunk
Left subclavian artery and vein
Subclavian axillary group
Left bronchomediastinal trunk
Pretracheal nodes
Central axillary group
Left internal thoracic trunk
Lateral axillary group
Subscapular axillary group
Pectoral axillary group
Brachial artery and veins and deep lymphatic vessels
Brachial node
Deep lymphatic vessels
Supratrochlear nodes
Deep cubital nodes
Radial node
Radial artery
Cephalic vein
Ulnar artery
Ulnar node
Radial node
Superf. inguinal nodes
Lymph vessels accompanying the palmar arches
Deep inguinal nodes
Lateral lymph vessels of the thumb
Deep lymphatic vessels
Lymphatic network
Lymph vessels passing to the network of the hand
Lymph vessels of the fingers
Femoral artery and vein with deep lymphatic vessels
Great saphenous vein
Popliteal nodes (in back of knee)
Small saphenous vein with lymph vessels
Anterior tibial artery and veins and lymph vessels
Posterior tibial artery and veins and lymph vessels
Anterior tibial node
Posterior tibial node
Peroneal artery and veins and lymph vessels
Great saphenous vein
Small saphenous vein
Peroneal artery and veins and lymph vessels
Posterior tibial artery and veins and lymph vessels
Dorsalis pedis artery and vein and lymph vessels
Dorsal venous arch
1
2
3
4
5
6
7
8
9
10
11
12
13
14
15
16
17
18
19
20
21
22
23
24
25
26
27
28
29
30
31
32
33
34
35


**Key**

1 Right brachiocephalic vein
2 Left brachiocephalic vein
3 Left common carotid artery
4 Anterior superior mediastinal nodes
5 Superior vena cava
6 Right cardiac lymph branch
7 Internal thoracic node
8 Right tracheobronchial nodes
9 Left tracheobronchial nodes
10 Right and left bronchopulmonary nodes
11 Internal thoracic lymph vessel ending in subclavicular nodes
12 Interpectoral nodes
13 Lymph vessels from deep part of breast
14 Posterior mediastinal nodes
15 Intercostal nodes and lymph vessels
16 Thoracic duct
17 Thoracic aorta
18 Descending right and left intercostal lymph trunks
19 Cisterna chyli
20 Intestinal trunk
21 Right and left lumbar trunks
22 Lumbar nodes
23 Testicular lymph vessels
24 Retroaortic node (lumbar nodes)
25 Preaortic node (lumbar nodes)
26 Common iliac nodes
27 Internal iliac artery and nodes
28 Sacral nodes
29 Lymph vessels to internal iliac nodes
30 Obturator vessels and nerve
31 Presymphysial node
32 Collecting lymph vessels from glans penis
33 Superficial lymph vessels from the penis
34 Lymph vessels from the scrotum
35 Lymph vessels of testis and epididymus

# Respiratory system

The respiratory system consists of the upper and lower respiratory tracts, the lungs, and the thoracic cage. Besides maintaining the exchange of oxygen and carbon dioxide in the lungs and tissues, the respiratory system helps to regulate the body's acid-base balance.

Any change in the respiratory system affects every other body system. Conversely, changes in other body systems may compromise the availability of oxygen. This section provides a brief overview of the respiratory system. For more detailed information on various structures of the respiratory system and gas exchange, see Part II, Head and neck, and Part V, Thorax.

## Upper respiratory tract

Structures of the upper respiratory tract include: nose, mouth, nasopharynx, oropharynx, laryngopharynx, and larynx. Besides warming and humidifying inhaled air, these structures provide for taste, smell, and the chewing and swallowing of food. In the upper respiratory tract, involuntary defense mechanisms – sneezing, coughing, gagging, and spasms – help to protect the respiratory system from infection and to prevent foreign body inhalation.

## Lower respiratory tract

The lower respiratory tract structures are the trachea, bronchi, and lungs. Bronchi branch into bronchioles, which in turn branch into lobules. The lobule includes the terminal bronchioles and alveoli. A mucous membrane containing hairlike cilia lines the lower tract. The constant movement of mucus by cilia cleans the tract and carries foreign matter upward for swallowing or expectoration. Functionally, the lower tract is subdivided into: conducting airways (the trachea and the primary, lobar, and segmental bronchi) and alveoli – the sites of gas exchange.

## Inspiration and expiration

Breathing involves two actions: inspiration, an active process, and expiration, a relatively passive one. Both actions rely on respiratory muscle function and the effects of pressure differences in the lungs.

### Normal respiration

During normal respiration, which is entirely passive, the external intercostal muscles aid the diaphragm, the major muscle of respiration. The diaphragm descends to lengthen the chest cavity; the external intercostal muscles – located between and along the lower borders of the ribs – contract to expand the anteroposterior diameter. This coordinated action pulls air into the lungs. When the diaphragm rises and the intercostal muscles relax, expiration occurs.

### Forced inspiration and active expiration

During exercise, when the body needs increased oxygenation, or in certain disease states that require forced inspiration and active expiration (such as pneumonia or respiratory failure), the accessory muscles of respiration also participate. The accessory muscles include:

- internal intercostals – on the inner surfaces of the ribs
- pectorals – in the upper chest
- sternocleidomastoids – on the sides of the neck
- scalenes – in the neck
- posterior trapezius – in the upper back
- abdominal rectus – in the anterior abdominal wall.

### External and internal respiration

Effective respiration has two components: gas exchange in the lungs, called external respiration, and gas exchange in the tissues, called internal respiration.

External respiration occurs through three processes:

- ventilation – gas transport into and out of the lungs
- pulmonary perfusion – blood flow from the right side of the heart, through the pulmonary circulation, and into the left side of the heart
- diffusion – gas movement through a semipermeable membrane from an area of greater concentration to one of lesser concentration.

All three processes are vital to maintaining adequate oxygenation and acid-base balance.

Internal respiration is the exchange of oxygen and carbon dioxide between blood capillaries and tissue cells. Internal respiration occurs only through diffusion.

## Acid-base balance

External and internal respiration together help maintain acid-base balance in the body. The systemic circulatory system carries oxygen-rich blood from the lungs – external respiration – to the tissues and exchanges oxygen for carbon dioxide produced by metabolism in body cells – internal respiration. Because carbon dioxide is more soluble than oxygen, it dissolves in the blood, where most of it forms bicarbonate (base) and the rest forms carbonic acid (acid).

### Respiratory responses

The lungs control levels of bicarbonate by breaking it down to carbon dioxide and water for respiratory excretion. In response to signals from the medulla, the lungs can change the rate and depth of breathing – and thereby the amount of carbon dioxide lost – to help maintain acid-base balance.

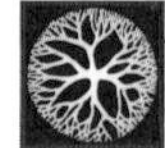

**Physiology**

In metabolic alkalosis – high blood bicarbonate levels – the rate and depth of ventilation decrease; the resulting retention of carbon dioxide increases carbonic acid levels. In metabolic acidosis – the result of excessive acid retention or bicarbonate – the rate and depth of ventilation increase to eliminate excess carbon dioxide, thus reducing carbonic acid levels.

Hypoventilation (reduced rate and depth of ventilation) leads to carbon dioxide retention and thereby causes respiratory acidosis.

Hyperventilation (increased rate and depth of ventilation) leads to less circulating carbon dioxide and causes respiratory alkalosis.

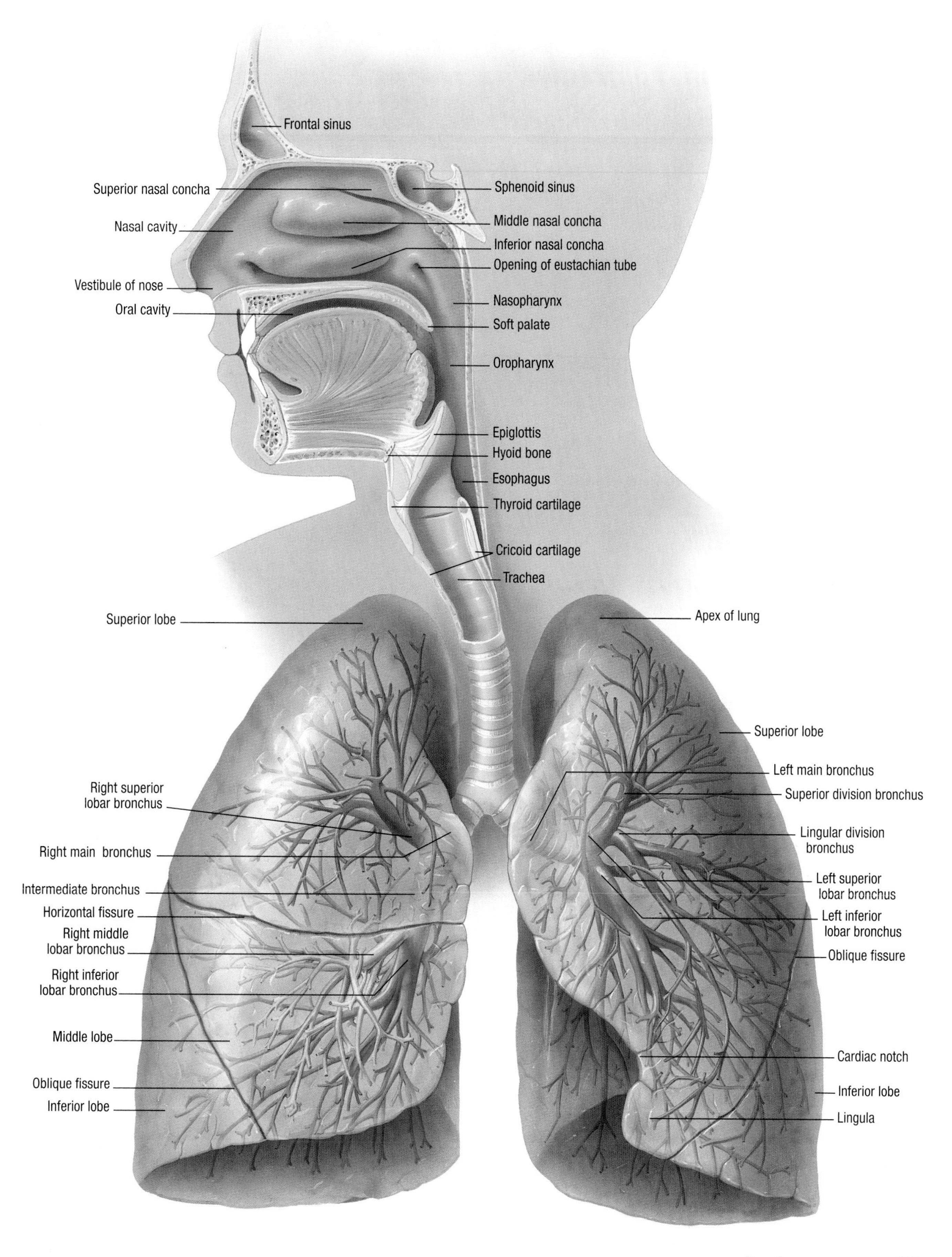
Frontal sinus
Superior nasal concha
Sphenoid sinus
Nasal cavity
Middle nasal concha
Inferior nasal concha
Opening of eustachian tube
Vestibule of nose
Oral cavity
Nasopharynx
Soft palate
Oropharynx
Epiglottis
Hyoid bone
Esophagus
Thyroid cartilage
Cricoid cartilage
Trachea
Superior lobe
Apex of lung
Superior lobe
Left main bronchus
Superior division bronchus
Right superior lobar bronchus
Lingular division bronchus
Right main bronchus
Left superior lobar bronchus
Intermediate bronchus
Horizontal fissure
Left inferior lobar bronchus
Right middle lobar bronchus
Oblique fissure
Right inferior lobar bronchus
Middle lobe
Cardiac notch
Oblique fissure
Inferior lobe
Inferior lobe
Lingula

# Gastrointestinal system

The GI system has two major components: the alimentary canal (also called the GI tract) and the accessory GI organs. Together, they serve two major functions:
- digesting food and fluid into simple chemicals that can be absorbed into the bloodstream and transported throughout the body
- eliminating waste products by excretion of feces.

This page provides a brief overview of the gastrointestinal system. For more detailed information on various structures and functions of the gastrointestinal system, see Part II, Head and neck, and Part VI, Abdomen.

## Alimentary canal

The alimentary canal is essentially a hollow muscular tube that begins in the mouth and extends to the anus. The structures of the alimentary canal include the:
- pharynx
- esophagus
- stomach
- small intestine
- large intestine
- rectum.

## Structures

The wall of the GI tract – from the lumen outward – consists of the following layers.
- mucosa – lines the lumen (also called the tunica mucosa)
- submucosa – surrounds the mucosa (also called tunica submucosa)
- tunica muscularis – surrounds the submucosa.
- visceral peritoneum – outer covering.

### Visceral peritoneum

The visceral peritoneum is the GI tract's outer covering. In the esophagus and rectum, it's also called the tunica adventitia; elsewhere in the GI tract, it's called the tunica serosa. The visceral peritoneum covers most of the abdominal organs and lies next to an identical layer – the parietal peritoneum – that lines the abdominal cavity.

The visceral peritoneum becomes a double-layered fold around the blood vessels, nerves, and lymphatics that supply the small intestine; it attaches the jejunum and ileum to the posterior abdominal wall to prevent twisting. A similar fold attaches the transverse colon to the posterior abdominal wall.

## Innervation

The submucosal plexus or the myenteric plexus (in the muscularis) stimulates transmission of nerve signals to the smooth muscle, which initiates peristalsis and mixing contractions. Parasympathetic stimulation via the vagus nerve (to most of the intestines) and the sacral spinal nerves (to the descending colon and rectum) increases gut and sphincter tone as well as the frequency, strength, and velocity of smooth-muscle contractions. Vagal stimulation also increases swallowing, visceral movement, and gastrin secretion. Sympathetic stimulation, by way of the spinal nerves from levels T6 to L2, slows peristalsis and inhibits GI activity.

## Accessory organs

Accessory organs – the liver, appendix, biliary duct system, and pancreas – contribute hormones, enzymes, and bile that are essential to digestion. They deliver their secretions to the duodenum through the hepatopancreatic ampulla, also called the papilla of Vater. The opening at the papilla of Vater is known as the sphincter of Oddi. There are also cells in the alimentary canal, specifically in the stomach and small intestine, that secrete hormones and enzymes. (See also Part VI, Abdomen.)

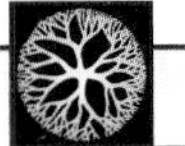

**Physiology**

### Digestion and elimination

Digestion starts in the oral cavity, where chewing, salivation, and swallowing all take place. As food passes through the esophagus, glands in the esophageal mucosal layer secrete mucus, which lubricates the bolus and protects the mucosal membrane from damage caused by poorly chewed foods. The bolus then enters the stomach.
- Stomach – The stomach's three major motor functions are storing food, mixing food with gastric juices to form chyme, and slowly parcelling chyme into the small intestine for further digestion and absorption.
- Small intestine – The small intestine performs most of the work of digestion and absorption. Here, intestinal contractions and various digestive secretions break down carbohydrates, proteins, and fats – actions that enable the intestinal mucosa to absorb these nutrients into the bloodstream (along with water and electrolytes) for subsequent use by the body.
- Large intestine – By the time chyme reaches the large intestine, it has been reduced to mostly indigestible substances. The bolus enters the large intestine where the ileum and cecum form the ileocecal pouch. The bolus then moves up the ascending colon, past the right abdominal cavity to the liver's lower border; crosses horizontally below the liver and stomach by way of the transverse colon; and travels to the iliac fossa through the descending colon.

From there, the bolus travels through the sigmoid colon to the lower midline of the abdominal cavity, then the rectum, and finally into the anal canal. The anus opens to the exterior through two sphincters. The internal sphincter is a thick, circular smooth muscle under autonomic control; the external sphincter is a ring of skeletal muscle under voluntary control.

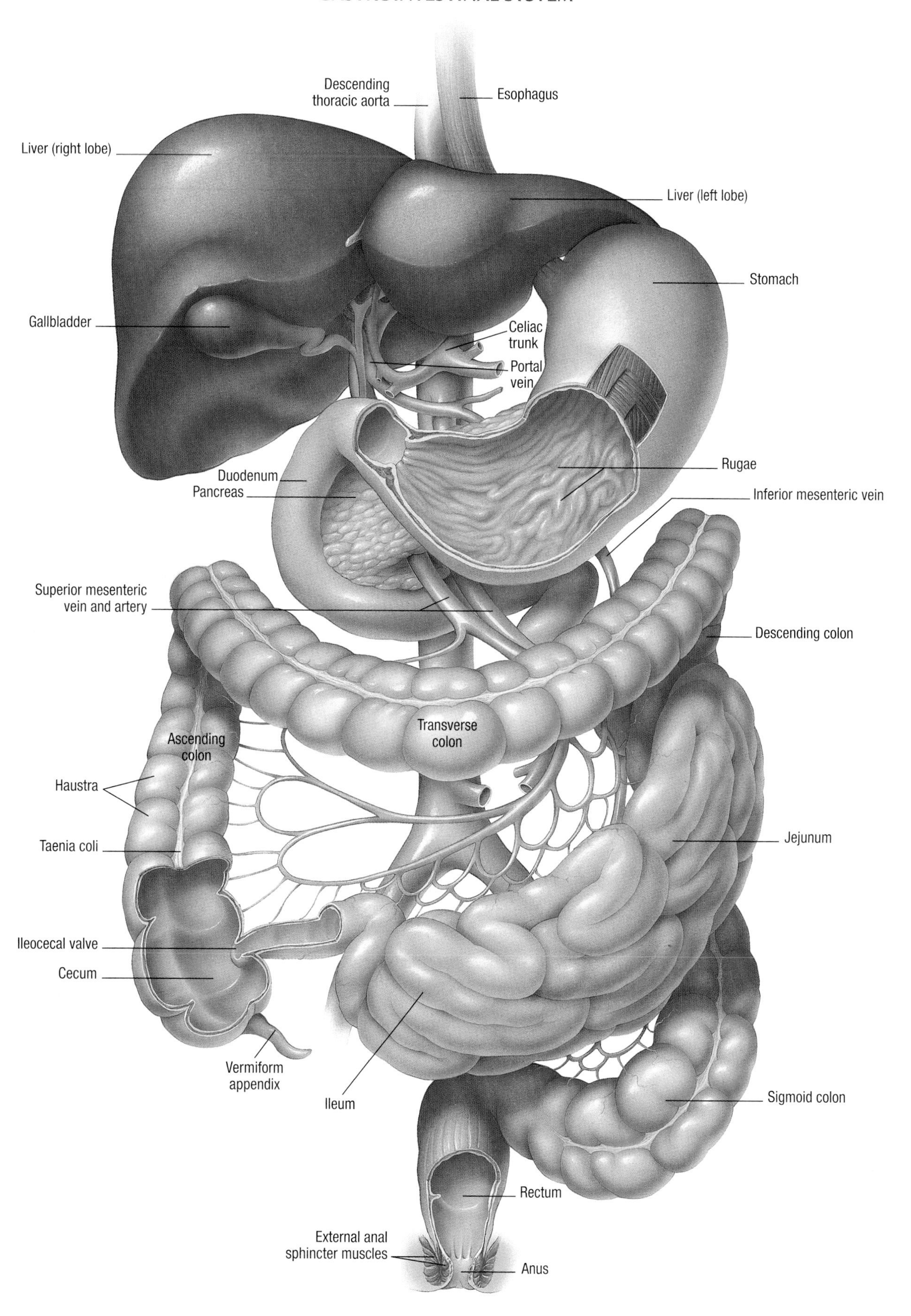
Descending thoracic aorta
Esophagus
Liver (right lobe)
Liver (left lobe)
Stomach
Gallbladder
Celiac trunk
Portal vein
Rugae
Duodenum
Pancreas
Inferior mesenteric vein
Superior mesenteric vein and artery
Descending colon
Transverse colon
Ascending colon
Haustra
Jejunum
Taenia coli
Ileocecal valve
Cecum
Vermiform appendix
Ileum
Sigmoid colon
Rectum
External anal sphincter muscles
Anus

# Urinary system

The urinary system consists of two kidneys, two ureters, the bladder, and the urethra. The ureters are fibromuscular tubes, which act as ducts, allowing urine to pass from the kidneys to the bladder. The bladder is a saclike urine-storage organ that lies anterior and inferior to the pelvic cavity and posterior to the symphysis pubis. The urethra is a small duct that channels urine outside the body from the bladder.

Together, these structures remove wastes from the body, help govern acid-base balance by retaining or excreting hydrogen ions, and regulate fluid and electrolyte balance. For more detailed information on various structures and functions of the urinary system, see Part VI, Abdomen, and Part VII, Pelvis.

## Kidney

The kidneys, bean-shaped organs located retroperitoneally, function to:

- eliminate wastes and excess ions as urine
- filter blood (regulating chemical composition and volume of blood)
- maintain fluid-electrolyte and acid-base balances
- produce erythropoietin (a hormone that stimulates red blood cell production) and enzymes (such as renin, which governs blood pressure and kidney function)
- convert vitamin D to a more active form.

### Nephron

The nephron, the functional unit of the kidney, mechanically filters fluids, wastes, electrolytes, acids, and bases into the tubular system. The nephron also selectively reabsorbs and secretes ions.

### Vasculature

Blood enters the kidney from the renal artery, which subdivides into several branches. Some of these branches distribute blood for filtration in the kidney; others nourish the kidney cells.

Of the blood brought to the kidney for filtration, about 99% returns to the general circulation through the renal vein. The remaining 1% undergoes further processing, resulting in urine-containing waste products that flow to the calyx and renal pelvis.

Blood enters and leaves the glomerular capillaries by two small blood vessels, the efferent and afferent arterioles. The glomerular capillaries act as bulk filters and pass protein-free and red blood cell–free filtrate to the proximal convoluted tubules.

## Urine formation

Urine formation is the result of three processes:

- glomerular filtration
- tubular reabsorption
- tubular secretion.

Normal urine contains sodium, chloride, potassium, calcium, magnesium, sulfates, phosphates, bicarbonates, uric acid, ammonium ions, creatinine, and urobilinogen. A few leukocytes and red blood cells (RBCs) and, in the male, some spermatozoa may enter the urine as it passes from the kidney to the ureteral orifice. Urine may also contain drugs if the person is taking drugs that undergo urinary excretion.

The kidneys regulate reabsorption and secretin in the nephrons, thus determining the composition of excreted urine.

## Hormones

Hormones help regulate tubular reabsorption and secretion. For example, antidiuretic hormone (ADH) acts in the distal tubule and collecting ducts to increase water reabsorption and concentrate urine. ADH deficiency decreases water reabsorption, causing dilute urine.

Another hormone, aldosterone, affects tubular reabsorption by regulating sodium retention and helping to control potassium secretion by epithelial cells in the tubules. When serum potassium levels rise, the adrenal cortex responds by increasing aldosterone secretion.

**Physiology**

By secreting the enzyme renin, the kidneys play a crucial role in blood pressure and fluid volume regulation. The renin-angiotensin system is an important homeostatic device for regulating the body's sodium and water levels and blood pressure. This system depends on feedback involving the juxtaglomerular apparatus in the glomeruli and the liver, lungs, and adrenal cortex.

Juxtaglomerular cells near each glomerulus secrete renin into the blood. The rate of renin secretion depends on the rate of perfusion in the renal afferent arterioles and on the serum sodium level. A low sodium load and low perfusion pressure (as in hypovolemia) increase renin secretion; a high sodium load and high perfusion pressure decrease it.

Renin acts throughout the body. In the liver, it converts angiotensinogen to the hormone angiotensin I. The lungs convert angiotensin I to angiotensin II, a potent vasoconstrictor. Angiotensin II acts on the adrenal cortex to stimulate production of the hormone aldosterone. Aldosterone, in turn, acts on the juxtaglomerular cells in the nephron to increase sodium and water retention and to stimulate or depress further renin secretion. This completes the feedback cycle that automatically restores homeostasis.

Renal hormones also regulate red blood cell production and calcium/phosphorus balance. In response to low arterial oxygen tension, the kidneys produce erythropoietin, which stimulates increased red blood cell production by the bone marrow.

To help regulate calcium/phosphorus balance, the kidneys filter and reabsorb approximately half of the body's unbound serum calcium and activate vitamin $D_3$, a hormone that promotes intestinal calcium absorption and regulates phosphate excretion.

## URINARY SYSTEM

### Anterior view

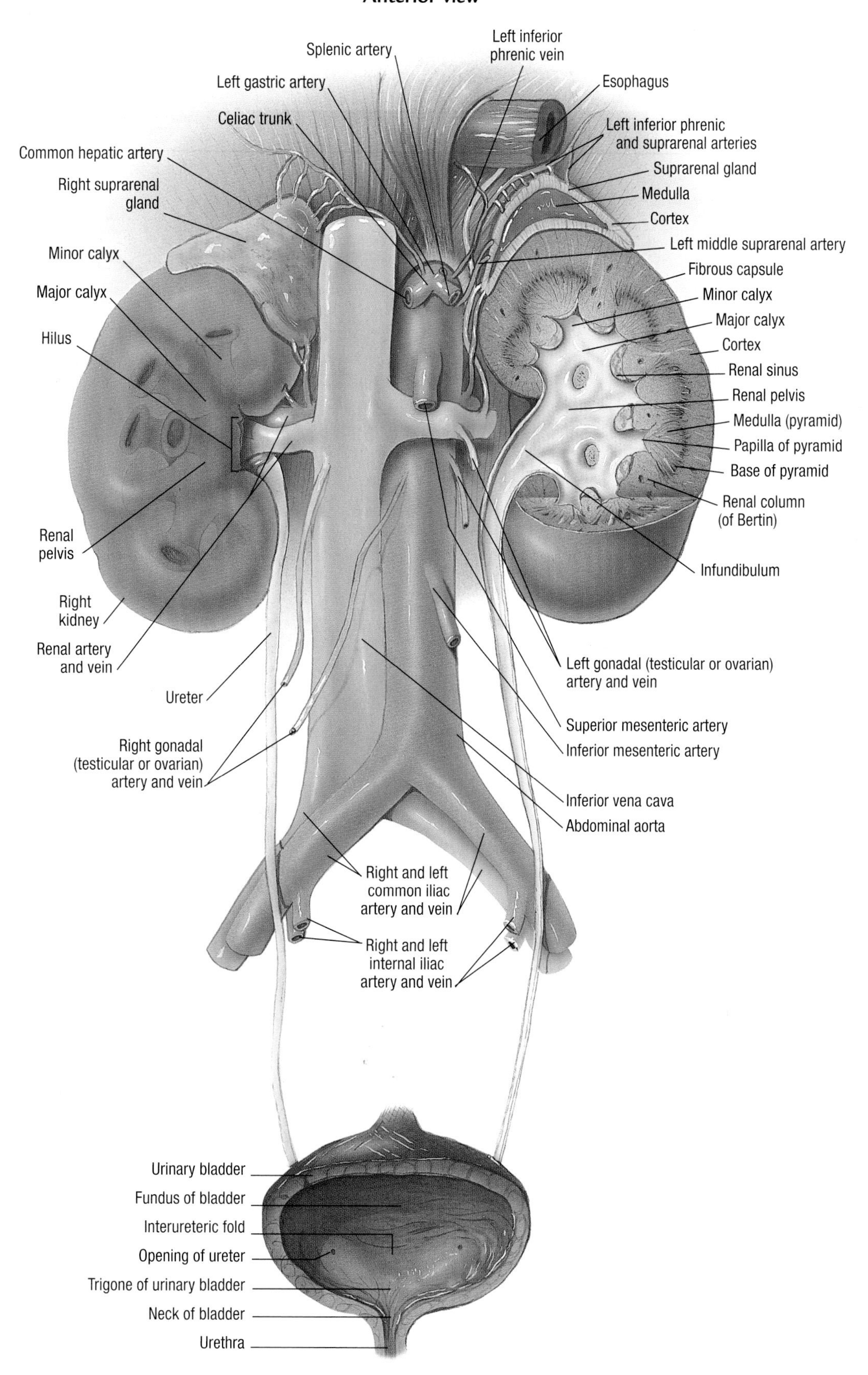

Splenic artery
Left inferior phrenic vein
Left gastric artery
Esophagus
Celiac trunk
Left inferior phrenic and suprarenal arteries
Common hepatic artery
Suprarenal gland
Right suprarenal gland
Medulla
Cortex
Left middle suprarenal artery
Minor calyx
Fibrous capsule
Major calyx
Minor calyx
Major calyx
Hilus
Cortex
Renal sinus
Renal pelvis
Medulla (pyramid)
Papilla of pyramid
Base of pyramid
Renal column (of Bertin)
Renal pelvis
Infundibulum
Right kidney
Renal artery and vein
Left gonadal (testicular or ovarian) artery and vein
Ureter
Superior mesenteric artery
Inferior mesenteric artery
Right gonadal (testicular or ovarian) artery and vein
Inferior vena cava
Abdominal aorta
Right and left common iliac artery and vein
Right and left internal iliac artery and vein
Urinary bladder
Fundus of bladder
Interureteric fold
Opening of ureter
Trigone of urinary bladder
Neck of bladder
Urethra

# Endocrine system

With the nervous system, the endocrine system regulates and integrates the body's metabolic activities. Glands, specialized cell clusters or organs, secrete hormones (chemical substances) in response to stimulation. The hormones are transported via the blood throughout the body; each one attaches to specific receptor proteins and triggers specific physiologic changes in a target cell.

## Endocrine glands

The major glands of the endocrine system collectively weigh less than 7 oz (200 g). They include:

- pituitary gland
- thyroid gland
- parathyroid glands
- adrenal glands
- pancreas
- thymus
- pineal gland
- gonads (ovaries and testes).

This page provides a brief overview of the glands of the endocrine system.

### Pituitary gland

The pituitary, a pea-sized gland, rests in the sella turcica, a depression in the sphenoid bone at the base of the brain. The pituitary has two main regions. The larger region, the anterior pituitary (adenohypophysis), produces at least six hormones:

- growth hormone (GH), or somatotropin
- thyrotropin, or thyroid-stimulating hormone (TSH)
- corticotropin
- follicle-stimulating hormone (FSH)
- luteinizing hormone (LH)
- adrenocorticotropin (ACTH).

The posterior pituitary – about 25% of the gland – is not an endocrine gland because it does not produce any hormones.

### Thyroid gland

The thyroid has two lobes that function as one unit to produce the hormones thyroxine ($T_4$), triiodothyronine ($T_3$), and calcitonin. Collectively referred to as thyroid hormone, $T_4$ and $T_3$ are the body's major metabolic hormones, regulating metabolism by speeding cellular respiration.

Calcitonin maintains the blood calcium level by inhibiting release of calcium from bone. Its secretion is controlled by the calcium concentration of the fluid surrounding thyroid cells.

### Parathyroid glands

The body's smallest known endocrine glands, the parathyroid glands work together as a single gland to produce parathyroid hormone (PTH). The main function of PTH is to help regulate the blood's calcium balance. This hormone adjusts the rate at which calcium and magnesium ions are removed from the urine and increases the movement of phosphate ions from the blood to urine for excretion.

### Adrenal glands

The adrenal glands, each lying atop a kidney, contain two distinct structures that function as separate endocrine glands – the medulla and the cortex.

The *adrenal medulla*, or inner layer of the adrenal gland, functions as part of the sympathetic nervous system, producing the catecholamines epinephrine and norepinephrine. Because these substances play important roles in the autonomic nervous system (ANS), the adrenal medulla is considered a neuroendocrine structure.

The outer layer, the *adrenal cortex*, forms the bulk of the adrenal gland. It has three zones, or cell layers: the outermost *zona glomerulosa*, the middle and largest *zona fasciculata*, and the innermost *zona reticularis*.

### Pancreas

The pancreas performs both endocrine (inward secretion) and exocrine (outward secretion) functions. Acinar cells, which make up most of the gland, regulate pancreatic exocrine function. The acinar cells secrete digestive enzymes, which flow through the pancreatic duct and enter the duodenum.

The endocrine cells of the pancreas are called the islet cells, or islets of Langerhans. Existing in clusters scattered among the acinar cells, the islets contain alpha, beta, and delta cells, which produce important hormones.

- alpha cells – glucagon, which raises the blood glucose level by triggering the breakdown of glycogen to glucose
- beta cells – insulin, which lowers the blood glucose level by stimulating the conversion of glucose to glycogen
- delta cells – somatostatin, which inhibits the release of growth hormone, corticotropin, and certain other hormones, and selectively inhibits the release of insulin and glucagon.

### Thymus

Located posterior to the sternum between the lungs, the thymus contains primarily lymphatic tissue. It reaches maximal size at puberty, and then starts to atrophy.

The thymus produces T cells, important in cell-mediated immunity; thus, its major role seems related to the immune system. However, the thymus also produces the peptide hormones thymosin and thymopoietin. Active in immunity, these hormones promote growth of peripheral lymphoid tissue.

### Pineal gland

The pinecone-shaped pineal gland lies at the back of the third ventricle of the brain. It produces the hormone melatonin, which has many widespread effects, including a possible role in the neuroendocrine reproductive axis.

### Gonads

The gonads are the ovaries in females and the testes in males. The ovaries produce ova (eggs) and the steroid hormones estrogen and progesterone.

The testes are paired structures that lie in an extra-abdominal pouch (scrotum). They produce spermatozoa and the hormone testosterone.

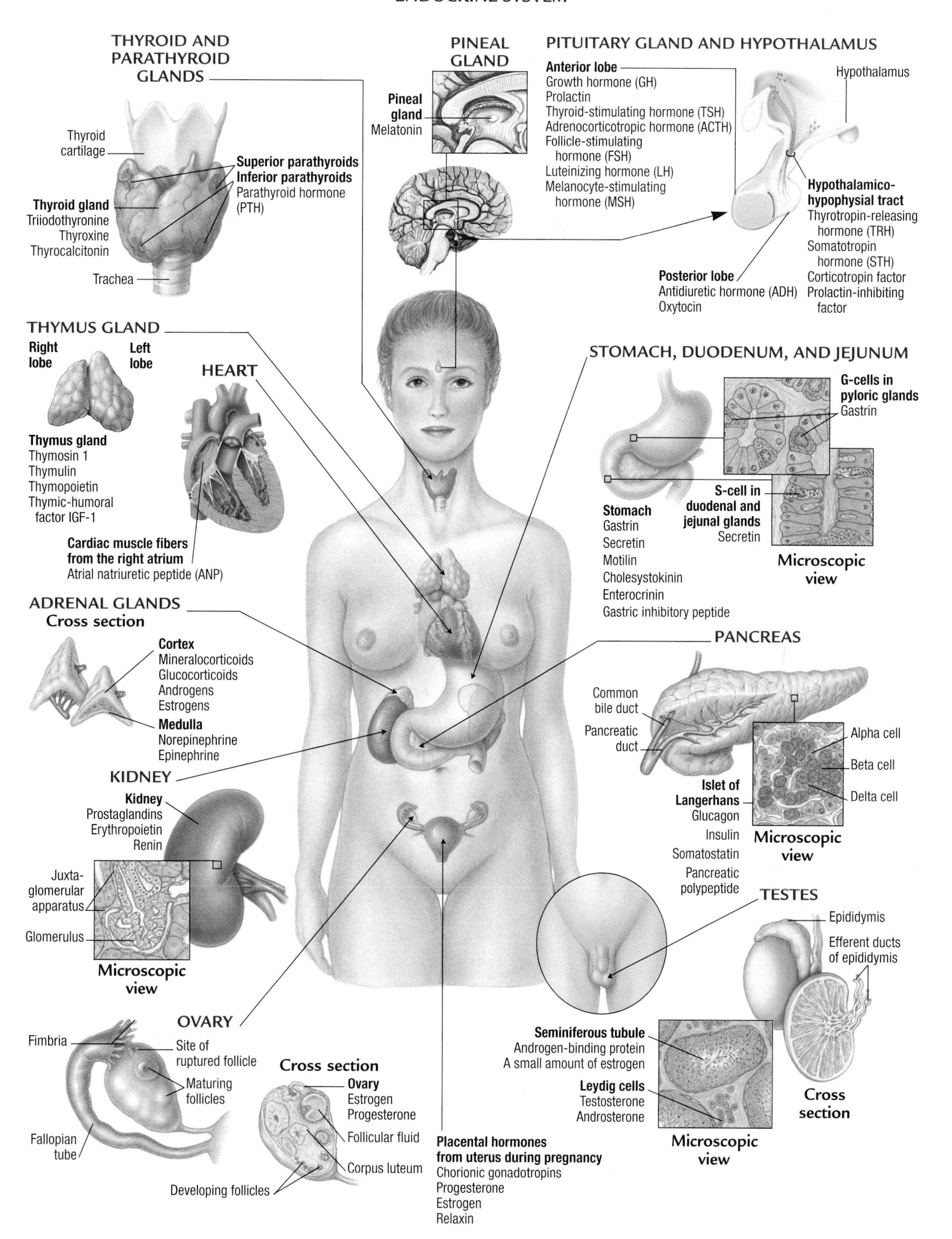
ENDOCRINE SYSTEM
THYROID AND PARATHYROID GLANDS
Thyroid cartilage
Superior parathyroids
Inferior parathyroids
Parathyroid hormone (PTH)
Thyroid gland
Triiodothyronine
Thyroxine
Thyrocalcitonin
Trachea
PINEAL GLAND
Pineal gland
Melatonin
PITUITARY GLAND AND HYPOTHALAMUS
Anterior lobe
Growth hormone (GH)
Prolactin
Thyroid-stimulating hormone (TSH)
Adrenocorticotropic hormone (ACTH)
Follicle-stimulating hormone (FSH)
Luteinizing hormone (LH)
Melanocyte-stimulating hormone (MSH)
Hypothalamus
Hypothalamico-hypophysial tract
Thyrotropin-releasing hormone (TRH)
Somatotropin hormone (STH)
Corticotropin factor
Prolactin-inhibiting factor
Posterior lobe
Antidiuretic hormone (ADH)
Oxytocin
THYMUS GLAND
Right lobe
Left lobe
Thymus gland
Thymosin 1
Thymulin
Thymopoietin
Thymic-humoral factor IGF-1
HEART
Cardiac muscle fibers from the right atrium
Atrial natriuretic peptide (ANP)
STOMACH, DUODENUM, AND JEJUNUM
G-cells in pyloric glands
Gastrin
S-cell in duodenal and jejunal glands
Secretin
Stomach
Gastrin
Secretin
Motilin
Cholesystokinin
Enterocrinin
Gastric inhibitory peptide
Microscopic view
ADRENAL GLANDS
Cross section
Cortex
Mineralocorticoids
Glucocorticoids
Androgens
Estrogens
Medulla
Norepinephrine
Epinephrine
PANCREAS
Common bile duct
Pancreatic duct
Alpha cell
Beta cell
Delta cell
Islet of Langerhans
Glucagon
Insulin
Somatostatin
Pancreatic polypeptide
Microscopic view
KIDNEY
Kidney
Prostaglandins
Erythropoietin
Renin
Juxta-glomerular apparatus
Glomerulus
Microscopic view
TESTES
Epididymis
Efferent ducts of epididymis
Seminiferous tubule
Androgen-binding protein
A small amount of estrogen
Leydig cells
Testosterone
Androsterone
Microscopic view
Cross section
OVARY
Fimbria
Site of ruptured follicle
Maturing follicles
Fallopian tube
Cross section
Ovary
Estrogen
Progesterone
Follicular fluid
Corpus luteum
Developing follicles
Placental hormones from uterus during pregnancy
Chorionic gonadotropins
Progesterone
Estrogen
Relaxin

# Integumentary system

The largest body system, the integumentary system includes the skin and its appendages —hair, nails, and certain glands. The integumentary system covers an area of about 11 to 21 square feet (1 to 2 $m^2$) and accounts for about 15% of body weight. The skin is considered an organ. Its many vital functions include:

- maintenance of body surface integrity
- protection of inner structures
- sensory perception
- regulation of body temperature and blood pressure
- synthesis of vitamin $D_3$ when stimulated by ultraviolet light
- excretion of sweat.

## Skin layers

Two distinct layers of skin, the epidermis and dermis, lie above a layer of subcutaneous fatty tissue (sometimes called the hypodermis).

### Epidermis

The outermost layer, the epidermis varies in thickness from less than 0.1 mm on the eyelids to more than 1 mm on the palms and soles. It consists of layers, or strata, listed from the outside inward: corneum, lucidum, granulosum, spinosum, basale.

**Clinical tip**

When you're trying to compare subtle temperature differences on your patient's skin, use the dorsal surfaces of your hands and fingers. They're the most sensitive to changes in temperature.

### Dermis

The skin's second layer, the dermis (also called the corium) is an elastic system that contains and supports blood vessels, lymphatic vessels, nerves, and epidermal appendages. Most of the dermis is composed of extracellular material called matrix. Matrix contains connective tissue fiber and provides elasticity and strength. The dermis has two layers:

- superficial papillary dermis — has fingerlike projections (papillae) that nourish epidermal cells
- lower or reticular layer — lies over a layer of subcutaneous fatty (adipose) tissue.

## Other structures

The skin contains several other relevant structures, including:

- basement membrane — collagenous membrane between the epidermis and dermis that holds them together
- Meissner's corpuscle — oval body in the dermis, thought to participate in tactile sensation
- Ruffini's corpuscle — oval capsule containing the ends of sensory fibers in the dermal papillae.

## Epidermal appendages

Epidermal appendages include the hair, nails, sebaceous glands, and sweat glands.

### Hair

Hairs are long, slender shafts composed of keratin. The expanded lower end of each hair forms a bulb, or root. On its undersurface, the root is indented by a hair papilla, a cluster of connective tissue and blood vessels.

Each hair lies within an epithelium-lined sheath called a hair follicle. A bundle of smooth-muscle fibers (arrector pili) extends from the dermis to the base of the follicle. When these muscles contract, the hair stands on end. Hair follicles also have a rich blood and nerve supply.

### Nails

Covering the distal surface of the end of each finger and toe, nails are specialized types of keratin. The nail plate, surrounded on three sides by the nail folds (cuticles), lies on the nail bed. The nail plate is formed by the nail matrix, which extends proximally for about ¼" (5 mm) beneath the nail fold.

The distal portion of the matrix shows through the nail as a pale crescent moon–shaped area, the lunula. The translucent nail plate distal to the lunula exposes the vascular nail bed, which imparts the characteristic pink appearance of the nails.

### Sebaceous glands

Sebaceous glands are present on all parts of the skin except the palms and soles. They are most prominent on the scalp, face, upper torso, and genitalia. Sebaceous glands produce sebum, a lipid substance, and secrete it into the hair follicle via the sebaceous duct. Sebum moves to the skin surface through the hair follicle opening. Sebum may help to waterproof the hair and skin, may promote the absorption of fat-soluble substances into the dermis, and may be involved in vitamin $D_3$ production. It may also have an antibacterial function.

### Sweat glands

The skin contains two types of sweat glands: eccrine and apocrine. The eccrine glands are present on most of the body, except for the lips, in the superficial layer of the dermis. These glands produce an odorless, watery fluid with a sodium concentration equal to that of plasma. A duct from the coiled secretory portion passes through the dermis, and the epidermis carries the fluid to the skin surface.

The apocrine glands are located chiefly in the axillary and anogenital areas near hair follicles. The coiled secretory portions of these glands lie deeper in the dermis than do those of the eccrine glands. An apocrine duct connects each gland to the upper portion of a hair follicle. Apocrine glands begin to function at puberty. However, they have no known biological function. As bacteria break down the fluids produced by these glands, body odor occurs.

**Age-related changes**

With aging, several changes occur in the skin:

- formation of facial lines
- decreased rate of skin cell replacement
- loss of skin elasticity
- decreased sweat gland output
- thinning hair and loss of hair pigment.

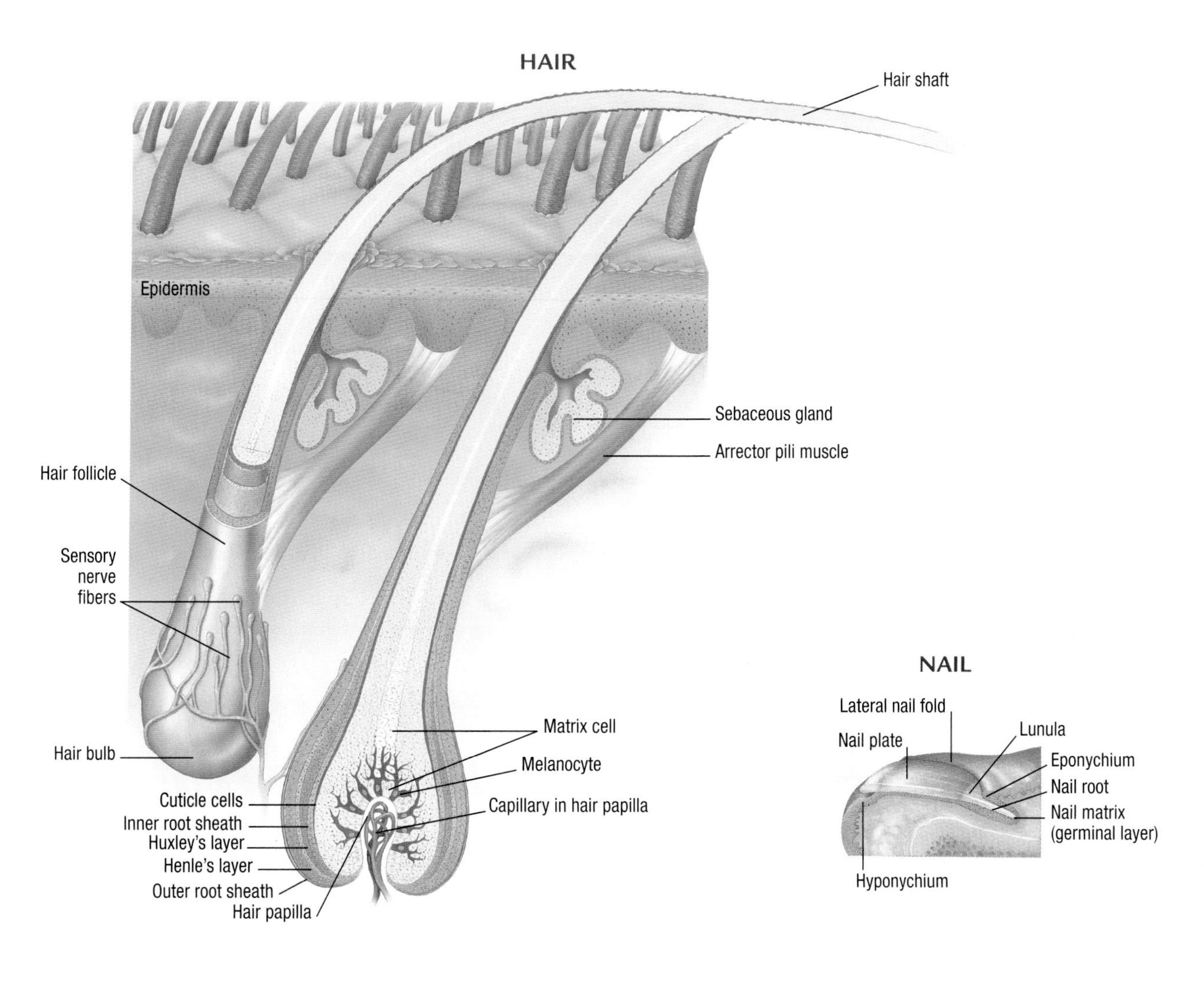
HAIR
Hair shaft
Epidermis
Sebaceous gland
Arrector pili muscle
Hair follicle
Sensory
nerve
fibers
Matrix cell
Hair bulb
Melanocyte
Cuticle cells
Capillary in hair papilla
Inner root sheath
Huxley's layer
Henle's layer
Outer root sheath
Hair papilla
NAIL
Lateral nail fold
Nail plate
Lunula
Eponychium
Nail root
Nail matrix
(germinal layer)
Hyponychium

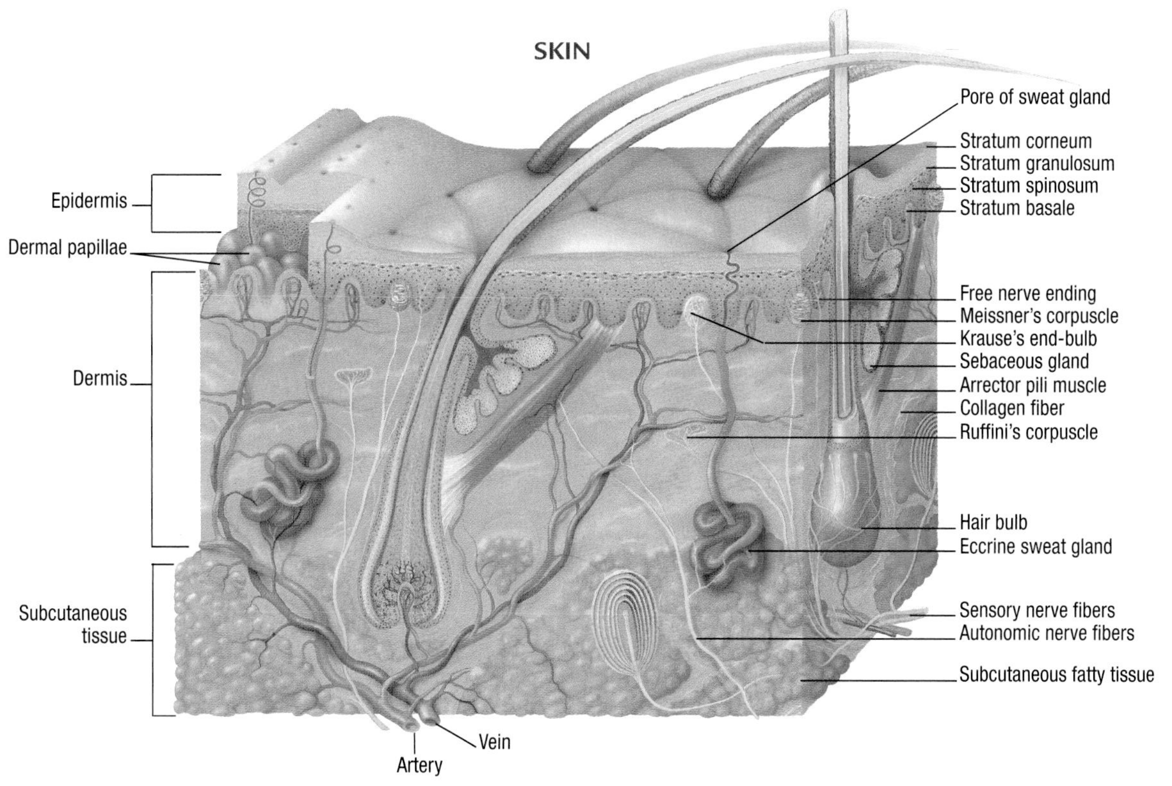
SKIN
Pore of sweat gland
Stratum corneum
Stratum granulosum
Stratum spinosum
Stratum basale
Epidermis
Dermal papillae
Free nerve ending
Meissner's corpuscle
Krause's end-bulb
Sebaceous gland
Dermis
Arrector pili muscle
Collagen fiber
Ruffini's corpuscle
Hair bulb
Eccrine sweat gland
Subcutaneous
tissue
Sensory nerve fibers
Autonomic nerve fibers
Subcutaneous fatty tissue
Vein
Artery

# APPENDICES, GLOSSARY, SELECTED REFERENCES, AND INDEX

## Appendix A

# Cranial nerves and their functions

Cranial nerves are described by type (sensory or motor) and function. Starting with the most anterior nerve, they are desig by Roman numerals and name.

| NERVE NUMBER AND NAME | TYPE AND FUNCTION |
|---|---|
| **I** (olfactory) | • Sensory: smell (olfaction) |
| **II** (optic) | • Sensory: vision |
| **III** (oculomotor) | • Motor: extraocular eye movement (superior, medial, and inferior lateral), pupil con striction, and upper eyelid elevation |
| **IV** (trochlear) | • Motor: extraocular eye movement (inferior lateral) |
| **V** (trigeminal) | • Motor: chewing, biting, and lateral jaw movements<br>• Sensory: transmitting stimuli from face and head, corneal reflex |
| **VI** (abducens) | • Motor: extraocular eye movement (lateral) |
| **VII** (facial) | • Sensory: taste receptors (anterior two-thirds of the tongue)<br>• Motor: facial muscle movement, including muscles of expression (those in the forel and around the eyes and mouth) |
| **VIII** (vestibulocochlear) | • Sensory: hearing, sense of balance |
| **IX** (glossopharyngeal) | • Sensory: sensations from throat; taste receptors (posterior one-third of tongue)<br>• Motor: swallowing movements |
| **X** (vagus) | • Motor: movement of palate and larynx, swallowing, gag reflex; activity of the thora and abdominal viscera, such as heart rate and peristalsis<br>• Sensory: sensations from throat, larynx, and thoracic and abdominal viscera (heart lungs, bronchi, and GI tract) |
| **XI** (spinal accessory) | • Motor: shoulder movement, head rotation |
| **XII** (hypoglossal) | • Motor: tongue movement |

# Appendix B

# Responses to autonomic nervous system stimulation

The parasympathetic and sympathetic divisions of the autonomic nervous system usually produce opposite responses, as shown in the examples below.

| EFFECTOR | PARASYMPATHETIC RESPONSE | SYMPATHETIC RESPONSE |
|---|---|---|
| **Eye** | | |
| Radial muscle of iris | • None | • Contraction (mydriasis) |
| Sphincter muscle of iris | • Contraction for near vision | • None |
| Lacrimal gland | • Stimulates tear secretion | • None |
| **Salivary glands** | | |
| Submandibular gland | • Increased secretion | • Decreased secretion |
| Sublingual gland | • Increased secretion | • Decreased saliva secretion |
| Parotid gland | • Increased secretion | • Decreased saliva secretion |
| **Heart** | | |
| | • Decreased rate and contractility | • Increased rate and contractility |
| **Lungs** | | |
| Bronchioles | • Contraction | • Relaxation |
| Mucous glands | • Stimulated secretion | • Inhibited secretion |
| **Stomach** | | |
| Motility and tone | • Increased | • Decreased (usually) |
| Sphincters | • Relaxation | • Contraction (usually) |
| **Intestine** | | |
| Motility and tone | • Increased | • Decreased |
| Sphincters | • Relaxation | • Contraction |
| **Urinary bladder** | | |
| Bladder muscle | • Contraction | • Relaxation |
| Trigone and sphincter | • Relaxation | • Contraction |
| **Skin** | | |
| Erector pili | • None | • Contraction |
| Sweat glands | • None | • Increased secretion |

*(continued)*

| EFFECTOR | PARASYMPATHETIC RESPONSE | SYMPATHETIC RESPONSE |
|---|---|---|
| **Adrenal medulla** | | |
| | • None | • Secretion of epinephrine and norepinephrine |
| **Liver** | | |
| | • None | • Glycogenolysis |
| **Pancreas (acini)** | | |
| | • Increased secretion | • Decreased secretion |
| **Adipose tissue** | | |
| | • None | • Lipolysis |
| **Juxtaglomerular cells** | | |
| | • None | • Increased renin secretion |
| **External genitalia** | | |
| | • Erection due to vasodilation | • Ejaculation of spermatozoa<br>• Reversed uterine peristalsis |

Appendix C

# Guide to skeletal muscles

This table groups muscles by anatomic region and provides their points of origin and insertion.

| MUSCLE | ORIGIN | INSERTION |
|---|---|---|
| **Head and neck** | | |
| ***Face*** | | |
| Buccinator | • Mandible (alveolar process)<br>• Maxilla | • Orbicularis oris<br>• Skin at mouth angle |
| Corrugator supercilii | • Frontal bone | • Skin of eyebrows |
| Depressor anguli oris | • Mandible below mental foramen | • Skin and muscles at mouth angle |
| Depressor labii inferioris | • Mandible between symphysis and mental foramen | • Skin and muscles of lower lip |
| Epicranius frontalis | • Aponeurotic structure of scalp | • Skin and muscles of scalp |
| Epicranius occipitalis | • Occipital bone | • Aponeurotic structure of scalp |
| Levator labii superioris | • Eye orbit (lower margin) | • Skin and muscles of upper lip<br>• Wing of nose |
| Mentalis | • Mandible near symphysis | • Skin of chin |
| Orbicularis oculi | • Frontal bone and maxilla<br>• Medial palpebral ligament | • Skin surrounding eyes and eyelids |
| Orbicularis oris | • Muscles surrounding mouth | • Skin surrounding mouth |
| Platysma | • Fascia over pectoralis major and deltoid muscles | • Mandible (lower border)<br>• Skin of cheek and neck |
| Procerus | • Nasal bone (lower portion)<br>• Lateral nasal cartilage (upper part) | • Skin between eyebrows |
| Risorius | • Fascia of masseter muscle | • Skin at mouth angle |
| Zygomaticus major | • Zygomatic bone | • Skin and muscles above mouth angle |
| Zygomaticus minor | • Zygomatic bone | • Skin and muscles above mouth angle |
| ***Muscles of mastication*** | | |
| Temporalis | • Temporal fossa | • Mandible (coronoid process and ramus) |
| Masseter | • Zygomatic arch | • Mandible (angle and ramus) |
| Medial pterygoid | • Sphenoid bone (lateral pterygoid plate) | • Mandible (internal surface) |
| Lateral pterygoid | • Lateral pterygoid plate (lateral surface)<br>• Sphenoid bone (great wing) | • Mandible (just below condyle) |
| ***Extrinsic muscles of the tongue*** | | |
| Genioglossus | • Mandible (internal surface) | • Near symphysis |
| Hyoglossus | • Hyoid bone (body and greater projection) | • Tongue (sides) |
| Styloglossus | • Temporal bone (styloid process) | • Tongue (sides) |

*(continued)*

| MUSCLE | ORIGIN | INSERTION |
|---|---|---|
| **Head and neck** *(continued)* | | |
| ***Neck*** | | |
| Sternocleidomastoid | • Sternum (manubrium)<br>• Clavicle (medial portion) | • Temporal bone (mastoid process) |
| Digastric | • Mandible (lower border)<br>• Temporal bone (mastoid notch) | • Intermediate tendon on hyoid bone |
| Stylohyoid | • Temporal bone (styloid process) | • Hyoid bone |
| Mylohyoid | • Mandible (inner surface from symphysis to angle) | • Hyoid bone |
| Geniohyoid | • Mandibular symphysis (inner surface) | • Hyoid bone |
| Sternohyoid | • Clavicle (manubrium and medial end) | • Hyoid bone |
| Sternothyroid | • Manubrium | • Larynx (thyroid cartilage) |
| Thyrohyoid | • Larynx (thyroid cartilage) | • Hyoid bone |
| Omohyoid | • Scapula (superior border) | • Hyoid bone |
| **Vertebral column** | | |
| Semispinalis thoracis<br>Semispinalis cervicis<br>Semispinalis capitis | • Vertebrae (transverse processes of all thoracic and seventh cervical) | • Vertebrae (spinous processes of second cervical through fourth thoracic)<br>• Occipital bone |
| Multifidi | • Ilium and sacrum (posterior surface)<br>• Vertebrae (transverse processes of lumbar, thoracic, and lower cervical) | • Vertebrae (spinous processes of lumbar, thoracic, and cervical) |
| Rotatores | • Vertebrae (transverse processes) | • Next superior vertebra (base of spinous process) |
| Interspinales | • Vertebrae (superior surfaces of all spinous processes) | • Next superior vertebra (inferior processes) |
| Scalene muscles | • Cervical vertebrae (transverse processes) | • Ribs (first and second) |
| Intertransversarii | • Vertebrae (transverse processes) | • Next superior vertebra (transverse process) |
| Splenius capitis<br>Splenius cervicis | • Vertebrae (spinous processes of upper thoracic and seventh cervical, from ligamentum nuchae) | • Occipital bone<br>• Temporal bone (mastoid process)<br>• Vertebrae (transverse processes of upper three cervical) |
| **Erector spine group** | | |
| Iliocostalis lumborum<br>Iliocostalis thoracis<br>Iliocostalis cervicis | • Sacrum (crest)<br>• Vertebrae (spinous processes of lumbar and lower thoracic)<br>• Iliac crests<br>• Rib angles | • Rib angles<br>• Vertebrae (transverse processes of cervical) |
| Longissimus thoracis<br>Longissimus cervicis<br>Longissimus capitis | • Vertebrae (transverse processes of lumbar, thoracic, and lower cervical) | • Next superior vertebra (transverse process)<br>• Temporal bone (mastoid process) |
| Spinalis thoracis<br>Spinalis cervicis | • Vertebrae (spinous processes of upper lumbar, lower thoracic, and seventh cervical) | • Vertebrae (spinous processes of upper thoracic and and cervical) |

| MUSCLE | ORIGIN | INSERTION |
|---|---|---|
| **Muscles of respiration** | | |
| Diaphragm | • Rib cage (inferior border)<br>• Xiphoid process<br>• Costal cartilages<br>• Vertebrae (lumbar) | • Central tendon of diaphragm |
| External intercostal muscles | • Ribs (inferior border)<br>• Costal cartilages | • Next inferior rib (superior border) |
| Internal intercostal muscles | • Ribs (inner surface)<br>• Costal cartilages | • Next inferior rib (superior border) |
| Subcostales | • Ribs (inner surface, near angles) | • Second or third inferior rib (inner surface) |
| Transversus thoracis | • Sternum (inner surface) | • Costal cartilages (inner surface) |
| **Muscles of the shoulder** | | |
| Trapezius | • Costal cartilages (inner surface)<br>• Occipital bone<br>• Ligamentum nuchae<br>• Vertebrae (spinous processes of seventh cervical and all thoracic) | • Clavicle (lateral third)<br>• Acromion process<br>• Scapula (spine) |
| Rhomboideus major | • Vertebrae (spinous processes of second through fifth thoracic) | • Scapula (vertebral border below spine) |
| Rhomboideus minor | • Vertebrae (spinous processes of seventh cervical and first thoracic) | • Scapula (vertebral border at base of spine) |
| Levator scapulae | • Vertebrae (transverse processes of upper four cervical) | • Scapula (vertebral border above spine) |
| Pectoralis minor | • Ribs (anterior surface of third through fifth) | • Scapula (coracoid process) |
| Serratus anterior | • Ribs (outer surface of first nine) | • Scapula (ventral surface of vertebral border) |
| Subclavius | • Rib (outer surface of first) | • Clavicle (inferior surface of lateral portion) |
| **Abdominopelvic cavity** | | |
| ***Abdominal wall*** | | |
| External abdominal oblique | • Ribs (external surface of lower eight) | • Iliac crest (anterior half) |
| Internal abdominal oblique | • Inguinal ligament<br>• Iliac crest<br>• Lumbodorsal fascia | • Linea alba<br>• Pubic crest<br>• Ribs (lower four) |
| Transversus abdominis | • Inguinal ligament<br>• liac crest<br>• Pubic crest<br>• Lumbodorsal fascia<br>• Ribs (costal cartilages of last six) | • Linea alba<br>• Pubic crest |
| Rectus abdominis | • Pubic crest | • Ribs (costal cartilages of fifth through seventh) |
| Quadratus lumborum | • Iliac crest<br>• Rib (lower border of twelfth) | • Rib (lower border of twelfth)<br>• Vertebrae (transverse processes of upper lumbar) |

*(continued)*

| MUSCLE | ORIGIN | INSERTION |
|---|---|---|
| **Abdominopelvic cavity** *(continued)* | | |
| ***Pelvic floor*** | | |
| Levator ani | • Pubic bone (inner surface of superior ramus)<br>• Lateral pelvic wall<br>• Ischium (spine) | • Coccyx (inner surface) |
| Coccygeus | • Ischium (spine)<br>• Coccyx<br>• Sacrospinous ligament | • Coccyx<br>• Sacrum |
| **Upper extremities** | | |
| ***Arm movement*** | | |
| Pectoralis major | • Clavicle (medial half)<br>• Sternum<br>• Ribs (costal cartilages of upper six)<br>• External oblique (aponeurosis) | • Humerus (greater tubercle) |
| Latissimus dorsi | • Vertebrae (spinous processes of lower six thoracic and all lumbar)<br>• Sacrum<br>• Ilium (posterior crest) | • Humerus (medial margin of intertubercular groove) |
| Deltoid | • Clavicle (lateral third)<br>• Acromion process<br>• Scapula (spine) | • Humerus (deltoid tubercle) |
| Supraspinatus | • Scapula (supraspinatus fossa) | • Humerus (greater tubercle) |
| Infraspinatus | • Scapula (infraspinatus fossa) | • Humerus (greater tubercle) |
| Subscapularis | • Scapula (subscapular fossa) | • Humerus (lesser tubercle) |
| Teres major | • Scapula (dorsal surface of inferior angle) | • Humerus (lesser tubercle) |
| Teres minor | • Scapula (axillary border) | • Humerus (greater tubercle) |
| Coracobrachialis | • Scapula (coracoid process) | • Humerus (medial surface of medial third) |
| ***Forearm movement*** | | |
| Biceps brachii | • Long head: scapula (supraglenoid tubercle)<br>• Short head: scapula (coracoid process) | • Radius (tubercle) |
| Brachialis | • Humerus (anterior surface of distal half) | • Ulna (coronoid process) |
| Triceps brachii | • Long head: scapula (infraglenoid tubercle)<br>• Lateral head: humerus (posterior surface, above radial groove)<br>• Medial head: humerus (posterior surface, below radial groove) | • Ulna (olecranon process) |
| Anconeus | • Humerus (lateral epicondyle) | • Ulna (lateral surface of olecranon process) |
| Brachioradialis | • Humerus (lateral supracondylar ridge) | • Radius (styloid process) |

| MUSCLE | ORIGIN | INSERTION |
|---|---|---|
| **Wrist, hand, and fingers** | | |
| ***Anterior superficial muscles*** | | |
| Pronator teres | • Humerus (medial epicondyle)<br>• Ulna (coronoid process) | • Radius (shaft, middle of lateral surface) |
| Flexor carpi radialis | • Humerus (medial epicondyle) | • Second and third metacarpals (ventral surface) |
| Palmaris longus | • Humerus (medial epicondyle) | • Palmar aponeurosis |
| Flexor carpi ulnaris | • Humerus (medial epicondyle)<br>• Olecranon process<br>• Ulna (posterior surface, proximal two-thirds) | • Third through fifth metacarpals |
| Flexor digitorum superficialis | • Humerus (medial eipcondyle)<br>• Ulna (coronoid process)<br>• Radius (anterior surface) | • Second through fifth fingers (ventral surfaces of middle phalanges) |
| ***Anterior deep*** | | |
| Flexor digitorum profundus | • Humerus (medial epicondyle and coronoid process)<br>• Interosseous membrane<br>• Ulna (ventral surface) | • Second through fifth fingers (distal phalanges, ventral surface of base) |
| Flexor pollicis longus | • Radius (ventral surface)<br>• Interosseous membrane | • Thumb (distal phalanges, ventral surface of base) |
| Pronator quadratus | • Ulna (distal ventral surface) | • Radius (distal ventral surface) |
| ***Posterior superficial*** | | |
| Extensor carpi radialis longus | • Humerus (lateral supracondylar ridge) | • Second metacarpal (dorsal surface of base) |
| Extensor carpi radialis brevis | • Humerus (lateral epicondyle) | • Third metacarpal (dorsal surface of base) |
| Extensor digitorum communis | • Humerus (lateral epicondyle) | • Second through fifth fingers (dorsal surfaces of phalanges) |
| Extensor digiti minimi | • Tendon of extensor digitorum communis | • Fifth finger (tendon of extensor digitorum communis, on dorsum) |
| Extensor carpi ulnaris | • Humerus (lateral epicondyle) | • Fifth metacarpal (base) |
| ***Posterior deep*** | | |
| Supinator | • Humerus (lateral epicondyle) | • Radius (proximal end, lateral surface of shaft) |
| Abductor pollicis longus | • Radius and ulna (posterior surface of middle portion)<br>• Interosseous membrane | • First metacarpal (base) |
| Extensor pollicis brevis | • Radius (posterior surface of middle portion)<br>• Interosseous membrane | • Thumb (base of first phalanx) |
| Extensor pollicis longus | • Ulna (posterior surface of middle portion)<br>• Interosseous membrane | • Thumb (base of last phalanx) |
| Extensor indicis | • Ulna (posterior surface of distal end)<br>• Interosseous membrane | • Second finger (tendon of extensor digitorum communis) |

*(continued)*

| MUSCLE | ORIGIN | INSERTION |
|---|---|---|
| **Wrist, hand, and fingers** *(continued)* | | |
| ***Intrinsic hand*** | | |
| Abductor pollicis brevis | • Flexor retinaculum<br>• Scaphoid<br>• Trapezium | • Thumb (proximal phalanx) |
| Opponens pollicis | • Flexor retinaculum<br>• Trapezium | • Thumb (metacarpal, lower border) |
| Flexor pollicis brevis | • Flexor retinaculum<br>• Trapezium<br>• First metacarpal | • Thumb (base of proximal phalanx) |
| Palmaris brevis | • Flexor retinaculum | • Hand (skin on ulnar border) |
| Adductor pollicis | • Capitate<br>• Second and third metacarpals | • Fifth finger (base of proximal phalanx) |
| Abductor digiti minimi | • Pisiform<br>• Flexor carpi ulnaris (tendon) | • Fifth finger (base of proximal phalanx) |
| Flexor digiti minimi brevis | • Flexor retinaculum<br>• Hamate | • Fifth finger (base of proximal phalanx) |
| Opponens digiti minimi | • Flexor retinaculum<br>• Hamate | • Fifth finger (metacarpal) |
| Lumbricales | • Tendons of flexor digitorum profundus | • Tendons of extensor digitorum communis |
| Dorsal interossei | • Metacarpals (adjacent sides) | • Second through fourth fingers (proximal phalanges) |
| Palmar interossei | • Second metacarpal (medial side)<br>• Fourth and fifth metacarpals (lateral side) | • Same finger (proximal phalanx) |
| **Lower extremities** | | |
| ***Femur (thigh) movement*** | | |
| Iliopsoas | • Psoas: major vertebrae (transverse processes and bodies of last thoracic and all lumbar)<br>• Iliacus: iliac crest and fossa | • Femur (lesser trochanter) |
| Gluteus maximus | • Ilium (posterior gluteal line)<br>• Sacrum and coccyx (posterior surfaces) | • Femur (gluteal tubercle)<br>• Iliotibial tract of fascia lata |
| Gluteus medius | • Ilium (outer surface, between posterior and anterior gluteal lines) | • Femur (lateral surface of greater trochanter) |
| Gluteus minimus | • Ilium (outer surface, between anterior and inferior gluteal lines) | • Femur (anterior surface of greater trochanter) |
| Tensor fasciae latae | • Iliac crest (anterior portion)<br>• Anterior superior iliac spine | • Iliotibial tract of fascia lata |
| Piriformis | • Sacrum (anterior surface) | • Femur (superior border of greater trochanter) |
| Obturator internus | • Obturator membrane (inner surface)<br>• Obturator foramen (bony margins) | • Femur (trochanteric fossa) |
| Obturator externus | • Obturator membrane (outer surface)<br>• Obturator foramen (bony margins) | • Femur (trochanteric fossa) |

| MUSCLE | ORIGIN | INSERTION |
|---|---|---|
| **Lower extremities** *(continued)* | | |
| ***Femur (thigh) movement*** | | |
| Gemellus superior | • Ischial spine | • Femur (greater trochanter) |
| Gemellus inferior | • Ischial tubercle | • Femur (greater trochanter) |
| Quadratus femoris | • Ischial tubercle | • Femur (shaft, just below greater trochanter) |
| Adductor magnus | • Pubis and ischium (inferior rami)<br>• Ischial tubercle | • Femur (linea aspera, adductor tubercle) |
| Adductor longus | • Pubis (crest and symphysis) | • Femur (linea aspera) |
| Adductor brevis | • Pubis (inferior ramus) | • Femur (linea aspera) |
| Pectineus | • Pubis (superior ramus) | • Femur (posterior surface, just below lesser trochanter) |
| Gracilis | • Symphysis pubis<br>• Pubic arch | • Tibia (medial surface, just below condyle) |
| ***Anterior compartment*** | | |
| Sartorius | • Anterior superior iliac spine | • Tibia (proximal medial surface, below tubercle) |
| Quadriceps femoris | • Rectus femoris: ilium (anterior inferior spine)<br>• Vastus lateralis: femur (linea aspera, greater trochanter)<br>• Vastus medialis: femur (linea aspera)<br>• Vastus intermedius: femur (anterior surface of shaft) | • Tibia (via patella and patellar ligament) |
| **Hamstring group** | | |
| Biceps femoris | • Long head: ischial tubercle<br>• Short head: linea aspera | • Fibula (lateral surface of head) |
| Semitendinosus | • Ischial tubercle | • Tibia (medial surface of proximal end) |
| Semimembranosus | • Ischial tubercle | • Tibia (medial surface of proximal end) |
| **Feet and toes** | | |
| ***Anterior*** | | |
| Tibialis anterior | • Tibia (lateral condyle, proximal two-thirds of shaft)<br>• Interosseous membrane | • Tarsal (first proximal two-thirds of cuneiform)<br>• Metatarsal (first) |
| Extensor hallucis longus | • Fibula (anterior surface of middle portion)<br>• Interosseous membrane | • Great toe (dorsal surface of distal phalanx) |
| Extensor digitorum longus | • Tibia (lateral condyle)<br>• Fibula (proximal three-fourths of anterior surface)<br>• Interosseous membrane | • Second through fifth toes (dorsal surfaces of phalanges) |
| Peroneus tertius | • Fibula (distal one-third of anterior surface)<br>• Interosseous membrane | • Fifth metatarsal (dorsal surface) |

*(continued)*

| MUSCLE | ORIGIN | INSERTION |
|---|---|---|
| **Feet and toes** *(continued)* | | |
| ***Lateral*** | | |
| Peroneus longus | • Base of fibula (proximal two-thirds of lateral surface) | • First metatarsal<br>• Medial cuneiform |
| Peroneus brevis | • Fibula (distal two-thirds) | • Fifth metatarsal (lateral side) |
| ***Posterior*** | | |
| Gastrocnemius | • Femur (medial and lateral condyles) | • Calcaneus (via Achilles tendon) |
| Soleus | • Fibula (posterior surface)<br>• Tibia (middle one-third) | • Calcaneus (via Achilles tendon) |
| Plantaris | • Femur (lower surface, above lateral condyle) | • Calcaneus (via Achilles tendon) |
| Popliteus | • Femur (lateral condyle) | • Tibia (proximal portion) |
| Flexor hallucis longus | • Fibula (lower two-thirds) | • Great toe (distal phalanx) |
| Flexor digitorum longus | • Tibia (posterior surface) | • Second through fifth toes (distal phalanges) |
| Tibialis posterior | • Tibia (posterior surface)<br>• Fibula (posterior surface)<br>• Interosseous membrane (posterior surface) | • Navicular bone<br>• All three cuneiforms<br>• Cuboid bone<br>• Second through fourth metatarsals |
| **Intrinsic foot** | | |
| ***Dorsal*** | | |
| Extensor digitorum brevis | • Calcaneus (lateral surface) | • Tendon of extensor digitorum longus |
| Dorsal interossei | • Adjacent metatarsal (bases)<br>• Second toe (both sides)<br>• Third and fourth toes (lateral side) | • Proximal phalanges |
| ***Plantar*** | | |
| Abductor hallucis | • Calcaneus | • Great toe (proximal phalanx), with tendon of flexor hallucis brevis |
| Flexor digitorum brevis | • Calcaneus<br>• Plantar aponeurosis | • Second through fifth toes (middle phalanges) |
| Abductor digiti minimi | • Calcaneus<br>• Plantar aponeurosis | • Fifth toe (proximal phalanges) |
| Quadratus plantae | • Calcaneus | • Tendons of flexor digitorum longus |
| Lumbricales | • Tendons of flexor digitorum longus | • Tendons of extensor digitorum longus |
| Flexor hallucis brevis | • Cuboid bone<br>• Lateral cuneiform | • Great toe (proximal phalanx) |
| Adductor hallucis | • Oblique head: second, third, and fourth metatarsals<br>• Transverse head: ligaments of metatarsophalangeal joints | • Great toe (proximal phalanx) |
| Flexor digiti minimi brevis | • Fifth metatarsal | • Fifth toe (proximal phalanx) |
| Plantar interossei | • Third through fifth metatarsals | • Same toe (proximal phalanx) |

Appendix D

# Prefixes and suffixes

## Prefixes

| | |
|---|---|
| a(n)- | absence, without |
| ab- | away from |
| acou- | hearing |
| acr(o)- | extremity, peak |
| ad- | toward |
| adeno- | gland |
| adipo- | fat |
| alb- | white |
| ambi- | on both sides |
| andr(o)- | male |
| angi(o)- | vessel |
| ankyl(o)- | crooked |
| ante- | before, forward |
| anti- | against |
| apo- | away from, separate |
| arteri(o)- | artery |
| arthr(o)- | joint |
| articul- | joint |
| auri- | ear |
| aut(o)- | self |
| bi- | two |
| bili- | bile |
| bio- | pertaining to life |
| blast- | embryonic state |
| blephar(o)- | eyelid |
| brachi(o)- | arm |
| brady- | slow |
| brom(o)- | stench |
| bronch(o)- | bronchus |
| bucc(o)- | cheek |
| calc- | heel |
| capit- | head |
| carcin(o)- | cancer |
| cardi(o)- | heart |
| caud- | tail |
| cephal(o)- | head |
| cerebr(o)- | cerebrum |
| cervic(i)(o)- | neck |
| cheil(o)- | lip |
| chol(e)(o)- | bile |
| chondr(o)- | cartilage |
| cili- | eyelid, eyelash |
| circum- | around |
| col(i)(o)- | colon |
| colp(o)- | vagina |
| con- | with |
| contra- | opposite |
| corpor- | body |
| cost(o)- | rib |
| crani(o)- | skull |

| | |
|---|---|
| cut- | skin |
| cyst(i)(o)- | bladder |
| cyt(o)- | cell |
| dacr(o)- | tears |
| dactyl(o)- | finger, toe |
| dendr(o)- | treelike |
| dent(i)(o)- | tooth |
| derm- | skin |
| dextr(o)- | right |
| dia- | across, through |
| digit(i)- | finger, toe |
| disc(i)(o)- | disk-shaped |
| dors(o)- | back |
| dys- | difficult, painful |
| ect(o)- | outside |
| end(o)- | inward |
| enter(o)- | intestine |
| ep(i)- | on, upon |
| eso- | within |
| esthe- | perception, feeling |
| ex(o)- | outside |
| extra- | outside of, beyond |
| fasci- | bundle |
| fibr(o)- | fiber |
| fil- | threadlike |
| fiss- | cleft, split |
| gastr(o)- | stomach |
| gloss(o)- | tongue |
| gluc(o), glyc(o)- | sweet |
| gon(o)- | semen, seed |
| hem(a)(o)- | blood |
| hepat(o)- | liver |
| histi(o), hist(o)- | tissue |
| hyp(o)- | below, under |
| hyper- | above, beyond |
| hyster(o)- | uterus |
| ile(o)- | ileum |
| ili(o)- | ilium, flank |
| infra- | beneath |
| inter- | between, among |
| is(o)- | equal, identical |
| ischi(o)- | hip |
| jejun(o)- | jejunum |
| juxta- | near |
| kerat(o)- | horny tissue, cornea |
| labio- | lips |
| laryng(o)- | larynx |
| latero- | side |
| leuk(o)- | white |
| lip(o)- | fat |
| lymph(o)- | lymph |

(continued)

mamm(o)- breast
mast(o)- breast
medi(o)- middle
mega- great, large
meta- beyond, change
metr(o)- uterus
mio- less
mito- threadlike
mono- one
my(o)- muscle
myel(o)- marrow, spinal cord
myx(o)- mucus
nas(o)- nose
ne(o)- new
nephr(o)- kidney
neur(o)- nerves
ocul(o)- eye
omphal(o)- navel
onych(o)- nail
ophthalm(o)- eye
orchi(o)- testes
oro- mouth
oss-, oste(o)- bone
ot(o)- ear
ov(i)(o)- egg, ova
par(a)- beside
pell- skin
peri- around
phako- lens
pharyng(o)- pharynx
phleb(o)- vein
pil(i)(o)- hair
pleur(o)- pleura, rib, side
plic- fold, ridge
pneum(o)- lung
pod(o)- foot
poly- much, many
post- behind, after
pre- before, in front
pro- before, in front
proct(o)- rectum
prote(o)- protein
proto- first
pseudo- false
pulmo(n)- lung
pyel(o)- pelvis (kidney)
pyr(o)- heat
ren(o)- kidney
reticul(o)- netlike
retr(o)- netlike
retr(o)- backward
rhin(o)- nose
sarc(o)- flesh
scler(o)- hard
scolio- crooked
sept(o)- septum
squam(o)- scale
sub- under
super- above, over
supra- above, upon
sym- union
syn- union
test- testicle
thorac(o)- chest
trache(o)- trachea
trans- across, through
trich(o)- hair
vas(o)- vessel
ven(i)(o)- vein
ventr(o)- belly, front
ventriculo- ventricle
vertebr(o)- vertebra
vesic(o)- bladder

## Suffixes

-algia pain
-ary pertaining to
-biosis life
-blast embryonic state
-cele hernia, tumor
-crine secretion
-cyte cell
-dorsal back
-ectomy surgical removal
-osm(o) smell
-plasia growth
-pnea breathing
-poiesis production
-por(o) passageway
-praxia movement
-rrhea fluid discharge
-stomy opening
-trichia hair
-trophy growth, nutrition
-tropic influential
-uria urine

## Appendix E

# Abbreviations and symbols

## Abbreviations

| | |
|---|---|
| a | arterial blood |
| A | alveolar gas |
| ABG | arterial blood gas |
| Ach | acetylcholine |
| ACTH | adrenocorticotropic hormone, adrenocorticotrophic hormone |
| ADH | antidiuretic hormone |
| ADP | adenosine diphospate |
| ANS | autonomic nervous system |
| AP | anteroposterior |
| ATP | adenosine triphosphate |
| AV | atrioventricular, arteriovenous |
| AVN | atrioventricular node |
| b | blood |
| BBB | blood-brain barrier |
| BUN | blood urea nitrogen |
| C | cervical vertebrae; carbon |
| CBC | complete blood count |
| CBF | cerebral blood flow |
| CN | cranial nerve |
| CNS | central nervous system |
| cm | centimeter |
| COPD | chronic obstructive pulmonary disease |
| CSF | cerebrospinal fluid |
| CVA | costovertebral angle; cerebrovascular accident |
| DNA | deoxyribonucleic acid |
| DIP | distal interphalangeal |
| DTR | deep tendon reflex |
| EAO | external abdominal oblique muscle |
| ECF | extracellular fluid |
| ENT | ear, nose, throat |
| FSH | follicle-stimulating hormone |
| GFR | glomerular filtration rate |
| GI | gastrointestinal |
| GU | genitourinary |
| HCl | hydrochloric acid |
| HEENT | head, eyes, ears, nose, throat |
| IAO | internal abdominal oblique muscle |
| ICF | intracellular fluid |
| ICP | intracranial pressure |
| IP | interphalangeal |
| IVC | inferior vena cava |
| L | lumbar vertebrae |
| LAD | left anterior descending artery |
| LES | lower esophageal sphincter |
| LLQ | left lower quadrant |
| LUQ | left upper quadrant |
| MAP | mean arterial pressure |
| MCL | midclavicular line |
| ml | milliliter |
| MP | metacarpophalangeal, metatarsophalangeal |
| Na | sodium |
| NSR | normal sinus rhythm |
| PIP | proximal interphalangeal |
| PMI | point of maximum impulse |
| PTH | parathyroid hormone |
| RAS | reticular activating system |
| RBC | red blood cell |
| RCA | right coronary artery |
| REM | rapid eye movement |
| RNA | ribonucleic acid |
| RLQ | right lower quadrant |
| ROM | range of motion |
| RUQ | right upper quadrant |
| SA | sinoatrial |
| SMA | superior mesenteric artery |
| SVC | superior vena cava |
| T | thoracic vertebrae |
| $T_3$ | 3,5,5′-triiodothyronine |
| $T_4$ | tetraiodothyronine (thyroxine) |
| TA | transverse abdominal muscle |
| TMJ | temporomandibular joint |
| TSH | thyroid-stimulating hormone |
| v | venous blood; vein; volt |
| WBC | white blood cell; white blood cell count |
| VC | vision, color; vital capacity |

## Symbols

| | |
|---|---|
| ā | before |
| @ | at |
| c̄ | with |
| L | left |
| R | right |
| ♀ | female |
| ♂ | male |
| ↑ | increase |
| ↓ | decrease |
| ? | questionable |
| p̄ | after |
| s̄ | without |
| x | times |
| i̇̄, i̇̄i̇̄ | one, two, etc |
| Ⓛ | left |
| Ⓡ | right |
| LA | left arm |
| RA | right arm |
| LL | left leg |
| RL | right leg |
| ∅ | none |
| ṡṡ, s̄s̄ | one-half |
| ∧ | diastolic blood pressure |
| ∨ | systolic blood pressure |
| 1° | primary, first degree |
| 2° | secondary, second degree |
| 3° | tertiary, third degree |
| āā | of each |
| ″ | inch |
| # | pound, number |
| ♏ | minim |
| ℞ | prescription |
| ʒ | dram |
| ℥, | ounce |
| = | equal to |
| ≅ | approximately equal to |
| ≈ | approximately |
| ȱ | no |
| 1:1 | one to one |
| + | positive |
| - | negative |
| +/- | may or may not be appropriate |
| > | greater than |
| ≥ | greater than or equal to |
| < | less than |
| ≤ | less than or equal to |

# Glossary

**Abdomen** The portion of the trunk between the thorax and the pelvis. It contains the major visceral organs: the lower part of the esophagus, the stomach, the intestines, the liver, the spleen, and the pancreas.

**abdominal aorta** The portion of the descending aorta that lies in the abdomen and conveys blood to body structures below the diaphragm. Its branches are the celiac, superior mesenteric, inferior mesenteric, middle suprarenals, renals, testicular, ovarian, inferior phrenics, lumbars, middle sacral, and common iliacs.

**abdominal aponeurosis** The point at which the tendons of the oblique and transverse muscles of the abdomen join.

**abducens nerve** One of a pair of motor nerves (the sixth cranial nerves) arising from the pons that control movement of the lateral rectus muscle of the eye. Also called nervus abducens.

**abduct** To move away from the body.

**accessory nerve** Either of a pair of cranial nerves supplying the muscles of the pharynx, larynx, soft palate, and sternocleidomastoid and trapezius muscles that permit speech, swallowing, and certain movements of the head and shoulders. Each nerve is rooted both in the sides of the medulla and in the cervical spine. Also called eleventh cranial nerve, nervus accessorius, spinal accessory nerve.

**accommodation reflex** The adjustment made by the eyes to focus on near objects, consisting of three actions that change the path of light rays reaching the retina: pupil constriction, ciliary muscle contraction that causes increased rounding of the lens, and convergence of the eyes. Also called ciliary reflex.

**acetabulum, pl. acetabula** The cup-shaped depression containing the head of the femur, at the juncture of the ilium, ischium, and pubis.

**Achilles tendon reflex** Plantar flexion of the foot resulting from contraction in response to a sharp tap on the gastrocnemius muscle at the back of the ankle. Absence of this reflex may occur in patients with diabetes mellitus or peripheral neuropathy. Also called ankle reflex, calcaneal tendon reflex.

**acinus, pl. acini** 1. A small, saclike structure, also called alveolus. 2. The part of the airway distal to the terminal bronchiole.

**acoustic nerve** Either of the pair of cranial nerves involved in hearing and in maintaining equilibrium. Each acoustic nerve is composed of the cochlear and vestibular nerves. Also called eighth cranial nerve.

**acromioclavicular articulation** The joint between the acromial end of the clavicle and the medial margin of the acromion of the scapula.

**acromion** The lateral triangular projection of the spine of the scapula. It articulates with the clavicle and forms the point of the shoulder. Also called acromial process.

**actin** One of the two protein components of muscle fibers; actin is active in muscle contraction and cellular movement and helps to maintain the shape of a cell.

**adduct** To draw a limb toward the median axis of the body.

**adenohypophysis** The anterior lobe of the pituitary gland, which secretes a number of hormones, including growth hormone, thyrotropin, adrenocorticotropic hormone, melanin-stimulating hormone, follicle-stimulating hormone, luteinizing hormone, prolactin, and beta lipotropin, as well as the neurotransmitter endorphin. Also called anterior pituitary.

**adipose** Of or pertaining to fat or fatty tissue.

**adrenal cortex** The portion of the adrenal gland that produces and secretes steroid hormones. The outer portion of the cortex is a deep yellow; the inner portion, dark red or brown.

**adrenal gland** One of the two triangle-shaped, ductless glands at the superior pole of each kidney. The gland secretes cortisol, androgens, and aldosterone (from the adrenal cortex) and the catecholamines epinephrine and norepinephrine (from the adrenal medulla).

**adrenergic** 1. Pertaining to sympathetic nerve fibers of the autonomic nervous system. 2. Of or relating to substances, such as drugs or hormones, that produce adrenergic effects.

**afferent** Moving toward the center; carrying impulses or substances toward the center. The term usually refers to nerves, but may also refer to arteries, veins, and lymphatics.

**ala, pl. alae** 1. Any structure resembling a wing. 2. The axilla (armpit).

**alveolar canal** One of the canals of the upper jawbone (maxilla) that serve as a passage for nerves and blood vessels to the upper teeth.

**alveolar duct** The branch of a bronchiole that leads to the alveolar sacs in the lung.

**alveolus** 1. Any small, saclike structure. 2. An air-filled cell in the lung. 3. The follicle of an alveolar or racemose gland. 4. One of the honeycomb-like depressions in the stomach wall.

**ampulla, pl. ampullae** A saclike swelling or pouch formed in a portion of a duct or canal.

**anal** Of or relating to the anus.

**anal canal** The terminal portion of the large intestine, situated between the rectal ampulla and the anus.

**anal crypt** A channel separating the rectal columns.

**anatomic snuffbox** A small hollow formed by tendons on the back of the hand near the radial aspect of the wrist, visible when the thumb is abducted and extended.

**ankle** 1. The joint formed by the talus, tibia, and fibula. 2. The bone known as the ankle bone, or talus. 3. The part of the leg where this joint is located.

**annular horn** Motor portion of gray matter

**annular ligament** A ligament that encircles a bone or other structure.

**anterior longitudinal ligament** The wide ligament that extends from the occipital bone and anterior tubercle of the first cervical vertebra to the sacrum. Also called anterior common ligament.

**anus** The external orifice at the lower end of the gastrointestinal tract through which feces exit the body.

**aorta** The main trunk of the systemic arterial circulation, having four parts: the ascending aorta, the arch of the aorta, and the two branches of the descending aorta (the thoracic portion and the abdominal portion).

**aortic valve** A valve at the opening between the left ventricle and the aorta, preventing the backflow of blood into the left ventricle. Also called the tricuspid valve, it is composed of three half-moon-shaped cusps, which open during systole and close during diastole.

**aperture** An opening or an orifice in a body structure.

**apex, pl. apices** The extreme end or tip of a body structure, such as the apex of the heart.

**apocrine gland** A sweat gland, prevalent under the arms, around the areolae of the breasts, and in the genital and anal areas, whose duct opens into the hair follicle above the sebaceous duct. The action of bacteria on apocrine gland secretions produces the characteristic odor of perspiration.

**aponeurosis, pl. aponeuroses** A flat, fibrous sheet of connective tissue or an expanded tendon that attaches muscle tissue to bone or other tissue. At some sites, the aponeurosis may also serve as fascia, binding muscles together.

**aqueous humor** The clear, watery liquid that fills and circulates in the posterior and anterior chambers of the eye. It is produced by the ciliary processes and is reabsorbed into the venous circulation.

**arachnoid** 1. Resembling a cobweb or spiderweb. 2. The arachnoid membrane, a delicate, fibrous tissue forming the middle of the three tissue layers that cover the brain and spinal cord.

**areola, pl. areolae** 1. A circular zone of differently pigmented tissue, such as that which surrounds the breast nipples (areola mammae) or a reddened area around a wheal or pustule. 2. In the eye, the part of the iris that surrounds the pupil.

**areolar gland** One of a number of sebaceous glands that protrude from the areolae on a woman's breasts and secrete a lubricating fluid during breast-feeding. Also called Montgomery's follicles or glands.

**areolar tissue** Tissue composed of collagenous and elastic fibers, connective tissue cells, and a semifluid matrix, which forms interstitial tissue, membranes surrounding blood vessels and nerves, and a portion of fascia. Also called fibroareolar tissue.

**arteriole** The smallest branch of the arterial circulation, which communicates with the network of capillaries.

**arteriovenous** Of or relating to both arteries and veins.

**artery** One of the blood vessels that carry oxygenated blood away from the heart and toward the periphery of the body.

**articular capsule** A sac of tissue that envelops a joint, composed of an inner synovial membrane and an outer fibrous membrane.

**articular disk** The plate of fibrocartilage found in certain joints that attaches to the articular capsule and separates the ends of two bones that do not fit together well.

**ascending aorta** One of the four main segments of the aorta.

**ascending colon** The portion of the colon (large intestine) that extends from the cecum to the right hepatic flexure, usually at the umbilical level.

**atlanto-occipital joint** A joint at the base of the skull between the occipital bone and the first cervical vertebra (atlas) that allows flexion of the neck toward the chest and side-to-side movements of the head.

**atlas** The first cervical vertebra. It articulates above with the occipital bone and below with the axis (second cervical vertebra), rotating around the toothlike projection of the axis.

**atrioventricular septum** One of two membranous structures in the heart that separate the atria from the ventricles.

**atrioventricular valve** A valve between an atrium and a ventricle that prevents backflow of blood from the ventricles. The mitral (bicuspid) valve is located between the left atrium and left ventricle; the tricuspid valve, between the right atrium and right ventricle.

**atrium of the heart** One of the two chambers that form the upper half of the heart. Deoxygenated blood from the superior vena cava, inferior vena cava, and coronary sinus drains into the right atrium; oxygenated blood from the pulmonary veins, into the left atrium. During diastole, blood empties from the atria into the ventricles.

**auricle** 1. The shell-like structure of the external ear; the pinna.

**autonomic** Of or relating to the autonomic nervous system.

**autonomic nervous system** The division of the nervous system that regulates the activity of cardiac muscle, smooth muscle, and glands. It is subdivided into the sympathetic nervous system, which accelerates the heart rate, narrows blood vessels, and raises blood pressure; and the parasympathetic nervous system, which decreases the heart rate, accelerates intestinal peristalsis and gland activity, and relaxes the sphincters.

**axilla, pl. axillae** The space below the shoulder joint where the upper part of the underside of the arm meets the side of the chest. Also called armpit.

**axillary vein** One of a pair of veins that continues from the basilic and brachial veins and becomes the subclavian vein at the outer border of the first rib.

**axis, pl. axes** 1. In anatomy, an imaginary line that passes vertically through the center of the body or a body part, used to provide a starting point for anatomic angles and other references. 2. The second cervical vertebra, upon which the atlas (the first cervical vertebra) rotates, permitting turning, extension, and flexion of the head.

**axon** The process of a neuron that conducts nervous impulses away from the nerve cell body.

**azygos vein** One of the seven thoracic veins. It arises from either the right ascending lumbar vein or the inferior vena cava, passes through the diaphragm at the aortic orifice, travels to the right of the vertebral column to the fourth thoracic vertebra, then rises ventrally over the root of the right lung and terminates in the superior vena cava.

**Ball-and-socket joint** A synovial joint in which the ball-shaped head of an articulating bone fits into a concave socket of another bone. This type of joint permits the widest range of motion.

**band** In anatomy, a bundle of fibers that binds one part of the body to another or that encircles a structure.

**Bartholin's duct** The major duct draining the sublingual salivary gland.

**Bartholin's gland** One of two small, mucus-secreting glands on either side of the vaginal opening.

**basal cell** A cell in the basal layer of the epidermis.

**basal ganglia** Interconnected islands of gray matter, such as the caudate nucleus, putamen, and pallidum, deep within the cerebral hemispheres and in the upper brain stem. Also called basal nuclei.

**base of the heart** The portion of the heart that lies opposite the apex and just below the second rib, forming the upper border of the heart. It consists primarily of the left atrium, part of the right atrium, and the proximal portions of the great vessels.

**base of the skull** The floor of the cranium, containing the anterior, middle, and posterior cranial fossae and numerous foramina.

**basement membrane** A sheet of fragile, noncellular tissue on which the basal surfaces of epithelial cells rest. Its two layers are the basal lamina and reticular lamina.

**basilar** Of or describing a base or a basal area.

**beta cells** 1. Insulin-producing cells that compose the bulk of the islets of Langerhans. 2. Basophilic cells of the anterior lobe (adenohypophysis) of the pituitary gland.

**biceps brachii** The long fusiform muscle of the upper arm on the anterior surface of the humerus, arising in two heads from the scapula.

**bicuspid** Having two cusps or points.

**bile** A bitter, yellow-green fluid secreted by the liver and stored in the gallbladder. Bile flows from the gallbladder through the common bile duct after a fatty meal; it emulsifies the fats, preparing them for further digestion and absorption in the small intestine.

**biliary duct** Common hepatic duct; conveys bile from the liver to the duodenum.

**biparietal diameter** The distance separating the protuberances of the two parietal bones of the skull.

**bladder** 1. A saclike organ that acts as a receptacle for secretions. 2. The urinary bladder.

**blepharon** Eyelid.

**blood vessel** Any part of the network of tubes that carry blood throughout the body.

**blood-brain barrier (BBB)** An anatomic-physiologic feature of the brain, thought to consist of the walls of capillaries in the central nervous system and surrounding glial membranes; it slows or prevents passage of most toxic substances from the blood into the central nervous system.

**bone** The connective tissue that forms the skeleton of most vertebrates, comprising the 206 bones of the human skeleton. This dense, slightly elastic tissue is composed of organic material (cells and matrix) and inorganic material (mineral components, chiefly calcium phosphate and calcium carbonate).

**bone marrow** The soft tissue that fills bone cavities. Yellow marrow consists mainly of fat cells. Red marrow, found in developing bone in the sternum, ribs, and vertebral bodies of adults, is the production site of erythrocytes and granular leukocytes.

**Bowman's capsule** The cup-shaped dilatation that forms the beginning of a renal tubule and surrounds the glomerulus. Also called glomerular capsule.

**brain** One of the two components of the central nervous system, encased in the cranium and continuous with the spinal cord.

**brain stem** The portion of the brain that connects the cerebral hemispheres to the spinal cord, consisting of the medulla oblongata, pons, and mesencephalon.

**broad ligament of the uterus** A fold of the peritoneum that supports the uterus from either side.

**bronchial tree** The trachea, bronchi, and their branches. The trachea divides to form the bronchi, which branch further into lobar and segmental bronchi, ultimately forming bronchioles and alveoli. Gas exchange takes place in the alveoli.

**bronchiole** A small subdivision of the bronchial tree that extends from the bronchi into the lobes of the lung.

**bronchus, pl. bronchi** Any of the larger air passages of the lungs, which convey inhaled air and exhaled waste gases.

**bulbourethral gland** One of two small glands next to the prostate that secrete the fluid component of the seminal fluid.

**bundle of His** A band of fibers arising in the atrioventricular (AV) node of the heart through which cardiac impulses are transmitted to the ventricles. After leaving the AV node, the bundle of His travels through the atrioventricular junction and beneath the endocardium of the right ventricle. At the upper end of the muscular part of the interventricular septum, it separates into right and left bundle branches, which descend as Purkinje fibers into the walls of the right and left ventricle, respectively. Also called atrioventricular bundle.

**bursa, pl. bursae** A fibrous, saclike cavity containing a viscid fluid, situated between certain tendons and the bones beneath them. The bursae act to prevent friction.

**Calcaneus** The largest tarsal bone, generally known as the heel bone. The calcaneus articulates proximally with the talus and distally with the cuboid.

**calvaria** The skull cap or superior portion of the cranium where the frontal, parietal, and occipital bones meet.

**capillary** Any of the tiny vessels (roughly 0.008 mm in diameter) that join the arterioles and venules. The walls of the capillary act as semipermeable membranes for the exchange of various substances between the blood and interstitial fluid.

**capillus, pl. capilli** A hair, especially of the scalp.

**capitulum, pl. capitula** A small, rounded eminence on a bone where it articulates with another bone.

**capsule** A part of the body that encloses an organ or other body part.

**cardiac muscle** Striated, involuntary muscle of the myocardium, or wall of the heart, containing dark intercalated disks where the cytoplasmic membranes of two cardiac fibers abut. Cardiac muscle moves blood through the heart and into blood vessels.

**cardiac plexus** A network of nerves surrounding the base of the heart.

**cardiac sphincter** A circle of muscle fibers located where the esophagus and stomach meet; another name for the lower esophageal sphincter.

**cardiovascular system** The network consisting of the heart, blood vessels, and lymphatics that serves as the body's transport system. Also called circulatory system.

**carotid body** A small neurovascular structure at the bifurcation of the carotid arteries that detects variations in blood oxygen levels and changes in blood pressure.

**carotid sinus** A dilated portion of the internal carotid artery, near the bifurcation of the common carotid artery.

**carpal tunnel** A passageway created by the carpal bones and the flexor retinaculum tendon, through which the median nerve and the flexor tendons pass.

**carpus** The wrist; the joint at which the arm and hand meet, composed of eight bones.

**cartilage** A fibrous connective tissue found in the joints, thorax, and various rigid tubes, such as the larynx, trachea, nose, and ear of the adult. In utero, cartilage serves as the temporary skeleton of the embryo, and then as the blueprint for bone development.

**cartilaginous joint** A slightly movable joint in which cartilage connects the bony surfaces. The cartilaginous joints include symphyses (such as the symphysis pubis) and synchondroses (such as the joints between the ribs and the sternum).

**cavernous sinus** One of two irregularly shaped, bilateral venous spaces in the head between the sphenoid bone and the dura mater.

**cecum** A cul-de-sac, or blind pouch, at the beginning of the large intestine.

**celiac trunk** A branch of the abdominal aorta that supplies the stomach, liver, pancreas, spleen, and duodenum.

**cell** The structural and functional unit of all living organisms. Every cell contains a nucleus, cytoplasm, and organelles enclosed by a cytoplasmic membrane. Cells may exist as independent units of life (as in bacteria and protozoans), or they may cluster into colonies or tissues (as in higher animals and plants).

**centimeter (cm)** The metric unit of measurement of length equal to one hundredth of a meter, or 0.3937 inches.

**central nervous system (CNS)** Collective term for the brain and spinal cord; one of the two main divisions of the nervous system.

**cerebellar artery** One of three arteries that supply blood to the cerebellum, medulla, pineal body, and midbrain.

**cerebellar cortex** The gray matter of the cerebellum that covers the white matter in the medullary core. It consists of an external molecular layer and an internal granular cell layer.

**cerebellum, pl. cerebellums, cerebella** The part of the brain behind the brain stem in the posterior cranial fossa, consisting of a middle strip (the vermis) and two lateral cerebellar hemispheres (lobes). The cerebellum plays an important role in coordination, regulation of muscle activity, and maintenance of muscle tone and balance.

**cerebral aqueduct** The narrow channel between the third and fourth ventricles in the midbrain through which the cerebrospinal fluid travels.

**cerebral cortex** A thin mantle of gray matter on the surface of each cerebral hemisphere, folded into gyri and separated by fissures. It is responsible for the higher mental functions (memory, thought, intellect, and language) as well as for the control and integration of voluntary movement and the special senses (hearing, smell, taste, and sight).

**cerebral hemisphere** Either of the two halves of the cerebrum. Partly divided by the longitudinal cerebral fissure, the cerebral hemispheres are joined medially at the bottom of the fissure by the corpus callosum.

**cerebrospinal fluid (CSF)** The plasmalike fluid contained within the four ventricles of the brain, the subarachnoid space, and the central canal of the spinal cord.

**cerebrum, pl. cerebrums, cerebra** The largest portion of the brain, in the uppermost part of the cranium.

**cervical** Of or pertaining to the neck or the region of the neck of any organ or structure.

**cervical canal** The canal within the uterine cervix, between the uterus and the vagina.

**cervical plexus** The network of nerves formed by the ventral branches of the first four cervical nerves.

**cervical vertebra** One of the upper seven bones of the spinal column that form the neck.

**chordae tendineae** Tendinous strings that extend from the cusps of the atrioventricular valves to the papillary muscles of the heart, thus preventing valve inversion.

**choroid** A thin, pigmented vascular membrane that covers the posterior five-sixths of the eye. The choroid supplies blood to the retina and conducts arteries and nerves to the anterior structures.

**chyle** The milky fluid product of digestion, consisting mainly of lymph and triglyceride fats in an emulsion. Chyle passes through lacteals, fingerlike projections in the small intestine, into the lymphatic system for transport to the venous circulation.

**chyme** The creamy, semifluid substance produced in the stomach by gastric digestion. It then passes through the pylorus into the duodenum for further digestion.

**cilia, sing. cilium** Tiny, hairlike projections on the surfaces of some cells, which produce motion or a current in a fluid.

**ciliary body** The thickened part of the vascular tunic of the eye, joining the choroid with the iris. It is made up of the ciliary crown, ciliary processes and folds, ciliary orbiculus, ciliary muscle, and a basal lamina.

**ciliary process** Any of about 80 tiny fleshy folds on the posterior surface of the iris, which secrete aqueous humor into the posterior chamber of the eye.

**circle of Willis** A vascular network at the base of the brain where the internal carotid, anterior cerebral, posterior cerebral, anterior communicating, and posterior communicating arteries interconnect.

**clavicle** A long, horizontal, f-shaped bone that forms the anterior portion of the shoulder girdle on either side. Also called collar bone.

**clavicular notch** One of the two oval depressions on either side of the superior end of the sternum, where the clavicles articulate with the sternum.

**coccyx** The small bone at the base of the spine, named for its resemblance to the bill of the cuckoo. It is formed by three to five rudimentary vertebrae that are joined together and connected to the sacrum by a disk of fibrocartilage.

**cochlea** A spiral-shaped bony structure of the inner ear, containing apertures for passage of the cochlear division of the acoustic nerve.

**collagen** A protein substance consisting of white, glistening, inelastic fibers. It is found in the skin, tendon, bone, cartilage, and other connective tissue.

**collateral** In anatomy, a small side branch, such as an arteriole or venule.

**collecting tubule** A channel that carries fluids from secretory cells, such as those in the kidney that funnel urine from the distal convoluted tubules into the renal pelvis.

**colon** The part of the large intestine that extends from the cecum to the rectum, consisting of four segments (ascending, transverse, descending, and sigmoid).

**common bile duct** The duct formed by the joining of the cystic duct and hepatic duct.

**common carotid artery** An artery that arises from the aortic arch and has cervical and thoracic portions. There are two common carotid arteries, the right and the left. The left is the longer of the two.

**concha** In anatomy, a part or structure of the body that resembles a shell, such as the outer ear.

**condyle** A rounded projection at the end of a bone that allows articulation with adjacent bones.

**condyloid joint** A synovial joint in which an ovoid head of one bone moves in an elliptical cavity of another. Condyloid joints permit all movement except axial rotation.

**conjunctiva** The mucous membrane that covers the inner surfaces of the eyelids and the anterior portion of the sclera.

**connective tissue** Tissue that supports and binds together body structures; it includes bone, cartilage, and adipose tissue. Connective tissue is composed of collagenous, reticular, and elastic fibers.

**coracoid process** The strong, curved projection of the superior border of the scapula, to which the pectoralis minor is attached. It overhangs the shoulder joint.

**cornea** The transparent, anterior portion of the eye.

**coronal suture** The juncture line between the frontal bone and the two parietal bones.

**coronary artery** One of two arteries that branch from the aorta and supply the heart muscle.

**coronary vein** A vein of the heart that channels blood from the capillary beds of the myocardium through the coronary sinus and into the right atrium.

**Corti's organ** A spiral structure on the basilar membrane in the cochlear duct of the ear that contains special sensory receptors for hearing. Also called spiral organ of Corti.

**costovertebral angle (CVA)** The angle formed on either side of the vertebral column, between the last rib and the lumbar vertebrae.

**Cowper's gland** Either of two pea-sized glands embedded in the urethral sphincter of the male, posterior to the membranous portion of the urethra. Also called bulbourethral gland.

**cranium** The skeleton of the head, encasing the brain. It is composed of eight bones—frontal, occipital, sphenoid, and ethmoid bones, and paired temporal and parietal.

**cricoid** A ring-shaped cartilage attached to the thyroid cartilage by the cricothyroid ligament at the level of the sixth cervical vertebra.

**cubitus** The joint between the arm and forearm; the bend of the arm.

**cul-de-sac** A blind pouch or cecum (such as the conjunctival cul-de-sac), or a tubular cavity closed at one end (such as the diverticulum).

**cuneiform bone** One of three cuneiform bones (anterior, medial, and lateral) of the foot, on the medial side of the tarsus, between the scaphoid bone and the first metatarsal.

**cuticle** 1. A layer that covers the free surface of an epithelial cell. 2. The narrow band of epidermis that extends from the wall of the nail onto the nail surface. Also called eponychium. 3. The sheath of a hair follicle.

## D

**Deep fascia** The most extensive of the three kinds of fascia, or fibrous connective tissue, of the body. The deep fasciae form a series of connective sheets and bands that hold the muscles and other structures in place. Constituting a continuous system, they split and merge in a network that is attached to the skeleton.

**deep tendon reflex (DTR)** Involuntary contraction of a muscle after sudden stretching resulting from sharp tapping on the muscle's tendon of insertion. DTRs include Achilles tendon, biceps, brachioradialis, patellar, and triceps reflexes. Also called myotatic reflex, tendon reflex.

**dendrite** The threadlike process that projects from the cell body of a neuron; it receives impulses and conducts them to the cell body.

**dense fibrous tissue** Connective tissue consisting of strong, compact, inelastic bundles of parallel collagenous fibers. Dense fibrous tissue provides structural support. Organized (regular) dense fibrous tissue includes the tendons, aponeuroses, and ligaments; unorganized (irregular) dense fibrous tissue includes the fascial membranes, dermis, periosteum, and capsules of organs.

**dermis** A layer of skin just below the epidermis, consisting of dense vascular connective tissue with blood and lymphatic vessels, nerves and nerve endings, glands, and hair follicles. Also called corium.

**descending aorta** The main portion of the aorta, consisting of the thoracic aorta and abdominal aorta. Continuing from the aortic arch into the trunk of the body, it branches to supply many parts of the body, such as the esophagus, lymph glands, ribs, and stomach.

**descending colon** The segment of the colon extending downward from the end of the transverse colon at the splenic flexure on the left side of the abdomen to the beginning of the sigmoid colon in the pelvic cavity.

**diaphragm** In anatomy, the musculofibrous partition that separates the abdominal and thoracic cavities. The concave surface of the diaphragm forms the roof of the abdominal cavity; the convex cranial surface forms the floor of the thoracic cavity. During inspiration, the diaphragm moves down and expands the volume of the thoracic cavity; during expiration, it moves up, reducing the volume.

**diaphysis** The shaft, or cylindrical portion, of a long bone between the epiphyses (ends), consisting of a tube of compact bone enclosing the medullary cavity.

**diencephalon** The portion of the brain between the telencephalon and mesencephalon. Forming the central core of the forebrain, it connects the cerebrum with the brain stem. The diencephalon includes the hypothalamus, thalamus, metathalamus, and epithalamus and most of the third ventricle.

**dilatator pupillae** An involuntary muscle innervated by nerve fibers from the sympathetic system that contracts the iris of the eye and induces pupil dilation. Radiating fibers in this muscle converge from the circumference of the iris toward the center, blending with fibers of the sphincter pupillae near the margin of the pupil.

**distal** 1. Farthest away. 2. Away from the point of origin, the midline, or a central part of the body.

**distal convoluted tubule** A convoluted portion of the renal tubule between the ascending loop of Henle and the collecting duct.

**dorsiflexion** Flexion or bending toward the extensor aspect of a limb, as in the upward bending of the hand on the wrist or the foot on the ankle.

**ductus epididymis** The duct, or tube, into which the coiled ends of the efferent ductules of the testis empty.

**duodenum** The first or proximal portion of the small intestine, between the pylorus and the jejunum.

**dura mater** The outermost, toughest, and most fibrous of the three membranes that surround the brain and spinal cord.

## E

**Ear** The sense organ for hearing, specialized for detecting sound and maintaining equilibrium.

**eccrine gland** One of two types of sweat glands in the dermal layer of the skin. Eccrine glands promote cooling through evaporation of their secretion — a clear substance with little or no odor that contains water, sodium chloride, and traces of albumin, urea, and other compounds.

**efferent** Moving away from the center toward the periphery, as certain arteries, veins, nerves, and lymphatics.

**ejaculatory duct** The passage through which semen enters the urethra; it is formed by the union of the ductus deferens with the excretory duct of the seminal vesicle.

**elbow** The joint that connects the arm and forearm; it is covered by a protective capsule associated with three ligaments and an extensive synovial membrane. The elbow joint accommodates the radioulnar articulation and permits flexion and extension of the forearm.

**endometrium** The inner mucous membrane lining of the uterus, which has three layers — stratum compactum, stratum spongiosum, and stratum basale.

**endothelium** The layer of epithelial cells that is derived from the mesoderm and lines the cavities of the heart, blood and lymph vessels, and serous cavities.

**endotracheal** Within or through the trachea.

**epicardium** The visceral portion of the pericardium. It envelops the heart and folds back upon itself to form the parietal portion of the serous pericardium. It is composed of a single sheet of squamous epithelial cells overlying delicate connective tissue.

**epicondyle** A projection or eminence on the surface of a bone above its condyle.

**epicranium** The entire scalp, including the integument, aponeuroses, and muscular sheets.

**epidermis** The surface, or outermost, layer of the skin.

**epididymis** One of a pair of elongated, cordlike structures whose tightly coiled duct stores and conveys sperm from the seminiferous tubules of the testes to the vas deferens.

**epidural** On or outside the dura mater.

**epidural space** The space between the dura mater of the brain or spinal cord and the walls of the vertebral canal.

**epigastric** Of or pertaining to the epigastrium, the upper middle region of the abdomen within the infrasternal angle.

**epiglottis** The lidlike, cartilaginous structure that overhangs the larynx and prevents food from entering the larynx and trachea during swallowing.

**epiphysis, pl. epiphyses** The articular end of a long bone, which is separated from the shaft by the epiphyseal plate until the bone stops growing, the plate is obliterated, and the shaft and head unite.

**epispadias** The urethral opening on the dorsum of the penis.

**epithelium** A continuous cellular sheet that covers the body's surface, lines body cavities, and forms certain glands. Types of epithelium include simple squamous, simple cuboidal, and stratified columnar.

**esophageal artery** The branch of the thoracic aorta that supplies the esophagus.

**esophagus** The muscular passage that extends from the pharynx —beginning at the cricoid cartilage, at the level of the sixth cervical vertebra — to the cardiac sphincter of the stomach.

**ethmoid bone** The light, spongy bone at the base of the cranium; it forms most of the walls of the superior part of the nasal cavity.

**eustachian tube** The channel, about 36 mm (1½″) long, that connects the tympanic cavity with the nasopharynx and adjusts air pressure in the inner ear to the external pressure. Also called auditory tube.

**exocrine gland** Any gland that opens onto the surface of the skin through a duct or ducts in the epithelium, as the sweat glands and sebaceous glands. Simple exocrine glands have one duct; compound glands have more than one duct.

**extension** Movement that increases the angle between two adjoining bones.

**external acoustic meatus** The S-shaped passage of the external ear that extends from the auricle to the tympanic membrane. Also called external auditory canal.

**external ear** The outer structures of the ear. Made up of the auricle, or pinna, and the external acoustic meatus, it funnels sound waves to the middle ear.

**extracellular fluid (ECF)** Body fluids outside the cells, including the interstitial fluid and blood plasma. Important components of ECF include protein, magnesium, potassium, chlorine, and calcium.

**extrapyramidal** Describing the tissues and structures of the brain outside the pyramidal tract, not running through the medullary pyramid — excluding the motor neurons, motor cortex, and corticospinal and corticobulbar tracts.

**extrapyramidal system** A complex of upper motor neurons that connects the basal ganglia, substantia nigra, and thalamic and subthalamic nuclei to each other and to parts of the cerebellum, cerebrum, and reticular formation. Working with the pyramidal system, it controls and coordinates movement and posture and integrates motor impulses originating in the cortex.

**extrapyramidal tracts** The tracts of those motor nerves outside the pyramidal tracts that run from the brain to the anterior horns of the spinal cord.

**eye** The organ of vision, within the bony orbit at the front of the skull and innervated by the optic nerve.

**Fallopian tube** One of a pair of tubes extending from the uterus and opening into the peritoneal cavity near the ovary.

**fascia, pl. fasciae** Sheets or bands of fibrous tissue that envelop the body under the skin and enclose muscles and muscle groups.

**fasciculus, pl. fasciculi, or fascile** A small bundle of fibers, usually of muscle, tendon, or nerve.

**femur** The thigh bone; the largest bone in the body, extending from the pelvis to the knee.

**fibrocartilage** Cartilage consisting of a dense matrix of collagenous fibers, for example, in the intervertebral disks.

**fibrous capsule** An anatomic structure made of fibrous elements that encloses an organ or body part.

**fibrous joint** An immovable joint in which the bones are connected by a continuous fibrous tissue, as in the sutures of the skull.

**fibrous tissue** Connective tissue consisting of closely woven elastic fibers, for example in tendons and ligaments.

**fibula** The outer bone of the lower leg, lateral to and smaller than the tibia.

**flexion** 1. Bending of a joint between two bones of the skeleton that decreases the angle between the bones. 2. Raising an arm or leg forward by movement at the shoulder or hip joint.

**fontanel** A soft spot, such as the spaces covered by tough membranes remaining between the bones of an infant's skull.

**foot** The distal extremity of the leg, consisting of the tarsus, metatarsus, phalanges, and related structures and tissue that envelops them.

**foramen magnum** Great foramen; a passage in the occipital bone that connects the spinal column and the cranial cavity; the spinal cord passes through it.

**foramen of Monro** The interventricular foramen; a passage between the lateral and third ventricles of the brain.

**fossa, pl. fossae** A usually longitudinal depression on the surface of a body part, especially on the end of a bone, as the olecranon fossa.

**frenum, pl. frenums, frena** A small fold of skin or mucous membrane that limits or restrains movement of a structure. Also called frenulum.

**frontal lobe** The anterior portion of the cerebral hemisphere, lying beneath the frontal bone and extending posteriorly to the central sulcus and inferiorly to the lateral fissure. It influences personality and is associated with the higher mental activities, such as planning, judgment, and conceptualizing.

**frontal sinus** One of a pair of small paranasal cavities in the frontal bone of the skull that communicates with the nasal cavity and is lined with a mucous membrane continuous with that of the nasal cavity.

**Gallbladder** A pear-shaped excretory sac that is lodged in a hollow on the surface of the liver; serves as a reservoir for bile; and contracts during digestion of fats, ejecting bile through the common bile duct into the duodenum.

**gingiva, pl. gingivae** The gums; portion of the oral mucosa that acts as supportive tissue for the teeth, covers the crowns of unerupted teeth, and encircles the necks of erupted teeth.

**gland** A group of specialized cells that secrete or excrete materials not related to ordinary metabolic needs.

**glans penis** The conical tip of the penis; an expansion of the corpus spongiosum that covers the head of the penis.

**gliding joint** A plane joint; a type of synovial joint that allows only gliding movements of opposing bones, as in the intermetacarpal joints.

**glottis, pl. glottises, glottides** The vocal apparatus, made up of the true vocal cords and the slitlike opening between them.

**gray matter** Nerve tissue composed of the cell bodies of the neurons. It does not include myelinated processes.

**Hard palate** The bony, front portion of the roof of the mouth.

**haversian canal** Any one of the small canals or channels forming the haversian system of compact bone.

**heart valve** One of the four structures at the openings between the chambers of the heart that control the flow of blood from one chamber to the next in one direction without backflow. The four valves are the atrial, mitral (bicuspid), pulmonary, and tricuspid.

**hepatic duct (common)** The duct formed by the right and left hepatic duct. The common hepatic duct joins with the cystic duct from the gallbladder to form the comon bile duct.

**hepatic portal vein** See portal vein.

**hilum, pl. hila** The area in an organ at which nerves and vessels enter and leave.

**hinge joint** A joint in which the convex end of one bone corresponds to the concave end of another bone, allowing motion in only one plane. The elbow is an example of a hinge joint. Also called ginglymus.

**hyaline cartilage** The smooth, translucent cartilage that covers the ends of bones at the joints. Also called true cartilage.

**hymen** A thin fold of membranous tissue that partly covers the vaginal opening.

**hyoid bone** A U-shaped bone suspended from the styloid processes of the temporal bones; it lies between the mandible and the larynx.

**Ileocecal valve** A cone-shaped structure with a star-shaped opening that lies between the ileum of the small intestine and the cecum of the large intestine. It allows the intestinal contents to move only in a forward direction.

**ileum** Distal portion of the small intestine, extending from the jejunum to the cecum.

**ilium, pl. ilia** The broad, flaring section of the hip bone that fuses with the ischium and pubis during early childhood to form the innominate bone.

**incus** One of the three small bones in the middle ear whose function is to conduct sound vibrations; the common name for this bone, because of its shape, is the anvil.

**inferior vena cava** The vein that drains blood from the lower body. It is formed by the two common iliac veins and terminates in the right atrium.

**inguinal** Of or relating to the groin.

**inguinal falx** The conjoined or conjoint tendon that forms the origin of the transverse and internal oblique abdominal muscles.

**inner ear** The area of the ear that consists of the vestibule, semicircular canals, and cochlea. Also called labyrinth.

**intracellular fluid (ICF)** The fluid contained inside tissue cells; accounts for 30% to 40% of body weight.

**iris** A circular diaphragm in the eye between the cornea and the lens. Composed of contractile tissue and perforated by the pupil, it regulates the entrance of light into the eye by contracting and dilating.

**Jaw** One of the two bony structures (mandible and maxilla) that hold the teeth and form the framework of the mouth.

**jejunum, pl. jejuna** The portion of the small intestine between the duodenum and the ileum.

**joint** An articulation between two or more bones. Joints are classified according to structure and movability as fibrous, cartilaginous, or synovial.

**Kidney** One of a pair of bean-shaped urinary organs on either side of the vertebral column in the retroperitoneal lumbar region of the abdomen.

**knee** A complex joint formed by the femur, tibia, and patella in which the thigh connects with the lower leg.

**Labia, sing. labium** 1. A fleshy edge or border. 2. The liplike folds of skin at the vaginal opening.

**labia majora, sing. labium majus** Two raised folds of adipose and connective tissue covered by skin, one on either side of the vagina outside the labia minora, that taper dorsally from the mons pubis to the perineum.

**labia minora, sing. labium minus** Two folds of skin between the labia majora that taper dorsally from the clitoris along both sides of the vagina.

**lacrimal apparatus** A network of eye structures (the lacrimal glands, upper and lower canaliculi, lacrimal sac, and nasolacrimal duct) that secretes and circulates tears from the surface of the eyeball.

**lambdoid suture** The line of juncture between the parietal and occipital bones of the skull (named for its resemblance to the Greek letter lambda).

**lamina, pl. laminae** 1. A thin, flat plate or layer. 2. The flattened part of the vertebral arch.

**large intestine** The portion of the gastrointestinal tract that extends from the pyloric opening of the stomach to the anus, consisting of the cecum, colon, rectum, and anal canal. Its primary functions are to absorb water and form feces.

**laryngeal nerve** Nerve with sensory and motor branches that supplies the laryngeal, pharyngeal, and tracheal structures.

**laryngeal veins** Inferior and superior veins of the larynx.

**larynx** A structure between the root of the tongue and the upper end of the trachea, just below and ventral to the lowest part of the pharynx. Composed of numerous cartilages and muscles, the larynx is the organ of voice, and it protects the airway from the entrance of liquids or solids during swallowing.

**latissimus dorsi** One of a pair of large, triangular muscles in the thoracic and lumbar region of the back. It extends the upper arm and adducts the upper arm posteriorly.

**left coronary artery** Either of a pair of branches from the ascending aorta, originating in the left posterior aortic sinus and branching into the left interventricular artery and the circumflex branch. It provides blood to both of the ventricles and to the left atrium.

**left pulmonary artery** The shorter and smaller of the two arteries that carry deoxygenated blood from the right side of the heart to the lungs. Rising from the pulmonary trunk, it typically has more separate branches than does the right pulmonary artery.

**left ventricle** The thick-walled, lower chamber of the heart that pumps blood through all the vessels of the body except those to and from the lungs.

**lens** The transparent, biconvex, crystalline lens of the eye, between the posterior chamber and the vitreous body.

**lesser omentum** A membranous fold of peritoneum that joins the lesser curvature of the stomach and the first part of the duodenum with the hepatic portal.

**lesser trochanter** A short, conical projection at the base of the neck of the femur; it is the insertion point for the tendon of the psoas major muscle.

**levator ani** One of a pair of muscles of the pelvic diaphragm that form the floor of the pelvic cavity and support the pelvic organs.

**Leydig's cells** Epithelioid cells of the interstitial tissue of the testis. These testosterone-secreting cells constitute the endocrine tissue of the testes. Also called interstitial cells, interstitial cells of Leydig.

**ligament** A flexible band of fibrous tissue, predominantly white, that supports and strengthens joints and connects bones or cartilages.

**limbic system** A group of brain structures below the cerebral cortex that are involved in feelings, motivation, and sexual arousal. The structures of the limbic system are the hippocampus, amygdaloid, dentate gyrus, gyrus fornicatus, archicortex, connections with the hippocampus, septal area, and medial area of the mesencephalon tegmentum.

**linea alba** The whitish median line in the anterior abdominal wall formed by the union of the aponeuroses of the three flat abdominal muscles. It extends from the symphysis pubis to the xiphoid process.

**linea arcuata** The tendinous band in the sheath of the rectus abdominis muscle below the umbilicus.

**liver** A large, dark red gland in the right side of the abdomen, immediately below the diaphragm and partially anterior to the stomach. Consisting of a right lobe, left lobe, caudate lobe, and quadrate lobe, the liver has many digestive, metabolic, and regulatory functions.

**lobar bronchus** An air passage that arises from the primary bronchus and passes into one of the lobes of the right or left lung.

**lobe** A relatively well-defined portion of an organ, such as the brain, liver, or lungs, that is bounded by sulci, fissures, or connective tissue.

**lobule** A small lobe.

**loculate** Divided into cavities or small spaces.

**loop of Henle** The long, U-shaped portion of a renal tubule, consisting of a descending limb and an ascending limb.

**loose fibrous tissue** A type of connective tissue consisting of elastic fibers, collagen fibers, and fluid-filled areolae.

**lumbar** Of or relating to the loins, or the part of the body between the thorax and the pelvis.

**lumbar nerves** The five pairs of spinal nerves that originate in the lumbar region.

**lumbar plexus** A network of spinal nerves formed by the ventral branches of the second to fifth lumbar nerves in the lumbar region of the back. (Some authorities include the first lumbar nerve.)

**lumbar vertebra** One of the five bones of the spinal column, between the thoracic vertebrae and the sacrum, that support the small of the back. Lumbar vertebrae have a body without facets and lack a foramen in the transverse process.

**lumbosacral plexus** A collective term for the lumbar and sacral nerve plexuses together.

**lumen, pl. lumina, lumens** The space enclosed by a vessel, duct, or other tubular or saclike organ.

**lung** One of a pair of spongy, highly elastic respiratory organs in the lateral cavities of the chest, separated from each other by the heart and mediastinal structures. The lungs lack muscle and are ventilated by respiratory movements.

**lymph** A colorless liquid found in the lymphatic system, into which it drains from the spaces between cells. Lymph resembles blood plasma, and consists mainly of water with dissolved salts and protein. It circulates through the lymphatic vessels, is filtered by the lymph nodes, and ultimately returns to the venous circulation.

**lymph node** One of many small masses of lymphoid tissue that filters out bacteria and other foreign particles to prevent them from entering the bloodstream and causing infection.

**lymphatic system** A specialized component of the circulatory system that produces, filters, and conveys lymph and produces various white blood cells. The lymphatic system includes a group of vessels called lymphatics that return the lymph to the blood; lymph nodes, along the paths of the collecting vessels; isolated nodules of lymphatic tissue, such as Peyer's patches in the intestinal wall; specialized lymphatic organs, such as the thymus, spleen, and tonsils; valves; and ducts.

## M

**Malleus** The outermost of the three auditory ossicles in the middle ear. Attached to the tympanic membrane, its club-shaped head articulates with the incus. Also called hammer.

**mammary gland** An accessory gland of the skin of female mammals that secretes milk in mature females. These glands are also present, although not functional, in children and in males.

**mandible** The bone of the lower jaw.

**manubrium** A general term for a handlelike structure or part; the uppermost portion of the sternum, which articulates with the clavicles and first two pairs of ribs.

**masseter** The thick, rectangular cheek muscle that raises the mandible and closes the jaw.

**mastoid process** A conical projection of the caudal, posterior portion of the temporal bone to which several muscles are attached, including the sternocleidomastoideus and splenius capitis. A hollow section contains air cells that are characterized by a large, irregular tympanic antrum in the superior anterior portion of the process.

**maxilla, pl. maxillae** One of a pair of large bones that form the upper jaw. It consists of a pyramidal body and four processes: the zygomatic, frontal, alveolar, and palatine.

**maxillary sinus** One of a pair of large spaces that form an air cavity in the maxillary body. See illustration on page 178.

**maxillary vein** One of a pair of deep facial veins, accompanying the maxillary artery and passing between the condyle of the mandible and the sphenomandibular ligament.

**meatus, pl. meatuses, meatus** An opening or passage in the body, such as the external acoustic meatus, which leads from the external ear to the tympanic membrane.

**medial malleolus** Bony prominence on the inside of the ankle.

**median aperture of fourth ventricle** An opening between the lower part of the roof of the fourth ventricle and the subarachnoid space.

**median plane** A vertical plane that separates the body into right and left halves and passes through the sagittal suture of the skull.

**mediastinal lymph node** One of the lymph nodes along the thoracic aorta that drains lymph from the esophagus, diaphragm, liver, and pericardium and sends lymph to the thoracic duct and inferior tracheobronchial nodes.

**mediastinum, pl. mediastina** Area separating the lungs. It contains the heart, vessels, trachea, esophagus, thymus, lymph nodes, nerves, and connective tissue.

**medulla, pl. medullas, medullae** The innermost part of a structure or organ.

**medulla oblongata** The most inferior part of the brain, continuing as the bulbous portion of the spinal cord just above the foramen magnum and separated from the pons by a horizontal groove.

**membrane** A thin layer of tissue that covers a surface, lines a cavity, or divides a space.

**meninges, sing. meninx** The three membranes (dura mater, pia mater, and arachnoid) that enclose the brain and the spinal cord.

**meniscus** Curved, fibrous cartilage in the knees.

**mesentery** A peritoneal fold attaching the stomach, small intestine, pancreas, and other abdominal organs to the dorsal body wall.

**metacarpus** The middle portion of the hand that consists of five slender bones numbered from the thumb side, metacarpals I through V.

**metatarsus** The middle portion of the foot that consists of five slender bones numbered from the medial side, metatarsals I through V

**midclavicular line** An imaginary vertical line on the front surface of the chest, passing through the midpoint of the clavicle and the center of the nipple, that divides each side of the anterior chest into two parts.

**middle ear** An irregularly shaped opening in the temporal bone containing the tympanic cavity with its auditory ossicles and the auditory tube, which carries air from the posterior pharynx into the middle ear.

**mitral valve** The valve connecting the left atrium and the left ventricle of the heart. Of the four heart valves, it is the only one with two cusps (or flaps) rather than three. Also called bicuspid valve.

**Montgomery's tubercle** The enlargement during pregnancy of sebaceous glands on the areola of the breast.

**motor area** The region of the cerebral cortex primarily involved in controlling or stimulating the contraction of voluntary muscles.

**mucous membrane** The tissue lining passages and cavities that communicate with the air.

**mucus** The viscous, slimy secretions of mucous membranes and glands, composed of mucin, white blood cells, water, inorganic salts, and exfoliated cells.

**muscle** A body tissue composed of fibers that are capable of contracting, thus causing movement of the parts and organs of the body.

**musculothoracic nerve** A branch of the brachial plexus, innervating the neck and axilla.

**myelin** A white, creamy substance composed primarily of lipids with some protein that forms an insulating sheath around various nerve fibers throughout the body.

**myocardium** The thick, contractile, middle layer of the heart wall composed of cardiac muscle cells.

**myosin** The most abundant protein in muscle tissue. The interaction between myosin and actin is responsible for the contraction and relaxation of muscle.

**Nail** A flattened, horny, elastic structure at the end of a finger or toe.

**nares** The two pairs of nasal openings, also called nostrils, which permit the inflow and outflow of air during breathing. Anterior nares are the external pair, also called nostrils. Posterior nares, also called choanae, allow the passage of air between the nose and the nasopharynx.

**nasal septum** The partition separating the nostrils; it is composed of bone and cartilage covered by mucous membrane.

**nasal sinus** One of many cavities in various skull bones that is lined with ciliated mucous membrane continuous with that of the nasal cavity.

**nasal turbinates** The ledges that form the sides and lower wall of each nasal cavity.

**nasion** The depression at the root of the nose below the eyebrows that indicates the frontonasal suture.

**nasopharynx** The region of the throat behind the nose, and extending from the posterior nares to the level of the soft palate.

**neck** A constricted section, such as the part of the body connecting the head and trunk.

**nephron** The functional unit of the kidney, it consists of the glomerulus, proximal and distal convoluted tubules, loop of Henle, and collecting duct.

**nerve** A bundle of fibers lying outside the central nervous system that connect the brain and spinal cord with various parts of the body.

**nerve trunk** White, glistening, cordlike bundle of nerves.

**nervous system** A system of extensive and intricate structures that activates, coordinates, and controls all bodily functions. It is divided into the central nervous system (brain and spinal cord) and the peripheral nervous system (cranial and spinal nerves).

**neuroglia** Cells and fibers that support nerve tissue.

**neuron** A cell of the nervous system, containing a nucleus within a cell body and extending one or more processes. Classification of neurons (for example, afferent, peripheral, sensory) is based on the direction in which they conduct impulses, location, and the kind of impulses they carry.

**nipple** A pigmented projection in the front of the mammary gland that contains outflow ducts from milk glands. Also called papilla mammae, mamilla, thelium.

**node** A small mass of tissue similar to a knot or swelling. Specific nodes include lymph nodes.

**nose** A specialized structure on the face that serves as a passageway for air to and from the lungs. It also serves as the organ of the sense of smell. The term includes the external nose and the nasal cavity. Also called nasus.

**nucha, pl. nuchae** The back, or nape, of the neck.

**nucleus, pl. nuclei** 1. The nucleus of a cell: an intracellular structure that contains the genetic codes for maintaining and reproducing that cell. 2. A group of central nervous system cells that support a common function, such as hearing or smell.

**nucleus pulposus** The pulpy, semifluid center of an intervertebral disk.

**Oblique fissure of the lung** 1. In the left lung, the groove marking the division between the upper and lower lobes. 2. In the right lung, the groove marking the division between the lower and middle lobes.

**obturator foramen** A large opening on the lower portion of the innominate bone. It is bordered by the pubis and ischium.

**occipital artery** Posterior part of the external carotid artery; divides into six branches that supply parts of the head and scalp.

**occipital bone** The bone in the lower back of the skull.

**occipital lobe** One of the five lobes of each cerebral hemisphere, at the back of the head and shaped like a three-sided pyramid.

**oculomotor nerve** One of a pair of cranial nerves that are essential for eye movement; they originate in the brain stem and supply extrinsic and intrinsic eye muscles, including the pupil.

**olecranon** The projection of the ulna that forms the point of the elbow. When the forearm is fully extended, it fits into the olecranon fossa of the humerus.

**olecranon bursa** The bursa of the elbow.

**olecranon fossa** The hollow in the posterior aspect of the humerus at the distal end. The olecranon of the ulna fits into it when the forearm is extended.

**olfactory nerve** One of two sensory nerves involved with the sense of smell.

**omental bursa** A peritoneal cavity behind the stomach, lesser omentum, and lower border of the liver and in front of the pancreas and duodenum. Also called lesser peritoneal cavity.

**optic disk** A small blind spot on the retina where the optic nerve enters the eye.

**optic nerve** One of two cranial nerves that arise in the retinal ganglion, traverse the thalamus, and connect to the visual cortex.

**orbicularis oculi** The palpebral, orbital, and lacrimal muscles, which move the eye.

**orbicularis oris** The muscle around the mouth; it includes fibers from other facial muscles, such as the buccinator, that insert into the lips, and adjacent muscles of the mouth.

**oropharynx** The division of the pharynx that extends from the soft palate to the hyoid bone and contains the palatine and lingual tonsils.

**osseous labyrinth** The internal ear's bony portion, which has three cavities: cochlea, semicircular canal, and vestibule. It transmits sound vibrations from the middle ear to the acoustic nerve.

**ossicles, auditory** The tiny bones of the middle ear—the incus, malleus, and stapes (commonly called the anvil, hammer, and stirrup)—that vibrate in response to sound waves and form an essential part of the sound conduction system from the tympanum to the inner ear.

**osteoblast** A cell active in the formation phase of bone growth

**osteoclast** A large, multinuclear cell originating in the marrow of growing bones and involved in the remodeling phase of bone growth

**ovary** One of two female gonads; they lie in the lower abdomen next to the uterus, in a fold of the broad ligament.

**Palate** The structure that separates the oral and nasal cavities and forms the roof of the mouth. The palate is divided into the soft (or fleshy) palate at the back of the mouth and the hard (or rigid) palate at the front of the mouth.

**palatine** Relating to the palate.

**palatine arch** The vault-shaped muscular structure that forms the soft palate and is located between the mouth and nasopharynx. An opening in it connects the mouth with the oropharynx; the uvula is suspended from the middle of the back edge of the arch.

**palatine bone** One of two bones of the skull that form the back of the hard palate, a portion of the nasal cavity, and the bottom of the eye orbit.

**palatine ridge** One of four to six ridges on the front surface of the hard palate.

**palatine tonsil** One of two small masses of lymphoid tissue between the palatopharyngeal and palatoglossal arches on either side of the fauces.

**palm** The ventral side, hollow or flexor surface, of the hand, between the wrist and the bases of the fingers.

**palmar aponeurosis** Ribbonlike fibrous tissue that runs from the tendon in the palm to the bases of the fingers. Also called volar fascia.

**palmar crease** One of several grooves that run across the palm of the hand to facilitate flexibility.

**pancreas** An elongated gland that extends transversely behind the stomach between the spleen and the duodenum; it secretes digestive enzymes, insulin, and glucagon.

**papilla, pl. papillae** A small, nipple-shaped elevation or projection. Examples of papillae include the conoid papillae of the tongue and the papillae of the dermis.

**paranasal** Near or alongside the nasal cavity – for example, the paranasal sinuses.

**paranasal sinuses** The mucosa-lined air cavities in cranial bones around the nose. They include the ethmoidal, frontal, maxillary, and sphenoidal sinuses, all of which communicate with the nasal cavity.

**parasympathetic** Of or referring to the craniosacral division of the autonomic nervous system, consisting of the ocular, bulbar, and sacral divisions. Primary function is regulating activities that conserve and restore body energy.

**parathyroid gland** One of several small glands, usually four, embedded in the posterior surface of the thyroid gland and attached to the lower edge. They secrete parathyroid hormone, a major regulator of calcium and phosphorous metabolism.

**parenchyma** The functional tissue of an organ as distinguished from its connective tissue or supporting framework.

**parietal bone** One of a pair of bones that form part of the top and sides of the skull. Each parietal bone adjoins five other bones: the opposite parietal, occipital, frontal, temporal, and sphenoid.

**parietal lobe** The upper central lobe of each cerebral hemisphere that occupies the parts of the surfaces that are covered by parietal bones.

**parietal lymph node** A lymph node in the walls of the thorax or along the larger blood vessels of the abdomen and the pelvis.

**parietal peritoneum** The serous membrane lining the abdominal and pelvic walls.

**parietal pleura** The serous membrane lining the walls of the thoracic cavity.

**parotid duct** A short tube that drains the parotid gland and empties into the mouth.

**parotid gland** The largest of the three main pairs of salivary glands, at the side of the face in front of and below the external ear.

**patella** The flat, triangular sesamoid bone at the front of the knee, within the tendon of the quadriceps muscle. Also called knee cap.

**patellar ligament** The central portion of the tendon of the quadriceps muscle behind the knee; it extends from the patella to the tibia.

**pelvic diaphragm** The caudal aspect of the body wall, made up of the levator ani and coccygeus muscles with the fasciae above and below them.

**pelvic inlet** The upper opening to the true pelvis, bounded by the sacral promontory, the arcuate lines of the ilia, and the symphysis pubis and pubic crest.

**pelvic outlet** The lower opening to the true pelvis, bounded by the pubic arch, the sacrotuberous ligaments, part of the ischium, and the coccyx.

**pelvis, pl. pelves** 1. The lower portion of the trunk, bounded by the two hip bones laterally and anteriorly and the sacrum and coccyx posteriorly; the upper border is known as the pelvic inlet; the lower border, or pelvic outlet, is closed by the levator ani and coccygeus muscles. 2. Any basinlike structure such as the renal pelvis.

**penis** The male external reproductive organ, which also contains the urethra and serves as the organ of urinary excretion; consists of the root, the corpora cavernosa, the corpus spongiosum, and the glans penis.

**pericardium, pl. pericardia** The sac surrounding the heart and the roots of the great vessels; made up of an outer layer of fibrous tissue and an inner serous layer.

**perilymph** The clear fluid in the space between the osseous labyrinth and the membranous labyrinth in the inner ear.

**perineal** Of or pertaining to the perineum.

**perineum** 1. The pelvic floor and its associated structures; located between the symphysis pubis and the coccyx, and bordered on the sides by the ischial tuberosities. 2. The body area between the thighs; bounded by the anus and scrotum in males and by the anus and vulva in females.

**periosteum** The fibrous vascular connective tissue covering the bones, consisting of an inner layer of collagenous tissue, a layer of fine elastic fibers, and an outer layer of dense vascular connective tissue; has bone-forming capability.

**peripheral nervous system** The portion of the nervous system made up of the motor and sensory nerves and ganglia outside the brain and spinal cord; consists of 12 pairs of cranial nerves, 31 pairs of spinal nerves, and their various branches.

**peritoneal cavity** The potential space between the parietal and visceral peritoneum.

**peritoneum** The extensive serous membrane that covers the abdominal wall and invests the viscera; divided into the parietal peritoneum and the visceral peritoneum.

**peroneal** Pertaining to the outside of the leg, to the fibula, or to the muscles and nerves found there.

**Peyer's patches** Elevated areas of lymphoid tissue in the terminal ileum near its junction with the colon; made up of lymph nodules packed closely together.

**phalanx, pl. phalanges** 1. Any one of the tapering bones that make up the fingers and toes. 2. Any of the plates that make up the reticular membrane of the organ of Corti.

**pharyngeal tonsil** A collection of lymphoid tissue on the posterior wall of the nasopharynx; when swollen, results in the condition known as adenoiditis.

**phrenic nerve** One of a pair of branches of the cervical plexus, arising from the fourth cervical nerve; primarily acts as the motor nerve for the diaphragm, but also provides sensory fibers to the pericardium.

**pia mater** The innermost of the three meninges covering the brain and the spinal cord; firmly adheres to both structures and carries a rich supply of blood vessels, which nourish the brain tissue.

**pineal gland** A small, flattened, cone-shaped structure suspended by a stalk in the epithalamus, between the superior colliculi and below the splenium of the corpus callosum; synthesizes and secretes melatonin in response to norepinephrine.

**piriformis** A flat, pyramidal muscle that lies almost parallel to the gluteus medius; originates in the ilium and the second to fourth sacral vertebrae and inserts into the greater trochanter of the femur; acts to rotate the thigh.

**pisiform bone** A small, pea-shaped bone in the proximal row of carpal bones; attached to the flexor retinaculum, the flexor carpi ulnaris, and the abductor digiti minimi.

**pivot joint** A uniaxial joint in which one bone rotates within a bony or ligamentous ring, as in the atlantoaxial joint.

**pleura, pl. pleurae** A delicate serous membrane enclosing the lung and lining the thoracic cavity, creating a potential space known as the pleural cavity.

**pleural cavity** The potential space between the visceral and parietal pleurae; contains a small quantity of fluid that acts as a lubricant. Also called pleural space.

**plexus, pl. plexuses** A network of intersecting nerves, blood vessels, or lymphatic vessels.

**plica, pl. plicae** A ridge or fold of tissue.

**plica semilunaris** The semilunar fold of the conjunctiva; a fold of mucous membrane at the inner corner of the eye.

**plicae transversales recti** Transverse folds of the rectum that support the weight of feces.

**point of maximal impulse (PMI)** The place on the chest where the left ventricular pulse is felt most strongly, usually in the fifth intercostal space just medial to the left midclavicular line.

**pons, pl. pontes** 1. The part on the ventral surface of the brain between the medulla oblongata and the mesencephalon. 2. A formation of tissue resembling a bridge that connects two parts of the same organ or structure.

**portal vein** A vein formed by the superior mesenteric and splenic veins behind the pancreas. It ascends anterior to the inferior vena cava, forms right and left branches in the liver and, within the liver, splits into branches resembling arteries that empty into the hepatic veins.

**posterior horn** Sensory portion of gray matter

**posterior longitudinal ligament** A thick, fibrous band that runs vertically along the entire length of the spine and connects the posterior surfaces of the vertebral bodies.

**prepatellar bursa** A bursa in the knee joint, between the quadriceps tendon and the lower portion of the femur.

**prepuce** The loose, retractable fold of skin that covers the glans penis. Also called foreskin.

**primary bronchus** One of two main air passages of the lungs. Branching from the trachea and conveying air to the lungs, the right and left primary bronchi are shaped differently (the right primary bronchus is wider and shorter); thus most foreign objects in the trachea usually enter the right bronchus.

**pronation** 1. The act of turning the forearm so the palm is downward and backward. 2. The act of lowering the medial margin of the foot during an outward rotation.

**psoas major** An abdominal muscle that flexes the thigh or trunk, originating from the lumbar vertebrae and fibrocartilages and the lower thoracic vertebrae and lumbar vertebrae.

**pubic symphysis** The pelvic joint, which consists of two pubic bones divided by a fibrocartilage disk and connected by two ligaments.

**pubis, pl. pubes** One of two pubic bones that, with the ischium and the ilium, form the hip bone.

**pudendal nerve** A branch of the pudendal plexus that arises from the second, third, and fourth sacral nerves. Passing between the piriformis and coccygeus muscles, the pudendal nerve leaves the pelvis through the greater sciatic foramen, passes over the spine of the ischium, and enters the ischiorectal fossa.

**pudendum, pl. pudenda** The external genitalia.

**pulmonary alveolus** A terminal air sac of the lung in which the exchange of oxygen and carbon dioxide takes place.

**pupil** The round opening at the center of the iris that allows light to enter the eye.

**Purkinje network** Muscle fibers that spread through the heart in a complex network to convey the impulses necessary for the near-simultaneous contraction of the right and left ventricles.

**pyloric sphincter** A muscular ring in the stomach that separates the pylorus from the duodenum.

**pylorus, pl. pylori, pyloruses** Distal portion of the stomach; empties into the duodenum.

**pyramidal tract** Two groups of fibers in the spinal cord in which motor impulses are conveyed to the anterior horn cells from the opposite side of the brain.

**Quadriceps femoris** The great extensor muscle of the anterior thigh that forms a large dense mass covering the front and sides of the femur; composed of the rectus femoris, the vastus lateralis, the vastus medialis, and the vastus intermedius.

**Radial artery** An artery in the forearm, beginning at the bifurcation of the brachial artery, passing in 12 branches to the forearm, wrist, and hand.

**radial nerve** The largest branch of the brachial plexus.

**radioulnar articulation** The pivot joint between the radial head, the ulna, and the annular ligament, which is responsible for the motion of the head of the radius in pronation and supination.

**rectum** The distal portion of the large intestine, beginning as a continuation of the sigmoid colon from the level of the third sacral vertebra and ending in the anal canal.

**rectus abdominis** A pair of anterolateral abdominal muscles that extend the length of the ventral aspect of the abdomen and are separated by the linea alba.

**reflux** A backward or return flow of a fluid.

**renal calyx** The first unit within the renal system carrying urine from the kidney cortex to the ureters and into the pelvis for excretion

**renal cortex** The outer layer of the kidney; contains about 1.25 million renal tubules, which remove bodily wastes as urine.

**renal medulla** The central portion of the kidney. A collection of 8 to 18 renal pyramids containing collecting ducts and portions of the loop of Henle.

**retina** A 10-layered, delicate, nervous tissue membrane of the eye that receives images and transmits impulses through the optic nerve to the brain.

**right coronary artery** One of two branches of the ascending aorta, arising in the right posterior aortic sinus, passing along the right side of the coronary sulcus, dividing into the right interventricular artery and a large marginal branch, and supplying the muscles of both ventricles, the right atrium, and the sinoatrial node.

**right hepatic duct** The duct that drains bile from the right lobe of the liver into the common bile duct.

**right pulmonary artery** The larger and longer of two arteries carrying blood from the heart to the lungs. It originates in the pulmonary trunk, bends to the right behind the aorta, and divides into two branches at the base of the right lung.

**right subclavian artery** A large artery that originates in the brachiocephalic artery. Arising from the right subclavian artery are several branches: the axillary, vertebral thoracic, and internal thoracic arteries and the cervical and costocervical trunks that pass blood to the right side of the upper body.

**rotary joints** Joints in which one bone pivots around a stationary bone. Also called pivot joints.

**rotation** The motion of a bone around its central axis, which may lie in a separate bone.

**ruga, pl. rugae** A wrinkle or fold such as the rugae of stomach: large folds of mucous membrane that form when the organ is empty.

**Sac** A pouch, or saclike structure; for example, the abdominal sac in the embryo eventually develops into the abdominal cavity.

**saccule** A small bag or pouch, as the laryngeal saccule.

**sacral** Relating to the sacrum.

**sacral foramen** One of several openings between the fused segments of the sacral vertebrae in the sacrum through which the sacral nerves pass.

**sacral plexus** A network of motor and sensory nerves formed by the lumbosacral trunk from the fourth and fifth lumbar and the first, second, and third sacral nerves.

**sacroiliac articulation** A joint in the pelvis formed by the joining of the sacrum and the ilium. Movement of the joint is limited.

**saddle joint** A joint that has two saddle-shaped surfaces, which are at right angles to each other, such as the carpometacarpal joint of the thumb. This type of joint allows no axial rotation but flexion, extension, adduction, and abduction are possible.

**sagittal** Relating to an imaginary straight line that extends from the front to the back through the median of the body (or a part of the body).

**sagittal suture** The immovable joint between the parietal bones. It runs along the midline of the skull from the coronal suture to the upper part of the lambdoid suture.

**saliva** A clear, somewhat viscous fluid that moistens and softens food and keeps the mouth moist. It is secreted by the salivary and mucous glands in the mouth and is composed of water, mucin, organic salts, and the digestive enzyme ptyalin.

**salivary gland** One of the six glands that secrete saliva in the mouth.

**scapula** The shoulder blade; a flat, triangular bone on the back of the shoulder. It forms the posterior part of the shoulder girdle, articulating with the humerus.

**sclera** The tough, white outer coating of the eyeball, extending from the optic nerve to the cornea.

**scrotum** The pouch of skin that contains the testes and their accessory organs.

**sebaceous gland** One of the many small glands in the skin that secrete sebum.

**sebum** A substance composed of keratin, fat, and cellular debris secreted by the sebaceous glands of the skin. Combined with sweat, sebum forms a moist, oily, acidic film that protects the skin against drying and bacterial and fungal infections.

**sella turcica** A saddle-shaped depression crossing the midline of the superior surface of the body of the sphenoid bone; it contains the pituitary gland.

**seminal duct** Any duct, such as the deferent duct or the ejaculatory duct, through which semen passes.

**semitendinosus** One of three posterior femoral muscles at the back and inner part of the thigh that flexes and rotates the leg and extends the thigh; remarkable for the great length of its tendon of insertion.

**sensorium** 1. The center in the brain that processes all sensations. 2. The complete sensory apparatus of the body.

**sensory** 1. Relating to the senses or sensation. 2. Referring to a structure, such as a nerve, that receives and transmits impulses from sense organs to the reflex and higher centers of the brain.

**sensory nerve** A peripheral afferent nerve that conducts sensory impulses from the sense organ receptors to the brain or spinal cord.

**serosa** Any serous membrane that lines the external walls of body cavities and secretes a watery exudate – the tunica serosa, for example.

**serous** Having the nature of serum; plasma without the clotting factors.

**serratus anterior** A thin muscle of the chest wall extending from the ribs under the arm to the scapula. It serves to draw the scapula forward and also to rotate the scapula to raise the shoulder.

**shoulder girdle** The clavicles and the scapulae; stabilizes the upper trunk and supports the arms.

**shoulder joint** The ball-and-socket articulation formed by the head of the humerus and the cavity of the scapula.

**sigmoid** 1. Resembling an S shape. 2. The sigmoid colon.

**sigmoid colon** Part of the colon describing an S-shaped curve between the pelvis and rectum.

**sigmoid mesocolon** A curved peritoneal fold that attaches the sigmoid colon to the pelvic wall.

**skeleton** The supporting framework of the body. It protects delicate structures, provides attachments for muscles, and allows body movement. The bone marrow serves as a major reservoir of blood and produces red blood cells.

**Skene's glands** Paraurethral glands of the female urethra that open just within the urethral orifice.

**skin** The tissue between the body and the external environment; the largest organ.

**skull** The bony structure of the head, consisting of the cranium and facial skeleton. The cranium contains and protects the brain.

**small intestine** Part of the gastrointestinal tract between the stomach and the iliocecal junction, extending for about 23′ (7 m); it is divided into the duodenum, jejunum, and ileum.

**soft palate** The posterior part of the palate, composed of mucous membrane, muscle fibers, and mucous glands.

**solar plexus** A complex network of nervous tissue surrounding the celiac roots and the superior mesenteric arteries at the first lumbar vertebra.

**sphenoid bone** The bone at the skull base, anterior to the temporal bones and the basilar part of the occipital bone; it resembles a bat with extended wings.

**sphenoidal fissure** A cleft between the great and small wings of the sphenoid bone.

**sphenoidal sinus** One of a pair of paranasal cavities lined with mucous membrane. A large sphenoidal sinus may extend into the roots of the pterygoid processes, the great wings, or occipital bone.

**sphincter** A muscle that surrounds a tube, duct, or orifice and alters the size of the lumen or orifice by contracting.

**sphincter pupillae** A muscle that expands the iris and narrows the pupillary diameter.

**spinal aperture** A large opening in a vertebra, formed by the vertebral body and its arch. Spinal apertures form the spinal canal.

**spinal canal** The cavity inside the vertebral column, through which the spinal cord passes.

**spinal cord** A long, almost cylindrical structure of nerve tissue, extending from the base of the skull to the upper lumbar region.

**spinal nerves** The nerves emerging from the spinal cord; there are 31 pairs (8 cervical, 12 thoracic, 5 lumbar, 5 sacral, and 1 coccygeal).

**spinal tract** An ascending or descending pathway for nerve impulses; in the white matter of the spinal cord.

**spleen** A soft, highly vascular, lymphatic organ between the stomach and diaphragm on the left side.

**splenius capitis** One of a pair of deep muscles of the back.

**sputum** Expectorated material coughed up from the lungs.

**stapes** The innermost auditory ossicle in the middle ear, resembling a stirrup; transmits sound vibrations from the incus to the internal ear.

**sternoclavicular articulation** The double-gliding joint between the sternum and clavicle.

**sternocostal articulation** The gliding joint of each rib cartilage and the sternum, excluding the first rib joint, in which the cartilage is directly connected with the sternum to form a synchondrosis.

**sternum** A long, flattened bone forming the middle part of the thorax. It supports the clavicles, articulates with the first seven pairs of ribs, and is made up of the manubrium, gladiolus, and xiphoid process.

**stomach** A saclike organ of digestion in the upper left abdomen.

**stratified epithelium** Squamous or columnar cells arranged in three or more layers.

**stratiform fibrocartilage** A fibrocartilage structure that forms a thin layer of grooves in the bone, through which certain muscles glide.

**stratum, pl. strata** A uniform, thick sheet or layer, usually occurring with other layers, as in the epidermis.

**stratum basale** The basal layer of the epidermis composed of a single layer of columnar cells.

**stratum corneum** Outer horny layer of the epidermis.

**stratum granulosum** The granular layer of the epidermis in which cells contain visible cytoplasmic granules that die, become keratinized, move to the surface, and flake away; it is absent in the palms and soles.

**stratum lucidum** A layer of the epidermis that occurs only in the thick skin of the palms and soles.

**stratum spinosum** A layer of the epidermis that is composed of several layers of prominent intercellular attachments.

**striated muscle** Voluntary skeletal muscles whose fibers are divided by transverse bands.

**stylohyoid ligament** Attachment of the tip of the styloid process of the temporal bone to the hyoid bone.

**stylohyoideus** One of four muscles of the styloid process of the temporal bone and the hyoid bone that lie anterior and superior to the posterior belly of the digastricus.

**subacromial bursa** The saclike, fluid-filled cavity separating the acromion and deltoid muscle from the point of insertion of the supraspinatus muscle and the greater tubercle of the humerus.

**subarachnoid** Located between the arachnoid and the pia mater.

**subarachnoid cisterns** Subarachnoid reservoirs containing cerebrospinal fluid.

**subclavian artery** A large artery at the base of the neck that supplies blood to the arms. It branches into the left and right subclavian arteries.

**subclavian vein** An extension of the axillary vein, which, together with the internal jugular, forms the brachiocephalic vein.

**subcutaneous** Under the skin.

**subdural** Between the dura mater and the arachnoid.

**sublingual** Under the tongue.

**sublingual gland** One of the pair of salivary glands under the tongue.

**submandibular duct** A passage for the secretion of saliva from the submandibular gland.

**submandibular gland** One of the salivary glands in the submandibular triangle, which extends from the digastricus to the stylomandibular ligament.

**subserous fascia** One type of fascia, beneath the serous membranes.

**substantia spongiosa ossium** Inner spongy layer of bone.

**superior mesenteric artery** Artery supplying blood from the heart to the small intestine and proximal half of the colon. It has the following branches: inferior pancreaticoduodenal, jejunal, ileal, ileocolic, right colic, and middle colic arteries.

**superior vena cava** The vein that drains blood from the head, neck, upper extremities, and chest. The second largest vein in the body, it empties deoxygenated blood into the right atrium of the heart.

**supraclavicular nerve** A branch of the cervical plexus, arising from the third and mostly the fourth cervical nerve. It has lateral and medial branches.

**suture** A fibrous joint in the skull, in which the bones are closely opposed.

**synapse** The area between two neurons or between a neuron and an effector organ, through which nerve impulses are transmitted.

**synchondrosis** An immovable joint. The surfaces of the bones that make up the joint are connected by cartilage.

**synovial bursa** A closed sac in the connective tissue between the muscles, tendons, ligaments, and bones that is filled with synovial fluid.

**synovial fluid** A clear, viscid fluid that lubricates the joints. Secreted within the bursae and tendon sheaths by the synovial membrane, it contains fat, albumin, mucin, and mineral salts.

**synovial joint** A joint connected by ligaments and lined with synovial membrane; articular cartilage covers the contiguous bony surfaces of this movable joint.

**synovial sheath** A tubular structure containing synovial fluid. Surrounding the tendon of a muscle, it helps the tendon glide through a fibrous or bony tunnel, such as under the flexor retinaculum of the wrist.

**Talus, pl. tali** The most proximal of the tarsal bones, which form the ankle joint.

**tarsal bone** One of the seven bones of the tarsus (ankle). The ankle includes the talus, calcaneus, cuboid, navicular, and the three cuneiform bones.

**tarsal gland** One of many sebaceous follicles between the tarsi and the conjunctiva of the eyelids.

**tarsometatarsal** Pertaining to the tarsus and the metatarsus of the ankle, especially at the instep of the foot, where the metatarsal bones meet the cuneiform and cuboid bones.

**tarsus, pl. tarsi** 1. The region of the junction between the foot and the leg. 2. One of the plates of connective tissue that form the framework of an eyelid.

**taste bud** One of the small organs of the gustatory (taste) nerve throughout the tongue and the roof of the mouth. They detect sweet, sour, and salty sensations.

**temporal artery** One of three cerebral arteries: superficial temporal artery, middle temporal artery, and deep temporal artery; supplies the temporal muscle, parotid gland, auricle, scalp, skin of the face, and masseter muscle.

**temporal bone** One of two large bones containing the hearing organs, forming part of the lateral surfaces and base of the skull and jaw.

**temporal lobe** The cerebral region below the lateral fissure.

**temporalis** The muscle that closes the jaws and retracts the mandible.

**temporomandibular joint** Formed by parts of the mandibular fossae of the temporal bone, the articular tubercules of the temporal and mandibular bones, the mandibular condyles, and five ligaments, this combined hinge and gliding joint connects the mandible to the temporal bone.

**temporoparietalis** One of two broad, thin muscles of the scalp that overlie the temporal fascia.

**tendo calcaneus** A powerful tendon at the back of the heel; the common tendon of the soleus and gastrocnemius muscles.

**tendon** A fibrous cord that connects a muscle to bone; it is generally covered in delicate fibroelastic connective tissue except at points of attachment.

**tensor fasciae latae** A muscle that arises from the iliac crest and iliac spine and extends to the deep fascia lata; one of 10 in the gluteal region. It flexes, abducts, and rotates the thigh.

**testis, pl. testes** One of two male gonads, which produce semen.

**thalamus, pl. thalami** Largest subdivision of the diencephalon. It relays all sensory stimuli (except olfactory stimuli) as they ascend to the cerebral cortex.

**thoracic aorta** The upper portion of the aorta, which supplies the heart, ribs, chest muscles, and stomach; it begins at the caudal border of the fourth thoracic vertebra and becomes the abdominal aorta at about the level of the diaphragm.

**thoracic arteries** The three arteries (lateral, superior, and internal) that distribute blood to the pectoral muscles, the mammary glands, the axillary area of the chest wall, the diaphragm, and the structures of the mediastinum and the anterior thoracic wall.

**thoracic duct** The main duct that carries lymph from the cisterna chyli to the junction of the left subclavian and left internal jugular veins.

**thoracic nerves** The nerves on each side of the thorax; include 11 intercostal nerves and one subcostal nerve.

**thoracic vertebra** Any of the vertebrae, designated T1 to T12, between the seventh cervical vertebra (C7) and the first lumbar vertebra (L1).

**thoracic visceral node** One of the three groups of lymph nodes that drain the thymus, pericardium, esophagus, trachea, lungs, and bronchi.

**thorax** The part of the body between the neck and the thoracic diaphragm, encased by the ribs. It contains the heart, lungs, and part of the gastrointestinal tract.

**thymus, pl. thymuses, thymi** A symmetric lymphoid organ in the mediastinum. It extends superiorly into the neck to the lower edge of the thyroid gland and inferiorly as far as the fourth costal cartilage. The thymus reaches maximum development at puberty and goes into decline throughout adulthood.

**thyrocervical trunk** A short, thick, arterial branch, deriving from the first portion of the subclavian arteries, close to the medial border of the scalenus anterior. It supplies blood to muscles and bones in the back, head, and neck.

**thyroid cartilage** The largest cartilage of the larynx; consists of two laminae fused at an acute angle in the middle line of the neck; forms the Adam's apple.

**thyroid gland** A vascular organ at the base of the throat. It has two lobes that are connected in the center by a narrow isthmus, and produces thyroxine, triiodothyronine, and the hormone calcitonin.

**tibia** The shin bone; the weight-bearing bone of the leg. The second longest bone of the body, it extends from knee to ankle and runs parallel to the fibula.

**tibiotarsal joint** The joint between the tibia and the tarsal bones.

**tonsil** A small, rounded mass of lymphoid tissue. The term is often used to designate the palatine tonsil.

**trachea** The windpipe; a tube composed of cartilage and membrane that extends from the larynx to the right and left bronchi.

**tragus, pl. tragi** The cartilaginous projection at the front opening of the ear.

**tunica adventitia** The fibrous elastic covering of blood vessels.

**tunica intima** The inner lining of a blood vessel.

**Ulna** The inner and larger bone of the forearm that lies parallel to the radius on the side opposite the thumb.

**ulnar artery** An artery arising near the elbow, it distributes blood to the forearm, wrist, and hand. This artery passes obliquely in a distal direction to become the superficial palmar arch.

**ulnar nerve** Originating in the medial and lateral cords of the brachial plexus, this nerve distributes to the muscles and the skin on the medial part of the hand and the ulnar side of the forearm.

**umbilical** Of or pertaining to the umbilicus or umbilical cord.

**umbilical cord** An elastic structure that connects the umbilicus and the placenta in the gravid uterus; it gives passage to the umbilical arteries and vein and transmits nourishment from the mother to the fetus.

**umbilical fissure** A groove on the visceral surface of the liver that separates its right and left lobes.

**umbilicus** The navel; site of attachment of the umbilical cord of the fetus.

**upper respiratory tract** The upper portion of the respiratory system, consisting of the nose, nasal cavity, ethmoidal air cells, frontal sinuses, sphenoidal sinuses, maxillary sinus, and larynx.

**ureter** The fibromuscular tube that conveys the urine from the kidney to the bladder.

**urethra** The membranous canal conveying urine from the bladder to the exterior of the body. About 1″ (2.5 cm) long in women, it is behind the symphysis pubis, anterior to the vagina. Much longer in men (about 8″ [20.5 cm]), it begins at the bladder and passes through the prostate gland, the tissue connecting the pubic bones, and the urinary meatus of the penis.

**urethral** Of or pertaining to the urethra.

**urinary bladder** The musculomembranous sac, in the anterior part of the pelvic cavity, that serves as a reservoir for urine; it then discharges it through the urethra.

**uvea** The vascular middle coat of the eye, consisting of the iris, the ciliary body, and the choroid.

**uvula, pl. uvulae** A pendular, fleshy mass hanging from the soft palate above the root of the tongue from the posterior border of the soft palate.

**Vacuole** 1. Any small space or cavity formed in the protoplasm of a cell. 2. A small space in the body, often consisting of secretions, fat, or cellular debris, that is surrounded by a membrane.

**vagal** Of or pertaining to the vagus nerve.

**vagina** The part of the female genitalia that connects the vulva to the cervix, in front of the rectum and behind the bladder.

**vagus nerve** The tenth cranial nerve. One of the nerves of the cranium that makes speech, swallowing, and many other bodily functions possible. Motor portion provides movement of swallowing and visceral muscles; sensory portion provides sensation from organs and position sense (proprioception).

**vascular** Of or pertaining to a blood vessel.

**vas deferens, pl. vasa deferentia** The excretory duct of the testis, which unites with the excretory duct of the seminal vesicle to form the ejaculatory duct.

**vasoconstriction** Narrowing of blood vessels, especially the arterioles and veins in the skin and abdominal viscera, leading to decreased blood flow; caused by contraction of vascular smooth muscle.

**vasodilation** Increased caliber or distention of blood vessels, particularly arterioles, caused by relaxation of vascular smooth muscle.

**vastus intermedius** A muscle in the center of the thigh; one of four of the quadriceps femoris.

**vastus lateralis** The largest of the four muscles that make up the quadriceps femoris; on the outside of the thigh, extending from the hip joint to the common quadriceps tendon; inserted in the patella; extends the leg.

**vastus medialis** One of the four muscles that make up the quadriceps femoris; on the inside of the thigh; inserting into the patella by the common quadriceps tendon; extends leg.

**vein** A blood vessel that conveys blood toward the heart.

**vena cava, pl. venae cavae** One of the two large venous trunks that return blood from the peripheral circulation and drain into the right atrium. The inferior vena cava receives blood from the lower extremities and from the pelvic and abdominal viscera; the superior vena cava, from the head, neck, upper extremities, and chest.

**ventricle** A cavity, such as those in the brain; or, the lower chambers of the heart.

**venule** Any of the small blood vessels that gather blood from the capillaries and anastomose to form veins.

**vertebra, pl. vertebrae** Any of the 33 bones of the spinal column: 7 cervical, 12 thoracic, 5 lumbar, 5 sacral, and 4 coccygeal; most consist of a body, an arch, a spinous process for muscle attachment, and pairs of pedicles and processes.

**vertebral artery** One of a pair of arteries that branch from the subclavian arteries and supply the muscles of the neck, the vertebrae and spinal cord, the cerebellum, and the interior of the cerebrum.

**vertebral body** The weight-supporting central portion of a vertebra.

**vertebral canal** The cavity inside the vertebral column through which the spinal cord passes; also called spinal canal.

**vertebral column** The backbone; the flexible structure made up of 33 vertebrae that extends from the base of the skull to the coccyx; forms a bony case for the spinal cord and the longitudinal axis of the skeleton.

**vesicouterine** Pertaining to the bladder and the uterus.

**vessel** Any tubular structure that conveys fluids, such as blood and lymph; primarily refers to the arteries, veins, and lymphatic vessels.

**vestibular** Of or pertaining to a vestibule.

**vestibular gland** 1. One of a pair of small secretory glands on each side of the vaginal orifice; also called Bartholin's gland, greater vestibular gland. 2. Any of the small mucous glands between the urethral and vaginal orifices; also called lesser vestibular gland.

**villus, pl. villi** A tiny projection on the surface of a mucous membrane, as in the arachnoid villi, chorionic villi, and intestinal villi; diffuses and transports fluids and nutrients.

**visceral peritoneum** The largest serous membrane in the body, which covers the viscera and holds it in position.

**visceral pleura** The serous membrane that surrounds the lungs and lines the fissures between the lobe.

**viscus, pl. viscera** An organ within a cavity, particularly the abdominal cavity.

**vitreous humor** The semigelatinous substance contained in the interstices of the stroma in the vitreous body, filling the cavity behind the lens of the eye.

**vitreous membrane** Not a true membrane, but a dense collection of collagen fibers in the posterior vitreous body.

**vomer** The flat bone forming the posterior and inferior part of the nasal septum.

**vulva** The external genitalia of the female. It includes the labia majora, labia minora, mons pubis, bulb of vestibule, vestibule of vagina, greater and lesser vestibular glands, and vaginal orifice.

**Wagner's corpuscle** A small sensory end organ that has a connective tissue capsule and tiny stacked plates. These special organs are found in the dermis of the hand and foot, forearm, lips, tongue, palpebral conjunctiva, and mammary papilla.

**white fibrocartilage** A mixture of tough, white fibrous tissue and flexible cartilaginous tissue. The white fibrous tissue dominates and is divided into circumferential fibrocartilage, connecting fibrocartilage, interarticular fibrocartilage, and stratiform fibrocartilage.

**white matter** A tissue consisting of myelinated and unmyelinated nerve fibers and embedded in neuroglia. It is the tissue surrounding the gray matter in the spinal cord.

**Xiphoid process** The smallest of three parts of the sternum, articulating caudally with the body of the sternum and laterally with the seventh rib.

**Yellow cartilage** The most elastic of the three main kinds of cartilage, consisting of a network of elastic fibers in a flexible, fibrous matrix. It is found in the external ear, the auditory tube, the epiglottis, and the larynx.

**Zona, pl. zonae** A zone or girdle encircling a region or area; an area with a specific boundary or characteristics.

**zona arcuata** An inner tunnel, the canal of Corti, which is located in the cochlea of the ear.

**zona fasciculata** Middle section of the cortex of the adrenal gland.

**zona glomerulosa** Outer section of the cortex of the adrenal gland.

**zona orbicularis** The orbicular zone of the hip joint in which circular fibers of the articular capsule form a ring around the neck of the femur.

**zona reticularis** Inner section of the cortex of the adrenal gland.

**zygomatic arch** The arch formed by the joint of the temporal process of the zygomatic bone and the zygomatic process of the temporal bone.

**zygomatic bone** One of the pair of quadrangular bones that form the prominence of the cheek and the lower part of the eye socket. It joins with the frontal bone, the maxilla, the zygomatic process of the temporal bone, and the sphenoid bone.

**zygomatic nerve** Sensory branch of the maxillary division of the trigeminal nerve that supplies the skin over the cheekbone and the temple.

**zygomatic process** Any of several strong, bony processes that form part of the zygomatic arch.

**zygomaticus major** One of the muscles of the mouth originating at the zygomatic bone in front of the temporal process and inserting into the corner of the mouth.

**zygomaticus minor** One of the muscles of the mouth originating at the malar surface of the zygomatic bone and inserting into the upper lip.

# Selected references


Abrahams, P.H., et al. *McMinn's Color Atlas of Human Anatomy,* 4th ed. London: Mosby International Limited, 1998.

Agur, A.M.R., and Lee, M.J. *Grant's Atlas of Anatomy*, 10th ed. Philadelphia: Lippincott Williams & Wilkins, 1999.

Hiatt, J.L., and Gartner, L.P. *Textbook of Head and Neck Anatomy,* 3rd ed. Philadelphia: Lippincott Williams & Wilkins, 2000.

Lutjen-Drecoll, E., and Rohen, J.W. *Atlas of Anatomy: The Functional Systems of the Human Body.* Baltimore: Williams & Wilkins, 1998.

Marieb, E.N. *Essentials of Human Anatomy and Physiology*, 6th ed. San Francisco: Benjamin Cummings, 2000.

McCracken, T.O. *New Atlas of Human Anatomy*. New York: Barnes & Noble, 1999.

Moore, K.L., and Dalley II, A.F. *Clinically Oriented Anatomy*, 4th ed. Philadelphia: Lippincott Williams & Wilkins, 1999.

Netter, F.H. *Atlas of Human Anatomy*, 2nd ed. East Hanover, New Jersey: Novartis, 1997.

Olson, T.R. *A.D.A.M. Student Atlas of Anatomy.* Baltimore: Williams & Wilkins, 1996.

Putz, R., and Pabst, R. (eds.), and Taylor, A.N. (trans. & ed.). *Sobotta Atlas of Human Anatomy*, 12th English ed., Vols. 1 and 2. Baltimore: Williams and Wilkins, 1997.

Schlossberg, L. (illus.), and Zuidema, G.D. (ed.). *The Johns Hopkins Atlas of Human Functional Anatomy*, 4th ed. Baltimore: Johns Hopkins University Press, 1997.

Seeley, R.R., et al. *Anatomy and Physiology*, 5th ed. Boston: McGraw-Hill, 2000.

Shier, D., et al. *Hole's Essentials of Human Anatomy and Physiology*, 7th ed. Boston: McGraw-Hill, 2000.

Snell, R.S. *Clinical Anatomy for Medical Students,* 6th ed. Philadelphia: Lippincott Williams & Wilkins, 2000.

Stern, J.T. Jr. *Core Concepts in Anatomy.* Philadelphia: Lippincott-Raven Publishers, 1997.

Thibodeau, G.A., and Patton, K.T. *Anatomy & Physiology,* with Student Survival Guide, 4th ed. Harcourt Health Sciences. St. Louis: Mosby, 1999.

Van De Graff, K.M., and Fox, S.I. *Concepts of Human Anatomy & Physiology,* 5th ed. Boston: William C. Brown/McGraw-Hill, 1999.

# Index



t refers to table.

t refers to table.

t refers to table.

t refers to table.

t refers to table.

t refers to table.

t refers to table.

t refers to table.

t refers to table.

t refers to table.

t refers to table.

t refers to table.

## WXY



## Z



t refers to table.